TECTONIC CONTROLS AND SIGNATURES
IN SEDIMENTARY SUCCESSIONS

Tectonic Controls and Signatures in Sedimentary Successions

EDITED BY L.E. FROSTICK
AND R.J. STEEL

SPECIAL PUBLICATION NUMBER 20 OF THE
INTERNATIONAL ASSOCIATION OF SEDIMENTOLOGISTS
PUBLISHED BY BLACKWELL SCIENTIFIC PUBLICATIONS
OXFORD LONDON EDINBURGH BOSTON
MELBOURNE PARIS BERLIN VIENNA

© 1993 The International Association
of Sedimentologists
and published for them by
Blackwell Scientific Publications
Editorial Offices:
Osney Mead, Oxford OX2 0EL
25 John Street, London WC1N 2BL
23 Ainslie Place, Edinburgh EH3 6AJ
238 Main Street, Cambridge
    Massachusetts 02142, USA
54 University Street, Carlton
    Victoria 3053, Australia

Other Editorial Offices:
Librairie Arnette SA
1, rue de Lille
75007 Paris
France

Blackwell Wissenschafts-Verlag GmbH
Düsseldorfer Str. 38
D-10707 Berlin
Germany

Blackwell MZV
Feldgasse 13
A-1238 Wien
Austria

First published 1993

Set by Semantic Graphics, Singapore
Printed and bound in Great Britain
at The Alden Press, Oxford

DISTRIBUTORS

Marston Book Services Ltd
PO Box 87
Oxford OX2 0DT
(*Orders*: Tel: 0865 791155
        Fax: 0865 791927
        Telex: 837515)

USA
Blackwell Scientific Publications, Inc.
238 Main Street
Cambridge, MA 02142
(*Orders*: Tel: 800 759-6102
        617 876-7000)

Canada
Oxford University Press
70 Wynford Drive
Don Mills
Ontario M3C 1J9
(*Orders*: Tel: 416 441-2941)

Australia
Blackwell Scientific Publications Pty Ltd
54 University Street
Carlton, Victoria 3053
(*Orders*: Tel: 03 347-5552)

A catalogue record for this title
is available from the British Library

ISBN 0-632-03745-8

Library of Congress
Cataloging in Publication Data

Tectonic controls and signatures in sedimentary successions/
    edited by L.E. Frostick and R.J. Steel.
        p.   cm.—(Special publication no. 20 of the
    International Association of Sedimentologists)
        Includes bibliographical references and index.
        ISBN 0-632-03745-8
        1. Sedimentation and deposition.   2. Geology,
    Stratigraphic.   3. Geodynamics.   I. Frostick, L.E.
    II. Steel, R.J.   III. Series:
    Special publication . . . of the International Association of
    Sedimentologists; no. 20.
    QE571.T35   1993
    551.3′03—dc20

Contents

Convergent Plate-Margin Basins

Preface

The papers in this Special Publication arise from the 13th International Sedimentological Congress in Nottingham in August 1991, theme C, dealing with sedimentation in a variety of tectonic settings. Sixteen of the talks presented originally are included here, together with eight additional papers which were invited to provide a better balance to the book.

Stratigraphers and sedimentologists who are presently describing and interpreting the infill of sedimentary basins are generally agreed that it is difficult to disentangle the signatures of tectonic processes from those of climate and eustatic sea-level change in the resultant rock succession. This problem is now being tackled in a number of ways. One of the avenues which we believe will *not* lead very far are the attempts to use *scale* of sequence or of motif (be it thickness or time) to assign unique control responsibility to either tectonics or eustacy or climate. Until better criteria are developed to distinguish between the roles played by the major variables, it is still most useful to document and interpret basin-fill architectures where we know,

from independent evidence which one of the main controls is likely to have been a *major* contributor. For this reason we have assembled the present collection of papers in which it can be claimed that the tectonic setting is fairly well established, and assumed that the tectonic factor has contributed to the resultant signatures.

The book is organized into three sections. The first section contains four papers where compressional and extensional tectonic controls have left their mark clearly on the geometry of the basin and on the architecture of the infill, and one case where Late Holocene earthquakes appear to have left a disappointingly ambiguous sedimentary record. The other two sections deal with a variety of basinal successions where the tectonic setting is respectively divergent and convergent plate margin.

RON J. STEEL
Bergen, Norway

LYNNE E. FROSTICK
Reading, UK

Spec. Publs Int. Ass. Sediment. (1993) **20**, 1–9

Tectonic signatures in sedimentary basin fills: an overview

L.E. FROSTICK* *and* R.J. STEEL†

*Postgraduate Research Institute for Sedimentology,
University of Reading, Reading RG6 2AB, UK; and
†Department of Geology, University of Bergen,
Alléget 41–5007, Bergen, Norway

ABSTRACT

The recognition of tectonic signatures within the sedimentary rock record gives important insights into the evolution of basins at a variety of scales. It can help to define both basin type and its shape by identifying the location and character of active faults. At this scale the sedimentary response to tectonism is encapsulated in the broad architecture of the basin fill and in the distribution of sedimentary facies in space and time. Evidence of tectonism is held within the geometry of the fill and within the sedimentary pile at the megasequence scale. At a more detailed scale, sediments can also yield data which can be used to infer the timing and rates of fault movement. Tectonic signatures at this scale can range from the character of individual sequences down to details of single beds, as is the case for some slumped deposits which can be connected with specific earthquakes (e.g. in the marginal lake sediments of Bogoria, Kenya) (McCall, 1967; Tiercelin, 1990).

Recognition of the role of tectonism in controlling sediment character gives the basin analyst the potential for forward modelling. Once the location and programme of fault activity is known, it is possible to predict cross-basin patterns of facies and how they are likely to change through time. Conversely, in situations where the patterns of fault activity are poorly understood, recognizable responses in the sedimentary fill can provide the basis for tectonic interpretation. Such predictions have economic implications since they can allow petroleum geologists to infer the likely locations of both source rocks and potential reservoirs and to establish the timing of any structural modification. This helps to confirm the juxtaposition of an oil source, migration pathway and trap at an apposite time. The identification of tectonic signatures is therefore central both to basin analysis and to the development of exploration strategies.

TECTONIC SIGNATURES

Tectonic activity affects the development of facies within a sedimentary basin almost exclusively via its control on topography. In a basin where all other influences are presumed constant there will be a critical value of subsidence which will result in a topographic expression for that basin. This value for subsidence will depend upon the rate at which sediment is supplied, that is the rate at which the accommodation provided by the evolving structure is filled. Where subsidence is equalled or outstripped by sediment supply, the surface expression of tectonic activity may be subdued. In this case, local geomorphological controls on sediment character become more important than tectonics in controlling facies development, although there will still be a strong control on the geometry of the basin fill and on the three-dimensional distribution of the facies (Alexander & Leeder, 1987). Above this critical value for subsidence a positive basin topography will develop, the character of which is a function of subsidence history. Patterns of sedimentation will, in effect, represent the losing battle being fought by the depositing sediment with the ever deepening basin through time. In some modern, active basins several kilometres of accommodation await filling (e.g. in parts of the East African Rift System). This raises the problem of whether sediments are responding directly to each structural development, or whether much of the basin fill post-dates tectonism and contains only a remote memory of prior events. The work of Blair (1987) underlines the fact that this problem exists, especially at the scale of individual fault movements. He points out that alluvial fans issuing from fault scarps are unlikely to be able to keep pace with the vertical movement on a fault while it is active. As a result, at these times, coarse sediment can only be found immediately adjacent to the scarp. In almost all basins the supply of sediment is such that coarse alluvium can only prograde towards the basin depocentre during tectonically relatively quiet periods. This contradicts the long held view of many sedimentologists that bands of coarse sediment within a succession can be related directly to bouts of fault activity.

On a large scale the general tectonic setting controls the size, shape, orientation and structural evolution of a basin, all of which exert an overall control on the sedimentary facies which can develop. This can be a direct effect, for example by controlling the location of source areas for clastic sediment, or an indirect influence, such as by controlling the intensity of wind–wave activity on beaches since basin margin orientation determines the relationship between shorelines and prevailing wind patterns. But when the effects of tectonics are analysed at a variety of scales it becomes evident that tectonics controls sediment character through its influence on geomorphology. If we analyse these controls, they can be classified into six different groups:

1 changes in the overall accommodation space available for filling within the basin;
2 changes in the direction of tilt and the location and size of the depocentre;
3 changes in the character and orientation of the basin margins and in overall basin size and shape (e.g. by backstepping of marginal faults);
4 changes in the gradient both at the margins of the basin through faulting, and within the basin (e.g. through tilting of the basin floor);
5 deflection of sedimentary systems where tectonically controlled morphological changes act as barriers to sediment transfer both into the basin from the hinterland and between various areas within the basin; and

6 changes in the rate of sediment supply due to uplift or erosion of the supplying hinterland.

The accommodation available within the basin at any one time will control the geometry and character of the sedimentary fill on a gross scale. Accommodation generally increases during bouts of fault activity and then decreases as sediments fill the available space. As most basins have a degree of asymmetry the large-scale effect is to produce a series of superimposed wedges of sediment, the size of which depends upon the magnitude of the changes in accommodation. There is also a more detailed effect on the character of the fill and on the distribution of facies. A rapid increase in accommodation deepens the basin and prevents penetration of clastic material. The facies which develop as a result become increasingly dominated by deeper water deposits, with a general upward-fining character, and in many cases by biogenic deposits. By contrast, a slowing of the rate of increase in basin accommodation to a point where sediment supply equals or exceeds it can produce sediment bypass and unconformities, as well as allowing clastic sediment to penetrate into the basin to produce a characteristic coarsening-upward sequences (Steel *et al.*, 1988). At the top of such a sequence there may be evidence of the basin becoming almost completely filled with sediment. There may be well developed soils (Wright & Allen, 1989) which suggest that the surface spent time in the soil-forming horizon, for example the Stotfield cherty rock in the Moray Firth which is a silicified calcrete developed during a prolonged Late Triassic period of tectonic quiescence (Frostick *et al.*, 1988). Basin-wide sandy alluvial or shallow-marine sediments both suggest filling of the basin and diminution of the effects of tectonic barriers (e.g. Steel, 1993). Intensive aeolian reworking of alluvial deposits can indicate that the basin was in a period of relative stasis, i.e. almost filled with sediment and with little or no tectonism. The sediment patterns reflecting accommodation changes, whether generated by tectonic or eustatic factors, are well documented in recent sequence stratigraphy literature (e.g. Posamentier *et al.*, 1988).

Changes in the tilt of the basin floor or in the position of the depocentre due to tectonism or volcanism will influence the character of the fill at several scales. It will change the overall geometry of the deposit, the locations and extent of deep and shallow water facies and the direction of clastic sediment transport. For example, in an arid zone continental basin, the location of the shores of a shallow, ephemeral and even a deeper perennial lake can shift many kilometres in response to faulting (Grove, 1986; Watkins, 1986; Tiercelin, 1990).

The size and shape of every basin evolves through time and as it evolves the character and distribution of facies will alter to reflect the changing morphology. Of particular importance are shifts in the loci of erosion and deposition which can result in a discrete distribution of intra-sequence unconformities. This is particularly evident on the margins of foreland basins where forward propagation of thrusts into the evolving sedimentary pile progressively shifts the erosive zone of an alluvial fan downstream resulting in what is known as a progressive unconformity (Riba, 1976).

Both the *direction* in which faults propagate and the *sense* of fault movement will affect the type of sequence which develops, especially at positions close to the basin margin. A basin margin fault which has a stable position during continuous subsidence creates a narrow but thick belt of coarse-grained facies (Fig. 1a(i)). A basin-margin which has limited (Fig. 1a(ii)) or repeated (Fig. 1a(iii)) back-faulting will commonly show a large-scale, irregular upward-fining pattern because of landward retreat of the marginal fans across a set of normal faults (Fig. 1a, after Heward, 1978). In foreland basins, where the marginal overthrusts migrate *into* the basinal area, the large-scale motif is commonly upward-coarsening, from deep marine up to alluvial facies, although there is commonly a basal, sandy transgressive package (Fig. 1b, from Steel *et al.*, 1985). Along oblique-slip basin margins, where point-sourced fans are gradually torn laterally from their formative river headwaters, the vertical architecture is more complex. Oblique-slip faulting combined with progradational fan trends produces a complex range of upward-fining, upward coarsening-to-fining, and upward-coarsening motifs (see Steel, 1988, for discussion). The most commonly occurring motif, however, is an asymmetric (in time and space) upward coarsening-to-fining one (Fig. 1c).

The way in which any given basin margin evolves depends largely on the character of the faulting. In extensional basins there has been much debate concerning the relative importance of faults connected to a crustal detachment zone and those which cut down through the crust without a detachment (Wernicke & Burchfiel, 1982; Crews & Mc-

Grew, 1990; Roberts *et al.*, 1991). Even within the proponents of detachment faults, there is controversy over the shape of the fault plane, is it listric or planar? Examples of these various fault geometries are given in Fig. 2. Schlische (1991) examined the sedimentary consequences of structural evolution for three different types of extensional faulting, detachment faulting, domino faults and steep faults, which grow through time (termed the fault growth model). He constructed models of the sedimentary fills of basins associated with these various fault types for three different rates of sediment supply, assuming that the supply did not vary during fill development. A synopsis of his results for one rate of sediment supply is given in Fig. 3. He concluded that onlap of younger sediments onto older deposits is a feature of the fault growth model and that, since onlap is widespread in extensional basins (see Table 1), this is likely to be the predominant type of faulting. The problem remains that it is feasible to develop onlapping relationships within any of the basins if sediment supply varies sufficiently. However, Schlische's, work is pioneering and points to ways of using the sedimentary patterns for inferring structure where this is not known from other, independent, evidence.

The gradients of the margins and floor of a basin, if maintained for long time periods, exert a fundamental control on patterns of erosion, transport and deposition. They control the mechanical energy available to the system for erosion, the extent to which clastic sediments can be reworked and also, through the steepness of any lake or sea shoreline, determine the effectiveness of wind–waves in sedi-

Table 1. Basins in which younger syn-rift strata onlap pre-rift rocks of the hanging wall block (After Schlische and Olsen, 1990.)

Basin	Location	Age
Richmond	Eastern USA	Triassic
Newark	Eastern USA	Triassic–Jurassic
Atlantis	Offshore eastern USA	Triassic–Jurassic?
Long Island	Offshore eastern USA	Early Mesozoic
Nantucket	Offshore eastern USA	Early Mesozoic
Fundy	Eastern Canada	Triassic–Jurassic
Hopedale	Labrador Margin	Mesozoic
Saglek	Labrador Margin	Mesozoic
North Viking	North Sea	Mesozoic
Tanganyika	East Africa	Cenozoic

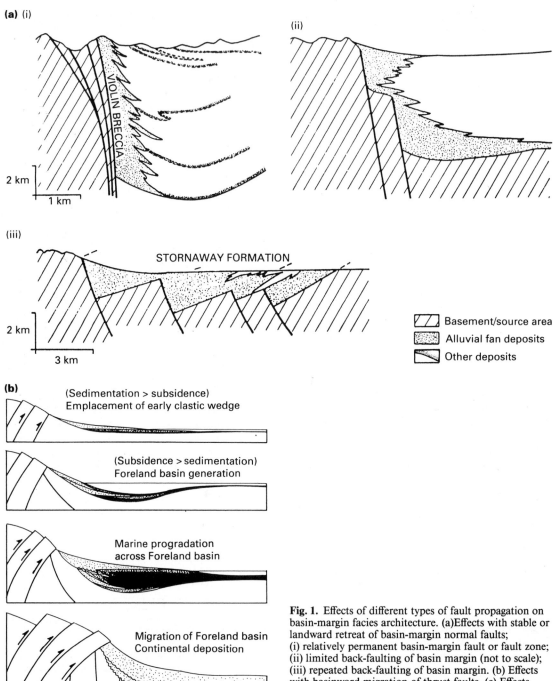

Fig. 1. Effects of different types of fault propagation on basin-margin facies architecture. (a)Effects with stable or landward retreat of basin-margin normal faults; (i) relatively permanent basin-margin fault or fault zone; (ii) limited back-faulting of basin margin (not to scale); (iii) repeated back-faulting of basin margin. (b) Effects with basinward migration of thrust faults. (c) Effects with oblique-slip movement of point-sourced, prograding alluvial fans. In (c) the resultant architecture is a complex of vertically and laterally asymmetric clastic wedges, with a coarsening-to-fining upward motif dominating.

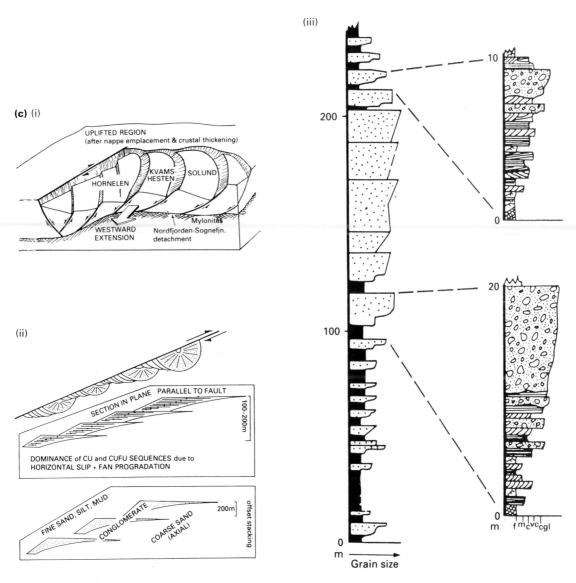

Fig. 1. (*continued*).

ment winnowing and the width of facies belts. Steep gradients produce river systems which can carry coarse material but which do not have the time to sort and round the sediment. Steep shores have narrow beach and shoreface facies belts which quickly pass offshore into deep water sediments. By contrast shallow gradients encourage extended periods of reworking and sorting within broader facies belts (Ingle, 1966; Galvin, 1968).

Rising topography both within and at the margins

of a basin will influence the quantity of clastic sediment which is supplied and the main locations at which it enters. In a rising hinterland, rivers must incise to keep pace with uplift and the sediment load they carry will increase as a result. But the rate at which this can occur depends on both geology and climate. Soft sedimentary strata offer little resistance to the erosive forces of incising streams irrespective of the degree of weathering whereas hard crystalline basement rocks can be removed

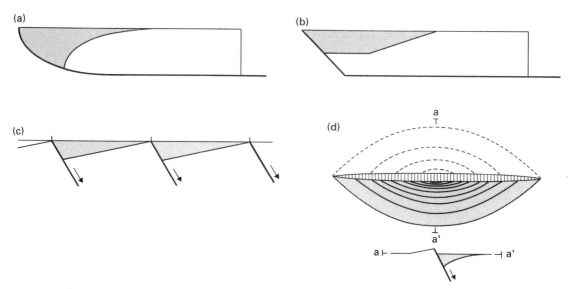

Fig. 2. Models for extensional fault systems. Stippled areas indicate accommodation space available for filling by sediments (a) Listric border fault; (b) linked fault; (c) domino fault; (d) fault growth model. (After Schlische, 1991.)

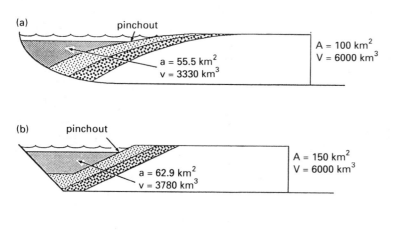

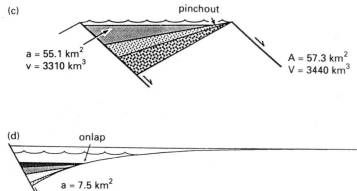

Fig. 3. Results of computer modelling of the sedimentary fill in extensional basins with different border fault character. Sediment supply in all cases approximates 3500 km³ per 5 My and the figures represent the situation after the passage of 15 My. (a) Listric border fault; (b) linked fault system; (c) domino fault system; (d) fault growth model. A, cross-sectional area; a, incremental area (increase in basin); V, volume of the basin; v, incremental volume of sediment deposited. (After Schlische, 1991.)

only after considerable weathering under humid conditions. It is interesting to compare the fills of two adjacent rift basins on the coast of Brazil, the Tucano and Reconcavo Basins. The marginal faults of the Tucano Basin cut into relatively easily eroded Palaeozoic sediments and the basin fill is dominated by thick sand sequences. The Reconcavo sits within Precambrian basement rocks which are resistant to erosion. As a result, fill sediments which accumulated during periods of fault activity are dominated by organic-rich shales. It is therefore the main oil exploration target in the area and contains a large proportion of the Brazilian known oil reserves.

Changing topography can deflect river drainage either into or away from the basin and can prevent or encourage sediment transfer within the basin. A classic case of river deflection can be seen in the East African Rift System where initial doming has diverted rivers away from the rift and footwall uplift has helped to maintain the diversion so that many basins are starved of clastic sediment (Frostick & Reid, 1989). In addition, in some areas the propagation of faults away from the basin depocentre has captured rivers, diverting them parallel to the rift and increasing starvation (Frostick & Reid, 1987; Fig. 4).

OTHER FACTORS INFLUENCING THE CHARACTER OF BASIN FILLS

Above a critical value of differential movement — depending on sedimentation rate — topography is generally controlled by the evolving structure. At a more detailed scale and on a shorter time-scale, other additional factors will influence facies development. These are: regional geology via rock erodibility and sediment supply (discussed above); climate — particularly in non-marine basins; and eustasy — particularly in marine basins (discussed by Bartsch-Winkler & Schmall, this volume, pp. 91–108.

Climate exerts a strong control on the character of the basin fill within lacustrine basins. Although tectonics exert a fundamental control on the potential depth of a lake, the hydrological balance (rainfall, runoff and evapotranspiration) will determine the size and chemistry of the lake which actually develops (Eugster, 1986; Tiercelin, 1990). Any climatic shifts will change this balance and result in either expansion or contraction of the lake which will, in turn, change the distribution of facies and the sequence signature. This has happened several times during the Pleistocene history of lakes in the East African Rift System (Street & Grove, 1976,

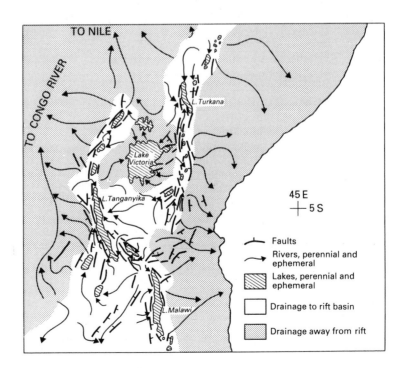

Fig. 4. Present-day river drainage in the East African Rift System.

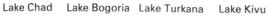

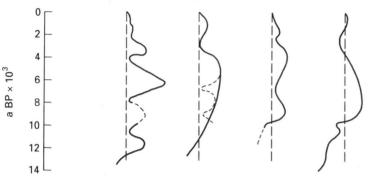

Fig. 5. Schematic diagram of lake level changes in some African lakes over the past 14 000 years. (After Talbot, 1988.)

1979). The latest major rise and fall occurred approximately 10 000 years ago and its effects are recorded in the sedimentary records of a number of the larger lakes (Fig. 5). Expanding lake conditions will leave a fining-upward signature which may be confused with the immediate effects of fault activity. A retreating lake, on the other hand, allows coarse sediment to prograde out into the basin giving a coarsening-upward sequence which could be misinterpreted as resulting from a post-faulting quiescent phase.

Differentiating between the effects of climate and tectonism is not easy, especially for short time-scale changes. Climatic fluctuations generally occur on time-scales of years to thousands of years (Street & Grove, 1979) and so are unlikely to control the development of megasequences. At sequence scale, other corroborative evidence of climatic change should always be sought. This can be obtained from soils, evidence of aeolian activity, evidence of vegetation change, for example from pollen, development of evaporites and other indicators of changing lake chemistry (Coulter, 1963; Cerling *et al.*, 1977; Eugster, 1980). This can be seen for instance in the deposits of Lake Lisan, the Pleistocene precursor to the present Dead Sea. Here the geochemical and isotope work of Katz and Kolodny (1977) has shown that a much more dilute saline lake existed during the wetter climatic phase of more than 8000 years BP.

SUMMARY

It must always be remembered that similar sedimentary signatures can arise under a variety of tectonic conditions and that corroborative evidence of any conclusions should always be sought. Most rules for inferring tectonic causes for sedimentary changes assume that all else remains constant. This is almost never the case and any scientist investigating tectonics and sedimentation must always take other controls (including allocyclic) into account.

REFERENCES

ALEXANDER, J. & LEEDER, M.R. (1987) Active tectonic control on alluvial architecture. In: *Recent Developments in Fluvial Sedimentology* (Eds Ethridge, F.G., Flores, R.M. & Harvey, M.D.) Soc. Econ. Palaeont. Mineral. Spec. Publ. **39**, 243–252.

BLAIR, T.C. (1987) Tectonic and hydrologic controls on cyclic alluvial fan, fluvial and lacustrine rift-basin sedimentation, Jurassic-lowermost Cretaceous Totos Santos Formation, Chiapas, Mexico. *J. Sediment. Petrol.* **57**, 845–862.

CERLING, T.E., HAY, R.L. & O'NEILL, J.R. (1977) Isotopic evidence for dramatic climatic changes in East Africa during the Pleistocene. *Nature* **267**, 137–138.

COULTER, G.W. (1963) Hydrogeological changes in relation to biological production in southern Lake Tanganyika. *Limnol. Oceanogr.* **8**, 463–477.

CREWS, S.G. & McGREW, A.J. (1990) Influences of structural style on rift basin morphology and non-marine sequence geometry. *Geol. Soc. Am. Abstr.* **22(7)**, A239.

EUGSTER, H.P. (1980) Lake Magadi Kenya: a model for rift valley hydrochemistry and sedimentation? In: *Sedimentation in the African Rifts* (Eds Frostick, L.E., Renaut, R.W., Reid, I. & Tiercelin, J.J.) Geol. Soc. London Spec. Publ. **28**, 177–189.

EUGSTER, H.P. (1986) Lake Magadi, Kenya and its precursions. *Dev. Sediment.* **28**, 195–232.

FROSTICK, L.E. & REID, I. (1987) Tectonic control of desert sediments in rift basins ancient and modern. In: *Desert Sediments: Ancient and Modern* (Eds Frostick, L.E. & Reid, I.) Geol. Soc. London Spec. Publ. **35**, 53–68.

FROSTICK, L.E. & REID, I. (1989) Is structure the main control of river drainage and sedimentation in rifts? *J. Afr. Earth Sci.* **8**, 165–182.

FROSTICK, L.E. REID, I., JARVIS, J. & EARDLEY, H. (1988) Triassic sediments of the Inner Moray Firth, Scotland: early rift deposits. *J. Geol. Soc.* **145**, 235–248.

GALVIN, C.J. (1968) Breaker type classification on three laboratory beaches. *J. Geophys. Res.* **73**, 1–9.

GROVE, A.T. (1986) Geomorphology of the African rift system. In: *Sedimentation in the African Rifts* (Eds Frostick, L.E., Renaut, R.W., Reid, I. & Tiercelin, J.J.) Geol. Soc. London Spec. Publ. **25**, 9–16.

HEWARD, A.P. (1978) Alluvial fan sequence and megasequence models: with examples from Westphalian D-Stephanian B coalfields, northern Spain. In: *Fluvial Sedimentology* (Ed. Miall, A.D.) Can. Soc. Petrol. Geol. Mem. **5**, 669–702.

INGLE, J.C. (1966) The movement of beach sand. *Dev. Sediment.* **5**, 221 p.

KATZ, A. & KOLODNY, Y. (1977) The geochemical evolution of the Pleistocene. Lake Lisan, Dead Sea System. *Geochim Cosmochim Acta* **41**, 1609–1626.

McCALL, G.J.H. (1967) *Geology of the Nakuru — Thompsons Falls, Lake Hannington Area.* Geological Survey of Kenya Report, 78p.

POSAMENTIER, H.W., JERVEY, M.T. & VAIL, P.R. (1988) Eustatic controls on clastic deposition. I — conceptual framework. In: *Sea-levl Changes: an Integrated Approach* (Eds Wilgus, C.K., Hastings, B.S., Kendall, C.G.St.C., Posamentier, H.W., Ross, C.A. & Van Wagoner J.) Soc. Econ. Paleont. Mineral., Spec. Publ. **42**, 109–124.

RIBA, O. (1976) Syntectonic unconformities of the Alto Cardener Spanish Pyrenees: a genetic interpretation. *Sediment. Geol.* **15**, 213–233.

ROBERTS, A.M., YIELDING, G. & FREEMAN, B. (1991) *The Geometry of Normal Faults.* Geol. Soc. London Spec. Publ., 56, 264 p.

SCHLISCHE, R.W. (1991) Half-graben basin filling models: new constraints on continental extensional basin development. *Basin Res.* **3**, 123–141.

SCHLISCHE, R.W. & OLSEN, P.E. (1990) Quantitative filling model for extensional basins with application to the early Mesozoic rifts of eastern North America. *J. Geol.* **98**, 135–155.

STEEL, R.J. (1988) Coarsening upward and skewed fan-bodies: symptoms of strike-slip and transfer fault movement in sedimentary basins. In: *Fan Deltas: Sedimentology and Tectonic Settings* (Eds Nemec, W. & Steel, R.J.) Blackie, Glasgow, pp. 75–83.

STEEL R.J. (1993) Triassic–Jurassic megasequence stratigraphy in the northern North Sea: rift to post-rift erolution. In: *Petroleum Geology of Northwest Europe* (Ed. Parker J.R.) Geol. Soc. London, p. 299–315.

STEEL, R.J., MAEHLE, S., NILSEN, H., RØE, S.L. & SPINNANGR, Å (1977) Coarsening-upward cycles in the alluvium of Hornelen Basin (Devonian), Norway: sedimentary response to tectonic events. *Geol. Soc. Am. Bull.* **88**, 1124–113.

STEEL, R.J., GJELBERG, J., HELLAND H.W., KLEINSPEHN, K., NØTTVEDT, A. & RYE-LARSEN, M. (1985) The Tertiary strike-slip basins and orogenic belt of Spitsbergen. In: *Strike-slip Deformation, Basin Formation and Sedimentation* (eds Biddle, K.T. & Christie-Blick, N.) Soc. Econ. Paleont. Mineral. Spec. Publ. **37**, 339–359.

STREET, F.A. & GROVE, A.J. (1976) Environmental and climatic implications to Quaternary lake-level fluctuations in Africa. *Nature* **261**, 385–390.

STREET, F.A. & GROVE, A.J. (1979) Global maps of lake level fluctuations since 30 000 year BP. *Quat. Res.* **12**, 83–118.

TALBOT, M.R. (1988) The origins of lacustrine oil source rocks: evidence from the lakes of tropical Africa. In: *Lacustrine Petroleum Source Rocks* (Eds Fleet, A.J., Kelts, K. & Talbot, M.R.) Geol. Soc. London Spec. Publ. **40**, 29–44.

TIERCELIN, J.J. (1990) Rift-basin sedimentation: responses to climate, tectonism and volcanism. Examples from the East African Rift. *J. Afr. Earth Sci.* **10**, 283–305.

WATKINS, R.T. (1986) Volcano-tectonic control on sedimentation in the Koobi Fora sedimentary basin, Lake Turkana. In: *Sedimentation in the African Rifts* (Eds Frostick, L.E., Renaut, R.W., Reid, I. & Tiercoelin, J.J.) Geol. Soc. London Spec. Publ. **25**, 85–95.

WERNICKE, B. & BURCHFIEL, B.C. (1982) Notes on extensional tectonics. *J. Struct. Geol.* **4**, 105–115.

WRIGHT, V.P. & ALLEN, J.R.L. (1989) *Palaeosols in Siliciclastic Sequences.* Postgraduate Research Institute for Sedimentology Short Course Notes O1, Reading Univenity 98 p.

Sedimentary Response to Tectonic Events

Spec. Publs Int. Ass. Sediment. (1993) **20**, 13–35

The interrelations of post-collision tectonism and sedimentation in Central Asia

M.E. BROOKFIELD

Land Resource Science, Guelph University, Guelph, Ontario N1G 2W1, Canada

ABSTRACT

During the Late Tertiary the northwest Himalaya and adjacent ranges underwent progressively increasing uplift associated with crustal thickening and differential erosion. At the same time the adjacent Tarim and Tadjik basins and the Himalayan foredeeps subsided accumulating thick piles of clastic sediments. The ages of these accumulations have been used to infer contemporary tectonism in the mountains. Radiometric and fission track dating together with faunal and floral studies and sediment budget and river drainage studies allow gross rates of uplift and subsidence to be compared with net rates for different areas. These studies show that the bulk of the derived clastics were eroded from the fronts of the Karakorum, Himalayan and Pamir allochthons during progressively increasing rates of uplift and thrusting from Early Miocene times onwards. Nevertheless, cooling Cenozoic climates, partly driven by uplift, have also contributed to increasing physical erosion and coarsening of the deposited sediments through time. Furthermore, each range shows an independent history of uplift and erosion within the framework of generally increasing uplift and erosion in Late Tertiary times. Contemporary deposition in adjacent basins is partly dependent on the timing of adjacent uplifts. But uplift and deposition also depend on erosion and deposition along the courses of major rivers — which have not remained constant. The thickness of sediment accumulating in marginal basins and the isostatic uplift of ranges depend on when the river changed its course, on when and how much temporary storage occurred within intermontane basins, and the time at which the intermontane barriers were breached and their sediments eroded. In some cases, e.g. the Indus, Tsangpo, Salween and Mekong, the rivers have even switched into different major oceans or basins. Coarse clastic sediment pulses cannot be used to infer increased tectonism in adjacent ranges.

INTRODUCTION

During the Late Tertiary, the mountain ranges and plateau of Central Asia (including the Himalaya, Karakorum, Pamir and Kun Lun ranges and Pamir and Tibetan plateaux) underwent progressive and increasing uplift associated with crustal thickening and differential erosion. At the same time, the adjacent northern and southern foredeeps subsided, accumulating thick clastic sedimentary sequences in northern (Tadzhik, Tarim) and southern (Indus, Ganges) basins (Fig. 1).

This area is one of the best places to study the interrelationships between tectonics and sedimentation. Mountain building is still ongoing forming the highest ranges and plateaux in the world. The architecture is relatively simple, at least in the

Himalaya, with comparable structures extending along strike for over 2000 km between the western and eastern syntaxes. Mountain building is clearly related in space and time to the collision of India and Asia, and the history of Indian plate movement is available from the magnetic anomalies of the Indian Ocean (Mattauer, 1986).

In addition, this area shows a complex interplay between indentation of Asia by northwestern India, lateral movement of large blocks and differential uplift and subsidence associated with changing patterns of tectonism in often closely adjacent crustal blocks. Figure 2 shows how the various Mesozoic sutures of the area have been deformed and disrupted by the indentation and how zones of

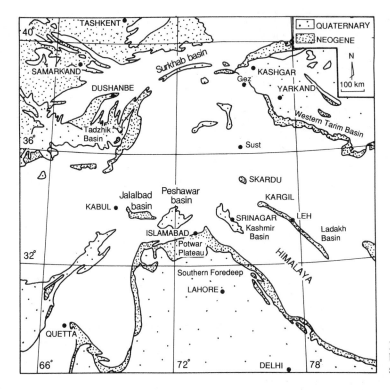

Fig. 1. Location map for sedimentary basins of the northwest Himalaya syntaxis and adjacent areas.

active thrust faulting rapidly pass laterally into zones of transcurrent motion.

A study of the interrelationships of sedimentation and tectonics has a bearing on the tectonic models used to explain the collision. For example, did the Indian Shield underthrust the entire Tibetan plateau (Powell, 1986)? Was limited underthrusting followed by indenting of India into Asia with Asian blocks being moved northwards and eastwards along major faults, particularly at the western and eastern syntaxes (Molnar & Tapponnier, 1975)? Or was most of the shortening due to internal deformation of the mountains and plateaux behind (England & Molnar, 1990)?

A particular problem is the time lag between initial Late Eocene collision of India and Asia, and the post-Miocene major uplifts of the Himalaya and adjacent ranges and Tibet (Tapponnier *et al.*, 1986). This lag may be related to the progressive telescoping of the northern Indian margin back to its original thickness along reversing normal faults, followed by progressive crustal thickening and nappe formation in continental basement and cover (Searle, 1986).

The purpose of this article is to relate uplift and erosion, and subsidence and deposition to the devel-

opment of the sedimentary basins and mountains formed (and destroyed) during the collision of India with Central Asia. Various lines of evidence for uplift of the Central Asian mountains indicate that major uplift did not start until the Early Miocene and progressively accelerated through the Late Cenozoic. This evidence also indicates a major break between simple compressive shortening during the Early Tertiary and Miocene and younger thrusting and lateral escape along very different trends to earlier ones.

The evidence includes: foredeep and intermontane basin stratigraphy and sedimentology; radiometric and fission track dating; sediment budgets of erosion of the mountains and deposition in foredeeps north and south of the mountains; and changing drainage patterns.

The results of this study illustrate that simplistic models of basin development in collision orogens are unlikely to be correct.

TECTONIC BACKGROUND

India commenced colliding with Asia in the Mid-Eocene around 50 Ma (Fig. 3). The early stages

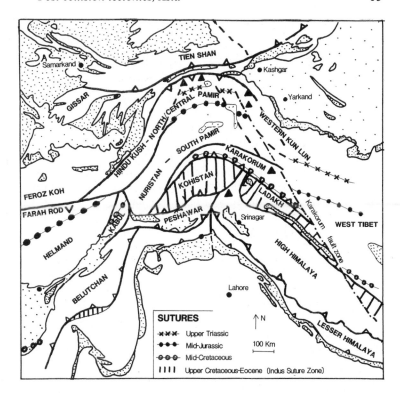

Fig. 2. Location map of tectonic units and sutures of the northwest Himalaya and adjacent areas.

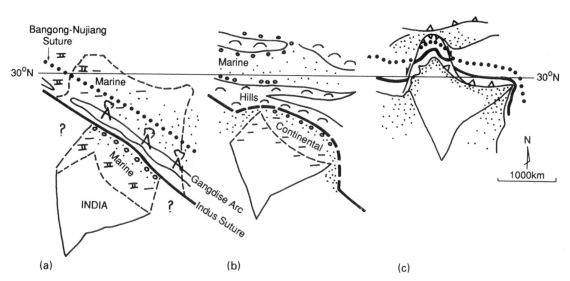

Fig. 3. Cartoons showing development of sedimentary basins during the Tertiary related to plate movements. Plate orientation and positions after Dewey *et al.* (1989). Lithofacies and basins from Krestnikov (1963) and personal observations. Present position of Himalayan frontal thrusts dashed. Line across page is 30°N datum for motions. (a) Mid-Eocene start of collision (around 50 Ma). Note that the marine basin *behind* the Gangdese–Ladakh magmatic arc is filled from the southeast by major rivers draining the Mesozoic orogens of Southeast Asia. (b) Late Oligocene completion of telescoping of the northern Indian margin and deformation of Tibet (30 Ma). (c) Present situation (0 Ma): main motion since the Miocene on the Main Central and associated thrusts and the North Pamir thrusts linked by strike-slip faults.

from Eocene to Early Miocene times involved restacking of the northern Indian passive margin back to its original thickness together with deformation and uplift of the Tibetan plateau and its marginal magmatic arcs (for discussion of these concepts see Dewey *et al.*, 1989). Sedimentary petrography shows that the southern foredeeps of the Himalaya were receiving volcaniclastic sediments from the Asian arcs as early as the Mid-Eocene (50 Ma) (Nicora *et al.*, 1987). During the following phase of around 30 Ma marine foredeeps existed to the south and north of the collision zone. These foredeeps were progressively filled with clastic sediments derived from the slowly uplifting areas to north and south. Marine conditions persisted *both north and south* of the collision zones into the Oligocene. The northern basins, including the Tarim and Tadzhik basins may have been underlain by oceanic crust in the Early Tertiary (Leith, 1985; Hsu, 1988).

By the Early Miocene internal deformation of the Tibetan and East Pamir plateaux had practically ceased (Dewey *et al.*, 1989) and both plateaux had started their progressively increasing Late Tertiary uplift. Miocene and younger movements in the south are concentrated on the Main Central Thrust (an intracrationic thrust now deforming the Indian Shield and its cover) and the Karakorum Thrust (a reactivated Mesozoic suture now carrying Asian continental crust over previously emplaced Cretaceous–Early Tertiary nappes) (Brookfield, in press). The junction between these zones roughly approximates to the Karakorum fault, a large right-lateral fault offsetting the Mesozoic sutures.

Miocene and younger movements in the north are concentrated along the North Pamir overthrusts and the large transcurrent faults along the edges of the Kun Lun and Hindu Kush mountain systems (Fig. 2). Structural data show a minimum of 470 km internal continental shortening at the northwest syntaxis (Coward & Butler, 1985). At least 120 km shortening occurred even within the Zanskar shelf portion of the northwest Indian passive margin (Searle, 1986). This Zanskar shelf synclinorium is obliquely truncated by the Asian Ladakh arc so a great deal of further shortening is also needed. These observations imply that the northern Indian passive margin was at least 600 km wide at initial collision, as shown on Fig. 3, and not 400 km wide, based on younger collision at 45 Ma (Dewey *et al.*, 1989). Figure 3 is also consistent with 1500 km of northward motion of Lhasa relative to Asia.

The uplifts, subsidences and lateral motions associated with these Miocene and younger movements provide a fascinating illustration of the complex interrelationships of tectonism and sedimentation in an active collision zone.

CONCEPTS

The components of uplift are defined as follows. Uplift is positive vertical elevation with respect to the geoid (basically mean sea-level). This normally refers to net uplift, combining the effects of gross (or total) uplift minus the effects of erosion. Thus: net uplift = gross uplift – erosion. Net uplift can be obtained by geodetic measurements which give present values of about 1 mm/year for the High Himalaya and Tibet (Chi-yen Wang & Yaolin Shi, 1982).

Gross uplift can be determined from sedimentary and fossil evidence, if erosion has not removed it. Thus, Cretaceous and Eocene marine limestones now occur at heights of over 4 km and therefore must have been uplifted by that amount. Eustatic sea-level changes amount to less than a couple of hundred metres at most, are insignificant compared to tectonic effects in this study, and are therefore ignored. Continental deposits characteristic of Oligocene and Miocene lowlands now occur high within the mountains and on the Tibetan plateau. A major problem is determining the original altitude at which these beds were deposited. Warmer conditions probably persisted to higher altitudes in the Mid-Tertiary (Molnar & England, 1990). Thus, the original altitude of deposition of continental (but not marine) beds may be underestimated.

Gross subsidence may also be inferred from the depth of burial of marine and lowland continental rocks below sea-level in the foredeep basins.

If erosion has stripped the sediments, then erosion rates can sometimes be found with radiometric and fission track dating as well as by working out the sedimentary budgets of erosion/deposition of the mountain belts — though a number of assumptions have to be made.

The following sections discuss some of the evidence for uplift and displacement of the Central Asian mountains and Tibet, and from this some idea of net uplifts in the late Tertiary are worked out. The time-scales used are from Repenning (1984) and Berggren *et al.* (1985).

NEOTECTONIC EVIDENCE

Seismic and gravity evidence indicate not only that the Indian plate is descending towards the north, but also that an oceanic slab is still descending *south* beneath the Pamir (Billington *et al.*, 1977). Any tectonic model of Himalayan orogenesis must account for *both*, not simply the northwardly subducting Indian plate. Geodetic measurements are available (mainly from the former USSR) for several of the tectonic blocks shown on Fig. 2.

At the junction of the North Pamir and Tien Shan, both horizontal and vertical movements have been measured over tens of years. Thus, the southern Tien Shan are advancing south at about 15–20 mm/year over the Surkhob basin: the vertical uplift is around 5–15 mm/year suggesting that the marginal Tien Shan fault is inclined at about 30° to the north (Pevnev *et al.*, 1975). On the south, the North Pamir are bulldozing Mesozoic–Cenozoic molasse of the Peter the Great Range northwards over the Surkhob depression (personal observation, 1990) with an uplift rate of between 3 and 11 mm/year (Finko & Enman, 1971). These vertical movements are accompanied by strong left-lateral displacement of between 10 and 15 mm/year on the western flanks of the Pamir (Kuchay & Trifonov, 1977). This is also an area where the Tadzhik basin and Gissar Range are moving southwestwards at comparable rates as they are squeezed out of the Pamir–Tien Shan junction (Bulanzhe *et al.*, 1980). The total convergence rate between the Tien Shan and Pamir is thus up to 35 mm/year. TInpost-glacial times, rivers have cut downwards at rates of up to 10 mm/year (Nikonov, 1970), presumably keeping pace with uplift.

Molnar (1984) reported rates of 6 mm/year differential uplift between the High Himalaya and the plains, and up to 10 mm/year across one major fault. Geodetic rates of uplift thus seem to be up to 15 mm/year both on the north and south, relating to measured horizontal convergence between India and Asia of up to 20 mm/year on the south and up to 35 mm/year on the north.

The present rate of plate convergence between India and Eurasia is about 50 mm/year (Molnar, 1984). Thus, only about 50% of the crustal shortening can be accommodated by convergence in the Himalayan frontal zone. Somewhat more (70%) can be taken up between the Tien Shan and Pamir. The residual crustal shortening either takes place north of the Tibetan plateau or is the result of eastward extrusion of Southeast Asia over the Pacific (cf.

Peltzer & Tapponnier, 1988). Miocene granites are related to intracontinental thrusting in the Himalaya and Karakorum and are the same age as anatectic melts generated by major left-lateral motion on the Red River fault zone in Yunnan (Scharer *et al.*, 1990a).

SEDIMENT BUDGET STUDIES

The amount of southward underthrusting on the Main Central Thrust of the Himalaya has been inferred, on structural and palaeomagnetic grounds, to be at least 650 km in Nepal, decreasing slightly westwards (Klootwijk *et al.*, 1985). This is consistent with the amazing facies contrasts between the Lesser Himalaya and High Himalaya stratigraphic sequences. These have almost no units in common, yet the Lesser Himalaya sequences are similar to those of the Indian shield (Brookfield, in press). The facies contrast requires the removal of at least half of a wide continental shelf along the Main Central Thrust. Thus, vertical uplift of the Himalaya is partly a consequence of continual southward displacement providing fresh rock for erosion at the Himalayan front. A wide internal segment of the northern Indian margin has been removed by erosion at the front of an intracontinental thrust (Mattauer, 1986). Similarly, a significant portion of the North Pamir record has been removed by erosion at the front of the North Pamir overthrusts. It is possible that the entire crustal block between the Pamir and Himalaya will be removed in this way, juxtaposing Asian and Indian continental crust. Thus, these mountain ranges might end up being represented only by sediment in the adjacent foredeeps.

The amount of material removed from the mountains can be determined by sediment budget studies, which also give the times at which this thrusting occurred. One can add up the amount of sediment delivered during successive time intervals to marginal sedimentary basins during development of the mountains. These give volumes of material eroded from the Himalaya, Pamir and adjacent ranges during each time interval. High grade metamorphic rocks are only exposed in the High Himalaya, Karakorum, Kun Lun and Pamir. There has been almost no erosion of the Tibetan plateau in the late Tertiary and the volume of exposed Lesser Himalaya pre-Tertiary rocks is quite small. Thus, the sediment delivered to marginal sedimentary basins

by the Indus, Ganges, Brahmaputra and associated
rivers and palaeorivers is almost entirely derived
from the High Himalaya, Pamir and adjacent
mountain ranges.

Only a small amount of clastic sediment was
delivered to both northern and southern foredeeps
in the Oligocene when both were still marine
(Fig. 3). The vast bulk of sediment has been transported into the foredeeps, particularly the Indus
and Bengal fans, with accelerating rates from Mid-
Miocene times onwards (Fig. 4). Though thick Late

Tertiary accumulations occur in the northern
Tadzhik and Tarim basins, the volume of these is,
despite their thickness, relatively small compared to
the volume in the southern fans.

The calculations are shown on Fig. 4 and tabu-
lated on Table 1. They show the volume of solid
sediment deposited in adjacent basins for the three
comparable periods; Oligocene (15 My), Early to
Mid-Miocene (10 My) and Late Miocene to Recent
(12 My). The dissolved loads of the Indus, Ganges
and Brahmaputra rivers are at least one-tenth

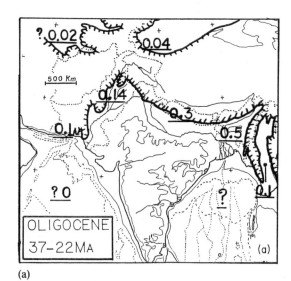

(a)

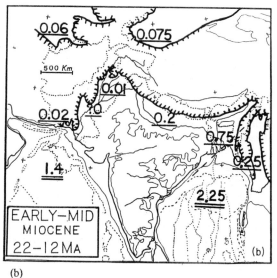

(b)

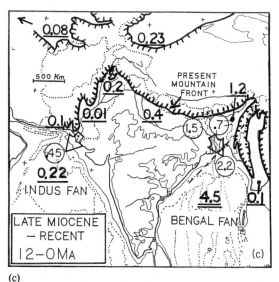

(c)

Fig. 4. Sediment budgets for : (a) the Oligocene; (b) the
Early–Mid-Miocene; (c) the Late Miocene to Recent
(circled values are measured Recent suspended sediment
loads of the main rivers, in 10^6 metric tonnes/year).
Underlined values are in 10^6 km^3 for individual basins.
Value in box is total sediment *volume* preserved for the
period. Time-scale from Berggren *et al.* (1985). Budget
data sources in Brookfield (in press). Note enormous
value for Bengal fan volume in (c).

Table 1. Sedimentary budgets for each basin (in 10^6 km^3)

Basin	Late Miocene–Recent (0–12 Ma)	Early to Mid-Miocene (12–22 Ma)	Oligocene (22–37 Ma)
Indus fan	0.22	1.4	?0
Makran	0.1	0.02	?0.1
Indus Valley	0.01	?0	?0
Northwest Himalaya	0.2	?0.01	0.14
Central–East Himalaya	0.4	0.2	0.3
Bengal Basin	1.2	0.75	0.5
Bengal fan	4.5	2.25	?
Nagaland	0.1	0.25	<1
Tadzhik	0.08	0.06	?0.02
Southwest Tarim	0.23	0.07	0.04
Totals	7.04	5.01	2.1
Rough correction for addition of dissolved load	7.74	5.5	2.3

weight of the solid load at present (Subramanian *et al.*, 1987). If this was true in the past then the volumes of rock eroded from the High Himalaya have to be increased as in Table 1. The correction gives total sediment volumes (in 10^6 km^3) of 2.3 (Oligocene), 5.5 (Early to Mid-Miocene) and 7.7 (Late Miocene to Recent).

The volume of rock eroded from the Himalaya can be determined with an average ratio of the densities of sedimentary (2.5) to crystalline rock (2.8). This gives an underestimate, since a significant part of the Himalaya is sedimentary rock. Volumes of crystalline rock eroded in the Oligocene, Early–Mid-Miocene and Late Miocene to Recent are, respectively, 2.0×10^6 km^3, 4.9×10^6 km^3 and 6.6×10^6 km^3. Since neither Tibet nor the Indo-Gangetic plain underwent significant erosion in the later Tertiary (Chang Chengfa *et al.*, 1986), all this material must have been removed from the High Himalaya and northwest syntaxis ranges.

If it is assumed that this material came from erosion of a crystalline slab (the rising High Himalaya and Karakorum, Pamir, etc.) and assumed that this slab detached at the base of the Indian continental crust, then one can calculate the width necessary for a 35-km-thick slab, 2500 km long, to supply these volumes (Fig. 5 and Table 2). The calculated widths show minimum southward displacements across the Main Central Thrust for successive periods.

However, it seems likely that less than half the thickness of the continental crust is involved in the

southward thrusting. That is, the crust delaminates at mid-crustal levels rather than at the base of the crust (Mattauer, 1986). Rocks exposed near the base of the High Himalaya slab, just above the Main

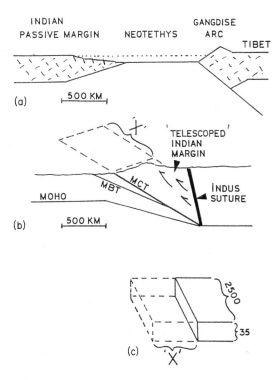

Fig. 5. Diagram showing method of calculating width (X) of block removed by erosion at different periods.

Table 2. Width (X) of 2500-km-long Himalayan slab required to account for sediment budgets at different thicknesses

	Southward displacement (X km) thickness[a]		Rate (mm/year)[b]	
	c.35 km[c]	*c*.20 km[d]	*c*.35 km[c]	*c*.20 km[d]
Late Miocene to Recent	75.4	132.0	6.3	11
Early to Mid-Miocene	56.0	98.0	5.6	9.8
Oligocene	22.8	40.0	1.5	2.6
Total	154.2	270.0		

[a] Width (southward displacement) $(X) = \dfrac{\text{total sediment volume}}{\text{length (2500 km)} \times \text{thickness}}$.

[b] Rate of thrusting $= \dfrac{\text{southward displacement (X)}}{\text{time}}$.

[c] Total thickness of Indian continental crust overthrust.

[d] Indian crust delaminated at mid-crustal levels.

Central Thrust, reached kyanite grade at temperatures of around 600°C (LeFort, 1986). This corresponds to pressures of around 5 kb, i.e. 20 km depth (Thompson & England, 1984). In that case the given values must be multiplied by 1.75 (Table 2).

The end result, for a 20-km-thick-slab, is that the total southward displacements on the Main Central Thrust and Northern Suture are: 40 km for the Oligocene, 98 km for the Early to Mid-Miocene, and 132 km for the Late Miocene to Recent. These values are in keeping with the increasing erosion rates derived from radiometric and fission track studies. The total southward displacement on the thrusts is thus around 270 km for the Oligocene to Recent and, more importantly, 230 km since the Early Miocene.

A consequence of these calculations is that most of the shortening in the Himalayan–Tibet area since the Mid-Miocene has taken place by thrusting along the Main Central Thrust and Northern Suture. Relatively little has been eroded from the front of the North Pamir. The calculated values are compatible with palaeomagnetic estimates of 440 km post-collision shortening between southern Tibet and India (Dewey *et al.*, 1989), thrust displacements in other orogens (Wu-ling Zhao & Morgan, 1985), and with concepts of Miocene extensional tectonics in the Tibetan plateau (Royden & Burchfiel, 1987). They are incompatible with 1000 km underthrusting of continental crust below the Tibetan plateau (Powell, 1986). There is also no need to indent Asia in post-Miocene times — which is in keeping with the known small, Late Tertiary displacements along the marginal Tibetan plateau faults (Chang Chenfa *et al.*, 1986).

Young micas aged between 18 and 0 Ma occur in Bengal fan sediments despite the absence of exposed lower crustal rocks in the Himalaya (Copeland & Harrison, 1990). TiO_2/Al_2O_3 ratios shows rapidly increasing stream gradients and hence increased erosion of sources feeding the fan (Schmitz, 1987). Both are explained with the concept of continual lateral, as well as vertical, erosion of the front of emplacing continental crust, but neither is compatible with simple vertical uplift. The vertical uplift in the Himalaya is only the most obvious component, since the more important horizontal component is mostly obliterated by erosion.

Thus, southward motion above the Main Central Thrust and erosion at the High Himalaya and Karakorum behind increased exponentially in the later Tertiary (and less markedly for northward thrusting of the Pamir). The onset of accelerating uplift corresponds with the Early to Mid-Miocene anatectic intrusions of the High Himalaya and Karakorum and with the start of accelerating erosion of the mountainous Himalaya, Karakorum, Pamir and Kun Lun margins of the Pamir and Tibetan plateau margins as determined by river gradients.

RIVER GRADIENTS

Studies of river gradients are an old tradition in the Himalaya (Burrard & Hayden, 1907) and are an important indicator of tectonic activity, though neglected until recently (Seeber & Gornitz, 1983).

The longitudinal profile of a river is sensitive to base-level changes due to ongoing processes of uplift and subsidence. River gradients can be used to infer disequilibrium in river systems which can then be related to tectonics (Seeber & Gornitz, 1983).

A river is referred to as 'graded' when gradient, width and depth of its channel are in equilibrium with discharge and load. Though a river can never be entirely graded due to fluctuations in discharge and load, changes in rock type and climate, and tectonic effects on its course, the longitudinal profile of a river can often approximate to a straight line in a semi-logarithmic plot. Thus, $H = C - K \ln L$ where H is elevation, L is the distance from the source, and C and K are constants. K, the slope of this idealized profile, is called the *gradient index*. K can be evaluated by: $K = H_i - H_j / \ln L_j - \ln L_i$ where i and j refer to two points along the river profile. K can be used to characterize any stretch of the river as well as the entire profile.

In this study the gradient index of various rivers is computed from their confluence to their source following the longest course. Tributaries are calculated separately from source to confluence with a larger river. The upper point 'i' is taken as the highest contour crossed by the stream below its source.

Except near the source, the gradient index of a stream can be approximated by the SL number; $SL = (\Delta H/\Delta L) \cdot L_m$ where $(\Delta H/\Delta L)$ is the slope of the reach and L_m is the distance from the source to the midpoint of the reach.

Doing these analyses allows the determination semi-quantitatively of the deviation of the river (and stretches of it) from an ideal graded profile; and the results can be used, hopefully, to infer the causes of the deviation.

Gornitz and Seeber (1981) have already done such analyses for Hazara in Pakistan and also for the main Himalaya Range (Seeber & Gornitz 1983) following the work of Burrard and Hayden (1907). Here, I have added the rivers of northernmost Pakistan, Afghanistan, Tadhjikstan and Xinjiang. The various rivers and their calculated gradients are shown on Figs 6 to 9.

For the Himalaya, the average topographic profile shows two distinct gradients. A relatively gentle slope across the Lesser Himalaya rises to 3 km at the junction with the High Himalaya. A steeper slope runs across the High Himalaya, levelling off at 5 km and extending across the Indus Suture into Tibet (Burrard & Hayden, 1907; Seeber & Gornitz, 1983). The High Himalaya topographic front is associated with a narrow belt of intermediate magnitude, thrust-associated earthquakes. The trace of the Main Central Thrust follows the topographic front. High river gradients are associated with this zone, suggesting continued major southward thrusting (Fig. 6). On the one hand, some of the steepness may be attributed to lithology since the Himalaya consist of crystalline rock in contrast to the low grade metamorphic and sedimentary rock of the Lesser Himalaya. On the other hand, the river courses are gentle even when they flow over crystal-

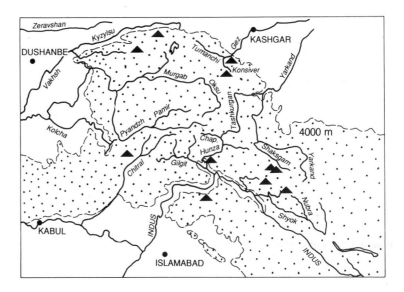

Fig. 6. Major rivers of the northwestern syntaxes.

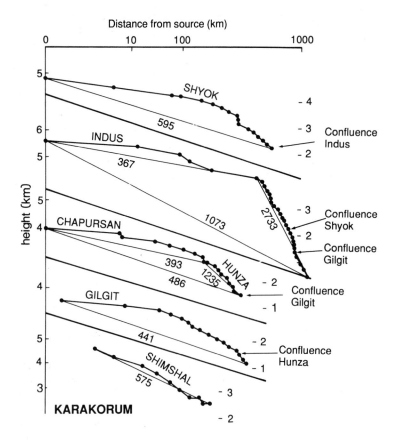

Fig. 7. River gradients in the Karakorum and Himalaya.

line rocks behind the High Himalaya and Karakorum. In any case, over long time periods the rivers would grade anyway.

No knick points are found in the Lesser Himalaya which is thus not a major active tectonic boundary. Similar steep gradients are found on all the rivers cutting across the Karakorum fault and associated strike-slip faults, and these are also seismically active.

The Indus, Tsangpo and Sutlej have steep profiles where they cut the Karakorum and Main Himalayan Range. But they have gentle profiles where they flow parallel to the ranges. Since all three rivers have sources very close together, the Sutlej may have captured the main longitudinal drainage of the Tibetan plateau along the present Indus–Tsangpo upper drainage. Uplift or local ice-cap formation in the Late Quaternary may have caused reversal of flow in both Indus and Brahmaputra rivers so that they now drain past the western and eastern syntaxes. The feeder rivers to the Brahmaputra and Indus tend to indicate flow of the main rivers in an opposite direction to their present flow towards the Sutlej. These tributary streams tend to join the Indus and Tsangpo with acute angles pointing *downstream* — the opposite to the normal situation; though in places they are now modifying their junctions.

In the northwest ranges (Karakorum, Pamir, Kun Lun, etc.), the steepest gradients mark the edge of the Pamir plateau — a Miocene offset of the western Tibetan plateau. It is noteworthy that the *northern* streams draining the North Pamir and Kun Lun have the steepest gradients suggesting that uplift is greater there than on the southern flanks of the ranges. This uplift is related to the underthrusting of the North Pamir by the southern Tien Shan. Most of the highest mountains are associated either with transcurrent faults with components of thrusting (K2, Muztagh Ata, Kongur) or with the northwardly directed thrusting in the Pamir (Pik Lenin, Pik Communism) (Fig. 6).

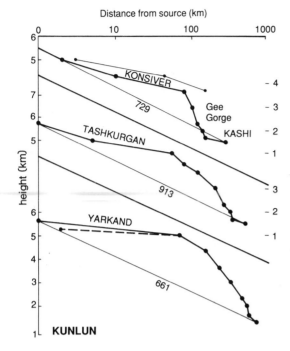

Fig. 8. River gradients in the Kun Lun.

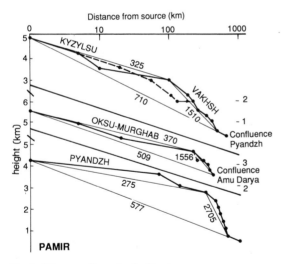

Fig. 9. River gradients in the Pamir.

The river gradient studies confirm the sediment budget studies. It seems that the steep edges of the North Pamir, Karakorum and High Himalaya are erosional in nature. The edges are related to erosion of the fronts of crystalline nappes during emplacement. The position of these fronts may be a delicate

balance between thrust and erosion rates. Erosion and tectonic activity may therefore be coupled, with erosion rates in certain areas determining the rate of tectonic activity (Jamieson & Beaumont, 1988). A consequence is that the area of the Pamir and Tibetan plateaux is shrinking by marginal removal of material. If the Indian Shield continues to underthrust to the north, it is possible that the entire superstructure of the marginal collision magmatic arcs will be removed by erosion, juxtaposing the shelf platform deposits of India, Tibet and Asia.

RADIOMETRIC AND FISSION TRACK DATING

Uplift rates in the Himalaya and Tibet have often been inferred from radiometric and fission track dating. In fact, radiometric and fission track dates give only cooling temperatures, and these can only be related to depth from the contemporary land surface by assuming an average geothermal gradient or by calculating the thermobarometry of the samples dated.

Erosion rates can be calculated from the depth of rock removed between successive closure temperatures of the systems. However, such erosion rates do not necessarily relate to uplift — the land surface could stay almost at sea-level as successive layers are eroded off the land surface. Nevertheless, in general terms there is an increasing logarithmic relationship between erosion and height in mountains (Schumm, 1963). The main problem in the calculation of uplift rates with this method is the choice of a suitable geothermal gradient. A common choice is an average geothermal gradient of 30°C/km. In many cases, e.g. the high-level Karakorum Miocene batholiths, the geothermal gradient assumed must be far too low. This problem may be overcome by thermobarometry of dated rocks or, where this is not available, by comparing igneous data with data from much older metamorphic rocks in the same tectonic block.

The cooling history of a rock can be determined by using various dating methods on different minerals. This follows from the fact that each datable phase within each isotopic system can be assigned a characteristic closure temperature. Thus, by using the $^{40}Ar/^{39}Ar$ method on hornblende and mica, and the fission track method on sphene, zircon and apatite, the thermal history of a rock can be sampled at roughly 500°C, 350–300°C, 250°C, 200°C and 100°C respectively (Zeitler, 1985).

Figure 10 shows radiometric dates and cooling curves for major tectonic units, based on closure temperatures of various minerals. The steepness of the curves is directly related to rates of cooling. By assuming an average geothermal gradient of 30°C/km, these curves by inference can be related to rates of erosion (Fig. 10). Also shown are cooling gradients possibly unrelated to erosion and due to reheating by high-level intrusions.

The following interesting points arise from these plots.

1 That recent rapid cooling (and erosion) is restricted to areas north of the Main Central Thrust and Karakorum Thrust (Fig. 10b,c) (cf. Scharer *et al.*, 1990b). This suggests — as do the facies distributions in Mesozoic–Tertiary deposits — that the

Main Boundary Thrust is a late, minor feature. Most of the post-Oligocene shortening along the northern Indian margin has taken place along the Main Central Thrust and Karakorum Thrust, linked by the strike-slip Karakorum fault zone (Fig. 1). The Pamir started uniform rapid uplift in the Late Oligocene, much earlier than uplift in the Karakorum and High Himalaya. The Karakorum and High Himalaya show slow cooling and erosion starting in the Late Eocene (initial collision), but accelerating markedly after the Miocene intrusions associated with the start of thrusting on the Main Central Thrust and Northern Suture (Mattauer, 1986) (Fig. 10b,c). These intrusions have been related to partial melting of continental crust associated with subduction of Indian continental crust (LeFort, 1986). The $^{40}Ar/^{39}Ar$ ages obtained by

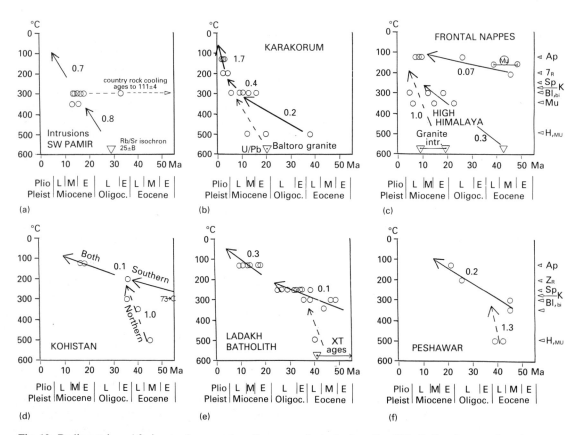

Fig. 10. Radiometric and fission track ages and cooling curves for tectonic units of Fig. 2. Rough rates of erosion (mm/year) are based on a geothermal gradient of 30°C/km. Dotted arrows refer to cooling after intrusion. Solid arrows may be cooling due to uplift and erosion (data sources in Brookfield, 1989). Circles are fission track minerals: upper case abbreviations indicate minerals used for K/Ar; lower case abbreviations indicate Rb/Sr. Crystallization ages are shown at base where available.

Maluski *et al.* (1988) indicate very strong tectonometamorphic and magmatic activity during the Miocene both in the upper part of the High Himalaya slab and further north in the Tethyan cover. Similar events occurred in the Karakorum (Brookfield, 1981; Coward *et al.*, 1986; Brookfield & Reynolds, 1990).

2 In Kohistan and Ladakh cooling (and erosion) appears to have been relatively constant after Late Cretaceous ophiolite emplacement (Fig. 10d,e). Higher apparent erosion rates in northern Kohistan and in places (dotted lines) in Ladakh are related to the rapid cooling of high-level intrusions (Zeitler, 1985). The Ladakh and Kohistan curves can be related to the start of emplacement of the Transhimalaya batholiths of the Asian margin over the northern Indian margin during and after collision. The frontal nappes of the High Himalaya also show slow erosion rates (Fig. 10c top).

3 Some blocks do not fit this simple scheme. The Peshawar block shows peak metamorphism at about 35.5 Ma, followed by major uplifts in the Late Oligocene–Early Miocene (Fig. 10f). This relates to the emplacement of a complex of overlapping nappes (Treloar & Rex, 1990) at an earlier period to their supposed equivalents in the High Himalaya in India. In fact, as noted by Yeats and Lawrence (1984), there seems to be no structural equivalent of the High Himalaya west of western syntaxis.

The conclusion from dating this is that initial uplift of the Ladakh and Kohistan arcs occurred shortly after initial collision and that areas to the south (Peshawar) were slowly squeezed southwards after initial collision, forming progressively southward migrating topographic fronts (Fig. 3).

During the Early Miocene, displacements and crustal shortening shifted to the narrow Main Central Thrust and Karakorum Thrust belts. Areas to the north did not start to undergo major erosion (related to uplift) until the Miocene.

A consequence of this is that east of the syntaxis, the earlier Tertiary structures have been destroyed by southward motion on the Main Central Thrust.

BASINS AND MOUNTAINS

The general conclusions based on neotectonics, sediment budgets, river gradients and radiometric dating must now be compared with actual stratigraphical, sedimentological and biological evidence for the evolution of individual basins and mountain ranges. Tentative interpretations are summarized on Figure 12. Data bases and detailed histories of the basins are presented in Brookfield (in press).

Northern foredeeps

The bulldozing of the Pamir into Central Asia has formed a narrow trough, the Surkhab basin, flanked by larger foredeep basins, the Tadzhik basin on the west and the southwest Tarim basin (Fig. 1).

Surkhab basin

The Surkhab basin is about 300 km long and 30 km wide and contains a very thick 3.5-km Upper Oligocene to Pleistocene section (Krestnikov, 1963). Little information is available for it.

Geodetic measurements show that it is actively subsiding at a rate of around 0.3 mm/year in places: Holocene fault escarpments reach a height of 13 m. South of the Surkhab fault, the land is rising at over 10 mm/year (Finko & Enman, 1971).

Tadzhik basin

The Tadzhik basin is a depression filled with 10–14 km of Mesozoic–Cenozoic deposits. It lies between the Late Palaeozoic Gissar–Alai range on the north and west, and the Pamirs and Hindu Kush on the east and south (Loziyev, 1976) (Fig. 1). The Hindu Kush and Pamirs are actively overthrusting the Tadzhik depression and the basin is closing rather than being displaced to the west (Abers *et al.*, 1988). The depression is thus the site of active crustal shortening by folding and faulting as well as tectonic and sedimentary loading of a subsiding foredeep.

Northwesterly directed folding and thrusting commenced after deposition of Palaeogene deposits. The Mesozoic–Palaeogene deposits are about 4–5 km thick and formed a continental back-arc basin behind magmatic arcs at the edge of the Asian plate (personal observation and unpublished data of 1990). The section is not a subsiding passive margin sequence (Leith, 1985). The last marine deposits in the basin are Oligocene red lagoonal gypsiferous clastic sediments of the Shurysay Formation, with marine bivalves and gastropods (Glikman & Ishchenko, 1967). This forms a sea-level datum for subsidence measurements. At Ariktau, south of Dushanbe, the thin Oligocene is now 2 km *below* sea-level.

The succeeding Neogene non-marine clastics are

between 3 and 7 km thick and mark the deformation of the Tadhzik depression and oblique overthrusting by the West Pamir Range on the east
(Leith, 1985). The history recorded by these clastic
sediments is summarized in detail elsewhere
(Brookfield, in press).

Rapid and exponentially increasing sedimentation commenced in the Early Miocene. Two pulses
occurred in the Early and Middle Miocene, followed by repeated and increasing pulses of clastic
sedimentation, alternating with erosion, in the
Pliocene. Rates commenced at 0.4 mm/year and
culminated in 1.4 mm/year. These sedimentation
rates differ little across the Tadzhik depression,
suggesting that the rates are a general feature of the
basin and not simply due to local fault-block subsidence.

In the Dushanbe area, two major pulses of coarse
conglomeratic sedimentation can be dated palaeomagnetically as late Gilbert (4 Ma) and Olduvai
(2 Ma).

The area is now under erosion with the incising of
the Amudarya river. In the Middle Quaternary,
there was intense valley downcutting, to depths of
1500 to 2000 m, in the mountains and several
hundred metres in the plains (Nikonov, 1970). This
may be related to the onset of glacial conditions and
consequent burial of sources and more variable and
intense discharge from glaciated areas closer to the
present lower reaches of the rivers. In post-glacial
times, retreat of glacial source areas caused sedimentation in the former gorges. The rivers usually
flow over Quaternary sediments and not over bedrock, even in alpine areas. Throughout the enormous Amudarya basin, there are only isolated
reaches on which rivers are incised into bedrock.

Southwest Tarim basin

The southwest Tarim basin lies between a buried
basement high and the Kun Lun Range to the south
(Fig. 1). Mesozoic isostatic subsidence allowed the
accumulation of more than 10 km of sediment in
the basin which was essentially filled by the Cretaceous; after which thin marginal marine and terrestrial red beds were deposited during the Cretaceous
and Palaeogene. As Hsu (1988) notes, it is likely
that at least parts of the Tarim basin are underlain
by oceanic crust. The basin was reactivated during
Neogene times, when very thick continental clastics
were laid down in the marginal troughs (Hsu, 1988).
Over 12 000 m of sediments have accumulated in

the deepest parts of this trough since the Miocene;
6400 m since the start of the Pliocene.

In Neogene times, the foredeep north of the Kun
Lun subsided to an enormous extent. A narrow
trough accumulated up to 3500 m of coarse clastic
sediment. The western Kum Lun has been rising at
equally remarkable rates, mostly since the Middle
Miocene. Marine Eocene rocks are now at 5000 m
(Kaz'min & Faradzhev, 1963). In the Early Miocene, uplift of the Kun Lun caused the formation
of a large lake in the Tarim basin into which coarse
alluvial fans fed from the south. Nevertheless, the
geomorphological contrast was relatively weak. The
lake was very shallow and the marginal fans relatively fine-grained. In the Middle to Late Miocene,
the Kun Lun increased its rate of uplift. A deep lake
was now fed by coarse fan deltas building directly
into deep water (Giu Dongzhou, 1987).

Much of the Neogene tectonic movement took
place along the Karakorum–Tashkurgan fault
zones — dextral strike-slip faults with up to 270 km
lateral displacement (Trifonov, 1978). Block uplifts
along these faults have formed the huge 6 to 7.5-km-
high-Kongur,-Muztagh Ata and Kung Ata Tagh massifs southwest of Kashgar, and K2 and associated
peaks in the Karakorum (personal observation,
1989). Uplift is still taking place. Norin (1932) noted
a post-glacial lake shoreline which had been tilted
considerably to the east in the last few thousand
years. Late Quaternary gravels are progressively
tilted up to 45° angles against the northern edge of
the Kun Lun. The Gez river plunges 2 km down a
rocky gorge after being ponded back into swamps
against the southern boundary fault of the Kun Lun
(personal observation, 1989).

Intermontane basins

Intermontane basins are those formed *within*
mountain ranges by faulting, erosion and *relative*
subsidence. They are formed by a combination of
processes. Most of the intermontane basins of the
Pamir, Karakorum and Himalaya are almost entirely Quaternary in age, are sites of only temporary
storage of sediment, and are mostly now being
eroded (Figs 1 & 11). Erosion during the great Late
Cenozoic mountain uplifts has in most cases removed any pre-Quaternary basins. Only along the
Indus Suture zone has Quaternary backthrusting
partially preserved Miocene basins. Post-Oligocene
sediments are sometimes preserved on the Tibetan
plateau.

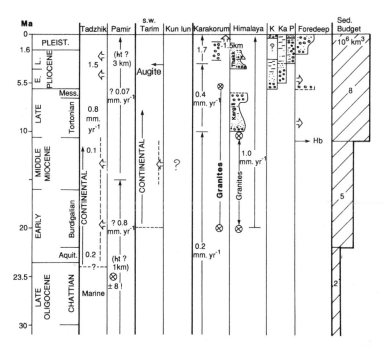

Fig. 11. Summary of Tertiary events for basins and ranges.

A convenient distinction in the mountains is between 'true' intermontane basins entirely within the mountains, and 'piggyback' basins formed by Late Cenozoic thrusting. The latter are now being carried southwards on the frontal nappes of the Himalaya.

'True' intermontane basins

Pamir. Marine deposits persist in the northern Pamirs until the Late Eocene (Moralev *et al.*, 1967). Starting in the Late Oligocene, at which time they were gentle rolling hills, the entire Pamirs started rising (Krestnikov, 1963). In the Central Pamir, Late Oligocene coarse conglomerates form the base of the succeeding volcaniclastic molasse sequence (Leonov & Nikonov, 1988).

This Early Neogene uplift formed the present Pamirs into a region of mainly erosion. Thus, few Neogene deposits are preserved, apart from the Quaternary deposits of recent intermontane valleys and depressions (Fig. 1).

The most intense uplift occurred in the northern and western Pamirs. Here, uplift, deep erosion and consequent isostatic adjustments are responsible for the development of steep-sided valleys, sharp-crested watersheds and the highest mountains. As in the High Himalaya, this is due to continual erosion at the front of active overthrusts. The eastern Pamirs have broader valleys and watersheds and have a greater component of simple vertical uplift.

At the end of the Pliocene and beginning of the Pleistocene, the northern and western Pamirs rose at a slower rate, resulting in greater lateral and less vertical erosion of valleys. Marginal to the range, in the Tadzhik depression, this was a time of valley filling. By the Pleistocene, the Pamirs as a whole had net uplifts of at least 1500–2000 m since the end of the Oligocene. This averages out to a rate of 0.12 mm/year. Belousov (1976) considered that increasingly rapid uplifts of the Middle–Late Pleistocene mark the stage that has determined the present structure and relief of the Pamirs. However, this is mainly based on palaeobotany which likely greatly overestimates the uplift (Molnar & England, 1990).

Another approach is to use gravity data to calculate Bouger and isostatic anomalies for the Pamirs (Artem'yev & Belousov, 1979). If there had been no erosion, the mean elevation of the Pamirs would be about 5.65 km based on the highest records of marine Eocene. About 4 km of this purely tectonic uplift must have occurred since the Early Miocene. This involves a thickening of the crust of about 30 km since the Eocene and more pertinently about 20 km since the Oligocene.

The cause of the latter increase is probably the underthrusting of the entire Pamir by Asian crust since the Early Miocene start of continental collision in the area (Leith, 1985).

The present mean elevation of the Pamirs is 4.25 km compared to the possible 5.65 km. The difference could be taken as the average erosional truncation compensated by isostasy. The 1.4-km height difference requires about 8.4 km of continental material to be removed.

The rather meagre radiometric dating data from the Pamir suggest Neogene intrusions starting in the Late Oligocene/Early Miocene followed by steady cooling until Recent times (Fig. 10).

Displacements on Recent faults in the Pamirs have been estimated from radiocarbon dating of offsets. Rates of displacement for three faults in the last 10 000 years range from 1 to 3 mm/year (Nikonov, 1981).

In general terms the Pamirs closely follow the Neogene histories recorded in the High Himalaya, Tibetan plateau and Karakorum. Progressively increasing uplift is associated with continental subduction and formation of anatectic granite magmas from about 20 Ma onwards (LeFort, 1986).

Karakorum. Studies of sediments are essentially restricted to two areas, the Indus valley at Skardu and the Hunza valley north of Gilgit (Fig. 1). Neither contains any pre-Quaternary deposits. The valley fill sequences at Sust, and elsewhere north of Hunza, seem to fit the Hunza sequence studied by Derbyshire *et al.* (1984) (personal observation, 1989).

In the *Skardu basin,* a thick clastic sequence with glacial deposits has been preserved on the northern side of the present wide valley. This Bunthang sequence has been described in some detail by Cronin and Johnson (1989). Above a basal coarse diamictite, the lower Bunthang sequence consists of a 50 m of micaceous sand passing up into 300 m of silty clays deposited in a glacial lake. The middle Bunthang sequence consists of alluvial fan boulder conglomerates and interbedded sandstones up to 190 m thick which thicken towards the mountain front to the northeast. The upper Bunthang sequence onlaps underlying inclined fanglomerates. It consists of up to 70 m of mixed sand, silt and clay deposited by longitudinal streams and associated lakes. The sequence pre-dates the last ice advance into the Skardu valley, since the deposits were eroded and sculptured by it.

No fossils have been found in the Bunthang

sequence. But the lower and upper sequence are reversely magnetized throughout (most of the middle sequence is too coarse for palaeomagnetic analysis). These thick bodies of reversely magnetized sediment may have been deposited during the Matuyama reversed chron (between 2.48 and 0.73 Ma). Provisionally accepting this date, then the sequence was deposited in the latest Pliocene–Middle Pleistocene (Repenning, 1984). However, in which part of this interval is unknown. Cronin and Johnson (1989) calculated that the entire sequence could have been deposited within 500 000 years.

At any rate, the entire sequence now towers up to 500 m above the present level of the Indus, which was actively aggrading until recently in the Skardu basin. This suggests a history comparable to that of the Pamir (despite the poor dating). Pliocene to Early Pleistocene deposition was followed by Middle to Late Pleistocene uplift and erosion and by Quaternary valley filling.

In the *Hunza valley*, a complex of eight glacial deposits are mostly younger than 150 000 years BP and no older deposits are preserved (Derbyshire *et al.*, 1984). The earliest widespread phase, attributed to the early Pleistocene, left isolated and deeply weathered erratics on summit surfaces above 4150 m. The succeeding phases occur at 2500 m (older than 139 000 years) and lower. These lower elevations reflect valley incision during the Pleistocene and are related to uplift of the Karakorum. If one *assumes* that the earlier glaciers reached around the same height, then early Pleistocene uplift of over 1500 m is indicated.

Himalaya. True intermontane basins have only been preserved in two areas. The first area is where the High Himalaya and its cover backthrust northward over the initial early to Mid-Tertiary foredeep along the Indus suture zone in Ladakh and Tibet. This narrow linear zone was subsequently filled with Late Tertiary clastic sediments from both sides. Details of this zone are only available for the Kargil basin in Ladakh and the area around Xigatse in southern Tibet.

The second area is where Quaternary sediments have been deposited in late-stage grabens crosscutting the Himalaya trend. The only reasonably well-studied example is the Thakkhola Basin in Nepal.

The *Kargil basin* forms the western end of the long elongate intermontane Late Tertiary basin along the Indus Suture zone (Fig. 1). It contains a

Miocene to Pliocene late orogenic clastic sequence (Brookfield & Andrews-Speed, 1984a,b). Only the Kargil section has been studied in any detail, and sufficiently to determine rough ages and environments for the intermontane clastic sediments.

Upper Miocene coarse clastic alluvial fan deposits pass upwards into braided and meandering stream deposits and finally into lake sediments, which are overlain by coarse alluvial fan deposits. Sediments were derived from both sides of the basin, from the southern Zanskar Range and the northern Ladakh Range. The basin was deformed and overthrust from the south by Indus Suture units probably in Pliocene times when terraces related to the present Indus formed (Fort, 1982). This ubiquitous Late Tertiary backthrusting north of the High Himalaya has destroyed much of the Late Tertiary intermontane lake record along the Indus Suture zone (Searle, 1983).

For the middle Upper Miocene section, angiosperm pollen suggest a temperate climate (Bhandari *et al.*, 1977) and the gastropod *Subzebrinus* a high altitude of at least 2000 m (Tewari & Dixit, 1972). A lake developed after this, possibly due to blocking of the ancestral Indus by large landslides — downcutting by the river would undoubtedly be capable of keeping pace with any gradual uplift. The change to semi-arid climates in the topmost units is possibly related to uplift of the High Himalaya to the south and the formation of the present topographic barrier. There is no precise fossil or palaeomagnetic dating of the deposits.

The *Thakkhola basin* lies between 3000 and 4000 m and developed as a graben at right angles to the Himalayan trend, in Late Tertiary times (Fig. 11). It is filled with more than 850 m of sediment and has been studied mainly by Fort and co-workers (Fort *et al.*, 1982a,b).

The Tetang Formation consists of two gravel/sand fluvial deposits with a lake sequence, including carbonates, between. The sediments were derived from the north and east. This formation is probably Pliocene. It contains dominant conifer pollen comparable to the Middle Siwalik floral zones II and III, and Pliocene ostracods. Yoshida *et al.* (1984) reported pollen assemblages now characteristic of 900–1800 m, i.e. over 2000 m lower than the present altitude.

The Thakkhola Formation lies unconformably on the underlying beds and consists of basal fluvial boulder conglomerates overlain by mixed fluvial/lacustrine beds. These sediments were derived from the south and west (Yoshida *et al.*, 1984) and probably reflect the rise of the High Himalaya. Since these units were deposited, the Kali Gandaki gorge has been cut, filled with sediment (Sammargaon Formation), and again recut (Fort *et al.*, 1982a). There is no good fossil dating of the Thakkhola basin sediments.

Cross-correlation of the palaeomagnetic reversals in the combined sections of the Tetang Formation with the standard scale indicates the Gauss to Early Matuyama epochs for the Tetang Formation (i.e. 3.4 to 2 Ma). The Thakkhola Formation was deposited during a mainly normal epoch — probably the Brunhes (less than 0.75 Ma) — and the younger Kali Gandaki gorge sediments in relatively recent times.

Although the dating of events in the Thakkhola basin is still rather imprecise, in general terms it seems to fit the Kashmir chronology. Like the Kashmir basin it probably started its development in the Late Pliocene. It also shows reversal of drainage and uplift in Pleistocene times.

Piggyback basins

These are basins carried passively on the backs of thrust sheets (Ori & Friend, 1984). They usually involve ponding of the former drainage and the formation of temporary lakes which then drained rapidly by overflow and downcutting. The Peshawar, Kashmir and Katmandu basins are examples, all now breached, which were carried southward and upwards during development of the Main Boundary Thrust during the Late Cenozoic (Burbank & Raynolds, 1984) and consequent uplift of the frontal Attock, Pir Panjal and Mahabharat Ranges of the Himalaya.

The present Potwar plateau is an incipient example, which developed in Plio-Pleistocene times as the Salt Range developed (Burbank & Raynolds, 1984).

Katmandu basin. The Katmandu basin, like the Kashmir basin in India, evolved in response to tectonic movements associated with southward thrusting of the Mahabharat Range on the Main Boundary Thrust and with climatic changes due to Quaternary glaciation. It has been described by Dongol (1985, 1987). The maximum exposed thickness of sediment is 280 m and the top of the sequence lies at only 1500 m.

The basal Tarebhir Gravel (60 m) consists of stream-flood and debris-flow breccias of alluvial fans, derived from the south (Mahabharat granite clasts), interfingering near the top with better sorted braided stream deposits with northern source petrographies.

The following three formations are found on three distinct terraces along the Bagmati river. The terraces are tilted slightly northwards indicating post-depositional relative uplift to the south.

Disconformably overlying the basal gravels is a 50-m-thick deltaic–lacustrine sequence of alternating sandstones, siltstones and mudstones with interbedded thin lignites. The lignites contain the only Quaternary vertebrate fauna recorded from the Katmandu basin. The Plio-Pleistocene fauna are elephant, rhinoceros, water buffalo, swamp deer, river hog and bush pig. These suggest environments varying from reed swamp to grass jungles to humid jungle to open woodland and light forest around the lake (Dongol, 1985).

Mudstones transitionally overlie the lignites and consist of 40 m of massive to laminated mudstones with occasional leaf impressions. They represent a more offshore, deeper euxinic lake environment. Northern exposures show more variable, shallower water conditions including rapid deposition from flooding streams. Pollen of dominantly pine, oak and alder changes upwards into dominantly pine, fir and spruce pollen with a sudden decrease in the percentage of oak (Yoshida & Igarashi, 1984). This suggests an initial subtropical moist climate changing to a much cooler and drier one at the top.

The overlying gravels consist of up to 50 m of massive stream-flood alluvial fan gravels interbedded with coarse feldspathic sandstone. The pebbles are all derived from the south indicating rejuvenation of the Mahabharat Range.

A dark, organic-rich clay is confined to the northern part of the basin and is the last and youngest deposit. It formed in a residual lake dammed behind a bedrock ridge.

Thus, the Katmandu basin shows two phases of uplift of the southern Mahabharat Range, due to thrust movements on the Main Boundary Thrust. The first ponded back a large lowland Pliocene lake in which lignites and mudstones accumulated. The upper part of this sequence is entirely normally magnetized and correlates with either the late Gauss (3–2.5 Ma) or Olduva (around 1.8 Ma) chrons (Yoshida & Igarashi, 1984). The mammal fauna suggest the earlier late Gauss interval. At any rate the climate cooled during deposition of these lake deposits.

This lake drained, probably to the east (Dongol, 1987), and was followed by a long period of non-deposition lasting the entire early and middle Pleistocene (Yoshida & Igarashi, 1984). Then, within the Brunhes normal chron (less than 0.74 Ma) rapid deposition of gravels, followed by local tectonic ponding and deposition of organic-rich clay. This could be caused tectonically or climatically.

Gross uplift of the Katmandu basin has been relatively small, certainly not exceeding 1500 m (its present height) since the Pliocene.

Kashmir basin. This basin developed as a lacustrine piggyback basin above the developing Pir Panjal Thrust sheet. Its late Cenozoic history has been summarized by Burbank and Johnson (1983).

The basin developed around 3–4 Ma, based on stratigraphic sequences in the basin and changes in Siwalik sedimentation south of the Pir Panjal. The stratigraphic sequence, well-dated palaeomagnetically, shows initial ponding of a lake, followed by rapid uplift of the northeastern margin around 3.5 Ma. Slower uplifts between 3.5 and 0.4 Ma were followed by rapid uplift of the Pir Panjal and northward tilting of the lake deposits. This last event involves a minimum of 1400–1700 m differential uplift between the southern and northern margins of the basin: the minimum uplift rate is thus about 4 mm/year. The total uplift is probably much greater. From 4 to 1.8 Ma, the conglomerates were shed mostly from the northeastern boundary fault scarps. After 1.8 Ma, the conglomerates were shed from the south and southwestern Pir Panjal margin.

Thus, the Kashmir basin fits the Late Pliocene and Late Quaternary uplift sequence of the Pamir, Karakorum and Kathmandu basins.

Peshawar basin. In the Peshawar basin, sedimentation began around 3 Ma. Before this, movement on the Attock Thrust had caused uplift of the ancestral Attock Range and generated a tectonic depression to the north. This ponded the fluvial systems which had previously drained across the area and localized sedimentation in the newly formed basin (Burbank & Raynolds, 1984). Intermontane basin sedimentation was terminated by accelerated uplift of the Attock Range after about 0.6 Ma. Subsequently, there was catastrophic flooding caused probably by landslide dams in the main Himalayan valleys to the north (Burbank & Tahirkheli, 1985).

The development of the Peshawar and Kashmir basins was not synchronous. The Kashmir basin seems to have formed at least 1 Ma earlier than the Peshawar basin (Burbank & Tahirkheli, 1985). Phases of active faulting on either side of the western syntaxis do not correlate.

Jalalabad basin. The Jalalabad and adjacent Lagman basins were formed by block faulting in the late Cenozoic and filled with continental clastics up to 650 m thick (Raufi & Sickenberg, 1973). The lowest unit consists of alternating conglomerates and sandstones in several cyclical sequences, and contains Upper Pliocene mammals and plants. The mammals are a low diversity fauna of rodent, camel and ox species, suggesting dryish savannah conditions.

These beds are disconformably overlain by conglomerates which thicken northwards and are probably of Pleistocene age. Pebble orientations show southerly and southwesterly current flow from the northern mountains. The entire sequence was then folded prior to the development of the present Kabul river system, with its three terraces. No palaeomagnetic work has yet been done in this basin.

Southern foredeep

The foredeep in front of the rising Himalaya developed progressively from Late Eocene times onwards as India collided with Asia. The early stages involved progressive telescoping of a northern Indian passive margin with sediments being deposited primarily in a marine foredeep in front of the progressively southward, moving Asian margin. Internal deformation of the Indian Shield and the start of uplift of the main Himalaya started in the Lower Miocene as reflected in the sedimentary sequences. The sediments of the Kasauli, Murree and Dharamsala Groups mark the start of continental sedimentation derived from the Himalaya. The change from marine to continental sedimentation, though naturally diachronous in different areas, occurred mostly between the Aquitanian and Burdigalian (Sahni & Chandra, 1980), i.e. around 20 Ma. The overlying Siwalik Group and related continental sediments record the following 20 Ma of uplift and denudation of the adjacent Himalaya (Johnson *et al.*, 1981).

This 20 My datum is a convenient base, since it also marks the start of movement on the Main Central Thrust and intrusion of granite and meta-

morphism in the High Himalaya (LeFort, 1986).

Only a brief outline of the Siwalik sedimentation is necessary here. From Early Miocene times onwards, the Siwaliks generally increase in coarseness and the grade of the source rocks also shows an increase. The variation in heavy mineral frequency distributions suggests increasing intensity of denudation — and hence uplift — with decreasing age (Chaudhri, 1975). The coarse-grained Siwaliks (Upper Pliocene to Pleistocene) have rock fragments of schist and gneiss occasionally exceeding 50% of the sandstone modal compositions. They contain increasing amounts of unstable detrital heavy minerals, including sillimanite. Weathering was thus mainly by mechanical disintegration and distance of transport was short (Chaudhri, 1975). The Upper Siwaliks mark a significant coarsening of the debris from the adjacent Himalaya and stretch the entire length of the range. If not markedly diachronous, they probably mark major uplifts of the Himalaya. However, the onset of Quaternary glaciation may also have been a factor (Molnar & England, 1990).

In the northwestern Himalaya, the topmost Siwalik is marked by the Boulder Conglomerate — a coarse braided stream fanglomerate — which sharply overlies underlying meandering stream deposits. In the Punjab, the base of this is dated palaeomagnetically as the end of the Olduvai event (around 1.7 Ma) and sedimentation becomes coarsely conglomeratic at the Jaramillo event (around 0.97 Ma) (Azzaroli & Napoleone, 1982). In Jammu, two different areas show conglomeratic sedimentation starting at the same Olduvai and Jaramillo events. The earlier Olduvai event at Jammu may be related to southward thrusting of the Pir Panjal (Ranga Rao *et al.*, 1988). This is confirmed by the synchronous change from a northerly to a southerly source for the Kashmir basin conglomerates (Burbank & Johnson, 1983). However, the Pleistocene also marks a sharp climatic cooling and this may have contributed to increased physical denudation in the mountains.

Around 9 Ma, an earlier phase of uplift to the north is recorded in the Potwar plateau and is marked by a doubling in the sedimentation rate from 0.22 mm/year to 0.49 mm/year (Johnson *et al.*, 1982). The Potwar plateau is an incipient piggyback basin which rose on the subhorizontal detachment surface thrusting the Salt Range over the Indian Shield (Burbank & Raynolds, 1984). It contains Miocene to Recent fluvial sediments.

Southward thrusting commenced around 2.1 Ma. Very rapid uplift is shown in the Soan syncline, where the 1.9-Ma-old Lei Conglomerate truncates earlier deposits. Here, at least 3000 m of structural relief was developed within 200 000 years (Burbank & Raynolds, 1984).

CONCLUSION

Uplift and subsidence rates and erosion and deposition rates, based on flora, radiometric and fission track data, sediment budget studies, stratigraphy and sedimentology, show a remarkable correspondence for the Late Cenozoic in Central Asia. The exponentially increasing rates of uplift for the Tibetan and Pamir plateaux and the Himalaya, Karkorum, Pamir and adjacent ranges are real. The exponentially increasing uplift is not an artifact due to climatic change (Mathur, 1984 Molnar & England, 1990). This places constraints on models for collision mountain belts.

The delay between initial India–Asia collision in the Late Eocene and the start of major uplift in the Early Miocene may be related to two factors. The first factor is the progressive telescoping of a thinned northern Indian passive margin back to an original continental thickness, with consequent telescoping and partial destruction of early foredeep basins. The second factor is compressional thickening of the Tibetan plateau crust, marked by simple deformation of all pre-Miocene sequences (Dewey *et al.*, 1989).

The Early Miocene start of exponential uplift of the Himalaya can be related to underthrusting of Indian Shield crust below the Himalaya along the Main Central Thrust and underthrusting of Asian crust below the Pamirs — and possibly Karakorum. However, since Tibet was not deformed, similar uplift of the plateau is more likely due to changes in the upper mantle as a consequence of the start of continental underthrusting. Possible causes are the detachment of the Indian lithosphere and consequent thermal equilibration below Tibet.

But, on a cautionary note, subsidence and uplift are not strictly synchronous among and between basins and ranges. Each basin or range, though generally following the increasing rates of the Late Cenozoic, has its own history (Fig. 12). Thus, the Kashmir basin started its development before the Peshawar basin and the Pamirs started a more gradual uplift before the Karakorum. The surface expression of deep underlying changes varies according to the position and orientation of the surface blocks, the local stresses acting on their edges, and the often unpredictable development of drainage patterns which differentially wear them down. The sudden influx of sediment into Pakistan in the Miocene suggests that the Indus river may have changed its course from northwesterly through the Tashkurgan valley to southerly across the Karakorum and Himalaya. This change would involve switching of major clastic deposition from the early Tadzhik and Surkhab basins into the present Indus basin. Headward erosion of the Sutlej may have beheaded this early Indus and its downcutting may have caused isostatic uplift around the sources of the Indus, Sutlej and Tsang Po. Reversal of drain-

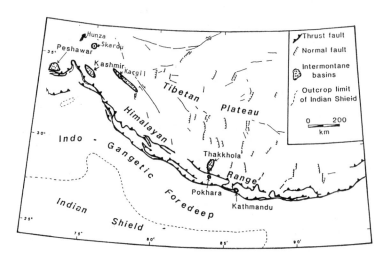

Fig. 12. Location map of Central Himalaya intermontane and piggyback basins. (From Burbank & Johnson, 1983.)

age shown by the Tsang Po indicates that it may represent the original headwaters of the early Indus rising in the eastern syntaxes of the Himalaya.

Detailed sedimentology may soon permit reconstructions of the palaeodrainages of these mighty rivers. Eventually it might even be possible to reconstruct the temporary and ever-changing palaeotopography of mountain ranges and sedimentary basins during the development of the India–Asia collision zone.

ACKNOWLEDGEMENTS

I thank Dr J. Alexander for criticism of the manuscript. Field work in northern Pakistan, India, southern USSR, western China and Tibet was founded by NSERC (Canada) and the National Geographic Society. I would like to thank the villagers, tribesmen and government officials in the various countries for hospitality, accommodation and assistance. I am particularly indebted to: the Tadzhik and Kirghiz shepherds in the North Pamir; the villagers of Hushe, Dansam and Khorkundus in the Karakorum and Honupattan and Omlung in Ladakh; and the monks of Lamayuru, Wanlah, Rumbok, Senli and Samehum in Ladakh and Zanskar.

REFERENCES

ABERS, G., BRYAN, C., ROECKER, S. & McCAFFREY, R. (1988) Thrusting of the Hindu Kush over the southeastern Tadjik basin, Afghanistan: Evidence from two large earthquakes. *Tectonics* 7, 41–56.

ARTEM'YEV, M.YE. & BELOUSOV, T.P. (1979) Isostasy and neotectonics of the Pamirs and southern Tien Shan. *Dokl. Akad. Nauk SSSR* 249, 51–54.

AZZAROLI, A. & NAPOLEONE, G. (1982) Magnetostratigraphic investigation of the Upper Sivaliks near Pinjor, India. *Riv. Ital. Paleont.* 87, 739–762.

BELOUSOV, T.P. (1976) Evolution of vertical movements in the Pamirs in the Pleistocene and Holocene. *Geotectonics* 10, 68–75.

BERGGREN, W.A., KENT, D.V., FLYNN, J.J. & VAN COUVERING, J.A. (1985) Cenozoic geochronology. *Geol. Soc. Am. Bull.* 96, 1407–1418.

BHANDARI, L.L., VENKATACHALAYA, B.S. & PATAP SINGH (1977). Stratigraphy, palynology and palaeontology of Ladakh Molasse Group. *Proc. 4th Colloq. on Indian Micropalaeontology and Stratigraphy 1974–75, Oil and Gas Commission and Institu. Petrol.*, Dehradun, pp. 127–133.

BILLINGTON, S., ISACKS, B.L. & BARAZANGI, M. (1977) Spatial distribution and focal mechanisms of mantle earthquakes in the Hindu Kush–Pamir region: a contorted Benioff zone. *Geology* 5, 699–704.

BROOKFIELD, M.E. (1981) Metamorphic distributions and events in the Ladakh range, Indus suture zone and Karakorum mountains. In: *Metamorphic Tectonites of the Himalaya* (Ed Saklani, P.S.) Today and Tomorrows Publ., New Delhi, pp. 1–14.

BROOKFIELD, M.E. (1989) Miocene to Recent uplifts of the northwestern Himalaya and adjacent areas. In: *Intermontane Basins: Geology and Resources* (Ed. Thanasuthipitak, T. and Ounchanum, P. Univ. Chinag Mai, Thailand, pp. 452–467.

BROOKFIELD, M.E. (*in press*) Miocene to Recent uplift and sedimentation in the NW Himalaya and adjacent areas. In: *Himalayas to the Sea: Geology, Geomorphology and the Quaternary* (Ed. Shroder, Jr., J.F.) Routledge Press, London.

BROOKFIELD, M.E. & ANDREWS-SPEED, C.P. (1984a) Sedimentology, petrography and tectonic significance of the shelf, flysch and molasse clastic deposits across the Indus Suture Zone, Ladakh, NW India. *Sediment. Geol.* 40, 249–286.

BROOKFIELD, M.E. & ANDREWS-SPEED, C.P. (1984b) Sedimentation in a high-altitude intermontane basin — the Wakka Chu Molasse (mid-Tertiary, northwestern India). *Bull. Indian Geol. Assoc.* 17, 175–193.

BROOKFIELD, M.E. & REYNOLDS, P.H. (1981) Late Cretaceous emplacement of the Indus suture ophiolitic melanges and an Eocene–Oligocene magmatic arc on the northern edge of the Indian plate. *Earth Planet. Sci. Letters* 35, 157–162.

BROOKFIELD, M.E. & REYNOLDS, P.H. (1990) Miocene ^{40}Ar/^{39}Ar ages from the Karakorum Batholith and Shyok Mélange, northern Pakistan, indicate late Tertiary uplift and southward displacment. *Tectonophysics* 172, 155–167.

BULANZHE, YU.D., GUSEVA, T.V., PEVNEV, A.K. & ULASHINA, S.A. (1980) Nature of Holocene horizonatal movements in the transition zone from the Pamirs to the Tien Shan. *Dokl. Akad. Nauk SSSR* 254, 23–25.

BURBANK, D.W. & JOHNSON, G.D. (1983) The Late Cenozoic chronologic and stratigraphic development of the Kashmir Intermontane Basin, Northwestern Himalaya. *Palaeogeogr., Palaeoclimatol., Palaeoecol.* 43, 205–235.

BURBANK, D.W. & RAYNOLDS, R.G.H. (1984) Sequential late Cenozoic structural disruption of the northern Himalaya foredeep. *Nature* 311, 114–118.

BURBANK, D.W. & TAHIRKHELI, R.A.K. (1985) The magnetostratigraphy, fission-track dating, and stratigraphic evolution of the Peshawar intermontane basin, northern Pakistan. *Geol. Soc. Am. Bull.* 96, 539–552.

BURRARD, S.G. & HAYDEN, H.H. (1907) *Geography and Geology of Himalayan Mountains and Tibet. Part 3. The rivers of the Himalaya and Tibet.* Govt. India Press, Calcutta, pp. 119–230.

CHANG CHENGFA *ET AL.* (26 AUTHORS) (1986) Preliminary conclusions of the Royal Society and Academia Sinica 1985 geotraverse of Tibet: *Nature* 323, 501–507.

CHAUDHRI, R.S. (1975) Sedimentology and genesis of the Cenozoic sediments of northwestern Himalayas (India). *Geol. Rundsch.* 64, 958–977.

CHI-YEN WANG & YAOLIN SHUI (1982) On the tectonics of the Himalaya and the Tibet plateau. *J. Geophys. Res.* 87, 2949–2957.

COPELAND, P. & HARRISON, T.M. (1990) Episodic rapid uplift

in the Himalaya revealed by $^{40}Ar/^{39}Ar$ analysis of detrital K-feldspar and muscovite, Bengal fan. *Geology* **18**, 354–357.

COWARD, M.P. & BUTLER, R.W.H. (1985) Thrust tectonics and the deep structure of the Pakistan Himalaya. *Geology* **13**, 417–420.

COWARD, M.P. *ET AL.* (8 AUTHORS) (1986) Collision tectonics in the NW Himalaya: In: *Collision Tectonics* (Eds Coward, M.P. & Ries, A.C.) Geol. Soc. London Spec. Publ. **19**, 203–219.

CRONIN, V. & JOHNSON, W.P. (1989) Chronostratigraphy of the late Cenozoic Bunthang sequence and possible mechanisms controlling base-level in Skardu intermontane basin, Karakorum Himalaya, Pakistan. *Geol. Soc. Am. Spec. Pap.*

DERBYSHIRE, E., LI JIJUN, PERROTT, F.A., XU SHUYING & WATERS, R.S. (1984) Quaternary glacial history of the Hunza Valley, Karakorum mountains, Pakistan. In: *International Karakorum Project* (Ed. Miller, K.J.) Cambridge University Press, Cambridge, vol. 2, pp. 456–495.

DEWEY, J.E., CANDE, S. & PITMAN III, W.C. (1989) Tectonic evolution of the India/Eurasia collision zone. *Eclogae Geol. Helv.* **82**, 717–734.

DONGOL, G.M.S. (1985) Geology of the Kathmandu fluviatile lacustrine sediments in the light of new vertebrate fossil occurrences. *J. Nepal Geol. Soc.* **3**, 43–57.

DONGOL, G.M.S. (1987) The stratigraphic significance of vertebrate fossils from the Quaternary deposits of the Kathmandu basin, Nepal. *Newsl. Stratigr.* **18**, 21–29.

ENGLAND, P. & MOLNAR, P. (1990) Right-lateral shear and rotation as the explanation for strike-slip faulting in eastern Tibet. *Nature* **344**, 140–142.

FINKO, YE.A. & ENMAN V.B. (1971) Present surface movements in the Surkhob fault zone. *Geotectonics* **5**, 330–334.

FORT, M. (1982) Apport de la teledetection la connaissance des formations superficielles et des structures dans le bassin de Leh (Valle de l'Indus, Himalaya du Ladakh). *Bull. Soc. Geol. France* **24**, 97–104.

FORT, M., FREYTET, P. & COLCHEN, M. (1982a) Structural and sedimentological evolution of the Thakkhola–Mustang graben (Nepal Himalayas). *Z. Geomorph, N.F., Suppl. Bd.* **42**, 75–98.

FORT, M., FREYTET, P. & COLCHEN, M. (1982b) The structural and sedimentological evolution of the Thakkhola–Mustang graben (Nepal Himalaya) in relation to the uplift of the Himalayan range. *Proc. Symposium on Qinghai-Xizang (Tibet) Plateau.* Science Press, Beijing, pp. 307–313.

GIU DONGZHOU (1987) Sedimentary models of gypsum-bearing clastic rocks and prospects for associated hydrocarbons west of the Tarim basin (China) in Miocene. In: *Evaporite Basins* (Ed. Peryt, T.M.) Springer Verlag, New York, pp. 123–132.

GLIKMAN, L.S. & ISHCHENKO, V.V. (1967) Marine Miocene sediments in Central Asia. *Dokl. Akad. Nauk. SSSR* **177**, 78–81.

GORNITZ, V. & SEEBER, L. (1981) Morphotectonic analysis of the Hazara arc region of the Himalayas, North Pakistan and northwest India. *Tectonophysics* **74**, 263–282.

HSU, K.J. (1988) Relict back-arc basins: principles of recognition and possible new examples from China. In: *New Perspectives in Basin Analysis* (Eds Kleinspehn,

K.L. & Paola, C.) Springer Verlag, New York, pp. 245–263.

JAMIESON, R.A. & BEAUMONT, C. (1988) Orogeny and metamorphism: a model for deformation and pressure–temperature–time paths with application to the central and southern Appalachians. *Tectonics* **7**, 417–445.

JOHNSON, G.D., REY, P.H., ARDREY, R.H., VISSER, C.F., OPDYKE, N.D. & TAHIRKHELI, R.A.K. (1981) Paleoenvironments of the Siwalik Group, Pakistan and India. In: *Hominid sites: their geologic settings* (Eds Rapp, G. Jr. & Vondra, C.F.) *Am. Assoc. Adv. Sci. Selected Symposium* **63**, 197–254.

JOHNSON, N.M., OPDYKE, N.D., JOHNSON, G.D., LINDSAY, E.H. & TAHIRKHELI, R.A.K. (1982) Magnetic polarity stratigraphy and ages of Siwalik Group rocks of the Potwar Plateau, Pakistan. *Palaeogeogr., Palaoeclimatol., Palaeoecol.* **37**, 17–42.

KAZ'MIN, V.G. & FARADZHEV, V.A. (1963) Tectonic development of the Yarkand sector of the Kun Lun Shan. *Internat. Geol. Review* **5**, 180–188.

KLOOTWIJK, C.T., CONAGHAN, P.J. & POWELL, C.McA. (1985) The Himalayan arc: large scale continental subduction, oroclinal bending and back-arc spreading. *Earth Planet. Sci. Lett.* **75**, 167–183.

KRESTNIKOV, V.N (1963) History of the geological development of the Pamirs and adjacent regions of Asia in the Mesozoic–Cenozoic (Upper Cretaceous–Quaternary). *Int. Geol. Review* **5**, 38–62.

KUCHAY, V.K. & TRIFONOV., V.G. (1977) A young left-lateral displacement in the Darvaz-Karakul fault zone. *Geotectonics* **11**, 218–226.

LEFORT, P. (1986) Metamorphism and magmatism during the Himalayan collision: In: *Collision Tectonics* (Eds Coward, M.P. & Ries, A.C.) Geol. Soc. London Spec. Publ. **19**, 159–172.

LEITH, W. (1985) A mid-Mesozoic extension across central Asia. *Nature* **313**, 567–570.

LEONOV, YU. G & NIKONOV, A.A. (1988) Problems of the neotectonic evolution of the Pamir–Tien Shan mountain complex. *Geotectonics* **22**, 178–187.

LOZIYEV, V.P. (1976) Present structure and types of local deformation in the south Tadzhik depression. *Geotectonics* **10**, 291–296.

MALUSKI, H., MATTE, P. & BRUNEL, M. (1988) Argon 39–argon 40 dating of metamorphic and plutonic events in the North and High Himalaya belts (southern Tibet–China). *Tectonics* **7**, 299–326.

MATHUR, Y.K. (1984) Cenozoic palynofossils, vegetation, ecology and climate of the north and northwestern Subhimalayan region, India: In: *The Evolution of the East Asian Environment* (Ed. White, R.O.) Centre of Asian Studies, Hong Kong, **2**, 504–549.

MATTAUER, M. (1986) Intracontinental subduction, crust-mantle décollement and crustal-stacking wedge in the Himalaya and other collision belts. In: *Collision Tectonics* (Ed. Coward, M.P. & Ries, A.C.) Geol. Soc. London Spec. Publ. **19**, 37–50.

MOLNAR, P. (1984) Structure and tectonics of the Himalaya: constraints and implications of geophysical data. *Ann. Rev. Earth Planet, Sci.* **12** 489–518.

MOLNAR, P. & ENGLAND, P. (1990) Late Cenozoic uplift of mountain ranges and global climate change: chicken or egg? *Nature* **346**, 29–34.

MOLNAR, P. & TAPPONNIER, P. (1975) Cenozoic tectonics of Asia: effects of a continental collision: *Science* **189**, 419–426.

MORALEV, V.M. SKOTARENKO, V.V. & FOKINA, N.A. (1967) Eocene paleogeography of the northern Pamirs. *Dokl. Akad. Nauk SSSR* **175**, 100–101.

NICORA, A. GARZANTI, E. & FOIS E. (1987) Evolution of the Tethys Himalaya continental shelf during Maastrichtian to Paleocene (Zanskar, India). *Riv. Ital. Paleont. Strat.* **92**, 439–496.

NIKONOV, A.A. (1970) Evolution of river valleys in the southern part of central Asia in the Anthropogene. *Dokl. Akad. Nauk. SSSR* **195**, 29–31.

NIKONOV, A.A. (1981) Dating of seismotectonic movements and old earthquakes in the mountains of Soviet Central Asia by means of radiocarbon analysis and archeologic data. *Dokl. Akad. Nauk. SSSR* **257**, 62–65.

NORIN, E. (1932) Quaternary climatic changes within the Tarim basin. *Geogr. Rev.* **22**, 591–598.

ORI, G.G. & FRIEND, P.F. (1984) Sedimentary basins formed and carried piggyback on active thrust sheets. *Geology* **12**, 475–478.

PELTZER, G. & TAPPONNIER, P. (1988) Formation and evolution of strike-slip faults, rifts and basins during the India–Asia collision: an experimental approach. *J. Geophys. Res.* **93 B12**, 15085–15117.

PEVNEV, A.K., GUSEVA, T.V., ODINEV, N.N. & SAPRYKIN, G.V. (1975) Regularities of the deformations of the earth's crust at the joint of the Pamirs and Tien-Shan. *Tectonophysics* **29**, 429–438.

POWELL, C. McA. (1986) Continental underplating model for the rise of the Tibetan Plateau. *Earth Planet. Sci. Lett.* **81**, 79–94.

RANGA RAO, A., AGARWAL, R.P., SHARMA, U.N. & BHALLA, M.S. (1988) Magnetic polarity stratigraphy and vertebrate palaeontology of the Upper Siwalik Subgroup of Jammu Hills, India. *J. Indian Geol. Soc.* **31**, 361–385.

RAUFI, F. & SICKENBERG, O. (1973) Zur Geologie und Palaeontologie der Becken von Lagman und Jalalabad. *Geol. Jb.* **B3**, 63–99.

REPENNING, C.A. (1984) Quaternary rodent biochronology and its correlation with climatic and magnetic stratigraphies. In: *Correlation of Quaternary Chronologies* (Ed. Mahaney, W.C.) Geo Books, Norwich, pp. 105–118.

ROYDEN,L.H. & BURCHFIEL, B.C. (1987) Thin-skinned N–S extension within the convergent Himalayan region: gravitational collapse of a Miocene topographic front. *Continental Extension Tectonics* (Eds Coward, M.P., Dewey, J.F. & Hancock, P.L.) Geol. Soc. London Spec. Publ. **28**, 611–619

SAHNI, A. & CHANDRA, M. (1980) Lower Miocene (Aquitanian–Burdigalian) palaeobiogeography of the Indian subcontinent. *Geol. Rundsch.* **69**, 824–848.

SCHARER, U., COPELAND, P., HARRISON, T.M. & SEARLE, M.P. (1990a) Age, cooling history, and origin of post-collisional leucogranites in the Karakorum batholith; a multi-system isotope study. *J. Geol.* **98**, 233–251.

SCHARER, U., TAPPONNIER, P., LACASSIN, R., LELOUP, P.H.,

ZHONG DALAI & JI SHAOCHENG (1990b) Intraplate tectonics in Asia: a precise age for large-scale Miocene movement along the Ailao Shan–Red River shear zone, China. *Earth Planet. Sci. Letters* **97**, 65–77.

SCHMITZ, B. (1987) The TiO_2/Al_2O_3 ratio in the Cenozoic Bengal abyssal fan sediments and its use as a paleostream energy indicator. *Mar. Geol.* **76**, 195–206.

SCHUMM, S.A. (1963) The disparity between present rates of denudation and orogeny. *U.S. Geol. Surv. Prof. Pap.* **454H**, H1–H13.

SEARLE, M.P. (1983) Stratigraphy, structure and evolution of the Tibetan–Tethys zone in Zanskar and the Indus suture zone in the Ladakh Himalaya. *Trans. Roy. Soc. Edin. Earth Sci.* **73**, 205–219.

SEARLE, M.P. (1986) Structural evolution and sequence of thrusting in the High Himalaya, Tibetan–Tethys and Indus suture zones of Zanskar and Ladakh, Western Himalaya. *J. Struct. Geol.* **8**, 923–936.

SEEBER, L. & GORNITZ, V. (1983) River profiles along the Himalayan arcs as indicators of active tectonics. *Tectonophysics* **92**, 335–367.

SUBRAMANIAN, V., SITASAWADI, R., ABBAS, N. & JHA, P.K. (1987) Environmental geology of the Ganga River basin. *J. Geol. Soc. India* **30**, 335–355.

TAPPONNIER, P., PELTZER, G. & ARMIJO, R. (1986) On the mechanics of the collision between India and Asia: In: *Collision Tectonics* (Eds. Coward, M.P. & Ries, A.C.) Geol. Soc. London Spec. Publ. **19**, 115–157.

TEWARI, B.S. & DIXIT, P.C. (1972) A new terrestrial gastropod from freshwater beds of Kargil, Ladakh, J & K State. *Bull. Indian Geol. Assoc.* **4**, 61–67.

THOMPSON, A.B. & ENGLAND, P.C. (1984) Pressure–temperature–time paths of regional metamorphism II: Their inference and interpretation using mineral assemblages in metamorphic rocks. *J. Petrol.* **25**, 929–955.

TRELOAR, P.J. & REX, D.C. (1990) Cooling and uplift histories of the crystalline thrust stack of the Indian Plate internal zones west of Nanga Parbat, Pakistan Himalaya. *Tectonophysics* **180**, 323–349.

TRIFONOV. V.G. (1978) Late Quaternary tectonic movements of western and central Asia. *Geol. Soc. Am. Bull.* **89**, 1059–1072.

WU-LING ZHAO & MORGAN, W.J. (1985) Uplift of the Tibetan plateau. *Tectonics* **4**, 359–369.

YEATS, R.S. & LAWRENCE, R.D. (1984) Tectonics of the Himalayan Thrust Belt in Northern Pakistan. In: *Marine Geology and Oceanography of Arabian Sea and Coastal Pakistan* (Eds Haq, B.U. & Milliman, J.D.) Van Nostrand Reinhold Co., New York, pp. 177–198.

YOSHIDA, M. & IGARASHI, Y. (1984) Neogene to Quaternary lacustrine sediments in the Kathmandu valley, Nepal. *J. Nepal Geol. Soc.* **4**, 73–100.

YOSHIDA, M., IGARASHI, Y., ARITA, K., HAYASHI, D. & SHARMA, T. (1984) Magnetostratigraphic and pollen analytic studies of the Takmar series, Nepal Himalaya. *J. Nepal. Geol. Soc.* **4**, 101–120.

ZEITLER, P. (1985) Cooling history of the NW Himalaya, Pakistan. *Tectonics* **4**, 127–151.

Spec. Publs Int. Ass. Sediment. (1993) **20**, 37–48

Distribution of Plio-Pleistocene and Modern coarse-grained deltas south of the Gulf of Corinth, Greece

M. SEGER *and* J. ALEXANDER

Department of Geology, University of Wales College of Cardiff,
PO Box 914, Cardiff, CF1 3YE, UK

ABSTRACT

Geomorphology is strongly controlled by surface deformation in areas of active tectonism and distinct 'tectonic geomorphologies' develop. There is a close relationship between geomorphology and processes of erosion and sedimentation. Sediment distribution patterns along the southern margin of the Gulf of Corinth are controlled by the inherited pre-rift geomorphology, the evolving tectonic geomorphology, and both Mesozoic basement and basin-fill lithologies (exposed Neogene sediments). Drainage basin characteristics have a particularly important control on sediment yield and the location and nature of coarse-grained deltas. Tectonic extension in the Gulf of Corinth Basin has been concentrated along WNW–ESE trending normal faults which define prominent topographical boundaries between the Hellenide Massif of the North Peloponnese, uplifted Neogene sediments and the seismically active Gulf of Corinth. A series of offlapping, Plio-Pleistocene 'Gilbert-type' deltas step down to the southern shore of the Gulf. Topsets are located at altitudes up to 1700 m and foresets are up to 700 m thick. Successive delta abandonment is attributed to uplift of the North Peloponnese and northward migration of faulting activity. Modern deltas are restricted to the west part of the area, whereas older deltas occur throughout the exposed Neogene basin fill. Differential movement between fault blocks coupled with regional uplift caused reversal of drainage across fault blocks, channel incision, development of new drainage basins on the emergent Neogene sediments and sediment starvation in the east of the basin.

INTRODUCTION

Sedimentation and erosion occur in response to local geomorphic features which, in areas of crustal extension, are controlled by the prevailing tectonism. Alexander and Leeder (1987), Leeder and Alexander (1987), Leeder and Gawthorpe (1987) and others have suggested that the distribution of sedimentary facies in extensional basins generally depends upon the effects of the prevailing tectonism through surface tilting, uplift and subsidence. Frostick and Reid (1989, 1990) described how the development of the East African Rift placed a structural control on basin geomorphology which affected sedimentation. Similar observations on the tectonic control of geomorphology and tectonic–geomorphological control of sedimentation have been presented for other basins (Hooke, 1972; Alexander & Leeder, 1990; and others). Ori (1989) discussed the relationship between

the sediment distribution and faulting in the Gulf of Corinth Basin. This paper expands on Ori's (1989) ideas and discusses the interaction of tectonic activity and drainage to show how they act together to control the distribution of coarse sediment in the Gulf of Corinth Basin.

The Gulf of Corinth is a prominent 30-km-wide and 120-km-long geomorphic feature of central Greece, separating the North Peloponnese from mainland Greece (Fig. 1). The Gulf of Corinth Basin has an areal extent of approximately 4000 km² of which the present Gulf occupies approximately 2300 km² (Fig. 2). Thick deposits of Pliocene to Recent sediments crop out over an area of approximately 1700 km² along the southern margin of the Gulf of Corinth. They form a narrow coastal strip 5–10 km wide between Patras and Xilokastro in-

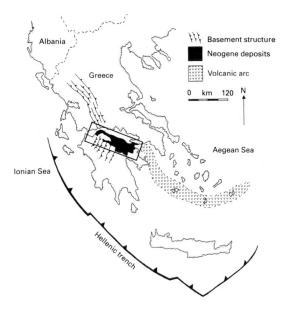

Fig. 1. Map of Greece to show the location of the study area and tectonic regions. The area in the box correspond to that illustrated in Fig. 4.

Pleistocene conglomerate bodies and their sedimentological characteristics have been investigated. Modern geomorphic features have also been mapped and analysed with particular attention given to drainage basin area, geological substrate, altitude and proximity to normal faults, in an attempt to explain the conglomerate distribution.

TECTONIC HISTORY AND BASIN STRUCTURE

The Gulf of Corinth region forms a tectonic boundary between the extending terrains of northern and central Greece and the uplifting Peloponnese landmass. The region has been undergoing crustal extension since the Late Miocene (Kelletat et al., 1976). Since the Late Miocene, tectonic deformation has been concentrated along a series of WNW–ESE trending normal faults which mostly downthrow to the north (Fig. 2). There are two fault sets, both comprising a series of 10–20-km-long fault segments which step irregularly from northwest to southeast. They cut the pre-extension, NNW–SSE, structural strike of the Hellenide Massif (Pe-Piper & Piper, 1984) (Figs 1 & 2). A presently seismically quiescent set of faults delineates the basin from the Hellenide Massif to the south. The more northerly set of faults is seismically active and controls the bathymetry of the Gulf of Corinth (Higgs, 1988). Synthetic and antithetic faults accommodate deformation in the Gulf (Brooks & Ferentinos, 1984; Ferentinos et al., 1988; Higgs, 1988). The Gulf of Corinth is presently undergoing NNE–SSW exten-

creasing to a 30-km-broad coastal plain between Kiato and Corinth (Fig. 2). Thus it seems that the Gulf of Corinth was of greater areal extent in the Pliocene than at present (Ori, 1989).

The basin-fill sediments are dominated by marls and conglomerates. The conglomerates represent about 35% of the exposed sediments and are mostly restricted to the southwest parts of the basin. In this study, the distribution of major Pliocene–

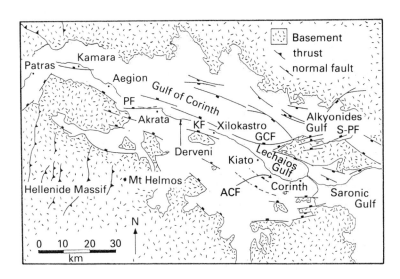

Fig. 2. Location map showing the area of the Gulf of Corinth Basin including the Gulf of Corinth and the Neogene outcrop. The major fault strands are: PF, Plantanos fault strand; KF, Kamara fault strand; ACF, Acro Corinth fault strand; GCF, Gulf of Corinth fault strand; SPF, Shinos-Psatha fault strand.

sion (Schroder & Kelletat, 1976; McKenzie, 1978; Jackson *et al.*, 1982; Vita-Finzi & King, 1985; Roberts & Jackson, 1991). The extension has been attributed to back-arc spreading behind the Hellenic Trench (Le Pichon & Angelier, 1979; Le Pichon, 1982) and consequent block-faulting and rotation associated with the south-southwest movement of central Greece relative to the Aegean Sea and the Hellenic Trench (Jackson & McKenzie, 1988; Taymaz *et al.*, 1991) (Fig. 1).

The seismically quiescent basin-margin faults are separated from the active faults by a 5–30-km-wide strip of Neogene sediments and Mesozoic basement (Fig. 2). Within this area, fault scarps define the northern side of a series of prominent basement inliers. The exposed fault scarps, for example, the Plantanos fault scarp near Aegion, the Kamara fault scarp near Xilokastro and Acro Corinth (Fig. 2), are multifaceted, corrugated and slicken-sided (Stewart & Hancock, 1988).

The simple model of the evolution of the Gulf of Corinth Basin presented in Fig. 3 was developed from consideration of published structural and tectonic interpretations (including: Schroder & Kelletat, 1976; McKenzie, 1978; Le Pichon & Angelier, 1979; Jackson *et al.*, 1982; Le Pichon, 1982; Vita-Finzi & King, 1985; Jackson & McKenzie, 1987, 1988; Roberts & Jackson, 1991; Taymaz *et al.*, 1991). The simple model illustrates how differential fault movement changed the topography and it is used below to explain the evolution of

drainage and facies distributions. An early phase of basin development was associated with displacement on both fault sets giving rise to a basin approximately twice the size of the present gulf. The onset of regional uplift in the North Peloponnese resulted in the confinement of fault activity to the north, and basin narrowing occurred. Subsequent differential movement between the fault sets resulted in uplift and exposure of Neogene sediments in a tilt block between the southern shore of the Gulf of Corinth and the Hellenide Massif. The hanging walls of the northern faults continued to subside creating the Gulf of Corinth and uplift continued in the south through the Quaternary. The nature of the uplift cannot be explained wholly by the mechanism of footwall uplift. White *et al.* (1987) suggest that the uplift is an isostatic response to lower crustal accretion of igneous material and note that the Gulf of Corinth Basin is at the western tip of the Aegean arc (Fig. 1). Collier (1990), Roberts and Jackson (1991) and Leeder *et al.* (1991) suggest that an element of the uplift may be due to underplating of the North Peloponnese area by sediment from the Hellenic Trench to the south.

BASIN GEOMORPHOLOGY

The basin area (Fig. 2) can be divided into three distinct physiographical provinces: the Gulf of Corinth, the strip of uplifted and exposed Neogene

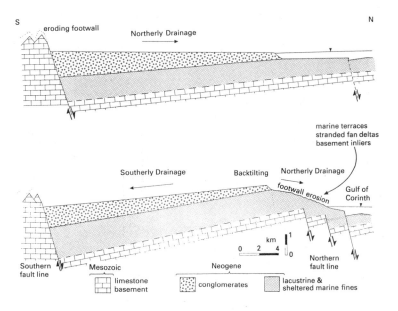

Fig. 3. Simple model of the evolution of the southern part of the Gulf of Corinth Basin.

sediments and the Hellenide Massif. The Gulf of Corinth is a physiographical trough whose bathymetry (maximum depth of 840 m) and coastal morphologies reflect the structural asymmetry of the controlling half-graben. The northern margin of the Gulf is characterized by gentle gradients (10°–20°) whereas most of the southern margin is steep (30°–40°). The Alkyonides Gulf which forms the eastern arm to the Gulf of Corinth has a similar morphology and is bounded to the south by the Shinos–Psatha Fault (Fig. 2). However, the Lechaios Gulf is a relatively shallow (c. 150 m) shelf located on the footwall of the Gulf of Corinth Fault at its eastern tip. The characteristics of the Gulf of Corinth are described by Heezen et al. (1966), Brooks and Ferentinos (1984) and Ferentinos et al. (1988).

The Hellenide Massif forms a prominent topographical high to the south of the Gulf and is the primary sediment source area for the Gulf of Corinth Basin. The Hellenide Massif comprises a series of NNW–SSE trending Alpine thrust nappes (Fig. 2) which control the trend of the component mountain ranges and wide valleys. The northern front of the Massif corresponds to the strike of the southern fault strands running some 150 km from the Gulf of Patras to the Saronic Gulf (Fig. 2). The relief and topographical expression of the Massif decreases from an average altitude of 1200 m (maximum of 2400 m at Mt Helmos) in the northwest to an average altitude of 500 m (maximum of 1000 m) in the southeast.

The strip of exposed Neogene sediments between the Hellenide Massif and the Gulf of Corinth is locally in excess of 2 km thick and is dominated by marls and conglomerates. The marls have a variable dip with a recorded average of 8° towards the southern basin margin. The area is characterized by rolling hills whose altitude rarely exceeds 900 m but locally reach 1700 m. West of Akrata the Neogene outcrop is restricted to a narrow, 5–10-km-wide coastal strip of conglomerates on the footwall of the Gulf of Corinth Fault (Fig. 2). In the east a 30-km-wide coastal plain is characterized by a series of 20 marine terraces giving the emergent coastline a 'staircase morphology' (Keraudren & Sorel, 1987).

MODERN AND PLIO-PLEISTOCENE COARSE-GRAINED DELTAS

Sedimentation over most of the floor of the Gulf of Corinth is dominated by fine-grained turbidities and sediment settling out from suspension in the water column (Brooks & Ferentinos, 1984). Coarse-grained sedimentation is limited to a fringe of deltas lining the coastline between Kamara and Xilokastro (Figs 2 & 4). These deltas are fed by rivers which drain into the Gulf from the North Peloponnese (Fig. 4). In the west between Kamara and Akrata the deltas are well developed forming a 5-km-wide coastal plain of coalescing or isolated deltas (Fig. 4). East of the River Krathis delta the Modern coastal plain narrows to 1–2 km with only small, poorly developed deltas present. No major Modern deltas exist southeast of Xilokastro (Fig. 4). The largest individual delta is the Phoenix delta with a land area of 7.5 km^2, while the smallest is the Krios delta with a land area of 1 km^2 (Fig. 4). The Megantis, Selinos, Kerintis and Vouraikos river deltas have coalesced to cover an area in excess of 40 km^2. The vertical distance from delta top to the floor of the Gulf is 400–500 m. The deltas prograde basinward over the fine-grained marine sediments, with coarse sediments reaching the fan toes in slumps, debris flows and turbidity currents (Ferentinos et al., 1988). Seismic sections across the Gulf show well-bedded facies (turbidite sequences) running into chaotic facies (debris flows and slumps) close to the fault on the southern side of the Gulf (Brooks & Ferentinos, 1984).

A series of large Plio-Pleistocene Gilbert-type delta deposits are located above the Gulf of Corinth between Patras and Kiato (Ori, 1989) (Fig. 4). These deposits form an extensive, coarse-grained, clastic apron and rest unconformably upon southward-tilted marls with varying angular discordance. Locally, as at Trikala, conglomerates rest unconformably on basement lithologies. Delta topsets are found at altitudes ranging from 3 to 1700 m, giving a 'staircase' morphology with successively younger deltas offlapping and stepping down towards the Modern coastal plain (Fig. 5).

The Evrostini delta (named eponymously) has an area of 6.8 km^2 (Fig. 6), Mt Mauron delta has an area of 4 km^2 and to the east, coalesced deltas cover an area of 35 km^2 (Fig. 4). They consist of weakly consolidated conglomerates which grade distally into sandy turbidites and marls (Ori, 1989). Foresets range from the c. 500-m-thick Evrostini delta to the 20–40-m-thick coalesced Mega Valtos, Souli and Krioneri deltas (Fig 4). The foreset dip orientations, smaller-scale palaeocurrent indicators and facies patterns demonstrate that sediment dispersal

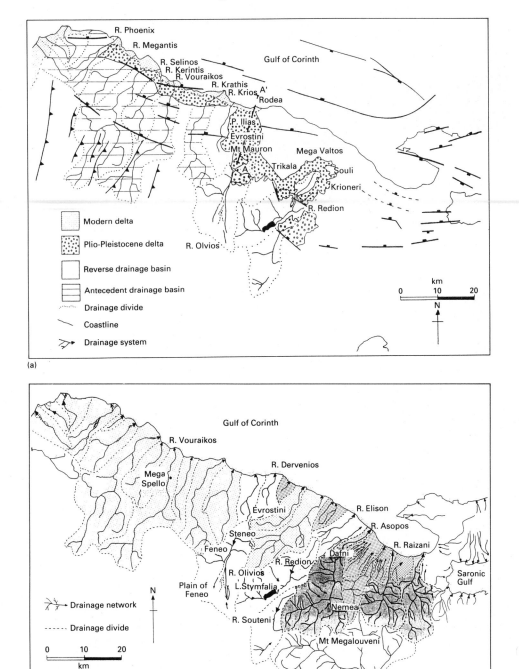

Fig. 4. Drainage patterns. (a) Map comparing the distribution of the major Plio-Pleistocene to Recent deltas and types of drainage basin. (b) Map to show major drainage courses, drainage basins and types of drainage basin.

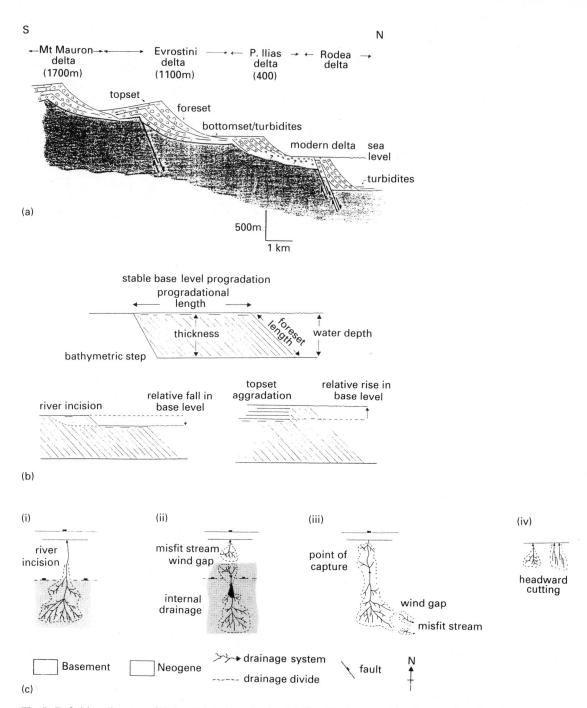

Fig. 5. Definition diagrams of deltas and drainage basins. (a) Simplified cross-section through the offlapping and stepping down Mt Mauron, Evrostini and Prophet Ilias deltas (modified from Ori, 1989). (b) Delta structure and its relation to changing base level. (c) Definition of the major drainage types identified in the North Peloponnese. (i) Antecedent (original) drainage. Drainage has maintained course across later tectonic tiltblock. Soft Neogene substrate aids river incision. (ii) Reverse drainage. Drainage abandons original course due to tectonic backtilting. New course is opposite to original. Hard basement substrate prevents river maintaining original course. (iii) Capture drainage. Reverse drainage has been captured back to original course by misfit stream. Soft Neogene substrate aids capture. (iv) Juvenile drainage. Drainage developing on emergent basin deposits.

Fig. 6. Photograph of Evrostini delta to illustrate the massive nature of the stranded Plio-Pleistocene fan deltas. Topsets crop out at 1109 m above present sea-level. Foresets are in excess of 500 m thick.

was generally from south to north with delta progradation towards northeast and east.

The location of the deltas in fault transfer zones (Ori, 1989) and in areas adjacent to major faults suggests a strong structural control on their formation, as has been recorded in other basins (Leeder & Gawthorpe, 1987; Colella, 1988; Gawthorpe & Collela, 1990). The size of the deltas decreases from the northwest to the southeast. The larger Prophet Ilias, Mt Mauron and Evrostini deltas (with foresets 400–700 m thick) are located in transfer zones between major onshore faults (Fig. 4). The growing tectonic topography provided the bathymetric differentiation required for Gilbert-type delta formation. There are no major faults onshore in the eastern part of the area and consequently topographical features were less pronounced and the deltas were smaller, between 20 and 40 m thick.

The presence of stacked, offlapping delta sequences suggests that base-level variation also had a strong control on delta formation. Rivers which fed the now subaerially exposed deltas, incised down as base level fell and fed younger, lower deltas (Fig. 5). A sequence of offlapping deltas has been identified, for example, south of Derveni (Fig. 2) where the Mt Mauron delta (topset altitude 1700 m) was replaced seaward by the Evrostini delta (altitude 1200 m) which was in turn replaced by the Prophet Ilias delta (altitude 200 m) (Fig. 4). The River Dervenios has incised into the Prophet Ilias delta and is now forming a delta at Rodea. Following Keraudren and Sorel (1987), each preserved delta topset represents a glacio-eustatic highstand and was preserved as a distinct feature because of uplift before the subsequent highstand.

The Modern and Plio-Pleistocene deltas are of similar magnitude, foresets are of similar length and the facies associations and facies geometries of the uplifted Plio-Pleistocene deltas are the same as those of the Modern deltas. Gilbert-type delta thickness is controlled by the depth of water into which the deltas prograde (Fig. 5). The foreset thicknesses of the Plio-Pleistocene deltas (for example the 700-m-thick Evrostini delta foresets; Fig. 6) indicate that at times the Plio-Pleistocene gulf had a similar bathymetry to the present Gulf of Corinth.

The Modern deltas do not have the same distribution as the Plio-Pleistocene deltas (Fig. 4a). In the west, Plio-Pleistocene deltas crop out above the Modern coastal plain of coalesced deltas; however, east of Akrata (Fig. 2) there are no major Modern deltas although there are stranded Plio-Pleistocene deltas (the Evrostini, Mt Mauron and Mega Valtos deltas for example). The Gulf of Corinth Fault controls the offshore bathymetry and has created a steep slope which would encourage Gilbert-type deltas to develop if coarse sediment was being supplied to the shore in this area; however, no Modern deltas are present. Therefore, the distribution of Modern deltas is not only attributed to a simple tectonic control but is also related to sites of sediment supply from rivers which drain the North Peloponnese.

DRAINAGE PATTERNS

Drainage into the Gulf of Corinth is predominantly from rivers which originate in the North Peloponnese. Areas to the north and east of the Gulf are characterized by rivers which flow into the Aegean or Ionian Seas (Fig. 1). Twenty-four major drainage basins can be identified in the North Peloponnese (Fig. 4b) and these can be classified into four types: (1) antecedent drainage basins, (2) reverse drainage basins, (3) capture drainage basins and (4) juvenile drainage basins (Fig. 5). Of the 24 major drainage basins, 12 are antecedent drainage basins, four are reverse drainage basins, two are capture drainage basins and six are large juvenile drainage basins. Of the four which are reverse drainage basins, three drain internally and one has been redirected into the Gulf of Argos. Antecedent drainage dominates the Gulf between Patras and Akrata. East of Akrata, towards Xilokastro, reverse and juvenile drainage basins dominate. Juvenile and capture drainage basins dominate between Xilokastro and Corinth (Fig. 4b). The distribution of different types of drainage basin reflects the effects of basin tectonism, substrate and relief.

1 *Antecedent drainage* occurs where a river has maintained its original direction of flow across later tectonic topography. River erosion and deposition is able to keep pace with the rate of uplift or subsidence so that the river maintains its course. The river incises through uplifted topographical features, such as fault footwalls, and erosion proceeds upstream by knick-point retreat. In the North Peloponnese, antecedent rivers flow through deep gorges in their lower reaches but have mature pre-uplift forms in their upper reaches above headward-eroding knick points. Such drainage basins are common in the western part of the North Peloponnese between Patras and Akrata (Fig. 4b). For example, the River Vouraikos has a mature upper drainage system with wide flood plains and sinuous rivers. It passes through knick points close to Mega Speilio Monastery, from whence the river flows through steep-sided gorges 800 m deep, to a narrow coastal plain where the river feeds a Modern delta. River incision is partly in response to the regional uplift in the North Peloponnese.

2 *Reverse drainage* develops when the flow direction along part of a river's course is reversed; in the study area this is caused by tectonic deformation of the river bed. Reverse drainage basin types consist of two opposing drainage elements: a reverse drainage component and a misfit drainage component. Water and sediment discharge in the original flow direction is greatly reduced by the reduction in catchment size and the resultant stream is termed a 'misfit' stream as its size is too small to explain the scale of the valley or alluvial plain through which it flows. The misfit stream incises and undergoes headward erosion in response to continued surface deformation. This misfit stream occupies a narrow, elongate basin exhibiting an immature dendritic drainage pattern with few tributaries. The point of gradient inflexion resulting from tectonic deforma-

tion between the two drainage elements is a dry valley and is termed a 'wind gap'. The reverse drainage element either becomes redirected to an adjacent drainage basin or drains internally due to topographical confinement. The reverse drainage basins tend to display dendritic drainage patterns (Fig. 5).

Four reverse drainage basins have developed in the North Peloponnese between Akrata and Kiato in response to the uplift and backtilting of the southern margin of the Gulf of Corinth Basin. A major reverse and misfit drainage basin is located south of the town of Derveni (Fig. 2). The misfit drainage basin of the River Dervenios has its source at an altitude of 1200 m above the village of Evrostini (Fig. 4b). The reverse drainage basin is drained to the south by the River Olvios which has cut down into Neogene sediments and locally into basement at Steneou. The River Olvios flows into the Plain of Feneo which is an internal drainage basin from which outflow is prevented by the surrounding topography of the Hellenide Massif. The reversal of drainage, the reworking of Neogene sediment and its transport to the south is responsible for infilling topographical lows and results in the flat nature of the Plain of Feneo. The two drainage elements are separated by a wind gap which is approximately 4 km long, 2 km wide and 200 m deep, with an altitude of 1400 m. This dry valley cuts through the topset of the Mt Mauron delta.

Lake Stymfalia and the River Redion, 20 km to the east of the Plain of Feneo, form another reverse drainage basin, with the River Elison forming the misfit component (Fig. 4b). A third reverse drainage area is drained by the River Souteni which is regarded as a stranded tributary of Lake Stymfalia.

3 *Capture drainage* in the Corinth area occurs when reverse drainage elements return to their original flow directions. Headward erosion by misfit streams through wind gaps extends into the area of reverse drainage and 'captures' the streams. Where a river has been redirected a series of wind gaps and misfit streams flowing in the reverse direction will be produced.

Capture drainage basins are recognized in the east of the Gulf of Corinth Basin where most of the drainage basins are developed on Neogene substrates. The Neogene deposits are easier to erode than basement lithologies and misfit streams have a faster rate of headward erosion. Consequently, river capture is more likely to occur in the area of Neogene outcrop than on basement lithologies. This contrast explains the west to east change from reverse drainage basins to capture drainage basins between Akrata and Kiato (Fig. 5) which mirrors the transfer of basin substrate from basement to Neogene lithologies (Fig. 2). The reverse drainage basins of Lake Stymfalia and Feneo have basement substrates whereas the capture drainage basins of the Raizani and Asopos rivers lie on Neogene deposits (Fig. 4b). The River Asopos has a well-developed dendritic drainage pattern; the upper reach drains the Plain of Nemea and the river debouches into the Gulf east of Kiato. The Plain of Nemea has a morphology and size comparable to the Plain of Feneos and is considered to have once been an area of reverse drainage; basin drainage was redirected south. This theory is supported by the presence of two wind gaps to the south of the town of Nemea which cut through Mt Megalovouni. The gradient inflexion point was located close to the village of Dafni, where the river begins to flow through steep-sided valleys with narrow flood plains until it reaches the coastal plain (Fig. 4b).

4 *Juvenile drainage* basins along the southern margin of the Gulf of Corinth consist of small incising and headward-eroding streams. These immature drainage systems have developed along the entire coastal plain. They are most common on the eastern coastal plain between Kiato and Ancient Corinth and form linear drainage features incising through uplifted marine terraces and basin marls (Fig. 4b).

The drainage can be split into two distinct regions reflecting the topographical expression which results from changing substrate lithology. In the area of high relief, between Patras and Xilokastro, drainage basins are elongate, exhibiting trellis drainage patterns (Fig. 4). The rivers have mature upper reaches resting on basement lithologies and immature lower reaches which rest on exposed Neogene deposits. The patterns of the upper reaches are interpreted as relics of pre-extension drainage and reflect the NNW–SSE structural grain of the Hellenide Massif. In the east, between Kiato and Corinth, the topographical expression of the Hellenic Massif is greatly reduced; drainage basins are of similar size to those to the west but exhibit dendritic drainage patterns (Fig. 4). Whole drainage basins have developed on the uplifted Neogene sediments. Their S–N orientation is similar to the structural grain of the Hellenide Massif and it is postulated that rivers originating in the Hellenide Massif superimposed their drainage orientation onto the emergent basin sediments.

These drainage basins, therefore, record the tectonic modification of fluvial geomorphology by uplift and tectonic tilting and also record the influence of drainage basin substrate.

FLUVIAL GEOMORPHOLOGY AND THE DISTRIBUTION OF CONGLOMERATE BODIES

Modern deltas are located at the mouths of major antecedent drainage basins (Fig. 4a). Plio-Pleistocene deltas are located throughout the Neogene basin. In the west, between Patras and Akrata, Plio-Pleistocene deltas are incised by the drainage systems which feed the modern deltas and, together with them, form an offlapping succession of deltas which step down to the shore of the Gulf of Corinth. To the east of Akrata, major Plio-Pleistocene deltas are located at the heads of reverse drainage basins (Fig. 4a). Mt Mauron, Evrostini and Ilias deltas mark the head of the River Olvios reverse drainage basin. Further to the east the coalesced deltas of Souli, Krioneti and Mega Valtos mark the head of the Lake Stymfalia reverse drainage basin. It is postulated that the rivers in these basins once flowed north providing sediment to prograding Gilbert-type deltas. Northward-directed palaeocurrents have been reported in Olvios and Redion river terraces (Roberts, 1988; Dufaure, 1975). Uplift and backtilting are believed to be responsible for encouraging drainage reversal leading to delta abandonment. With no major rivers draining into the Gulf the formation of deltas along the coast line is severely restricted as is demonstrated by the present delta distribution. Their absence in areas of capture drainage is partly explained by the drainage basin substrate lithology, as the marls and turbidities, when eroded, provide little coarse-grained material. A more important control on the absence of Modern deltas along part of the southern shore of the Gulf of Corinth is the area of the drainage basin (or part of it) which is sourced in the Hellenide Massif.

EVOLUTION OF BASIN GEOMORPHOLOGY, DRAINAGE PATTERNS AND COARSE-GRAINED DELTA DISTRIBUTION

The evolution of the drainage in the North Peloponnese is represented by a series of block diagrams in Fig. 7. The pre-extension topography had a profound effect on the development of drainage. The

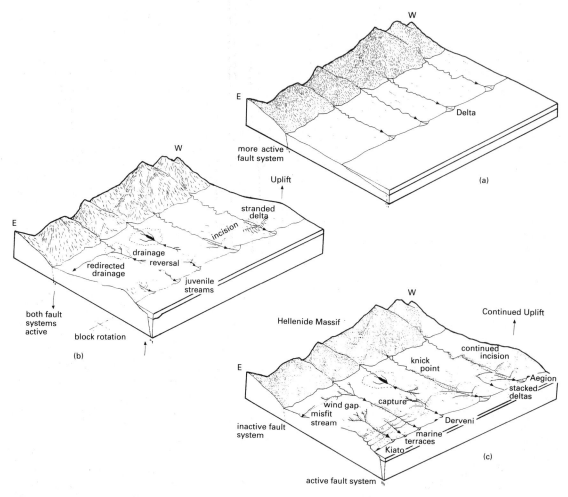

Fig. 7. Schematic block diagrams illustrating the tectono-geomorphic evolution of the southern margin of the Gulf of Corinth. (a) Plio-Pleistocene, (b) Pleistocene, (c) present day.

NNW—SSE structural strike of the Hellenide Massif determined the orientation of the early drainage basins. With the onset of extensional faulting, sites of river input to the young extensional basin were inherited from the Hellenide Massif. The distribution of Plio-Pleistocene deltas corresponds to the drainage basins originating in the Massif (Fig. 7a).

Changes in the configuration of surface deformation associated with the onset of regional uplift, and basin narrowing, caused the forced regression of the southern shorelines of the Gulf of Corinth. This led to the superimposition of south–north drainage upon the coastal fringe of Neogene deposits as it became exposed rapidly. Subaerial exposure of the Neogene basin sediments allowed the development

of new juvenile drainage systems. Backtilting of the uplifted basin fault blocks led to the reversal of drainage in the eastern half of the basin. In the west, the well-developed drainage basins of the Hellenide Massif were able to incise into their own deposits as they were uplifted, allowing the rivers to maintain their course towards the north. The developing antecedent drainage was aided by the high relief and more established drainage basins of the Hellenide Massif. The decrease in topographical relief of the Massif, absence of inherited drainage basins to the east, and the development of younger drainage basins on the uplifted Neogene sediments resulted in rivers which were unable to downcut fast enough to keep pace with surface deformation. This resulted in reversal and redirection of drainage. Drainage reversals led to the abandonment of Plio-Pleistocene deltas east of Akrata. Along the uplifting coastal plain juvenile streams cut headward through the series of marine terraces (Fig. 7b).

With continued uplift, antecedent rivers continued to cut down, causing the formation of deeply incised gorges and cliffs. These rivers now supply sediment to the Modern apron of deltas. Between Akrata and Xilokastro, reverse drainage basins are located on hard Mesozoic basement lithologies which are resistant to erosion and, as a result, river incision is unable to keep pace with uplift and capture of the reverse drainage cannot take place. East of Kiato, juvenile and misfit streams rapidly incised into the uplifted Neogene sediments allowing the capture of reverse and redirected drainage systems. Older juvenile systems have begun to develop more mature dendritic drainage patterns and a younger series of juvenile systems is developing on the emergent coastal plain. Capture of drainage has left a series of dry valleys and misfit streams (Fig. 7c).

CONCLUSIONS

The nature and distribution of Plio-Pleistocene to Recent sediments exposed along the southern margin of the Gulf of Corinth indicate that the extensional Gulf of Corinth Basin was formerly almost twice the width of the Modern Gulf of Corinth. It is suggested that a major basin-narrowing event caused the emergence of Neogene sediments and that this event marked the onset of regional uplift in the North Peloponnese.

The distribution of Neogene conglomerate bodies and Modern deltas reflects the control that tectonic deformation, drainage basin substrate and relief have on fluvial geomorphology. All of the major Plio-Pleistocene and Modern deltas were constructed by rivers with mature upper reaches located in the Hellenide Massif. The rivers which drain only, or mostly, areas of Neogene sediments have not built major deltas as erosion of the young basin sediments produces only minor quantities of coarse sediment. This demonstrates the strong control of catchment lithology and, therefore, sediment supply on conglomerate distribution.

The earliest sites of coarse clastic deposition, which occur along the length of the Gulf of Corinth Basin, were controlled by the position of alluvial valleys developed on the thrust sheets of the Helle-

nide Massif. Regional uplift of the North Peloponnese caused Neogene sediments to emerge and rivers to incise. Old drainage basins were superimposed on the newly emergent coastal strip and juvenile drainage systems grew. The uplift caused rivers to cut down through their delta deposits and build new, topographically lower, deltas. Consequently many of the conglomerate bodies at lower altitudes are younger than those at higher altitudes, the reverse of what is commonly assumed in ancient extensional basin-fills.

Surface tilting, associated with block-faulting, modified the developing drainage patterns and led to drainage reversal where river incision could not keep pace with surface deformation. Where drainage reversal has occurred, stranded deltas have been found which were constructed by northward-flowing rivers where now only small streams, or no streams, flow in that direction. Headward erosion of the misfit component of drainage systems has allowed capture of some of the reverse components and the return of a major northward-flowing river.

The stepped distribution of deltas suggests that sea-level fluctuation, superimposed on basin subsidence and uplift, controlled conglomerate body thickness by controlling the altitude of delta topsets.

ACKNOWLEDGEMENTS

Thanks to Mobil, NERC and BSRG for contributing towards the cost of attending the 13th International Sedimentology Conference at Nottingham where this paper was presented. Thanks to Gary Nichols, Mike Leeder and an anonymous referee for comments on earlier versions. MJS acknowledges his NERC studentship. Permission to work in central Greece was granted by IGME, Athens.

REFERENCES

ALEXANDER, J. & LEEDER, M.R. (1987) Active tectonic control of alluvial architecture. In: *Recent Developments in Fluvial Sedimentology* (Eds Ethridge, F.G., Flores, R.M. & Harvey, M.D.) S.E.P.M. *Spec. Publ. Soc. Sedim. Geol.* **39**, 243–252.

ALEXANDER, J. & LEEDER, M.R. (1990) Geomorphology and surface tilting in an active extensional basin, SW Montana, USA. *J. Geol. Soc. Lond.* **147**, 1–7.

BROOKS, M. & FERENTINOS, G. (1984) Tectonics and sedimentation in the Gulf of Corinth and the Zakynthos and Kefallinia channels, Western Greece. *Tectonophysics* **101**, 25–54.

COLLIER, R.E.LL. (1990) Eustatic and tectonic controls upon quaternary coastal sedimentation in the Corinth Basin, Greece. *J. Geol. Soc. Lond.* **147**, 301–314.

COLELLA, A. (1988) Pliocene—Holocene fan deltas and braid deltas in the Crati Basin, southern Italy: a consequence of varying tectonic conditions. In: *Fan Deltas: Sedimentology and Tectonic Settings* (Eds Nemec, W. & Steel, R.J.) Blackie & Son, London, pp. 50–74.

DUFAURE, J.J. (1975) Le relief du Peloponnese. Unpublished thesis, University of Paris, 1422 pp.

FERENTINOS, G., BROOKS, M. & DOUTSOS, T. (1985) Quaternary tectonics in the Gulf of Patras, western Greece. *J. Struct. Geol.* **7**, 713–717.

FERENTINOS, G., PAPATHEODOROU, G. & COLLINS, M.B. (1988) Sediment transport processes on an active submarine fault escarpment: Gulf of Corinth, Greece. *Mar. Geol.* **83**, 43–61.

FROSTICK, L. & REID, I. (1989) Is structure the main control of river drainage and sedimentation in rifts? *J. Afr. Earth Sci.* **8**, 165–182.

FROSTICK, L. & REID, I. (1990) Structural control of sedimentation patterns and implications for the economic potential of the East African Rift basins. *J. Afr. Earth Sci.* **10**, 307–318.

GAWTHORPE, G.L. & COLELLA, A. (1990) Tectonic controls on coarse-grained delta depositional systems in rift basins. In: *Coarse-grained Deltas* (Eds Collella, A. & Prior, D.B.) Int. Ass. Sediment. Spec. Publ. **10**, 113–127.

HEEZEN, B.C., EWING, M. & JOHNSON, G.L. (1966) The Gulf of Corinth floor. *Deep-Sea Res.* **13**, 381–411.

HIGGS, W.G. (1988) Syn-sedimentary structural controls on basin deformation in the Gulf of Corinth, Greece. *Basin Res.* **1**, 154–165.

HOOKE, R.LE B. (1972) Geomorphic evidence for Late-Wisconsin and Holocene tectonic deformation, Death Valley, California. *Geol. Soci. Am. Bull.* **83**, 2073–2098.

JACKSON, J.A. & McKENZIE, D. (1987) Active normal faulting and crustal extension. In: *Continental Extensional Tectonics* (Eds Coward, M.P., Dewey, J.F. & Hancock, P.L.) Geol. Soc. London Spec. Publ. **28**, 3–17.

JACKSON, J.A. & McKENZIE, D. (1988) Rates of active deformation in the Aegean Sea and surrounding regions. *Basin Res.* **1**, 121–128.

JACKSON, J.J., GAGNEPAIN, J., HOUSEMAN, G., KING, G.C.P., PAPADIMITRIOU, P., SOUFLERIS, C. & VIRIEUX, J. (1982) Seismicity, normal faulting and the geomorphological development of the Gulf of Corinth (Greece): the Corinth earthquakes of February and March 1981. *Earth Planet. Sci. Lett.*, **57**, 377–397.

KELLETAT, D., KOWALCZYK, G., SCHRODER, B. & WINTER, K. (1976). A synoptic view on the neotectonic development of the Peloponnesian coastal regions. *Z. Dtsch. Geol. Ges.* **127**, 447–465.

KERAUDREN, B. & SOREL, D. (1987) The terraces of Corinth (Greece) — a detailed record of eustatic sea level variations during the last 500 000 years. *Mar.Geol.* **77**, 99–107.

LEEDER, M.R. & ALEXANDER, J. (1987) The origin and tectonic significance of asymmetrical meander belts. *Sedimentology* **34**, 217–226.

LEEDER, M.R. & GAWTHORPE, R.L. (1987) Sedimentary models for extensional tilt block/half graben basins. In: *Continental Extensional Tectonics* (Eds Coward, M.P., Dewey, J.F. & Hancock, P.L.) Geol. Soci. London Spec. Publ. **28**, 139–152.

LEEDER, M.R., SEGER, M.J. & STARK, C.P. (1991) Sedimentology and tectonic geomorphology adjacent to active and inactive normal faults in the Megara Basin and Alkyonides Gulf, central Greece. *J. Geol. Soc. London* **148**, 331–343.

LE PICHON, X. (1982) Land-locked oceanic basins and continental collision: the eastern Mediterranean as a case example. In: *Mountain Building Processes* (Ed. Hsu, K.) Academic Press, New York, pp. 201–211.

LE PICHON, X. & ANGELIER, J. (1979) The Hellenic arc and trench system: a key to the neotectonic evolution of the eastern Mediterranean area. *Tectonophysics* **60**, 1–42.

McKENZIE, D. (1978) Active tectonics of the Alpine–Himalayan belt: the Aegean Sea and surrounding regions. *Geophys J. R. Astron. Soc.* **55**, 217–254.

ORI, G.G. (1989) Geological history of the extensional basin of the Gulf of Corinth (?Miocene–Pleistocene), Greece. *Geology* **17**, 918–921.

PE-PIPER, G. & PIPER, D.J.W. (1984) Tectonic setting of the Mesozoic Pindos basin of the Peloponnese. In: *The Geological Evolution of the Eastern Mediterranean* (Eds Dixon, J.E. & Robertson, A.H.F.) Geol. Soc. London Spec. Publ. **17**, 563–567.

ROBERTS, S. (1988) Active normal faulting in Central Greece and Western Turkey. PhD Thesis, University of Cambridge.

ROBERTS, S. & JACKSON, J.A. (1991) Active normal faulting in central Greece: an overiew. In: *The Geometry of Normal Faults* (Eds Roberts, A.M., Yielding, G. & Freeman, B.) Geol. Soc. London Spec. Publ. **56**, 125–142.

SCHRODER, B. & KELLETAT, D. (1976) Geodynamical conclusions from vertical displacement of Quaternary shorelines in the Peloponnesos/Greece. *Neues Jahrb. Geol. Palaeont. Abh.* **150**, 174–186.

STEWART, I.S. & HANCOCK, P.L. (1988) Normal fault zone evolution and fault scarp degradation in the Aegean region. *Basin Res.* **1**, 139–153.

TAYMAZ, T., JACKSON, J.A. & McKENZIE, D. (1991) Active tectonics of the north and central Aegean sea *Geophys J. Int.* **106**, 433–490.

VITA-FINZI, C. & KING, G.C.P. (1985) The seismicity, geomorphology and structural evolution of the Corinth area of Greece. *Philos. Trans. R. Soc. London* **A314**, 379–407.

WHITE, R.S., SPENCE, G.D., FOWLER, S.R., McKENZIE, D.P., WESTBROOK, G.K. & BOWEN, A.N. (1987) Magmatism at rifted continental margins. *Nature* **330**, 439–444.

Spec. Publs Int. Ass. Sediment. (1993) **20**, 49–65

Some controls on sedimentary sequences in foreland basins: examples from the Alberta Basin

D.J. CANT *and* G.S. STOCKMAL

Geological Survey of Canada, Institute of Sedimentary and Petroleum Geology, 3303 33rd St. NW, Calgary, Alberta T2L 2A7, Canada

ABSTRACT

The stratigraphy and sedimentology of foreland basin clastic wedges is believed to be controlled largely by the spatial and temporal effects of 'time lag' between the initiation of flexure-induced subsidence and the deposition of large amounts of sediment generated from the overthrust belt. The initial accretion of a terrane to an older continental margin, thereby converting the margin to a compressional orogen, should be associated with a relatively large time-lag effect because initial loading by overthrusting takes place on the continental slope, in the absence of significant early subaerial relief across the thrust belt. The large time-lag effect results in a succession grading from deep-water to shallow-water sediments in the early foreland basin.

Clastic wedges resulting from later overthrusting events are generally expected to show considerably reduced time-lag effects, with composite transgressive sequences succeeded by highstand regressive sequences, both composed mainly of non-marine, marginal-marine and shallow-marine sediments of varied lithologies. This reduced time-lag effect may reflect a relatively continuous background supply of orogen-derived sediment readily available to fill accommodation space as it is created. In extreme cases, some clastic wedges may show essentially no time-lag effect, lacking basal transgressive or deep-water sediments.

As well as orogenesis, other factors also condition foreland successions. In the Alberta Basin, the effects of some base-level change events (eustatic?) are more dramatic in cratonward areas where tectonic subsidence rates are lower. Valley-fill and lowstand deposits reflecting base-level drops are more common and easily recognizable toward that side of the basin. The structures and lithologies of the pre-existing shelf/miogeoclinal succession also affect thicknesses and facies of foreland units through their influence on relief of the basal unconformity and synsedimentary salt dissolution.

INTRODUCTION

Research on the origin and evolution of foreland basins has made strong advances in the last decade, principally due to developments in understanding the mechanics of sedimentary basin formation, as well as in stratigraphic and subsidence modelling. Price (1973) suggested that subsidence of the Alberta foreland basin reflected lithospheric flexure due to the weight of the adjacent overthrust belt. Building on this general hypothesis, Beaumont (1981), Jordan (1981), Quinlan and Beaumont (1984), Stockmal *et al.* (1986), Stockmal and Beaumont (1987), Flemings and Jordan (1989, 1990), and others, using a variety of mechanical descriptions for lithospheric flexure, have modelled fore-land subsidence and resulting stratigraphic thick-nesses of foreland basin fill. These geodynamic models quantitatively show that lithospheric flexure caused by overthrust loading, coupled with the resulting topographical expression across the thrust belt, is an adequate mechanism for generating the foreland basin and the stratigraphy of the deposits within it.

Qualitative geological interpretations utilizing predictions arising from the concept of lithospheric flexure have progressed in parallel with the quanti-tative modelling. Among the earliest such efforts was that of Jacobi (1981) who interpreted an uncon-formity in western Newfoundland separating the

Middle Ordovician Appalachian carbonate platform from the overlying foreland basin sequence to be the result of uplift due to peripheral bulge migration, followed by subsidence. Many others, including Quinlan and Beaumont (1984), Tankard (1986), Heller *et al.* (1988) and Cant and Stockmal (1989) have also used concepts stemming from quantitative foreland model studies to interpret stratigraphical features. Here we discuss some aspects of foreland clastic wedges in the context of sequence stratigraphical concepts, and consider possible interpretations in terms of flexural models for the lithosphere.

The tectonic setting of foreland basins ensures that the evolution of the associated orogen is a primary factor affecting the stratigraphy and sedimentology of the fill because of its influence on sources and rates of sediment supply as well as on subsidence. Some stratigraphic effects in the sedimentary fill, however, are better attributed to other factors: i.e. valley fills resulting from base-level changes, some of which seem best interpreted as eustatic in origin, and thickness and facies anomalies resulting from the influence of pre-existing structures and strata in the basin. Because all such factors are operative simultaneously, superimposed on a 'background' of flexural tectonic subsidence, separation of their effects may not be possible in many cases, and considerable future research will be required to evaluate the relative importance of each. We discuss general principles, in many cases derived indirectly from model studies, and show some examples from the Alberta Basin.

In any foreland basin, many unanswered questions about the stratigraphy remain, especially with regard to relatively small-scale variations. The broad-scale synthetic stratigraphy generated by geodynamic models of lithospheric flexure has been used to develop generalized hypotheses concerning finer-scale details of foreland sequences. Scale, however, remains a principal problem in utilizing the results of geodynamic models: the models of Beaumont (1981), for example, which focused on the lithospheric response to thrust loading, produced a remarkable match to the observed cross-sectional stratigraphy of the Alberta Basin on a basin-wide scale. However, on the scale of an individual clastic wedge or depositional sequence, other factors such as eustasy, autocyclic sedimentary processes, climate, local variations in type and amount of sediment supply, inherited conditions, in-plane stress and bending moments acting on the lithosphere,

and even dynamic forces in the asthenosphere, may have to be included. Attempts to model the effects of some of these variables have been made (e.g. Jervey, 1988; Flemings & Jordan, 1989, 1990; Mitrovica *et al.*, 1989; Sinclair *et al.*, 1991; Beaumont *et al.*, 1992); future modelling efforts will likely be directed toward understanding their combined effects.

For the stratigrapher/sedimentologist, the most important use of modelling may be to establish multiple hypotheses to explain an observation in terms of the factors mentioned above, and to constrain interpretations. Comparative studies of different basins will also enable researchers to establish the degree of variability of foreland basin fill in different settings and possibly the reasons for these variations. Future research on foreland basins will surely incorporate model results, enabling further refinements.

THE DEEP-TO-SHALLOW-WATER SUCCESSION IN FORELAND BASINS

A common theme in descriptions of foreland basin-fill sequences is the deep-to-shallow-water vertical succession, first noted in the Alpine foreland as the 'synorogenic' Flysch succeeded by the 'post-orogenic' Molasse (Bertrand, 1897). This type of sequence has been described in many basins by many authors (e.g. Covey, 1986; Ricci Lucchi, 1986; Caron *et al.*, 1989). Crook (1989) provides a description of an ongoing suturing event in New Guinea and the sedimentology of the resulting basin fill, which grades from turbidites to non-marine deposits. Cant and Stockmal (1989), working from the accretion model of Stockmal *et al.* (1986) and incorporating the concepts of peripheral bulge migration and basin 'rebound' (regional uplift), derived an idealized shallowing-upward unconformity-bounded sequence (Fig. 1) for foreland basin clastic wedges. We suggest that a fundamental control on the facies succession within the sequence is the effect of 'time-lag' between the onset of flexural subsidence and the subsequent higher rates of sediment supply to the basin.

Devlin *et al.* (1990) proposed an important model which also incorporates the concept of time-lag between subsidence and sedimentation, but for basins which lack deep-water sediments. In the non-marine to marginal-marine Wyoming foreland basin, the basal portion of a clastic cycle consists of

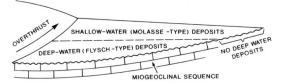

Fig. 1. Diagram illustrating typical foreland basin fill resulting from initial terrane accretion at a former passive continental margin. As the leading edge of the terrane overrides the continental slope, foreland flexural subsidence begins before much sediment is shed from the overthrust belt. Deep-water conditions in the foreland therefore result. Continued convergence pushes the overthrust belt over the slope break, creates topography in the juvenile thrust belt, and substantial sediment is shed into the basin. Deep-water 'flysch-type' deposits accumulate, onlapping the basal unconformity. As overthrusting proceeds, the sediment supply exceeds subsidence, resulting in shallow-marine and non-marine deposits. 'Rebound' during erosional lightening of the thrust load, during relative tectonic quiescence, can result in the upper unconformity. (Modified from Cant and Stockmal, 1989.)

transgressive depositional systems tracts, followed by progradational highstand systems tracts (Fig. 2). These authors related the transgressive-to-regressive pattern of systems tract development to high subsidence rates during tectonic activity followed by lower rates during tectonic quiescence. This model is fundamentally an application and expansion of the time-lag concept operating within a foreland which is 'predisposed' to shallow-water or non-marine environments. Its utility in analysing the shallow-water clastic wedges in the Alberta foreland basin is discussed below.

Fig. 2. A typical clastic wedge resulting from a subsequent deformation (accretion?) event, when the foreland basin is predisposed to shallow-marine and non-marine facies. The early phase when subsidence exceeds sedimentation is reflected by the transgressive depositional systems tract; the succeeding phase when sedimentation overwhelms subsidence, by the highstand, prograding depositional systems tract. The basic concepts were presented from the Wyoming foreland basin by Devlin *et al.* (1990).

The transition from passive margin to foreland basin — subsidence phase

Some foreland sequences begin essentially with deep-water deposits, without any apparent preservation of transgressive deposits reflecting the subsidence-dominated phase of the basin development. In young basins at the edge of a continental block, this may reflect original oceanic crust at oceanic depths caught up in foreland basin development (Crook, 1989). In other basins on continental crust with underlying shallow-water sequences, it appears that almost complete sediment starvation during early subsidence resulted in the absence of transgressive sedimentation (Ricci Lucchi, 1986).

Recently, Allen *et al.* (1990) and Sinclair *et al.* (1991) have documented a basal transgressive phase in part of the northern Alpine foreland basin that is composed of a basal thin glauconitic sandstone, then shallow-water nummulitic limestones, followed by pelagic deeper-water globigerina shales and marls, below the Flysch. Subsidence was initially slow enough so that some deposition occurred during the deepening phase; however, the types of lithologies indicate that only limited amounts of detrital sediments were being supplied from the overthrust belt, and largely biogenic deposits accumulated. In another case, significant deposition of limestone occurred during this transgressive phase. In the Ordovician of western Newfoundland, Jacobi (1981) suggested that up to 350 m of deepening-upward platform-to-slope carbonates (Lower Llanvirnian Table Head Group; Stenzel *et al.*, 1990), accumulated at the base of the foreland sequence, immediately above the interpreted peripheral bulge unconformity.

We consider the conversion of a passive margin to a foreland basin to coincide with the change in the mechanism of subsidence, i.e. from passive margin thermal decay to foreland basin flexure, rather than with any change in properties of the sediments themselves, such as their petrology or palaeocurrent patterns. The time-lag effect exists because flexural subsidence may begin significantly earlier than arrival of sediment supplied from the orogen. In this framework, the direction of sediment supply of a deep-water deposit is therefore not a critical criterion for determining the time of transition from a passive margin to a foreland basin. We suggest that clastic sequences bracketing the transition, but supplied from the craton, cannot necessarily be excluded from belonging to the foreland subsidence regime on the basis of palaeoflow directions or petrographic provenance. Clearly, the distinction between foreland basin units and pre-foreland units may be difficult to make, particularly in cases where the pre-existing sediments were accumulating in deep water (e.g. Caron *et al.*, 1989; Crook, 1989).

Deep-water facies

Deep-water foreland basin fill has been described by many authors, especially from Alpine forelands (Bertrand, 1897; Ricci Lucchi, 1986; Caron *et al.*, 1989) where many classic sedimentological studies of turbidites (Bouma, 1962) and submarine fan deposits (Mutti & Ricci Lucchi, 1972) were conducted. Covey (1986) and Cant and Stockmal (1989) attributed the deep-water facies to low rates of sediment supply during the early stage of accretion and thrusting, so that flexure-induced subsidence exceeded sediment accumulation and deep-water conditions (bathyal or abyssal) prevailed. This time-lag effect is implicit in the archetypal geodynamic models of Stockmal *et al.* (1986) for the accretion of a terrane to a continental margin through A-type subduction. A time-lag effect is expected because at an early stage in orogenesis, the overthrust belt (colliding accretionary wedge) structurally overlies the continental slope and lacks significant topographical expression above sea-level. Therefore, relatively little sediment is supplied even though the weight of the belt has begun to flexurally depress the early foreland basin. This situation, where an overthrust wedge impinges upon a former passive margin (facing either an ocean or a back-arc basin), is therefore predisposed to deep-water conditions (Covey, 1986; Cant & Stockmal, 1989). As the overthrust wedge structurally thickens and advances across the shelf–slope transition of the former passive margin, producing significant subaerial relief, sediment supply rates should increase dramatically. We suggest that the effect of time-lag between the onset of subsidence and the subsequent period of increasing rates of sediment supply is a fundamental control of foreland facies and sequences, as discussed below.

Are deep-water facies always formed?

As noted above, basal deep-water sediments are characteristic of many foreland basins around the world Hiscott *et al.*, 1986; (Ricci Lucchi, 1986; Caron *et al.*, 1989). However, many other foreland basins apparently lack facies reasonably interpreted as deep-water (deposited below the shelf break): e.g. Indo-Gangetic plains (Johnson *et al.*, 1986); Timor–Tanimbar Trough (Audley-Charles, 1986); northern Argentina foreland (Beer & Jordan, 1989); the western Interior Basin of North America (Dickinson *et al.*, 1988; Cant & Stockmal, 1989).

In the Western Canadian foreland basin, marine shales of the Jurassic Fernie Group are in the most appropriate stratigraphical position, near the base of the foreland basin succession, for deep-water facies to occur. Although some thin sandstone beds in this unit have been interpreted as turbidites in the past (Hamblin & Walker, 1979), no suggestion of submarine fan facies or even slope facies has ever been made. The sandstone beds can be interpreted as flood-deposit plumes from fluvial distributaries, wind-forced shelf currents, or storm deposits which perhaps gained enough inertia or density contrast to the surrounding water mass to flow beyond storm-wave base (A.P. Hamblin, personal communication, 1990). These pass gradationally upwards with no major changes in lithology into hummocky cross-stratified sandstones interpreted as deposited above storm-wave base (Hamblin & Walker, 1979). Also, no deep-marine interpretations of Jurassic sediments have been made on the basis of palaeontology (Hall, 1984; Stronach, 1984). The possibility exists, however, that the surviving outcrops are located too far cratonward to show deep-water sediments, and that now-cannibalized Jurassic abyssal and bathyal deposits once existed to the west. Similarly, by analogy with the Taiwan foreland basin, Covey (1986) has attributed the lack of deep-water facies in some foreland basins simply to lack of preservation because of ongoing overthrusting and cannibalization of early deposits.

All other factors being equal, the relative strengths of the lithosphere underlying flexural foreland basins can be estimated by considering the widths of the basins. For a given tectonic load, a wide but shallow basin indicates a strong lithosphere, whereas a narrow but deep basin indicates a weak lithosphere (e.g. Molnar & Lyon-Caen, 1988). For example, in their models of the Alberta and Molasse basins, Stockmal and Beaumont (1987) required strong and weak lithosphere, respectively, to match the observed large-scale stratigraphical thicknesses. A margin underlain by weak lithosphere may be more favourably disposed to the development of an early deep-water facies than one underlain by a strong plate, but once the overthrust wedge has overridden the former passive margin hinge line and much of the 'predisposition' toward relatively deep-water facies has been consumed, the differences begin to diminish (Stockmal *et al.*, 1986).

Shallow-water facies

The progressive overriding of the continental margin by the overthrust wedge results in high sediment supply rates to the foreland and filling of the basin to shallow-marine or non-marine conditions. In Alpine forelands, this is the Molasse phase, with common shallow-marine (Allen & Homewood, 1984) and non-marine clastic deposits (Bersier, 1959). In basinal areas remote from sediment input points, shallow-water to shoreline carbonates and evaporites may accumulate, as in the modern Arabo-Persian Gulf foreland (Purser, 1985). In these later stages of basin filling, cratonward migration of the overthrust wedge in many cases incorporates large amounts of older passive margin and foreland sediments that are cannibalized and recycled into younger foreland deposits.

A foreland basin can be defined as 'filled' if the space up to the crest of the peripheral bulge is occupied by sediment (Flemings & Jordan, 1989). Overfilled and underfilled basins also occur. Note that this does *not* imply that filled or overfilled basins contain only non-marine deposits or that underfilled basins contain only marine deposits. Facies are determined by the relationship of the sediment surface to sea-level, not to the crest of the peripheral bulge.

The Canadian Cordillera is a collage of terranes which accreted to the western continental margin of North America from Mid-Jurassic to Early Tertiary times (e.g. Monger & Price, 1979; Monger *et al.*, 1982; Gabrielse & Yorath, 1989). Cant and Stockmal (1989) hypothesized that the initial Mid-Jurassic accretion of the Intermontane superterrane occurred by overthrusting of the leading edge of the terrane onto the continental slope, and subsequent accretion events perhaps affected the foreland by displacing the intervening Intermontane and other accreted terranes further toward the craton. This indirect mechanism was proposed because the later-accreted terranes are too far from the basin (beyond the flexural half-wavelength of the lithosphere) to contribute directly, through flexural loading, to the later Cretaceous and Tertiary foreland subsidence

events. Compression of earlier-accreted, inboard terranes and displacement of the overthrust belt created by the earlier accretions and shortening events seem necessary to account for the subsidence of the foreland so far from the active convergent margin. Direct evidence for reactivation of the overthrust belt is lacking, as only the latest events have as yet been documented.

In such cases of orogenic loading following initial accretion, the effect of time-lag between load-induced foreland subsidence and the arrival of sediment generated by this phase of thrusting may be minor or even non-existent if sediment was being continuously shed from the pre-existing thrust belt. Therefore, once the former passive margin has been buried tectonically by the overthrust terranes and by foreland sedimentation, additional foreland depositional episodes will be predisposed toward shallow-water and non-marine conditions (Cant & Stockmal, 1989; Stockmal *et al.*, 1992). In this context, a non-marine to marginal- and shallow-marine clastic wedge such as the Mannville Group (discussed below) is logically attributable to a deformation (accretion?) event subsequent to the initial accretion on the former passive margin.

During a period of no thrust-loading and therefore no first-order formation of accommodation space in the basin, material shed into it may be bypassed to other basins. In the Tertiary, sediment from the North American Cordillera apparently passed through the foreland basin and contributed to major clastic wedges along the passive Gulf Coast margin of Texas and Louisiana.

EXAMPLES FROM THE ALBERTA FORELAND BASIN

Figure 3 presents a generalized stratigraphy of the Alberta foreland basin (Cant & Stockmal, 1989; Stockmal *et al.*, 1992). We interpret here three unconformity-bounded units using the concepts of foreland sedimentation discussed above. Previously, analyses of the basin fill have generally focused on discrete, coarsening-upward clastic wedges (e.g. Eisbacher *et al.*, 1974; Cant & Stockmal, 1989). In light of the time-lag concept, however, the critical relationship between subsidence and sedimentation becomes the focus: is the basin subsiding faster than it is filling, or vice versa? A useful approach is to subdivide the fill into depositional systems tracts, whether transgressive or

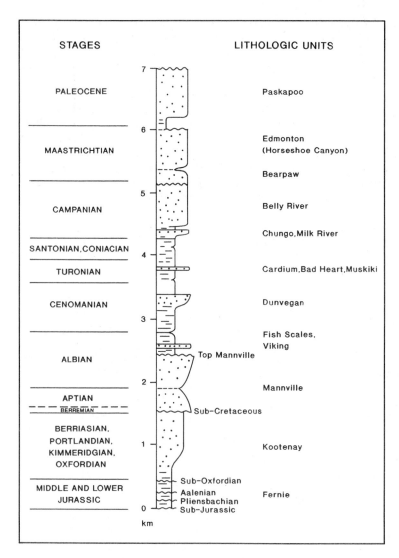

Fig. 3. Simplified stratigraphy of the Alberta foreland basin. Thicknesses of units are representative of the western part of the basin. Major unconformities are named beside the lithologic column. The stratigraphic names are those used over the largest areas in the basin. In contrast to many other foreland basins, subsidence and sedimentation persisted over a relatively long period of time from at least the Mid-Jurassic to the Early Tertiary. Note the Upper Albian to Santonian shaley interval interrupted only by the Dunvegan clastic wedge, and thinner units of sandstone (Viking, Cardium, etc.) interpreted to be mainly the result of base-level fluctuations. The Jurassic nomenclature is simplified; the Green Beds and the Passage Beds mentioned in the text lie immediately above the sub-Oxfordian unconformity.

highstand regressive, thus using the fundamental principles of sequence stratigraphy (Posamentier *et al.*, 1988), without necessarily embracing the purely eustatic hypothesis.

Fernie–Kootenay Wedge. The stratigraphy and palaeogeography of the Jurassic to earliest Cretaceous Fernie–Kootenay sequence are not well known, in part because of limited outcrop in the deformed belt and its restricted eastward extent in the subsurface of the Plains. However, the Fernie–Kootenay succession is bounded by unconformities, and shows the classic shallowing-upward pattern from deep- shelf to non-marine deposits. Compared

to the Alpine Flysch-to-Molasse sequence, this pattern can be interpreted to reflect a lesser but still substantial time-lag effect, with initial rates of subsidence higher than rates of sediment supply.

The base of the foreland sequence (as defined above) may coincide with the base of the entire Fernie Formation which rests unconformably on Triassic and older rocks, or it may lie at one of the unconformities within the Fernie (Fig. 3). For example, the sub-Oxfordian unconformity may reflect uplift across the peripheral bulge and therefore represent the base of the foreland succession. Above the unconformity are the Green Beds, a 15-m-thick, glauconitic, belemnite-bearing silty shale which is believed to have been deposited in relatively shallow water (Frebold, 1957); they are succeeded by deep-shelf shales of the Passage Beds which grade up into the mostly non-marine Kootenay Group (Hamblin & Walker, 1979). The Green Beds can be regarded as the basal transgressive phase, followed by the regressive highstand Passage Beds to Kootenay succession. The distribution of the Green Beds in the more proximal (southern) parts of the basin suggests shallow-water accumulation — the longitudinal facies distribution of the later foreland appears to have been already established. However, as noted above, palaeoflow direction by itself is not necessarily a reliable criterion for identifying the base of the foreland sequence; the lower part of the Fernie, exhibiting multiple unconformities, cannot presently be excluded from the foreland basin (tectonic) subsidence regime.

Mannville Wedge. The Aptian to Middle Albian Mannville Group clastic wedge differs markedly from the Fernie–Kootenay wedge in that the proportion of coarse sediment is much higher, and it is composed largely of non-marine and marginal-marine sediments (Fig. 4). It should be noted that the ages of the stratigraphically lowest conglomerates and sandstones are questionable, whether entirely Late Aptian or as old as Barremian in part. The Mannville is unconformity-bounded over most of the basin, lying on rocks as old as Devonian. In the westernmost part of the preserved Mannville (called the Blairmore Group in outcrop), it is more

difficult to demonstrate a stratigraphical break between it and the upper part of the underlying Kootenay clastic wedge (Ricketts & Sweet, 1986). This possible absence of an erosional break can be interpreted to result from visco-elastic relaxation of the foreland plate (cf. Quinlan & Beaumont, 1984; Beaumont *et al.*, 1988) following Jurassic overthrusting, causing the basin to narrow as well as deepen (subside) adjacent to the orogen. The upper unconformity is short-lived, and is essentially a disconformity over most of the basin (Stelck & Kramers, 1980), extending at least to the Saskatchewan–Manitoba border.

The Lower Mannville is relatively thin (Fig. 5), non-marine over most of the basin, and grades northwestward to shallow-marine deposits. The unit comprises a transgressive parasequence set, with thick estuarine sandstones gradually backstepping southward over non-marine deposits (Fig. 4). The Middle Mannville is defined as the interval in which the transgression rapidly accelerated southward (Fig. 4), with a single shoreline-transgressive shelf sandstone (Bluesky Formation) laid down over the northern half of the basin. The Upper Mannville comprises a highstand systems tract with strong northwestward progradation (Fig. 4). In the southern part of the basin, non-marine deposits lie directly on the transgressive Middle Mannville. Farther north, the basal transgressive–regressive sequences consist of coarsening-upward shallow-marine shales which grade up to shelf and shoreline sandstones. The northwestward-prograding shore-

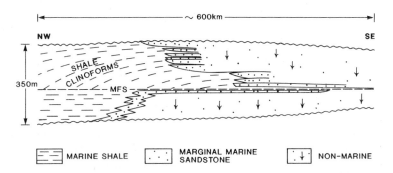

Fig. 4. Stratigraphy and regional facies of the Aptian–Albian Mannville clastic wedge, illustrated on a southeast (proximal) to northwest (distal) cross-section. The Lower Mannville is the dominantly non-marine unit, the Middle Mannville is the marginal-marine to shoreline sandstone immediately below the maximum flooding surface (MFS), and the Upper Mannville is the northwest-prograding wedge. The Lower Mannville is transgressive, with southward-stepping shoreline and estuarine sandstones. An abrupt transgression drove the shoreline from central Alberta hundreds of kilometres south, with marginal-marine shales of the MFS extending south across the international border. The Upper Mannville is a highstand prograding sequence which drove the shore zone back to north-central Alberta. This diagram should be compared with Figure 2.

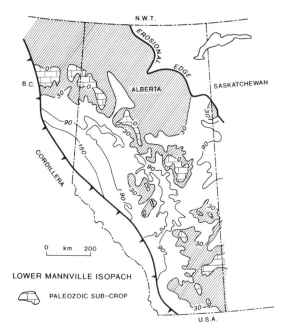

Fig. 5. Isopach map of the Lower Mannville. Visible are the westward thickening, caused by asymmetric subsidence characteristic of foreland basins, and the linear thickness trends caused by the subcrop of Palaeozoic miogeoclinal carbonates. The orientation of these ridges is probably a function in part of Mesozoic foreland tectonics. Thicknesses in metres. Hatched area is where the Lower Mannville is thinnest (0–30 m).

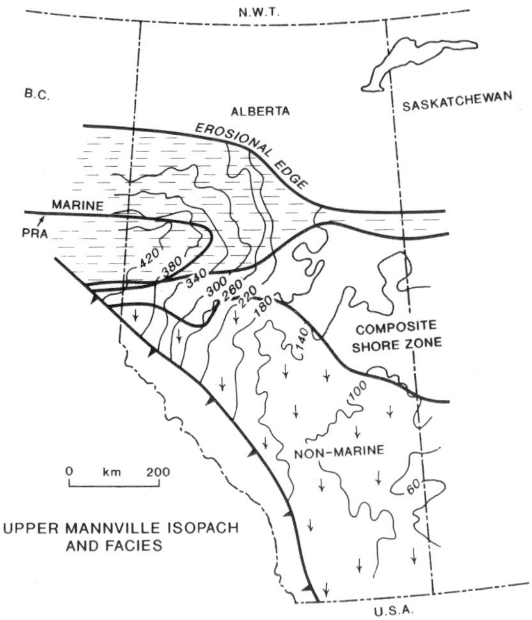

Fig. 6. An isopach map of the Upper Mannville, approximately the highstand depositional systems tract. The large thickness anomaly in the western portion of the area results from more rapid subsidence of the Peace River Arch (PRA), a Palaeozoic crustal structure. The generalized facies distribution of the higher part of the Upper Mannville is shown. Eroded area includes Upper Mannville behind well casings. Thicknesses in metres.

lines in the lower three parasequences progressively migrated through south-central Alberta until one regression abruptly drove the shore zone some 300 km north to the margin of the subsiding Peace River Arch (Fig. 6). This rapid progradation implies that rates of sediment supply vastly outpaced rates of formation of accommodation space (change in sea-level + subsidence). Traced far to the north into the area of marine shales, each individual transgressive–regressive sequence successively downlaps onto the transgressive Middle Mannville (Fig. 4).

A north–south section through the Mannville wedge fits the model of Devlin *et al.* (1990), with a basal transgressive phase followed by a major highstand regressive phase of somewhat greater thickness (Fig. 2). This stratigraphical pattern can be interpreted to reflect a much reduced time-lag effect between subsidence and sedimentation, as compared to the Alpine foreland fill or the Jurassic clastic wedge, due to the predisposition of the

foreland to shallow-water or non-marine conditions during deformation events subsequent to initial accretion. An important petrographic change from relatively clean, quartz-dominated sandstones to much less mature, highly lithic sandstones occurs at the Middle Mannville (Cameron, 1965; Wightman & Berezniuk, 1986) further supporting the hypothesis that the Middle Mannville 'inflection point' of the sequence represents the arrival of a new pulse of sediment from the source area.

Belly River–Edmonton Wedges. These two clastic wedges are not as distinctly separable as those underlying them, perhaps because the deformation/loading events in the Late Cretaceous occurred with little time separation between them. The Campanian Belly River wedge is separated lithologically from the overlying mostly Maastrichtian, Edmonton wedge by the Bearpaw Shale (Fig. 7), the result of a transgression which advanced from the east-southeast toward the overthrust belt. The Bearpaw

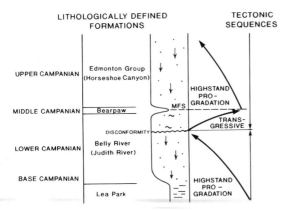

LITHOLOGICALLY DEFINED FORMATIONS

TECTONIC SEQUENCES

Fig. 7. The lithostratigraphy of the Campanian Belly River and Edmonton clastic wedges. The disconformity in the upper Belly River unit (Eberth, 1990) can be used to divide the section into tectonic sequences similar to those illustrated in Fig. 2, as shown on the right side of the diagram. The Bearpaw Shale which separates the major clastic wedges becomes the maximum flooding surface (MFS), within the tectonic sequence.

transgression did not reach the northwestern part of the basin where the undivided Belly River–Edmonton unit is termed the Brazeau Formation.

Within the upper Belly River Formation, a disconformity exists with some cobble conglomerates resting on it (Eberth, 1990), dividing non-marine sediments below from marine-influenced deposits above. Sandstones above the disconformity are also coarser and petrographically less mature (Eberth, 1990). As shown in Fig. 7, that part of the Belly River Formation above the disconformity can be regarded as the transgressive phase of the next tectonic sequence, and the Bearpaw Shale represents the maximum flooding surface, with the highstand or regressive sequence above. An important implication is that the lithologically defined formations are not equivalent to units defined on a sequence basis and, therefore, may not be as useful for stratigraphical analysis of basin tectonics.

VARIATIONS IN TIME-LAG EFFECT

The inferred time-lag effects between initiation of subsidence and subsequent sedimentation of clastic wedges in different foreland basins provides a useful conceptual linkage between these basins. We emphasize that the effect we are describing is the sedimentary result of variation of the relative rates of subsidence and sedimentation in the foreland: i.e. the effect of the time interval or 'lag' between production of accommodation space and the arrival of sediment to fill that space. The magnitude of the time-lag effect in the sedimentary record is not necessarily a measure of, nor proportional to, the absolute time-lag itself; nearly identical stratigraphical effects could occur in two foreland sequences, one deposited with high rates of subsidence and sedimentation and a short absolute time-lag, the other with low rates of the two parameters and a long absolute time-lag.

The magnitude of the time-lag stratigraphical effect depends on a variety of factors including the following.

1 The tectonic setting at the continental margin; i.e. was this subsidence/sedimentation event associated with the initial or a subsequent deformation or accretion event?

2 The period of 'tectonic quiescence' between consecutive loading events (considered here as equivalent to a period of no or simply reduced orogenic deformation), which influences the magnitude of erosion (load removal) across the existing overthrust belt.

3 The rheology and strength of the lithosphere underlying the continental margin; i.e. elastic or visco-elastic, relatively strong or weak.

4 The position of the sequence with respect to the flexure-inducing load and to the flexural bulge (cf. Flemings & Jordan, 1989, 1990).

5 The spatial partitioning of thrust deformation within the orogen ('in sequence' versus 'out-of-sequence' patterns).

6 The efficiency of sediment transport to the foreland.

The last two factors may involve the development of temporary sites of sediment storage such as minor basins developed on thrust sheets which are later cannibalized, causing fluctuations in coarse sediment supply. Ricci Lucchi (1986) invoked trapping of sediment in piggy-back basins on tops of thrust sheets as the mechanism which allowed subsidence of the Apennine Basin to relatively deep water. Other less well understood but significant variables such as climate are probably important as well (see Beaumont *et al.*, 1992).

Figure 8 depicts examples of foreland basin clastic sequences which we interpret to show different degrees of the stratigraphical time-lag effect. At one extreme, a large effect gives rise to a sediment-

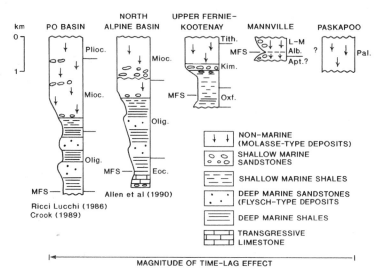

Fig. 8. Foreland clastic wedges, ordered according to the time-lag effect between the initiation of subsidence and the influx of large amounts of sediment. In basins which initiate at oceanic depths (Crook, 1989) or have initial rapid subsidence without sedimentation (Ricci Lucchi, 1986), a large time-lag effect results in deep-water conditions at the base of the clastic wedge, with the maximum flooding surface virtually at its base. The North Alpine sequence shows a minor transgressive sandstone and limestone sequence (Allen *et al.*, 1990; Sinclair *et al.*, 1991). The Fernie–Kootenay sequence shows relatively shallow-water deposits in a thinner sequence, with minor transgressive sediments. The Mannville is composed of shallow-marine and non-marine facies in a transgressive followed by a highstand succession reflecting a limited time-lag effect. The Paskapoo wedge (Fig.) is entirely non-marine where it is preserved, and may be the result of essentially no time-lag effect, but the unit is not well understood at present.

starved basin and deep-water conditions. The maximum flooding surface is essentially at the base of the section. Progressively decreasing magnitudes of the effect can be expected to result in shallower water sequences, with basal transgressive systems tracts. At the opposite end of the spectrum, sequences may show either no trend or reflect the influence of other variables such as eustatic sea-level change. Note that the absolute period of the time-lag may be very difficult to constrain since biostratigraphic resolution in many cases is inadequate. However, one of the best constrained sub-modern foreland sequences occurs in Taiwan (Covey, 1986; Teng, 1990). The transgressive phase lasted from approximately 5 Ma to 3 Ma, with the 500-m to 600-m section grading upward from coastal to outer offshore deposits (Teng, 1990, his Fig. 7). Above this lies 3 km of highstand, regressive sediments which infilled the basin.

DIRECTIONS OF FILL

Foreland basins are commonly filled longitudinally.

The Alberta foreland basin (Eisbacher *et al.*, 1974; Rahmani & Lerbekmo, 1975; Cant & Stockmal, 1989), the Apennine Basin (Ricci Lucchi, 1986), the Molasse Basin (Homewood *et al.*, 1986), the Indo-Gangetic Basin (Eisbacher *et al.*, 1974), and the modern Arabo-Persian Gulf (Purser, 1985) are all dominantly filled by sedimentary systems which emerge from their orogens, and then turn parallel to their trends. Because of progressive cannibalization of the orogenward edge of the basin by structural advance of the overthrust belt, most preserved deposits show only longitudinal palaeoflows and facies distributions (Figs 4 & 6).

In the Alberta Basin, most exceptions to longitudinal palaeoflow are coarse-grained units, either very proximal non-marine conglomerates, or sandstones and conglomerates reflecting base-level drops within the upper Albian to Santonian shallow-marine shales (Fig. 3). These coarse packages appear to have been shed directly from the orogen, with northeastward progradation (e.g. Plint *et al.*, 1986). The dominant slope in the Alberta foreland basin, however, along the basin axis, has been interpreted by us to be the result of differential tectonic loading along the orogen (Cant & Stockmal, 1989), largely because of the long time-spans over which the relatively consistent drainage directions persisted (75–80 My to the northwest, followed by 35–40 My to the southeast).

Note that the longitudinal fill of the Alberta Basin and other forelands causes some confusion in the usage of the terms proximal and distal. They refer to the relative distance from the source of the sediment, but this is assumed in many cases to be synonymous with the relative distance from the orogen. However, with longitudinal flow, proximal and distal in the case of the Mannville Group, for example, refer to relative distance northwest or southeast parallel to the orogen. It is necessary therefore to distinguish these terms from orogenward and cratonward (Fig. 9). The depositional topographical gradients lie in the longitudinal proximal–distal direction, whereas tectonic subsidence gradients extend in the transverse cratonward–orogenward direction. This results in palaeogeographies where deepest water conditions lie distally away from the major sediment input points, but nearly adjacent to the orogen (Fig. 9).

During periods when the overthrust load on the lithosphere is being lightened by erosion across the orogen (the 'destructive' state of Jamieson & Beaumont, 1988), the basin undergoes 'epeirogenic' uplift, with the greatest magnitudes of uplift and erosion near the orogen. For this reason, transverse patterns of drainage occur in some modern forelands undergoing long-period rebound (e.g. the present-day interior of western Canada), suggesting that unconformities formed by epeirogenic uplift might show transverse patterns of erosion.

These generalizations regarding directions of drainage are strictly applicable only to simple, flexurally depressed forelands. In deformed foreland basins with uplifted blocks which trend at a variety of angles relative to the orogen, drainage directions are controlled by the trends of these structures, as in the Laramide foreland of the USA (Dickinson *et al.*, 1988). In addition, where piggyback basins are developed on the backs of thrust sheets, as in the southern French and Italian Alpine forelands, local structures determine the local drainage patterns (Ricci Lucchi, 1986).

ADDITIONAL INFLUENCES ON FORELAND CLASTIC SEQUENCES

Base-level changes and inherited topographical/structural features are also important factors affecting foreland sedimentation. These factors are often related to or tempered by foreland and overthrust belt tectonics.

Base-level fluctuations

In recent years, sedimentologists have come to appreciate the role of changing base level in af-

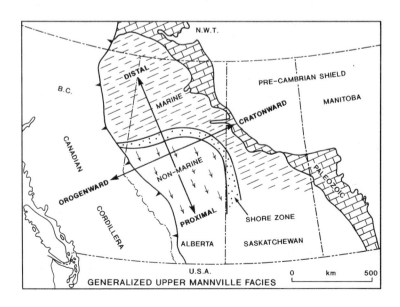

Fig. 9. In many studies, areas near the orogen are assumed to be proximal. Where longitudinal palaeoflow occurred, proximal–distal and orogenward–cratonward directions are almost at right angles, as illustrated here using facies distributions from the higher part of the Upper Mannville. General palaeoflow directions in the Cenomanian Dunvegan and later clastic wedges show the reverse trend, heading toward the southeast, thus also reversing the proximal–distal directions.

fecting some sedimentary sequences and facies relationships. The debate about the origin of base-
level changes is focused on the role of eustasy versus
tectonic causes such as intra-plate stresses (e.g.
Cloetingh, 1986; Karner, 1986). As pointed out by
several authors (e.g. Kendall & Lerche, 1988), no
absolute criteria are known to distinguish unambig-
uously the role of eustasy from that of tectonic
movements in the creation of base-level change.
The base-level fluctuations discussed here have
magnitudes exceeding 30 m, and periods less than
the limit of biostratigraphical resolution, presumably about 1–2 My. They occurred several hundreds
of kilometres from the overthrust belt, during periods of general subsidence and sedimentation in the
foreland.

The Cretaceous eustatic sea-level curve of Haq *et
al.* (1987) shows two distinct frequencies of variation: the longer frequency is on the scale of tens of
millions of years with amplitudes of 200–350 m
and can be accounted for by variations in volume of
the mid-ocean ridges (Kominz, 1984); the shorter
variations occur with periods generally less than
two million years and estimated amplitudes of up to
c. 100 m. Although the accuracy of this 'eustatic'
curve is debatable (Miall, 1991), some very prominent, short-term, high-rate eustatic changes on it
correlate in time with rapid base-level changes in
the Alberta foreland succession (e.g. Late Turonian
drop and rise). Given present uncertainties associated with interregional or intercontinental tectonic
mechanisms which might account for these base-
level changes (e.g. in-plate stresses), these correlatable base-level variations may be more readily
interpreted as eustatic, superimposed on the 'background' of foreland subsidence and sediment-
supply regime. Most base-level changes in the
Alberta foreland, however, show no definite correlation to any global curves.

During deposition of a clastic wedge, base-level
drops cause incision of valleys into non-marine and
marginal-marine sediments: streams are steepened,
and erosion results. Although controls on valley
formation, as opposed to development of a flat
erosion surface, are not well understood, examples
of rapid base-level fall favour development of valleys; for example, the incision of valleys during the
rapid Pleistocene sea-level fall, and excavation of
the Grand Canyon in the rapidly uplifted Colorado
Plateau area. Valleys are identified in the subsurface
where regional units are replaced by anomalous
deposits (Fig. 10) with narrow, linear shapes (Gross,

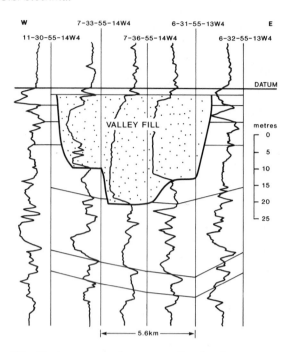

Fig. 10. A gamma-ray log cross-section showing an
incised valley filled by probable estuarine sandstone in
the topmost Mannville. These linear features interrupt a
regionally consistent pattern of sandstone–shale
shoreline facies successions. In many cases, the valleys
are filled by varying proportions of marine muds, but
controls on the lithology of the fill are not well
understood.

1980). The infill of the valleys is not uniform; it may
consist of varying proportions of sand and shale
from 0 to 100% of either component. These valleys
cut through non-marine areas, as well as the associated shoreline deposits, and are commonly filled
during the subsequent transgression by estuarine or
marginal-marine deposits. Infill by these transgressive facies is one criterion useful in some cases for
differentiating valleys from large contemporaneous
channels.

Middle Mannville valleys occur across the entire
basin (Fig. 11) as documented by many local studies
(e.g. Gross, 1980; Wood & Hopkins, 1989). As
discussed above, the Middle Mannville contains the
maximum flooding surface for the wedge, reflecting
the change from the subsidence-dominated tectonic
phase to the sedimentation-dominated phase, and
coincides with the petrographic change mentioned
above. The timing of this composite episode of
base-level change (events 1 and 2 on Fig. 11)

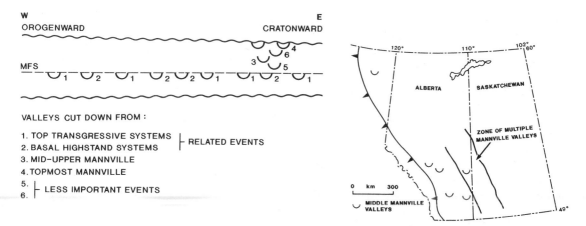

Fig. 11. The gross stratigraphy and regional distribution of valley fills in the Mannville. The Middle Mannville valleys (events 1 and 2), which occur slightly above and below the maximum flooding surface, range west to near the thrust belt. The Upper Mannville valleys (events 3–6) are restricted to eastern Alberta and western Saskatchewan.

suggests that it may be related to tectonic events rather than eustatic ones, although no specific mechanism is proposed.

Upper Mannville incision events (Fig. 11, events 3 to 6) present in eastern Alberta are not recognizable closer to the orogen. This pattern is interpretable as resulting from eustatic drops combined with the foreland subsidence pattern: the higher rates of tectonic subsidence in the western part of the basin inhibited valley formation during each short-term eustatic fall. On a detailed scale, the localization of valleys within the cratonward area of the basin seems to have been partly localized by dissolution of a Devonian salt during the Early Cretaceous (Wightman, 1990). The resulting ongoing local subsidence was apparently sufficient to help maintain the courses of several generations of river systems and their incised valleys.

As discussed above, the distribution of these Upper Mannville valleys can be interpreted in terms of a eustatic fall superimposed on the foreland subsidence. However, short-period in-plane compressions could possibly produce a similar distribution by raising the margin of the basin and depressing the deeper part. At the moment, definitive criteria to distinguish the alternatives are lacking.

Inherited features

Patterns of foreland subsidence and infilling are also affected by structural and sedimentary features directly inherited or partially modified from the succession below. In some cases, regional structures that developed during the pre-foreland history of the basin continue to respond to orogenic events in a different fashion or at different rates than other parts of the foreland. The Peace River Arch of north-central Alberta (Fig. 6), a crustal-scale structure formed in the Late Precambrian to Early Palaeozoic and oriented across the foreland, subsided differentially during deposition of the Upper Mannville wedge. Because of the higher rate of subsidence across the Arch, it controlled the thickness variations of foreland units across it and, in combination with the northward direction of fill, the transitions from non-marine to marine facies (Fig. 6) in several Upper Mannville trangressive–regressive sequences (Cant, 1984). The shorelines in several of these are stacked at the southern margin of the structure. Arch subsidence also prevented uplift and unconformity development on the top of the Mannville clastic wedge, thus modifying the regional stratigraphy shown in Figs 3 and 4.

The lithologies of the underlying platform succession also affected foreland deposition because of differential erosion of various units during long-period unconformity development. This created topography which affected thickness patterns and depositional facies in the Lower Mannville. Ridges estimated to be originally over 100 m high (Jackson, 1984) are present where resistant carbonates subcrop, separated by valleys underlain by less resistant units. The carbonate ridges were exposed during deposition of half the Mannville succession, and controlled variations in its thickness over much of the basin (Fig. 5). These ridges supplied small amounts of sediment and also localized shoreline sand bodies.

As mentioned above, solution of evaporitic units in the Palaeozoic shelf succession has occurred, likely in response to the water-flow pattern in the foreland. Dissolution of Palaeozoic evaporites in the Cretaceous has caused locally increased subsidence rates, with effects on foreland thickness patterns and facies. Specifically, dissolution of Devonian salt in the Early Cretaceous caused localization of incised valleys in the Mannville (Fig. 11) as well as a dip reversal with eastward slopes in Saskatchewan where the salt has been removed (Wightman, 1990). Both the salt edge and the Palaeozoic ridges in the Alberta Basin imposed local control of the overall north-northwest basin-fill pattern. However, because these elements themselves were uplifted and tilted by foreland flexure, they trend roughly parallel to the basin axis.

DISCUSSION

Comparison of foreland sequences from a variety of basins underscores the importance of tectonic controls. Specifically, the 'time-lag' stratigraphical effect is interpreted to be a primary control, differently expressed in time and space in different basins. We must be cautious in the application of this concept, however, because similar stratigraphical successions do not necessarily have identical causes. For example, the pattern of transgressive succeeded by regressive highstand systems tracts observed in the Mannville Group and other units is not a pattern unique to tectonic control. It can be formed by, and is believed by some authors to be largely controlled by, eustatic variation (e.g. Posamentier et al., 1988).

Eustasy can be interpreted to play an important role in foreland basins in certain circumstances. For example, in the Alberta Basin the Upper Albian to Cenomanian shales (Fig. 3) have sequences which apparently show transgressive deepening up to maximum flooding surfaces, followed by regressive shallowing up to unconformities cut during base-level drops. One of these is illustrated in Fig. 12, showing the transgressive portion of the Dunvegan Formation (as defined by Bhattacharya, 1988) and the Kaskapau Formation shales. The Second White

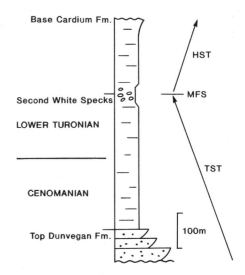

Fig. 12. The Cenomanian to Lower Turonian sequence in the Alberta Basin is composed almost entirely of the Kaskapau Shale, along with the topmost transgressive portion of the Dunvegan Formation (Bhattacharya, 1988). Note that the greatest thickness of the Dunvegan clastic wedge lies in the highstand depositional systems tract of the previous sequence. The Second White Specks are composed of coccolith-bearing shales, and contain the maximum flooding surface in this sequence.

Specks is a shale with scattered coccoliths which grades to a shaley limestone in places, and is interpreted to represent the maximum flooding surface of this sequence. This unit is believed by Caldwell (1984) to correlate with the peak transgression of the USA. Greenhorn cycle of Kauffman (1969) of the Western Interior Basin, The Greenhorn cycle is itself interpreted to be produced by eustatic variation (Kauffman, 1969). This correlation, of course, simply shifts the scale of the discussion to whether these interregional cycles are eustatic or not. However, it seems possible that eustasy may well play a role in controlling at least the timing of the maximum transgression in some shaley successions.

The presence of hundreds of metres of shale in the middle of the Alberta succession (Upper Albian to Santonian, Fig. 3) seems to be uncommon in the fill of foreland basins. The thin intercalations of sandstone and conglomerates in these shales all lie on, or are associated with, unconformities, and are believed to involve base-level drops. The most general explanation for the thick shale succession is that subsidence rates overwhelmed rates of coarse sediment supply, trapping it near the unpreserved basin margin. More specific explanations, such as trapping in marginal basins, or regional out-of-sequence structural development trapping within the overthrust belt can be modelled, but a confident solution must await additional geological constraints.

ACKNOWLEDGEMENTS

This paper benefited from the careful and thoughtful reviews of C. Busby-Spera, W. Helland-Hansen, and K.G. Osadetz. Geological Survey of Canada Contribution No. 26191.

REFERENCES

ALLEN, P.A. & HOMEWOOD, P. (1984) Evolution and mechanics of a Miocene tidal sandwave complex. *Sedimentology* 21, 63–81.

ALLEN, P.A., CRAMPTON, S. & SINCLAIR, H. (1990) The inception and early evolution of the north Alpine foreland basin, Switzerland. *Abstracts of 13th Internat. Sediment. Congress*, Nottingham, pp. 16–17.

AUDLEY-CHARLES, M.G. (1986) Timor–Tanimbar trough: the foreland basin of the evolving Banda orogen. In: *Foreland Basin* (Eds Allen, P.A. & Homewood, P.) *Int. Ass. Sediment. Spec. Publ.* 22, 91–102.

BEAUMONT, C. (1981) Foreland basins. *Geophys. J. R. Astron. Soc.* 55, 291–329.

BEAUMONT, C., QUINLAN, G. & HAMILTON, J. (1988) Orogeny and stratigraphy: numerical models of the Paleozoic in the Eastern Interior of North America. *Tectonics* 7, 389–416.

BEAUMONT, C., FULLSACK, P. & HAMILTON, J. (1992) Erosional control of active compressional orogens. In: *Thrust Tectonics* (Ed. McClay, K.). Chapman & Hall London, p. 1–18.

BEER, J.A. & JORDAN, T.E. (1989) The effects of Neogene thrusting on deposition in the Bermejo basin, Argentina. *J. Sedimeut. Petrol.* 59, 330–345.

BERSIER, A. (1959) Séquences détritiques et divagations fluviales. *Ecolog. Geol. Helv.* 51, 854–893.

BERTRAND, M. (1897) Structure des Alpes françaises et recurrence de certaines facies sedimentaires. *VIe International Congress* (Zurich), pp. 161–177.

BHATTACHARYA, J. (1988) Autocyclic and allocyclic sequences in river- and wave-dominated deltaic sediments of the Upper Cretaceous Dunvegan Formation, Alberta: Core examples. In: *Sequences, Stratigraphy, Sedimentology: Surface and Subsurface* (Eds James, D.P. & Leckie, D.A.) Can. Soc. Petrol. Geol. Mem. 15, 25–32.

BOUMA, A.H. (1962) *Sedimentology of some Flysch Deposits — a Graphic Approach to Facies Interpretation.* Elsevier Publishing Co., Amsterdam, 168 pp.

CALDWELL, W.G.E. (1984) Early Cretaceous transgressions and regressions in the southern interior plains. In: *The Mesozoic of Middle North America* (Eds Stott, D.F. & Glass, D.J.) an. Soc. Petrol. Geol. Mem. 9, 173–203.

CAMERON, E.M. (1965) Application of geochemistry to stratigraphic problems in Lower Cretaceous of Western Canada. *Am. Assoc. Petrol. Geol. Bull.* 49, 62–80.

CANT, D.J. (1984) Development of shoreline–shelf sand bodies in a Cretaceous epeiric sea deposit. *J. Sediment. Petrol.* 54, 541–556.

CANT, D.J. & STOCKMAL, G.S. (1989) The Alberta foreland basin: relationship between stratigraphy and Cordilleran terrane-accretion events. *Can. J. Earth Sci.* 26, 1964–1975.

CARON, C., HOMEWOOD, P. & WILDI, W. (1989) The original Swiss Flysch: A reappraisal of the type deposits in the Swiss Prealps. *Earth Sci. Rev.* 26, 1–45.

CLOETINGH, S. (1986) Intraplate stresses: a new tectonic mechanism for fluctuations of relative sea level. *Geology* 14, 617–620.

COVEY, M. (1986) The evolution of foreland basins to a steady state: evidence from the western Taiwan foreland basin. In: *Foreland Basins* (Eds Allen, P.A. & Homewood, P.) *Int. Ass. Sediment. Spec. Publ.* 22, 77–90.

CROOK, K.A.W. (1989) Suturing history of an allochthonous terrane at a modern plate boundary traced by flysch to molasse facies transitions. *Sediment. Geol.* 61, 49–79.

DEVLIN, W.J., RUDOLPH, K.W., EHMAN, K.D. & SHAW, C.A. (1990) The effect of tectonic and eustatic cycles on accommodation and sequence stratigraphic framework in the southwestern Wyoming foreland basin. *Abstracts of 13th Internat. Sediment. Congress*, Nottingham, p. 131.

DICKINSON, W.R., KLUTE, M.A., HAYES, M.J., et al. (1988) Paleogeographic and paleotectonic setting of Laramide sedimentary basins in the central Rocky Mountain region. *Geol. Soc. Am. Bull.* 100, 1023–1039.

EBERTH, D.A. (1990) Tectonic, sedimentologic and paleontologic significance of a disconformity in the Upper Judith River Formation (Campanian) of south central Alberta. *Abstracts of 13th Internat. Sediment. Congress*, Nottingham, p. 150.

EISBACHER, G.H., CARRIGY, M.A. & CAMPBELL, R.B. (1974) Paleodrainage pattern and late-orogenic basins of the Canadian Cordillera. In: *Tectonics and Sedimentation* (Ed. Dickinson, W.R.) Soc. Econ. Paleontol. Mineral. Spec. Publ. 22, 143–166.

FLEMINGS, P.B. & JORDAN, T.E. (1989) A synthetic stratigraphic model of foreland basin development. *J. Geophys. Res.* 94, 3851–3866.

FLEMINGS, P.B. & JORDAN, T.E. (1990) Stratigraphic modelling of foreland basins: Interpreting thrust deformation and lithosphere rheology. *Geology* 18, 430–434.

FREBOLD, H. (1957) The Jurassic Fernie Group in the Canadian Rocky Mountains and foothills. *Geol. Surv. Can. Mem.* 287, 197 ʃp

GABRIELSE, H. & YORATH, C.J. (1989) DNAG #4: the Cordilleran Orogen in Canada. *Geosci. Can.* 16, 67–83.

GROSS, A.A. (1980) Mannville channels in east-central Alberta. In: *Lloydminster and Beyond: Geology of Mannville Hydrocarbon Reservoirs* (Eds Beck, L.S., Christopher, J.E., & Kent, D.M.) Saskatchewan Geol. Soc. Spec. Publ. 5, 33–62.

HALL, R.L. (1984) Lithostratigraphy and biostratigraphy of

the Fernie Formation (Jurassic) in the southern Canadian Rocky Mountains. In: *The Mesozoic of Middle North America* (Eds Stott, D.F. & Glass, D.J.) Can. Soc. Petrol. Geol. Mem. **9**, 233–247.

HAMBLIN, A.P. & WALKER, R.G. (1979) Storm-dominated shallow marine deposits in the Fernie–Kootenay (Jurassic) transition, southern Rocky Mountains. *Can. J. Earth Sci.* **16**, 1673–1689.

HAQ, B.U., HARDENBOL, J. & VAIL, P.R. (1987) Chronology of fluctuating sea levels since the Triassic. *Science* **235**, 1156–1167.

HELLER, P.L., ANGEVINE, C.L., WINSLOW, N.S. & PAOLA, C. (1988) Two-phase stratigraphic model of foreland-basin sequences. *Geology* **16**, 501–504.

HISCOTT, R.N., PICKERING, K.T. & BEEDEN, D.R. (1986) Progressive filling of a confined Middle Ordovician foreland basin associated with the Taconic Orogeny, Quebec, Canada. In: *Foreland Basins* (Eds Allen, P.A. & Homewood, P.) *Int. Ass. Sediment. Spec. Publ.* **8**, 309–326.

HOMEWOOD P., ALLEN, P.A. & WILLIAMS, G.D. (1986) Dynamics of the Molasse Basin of western Switzerland. In: *Foreland Basins* (Eds Allen, P.A. & Homewood, P.) Int. Ass. Sediment. Spec. Publ. **8**, p. 199–214.

JACKSON, P.C. (1984) Paleogeography of the Lower Cretaceous Mannville Group of western Canada. In: *Elmworth — Case Study of a Deep Basin Gas Field* (Ed. Masters, J.A.). Am. Assoc. Petrol. Geol. Mem. **38**, 49–78.

JACOBI, R.D. (1981) Peripheral bulge — a causal mechanism for the Lower/Middle Ordovician unconformity along the western margin of the northern Appalachians. *Earth Planet. Sci. Lett.* **56**, 245–251.

JAMIESON, R.A. & BEAUMONT, C. (1988) Orogeny and metamorphism: a model for deformation and pressure–temperature–time paths with applications to the central and southern Appalachians. *Tectonics* **7**, 417–445.

JERVEY, M.T. (1988) Quantitative geological modelling of siliciclastic rock sequences and their seismic expression. In: *Sea Level Changes: An Integrated Approach* (Eds Wilgus, C.K., hastings, B.S., Posamentier, H., Van Wagoner, J., Ross, C.A. & Kendall, C.G.St.C.). Soc. Econ. Paleontol. Mineral. Spec. Publ. **42**, 47–70.

JOHNSON, G.D., RAYNOLDS, R.G. & BURBANK, D.W. (1986) Late Cenozoic tectonics and sedimentation in the northwestern Himalayan foredeep: I. Thrust ramping and associated deformation in the Potwar region. In: *Foreland Basins* (Eds Allen, P.A. & Homewood, P.) Int. Ass. Sediment. Spec. Publ. **8**, 273–291.

JORDAN, T.E. (1981) Thrust loads and foreland basin evolution, Cretaceous, western United States. *Am. Assoc. Petrol. Geol. Bull.* **65**, 2506–2520.

KARNER, G.D. (1986) Effects of lithospheric in-plane stress on sedimentary basin stratigraphy. *Tectonics* **5**, 573–588.

KAUFFMAN, E.G. (1969) Cretaceous marine cycles of the Western Interior. *Mountain Geol.* **6**, 227–245.

KENDALL, C.G.St.C. & LERCHE, I. (1988) The rise and fall of eustasy. In: *Sea-Level Changes — An Integrated Approach* (Eds Wilgus, C.K., Hastings, B.S., Posamentier, H., Van Wagoner, J., Ross, C.A. & Kendall, C.G.St.C.) Soc. Econ. Paleontol. Mineral. Spec. Publ. **42**, 3–17.

KOMINZ, M.A. (1984) Ocean ridge volumes and sea-level

change — an error analysis. In: *Interregional Unconformities and Hydrocarbon Accumulation* (Ed. Schlee, J.S.) Am. Assoc. Petrol. Geol. Mem. **36**, 108–128.

MIALL, A.D. (1991) Stratigraphic sequences and their chronostratigraphic correlation. *J. Sediment. Petrol.* **61**, 497–505.

MITROVICA, J.X., BEAUMONT, C. & JARVIS, G.T. (1989) Tilting of continental interiors by the dynamical effects of subduction. *Tectonics* **8**, 1079–1094.

MOLNAR, P. & LYON-CAEN, H. (1988) Some simple physical aspects of the support, structure, and evolution of mountain belts. *Geol. Soc. Am. Spec. Pap.* **218**, 179–207.

MONGER, J.W.H. & PRICE, R.A. (1979) Geodynamic evolution of the Canadian Cordillera — progress and problems. *Can. J. Earth Sci.* **16**, 770–791.

MONGER, J.W.H., PRICE, R.A. & TEMPLEMAN-KLUIT, D.J. (1982) Tectonic accretion and the origin of the two major metamorphic and plutonic welts in the Canadian Cordillera — progress and problems. *Geology* **10**, 70–75.

MUTTI, E. & RICCI LUCCI, F. (1972) Le torbiditi dell'Apennino settentrionale: introduzione all'analisi di facies. *Mem. Soc. Geol. Ital.* **11**, 161–199.

PLINT, A.G., WALKER, R.G. & BERGMANN, K.M. (1986) Cardium Formation 6. Stratigraphic framework of the Cardium in subsurface. *Bull. Can. Petrol. Geol.* **34**, 213–225.

POSAMENTIER, H.W., JERVEY, M.T. & VAIL, P.R. (1988) Eustatic controls on clastic deposition I — Conceptual framework. In: *Sea-Level Changes — An Integrated Approach* (Eds Wilgus, C.K., Hastings, B.S., Posamentier, H., Van Wagoner, J., Ross, C.A. & Kendall, C.G.St.C.) Soc. Econ. Paleont Mineral. Spec. Publ. **42**, 109–124.

PRICE, R.A. (1973) Large-scale gravitational flow of supracrustal rocks, southern Canadian Rocky Mountains. In: *Gravity and Tectonics* (Eds De Jong, K.A. & Scholten, R.A.) Wiley Interscience, New York, pp. 491–502.

PURSER, B.H. (1985) The Arabo-Persian Gulf: a modern foreland basin. *Programme and Abstracts, Internat. Assoc. Sediment. Symposium on Foreland Basins*, Fribourg, p. 105.

QUINLAN, G.M. & BEAUMONT, C. (1984) Appalachian thrusting, lithospheric flexure, and the Paleozoic stratigraphy of the eastern interior of North America. *Can. J. Earth Sci.* **21**, 973–996.

RAHMANI, R.A. & LERBEKMO, J.F. (1975) Heavy mineral analysis of Upper Cretaceous and Paleocene sandstones in Alberta and adjacent areas of Saskatchewan. In: *The Cretaceous System in the Western Interior of North America.* (Ed. Caldwell, W.G.E.) Geol. Assoc. Can. Spec. Pap. **13**, 607–632.

RICCI LUCCHI, F. (1986) The Oligocene to Recent foreland basins of the northern Apennines. In: *Foreland Basins* (Eds Allen, P.A. & Homewood, P.) Int. Ass. Sediment. Spec. Publ. **8**, 105–140.

RICKETTS, B.D. & SWEET, A.R. (1986) Stratigraphy, sedimentology, and palynology of the Kootenay–Blairmore transition in southwestern Alberta and southeastern British Columbia. *Geol. Surv. Can. Pap.* **84-15**, 41 pp.

SINCLAIR, H.D., COAKLEY, B.J., ALLEN, P.A. & WATTS, A.B. (1991) Simulation of foreland basin stratigraphy using a diffusion model of mountain belt uplift and erosion: an example from the central Alps, Switzerland. *Tectonics* **10**, 599–620.

STELCK, C.R. & KRAMERS, J.W. (1980) *Freboldiceras* from the Grand Rapids Formation of north-central Alberta. *Bull. Can. Petrol. Geol.* **28**, 509–521.

STENZEL, S.R., KNIGHT, I. & JAMES, N.P. (1990) Carbonate platform to foreland basin: revised stratigraphy of the Table Head Group (Middle Ordovician), western Newfoundland. *Can. J. Earth Sci.* **27**, 14–26.

STOCKMAL, G.S. & BEAUMONT, C. (1987) Geodynamic models of convergent margin tectonics: The southern Canadian Cordillera and the Swiss Alps. In: *Sedimentary Basins and Basin-Forming Mechanisms* (Eds Beaumont, C. & Tankard, A.J.) Can. Soc. Petrol. Geol. Mem. **12**, 393–411.

STOCKMAL, G.S., BEAUMONT, C. & BOUTILIER, R. (1986) Geodynamic models of convergent margin tectonics: transition from rifted margin to overthrust belt and consequences for foreland basin development. *Am. Assoc. Petrol. Geol. Bull.* **70**, 181–190.

STOCKMAL, G.S., CANT, D.J. & BELL, J.S. (1992) Relationship of the stratigraphy of the Western Canada Foreland Basin to Cordilleran tectonics: insights from geodynamic models. Amer. Assoc. Petrol. Geol. Mem. **55**, p. 107–124.

STRONACH, N.J. (1984) Depositional environments and cycles in the Jurassic Fernie Formation, southern Canadian Rocky Mountains. In: *The Mesozoic of Middle North America* (Eds Stott, D.F. & Glass, D.J.) Can. Soc. Petrol. Geol. Mem. **9**, 43–68.

TANKARD, A.P. (1986) On the depositional response to thrusting and lithospheric flexure: examples from the Appalachian and Rocky Mountain basins. In: *Foreland Basins* (Eds Allen, P.A. & Homewood, P.) Int. Ass. Sediment. Spec. Publ. **8**, 369–394.

TENG, L.S. (1990) Geotectonic evolution of late Cenozoic arc–continent collision in Taiwan. *Tectonophysics* **183**, 57–76.

WIGHTMAN, D. (1990) Salt solution and its effect on the western Canada Basin. *National Conference on the Earth Sciences* Banff, Alberta.

WIGHTMAN, D. & BEREZNIUK, T. (1986) Resource characterization and depositional modelling of the Clearwater Formation, Cold Lake Oil sands deposit, east-central Alberta. In: *Proceedings of the 1986 Tar Sands Symposium* (Eds Westhoff, J.D. & Marchant, L.C.) United States Department of Energy, Morgantown, West Virginia, pp. 20–45.

WOOD, J.M. & HOPKINS, J.C. (1989) Reservoir sandstone bodies in estuarine valley fill: Lower Cretaceous Glauconitic Member, Little Bow Field, Alberta, Canada. *Am. Assoc. Petrol. Geol. Bull.* **73**, 1361–1382.

Spec. Publs Int. Ass. Sediment. (1993) **20**, 67–88

Stratigraphy and sedimentary development of the Sant Llorenç del Munt fan-delta complex (Eocene, southern Pyrenean foreland basin, northeast Spain)

M. LÓPEZ-BLANCO

Dept. GDGP, Universitat de Barcelona, 08071 Barcelona, Spain

ABSTRACT

The Bartonian–Priabonian Sant Llorenç del Munt fan-delta complex is located on the southeast margin of the South Pyrenean foreland basin. During the deposition of this complex the basin margin was controlled by a series of sinistral strike-slip faults that, under transpressive conditions, developed some thrust structures.

The overall stratigraphical evolution of the fan-delta complex shows a large-scale progradation of the system to the northwest (the regressive part of a 2nd-order sequence). Fundamental sequences (4th order) of 50 000–200 000-year duration can be defined and these contain regressive, transgressive and occasional low-stand components. The stacking pattern of fundamental sequences defines composite sequences (3rd order, of about 800 000 years duration) which also have clear transgressive and regressive components. Fundamental sequences can be further subdivided into 5th-order sequences.

The deposition of the complex and the development of four orders of cyclicity is related to changes in relative sea-level and clastic supply, with both of these factors probably controlled largely by tectonic movements along the margin of this foreland basin, and climatic changes.

INTRODUCTION AND GEOLOGICAL SETTING

The Eocene Sant Llorenç del Munt fan-delta complex is located in the northeastern part of Spain, 40 km northwest of Barcelona, and comprises part of the Palaeogene fill of the South Pyrenean foreland basin. This basin has a triangular shape and is bounded by three main mountain ranges (the Pyrenees to the north, the Iberian chain to the southwest and the Catalan Coastal Ranges to the southeast). The Sant Llorenç del Munt complex is adjacent to the southeastern margin, close to the Catalan Coastal Ranges (Fig. 1).

The structure of the Catalan Coastal Ranges is characterized by a series of old basement faults that have moved during the Late Palaeozoic (Anadón *et al.*, 1985b), Mesozoic (Esteban & Robles, 1976; Anadón *et al.*, 1979; Marzo, 1980) and reactivated in the Tertiary (Fontboté, 1954; Anadón *et al.*, 1985b; Guimerà, 1988). There are two main fault systems.

1 Nearly vertical, basement-involved strike-slip faults that constitute a right-stepping, én echelon array with an average northeast to southwest orientation.

2 Transverse northwest-striking basement faults with smaller displacements (Anadón *et al.*, 1985b; Guimerà, 1988).

During the Alpine orogeny (Palaeogene) some of these faults (now located near the northwest boundary of the Catalan Coastal Ranges) were reactivated as strike-slip faults that under local transpressive conditions generated a series of southwest to northeast oriented thrusts and folds. These compressional structures show a northwest vergence and involve Palaeozoic, Mesozoic and also Tertiary materials. A product of this Palaeogene structuring was the emplacement of 'els Brucs' and 'les Pedritxes' thrust sheets, linked to the activity of one of the most important faults of the area ('Vallés-Penedés' fault), located in the source area of the Sant Llorenç del Munt fan delta (Fig. 2). These

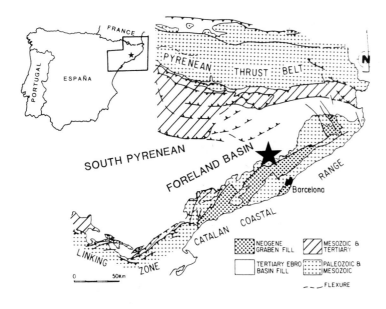

Fig. 1. Location of the Sant Llorenç del Munt fan-delta complex on the Iberian peninsula, near the southeastern border of the South Pyrenean foreland basin. (Modified from Anadón *et al.* 1989.)

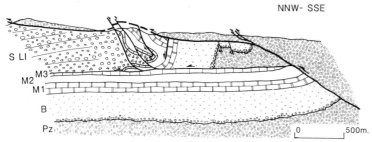

Fig. 2. Cross-section (with no vertical exaggeration) of the northwestern border of the Catalan Coastal Range in the Sant Llorenç del Munt area. Pz, Palaeozoic; B, Buntstandstein; M1, Lower Muschelkalk; M2, Middle Muschelkalk; M3, Upper Muschelkalk; S Ll, Sant Llorenç del Munt palaeogene conglomerates.

thrust sheets have almost 5 km of northwestwards displacement. The Mesozoic materials (mostly Triassic) overthrusted by these sheets, make up a complex structure showing important flexures, minor scale folds and, in some areas, complex thrust sequences with northwest vergence (Fig. 3). The compressional structuring of the Catalan Coastal Ranges took place from the Ilerdian–Cuisian at least stages to the Middle Oligocene, and was the product of a north–south compression due to the convergent motion of the European and African plates (Guimerà, 1988).

Thus, during the Palaeogene, a zone of high relief Palaeozoic and Mesozoic rocks was created at the southwest of the NE–SW faults (the Catalan Coastal Ranges) in front of a depressed area (the South Pyrenean foreland basin). The uplift and erosion of this high relief zone resulted in the formation of a series of alluvial fan complexes along the southeastern border of the South Pyre-

nean foreland basin (see Fig. 3). A major transgression during the Middle Eocene (Biarritzian) converted most of those alluvial fans, including the

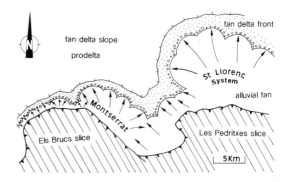

Fig. 3. Palaeogeographical sketch-map of the Sant Llorenç del Munt area during the Late Eocene. (Modified from Marzo & Anadón 1988.)

Sant Llorenç del Munt, into fan deltas (Anadón *et al.*, 1985a).

Stratigraphically the studied deposits are Bartonian and Priabonian in age, and they correspond to the Santa María Group (Pallí, 1972), the Sant Llorenç del Munt conglomerates (Anadón, 1978) and their distal equivalents.

The primary purpose of this paper is to document the facies distribution of the different environments that form the Sant Llorenç del Munt fan-delta complex, to develop a useful allostratigraphical subdivision of the series, and to reconstruct the palaeogeographical evolution of the system.

FACIES ANALYSIS

The studied deposits consist of an alternation of detrital fan-delta and carbonate-platform deposits (Anadón *et al.*, 1985a; Maestro, 1987; Travé, 1988).

Fan delta

The clastic fan-delta sediments are the volumetrically most abundant facies (more than 90%) in this complex. They can be subdivided from proximal (in the southeast) to distal facies (in the northwest) as:
1 alluvial fan;
2 fan-delta front;
3 fan-delta slope (Fig. 4).

Alluvial fan

The alluvial fan deposits (mostly conglomerates, sandstones and red clays) crop out in the most proximal areas of the basin (in the southeast near the basin-margin faults). Two main parts have been distinguished within the alluvial fan: proximal and distal (see Fig. 4).

The proximal alluvial fan was located close to the source area and its deposits are characterized by generally massive conglomeratic facies deposited by sediment gravity flows and less frequently by braided channels. The sediment gravity flow facies have been interpreted to be deposited by the deceleration of catastrophic flash-flooding events with high sediment content. The flows came through canyons from the source area, and spread out onto the fan surface (Marzo & Anadón, 1988). These energetic flows and strong tractive currents imply nearby slopes.

The distal alluvial fan is characterized by conglomeratic and sandy channel-fill deposits, surrounded by red floodplain clays and sandstones. These facies have been interpreted to be deposited by stream and sheet flows. Distributary channels were filled with sandstones and conglomerates while sheet sandstones and clays were deposited on the floodplain. In the distal alluvial fan, the original steep slope from the proximal zone had been reduced and the deposits are finer grained, from less energetic currents.

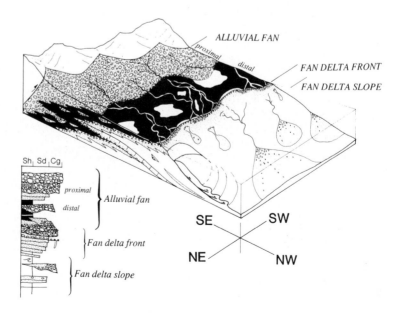

Fig. 4. Fan-delta facies associations. Block diagram showing the spatial distribution of the different facies belts within the Sant Llorenç del Munt fan delta. Idealized schematic columnar section (without vertical scale) showing the progradation of the Sant Llorenç del Munt fan delta.

Fan-delta front

The fan-delta front deposits form a facies belt located in the transitional zone between subaerial and submarine environments. This coarse-grained belt composed of sandstones and conglomerates passes laterally to finer sediments both landwards (subaerial distal alluvial-fan clays) and basinwards (marine prodelta marls) (Fig. 4). These facies, generally related with regressive episodes, usually show coarsening-and-thickening-upward sequences of sandstones and conglomerates (not thicker than 20 m) (Figs 5a & 5b). Also apparent is a large-scale, low-angle cross stratification linked to progradation. Sometimes, the fan-delta front deposits show badly defined fining-upward sequences (exceptionally 8 m thick) with subhorizontal stratification. These are aggrading or retrograding sequences, generally related to transgressive episodes.

The fan-delta-front facies were originally deposited as mouth bars, and then subsequently reworked and modified by coastal currents and organic reworking.

The mouth-bar facies in the most proximal zones consists of conglomerates with a sandy matrix and a diffuse horizontal or low-angle cross-stratification. These conglomerates pass distally to generally tabular sandstones with trough cross-stratification, horizontal and ripple lamination.

These deposits were laid down in front of distributary channels where their fluvial flows were expanded and diluted. The deposits were laid down from non-cohesive debris flows in the proximal areas that may have evolved to turbiditic and tractive currents in the distal parts (Marzo & Anadón, 1988).

Wave reworking often modified the original depositional character of the mouth-bar deposits. This reworking resulted in:
1 lags of the coarsest grain sizes (generally conglomerates) produced by the 'washing' out of finer grained sediments;
2 cross-stratification, laminations or imbrications (Fig. 6) which are products of the migration of bedforms on the shoreface, induced by currents parallel to the coast (SW–NE) or from sea to land (NW–SE);
3 some foreshore-parallel and low-angle laminations and the presence of wave-ripples.

(a)   (b)

Fig. 5. Regressive fan-delta front facies. Notice the coarsening-upward trend (a) from clays to sands, and (b) from sands to gravel.

Fig. 6. Proximal fan-delta-front facies. The lower part shows proximal mouth-bar (debris flow) deposits. The upper part is the product of the reworking of previous deposits by coastal currents (notice the pebble imbrication that marks a landward directed current).

Wave reworking occurs mostly in the distal parts of the mouth bars, or is linked to episodes of low sediment input which may be related to the abandonment of the feeder channels.

Organic activity also modified some of the original mouth-bar deposits and the wave-reworked sediments. This organic reworking is recognized by the presence of marine bioturbation of varying intensity such as simple burrows, bored clasts (Fig. 7) or even the complete homogenization of the sediment. Organic reworking may be associated with episodes of low sediment supply, as is wave reworking, but it is more important during transgressive stages.

Fan-delta slope

The fan-delta slope is the most distal facies belt of the complex (Fig. 4) and is generally characterized by a steeper depositional dip of the bedding compared to the delta-front and alluvial deposits. This facies belt consists principally of marly prodelta deposits. These are intercalated with sandstones and conglomerates deposited from sediment gravity flows, and also with slumped and slid blocks. The resedimentation becomes more frequent where the prograding fan delta reaches the slope break formed by the underlying prograding unit.

The recognized sediment gravity flow deposits (from proximal to distal) are composed of the following.

Fig. 7. Vertical view of a pebble lag, product of the wave reworking of mouth-bar deposits. Notice the presence of abundant bored clasts indicating a later organic reworking.

1 Single and multistorey sandy and conglomeratic channel-fill facies deposited by high-density channellized turbidites of type R3, S1 and S3 (Lowe, 1982); these facies have been interpreted as deposited by high-density turbidity currents inside small (a few metres wide) channels or gullies incised on the fan-delta slope.

2 Clays, silts and tabular sandstone beds laterally related to the channel-fill facies: these are interpreted as overbank and/or levee deposits laden from S1, S2, S3, R2 and R3 high-density turbidity currents (Lowe, 1982).

3 Marls and generally tabular or convex-up (from a few centimetres to 2 m thick) bodies of sandstone and conglomerate corresponding to high-density (R, R2, R3, S1, S2, or S3 of Lowe, 1982) and low-density (Ta-b or Ta-c of Bouma, 1962) turbidity currents; these facies are interpreted as deposited on small turbiditic apron lobes tens (occasionally hundreds) of metres wide fed by the above-described channels.

Slide and slump processes have also been recognized in the fan-delta slope and fan-delta front. Extensional zones with structures such as normal (and listric) faults, rollover anticlines and boudinage can be distinguished. These extensional features are generated in the head zone of the slumps (Lewis, 1971). Compressional features linked to slump movement have also been observed. These structures are thrusts, horses, duplexes, schistosity and downslope vergent folds. They are generated on the toe zone of the slump (Lewis, 1971) where its movement is being stopped, generally due to a decrease in slope.

Carbonate platform

The carbonate platform deposits comprise less than 10% of the Sant Llorenç del Munt complex. In some cases, these facies can be laterally traced into transgressive (wave- and organic-reworked) fan-delta-front sediments. In general, the carbonate facies directly overlie conglomeratic lags (produced by wave reworking of previous detrital deposits) and show a transgressive tendency (Fig. 9). Where the carbonate deposits are best developed (thick carbonate successions), they show a transgressive-regressive tendency.

Two major facies associations have been distinguished within the carbonate platform deposits: bioclastic bars and reefs.

Bioclastic bars

Bioclastic bars are the dominant facies within the carbonate platform deposits and usually represent the transgressive part of the transgressive–regressive carbonate successions. They are represented generally by limestone bodies from 0.1 to 3 m in thickness and have a lateral extent from tens of metres to 2.5 km in a dip section. Five main facies belts orientated parallel to the palaeocoastline have been distinguished (Fig. 8a). From shallow to deeper areas these consist of: very bioturbated marly and conglomeratic sandstones with some fauna of *Nummulites* and bivalves (1) interpreted as transgressive fan-delta-front deposits reworked by waves and organisms. These reworked fan-delta front facies pass basinwards into true bioclastic bar facies composed of Foraminifera packstones (Fig. 9), interpreted as 'in situ' accumulations of fauna (*Nummulites* (2) in the shallower parts and discocyclines (3) in the deeper parts) (Aigner, 1982; Serra-Kiel, 1982). Packstones may be interbedded with grainstones of *Nummulites* and discocyclines (4), interpreted as the product of the reworking by coastal currents of the 'in situ' fauna accumulations into small shoals. Towards the basin the discocycline packstones pass into discocycline marls (5).

Reefs

This facies can be traced laterally and vertically into bioclastic bars. Reefs were formed over a pebbly or a previously colonized substratum. The framework organisms were either red algae or corals. Two main types of reef have been differentiated: coastal and barrier reefs.

Coastal reefs were located in the shallowest part (southeast) of some carbonate platforms, just over wave-reworked conglomerates (Figs 8b & 10) and were generally deposited during the maximum transgressive stages of the carbonate successions. The core of the reef is composed of red algae and coral boundstones, passing basinwards into rudstones (product of the destruction of the reef). The coastal-reef facies generally passes into the bioclastic-bar facies towards the basin.

The barrier-reef facies is generally located in the uppermost part of the carbonate platform deposits and sometimes shows a regressive character with respect to the previous carbonate platform deposits, representing the regressive part of some

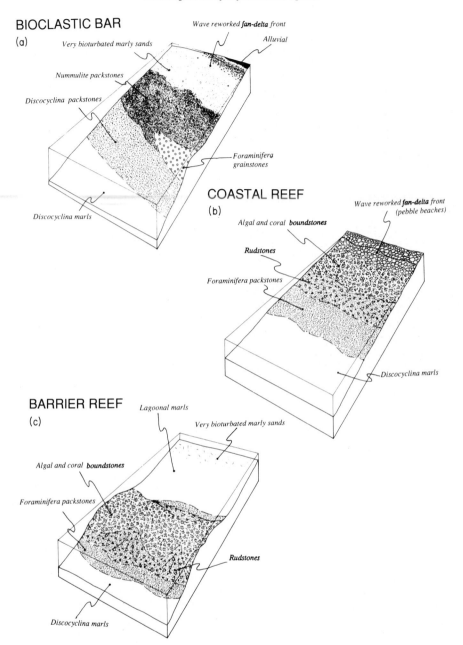

Fig. 8. Carbonate platform facies associations. Block diagrams showing the spatial distribution of the different facies belts within (a) bioclastic bars, (b) coastal reefs, (c) barrier reefs.

transgressive–regressive carbonate successions. The core of the reef is composed of coralline and algal boundstones. These bioconstructions formed barriers that protected a lagoonal zone to the southeast where marly sediments and bioclastic-bar facies

have been deposited (Fig. 8c). Towards the northwest there was a reef slope where rudstones, packstones, and discocyclina marls were deposited from proximal (shallower) to distal (deeper) parts.

Carbonate platform deposits are linked with

Fig. 9. Bioclastic bar facies showing a deepening-upward sequence. From base to top: wave-reworked conglomeratic deposits (pebble lag); 1 m thick, very bioturbated sands with shallow-marine fauna; Foraminifera packstone.

periods of low clastic sediment supply or when clastic detritus was restricted to the proximal areas of the system during relative sea-level rise. These transgressive episodes flooded the distal portion of the alluvial fan, generating an extensive shallow-water zone, a favourable environment for the development of the shallow-marine fauna and the generation of carbonate deposits.

STRATIGRAPHY

Because of the great variety of facies within the Sant Llorenç del Munt complex and their rapid lateral changes, a classical lithostratigraphical subdivision would not be useful. Thus, a stratigraphical subdivision based on allostratigraphical units is presented. These allostratigraphical units are defined as mappable stratiform bodies of sedimentary rock

that are defined and identified on the basis of their bounding discontinuities (NACSN, 1982).

The subdivision of sedimentary basin fills has been developed by several authors. The two most important schools are the Exxon group (Brown & Fisher, 1977; Mitchum *et al.*, 1977; Vail *et al.*, 1984; Vail, 1987; Van Wagoner *et al.*, 1987; Haq *et al.*, 1988; Mitchum & Van Wagoner, 1990) with depositional sequences, and Galloway (1989a, b) with genetic stratigraphical sequences.

The Exxon group has developed the methodology of sequence stratigraphy (the study of the relationships between beds or bedsets genetically related and chronostratigraphical units). The basic stratigraphical unit of sequence stratigraphy is the depositional sequence, defined by Mitchum *et al.* (1977) as a relative conformable succession of genetically-related strata bounded by unconformities or their correlative conformities. Depositional sequences can be subdivided into systems tracts which are linkages of contemporaneous depositional systems (Brown & Fisher, 1977) interpreted to be deposited during a specific portion of one complete cycle of eustatic fall and rise of sea-level (Vail, 1987). There are lowstand, transgressive, highstand and shelf-margin systems tracts. Systems tracts can be subdivided into parasequences which are the fundamental building blocks of sequences. A parasequence is defined as a relatively conformable succession of genetically-related beds or bedsets bounded by marine flooding surfaces and their correlative surfaces. Systems tracts are recognizable by their position within the sequence, the stacking pattern of the parasequence sets, and their depositional environments and facies (Van Wagoner *et al.*, 1990).

Sequence boundaries are surfaces formed during relative falls of sea-level. There are two types of sequence boundary: Type I sequence boundaries (Vail, 1987) are interpreted to occur when sea-level falls below the shelf break producing subaerial exposure and laterally equivalent submarine erosion, onlap of overlying coastal strata onto the sequence boundary, and an abrupt basinward shift of facies. Type II sequence boundaries (Vail, 1987) occur when sea-level does not fall below the depositional shoreline break, resulting in a downward shift of coastal onlap, minor subaerial erosion and exposure on the shelf, and vertical change in parasequence stacking pattern.

Mitchum and Van Wagoner (1990) develop slightly further the sequence stratigraphy model.

Fig. 10. Coastal reef deposits on top wave-reworked conglomeratic deposits.

Their work defines composite sequences as depositional sequences whose building units are high-frequency depositional sequences (instead of parasequences). They also define sequence sets as equivalent (in meaning) to systems tracts, but defined by the stacking pattern of high-frequency depositional sequences.

Depositional sequences and their components are a response to the interaction between eustasy, subsidence and sediment supply rates (Vail *et al.*, 1984; Vail, 1987; Van Wagoner *et al.*, 1987; Haq *et al.*, 1988) and are generated between two relative sea-level falls. The Exxon model ties the boundary to a specific cause like eustatic fall in sea-level because they assume that the rates of sediment supply and subsidence are constant.

Other methodology has been developed by Galloway (1989a) based on the genetic stratigraphical sequences of Frazier (1974).

Genetic stratigraphical sequences are the sedimentary product of a depositional episode, deposited during a period of regional palaeogeographical stability. They are packages of sediments recording a significant episode of basin-margin outbuilding and basin filling bounded by periods of widespread basin-margin flooding. Thus, the genetic stratigraphical sequences are bounded by maximum-flooding surfaces. These maximum-flooding surfaces are hiatal surfaces that reflect major reorganizations of the depositional systems. Maximum-flooding surfaces are recognized as the result of

three interacting factors (sediment influx, tectonic subsidence or uplift, and eustasy).

Depositional episodes and genetic stratigraphical sequences consist of three families of elements: offlap components, onlap or transgressive components, and bounding surfaces reflecting maximum marine flooding.

Stratigraphical subdivision

In the Sant Llorenç del Munt complex, transgressive–regressive allostratigraphical units of several scales, orders and frequencies have been distinguished. Going from more to less frequent, they have been termed fundamental sequences (López-Blanco & Marzo, 1991) and composite sequences (López-Blanco, 1991).

The fundamental sequences are the highest order, minor scale, laterally persistent and mappable transgressive–regressive sequences. Their thickness ranges from 3 to 80 m and may be subdivided into three parts (Fig. 11).

1 The first part is a basal transgressive part located between a transgressive surface and a maximum-flooding surface, consisting of carbonate platform facies and their lateral equivalents.

2 The second is an upper regressive part composed of prograding fan-delta deposits showing coarsening-and-thickening-upward sequences which overlie the maximum-flooding surface.

3 Pebble lags and occasionally coastal, turbiditic

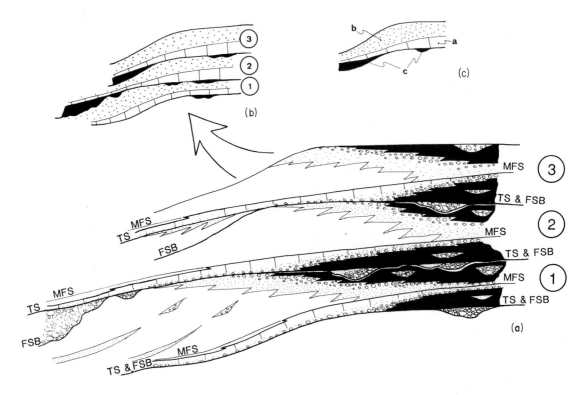

Fig. 11. (A) Diagram showing three types of fundamental sequences, their key surfaces (TS, transgressive surface; FSB, fundamental sequence boundary; MFS = maximum flooding surface), and their facies distributions (the symbols are similar to those in Fig. 4). (B) simplified diagram of (A). (C) Components of a fundamental sequence: a, transgressive part; b, regressive part; c, 'lowstand' deposits.

and alluvial channel-fill deposits may be related with deposition during relative sea-level falls. These deposits overlie part of the regressive deposits, and at the same time are overlain by the transgressive ones. They are deposited immediately above the fundamental sequence boundary (Fig. 11).

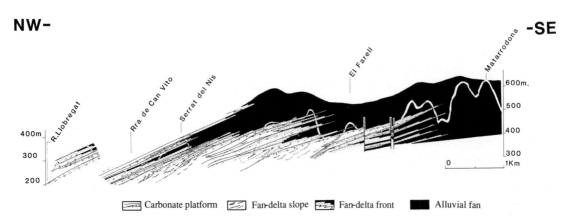

Fig. 12. Geological cross-section of the Sant Llorenç del Munt complex parallel to the main transport direction.

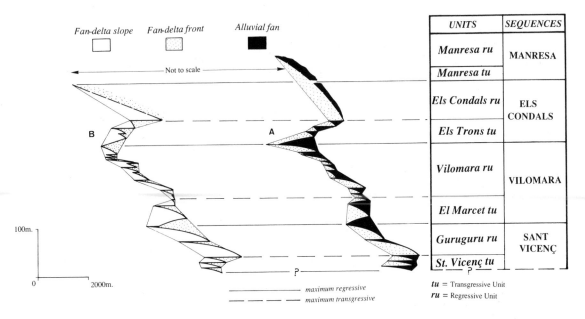

UNITS	SEQUENCES
Manresa ru	MANRESA
Manresa tu	
Els Condals ru	ELS CONDALS
Els Trons tu	
Vilomara ru	VILOMARA
El Marcet tu	
Guruguru ru	SANT VICENÇ
St. Vicenç tu	

tu = Transgressive Unit
ru = Regressive Unit

Fig. 13. Variations in the horizontal position of the coastline (A) and of the external margin of the fan-delta-front facies (B) in the section of Fig. 12. Location of the levels of maximum regression and maximum transgression and the position of units and composite sequences. Each little wedge represents a fundamental sequence.

The fundamental sequence boundaries are time surfaces separating transgressive deposits (1) from regressive deposits (2), except when interval (3) is present. The stacking pattern of fundamental sequences has been studied along an 8-km-long cross-section parallel to the main transport direction (Fig. 12). In addition, two graphs have been constructed (Fig. 13) in which the thickness of each fundamental sequence (measured above the maximum extent of the mouth-bar facies of the underlying fundamental sequence) has been related to the magnitude of the horizontal displacement of the coastline (boundary between alluvial fan and fan-delta-front deposits) and the external margin of the fan-delta front (transition from fan-delta-front to fan-delta-slope deposits).

On the basis of these two graphs, 11 different units have been defined (Figs 13, 14 & 15). The units are sets of fundamental sequences which display a common stacking pattern. There, the transgressive units are sets of fundamental sequences showing aggradational to retrogradational stacking patterns and the regressive units are sets of fundamental sequences which display a clear progradational stacking pattern.

The base of a transgressive unit coincides with the top of the previous regressive unit and with the top of the uppermost fundamental sequence belonging to the preceding regressive unit. This surface marks the change from regressive to transgressive stacking patterns, and is therefore a maximum regression surface.

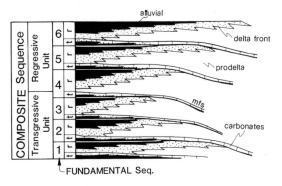

Fig. 14. Diagram showing the definition of the composite sequences, transgressive and regressive units, and fundamental sequences with their transgressive and regressive components (1, 2, 3, 4, 5 and 6 are fundamental sequences; mfs, maximum-flooding surface; t, transgressive part; r, regressive part).

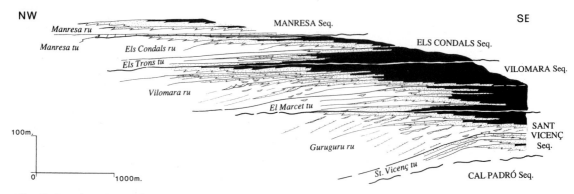

Fig. 15. Location of the defined units and composite sequences along the Fig. 12 cross-section.

The base of a regressive unit coincides with the top of the previous transgressive unit, but in this case it is given by a flooding surface (maximum-flooding surface) that divides a fundamental sequence in two. Thus, a fundamental sequence is divided between two different units, with its transgressive part belonging to the transgressive unit and its regressive part to the regressive unit.

The composite sequences (Figs 13, 14 & 15) have been defined as transgressive–regressive sequences composed of couplets of units (a transgressive unit at the base followed by a regressive unit at the top). The composite sequence boundaries are time surfaces separating transgressive units (the basal term of the sequence) from regressive units of the preceding sequences. Thus, these bounding surfaces have the same field expression as the fundamental sequence boundaries because the composite sequence boundaries are also fundamental sequence boundaries.

The main characteristics of the well-studied composite sequences and units are presented in Table 1.

Cyclicity

In the eastern part of the South Pyrenean foreland basin four main orders of transgressive–regressive cyclicity have been differentiated. The most extensive cycle has a time span of about 6 My, so it is similar in scale to 2nd-order cycles (Vail *et al.*, 1977; Haq *et al.*, 1988). The more frequent cycles are of 3rd, 4th and 5th order. The above-mentioned authors link the transgressive–regressive cycles only to global changes. In the Sant Llorenç del Munt complex the transgressive–regressive cycles are also related to variations in sediment supply and subsidence, both partly controlled by tectonism.

The 5th-order cycles are represented by 1–15-m-thick transgressive–regressive sequences, impersis-

Table 1. Main characteristics of some of the defined composite sequences. TU, transgressive unit; RU, regressive unit

	Sant Vicenç	Vilomara	Els Condals
Average thickness	100 m	169 m	130 m
TU average thickness	35 m	57 m	50 m
RU average thickness	65 m	112 m	80 m
Retreat of the coastline on the TU	1900 m	1000 m	4400 m
Advance of the coastline on the RU	2850 m	4615 m	3000 m
Advance–retreat	950 m	3615 m	– 1400 m
No. of fundamental sequences in the TU	3	3	7
No. of fundamental sequences in the RU	2	10	?
Advance–retreat	1.5	4.6	0.68
RU thickness/TU thickness	1.85	1.96	1.6

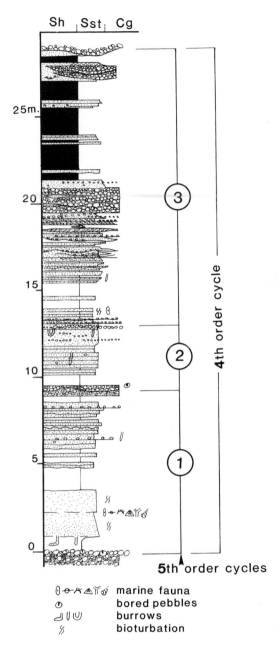

Fig. 16. Columnar section showing the field expression of the 4th- (fundamental sequences) and 5th-order cycles.

with some *Nummulites* and bivalves and wave-reworked facies) that is overlain by a coarsening-and-thickening-upward sequence of mouth-bar facies (Fig. 16). These cycles may be related to the migration and abandonment of the mouth bars in response to autocyclic (channel shifting) or less frequent allocyclic processes (relative sea-level changes and sediment supply variations).

The 4th-order cycles (Fig. 16) are represented by the previously defined fundamental sequences and have an approximate duration of between 50 000 and 200 000 years. This 4th-order cyclicity is believed to have been controlled principally by allocyclic processes because of its lateral continuity (km). The origin of these cycles would be the activation or abandonment of the fan-delta system in response to eustato-climatic variations (fig. 17) or tectonic activity (Figs 18, 19 & 20) on the active basin margin, controlling subsidence and sediment supply.

The 3rd-order cycles are represented by the composite sequences. These cycles are from 70 to 180 m thick and were deposited during intervals of approximately 800 000 years. As with the 4th-order cycles, interaction between larger-scale eustatic changes, variations of subsidence, and sediment supply may have controlled the 3rd-order cycles.

The 2nd-order cyclicity is represented by the Milany sequence of Puigdefàbregas *et al.* (1986). The studied area comprises part of the regressive term of this sequence which has been interpreted as forming in response to tectonic activity in the foreland basin (Puigdefàbregas *et al.*, 1986).

THE STRATIGRAPHICAL SUBDIVISION COMPARED TO OTHER MODELS

The stratigraphical subdivision detailed above is, in the following, compared with depositional sequences (Mitchum *et al.*, 1977), parasequences (Van Wagoner *et al.*, 1987) and composite sequences (Mitchum & Van Wagoner, 1990) of the Exxon model and with genetic stratigraphical sequences (Galloway, 1989a).

Fundamental sequences versus depositional sequences

As described above, the fundamental sequences are made up of two main parts: a lower transgressive

tent laterally, and mostly observed in prograding fan-delta-front deposits. They generally consist of a thin, basal reworked interval (bioturbated facies

CLIMATE

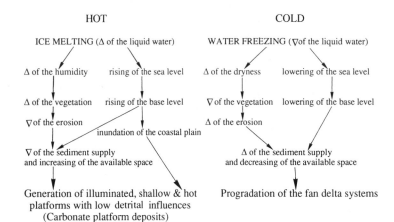

Generation of illuminated, shallow & hot
platforms with low detrital influences
(Carbonate platform deposits)

Prograadation of the fan delta systems

Fig. 17. Possible climatic influence
on the sedimentary cycles.

carbonate part, and an upper regressive fan-deltaic part.

The transgressive part is limited at the base by a flooding surface (transgressive surface) and at the top by a maximum-flooding surface. These bounding surfaces have generally clear expressions in the field, where they are related to transitional (coastal) facies. Towards deeper and shallower parts of the system, they are sometimes difficult to recognize because these surfaces may not separate different types of facies (Fig. 11). The transgressive part usually also shows an obvious transgressive character. Where it is composed of several carbonate layers, the beds often display an onlapping, or

TECTONICS

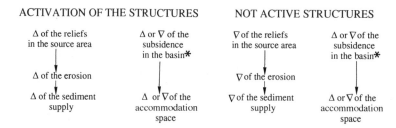

Fig. 18. Possible tectonic influence on the sedimentary cycles. (*) The increasing or decreasing rate of subsidence at a fixed point within the basin is dependent upon the type of deformation and structures that control the basin margin. If the tectonic activity is represented by a simple tilting, the rate of subsidence (or uplift) depends on the location with respect to the tilt axis (see Fig. 19c). At one side of this axis there is an uplift and an increase in relief; at the other side there is a relative sinking, giving a major rate of subsidence. The position of the tilting axis with respect to the coastline will control the movement of relative sea-level, generating relative transgressions in some cases (see Fig. 19(a) 3 & (b) 3), or regression in others (see Fig. 19(a) 1 & (b) 1). If the tectonic activity is represented only by relative movements along a major fault (see Fig. 20), then there will be an increasing amount of subsidence in the basin, accompanied by an increasing amount of uplift in the source area. Depending on the relations between the variations of accommodation space and the volume of sediment supply, there will be transgressive or regressive tendencies. A decrease in accommodation space (relative regression) and increase of the sediment supply result in *detrital progradation*. An increase in the accommodation space (relative transgression) and a decrease of the sediment supply results in *detrital retrogradation* (or deposition of carbonate platform facies).

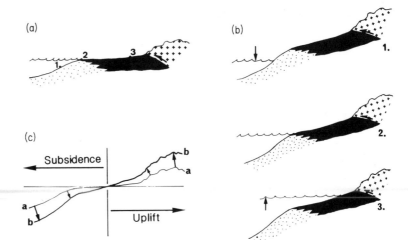

Fig. 19. Results of tilting on coastal sedimentation. (a) Initial stage (1, 2 & 3 represent the possible axes of tilting). (b) Situation after tilting (with 1, 2 & 3 representing the three different results depending on the different tilt axis locations shown in (a). (c) Scheme showing how subsidence and uplift rates are increased on either side of the axis.

backstepping, disposition just above the transgressive surface. These characteristics suggest that the transgressive parts of the fundamental sequences are similar to the transgressive systems tract of the depositional sequences (Brown & Fisher, 1977; Vail, 1987).

The regressive part of the fundamental sequences are fan-delta deposits showing a marked prograding pattern located just above the maximum-flooding surface, where their clinoforms display a downlap relationship. These characteristics coincide with the highstand systems tract (Brown & Fisher, 1977; Vail, 1987) of the Exxon model.

In the Sant Llorenç model the sequence boundary would generally be over the regressive part and below the transgressive, and is coincident with the transgressive surface. The sequence boundary is usually marked by a lag of coarse-grained material

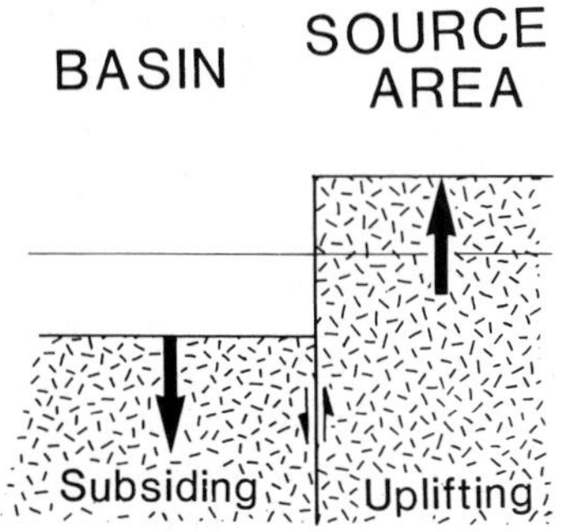

BASIN **SOURCE AREA**

Fig. 20. Diagram of a basin margin controlled by vertical movements along a major fault.

(pebbles and cobbles) located immediately above a truncation surface cutting the progradation surfaces of the preceeding fundamental sequence. These lags are interpreted as having been generated during the maximum regression (even a lowstand period) (Fig. 21), on areas having occasional subaerial exposure but without large-scale incision of ravines. Later, during the following transgressive episode these erosive surfaces and pavements were reworked by waves and organisms and the resulting deposit is generally a zone with clasts and coarse detrital material near the base of the carbonate facies (Figs 9 & 21).

Most of these surfaces reflect an abrupt shift of facies basinwards, are associated with regionally recognizable surfaces and indicate an important period of (even subaerial) erosion associated with a relative base-level fall. So, these surfaces can be considered similar to type I unconformities or type I sequence boundaries (Vail, 1987).

The type I depositional sequences (Vail, 1987) of the Exxon model have a lowstand systems tract (Brown & Fisher, 1977; Vail, 1987) that is not commonly observed in our fundamental sequences. Lowstand deposits should be located between a type I unconformity and the transgressive surface, but in the Sant Llorenç del Munt complex both of these surfaces are generally coincident. So, if there are any lowstand deposits they would be the lag of pebbles and cobbles mentioned above.

Only in a few cases have deposits similar to lowstand systems tracts been distinguished. Channellized turbiditic facies have been recognized.

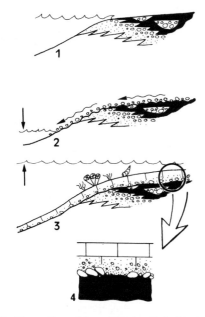

Fig. 21. Generation of the pebble lags linked to type I unconformities. (1) High relative sea-level stage, (2) lowering of the relative sea-level, (3) relative transgressive stage, (4) detail of the pebble lags and the transgressive part of a fundamental sequence.

Because of their location inside the sequence (see the fundamental sequence 2 on Fig. 11), their thickness, and their scale of incision, they may be interpreted as feeder channels of a submarine fan (Van Wagoner *et al.*, 1987) developed during a lowstand period. The fan deposits would be in the subsurface basinwards and do not crop out.

Occasionally, isolated deposits of prograding fandelta-front facies, separated from the main fandelta front deposits, have been recognized (see the fundamental sequence 3 on Fig. 11). Because of their location inside the sequences and their isolation from the main prograding body, these deposits could be considered as lowstand prograding wedges deposited during a stillstand in a relative lowstand period (Van Wagoner *et al.*, 1987; Posamentier & Chamberlain, 1990).

Sometimes there is an abundance of channelized alluvial deposits close to the sequence boundary. Perhaps these are incised valley fill deposits (Van Wagoner *et al.*, 1987), but there is no clear evidence for that. These deposits could be related only to the last stages of progradation of the system, when the aggradation rate is very low and there is much

bypassing, with predominance of coarse channellized alluvial deposits.

If some shelf-margin systems tract (Brown & Fisher, 1977; Vail 1987) deposits were present, they would be located between a type II unconformity and the transgressive surface. These type II surfaces do not mark any type of erosion linked to the rejuvenation of the fluvial systems, nor submarine erosion nor downward shifting of the facies belts. Thus, the distinction between the latest part of the highstand systems tract and any shelf-margin systems tract deposits within the regressive parts of the fundamental sequences would be very difficult to make. If any shalf-margin systems tract deposits exist they will correspond with the last episodes of progradation of the fan delta. If there is a shelf-margin systems tract, then the pebble lags overlying the transgressive surface would have an origin linked only to the transgressive reworking, or are related to a later, major sea-level fall (generating a type I unconformity after the deposition of the shelf-margin systems tract).

Therefore, it seems that fundamental sequences may compare with depositional sequences of the Exxon model (the fundamental sequences would be 4th-order depositional sequences). The main problem with the fundamental sequences is that they cannot be subdivided into parasequences like the Exxon model depositional sequences (Van Wagoner *et al.*, 1987). Fifth-order cycles have been observed, but they cannot be considered as parasequences because they include regressive and transgressive deposits and because of their scarce lateral continuity. So perhaps the fundamental sequences are equivalent (only as building blocks) to the Exxon model 4th-order parasequences (Vail *et al.*, 1977; Haq *et al.*, 1988).

Fundamental sequences versus parasequences

There are noticeable conceptual differences between fundamental sequences and parasequences:
1 Parasequences (Van Wagoner *et al.*, 1987) are conformable successions of genetically-related beds or bedsets bounded by a marine-flooding surface and their correlative surfaces, expressed by shallowing-upward prograding successions bounded by marine-flooding surfaces. Fundamental sequences show coarsening-and-shallowing-upward prograding successions (the regressive part), however they also contain a transgressive part composed of carbonate-platform deposits and their

lateral equivalents. Therefore, the transgressive deposits of the fundamental sequence are a critical point of contrast with the parasequence; these transgressive deposits are replaced only by a flooding surface in the latter.

2 Fundamental sequences contain components resembling different systems tracts (Brown & Fisher, 1977; Vail, 1987), whereas parasequences cannot be subdivided.

3 Parasequences are bounded by marine-flooding surfaces whereas fundamental sequence boundaries are expressed as maximum regressive surfaces.

Common points between fundamental sequences and parasequences are the scale (4th order), and that both can be used (through their stacking pattern) as fundamental building blocks to define larger-scale sequences.

Fundamental sequences can thus be considered similar to the Exxon model depositional sequences (Mitchum *et al.*, 1977; Van Wagoner *et al.*, 1987) or high-frequency sequences (Mitchum & Van Wagoner, 1990) but there is an important difference. Fundamental sequences are not divisible into parasequences.

Fundamental sequences versus genetic stratigraphical sequences

In the fundamental sequences the components of the model of Galloway (1989a) can be easily distinguished. Offlap components that include (1) sandy fluvial, delta plain, and bay–lagoon facies that reflect aggradation of the coastal plain, (2) sandy progradational deposits of the shore–zone, and (3) mixed aggradational–progradational slope deposits (Galloway, 1989a) would correspond to the prograding fan-delta deposits of the prograding regressive part of the fundamental sequence.

Onlap components consist of (1) reworked shore-zone and shelf facies deposited during and soon after shoreline retreat and (2) an apron of gravitationally resedimented, upper-slope and shelf-margin deposits at the toe of the slope (Galloway, 1989a). The transgressive part of the fundamental sequence would be represented by the reworked facies (1) of Galloway (1989a), while term (2) of his onlap components would be represented by cases like the (c component of the fundamental sequence 2 on Fig. 11).

The major differences between the fundamental and the genetic stratigraphical sequences are that fundamental sequences are bounded by the surface of maximum regression (they are transgressive–regressive sequences) while the genetic stratigraphical sequences are bounded by maximum-flooding surfaces (they are regressive–transgressive sequences).

Composite sequences versus Exxon and Galloway models

The fundamental sequences have been considered as working tools, equivalent in significance and scale to parasequences, in order to define major-scale sequences. This method of defining larger sequences by using the stacking pattern of the minor sequences has been utilized by Mitchum and Van Wagoner (1990). They first define parasequences (5th order), next systems tracts, then high-frequency depositional sequences (4th order) and sequence sets (similar to the systems tracts but instead of being made by piling of parasequences they are constructed by depositional sequences), and finally composite sequences (3rd order).

Depositional and composite sequences (Mitchum *et al.*, 1977; Van Wagoner *et al.*, 1987; Mitchum & Van Wagoner, 1990), and genetic stratigraphical sequences (Galloway, 1989a) are thus defined by joining elementary units like parasequences (Van Wagoner *et al.*, 1987), depositional sequences (Mitchum & Van Wagoner, 1990) or genetic stratigraphical sequences (Galloway, 1989a) into sets with similar stacking patterns like systems tracts (Brown & Fisher, 1977; Vail, 1987), sequence sets (Mitchum & Van Wagoner, 1990) or components (Galloway, 1989a).

In the Sant Llorenç del Munt fan-delta complex the composite sequences have been defined using the stacking patterns of fundamental sequences. Transgressive units and regressive units were thus constructed.

Transgressive units of this study are equivalent to the onlap components of Galloway (1989a) and to transgressive systems tracts (Brown & Fisher, 1977; Vail, 1987) or the transgressive sequence sets of Mitchum and Van Wagoner (1990) (Fig. 22).

Regressive units are equivalent to the offlap components of Galloway (1989a) and mainly to the highstand systems tracts or highstand sequence sets (Brown & Fisher, 1977; Vail, 1987; Mitchum & Van Wagoner, 1990) of the Exxon model. The lowstand systems tracts (or lowstand sequence sets), if present in the Sant Llorenç del Munt deposits, do not crop out (they would be in the subsurface) and

M. LÓPEZ BLANCO 1991		EXXON	GALLOWAY 1989
SEQUENCE	UNIT	CS	GSS
MANRESA	Manresa RU	HSS	Offlap comp.
	Manresa TU	TSS	Onlap comp.
ELS CONDALS	Els Condals RU	HSS	Offlap comp.
	Els Trons TU	TSS	Onlap comp.
VILOMARA	Vilomara RU	HSS	Offlap comp.
	El Marcet TU	TSS	Onlap comp.
SANT VICENÇ	Guruguru RU	HSS	Offlap comp.
	Sant Vicenç TU	TSS	Onlap comp.
CAL PADRO	La Creueta RU	HSS	Offlap comp.
	Cal Padró TU	TSS	Onlap comp.
RELLINARS	Rellinars RU	HSS	Offlap comp.

——————— Sequence boundary

------------- Maximum flooding surface

Fig. 22. Comparison of the Sant Llorenç del Munt units and composite sequences with the sequence sets and composite sequences of Exxon group (Mitchum & van Wagoner, 1990), components and genetic stratigraphical sequences (Galloway, 1989a). GSS = Genetic stratigraphical sequences; CS = composite sequences; HSS = highstand sequence sets; TSS = transgressive sequence set; TU = transgressive unit; RU = regressive unit.

have not been distinguished. The shelf-margin systems tract, if it exists, is not easily recognizable but it would correspond to the latest (uppermost) and most strongly prograding fundamental sequences of the regressive units.

The transgressive to regressive inflection is represented by the maximum-flooding surface that marks the limit between the transgressive and the regressive units within the Sant Llorenç del Munt composite sequences. That surface is equivalent to the maximum-flooding surface on top of the transgressive systems tract (Brown & Fisher, 1977; Vail, 1987) or sequence set (Mitchum & Van Wagoner, 1990) and the genetic stratigraphical sequence boundary (Galloway, 1989a).

The regressive to transgressive inflection is marked by the maximum regression surface that is the sequence boundary in the Sant Llorenç del Munt composite sequences. This surface corresponds to the change from offlap to onlap components (Galloway, 1989a) and in the Exxon model to the transgressive surface. If there are type I sequences (Vail, 1987; Mitchum & Van Wagoner, 1990) in the Sant Llorenç del Munt complex, the transgressive surfaces coincide at outcrop with the type I sequence boundary. If there are type II sequences, the type II unconformity would be inside our regressive units and these units would be divided between two different depositional sequences 'sensu' Mitchum and Van Wagoner (1990). The relationships between our sequences and other models are shown in Fig. 22.

PALAEOGEOGRAPHICAL EVOLUTION

The studied stratigraphical succession is composed of an alternation of deposits reflecting dominance of alluvial activity (progradation of the fan-delta system) and reflecting scarce alluvial activity (creation of shallow-water carbonate platforms and reefs over the abandoned fan-delta deposits). The fundamental sequences reflect this alternation.

The overall palaeogeographical evolution of the zone reflects a migration of the coastline and the different facies belts towards the northwest. This general regressive situation was punctuated by a series of transgressive pulses (high-frequency transgressions), reflected by the transgressive part of the fundamental sequences as well as higher period transgressions reflected by the transgressive units of the composite sequences. Figure (see pp. 86–7) 23 shows the loction of the different sedimentary environments (alluvial fan, fan-delta front and fan-delta slope) at the top of the regressive part of the uppermost fundamental sequence of each unit.

CONCLUSIONS

Two major sedimentary systems have been defined within the Sant Llorenç del Munt succession. These systems alternated in time and consisted principally of either prograding fan deltas or carbonate platforms (and their lateral equivalents). The prograding fan deltas were developed during periods of important alluvial activity and consisted of alluvial fan, fan-delta-front and fan-delta-slope deposits. The proximal alluvial fan was dominated by steep slopes and sediment gravity flow deposits, while distal parts of the alluvial fan were characterized by channelized streams and floodplain deposits. The fan-delta front was constructed by the building, migration and later reworking (by coastal currents or organic activity) of mouth bars. The fan-delta slope is composed principally of blue-grey marls which may be interbedded with sediment gravity flow deposits, slumps and slides. The carbonate platforms were built during periods of low alluvial activity (relative transgressive episodes) and onlap the fan-delta deposits. These platforms were shallow, close to the coast and sometimes developed coastal and barrier reefs.

There are at least three orders of transgressive–regressive cyclicity which are linked to the interaction between changes in relative sea-level, variations in clastic input and the shifting of the different deltaic lobes. These parameters would be linked to autocyclic (shifting) and allocyclic processes (tectonic activity on the basin margin and climatic variations). The most easily applicable stratigraphical model is the Galloway model, mainly because of its simplicity, and because generally the maximum-flooding surfaces are the easiest to recognize in the field. Nevertheless, a new stratigraphical subdivision has been developed for the Sant Llorenç del Munt system. Fundamental sequences are the major-order or minor-scale, laterally persistent and mappable transgressive–regressive sequences in which a transgressive and a regressive part (and occasionally isolated lowstand deposits) can be distinguished. Fundamental sequences are equivalent in architecture to depositional sequences (Mitchum *et al.*, 1977; Vail, 1987), and high-frequency sequences (Mitchum & Van Wagoner, 1990), and are equivalent to parasequences in scale and utility (by their ability to build major sequences). The units are sets of fundamental sequences showing a common stacking tendency and are similar to both the components of Galloway (1989a), the systems tracts of Exxon (Brown & Fisher, 1977; Vail, 1987), and the sequence sets of Mitchum and Van Wagoner (1990) but there are difficulties in recognizing lowstand systems tracts and/or shelf-margin systems tracts. Composite sequences are transgressive–regressive sequences made by couplets of units (a transgressive unit at the base followed by a regressive unit at the top), equivalent to the composite sequences of Mitchum and Van Wagoner (1990). (There are, however, differences between the two types of composite sequence because of our inability to differentiate between highstand systems tracts and lowstand or shelf-margin systems tracts within the regressive units of the Sant Llorenç model.)

Within the Sant Llorenç del Munt complex parasequences as defined above do not exist. The high-frequency sequences are composed of regressive and transgressive deposits. The recognition of key surfaces at a fundamental sequence scale (maximum-flooding surface, sequence boundary, transgressive surface) is not easy, because they are not generally accompanied by appreciable lithological changes. In detail, fundamental sequence and composite sequence boundaries often have the same field expression. A regional study is necessary to determine whether or not the fundamental sequence boundary corresponds with a composite sequence boundary.

The coastline and the associated sedimentary belts migrated towards the northwest during the studied period. The Sant Llorenç del Munt succession displays an overall regressive tendency which was punctuated by transgressive pulses of variable magnitude.

ACKNOWLEDGEMENTS

I wish to thank Mariano Marzo for his help, supervision, support and discussions during the development of this study. Also I wish to thank Beverly Burns and Ole J. Martinsen for their careful review and helpful criticism of the preliminary version of this paper, Deborah South and Ron Steel for the final corrections and the people from the Department de Geologia Dinàmica Geofísica i Paleontologia of the University of Barcelona and the Geologisk Institutt, Avd. A of the University of Bergen who helped me. This work was financed by the Servei Geologic de Catalunya and DGICYT projects PB91-0805 and PB91-0801.

REFERENCES

AIGNER, T. (1982) Event stratification in *Nummulite* accumulations and in shell beds from the Eocene of Egypt. In: *Cyclic and Event Stratification* (Eds Einsele, G. & Seilacher, A.) Springer-Verlas, Berlin; 248–268.

ANADÓN, P. (1978) *El Paleógeno continental anterior a la transgresión (Eocen medio) entre los rios Gaià y Ripoll (prov. de Tarragona y Barcelona).* Doctoral Thesis, Universitat de Barcelona, 267 pp.

ANADÓN, P., COLOMBO, F., ESTEBAN, M. & MARZO, M. (1979) Evolución tectonoe-stratigráfica de los catalánides. *Acta Geol. Hisp.* **11**, 73–78.

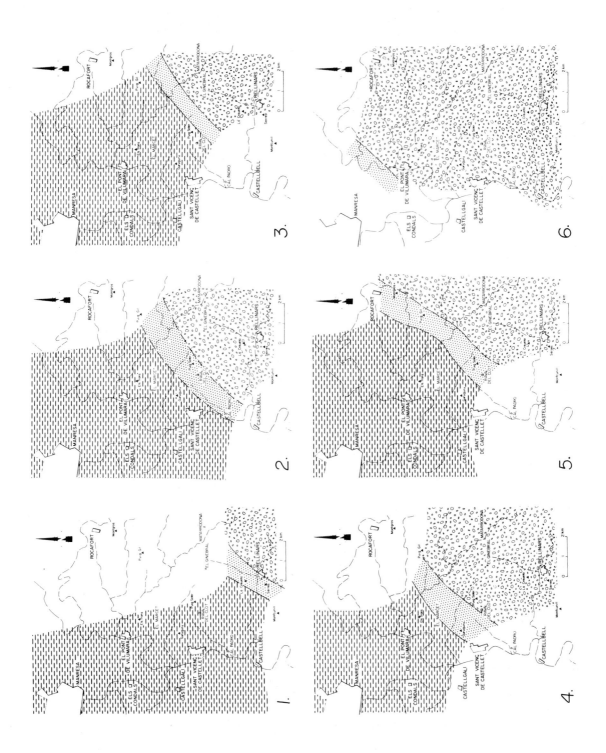

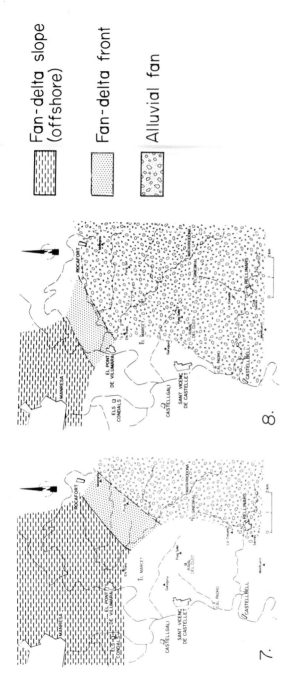

Fig. 23. Palaeogeographical maps corresponding to the last stages of progradation of the uppermost fundamental sequence of each defined unit. (1) Cal Padró Tu, (2) La Creueta Ru, (3) Sant Vicenç Tu, (4) Gurugurú Ru, (5) El Marcet Tu, (6) Vilomara Ru, (7) Els Trons Tu, (8) Els Condals Ru. The Manresa sequence (see Figs 13 & 15) has not been represented because we cannot control its transgressive unit; and because the last stages of its regressive unit are represented by alluvial fan deposits, the palaeocoastline being out of the map (northwest of Manresa city).

ANADÓN, P., MARZO, M. & PUIGDEFÀBREGAS, C. (1985a) The Eocene fan-delta of (SE Ebro basin, Spain). *IV Europ. Ref. Meet., Int. Assoc. Sedimentologists,* pp. 109–146.

ANADÓN, P., CABRERA, L., GUIMERÀ, J. & SANTANACH, P. (1985b) Paleogene strike-slip deformation and sedimentation along the southeastern margin of the Ebro basin. In: *Strike-Slip Deformation, Basin Formation and Sedimentation* (Eds Biddle, K.T. & Christie-Blick, N.) Spec. Publ. Soc. Econ. Paleont. Mineral. **37**, 303–318.

ANADÓN, P., CABRERA, L., COLLDEFORNS, B., COLOMBO, F., CUEVAS., J.L. & MARZO, M. (1989) Alluvial fan evolution in the SE Ebro basin: response to tectonics and lacustrine base level changes. *4th International Conference on Fluvial Sedimentology. Excursion Guidebook* 9 (Eds Marzo, M. & Puigdefàbregas, C.).

BOUMA, A.M. (1962) *Sedimentology of some Flyschs Deposits: a Graphic Approach to Facies Interpretation.* Elsevier, Amsterdam, 168 pp.

BROWN, L.F. & FISHER, W.L. (1977) Seismic–stratigraphic interpretation of depositional system: examples from Brazil rift and pull-apart basins. In: *Seismic Stratigraphy—Applications to Hydrocarbon Exploration* (Ed. Payton, C.E.) Am. Assoc. Petr. Geol. Mem. **26**, 213–248.

ESTEBAN, M. & ROBLES, S. (1976) Sobre la paleogeografía del Cretácico inferior de los Catalánides entre Barcelona y Tortosa. *Acta Geol. Hisp.* **14**, 242–270.

FRAZIER, D.E. (1974) Depositional episodes: their relationship to the Quaternary stratigraphic framework in the northwestern portion of the Gulf Basin. *Univ. Texas at Austin, Bur. Econ. Geol., Geologic Circular,* **74-1**, 28 pp.

FONTBOTÉ, J.M. (1954) Las relaciones tectónicas de la depresión del Vallés-Penedes con la Cordillera Prelitoral y con la Depresión del Ebro. *Tomo homenaje al Prof. E. Hernández Pacheco. R. Soc. Esp. Hist. Nat.,* pp. 281–310.

GALLOWAY, W.E. (1989a) Genetic Stratigraphic Sequences in Basin Analysis I: Architecture and genesis of flooding-surface bounded depositional units. *Am. Assoc. Petrol. Geol. Bull.* **73**, 125–142.

GALLOWAY, W.E. (1989b) Genetic stratigraphic sequences in Basin Analysis II: applications to northwest Gulf of Mexico Cenozoic basin. *Am. Assoc. Petrol. Geol. Bull.* **73**, 143–154.

GUIMERÀ, J. (1988) Estudi estructural de l'enllaç entre la serralada Ibèrica i la serralada costanera catalana. Doctoral thesis, Universitat de Barcelona, 600 pp.

HAQ, B.U., HARDENBOL, J. & VAIL, P.R. (1988) Mesozoic and Cenozoic chrono-stratigraphy and cycles of sea-level change. In: *Sea-level Change; an Integrated Approach* (Eds Wilgas, C.K.,Hastings, B.S. Posamentier, H., Van Wagoner, J., Ross, C.A. & Kendall, C.G.St.C.) Soc. Econ. Paleontol. Mineral. Spec. Publ. **42**, 71–108.

LEWIS, K.B. (1971) Slumping on a continental slope inclined at 1 deg.-4deg. *Sedimentology* **16**, 97–110.

LÓPEZ-BLANCO, M. (1991) *Estratigrafía y sedimentología del sector occidental del abanico costero de Sant Llorenç del Munt al este de Sant Vicenç de Castellet (Eoceno, Cuenca de antepaís surpirenaica).* MSc thesis, Universitat de Barcelona, 135 pp.

LÓPEZ-BLANCO, M. & MARZO, M. (1991) Estratigrafía secuencial del abanico costero de Sant Llorenç del Munt (cuenca del Ebro NW de Espana). *I. Congreso del grupo espanol del terciario. Abstracts book*, pp. 182–187.

LOWE, D.R. (1982) Sediment gravity flows: II. Depositional models with special reference to the deposits of high-density turbidity currents. *J. Sediment. Petrol.* **52**, 279–297.

MAESTRO, E. (1987) Estratigrafía i facies del complex deltaic (fan delta) de Sant Llorenç del Munt (Eocè mig-superior. Catalunya). Doctoral thesis, Universitat Autonoma de Barcelona, 299 pp.

MARZO, M. (1980) El Buntsandstein de los catalanides, estratigrafía y procesos de sedimentación. Doctoral thesis, Universitat de Barcelona. 322 pp.

MARZO, M. & ANADÓN, P. (1988) Anatomy of a conglomeratic fan-delta complex: the Eocene Montserrat Conglomerate, Ebro Basin, northeastern Spain. In: *Fan Deltas: Sedimentology and Tectonic Settings* (Eds Nemec, W. & Steel, R.J.). Blackie, Glasgow, pp. 318–340.

MITCHUM, R.M., VAIL, P.R. & THOMPSON, S. (1977) Seismic stratigraphy and global changes of sea level, part 2; the depositional sequence as a basic unit for stratigraphic analysis. In: *Seismic Stratigraphic Applications to Hydrocarbon Exploration* (Ed. C.W. Payton) Am. Assoc. Petrol. Geol. Mem. **26**, 53–62.

MITCHUM, R.M. & VAN WAGONER, J.C. (1990) High frequency sequences and eustatic cycles in the gulf of Mexico basin. *G.C.S.S.E.P.M. Foundation Eleventh Annual Research Conference Program and Abstracts*, pp. 257–267.

MITCHUM, R.M., SANGREE, J.B., VAIL, P.R. & WORNARDT, W.W. (1990) Sequence stratigraphy in late Cenozoic expanded sections, Gulf of Mexico. *Gulf Coast Section-Soc. Econ. Paleont. Mineral, 11th Ann. Res. Conf., Program and Abstracts*, pp. 237–256.

NORTH-AMERICAN COMMISSION ON STRATIGRAPHIC NOMENCLATURE (NACSN) (1982). North American Stratigraphic Code. *Am. Assoc. Petrol. Geol. Bull.* **67**, 841–875.

PALLÍ, L. (1972) *Estratigrafía del Paleogeno del Empordà y sus zonas limítrofes.* Doctoral thesis, Universitat Autonoma de Barcelona, 477 pp.

POSAMENTIER, H.W. & CHAMBERLAIN, C.J. (1990) Joarcam field, Alberta: a lowstand shoreline in a sequence stratigraphic framework. *13th Congress of Int. Assoc. Sedimentologists, Abstracts of papers*, p. 433.

PUIGDEFÀBREGAS, C., MUÑOZ, J.A. & MARZO, M. (1986) Thrust belt development in the eastern pyrenees and related depositional sequences in the southern foreland basin. *Int. Assoc. Sediment. Spec. Publ.* **8**, 229–246.

SERRA-KIEL, J. (1982) Contribució a la paleobiologia dels *Nummulites. Buttl. Inst. Cat. Hist. Nat.* **48**, 19–29.

TRAVÉ, A. (1988) *Estratigrafèa sedimentologia dels diposits deltaics de l'Eocè mitjà-superior al sector de Manresa.* MSc thesis, Universitat de Barcelona, 85 pp.

VAIL, P.R. (1987) Seismic stratigraphy interpretation using sequence stratigraphy, Part 1: seismic stratigraphy interpretation procedure. In: *Atlas of Seismic Stratigraphy* (Ed. Bally, A.W.) Am. Assoc. Petrol. Geol. Stud. Geol. **27**, 1–10.

VAIL, P.R., MITCHUM, R.M. & THOMPSON, S. (1977) Seismic stratigraphy and global changes of sea level, part 3, relative changes of sea level from coastal onlap. In: *Seismic Stratigraphic Applications to Hydrocarbon Exploration* (Ed. Payton, C.W.) Am. Assoc. Petrol. Geol., Mem. **26**, 63–97.

VAIL, P.R., HARDENBOL, J. & TODD, R.G. (1984) Jurassic unconformities, chrono-stratigraphy, and sea level changes from seismic stratigraphy and bio-stratigraphy. In: *Interregional Unconformities and Hydrocarbon Accumulation* (Ed. Schlee, J.S.) Am. Assoc. Petrol. Geol., Mem. **36**, 129–144.

VAN WAGONER, J.C., MITCHUM, R.M., POSAMENTIER, H.W. & VAIL, P.R. (1987) Seismic stratigraphy interpretation using sequence stratigraphy, Part 2: key definitions of sequence stratigraphy. In: *Atlas of Seismic Stratigraphy* (Ed. Bally, A.W.), Am. Assoc. Petrol. Geol., Stud. Geol. **27**, 11–14.

VAN WAGONER, J.C., MITCHUM, R.M., CAMPION, K.M. & RAHMANIAN, V.D. (1990) Siliciclastic sequence stratigraphy in well logs, cores and outcrops: concepts for high-resolution correlation of time and facies. *Am. Assoc. Petrol. Geol., Methods in Exploration Series* **7**, 55 pp.

Spec. Publs Int. Ass. Sediment. (1993) **20**, 91–109

Evidence for Late Holocene relative sea-level fall from reconnaissance stratigraphical studies in an area of earthquake-subsided intertidal deposits, Isla Chiloé, southern Chile

S. BARTSCH-WINKLER *and* H.R. SCHMOLL

US Geological Survey, Federal Center, Mail Stop 905, Denver, Colorado, USA

ABSTRACT

At Río Pudeto and Quetalmahue, two estuaries along the northern shore of Isla Chiloé that subsided as much as 2 m in the great 1960 earthquake, reconnaissance stratigraphical studies reveal evidence of a regressive, nearshore marine sequence. The intertidal deposits include a peat-bearing, high-intertidal marsh sequence as thick as 1.4 m overlying shell- and foraminifera-bearing silt and clay layers presumed to represent a deeper water, low-intertidal environment.

Stratigraphy indicates a relative sea-level fall since about 5000 years BP as evidenced by radiocarbon ages that constrain the peat-bearing sequences. Locally, low-intertidal silt and clay overlie high-intertidal peat layers, but such minor transgressions cannot be correlated from site to site. At Río Pudeto, the youngest foraminifera-bearing silt deposit is no younger than 1200 years BP. The youngest age of shells at Quetalmahue is about 2600 years BP. The oldest peat-bearing deposits that are not overlain by silt deposits are about 1350 years BP at Río Pudeto, and as old as 4900 years BP at Quetalmahue. At Río Pudeto, peat-bearing deposits, which are overlain by silt and clay, range in age from 760 to 5430 years BP, and at Quetalmahue from 290 to 5290 years BP. A beach terrace on the northwest coast of the Isla is estimated to have been emergent since 1150 + 130 years ago.

Although some relatively abrupt transgressions may be due to sudden coseismic subsidence, data are not sufficient to document regional subsidence during individual plate-interface earthquakes. Seven earthquakes in south central Chile since 1520, especially those that occurred in 1575, 1737, and 1837, are thought to have been of a magnitude comparable to that of the 1960 earthquake. Although the sedimentological effects of the 1960 earthquake on the intertidal zone were dramatic, only limited evidence of possible historic earthquakes is found on Isla Chiloé and nearby islands; the ages and displacements of these earthquakes are indeterminable.

Dead forests still mark some locations that subsided into the intertidal zone during the 1960 earthquake, particularly at Río Pudeto and southern coastal Chiloé. There is little evidence of post-1960 growth in any of these subsided areas. Tree-ring counts and tree-diameter measurements provide evidence that these trees survived the 1837 earthquake, and probably survived the 1737 earthquake, strongly suggesting that these earthquakes were of smaller magnitude than the 1960 event, or that the epicentre locations were further removed than the 1960 epicentre from Isla Chiloé, and that earthquake-induced relative sea-level changes differed from those occurring in 1960.

INTRODUCTION

The Chilean coast lies along the convergent margin between the subducting oceanic Nazca plate and the overriding continental South American plate. The region east of the Peru–Chile trench is characterized by frequent intraplate earthquakes whose hypo-centre locations deepen progressively to the east along the plate boundary, extending from the trench to the Andean volcanic chain (Plafker & Savage, 1970). The Chilean earthquake of May 1960, the largest earthquake of this century, was a result of movement along these two plates. The initial and culminating shock epicentres and the many aftershocks occurred in a coastal belt. The initial shock occurred on the Arauco Peninsula near Concepción about 600 km north of the study area (Fig. 1a), and the culminating shock occurred 33 h later, 140 km southwest of the initial and shock beneath the ocean (Plafker & Savage, 1970). These shocks produced parallel zones of coseismic subsidence and uplift along the coast of south–central Chile, reflecting as much as 20 m of slip on the megathrust between the Nazca and South American lithospheric plates (Plafker & Savage, 1970).

The earthquake caused elevation changes along the coast over an approximate length of 1000 km and width of 200 km between latitudes 37°S and 48°S (Saint-Amand, 1963; Plafker & Savage, 1970). From measurements made eight years after the

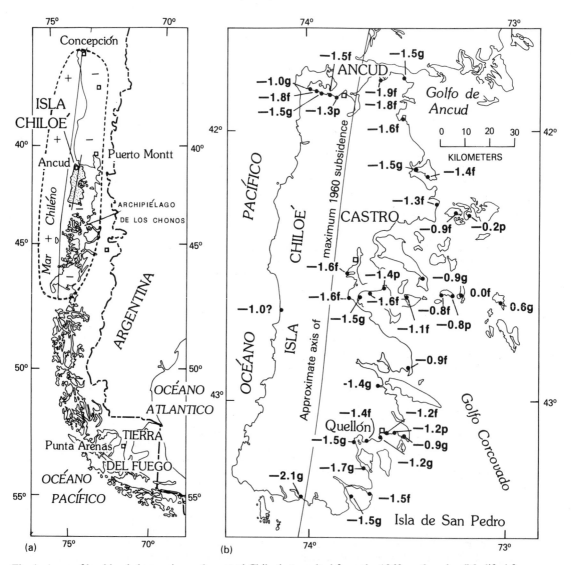

Fig. 1. Areas of land-level changes in south–central Chile that resulted from the 1960 earthquake. (Modified from Plafker & Savage, 1970.) (a) Outline of the focal region of the 1960 earthquake sequence; areas that underwent coseismic uplift (+) and subsidence (–) are indicated. (b) Approximate location of the axis of maximum subsidence on Isla Chiloé and measurement, in metres, of coastal subsidence at some locations by George Plafker in 1968. Estimated accuracy: f, fair, ±0.4 m; g, good, ±0.2 m; p, poor, ±0.6 m (Plafker & Savage, 1970, table 1).

earthquake, Plafker and Savage (1970) concluded that there were parallel zones of subsidence on land along the Chilean coast and uplift mostly beneath the Pacific Ocean nearest the trench (Fig. 1a); these parallel zones are thought to result from release of pre-earthquake stress in the upper South American plate.

Isla Chiloé was strongly shaken by the May 1960 earthquake. The Isla and nearby islands are within the 1960 zone of coseismic subsidence; the axis of maximum subsidence bisects Isla Chiloé from north to south. Shorelines, marshes, river mouths, and coastal forests on the Isla were suddenly tectonically lowered by as much as 2.1 m into the intertidal and subtidal zones.

In the seismically active region of southern coastal Chile, estuarine stratigraphy might be an important means of recording great subduction-zone earthquakes if there is preservation by subsequent burial of marshes and forests that have been tectonically lowered into the tidal zone and/or anomalous sandy material that may have resulted from liquefaction of coastal sediments or tsunami deposition. Such information could be applied to similar data being gathered to aid seismic-risk assessment (especially recurrence intervals for great earthquakes) in southern Alaska and the Pacific Northwest of southern Canada and western coastal United States.

The purpose of this study was to determine, by reconnaissance examination of intertidal stratigraphy and conventional radiocarbon dating, the essence of, and change in, relative sea-level during Late Holocene time. Additionally, we investigated the feasibility of more clearly defining past earthquakes using stratigraphical evidence. Specifically, this included determining whether (1) previous historic earthquakes are recorded by abruptly buried intertidal (peat) layers, (2) the extent of individual earthquakes could be mapped by identification, dating, and correlation of buried marsh surfaces, (3) the frequency of large earthquakes could be determined, (4) pre-, co-, or post-seismic earth movements could be inferred, and (5) stratigraphical evidence of tsunamis was preserved. Similar investigations using stratigraphical evidence to determine the character of past earthquakes have been described from parts of North America, including southern Alaska (Plafker & Rubin, 1978; Plafker, 1987; Bartsch-Winkler, 1988; Plafker, *et al.*; 1991; Bartsch-Winkler & Schmoll, 1992) and the northwestern United States and southwestern Canada (Atwater, 1992; Clague & Bobrowsky, 1990; Darienzo & Peterson, 1990; Nelson, 1991a, b).

In the vicinity of Isla Chiloé, historical earthquakes are recorded for 1575, 1737, and 1837 (Lomnitz, 1970). These historical earthquake events were thought to have been accompanied by land-level changes like those that have been documented for subduction-zone earthquakes (Plafker, 1969; Plafker & Savage, 1970; Nishenko, 1985), but they may not have had the same sense or amounts of movement as the 1960 earthquake (Muir-Wood, 1989). Evidence now indicates that the 1960 earthquake was not typical of past earthquakes that have occurred in central Chile (Barrientos & Ward, 1990). Large earthquakes (about 7–8 magnitude) apparently occur more frequently along the Nazca–South American plate boundary, and geographically overlap great earthquake events (>8 magnitude), which occur only once in four or five earthquake cycles (Barrientos & Ward, 1990).

Post-seismic sea-level changes have been measured by tide gauges at Puerto Montt (Fig. 1a) at the landward edge of the 1960 belt of coseismic subsidence, and indicate a relative lowering of sea-level (or emergence) of 3 cm/year from 1965 to 1970, which may be due to post-tectonic uplift (Wyss, 1978). But many of the tide-killed forests and marshes in Río Pudeto and other estuaries on Isla Chiloé are still submerged, with little evidence of shoreline progradation or land uplift .

ESTUARINE STRATIGRAPHICAL STUDIES OF LATE HOLOCENE TECTONIC LAND-LEVEL CHANGE

The extent of earthquake-caused deformation on sedimentary deposits is dependent on the earthquake magnitude and lessens with distance from the earthquake epicentre. Thus, stratigraphical and other geological evidence on historic land deformation along the Chilean coastline could theoretically provide some evidence of the magnitude and epicentre location for past earthquakes. Stratigraphical and sedimentological features that might provide evidence for earthquake-caused lowering of land level include: (1) numerous core sections containing abrupt regional upward change from peat to silt at similar stratigraphical horizons, indicating a sudden widespread change in depositional setting from marsh or forest to lower-intertidal or subtidal environment; (2) fossil evidence from numerous cores indicating a reversal from marsh to lower-intertidal to subtidal environments at like stratigraphical positions that might indicate sudden changes in relative sea-level; (3) many similar radiocarbon ages on the upper parts of peat layers with abrupt contacts that could indicate the timing of sudden regional subsidence; (4) the presence of liquefaction features, including convoluted or mixed bedding, and compacted layers at similar stratigraphical positions that might have been earthquake-induced; and (5) common occurrences of sand layers overlying correlatable peat-bearing beds in a given estuary that might indicate deposition by tsunami.

Methods

This reconnaissance study was conducted during the austral summer months of January and February, 1989. Field observations, hand coring, and measurement of stratigraphical sections in intertidal channel cuts and in marshes were undertaken during times of low tide when there was maximum exposure. Range in tide level for these months (highest high tide minus lowest low tide) was 0.9–

6.5 m. Hand-coring techniques were most successful in areas of peat, silt and clay, especially on the northern part of the Isla at Río Pudeto estuary and at a small unnamed estuary herein called Estero Quetalmahue (Figs 2, 3 & 4). On eastern and southern Isla Chiloé and adjacent Isla Quinchao, sections were mostly measured on sand beaches with thin, overlying marsh zones (Fig. 5). Several locations on the southern coast of the Isla that still show dead forests killed by earthquake-induced inundation in 1960 were visited, but they had few layered peat deposits. In order to learn whether forests submerged in the 1960 event were sufficiently old to have survived the 1837 earthquake, we collected two tree sections and tree-diameter data from pre-1960 remnant forests in several intertidal areas.

To locate marsh localities, pre- and post-1960 earthquake vertical and oblique black-and-white photographs taken by the US Air Force and the Chilean Air Force were used in addition to topographical maps at scales of 1:250 000, 1:50 000, and 1:25 000 obtained from the Instituto Geográfico Militar, Santiago.

Table 1. Radiocarbon results on samples from Isla Chiloé, Chile

Core	Sample	Laboratory	Date	±	Material dated	Depth (m)
Quetalmahue						
6	1	Beta-32377	Modern peat			1.0
10	2	Beta-36455	910	60	Silty peat	1.1
13	3	Beta-31889	3210	140*	Shells	1.65
14	4	Beta-31890	290	70*	Peat	0.25
14	5	Beta-31463	910	130*	Peat	0.75
14	6	Beta-31891	2660	150*	Organic silt	1.2
14	7	Beta-31892	5290	90*	Organic silt	2.6
16	8	Beta-31893	4890	90*	Peat	0.95
Río Pudeto						
1	9	Beta-31465	5430	180*	Peat	2.55
1	10	Beta-32373/ ETH-5672	6365	80*	Shell	4.75
2	11	Beta-31894	3520	80*	Peat	0.7
2	12	Beta-31895	5400	210*	Peat	1.3
5	13	Beta-32374	1330	80*	Peat	0.6
6	14	Beta-32375	720	60	Peat	0.6
6	15	Beta-32376	1170	140	Peat	1.1
20	16	Beta-32372	1370	80*	Peat	0.6
20	17	Beta-31464	760	80*	Peat	0.75
Other Isla Chiloé Locations						
19	1	Beta-31466	380	70	Peat	0.5
20	2	Beta-31467	530	70	Peat	0.7

*Extended counting method. ª, Accelerator method.

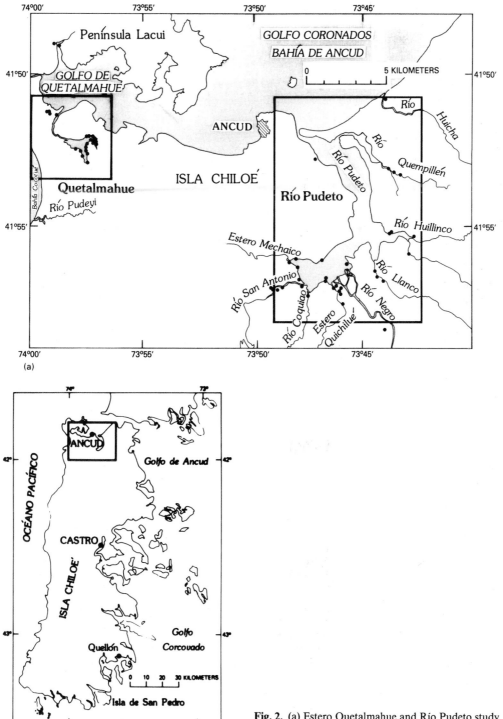

Fig. 2. (a) Estero Quetalmahue and Río Pudeto study areas, northern Isla Chiloé, southern Chile. Dots are station localities. (b) Map showing location of (a).

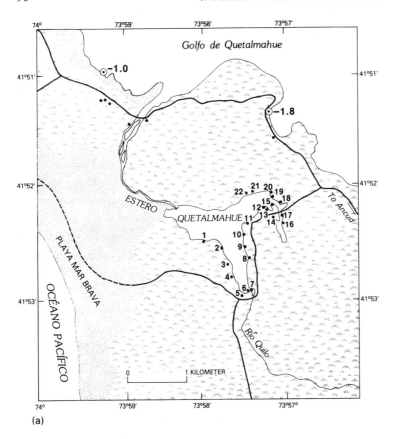

Fig. 3. (a) Numbered core sites and (b) measured sections (see facing page), Estero Quetalmahue, northern Isla Chiloé. Horizontal line at base of core section indicates that no further penetration by hand coring was possible. Asterisks indicate approximate depth of the 1960 layer. Circled dots indicate subsidence measurements of Plafker and Savage (1970, table 1).

(a)

Laboratory analyses

Radiocarbon dates reported have not been corrected to calendar years (see Table 1). Laboratory pre-treatment of radiocarbon samples included removal of obvious modern rootlets by picking or flotation in hot water, hot acid dispersion to rid the sample of carbonates, rinsing to neutrality in hot distilled water, gentle drying, and burning in a vacuum system. The samples underwent benzene synthesis and counting; small samples were given extended counting time. In one sample (core 1, sample 10), the ^{14}C was counted by accelerator mass spectrometry (AMS). Pre-treatment of this sample included washing, physical scrubbing and abrasion, acid etching, and conversion to CO_2. The CO_2 was purified and reacted with hydrogen on cobalt catalysts to produce graphite. Finally, the graphite was applied to copper targets and measurements made (in triplicate) by the Eidgenossiche Technische Hochschule University, Zurich, Switzerland.

Previous work on historic land-level change

Whereas the convergent margin of coastal Chile is known to have experienced mainly (historical) large earthquakes (e.g. Darwin, 1851), little is known of the magnitudes of past earthquakes along the Peru–Chile trench or the extent of land deformation associated with them, and nothing is known about the location of their epicentres. Charles Darwin (1851) noted evidence of shoreline change at many localities along the Chilean coastline, including the northern and eastern shoreline of Isla Chiloé and areas to the south, including Archipiélago de los Chonos (latitude 43°30′–46°00′S; Fig. 1a), and attributed these land-level changes to the 1837 and previous earthquakes. Specifically, Darwin (1851) attributed 2.4 m of uplift on Isla Lemu (west side of the Archipiélago de los Chonos) to have resulted from the 1837 earthquake. Other evidence (Brüggen, 1950) attributed coastal submergence in the area east of the Archipiélago to the same

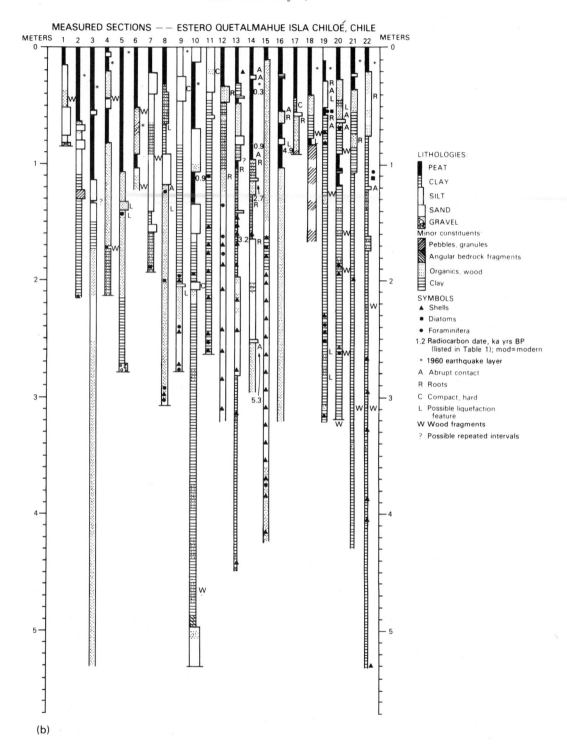

(b)

Fig. 3. (*continued*)

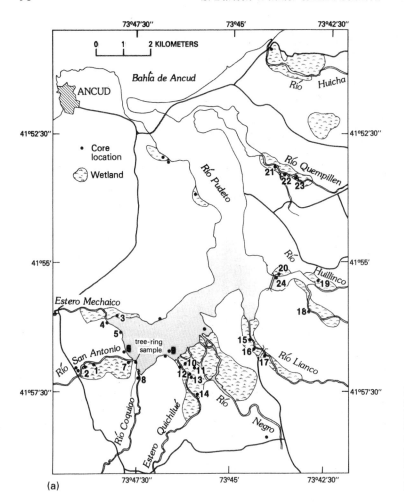

Fig. 4. Numbered core sites (a) and measured sections (b), Río Pudeto, northern Isla Chiloé. Horizontal line at base of core section indicates that no further penetration by hand coring was possible. Asterisks indicate approximate depth of the 1960 layer.

earthquake, prompting Plafker and Savage (1970) to suggest that historic land movement during earthquake events in this region was the same as that which occurred during the 1960 event. Using tree-ring information along the southern coastline south and east of Isla Chiloé, Muir-Wood (1989) concluded that the magnitude of the 1837 earthquake in Chile was comparable to that of the 1964 Alaskan earthquake, the second largest earthquake in the worldwide instrumental record. Further, he states that the belt of subsidence for the 1837 event was located further east than for the 1960 event, resulting in only very small amounts of subsidence along the eastern part of the Isla and no submergence, and possible uplift, of the western part of the Isla, including the Río Pudeto and Quetalmahue estuaries.

We noted on topographical maps that the terms 'estero' and 'rio' were variously used, and we questioned whether this usage had pertinence to sealevel changes. Pre- and post-earthquake low-tide, vertical and oblique aerial photographs in the Río Pudeto area indicate the extent of alteration of nearshore areas due to subsidence (Fig. 6). The estuary is now tidal; prior to the earthquake, it was mainly a river. Note also the designation of Estero Mechaico — now an estuary, but prior to 1960 a river. Quetalmahue estuary was only a small river influenced little by tides prior to the earthquake (Fig. 7). Alternation in designation of these river or estuarine systems on maps (names derived from common usage in this island with at least a 400-year-long record of European agrarian occupation), with no apparent link to current physiography, may

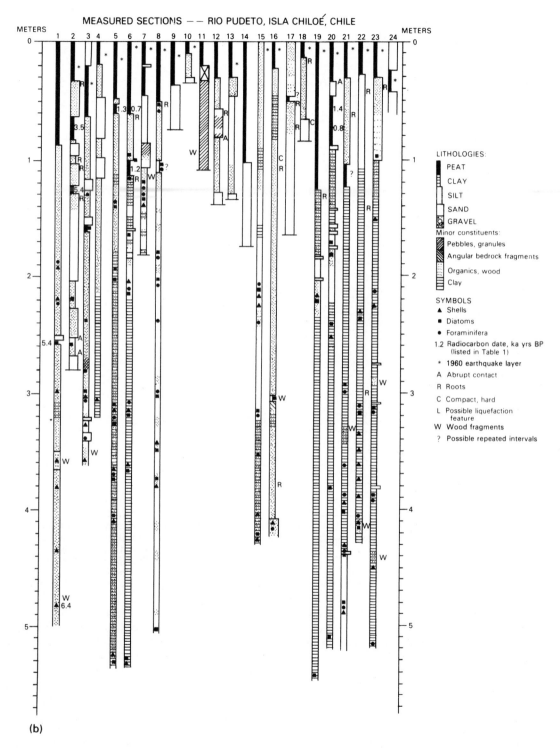

(b)

Fig. 4. (*continued*)

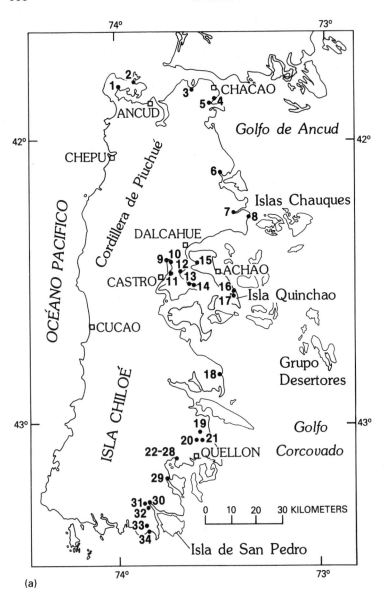

Fig. 5. Numbered core sites, (a) station locations, and measured sections (b) elsewhere on Isla Chiloé and nearby islands. Asterisks indicate approximate depth of the 1960 layer.

stem from a complicated history related to land-level changes resulting from earthquakes. We could find no conclusive evidence that the terms had any relevance to sea-level changes, but this interesting aspect of physiographical terminology may warrant further investigation of local historical records.

Peat and silt stratigraphy

The stratigraphy generally consists of peat deposits typically at and near the surface and dominantly silt deposits at greater depth (Figs 3–5). Peat and peat-bearing sediments, including both surface peats and relatively thin beds that are overlain by silt, are generally confined to the upper 1 m of the sections. The peat beds locally include intervening thin layers of silt and sand. (Some of the thin peat beds recorded at depth may be repeated during hand coring, especially those occurring at depths of about 1 and 2 m.) Typically, silt beds are more peat-rich in the upper sequences and contain increasing numbers of foraminifera and fragments of macroshells with increasing depth.

The sediments are primarily in the silt and clay

MEASURED SECTIONS − − LOCATIONS ON ISLA CHILOE AND NEARBY AREAS

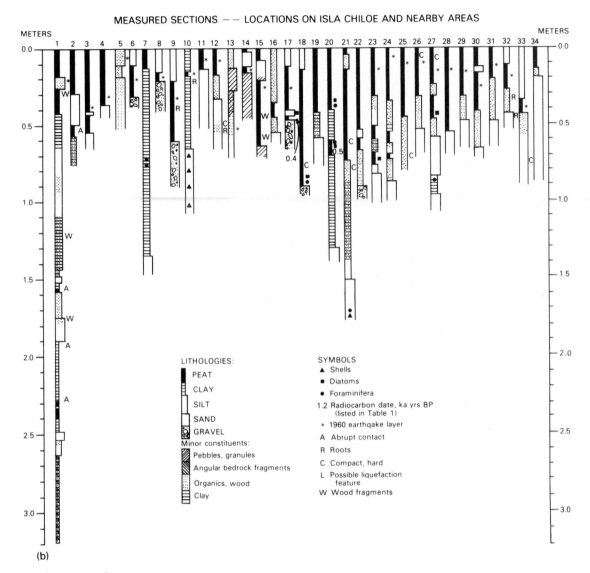

Fig. 5. (*continued*)

size-range, well- to moderately well-sorted, containing minor amounts of sand and gravel, and varying amounts of organic matter. Bedding is commonly thinly laminated to massive, and contains rare contorted intervals. Grain lithologies include metamorphic rock, volcanic and plutonic quartz, pumice, and mica. Peat deposits may contain some silt and clay. Whole shells and shell fragments, spicules, calcareous foraminifera, insect parts, seeds and diatoms provide evidence of deposition within nearshore and marsh environments. However, in the acidic environment of the peat marsh, much of

the calcareous material (shells and foraminifera) was post-depositionally altered to gypsum. Despite this alteration, we infer that sediments containing shells, spicules, diatoms and foraminifera (even though calcareous species may no longer be present in some cores due to diagenesis) are probably of deeper water (nearshore subtidal to low-intertidal zone) origin than those containing much plant and insect debris, which are thought to have been deposited near or within a marsh (high-intertidal zone).

The amount of post-1960 earthquake deposition

(a)

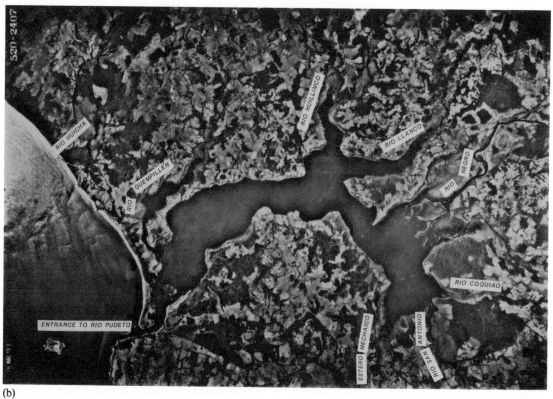

(b)

Fig. 6. Pre- and post-earthquake photographs showing change in physiography of Río Pudeto estuary resulting from the 1960 earthquake. (a) Oblique aerial photograph of Río Pudeto taken in 1944. (b) Vertical aerial photograph of Río Pudeto taken in 1961.

Fig. 7. Low-tide photograph of Estero Quetalmahue taken in February 1989 looking south-southwest near station 10 (Fig. 3) showing extent of present-day flooding by tidewater in the upper (southeastern) end of the estuary. Note the presence of pre-1960 fenceline and Río Quilo channel.

is variable on Isla Chiloé, but commonly is less than 20 cm. Section 6 at Quetalmahue (Fig. 3) indicates as much as 70 cm of post-earthquake deposition. In places where peat accumulation continued following the 1960 earthquake, the peat deposit presumed to contain the 1960 layer was usually recognizable as more dense sand only locally overlain by thin lenses of silt or organic silt. Typically, the presumed 1960 horizon (marked by submerged supratidal marsh vegetation) is undergoing erosion in the low-intertidal zone, and marsh aggradation is occurring in the high-intertidal zone.

Possible liquefaction features were noted exclusively at Estero Quetalmahue, where well-sorted silt and sand are the dominant lithologies, but where an irregular and convoluted mixture of grain sizes occurs in some intervals. In all areas, compacted intervals, which might indicate consolidation by earthquake shaking, were also noted, as well as the presence of wood fragments, roots, and abrupt changes from one lithology to another that might signify sudden changes in relative sea-level. The possible liquefaction features are not constrained to correlatable layers at any given region or within any depth range. Thus, because there is no correlation of these features from core to core or by age or stratigraphical position, we could not unequivocally relate them to regional earthquake occurrence. Liquefaction is not unusual in fluvial or estuarine set-

tings in fine sand and silt grain sizes, and may be a result of channel undercutting and slumping (Ovenshine *et al.*, 1976), wave-induced liquefaction (Dalrymple, 1979), and cyclic wetting and drying in the intertidal zone (Doe & Dott, 1980). The compacted intervals might result from normal compaction processes, the abrupt changes in lithology might be due to cut-and-fill and other transport processes, and the organic material might be redeposited.

As described previously, the most convincing evidence of past regionally induced subsidence, possibly due to an earthquake, is provided by peat beds that have been instantly buried by silt deposits, that are traceable across at least parts of one or more embayments, and that can be correlated by radiocarbon dating. However, the regional extent of such peat beds cannot be established adequately from our data. Instead, our sections indicate a complex stratigraphy that exhibits little recognizable continuity of individual beds and, especially as indicated by disparate ages of peat beds at comparable depths, that is probably dominated by processes of tidal channel cut-and-fill. At Quetalmahue (Fig. 3), only nine of the 22 sections contain buried peat. In three of the sections (10, 14, and 16), radiocarbon dates of about 900 and 5000 years are widely disparate. At Río Pudeto (Fig. 4), only seven of the 24 sections contain buried peat. Three of these sections (5, 6, and 20) provide evidence for two correlatable peat beds: (1) thin beds at disparate depths that date about 750 years BP in sections 6 and 20; and (2) somewhat thicker beds at roughly comparable depths in sections 5, 6 and 20 that date about 1250 years BP. The correlatable date from section 20 is equivocal, however, because it is not in chronological sequence with respect to a younger date stratigraphically lower in the peat section. At other localities (Fig. 5), 10 of the 34 sections contain buried undated peats. Dates from sections 17 and 20 yielded approximately similar ages of about 450 years BP.

In summary, 10 of the 80 stratigraphical sections we measured or cored have yielded peats whose approximate times of burial, using conventional radiocarbon techniques, are 450, 750, 900, 1250, and 5000 years BP. Because of the limited number of peat beds, their scattered geographical position, the probable lengthy time period between peat deposition and burial, and the disparate stratigraphical setting, we cannot establish that any of these peat-burial events are related to earthquake-caused subsidence, although it is possible that some could have resulted from such a cause. Even if the events are earthquake-caused, the five events scattered over more than 5000 years do not yield any information about the recurrence of great earthquakes in the region.

However, the radiocarbon dates could record the general progression from shallow water to marsh conditions and relative fall in Late Holocene sea-level, as evidenced by the general lithological change from silt to peat deposition recorded in almost all stratigraphical sections. At Quetalmahue estuary, the transition from shallow water to marsh conditions, marked by the uppermost shell- and foraminifera-bearing sediment in the core, ranges in depth from 0.2 m (core 13) to >2 m (cores 9 and 21); stratigraphical relations and a few radiocarbon dates indicate that the transition occurs at > 2700 years BP (core 14) but <5300 years BP (core 14) (Figs 3 & 4; Table 1). At Río Pudeto, this shallow water to marsh transition, using shells and foraminifera, ranges in depth from 0.5 (core 8) to about 3.0 m (cores 4 and 21); stratigraphical relations and radiocarbon dates indicate that the transition occurs between about 800 and 5400 years BP. The oldest peat-bearing deposits that are not overlain by silt deposits thought to represent deeper-water intertidal depths are as old as 4900 years at Quetalmahue (section 16) and about 5400 years old in Río Pudeto (section 2).

The 5000-year emergence may have created a regional unconformity during relative sea-level fall that is apparent in nearly all sections. While we have only a limited number of radiocarbon dates from which to make conclusions, it seems apparent that this unconformity occurs at widely spaced time intervals and at widely different depths in the cores. Due to facies changes in both estuaries, and assuming a progradational model wherein the more distal, higher regions of estuaries are filled in first, it is probable that the youngest marsh sequences occur in those stratigraphically thicker estuarine deposits closer to the mouth. Thus, such variable age results using stratigraphical indicators are critically dependent on the core site location within the estuary. In addition, it is possible that past earthquakes produced small unrecognized unconformities as a result of sudden submergence of marshes.

Tree-ring and tree-diameter analyses

Evidence for subsidence caused by earthquakes as recent as 1837 and 1737 would be difficult to ascertain directly by means of radiocarbon dating due, primarily, to the lack of sensitivity in the radiocarbon cycle. However, data from inundated forests on the Isla and nearby areas can provide evidence of prior earthquake-caused subsidence. Areas that subsided during the 1960 earthquake are marked by the presence of submerged forests killed by tidal inundation in 1960, particularly in Río Pudeto and southern coastal Chiloé. There is minimal evidence of post-1960 forest regrowth in all of these subsided areas in the 30 years since the earthquake. At Río Pudeto and elsewhere on Isla Chiloé and in the vicinity, former forests are still inundated by several metres of tidewater even at low tide, evidence for a slow sedimentation rate.

Diameters of the largest tree trunks were measured 1 m from their bases in the high-intertidal zone in order to estimate the ages of presumably the oldest trees affected by subsidence in 1960. A tree-ring count on a Coigüe (Nothofagus) tree located at the mouth of Río San Antonio (Río Pudeto estuary) that was 35 cm in diameter shows that the tree was more than 105 years old when it was killed by inundation in 1960 (David K. Yamaguchi, Institute of Arctic and Alpine Research, University of Colorado, Boulder, USA, oral communication, July 1990). In the same vicinity, Coigüe trees were measured to have diameters of as much as 73 cm; thus, it is likely that the larger trees were more than twice as old as the tree analysed and were growing at the time of the 1837 earthquake. Trees grow more slowly as they age, and the rings become more finely spaced. Trees with twice the diameter are more than twice as old (David Yamaguchi, oral communication, July 1990). It is likely that these larger-diameter trees also survived the 1737 earthquake. From this evidence, we conclude that in this area of Río Pudeto the forest was not affected by subsidence in 1837 or 1737 to the extent that it was in 1960. These historic earthquakes were probably of smaller magnitude (Barrientos & Ward, 1990) or possibly the epicentre locations were further from the Isla or the pattern of coseismic land deformation did not conform to that resulting from the 1960 event.

Tsunamis

Tsunamis occurred along the coastline of Chile shortly after the 1960 earthquake, and tide gauges recorded abnormalities. Most tsunamis were so small that they generally went unnoticed by the populace (Sievers *et al.*, 1963). However, on Isla Chiloé, all ports were rendered inoperable as a result of tsunami devastation. Changes in the level of the sea were immediate, and were the first evidence of land-level change, noted at virtually all small towns on and in the vicinity of the Isla, including Castro, Ancud, and Quellón (Fig. 1) (Sievers *et al.*, 1963). As a result of the subsidence that resulted from the 1960 earthquake, relative sea-level rose approximately 1.8 m. There were also immediate post-earthquake changes in the timing (with respect to the tide at Puerto Montt) and amplitude of the tide (Galli & Sanchez, 1963a). At Ancud, three waves were associated with the tsunami, the third being the highest (estimated 6–7 m in height) and the most destructive (Galli & Sanchez, 1963b). The tsunami that devastated the port area at Ancud is reported to have entered from the northeast across Península Lacui, across Bahía de Ancud, and into Río Pudeto, and possibly to have progressed to the west into Golfo de Quetalmahue (José A. Ulloa c., Ancud, Chiloé, oral communication, 1989). The low sand-bar entrance to Río Pudeto was permanently altered, improving access to the estuary by boats; such alteration may have been due to erosion by the tsunami. However, according to Codignotto (1983), this sand bar underwent considerable tectonically-induced compaction. There is no evidence for tsunami-laid sand or silt in our Río Pudeto cores (Fig. 4), although there are eye-witness accounts of the occurrence of a tsunami at many of our sites.

A tsunami that may have been as much as 10 m in amplitude entered the Quetalmahue estuary, either from the Pacific Ocean via Mar Brava west of Quetalmahue or from the east into Golfo de Quetalmahue and into the river that connects the estuary to Golfo de Quetalmahue (José A. Ulloa C., oral communication, 1989). Eye-witnesses describe a general rise in sea-level immediately after the earthquake, similar to a normal rise in the tide rather than a discrete wave, that covered the Quetalmahue region. At Quetalmahue, coarser grained deposits (silt to fine sand) are more commonly interbedded in the peat layers, and in several locations directly overlie peat beads (Fig. 3). In this reconnaissance study, we did not map the silt and sand interbeds in detail. However, we suggest that, in order to determine the feasibility of these being tsunami-laid sands, further detailed stratigraphical work is warranted in Estero Quetalmahue to determine (1) the location of these sand and peat layers with respect to geographical features and the 1960 peat layer, (2) possible internal bedding features in the sands that might determine direction of transport, and (3) the location and direction of the tsunami that occurred in this area.

Possible evidence for Late Holocene uplift on Isla Chiloé

Along the northwest coast of Isla Chiloé from Ancud to Bahía Cocotué (Fig. 2), a persistent uplifted strandline was noted at an elevation between about 1.5 and 2 m above present extreme high tide by George Plafker, Sergio E. Barrientos, and Steven N. Ward (written communication, 1990). This strandline is marked by sea caves (some of which were first reported near Ancud by Charles Darwin (1851)), by sea stacks such as the gigantic Roca Rühn in Bahía Cocotué, and by abandoned sea cliffs. On the west side of Isla Chiloé, abandoned sea cliffs estimated as high as 25 m occur in many coves and bays where they are separated from extreme high tides by a prograding strip of beach, dune, colluvial, and alluvial deposits extending to tens of metres wide. Trees (Alerce) growing on the terrace deposits and abandoned cliffs are as much as 85 cm in diameter and estimated to be more than 100 years old by local residents. Most of the sea cliffs are only slightly degraded and there is no perceptible soil development on the terrace deposits, which suggests that they are relatively young Holocene features.

According to Plafker, Barrientos, and Ward (written communication, 1990), a stream cut at Río

Pudeyi in Bahía Cocotué (Fig. 2) exposes the following section of unconsolidated deposits (from bottom to top): (unit 1) 150 cm of structureless, very fine to fine-grained dune(?) sand with sparse scattered pebbles in the upper part grading upwards into (unit 2) 20–30 cm of dark-brown, carbonaceous sand containing scattered pebbles, cobbles, and charcoal, and local lenses of gravel or charcoal of probable alluvial origin; (unit 3) 10 cm of dark-brown sandy peat, containing sparse logs as much as 20 cm in diameter and 4 m long at the base, of probable flood plain origin; (unit 4) 8 cm of fine-grained alluvial or wind-deposited sand; (unit 5) 7 cm of chocolate-brown to yellow-brown, homogeneous fresh-water peat; and (unit 6) as much as 95 cm of very fine to fine-grained sand with local, faint stratification that is probably wind-deposited. The oldest and lowest-dated unit containing charcoal remains was about 10 cm above extreme high tide level in an area that subsided an estimated 1–1.5 m during the 1960 earthquake (Plafker & Savage, 1970).

Four samples of charcoal, wood, and peat from the Río Pudeyi section were dated by ^{14}C gas-count by the US Geological Survey (Meyer Rubin, written communication to George Plafker, 1989). The age in radiocarbon years of charcoal from near the base of unit 2 is 1150 BP ± 130, wood from one of the logs in unit 3 is 850 BP ± 110, and samples of peat from the upper 2 cm of units 3 and 5 are 300 BP ± 100 and 280 BP ± 100 respectively. Together with the stratigraphy, these age data indicate to Plafker, Barrientos and Ward (written communication, 1990) that the terrace on which the Río Pudeyi alluvial and dune-sand deposits accumulated has been emergent since at least 1150 ± 130 years ago.

Discussion

There is no indication in the subsurface sediments in the Río Pudeto and Quetalmahue estuaries, or the other sections measured on Islas Chiloé and Quinchao, of any systematic pattern of separation of interbedding of peat and clastic sediments, as would be expected if marshes were repeatedly submerged over a large area. Other factors, such as facies changes, tide range, changes in tide range, or cutting and filling of channels, could produce such layering as is apparent in the core data from Isla Chiloé. These processes may or may not be a result of sudden physiographical alteration or subsidence

of the land surface; if they are, they mask the effects of potential tectonic subsidence in the stratigraphical record. Other complications include a variation in the amount of compaction in any given location that might have been caused by earthquake shaking because of varying grain size, porosity, degree of saturation of sediments, and other factors relating to wetland setting. Sedimentation rates in the Río Pudeto and Quetalmahue estuaries are low, so that there is no rapid burial and, thus, no ubiquitous separation of peat layers by intertidal sediment.

Stratigraphical evidence from the eastern and southern coastline of Isla Chiloé and Isla Quinchao indicates that a surficial peat layer, which in some locations contains silty beds, is resting on gravel, sand, or Pleistocene bedrock. At Río Yaldad (sections 22–24, Fig. 5), there are apparently two peat layers, but we do not have enough dates to test for possible correlation. These dual peat layers are a local occurrence within the eastern Río Yaldad estuary, and their proximity to the mouth of Río Yaldad at the head of the estuary suggests that they could have resulted from lateral migration of anastomosing channels, including processes of channel cut-and-fill, overbank deposition, and marsh aggradation.

CONCLUSIONS

Although the 1960 earthquake produced as much as 2 m of land subsidence (relative sea-level rise) along the shoreline of Isla Chiloé , and despite the known occurrence of past large earthquakes along the southern coast of Chile thought to have produced the same pattern of deformation of the land surface, the peat stratigraphy in this reconnaissance study of the Late Holocene deposits indicates net relative fall of sea-level from about 5000 years ago. Peat occurs to a depth of 1.5 m or less in estuaries on northern Isla Chiloé, and is less than about 5000 years old. Below about 1.5 m depth, clastic shell- or foraminifera-bearing sediments predominate, probably indicating deeper water (higher relative sea-level) prior to about 5000 years BP. Our core and stratigraphical data provide new information on relative past sea-level in the region; none of the evidence suggests that land subsidence exceeded 2 m (the maximum amount of subsidence caused by the 1960 earthquake) during the Late Holocene. Thus, even if such subsidence occurred during this time, some combination of residual isostatic rebound and regional uplift more than compensated for the subsidence and there has been a net regression during the Late Holocene.

We have no evidence that abrupt relative sea-level changes are regional across the Isla because we cannot trace any single event in the core stratigraphy. There is no apparent consistency from core to core of stratigraphical indicators such as interbedded peat and silt deposits with abrupt or peat-to-silt transitions, nor are ubiquitous tsunami-laid sand layers present in areas where tsunamis are known to have occurred. Limited numbers of ^{14}C ages do not suggest correlation of the tops of peat beds from core to core as might be expected if marshes were suddenly submerged over a large area.

On the contrary, the presence of thick and ubiquitous sequences of peat with local thin intercalations of silt that compose the uppermost part of most cores indicates probable gradual emergence. Evidence from a Late Holocene terrace at Río Pudeyi suggests emergence since at least 1150 years ago. Present-day tide-gauge records at Puerto Montt (Fig. 1) show a relative fall in sea-level since the earthquake. It appears certain from tree-ring evidence that in neither the 1837 nor the 1737 earthquakes was there subsidence here of a similar magnitude to the 1960 earthquake.

Along the west coast of Chile, evidence suggests that post-glacial relative sea-level has been falling during Holocene time (Walcott, 1972). Studies along the north coast of Chile (Paskoff, 1978, 1980) described terraces containing marine shells dated at 3980 years BP, and elevated beaches in central coastal Chile dated at 3700 and 4400 years BP. During the past 6000 years, 38 m of uplift on Isla Mocha, south-central Chile (30°28′S, 73°55′W), has been attributed to both coseismic and aseismic factors (Nelson *et al.* (1992)). Stratigraphical studies by Atwater and Vita-Finzi (1989) and Atwater *et al.* (1992) near Río Maullin, 20 km north of Isla Chiloé on the mainland, also indicate net Late Holocene emergence of intertidal deposits during the past 5000 years. Pleistocene and Holocene emergence has also been recorded on the Patagonian coast in beach deposits as high as 12 m (Codignotto, 1983).

On the west coast of Chile where numerous large interplate earthquakes have occurred in Holocene time, both seismic and interseismic tectonism must play a significant role in relative sea-level change. Our data suggest that the subsidence produced on Isla Chiloé and nearby islands by the 1960 earthquake was an unusual event rather than typical of

most previous events that affected the same area. This conclusion supports the interpretations of other workers. Barrientos and Ward (1990), based on geophysical modelling, suggest that the 1960 event was not typical of large earthquakes in central Chile, which have a recurrence cycle of about 128 years. According to Barrientos and Ward (1990), great 1960-type events may rupture the plate boundary only once in every 4–5 cycles (approximately every 500–650 years), and large earthquakes occur more frequently and geographically overlap the great events, filling in between the locked zones (Barrientos & Ward, 1990). Plafker (1972) earlier concluded that the recurrence interval for great earthquakes such as the 1960 event is longer than for earthquakes of lesser magnitude, based on measurements of the rate of strain accumulation and the size of the 1960 tsunami relative to other historic tsunamis.

ACKNOWLEDGEMENTS

Our work on Isla Chiloé was facilitated by Carolina Villagrán (University of Chile, Santiago) and Fred Goodell (Inter-American Geodetic Survey, Santiago). Many interested residents of Isla Chiloé allowed access to their property and equipment, and provided eye-witness accounts of local events that took place in the intertidal zones during and after the 1960 earthquake. José A. Ulloa C. (Ancud, Chiloé), and Gary R. Winkler, US Geological Survey, provided assistance in the field. David K. Yamaguchi (Institute of Arctic and Alpine Research, University of Colorado) identified the tree-ring counts reported herein. George Plafker (US Geological Survey) kindly provided unpublished information on the terrace deposits at Bahía Cocotué examined in February 1989 by himself, Sergio E. Barrientos (University of Chile, Santiago) and Steven N. Ward (University of California at Santa Cruz). We thank reviewers Alan R. Nelson and Brian F. Atwater, US Geological Survey, and two anonymous reviewers for many helpful comments.

REFERENCES

Atwater, B.F. (1992) Geologic evidence for earthquakes during the past 2000 years aong the Copalis River, southern coastal Washington. *J. Geophys. Res.* **97** no. B2, pp. 1901–1919.

ATWATER, B.F. & VITA-FINZI, C. (1989) Net Late Holocene emergence in the subsidence belt of the giant 1960 earthquake, southern Chile [abs.]: *EOS, Trans. Am. Geophys. Union.* **70**, 1330–1331.

ATWATER, B.F., JIMÉNEZ, H.N., & VITA-FINZI, O. (1992) Net lateHolocene emergence despite earthquake-induced submergence, south-central Chile. In: *Neotectonic Aspects of the Evolution of Quaternary Coasts* (Eds Ota, Y. Nelson, A.R., & Berryman, K.) Special Issue, Quaternary International, **15/16**, pp. 77–85.

BARRIENTOS, S.E., & WARD, S.M. (1960) Chile earthquake; inversion for slip distribution from surface deformation. *Geophys. J. Int.* **103**, pp. 589–598.

BARTSCH-WINKLER, S. (1988) Cycle of earthquake-induced aggradation and related tidal channel shifting, upper Turnagain Arm, Alaska, USA *Sedimentology* **35**, 621–628.

BARTSCH-WINKLER, S. & SCHMOLL, H.R. (1992) Utility of radiocarbon-dated stratigraphy in determining late Holocene earthquake recurrence intervals, upper Cook Inlet region, Alaska. *Geol. Soc. Am. Bull.* **104**, pp. 684–694.

BRÜGGEN, J.E. (1950) *Fundamentos de la Geología de Chile.* Instituto Geográphico Militar Santiago, 374 p.

CLAGUE, J.J. & BOBROWSKY, P.T. (1990) Holocene sea level change and crustal deformation, southwestern British Columbia. *Geol. Survey Can. Paper 90–1E*, 245–250.

CODIGNOTTO, J.O. (1983) Depósitos elevados y /o de acrecion Pleistoceno-Holoceno en la costa fueguino-patagónica. *Actas, Simp. osc. del nivel del mar durante el último hemiciclo deglacial en la Argentian, UNMDP, Mar del Plate*, 12–26.

DALRYMPLE, R.W. (1979) Wave-induced liquefaction: a modern analogue from the Bay of Fundy: *Sedimentology* **26**, 835–844.

DARIENZO, M.E. & PETERSON, C.D. (1990) Episodic tectonic subsidence of Late Holocene salt marshes, northern Oregon central Cascadia margin: *Tectonics* **9**, 1–22.

DARWIN, C. (1851) Geological observations on South America. In: *Geological Observations on Coral Reefs, Volcanic Islands, and on South America – being the Geology of the Voyage of the Beagle, under the Command of Captain Fitzroy, R.N., during the Years 1832 to 1836, Part III.* London, Smith, Elder & Co., pp. 1–29.

DOE, T.W. & DOTT, JR., R.H. (1980) Genetic significance of deformed cross-bedding – with examples from the Navajo and Weber Sandstones of Utah. *J. Sediment. Pet.* **50**, 793–812.

GALLI OLIVER, C. & SANCHEZ REJAS, J. (1963a) Relation between the geology and the effects of the earthquakes of May 1960 in the city of Castro and vicinity, Chiloé. *Seismol. Soc. Am. Bull.* **53**, 1263–1271.

GALLI OLIVER C. & SANCHEZ REJAS, J. (1963b) Relation between the geology and the effects of the earthquakes of May 1960 in the city of Ancud and vicinity, Chiloé. *Seismol. Soc. Am. Bull.* **53**, 1273–1280.

LOMNITZ, C. (1970) Major earthquakes and tsunamis in Chile during the period 1535 to 1953: *Geolog. Rundschau* **59**, 938–960.

MUIR-WOOD, R. (1989) The November 7th 1837 earthquake in southern Chile [abs.]. *Seismol. Res. Letters* **60**, 8.

NELSON, A.R. (1991a) Holocene tidal-marsh statigraphy in south–central Oregon – Evidence for localized sudden submergence in the Cascadia subduction zone: In: *Quaternary Coasts of the United States: Lacustrine and Marine Systems* (Eds Fletcher, C.P. & Wehmiller, J.F.) *SEPM Soc. Sediment. Geol. Spec. Publ.* **48**, pp. 287–306.

NELSON, A.R. (1991b) Discordant ^{14}C ages from buried tidal-marsh soils in the Cascadia subduction zone, southern Oregon coast. *Quat. Res.* **38**, pp. 74–90.

NELSON, A.R. & MANLEY, W.F. Holocene coseismic and aseismic uplift on Isla Mocha, south-central Chile: In: *Neotectonic Aspects of the Evolution of Quaternary Coasts*(Eds Ota, Yoko, Nelson, A.R. & Berryman, Kelvin) special issue, *Quat. Internat.* **15/16** pp. 61–76.

NISHENKO, S. (1985) Seismic potential for large and great interplate earthquakes along the Chilean and Southern Peruvian margins of South America – a quantitative reappraisal. *J. Geophys. Res.* **90**, 3589–3625.

OVENSHINE, A.T., LAWSON, D.E. & BARTSCH-WINKLER, S.R. (1976) The Placer River Formation: intertidal sedimentation caused by the Alaska earthquake of March 27, 1964. *Jour. Res., U.S. Geol. Survey* **4**, 151–162.

PASKOFF, R. (1978) Donnés à propos des variations rélatives du niveau de l'Océan Pacifique au cours de l'Holocène sur la côte du Chili du Nord et du Centre. *Proceedings 1978 International Symposium on Coastal Evolution in the Quaternary*, Sao Paulo, Brazil, pp. 449–452.

PASKOFF, R. (1980) Late Cenozoic crustal movements and sea level variations in the coastal areas of Northern Chile. In: *Earth Rheology, Isostacy, and Eustacy* (Ed. N.A. Mømer) John Wiley, Chichester, pp. 487–495.

PLAFKER, G. (1969) Tectonics of the March 27, 1964, Alaska earthquake. In: *The Alaska earthquake, March 27, 1964 – Regional Effects. U.S. Geol. Survey Prof. Pap. 543–1*, 74 pp.

PLAFKER, G. (1987) Application of marine-terrace data to paleoseismic studies. In: *Proceedings of Conference XXXIX – Directions in Paleoseismology.* (Eds: Crone, A.J. & Omdahl, E.M.) *U.S. Geol. Survey Open-file 87–673*, pp. 146–156.

PLAFKER, G. & RUBIN, M. (1978) Uplift history and earthquake recurrence as deduced from marine terraces on Middleton Island, Alaska. In: *Proceedings of Conference VI, Methodology for Identifying Seismic Gaps and Soon-to-Break Gaps. U.S. Geol. Survey Open-file 78–943*, pp. 687–721.

PLAFKER, G. & SAVAGE, J.C. (1970) Mechanism of the Chilean earthquakes of May 21 and 22, 1960. *Geol. Soc. Am. Bull.* **81**, 1001–1030.

PLAFKER, G., LAJOIE, K.R., AND RUBIN, M. (1992) Determining recurrence intervals of great subduction zone earthquakes in southern Alaska by radiocarbon dating. In: *Radiocarbon After Four Decades: An Interdisciplinary Perspective.* (Eds Taylor. R.E. Long, A. & Kra, R.S.)

SAINT-AMAND, P. (1963) Los terremotos de Mayo – an eyewitness account of the greatest catastrophe in recent history, Chile 1960. China Lake, California, Michelson Laboratory, U.S. Naval Ord. Test Station Tech. Art. 14, 39 p.

SIEVERS C., H.A., VILLEGAS C.G. & BARROS, G. (1963) The seismic sea wave of 22 May 1960 along the Chilean coast. *Bull. Seismol. Soc. Am.* **53**, 1125–1190.

WALCOTT, R.I. (1972) Past sea levels, eustasy, and deformation of the earth. *Quat. Res.* **2**, 1–14.

WYSS, M. (1978) Local changes of sea level before large earthquakes in South America. *Bull. Seismol. Soc. Am.* **66**, 903–914.

Divergent
Plate-Margin Basins

Spec. Publs Int. Ass. Sediment. (1993) **20**, 111–128

Sedimentation in divergent plate-margin basins

L.E. FROSTICK* *and* R.J. STEEL†

* Postgraduate Research Institute for Sedimentology,
University of Reading, Reading RG6 2AB, UK; and
† Department of Geology, University of Bergen,
Allégt 41-5007, Bergen, Norway

ABSTRACT

Divergent plate-margin basins are dominantly extensional in character and commonly form initially in either mantle-generated or lithosphere-generated rift zones. Individual basins frequently undergo an evolution from syn-rift phase, with active block rotation, to a post-rift phase where more broadly based subsidence is controlled by thermal cooling. The region may eventually become a new passive continental margin. The depositional infill and resulting sedimentary architecture can be most problematic in the syn-rift phase where it is controlled largely by the interplay of fault-block geometry, rates of fault movement and drainage area characteristics, as well as climatic and sea-level changes.

Within non-marine half-grabens, with the Pangean Triassic basins as prime examples, there is a generally good documentation of infill architecture, although there is still some uncertainty about the controls on the relative abundance of footwall-, hanging-wall-and axially-derived sediments. In marine half-grabens, where a sediment gravity flow association commonly dominates, there is more uncertainty about infill architecture because of the variable balance between limited sediment supply, sea-level change and fault-block tilt rates. Except in half-grabens adjacent to large and easily eroded hinterlands the volume of footwall-derived sediments is commonly overestimated. Examples of post-rift and rift–drift transition successions are discussed from the northern North Sea and Red Sea areas.

INTRODUCTION

Basins in divergent plate-margin settings are of considerable interest to Earth and environmental scientists. Ancient examples of such basins are of great economic importance since they contain significant accumulations of fossil fuels, notably oil and gas, as well as quantities of coal and other resources such as heavy metals (Frostick & Reid, 1990). In addition many present-day basins are sites of great ecological interest, some of the East African Rift lakes, for example, have unique faunas as a result of isolation and adaptation to changing water chemistry. Lakes Magadi and Natron are extreme examples of this (Eugster, 1980, 1986; Vincens & Casanova, 1987). As a result the scientific literature on these basins is vast and even that devoted just to geological aspects is extremely extensive. In this context it is interesting to note that there are still many controversies raging about almost every aspect of such basins, including their mode of formation, evolution and style of sedimentary fill. This paper focuses on the sedimentary infill, emphasizing the architecture of these deposits and how the variety of fill-types form in probable evolutionary trends in tectonics and structure.

TECTONIC EVOLUTION FROM RIFT TO OCEAN

An evolutionary series of basins is thought to exist between continental rifts and passive margins. This idea was first considered as part of the general theory of plate tectonics developed during the early 1960s by Hess (1962), Dietz (1963) and Wilson (1966) amongst others. Since this time the idea has been refined and modified by several authors, nota-

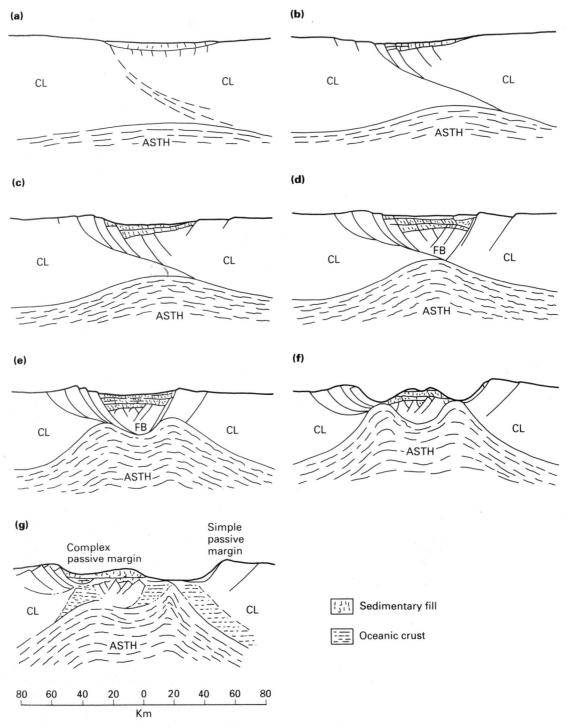

Fig. 1. Evolutionary sequence from rift to passive margin proposed by Rosendahl (1987). (a) Stage 1, sag basin (stretch trough). (b) Stage 2, juvenile half-graben. (c) Stage 3, mature half-graben. (d) Stage 4, facing half-graben. (e) Stage 5, alternating boundary fault activity. (f) Stage 6, juvenile rift basin. (g) Stage 7, successful spreading system. ASTH, asthenosphere; CL, continental lithosphere.

bly by Lowell and Genik (1972), Bally (1982) and Rosendahl (1987). The theory suggests that continental rifts evolve through a stage, designated the juvenile ocean phase, into fully fledged passive margins (Fig. 1). Evidence which supports the ideas encapsulated in this evolutionary sequence can be seen on many seismic sections of passive margins. Here it is often possible to identify superimposed packages of sediment which can be ascribed to the successive phases of development (i.e. rift, rift–drift, passive margin). An example of this is shown in Fig. 2, where interpretation of the structure on the Brazilian margin of the Atlantic Ocean shows all of the predicted features.

However, some margins are less complex and show little or no evidence of the rift to drift phase of basin development. In addition not all continental rifts evolve into passive margins. Rifts such as the Rhine and the North Sea have effectively died before reaching the stage at which sea floor spreading is initiated. These variations can largely be explained by the observation that the evolution of a rift can stop at any point up to and including the intrusion of ocean floor material. As a result there are ancient representatives of both immature and mature rifts as well as young oceans and passive margins. In addition, the structural evolution of these areas dictates a strong asymmetry which often leads to the preservation of the earlier rift and rift–drift packages on only one margin of the ocean.

The sedimentary evolution of rifted areas reflects a complex interplay between the effects of faulting, uplift and subsidence as well as the modifying influences of climate and eustacy. As a consequence

basin fills are extremely varied and generalization becomes difficult. Sedimentary models, for example for rift basins, are many and varied (see Leeder & Gawthorpe, 1987; Cohen, 1989; Frostick & Reid, 1990 for just a few examples). But even for rifts it is possible to identify some common themes, some features which, if not diagnostic, are almost always present. This effort to make some generalization has long had some practical importance, for example for the estimation of natural resources through the *prediction* of location and geometry of potential reservoir and source-rock lithosomes within complex rifted terranes.

CONTINENTAL RIFTS

There are several hundred continental rifts distributed around the globe (Burke, 1978; Easton,1983; Ramberg & Morgan, 1984) and they occur on almost all continents (Rosendahl, 1987). The structure and sedimentary fill of these basins has been the subject of geological interest for at least the whole of this century with the spectacular East African Rift acting as a special focus for attention. Indeed most of the early structural models were derived almost exclusively from African data (Gregory, 1921; Dixey, 1945a, b). But more recently a study of the variability of both contemporary and ancient rifts has allowed us to place the East African Rift in its proper context, as a relatively extreme example in a range of rift types.

It has become generally accepted that there are two broad categories of rift. These have been termed

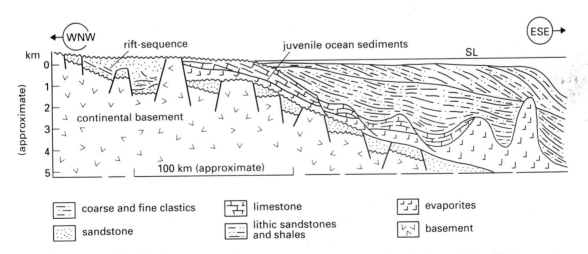

Fig. 2. Interpretation of a seismic section across the Brazilian continental margin. (After Ponte *et al.*, 1980.)

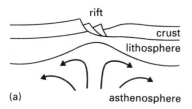

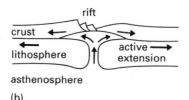

Fig. 3. (a) Mantle-generated rift; (b) lithosphere-generated rift. (After Turcotte & Emerman, 1983.)

either mantle-generated or active rifts and lithosphere-generated or passive rifts (Burke & Dewey, 1973; Sengor & Burke, 1978; Condie, 1982; Fig. 3). Mantle-generated rifts are formed as a consequence of mantle upwelling which causes both doming and volcanicity. By contrast lithosphere-generated rifts owe their origin to intraplate stresses causing extensional failure of the lithosphere. In the latter case additional mantle heat sources are unnecessary for failure to occur, although it will occur preferentially in warmer and younger crust. These two types of rift have contrasting tectonic histories and sedimentary fills as a function of their different origins.

Mantle-generated rifts

In this type, rifting is a response to doming, arching,

and uplift on a regional scale. The proposed causes of uplift are the presence of mantle plumes or hot spots (Sengor & Burke, 1978; Morgan & Baker, 1983; Keen, 1985) which may occur within a general plate setting that does not have to be extensional. The main diagnostic features of this type of rift are:

1 early doming leading to diversion of river drainage away from the site of rifting; and
2 high heat flow accompanied by early and extensive volcanic activity.

Lithosphere-generated rifts

In lithosphere-generated rifts, subsidence, not uplift, is the first expression of rifting. Any doming which occurs is subsequent to this and is the result

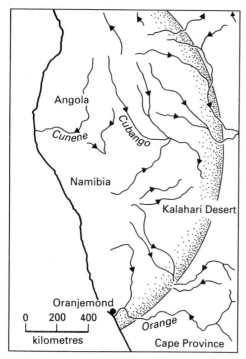

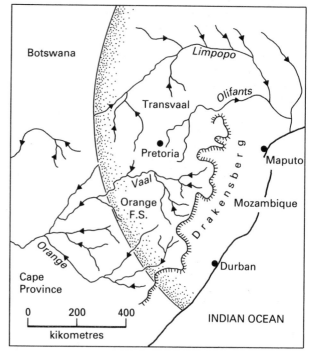

Fig. 4. Radical drainage patterns over domed areas interpreted as being due to rifting of the Atlantic and Indian Ocean margins of Afric.a (After Cox, 1989.)

of later thermal events. The cause of subsidence is in-plate extension causing stretching and thinning of the lithosphere. The diagnostic features of this type of rift are:

1 the early development of a sag basin trapping river drainage;

2 no initial uplift, although there may be doming later as the asthenosphere rises; and

3 later and/or sparse volcanism.

The contrasting patterns of structural development exhibited by the two types of rift result in distinctive patterns of sedimentary fill. Mantle-generated rift fill is floored by a marked erosional unconformity associated with the initial uplift. Radial drainage patterns away from the incipient rift may be formed (Cox, 1989; Frostick & Reid, 1989a; Fig. 4) and preserved even after the area becomes a passive margin. Early rift deposits show evidence of clastic sediment starvation. Frequently this takes the form of organic-rich lake deposits, as in Lake Tanganyika (Cohen, 1989). In some basins volcaniclastic sediments will dominate and compensate for the lack of alternative clastic material. By contrast, in lithosphere-generated rifts the early subsidence attracts river drainage and its accompanying clastic sediment. The initial sag becomes filled with fluvial and lake deposits which extend beyond the margins of the final rift. However, later infill-phases may show signs of clastic sediment starvation, as footwall uplift diverts rivers away from the evolving fault scarps.

Some rifts appear to fall easily into one or other of these categories, for example the East African Rift with its abundant volcanism and doming (Karson & Curtis, 1989) is apparently a good example of a mantle-generated rift. The Reconcavo Rift in eastern Brazil has all the sedimentary features of a lithosphere-generated rift, with extensive Jurassic sands beneath the main Cretaceous rift deposits. But many rifts are equivocal, with some structural features diagnostic of both types.

RIFT STRUCTURE AND THE SEDIMENTARY FILL

Subsidence rates and sedimentation

The tectonic significance of large-scale sedimentary architecture in actively subsiding basins has become better understood during the last decade. The spatial relationship between coarse-grained aprons of sediment and basin margins has long been clear, but the genetic relationship between the geometry (or progradational character) of such clastic wedges and changing rates of tectonic activity has been less obvious. As has been documented in thick synextensional successions by Steel *et al.* (1988) and from rift, pull-apart and foreland settings by Blair & Bilodeau (1992) and by Heller & Paola (1989), the most highly progradational levels within clastic wedges correspond to times of minimum tectonic activity along the basin margins. Conversely, the finest grained intervals (representing floodbasin, lacustrine and shaley marine deposits) of clastic wedges commonly coincide with the highest rates of differential subsidence. These relationships have also become clear in the light of the development of sequence stratigraphy concepts (e.g. Jervey, 1988; Posamentier *et al.*, 1988), where high rates of subsidence and base-level rise provide abundant accommodation space and cause coarse-grained facies to develop in narrow but thick belts adjacent to the basin margins. Low subsidence rates, or periodic slight uplift of the basin-margin regions result in a reduction in the amount of accommodation and higher rates of progradation. Assuming a reasonably continuous sediment supply, these theories allow a more predictive approach to the understanding of the architecture of clastic wedges and of the large-scale cyclothems developed within them. This approach has been applied to understanding the cyclic development of clastic wedges (5–15 My duration) in the extensional northern North Sea Mesozoic basin (Steel, 1993).

Rift geometry and sedimentation

Whatever the origin of individual rifts, there are structural features common to all. The basic structural element is now thought to be the half-graben (Gibbs, 1984; Rosendahl, 1987), not a full graben as was previously favoured. The half-graben segments are separated, along their length direction, by zones variously known as accommodation zones and transfer zones or faults. The single, major fault zones which control the asymmetrical basins are referred to as border faults and in adjacent segments they may reverse polarity, that is flip from one side of the rift axis to the other (Fig. 5) (see also Hamblin & Rust, 1989).

The geomorphological consequences of rift structure are a series of elongate tilted sub-basins ranging in size from 60 to 80 km long and 30 to 60 km wide (Rogers & Rosendahl, 1989) with depocentres asymmetrically placed close to the major control-

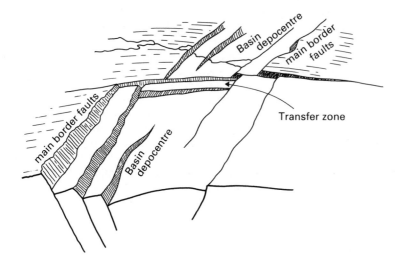

Fig. 5. General model for half-graben development in continental rifts. (After Frostick & Reid, 1990.)

ling border fault. The faulted margins of the individual sub-basins are steep, and backtilting on fault blocks can combine with footwall uplift to produce drainage gradients *away* from the main rift depocentre. However, the unfaulted or ramping margin of the individual sub-basins generally exhibits a relatively gentle gradient, although small faults and monoclines may locally disrupt this. At each end of the rift segments there are topographically high areas where basement rocks are shallow or outcropping. These cross-rift structures are the transfer zones and they tend to isolate one segment from another for the period during which they are active (Frostick & Reid, 1990).

NON-MARINE RIFT BASINS

The sedimentary fill of most active non-marine rifts reflects the above general structural framework. The geometry is generally wedge-shaped with stratal packages successively onlapping either basement rocks or older sediments (Schlische, 1989b). Hiatuses in deposition can be attributed either to tectonic or climatic changes and, at times, these are difficult to distinguish (Frostick & Reid, 1991). Careful mapping (e.g. with three-dimensional seismic control) may show angular unconformities between stratal packages, strongly suggesting the role of tectonic tilting *during* infill. In spite of the

difficulties in ascribing specific causes to particular breaks or changes in the pattern of sedimentation, the documentation of hiatus architecture is a critical facet of rift-basin analysis (see Underhill, 1991, for an example in marine, syn-rift fill).

The main environmental components of present-day, active, non-marine rifts are lakes, lake shores of a variety of types, deltas, rivers (either perennial or ephemeral), and fans. These environments are well documented in a number of rifts in widely different geographical locations, for example in East Africa (Cohen, 1989; Frostick & Reid, 1990), in the Basin and Range province of the Western USA (Myers & Hamilton, 1964; Anderson *et al.*, 1983), and in Israel (Garber *et al.*, 1987; Frostick & Reid, 1989b). The sedimentary fill in ancient continental rifts, such as the Triassic rift of the North Sea (Frostick *et al.*, 1988; Steel & Ryseth, 1990) and the Cretaceous Reconcavo Rift of Brazil, provides us with the result of the complex interplay between tectonism and climate as it affects the distribution and size of lakes, deltas, rivers and fans (Frostick *et al.*, 1988, 1992; Schlische & Olsen, 1990; Steel & Ryseth, 1990). The details of *how* these sedimentary environments and their facies react to changes in the rates and style of tectonic movement and climate are as yet poorly understood, though there is progress being made at this level too (e.g. Leeder & Alexander, 1987; Leeder *et al.*, 1988).

Transverse and axial infill

Although there is general agreement about the overall geometry of non-marine syn-rift fill there is still discussion about the facies distribution that is likely to develop within it. The main areas of disagreement concern the importance of fault-scarp fans and significance of axial drainage systems. Leeder and Gawthorpe (1987) have produced a rift model where axial drainage dominates and drainage across the ramp is subsidiary (Fig. 6). They and others (Hunt & Mabey, 1966; Hooke, 1972) also suggest that footwall-derived alluvial fans can be a major feature. In contrast, Frostick and Reid (1987a, b; 1990) have shown that the major fluvial input to the syn-rift basin can commonly cross the ramping margin and that footwall fans can be very limited in extent. The work of Blair and Bilodeau (1988) also supports the idea that fans are restricted in their basinward extent (though not necessarily in their volumetric importance) during periods of fault activity. It seems likely, as is the case of *marine* syn-rift infill, that the significant development of footwall-derived fans requires the presence of cross faults, fault offsets or other structural complications along the crestal area of the tilted block, in order to provide a focus for footwall-derived dispersal of eroded material.

Frostick and Reid (1989a) have argued that the establishment of axial drainage systems is inhibited by the presence of active cross-rift transfer zones. They also point out that backtilting and uplift of footwall blocks prevents all but the smallest of rivers from crossing the fault scarp. However, this segmentation of the rift will diminish and finally disappear once active rifting ceases. It is at this time, during thermal subsidence, that axial drainage will develop and may dominate, as in the Niger–Benue system in west Africa (Frostick & Reid, 1989a). It is also at this time that footwall-derived fans have the opportunity to develop greater radial extent as the topography on the fault scarp diminishes.

Frostick and Reid have based their models on experience from the East African Rift, and propose three different categories of rift based on the original crustal thickness and strength which control the complexity of the resulting structure (Fig. 7). Areas of thin crust allow the development of complex fault patterns, producing only shallow, small basins which are often filled with extensive volcanic deposits. In areas of moderate original crustal thickness faults are fewer though still quite complex. Volcanic material is still important but does not dominate the fill. In thick crustal areas only one, simple, border fault develops in each rift segment and the simple half-graben structure is filled with sediments which may include thick successions of non-clastic deposits as a result of clastic sediment starvation.

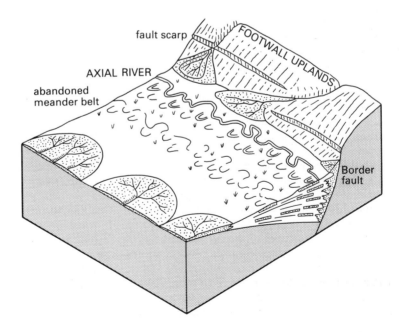

Fig. 6. A model for continental rift fill dominated by axial drainage. (After Leeder & Gawthorpe, 1987.)

(a)

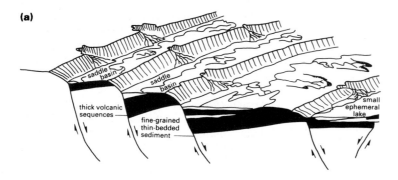

(b)

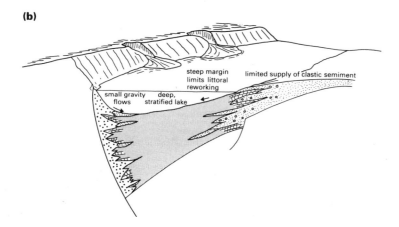

Fig. 7. Models for the sedimentary fills of continental rifts with different degrees of structural complexity. (After Frostick & Reid, 1990.) (a) Baringo-type model: volcanics + sediment supply ⩾ subsidence; (b) Tanganyika-type model: subsidence ≫ clastic sediment supply; (c) Turkana-type model: subsidence ≃ clastic sediment supply; (d) possible evolution in types.

Triassic examples

The Triassic system of North Africa, Europe and eastern North America offers some of the best examples of ancient, non-marine, rift basins where both transverse and longitudinal infilling alluvial systems vary in importance in different basins and stages of development (Manspeizer, 1988). A representative Triassic example is the Newark Basin of eastern North America which accumulated more than 7 km of syn-rift strata during the extension and initial break-up of the Pangean continent. The structural complexity and asymmetry, as well as the consequent asymmetrical distribution of fan, lacustrine, fluvial and deltic facies within the basin are shown in Fig. 8.

There are a number of syn-rift basins from the European Triassic where complex patterns of interfingering facies have been described. The Hebridean basins contain footwall-derived progradational clastic wedges which contrast strongly with a retrogradational architecture for the sandy alluvial sys-

tems on the hanging-wall slopes of the same basins (Steel, 1977). In the Inner Moray Firth sub-basin, on the other hand, hanging-wall alluvial drainage dominated, probably controlled by syndepositional roll-over structuring on this side of the rift (Frostick et al., 1988). Aspects of syn-rift sub-basins within the larger northern North Sea Triassic basin are also now emerging (Lervik et al., 1989; Frostick et al., 1992) but details are still fragmentary. Steel and Ryseth (1990) have explored the problem of distinguishing between patterns of sedimentation during periods of active extension and thermal subsidence in the Triassic succession of the northern North Sea. They highlight the disruptive effects of faulting during the deposition of the Teist Formation, making it impossible to correlate between sub-basins (Fig. 9). This is the only part of the Triassic succession in this area that shows evidence of true crustal extension, block rotation and syn-rift infill. However, in the Middle and Upper Triassic (Lomvi and Lunde Formations) the sediments span the whole width of the northern North Sea basin and are

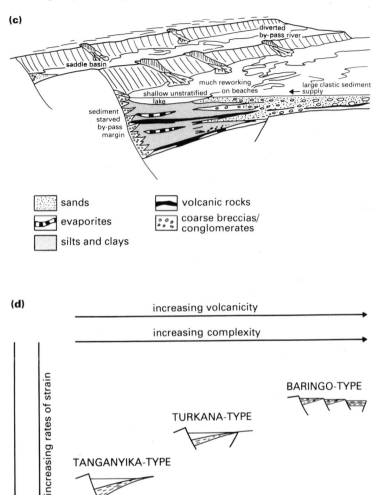

Fig. 7. (*continued*)

therefore classed as post-rift strata. Steel and Ryseth (1990) identify a number of coarsening-upward sequences in these later Triassic deposits that occur at a variety of scales. These suggest at least three major periods of coarse material progradation across the basin depocentre, each of which may be attributed to episodes of slowing in the rateof thermal subsidence during the post-rift period.

Fig. 8. Structural and facies-infill asymmetry of the Newark Basin. (From Manspeizer, 1988.)

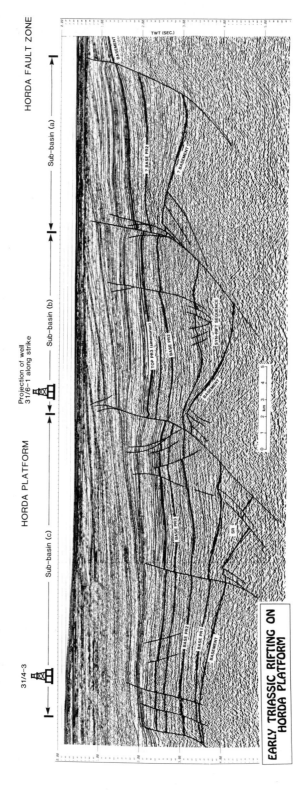

Fig. 9. Seismic cross-section showing Triassic syn-rift and post-rift configuration on the Horda Platform of the northern North Sea. (From Steel & Ryseth, 1990.)

The overall distribution of alluvial sedimentary systems will always respond to bouts of fault activity. Strong downfaulting prevents the basinward penetration of river-borne coarse sediment whereas periods of relative calm will allow the same rivers to prograde out into the basin and produce a characteristic coarsening-upward signature within the sequence. However, inferring tectonic controls must be approached with caution, especially at a small scale, since climatic change can produce a similar effect within a lake basin. Decreased rainfall can cause lakes to dry up, allowing river systems to spread coarse material over fine-grained lake sediments.

MARINE RIFT BASINS

Topography and facies distribution

Marine rift basins differ from non-marine ones in the character and architecture of their sedimentary infill. This architecture and the general predictability of sands and shales are important to hydrocarbon exploration; marine rift basins are well-known for being oil- and gas-prolific.

The onset of extension in an area usually involves break-up into a series of elongate sub-basins. If this occurs under conditions of relatively high sea-level, the sub-basins nearest to the hinterlands tend to trap most of the coarse clastic deposits, where the more axial sub-basins are commonly shale-filled or are underfilled. Commonly, there would be only limited sediment supply from exposed crestal areas of rotated blocks, and adjacent basinal areas would be covered by deep water. This situation with a high, but dynamic, sea-level, with a sediment supply area which is small (assuming a sub-basin far from the hinterland) but which will vary with sea-level, with shallow-marine sedimentary environments high on the footwall block but much deeper environments on the adjacent hanging-wall, leads to far less predictable facies distribution than is the case for non-marine rifts. Because relative sea-level is dependent on local tectonic movements, and because rates of uplift/subsidence may periodically far outstrip denudation and sediment supply, conventional sequence stratigraphy concepts are of limited value.

Facies prediction

In oil and gas exploration in marine rifts, one of the main tasks is to predict and map potential reservoirs in submarine aprons and fans, as well as adjacent shaley source rocks. Prediction of sand and shale distribution in such basins is usually approached by:

1 determination of the approximate sea-level position in relation to tilt-block crests;

2 estimation of the volume and lithology of sediment eroded from the crests of blocks;

3 reconstruction of the timing (rates of generation) and scale of rift topography in relation to sea-level behaviour;

4 tracing of potential sediment transport paths, using detailed structural/topographical maps, both downdip on either flank of the crestal zone and eventually by strike-feeding along the shallow water zones or axial transport in deeper areas;

5 reconstruction of depositional geometries, based on the above points, both in shallow and deeper water areas around the rifted topography.

This procedure demands a reasonable palaeo-geomorphological reconstruction of the rift topography, and a sedimentological approach to sequence stratigraphy is far more essential here than in post-rift basinal situations. The complexity in the work routine and the resultant uncertainty illustrates the variability of the sedimentary architecture in marine rift basins.

North Sea marine rift architecture

Figure 10 illustrates the intensive block-faulting and rotation associated with the Late Jurassic northern North Sea rift systems where the Viking Rift, the Moray Firth Rift and the Central Rift are the main components. This rifting event was initiated in the Mid–Late Jurassic, climaxed in the Early–Mid-Volgian and gradually abated in the Early Cretaceous (Surlyk, 1990). The sedimentary architecture produced by the sandstones and shales in such marine rift basins is still poorly understood, but examples from East Greenland (Wollaston Forland Group; Surlyk, 1990) and from the South Brae field (South Viking Graben) (Turner *et al.*, 1987) show conglomeratic and sandy sediment gravity flows to be the dominant facies and large-scale (up to 300 m), upward-fining sequences to be a key element of the vertical organization. The latter motifs appear to be caused by footwall-generated sediment discharge into a depocentre subject to continuous tilting and deepening (Surlyk, 1978 1990). Examples of this conglomerate–sandstone–

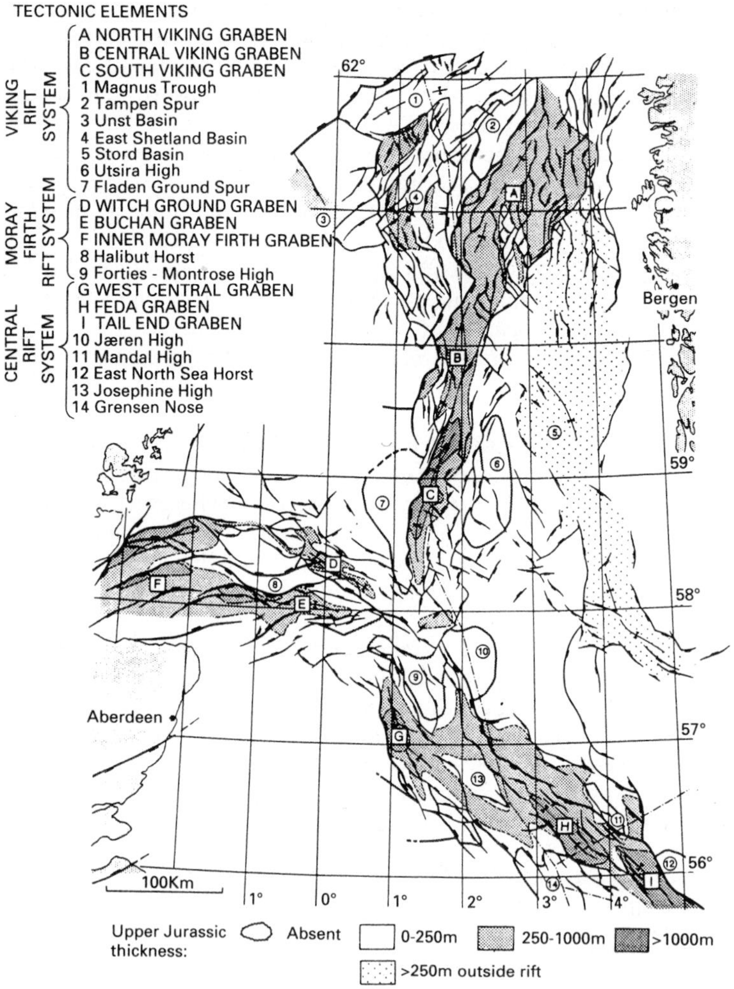
TECTONIC ELEMENTS
VIKING RIFT SYSTEM
A NORTH VIKING GRABEN
B CENTRAL VIKING GRABEN
C SOUTH VIKING GRABEN
1 Magnus Trough
2 Tampen Spur
3 Unst Basin
4 East Shetland Basin
5 Stord Basin
6 Utsira High
7 Fladen Ground Spur
MORAY FIRTH RIFT SYSTEM
D WITCH GROUND GRABEN
E BUCHAN GRABEN
F INNER MORAY FIRTH GRABEN
8 Halibut Horst
9 Forties - Montrose High
CENTRAL RIFT SYSTEM
G WEST CENTRAL GRABEN
H FEDA GRABEN
I TAIL END GRABEN
10 Jæren High
11 Mandal High
12 East North Sea Horst
13 Josephine High
14 Grensen Nose
62°
59°
58°
57°
56°
Bergen
Aberdeen
100Km
1°
0°
1°
2°
3°
4°
Upper Jurassic thickness:
Absent
0-250m
250-1000m
>1000m
>250m outside rift

Fig. 10. Fault patterns of the Late Jurassic rifting events of the northern North Sea. (From Spencer & Larsen, 1990.)

shale architecture with upward-fining motifs on different scales are illustrated from South Brae fans in Fig. 11 (facing p. 124). The Late Jurassic marine rift of the Inner Moray Firth area shows a broadly similar large-scale architecture (Underhill,1991), and this can be seen in the onshore-exposed succession of conglomerates along the Helmsdale Fault (Pickering, 1984; Theriault, 1992). Models of marine tilt-block architecture are shown in Fig. 12.

POST-RIFT INFILL

Terminology

That part of the basin infill which tends to overtop the crests of the fault-blocks, eventually filling the entire width of a broader basin, is normally referred to as the post-rift infill and is believed to accumulate during the thermal cooling phase of the basin's evolution. There is often some uncertainty within the *upper* parts of the syn-rift infill as to whether this has been passively deposited in an already established topography, or was deposited during continued extension. Seismic resolution is not always good enough to determine this. In these cases it is convenient practice to use descriptive criteria and to refer to the entire infill between adjacent rotated blocks as syn-rift. The opposite problem also arises within the post-rift part of the infill. Reactivation of the older tilt-blocks commonly allows some differential subsidence and local 'growth' of strata against reactivated fault-lines. Despite this, the strata are referred to as post-rift. The development of episodic

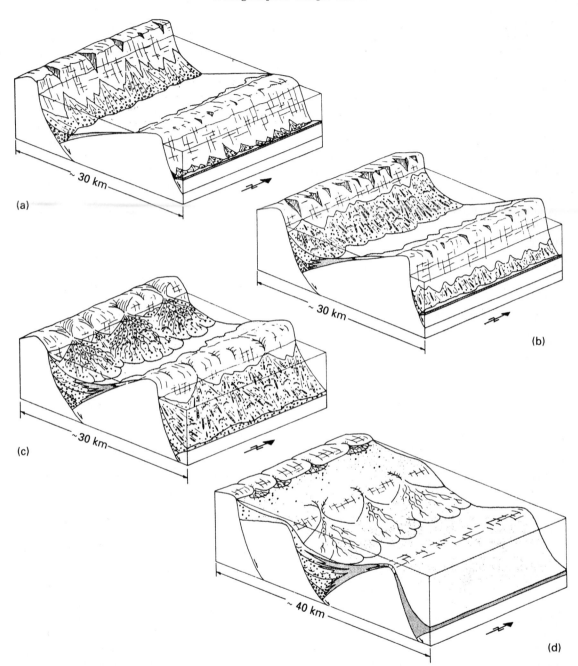

Fig. 12. Models of syn-rift infill by turbiditic depositional systems of the Wollaston Forland Group of East Greenland. The architecture is largely controlled by increasing relative sea-level (a–d) and by initial strong tilting (a–c) and later abate in tilting (d) (from Surlyk, 1990).

rift to post-rift or rift–drift couplets is well shown during the break-up of the Pangean supercontinent (Manspeizer, 1988).

North Sea Examples

The post-Valanginian succession of the northern North Sea provides one of the best known examples of strata deposited during a thermal cooling phase subsequent to Late Jurassic rifting. These post-rift strata overtop the earlier rift topography and progressively onlap across the entire northern Northern Sea Basin (Gabrielsen *et al.*, 1990).

The problem of distinguishing between patterns of sedimentation during the latest Permian–Early Triassic phase of crustal extension (Badley *et al.*, 1988; Roberts *et al.*, 1990) and the subsequent post-rift (Mid-Jurassic to Late Jurassic) succession in the northern North Sea was taken up by Steel (1993). The disruptive effects of Early Triassic extension and block rotation causes problems in attempting well-to-well correlation of Teist Formation and older strata, whereas the overlying post-rift Lomvi, Lunde and Statfjord Formations of Middle and Late Triassic age can be followed easily across the entire basinal area (Fig. 9). The post-rift succession has been divided, nevertheless, into a series of nine large-scale regressive to transgressive clastic wedges or megasequences each 200–1000 m in thickness and of 6–18 million years in duration. The megasequences were apparently initiated during periodic increases in the rate of thermal subsidence in the post-rift basin. Individual clastic wedges then developed, reaching their peak progradation during intervals of minimum subsidence rate or even slight uplift (Steel, 1993).

THE RIFT-DRIFT TRANSITION: JUVENILE OCEAN PHASE

Some continental rifts remain as rifts for long periods of geological time but evolve no further, for example the Rhine Rift will never evolve into an ocean. This is also the case for many rifts linked to the early development of the present-day oceans, the Reconcavo and Benue Rifts on either side of the Atlantic in Brazil and Nigeria respectively have both failed as proto-oceans and have become what are termed aulacogens.

In a few cases the separation of the continent is successful and a new ocean is born. The Red Sea–Gulf of Aden system is a present-day example of this stage of development. The separation in this case is now more than 200 km, Juvenile ocean-floor material is currently being intruded into the sea floor and estimates of the rate of spreading vary but it probably exceeds (Courtillot *et al.*, 1987). In evolving to this point the rift passes through various phases of structural development which are reflected in the character of the fill.

A synopsis of the possible stages in the evolution from rift to ocean is shown in Fig. 1. The first departure from the normal rift character is often the development of a second border fault on the opposite side of the basin from the original fault scarp. This has the effect of cutting off the supply of clastic sediment across the ramping side of the rift and sedimentation becomes dominated by biogenic deposits, often including organic-rich muds which can act as petroleum source rocks. Fault activity may switch from one margin to the other many times during this phase of evolution, producing many unconformities within the sedimentary pile.

The area continues to subside and extend, but access to clastic material is restricted by uplift on both margins which diverts drainage away from the basin. This is evident from an analysis of the river systems around the Red Sea–Gulf of Aden where rivers heading within 2 km of the Sea are diverted away into the Nile drainage (Fig. 13). At this stage the area may become marine either permanently or intermittently. But the sea is still very narrow and tectonically active and a barrier can easily develop in the narrow marine corridor blocking access to sea water. This leads to the deposition of thick sequences of evaporites. These, and the importance of organically sourced deposits, are the characteristic sedimentary features of this phase of development.

The Red Sea offers excellent examples of more-or-less typical juvenile ocean sequences in its outcrops and many oil exploration wells (Purser *et al.*, 1987; Montenat *et al.*, 1988). Most of the onshore deposits are marginal to the main basin depocentre and as a result thick evaporites and organic shales are not normally seen. What are seen are thick sequences of carbonates with some anhydrites but with only few, spatially and temporally restricted, beds of coarse alluvium (Purser *et al.*, 1987; Fig. 14). Offshore there are several wells which penetrate both thick salt deposits and organic shales. The salt acts as a capping deposit for some of

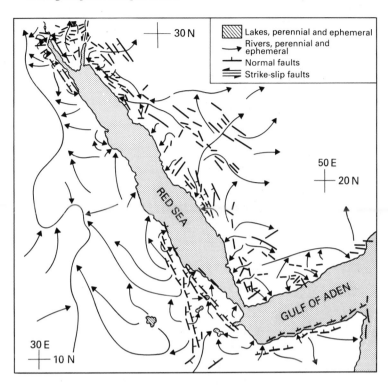

Fig. 13. Present-day patterns of river drainage in the Red Sea–Gulf of Aden region showing the influence of uplifted basin margins diverting drainage *away* from the depocentre. (After Frostick & Reid, 1989a.)

the offshore oil reservoirs but the sources occur deeper in the succession, since the major reservoir is the Nubian Sandstone.

The marginal deposits in parts of the Red Sea have been described by Purser and Hotzl (1988). They infer that the main rifting commenced in the Miocene and the earliest deposits are clastic, syn-rift sediments. Interbedded conglomerates, sands and silts are mainly poorly sorted and show evidence of deposition from flash floods. At the top of the sequence there are lake deposits, silts and fine sands, with spectacular slumps suggesting tectonic instability (Purser & Hotzl, 1988). There are limited basalt flows close to the base of the sequence which have been K/Ar dated at 25 Ma.

These syn-rift sediments have been faulted, tilted and eroded prior to the deposition of the succeeding marine sediments. These vary in age along the length of the rift controlled by the direction of the marine transgression and on the very varied pre-transgression topography. In general they are Burdigalian in age and represent a period of strong subsidence following active rifting. They comprise marine siliciclastic sands, bioclastic and oolitic carbonate sands, reefs and stromatolites. On both sides of the Red Sea there is evidence of synsedimentary tilting towards the subsiding rift axis. The onset of evaporite development occurred in the Middle Miocene, generally in the Langhian (Thiriet *et al.*, 1986). They are horizontally bedded deposits generally with no evidence of sabkha deposition. It is thought that some of the deposits may have developed by sulphatization of pre-existing carbonates, although the relative importance of these and primary deposits is not known. Further offshore, shales, marls and salts dominate the post-rift sequence. The salts began developing at a much earlier stage but the main salts are still of Langhian age.

Later, Plio-Pleistocene to Recent sediments are a mixture of restricted siliciclastics, often trapped by strong uplift on the rift margins, reefs and other carbonate deposits with Indian Ocean faunas (Purser & Hotzl, 1988). True oceanic conditions have therefore been established and the scene is set for both shores of the Red Sea to transform into passive margins.

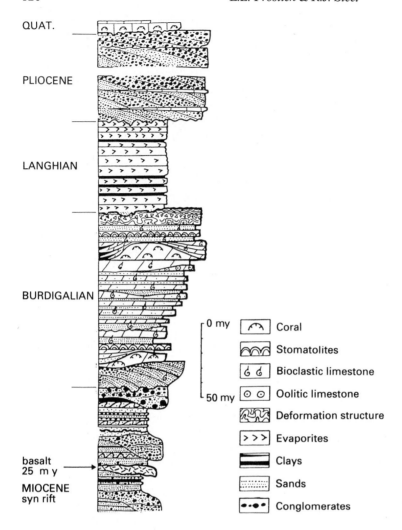

QUAT.

PLIOCENE

LANGHIAN

BURDIGALIAN

┌ 0 my

50 my

basalt
25 m y

MIOCENE
syn rift

⌒	Coral
⌒⌒	Stomatolites
6 6	Bioclastic limestone
⊙ ⊙	Oolitic limestone
〰	Deformation structure
> > >	Evaporites
▬	Clays
⋯	Sands
•-•-•	Conglomerates

Fig. 14. Juvenile-ocean
(post-Miocene) sedimentary
sequence on the margin of the Red
Sea. (After Purser & Hotzl, 1988.)

REFERENCES

Anderson, R.E., Zoback, M.L. & Thompson, G.A. (1983)
Implications of selected subsurface data on the struc-
tural form and evolution of some basins in the northern
Basin and Range province, Nevada and Utah. *Bull.
Geol. Soc. Am.* **94**, 1055–1072.

Badley, M.E., Price, J.D., Rambach Dahl, C. & Agdestein, T.
(1988) The structural evolution of the northern Viking
Graben and its bearing upon extensional models of
basin formation. *J. Geol. Soc. London* **145**, 455–472.

Bally, A.W. (1982) Musings over sedimentary basin evolution. *Philos. Trans. R. Soc. London* **A305**, 325–338.

Blair, T.C. & Bilodeau, W.L. (1988) Development of
tectonic cyclothem in rift, pull-apart and foreland basins: sedimentary response to episodic tectonism. *Geology* **16**, 517–520.

Burke, K. (1978) Evolution of continental rift systems in
the light of plate tectonics. In: *Tectonics and Geophysics

of Continental Rifts* (Eds Ramberg, I.B. & Neumann,
E.R.) Reidel, Dordrecht.

Burke, K. & Dewey, J.F. (1973) Plume generated triple
junctions: key indicators in applying plate tectonics to
old rocks. *J. Geol.* **81**, 406–433.

Cohen, A. (1989) Facies relationships and sedimentation
in large rift lakes: examples from Lakes Turkana and
Tanganyika. *Palaeogeogr. Palaeoclimatol. Palaeoecol.
Int. Assoc. Sedimentol. Spec. Publ. 13;* **70**, 65–80.

Condie, K.C. (1982) *Plate Tectonics and Continental Drift.*
Pergamon, Oxford.

Courtillot, V., Armijo, R. & Tapponnier, P. (1987) Kinematics of the Sinai triple junction and a two phase model of
Arabia–Africa rifting. In: *Continental Extensional Tectonics* (Eds Coward, M.P., Dewey, J.F. & Hancock, P.L.)
Geol. Soc. Londer Spec. Publ. **28**, 559–573.

Cox, K.G. (1989) The role of mantle plumes in the
development of continental drainage patterns. *Nature*
342, 873–876.

DIETZ, R.S. (1963) Collapsing continental rises, an actualistic concept of geosynclines and mountain building. *J. Geol.* **71**, 314–333.

DIXEY, F. (1945a) On the faulting of rift valley structures. *Geol. Mag.* **82**, 377–385.

DIXEY, F. (1945b) Erosion and tectonics in the East African rift system. *Q. J. Geol. Soc. London* **102**, 339–388.

EASTON, R.M. (1983) Crustal structure of rifted continental margins: geological constraints from the Proterozoic rocks of the Canadian Shield. *Tectonophysics* **94**, 371–90.

EUGSTER, H.P. (1980) Lake Magadi, Kenya and its precursors. In : *Hypersaline Brines and Evaporites* (Ed. Nissenbaum, A.) Developments in Sedimentology **28**, 195–232.

EUGSTER, H.P. (1986) Lake Magadi Kenya a model for rift valley hydrochemistry and sedimentation? In: *Sedimentation in the African Rifts* (Eds Frostick, L.E., Renaut, R.W., Reid, I. & Tiercelin, J.J.) Geol. Soc. Spec. Publ. **25**, 177–189.

FAERSETH, R. & PEDERSTEN, K. (1988) Regional sedimentology and petroleum geology of marine Late Bathonian–Valanginian sandstones in the North Sea. *Mar. Petrol. Geol.* **5**, 17–33.

FROSTICK, L.E. & REID, I. (1987a) A new look at rifts. *Geol. Today* **3**, 122–126.

FROSTICK, L.E. & REID, I. (1987b) Tectonic control of desert sediments in rift basins ancient and modern. In: *Desert Sediments: Ancient and Modern* (Eds Frostick, L.E & Reid, I.) Geol. Soc. London Spec. Publ. **35**, 53–68.

FROSTICK, L.E. & REID, I. (1989a) Is structure the main control of river drainage and sedimentation in rifts? *J. Afr. Earth Sci.* **8**, 165–182.

FROSTICK, L.E. & REID, I. (1989b) Climatic versus tectonic controls of fan sequences: lessons from the Dead Sea, Israel. *J. Geol. Soc. London* **146**, 527–538.

FROSTICK, L.E. & REID, I. (1990) Structural control of sedimentation patterns and implications for the economic potential of the East African Rift Basins. *J. Afr. Earth Sci.* **10**, 307–318.

FROSTICK, L.E., REID, I., JARVIS, J. & EARDLEY, H. (1988) Triassic sediments of the Inner Moray Firth, Scotland: early rift deposits. *J. Geol. Soc. London* **145**, 235–248.

FROSTICK, L.E., LINSEY, T. & REID, I. (1992) Tectonic and climatic control of Triassic sedimentation in the Beryl Basin, northern North Sea. *J. Geol. Soc. London* **149**, 13–26.

GABRIELSEN, R., FAERSETH, R.B., STEEL, R.J., IDIL, S. & KLØVJAN, O.S. (1990) Architectural styles of basin fill in the northern Viking Graben. In : *Tectonic Evolution of the North Sea Rifts* (Eds Blundell, D.J. & Gibbs, A.D.) Oxford University Press, Oxford, pp. 158–179.

GARBER, R.A., LEVYT, Y. & FRIEDMAN, G.M. (1987) The sedimentology of the Dead Sea. *Carbonates and Evaporites* **2**, 43–57.

GIBBS, A.D. (1984) Structural evolution of extensional basin margins, *J. Geol. Soc. London* **141**, 609–620.

GREGORY, J.W. (1921) *The Rift Valleys and Geology of East Africa*. London, Seeley, 479 pp.

HAMBLIN, A.P. & RUST, B.R. (1989) Tectono-sedimentary analysis of alternate polarity half-garben basin-fill successions: Late Devonian–Early Carboniferous Horton Group, Cape Breton Island, Nova Scotia. *Basin Research* **2**, 239–255.

HELLER, P.L. & PAOLA, C. (1992) The large-scale dynamics of grain-size variation in alluvial basins, 2: Application to syntectonic conglomerate. *Basin Res.* **4**, 91–102.

HESS, H.H. (1962) History of ocean basins. In: *Petrological Studies: a Volume in Honor of A.F. Buddington* (Eds Engel, A.E.J., James, H.L. & Leonard, B.F.) Geological Society of America, Boulder, Colorado, pp. 599–620.

HOOKE, R. LE B. (1972) Geomorphic evidence for Late Wisconsian and Holocene tectonic deformation, Death Valley, California. *Bull. Geol. Soc. Am.* **83**, 2073–2098.

HUNT, C.B. & MABEY, D.R. (1966) Stratigraphy and structure, Death Valley, California. *Prof. Pap. U.S. Geol. Surv.* A494.

JERVEY, M.T. (1988) Quantitative geological modelling of siliclastic rock sequences and their seismic expression. In: *Sea Level Changes: an Integrated Approach.* (Eds Wilgus, C.K., Hastings, B.S., Kendall, C.G.St.C., Posamentier, H., Ross, C.A. & Van Wagoner, J.) oc. Econ. Paleontol. Mineral. Spec. Publ. **42**, 47–70.

KARSON, J.A. & CURTIS, P.C. (1989) Tectonic and magmatic processes in the Eastern Branch of the East African Rift and implications for magmatically active continental rifts. *J. Afr. Earth Sci.* **8**, 431–453.

KEEN, C.E. (1985) The dynamics of rifting: deformation of the lithosphere by active and passive driving forces. *Geophys. J. R. Astron. Soc.* **80**, 95–120.

LEEDER, M.R. & ALEXANDER, J. (1987) The origin and tectonic sign of asymmetrical meander belts. *Sedimentol* **34**, 217–226.

LEEDER, M.R. & GAWTHORPE, R.L. (1987) Sedimentary models for extensional tilt-block (half-graben) basins. In: *Continental Extensional Tectonics* (Eds Coward, M.P., Dewey, J.F. & Hancock, P.C.) Geol. Soc. London Spec. Publ. **28**, 139–152.

LEEDER M.R., ORO D.M. & COLLIER, R. (1988) Development of alluvial fans and fan deltas in neotectonic extensional settings: inplications for the interpretation of basin fills. In: *Fan Deltas: Sedimentology and Tectonic settings* (Eds Nemec. W. & Steel, R.J.) Blackie, Glasgow, pp. 173–183.

LERVIK, K.S., SPENCER, A.M. & WARRINGTON, G. (1989) Outline of Triassic stratigraphy and structure in the central and northern North Sea. In: *Correlation in Hydrocarbon Exploration* (Ed. Collinson, J.D.) Norwegian Petroleum Society, Graham & Trotman, London.

LOWELL, J.D. & GENIK, G.J. (1972) Sea floor spreading and structural evolution of the southern Red Sea. *Bul. Am. Assoc. Petrol. Geol.* **56**, 247–259.

MANSPEIZER, W. (1988) Triassic–Jurassic rifting and opening of the Atlantic: an overview. In: *Triassic–Jurassic Rifting* (Ed. Manspeizer, W.) Developments in Geotectonics **22**, Elsevier, Amsterdam, pp. 41–81.

MONTENAT, C., D'ESTEVOU, P.O., PURSER, B., BUROLLET, P-F., JARRIGE, J-J., ORSZAG-SPERBER, F., PHILOBBOS, E., *et al.* (1988) Tectonic and sedimentary evolution of the Gulf of Suez and the northwestern Red Sea. *Tectonophysics* **153**, 161–177.

MORGAN, W.J. & BAKER, B.H. (1983) Introduction to processes of continental rifting. *Tectonophysics* **94**, 1–10.

MYERS, W.B. & HAMILTON, W. (1964) Deformation associated with the Hebgen Lake earthquake of August 17, 1959. *Prof. Pap. U.S. Geol. Surv.* **435**, 55–98.

PICKERING, K.T. (1984) The Upper Jurassic 'Boulder Beds'

and related deposits: a fault-controlled submarine slope, NE Scotland. *J. Geol. Soc. London* **141**, 357–74.

POSAMENTIER, H.W., JERVEY, M.T. & VAIL, P.R. (1988) Eustatic controls on clastic deposition. I. Conceptual framework. In: *Sea-level Changes: an Integrated Approach* (Eds Wilgus, C.K., Hastings, B.S., Kendall, C.G. St.C., Posamentier, H., Ross, C.A. & Van Wagoner, J.) Soc. Econ. Paleontol. Mineral. Spec. Publ. **42**, 109–124.

PURSER, B.H. & HOTZL, C.H. (1988) The sedimentary evolution of the Red Sea rift: a comparison of the northwest (Egyptian) and northeast (Saudi Arabia) margins. *Tectonophysics* **153**, 193–208.

PURSER, B.H., ORSZAG-SPERBER, F. & PLAZIAT, J-C. (1987) Sedimentation and rifting: the Neogene series of the NW Red Sea (Egypt). *Notes et Memores No. 21.Total Compagnie Francais des Petroles.*

RAMBERG, I.B. & MORGAN, P. (1984) Physical characteristics and evolutionary trends in continental rifts. *Tectonics* **7**, 165–216.

ROBERTS, A.M., YIELDING, G. & BADLEY, M.E. (1990) A kinematic model for the orthogonal opening of Late Jurassic North Sea rift system, Denmark–mid Norway. In: *Tectonic Evolution of the North Sea Rifts* (Eds Blundell, D.J. & Gibbs, A.D.) Oxford University Press, Oxford, pp. 180–199.

ROGERS, J.J.W. & ROSENDAHL, B.R. (1989) Perceptions and issues in continental rifting. *J. Afr. Earth Sci.* **5**, 137–142.

ROSENDAHL, B.R. (1987) Architecture of continental rifts with special reference to East Africa. *Ann. Rev. Earth Planeta. Sci.* **15**, 445–503.

SCHLISCHE, R.W. (1991) Half-graben basin filling models: new constraints on continental extensional basin development. *Basin Res.* **3**, 123–141.

SCHLISCHE, R.W. & OLSEN, P.E. (1990) Quantitative filling model for extensional basins with application to the early Mesozoic rifts of eastern North America. *J. Geol.* **98**, 135–155.

SENGOR, A.M.C. & BURKE, K. (1978) Relative timing of rifting and volcanism on earth and its tectonic implications. *Geophys. Res. Lett.* **5**, 419–421.

SPENCER, A.M. & LARSEN, V.B. (1990) Fault traps in the northern North Sea. In: *Tectonic Events Responsible for British Oil and gas Resources* (Eds Hardman, R.F.P. & Brooks, T.) Geol. Soc. London Spec. Publ. **55**, 281–298.

STEEL, R.J. (1977) Triassic rift basins of north west Scotland: their configuration, infilling and development. In: *Proceedings Northern North Sea Symposium* (Eds Finstad, K.G. & Selley, R.C.) Norwegian Petroleum Society, Oslo.

STEEL, R.J. (1988) Coarsening-upward and skewed fan bodies: symptoms of strike-slip and transfer fault movement in sedimentary basins. In: *Fan Deltas, Sedimentology and Tectonic Settings* (Eds Nemec W., & Steel R.J.) Blackie, Glaygow, pp. 75–83.

STEEL, R.J. (1993) Triassic–Jurassic megasequence stratigraphy in the northern North Sea: Rift to post-rift evolution. In: *Proceedings of the 4th Conference on Petroleum Geology of North West Europe* (Ed. Parker, J.R.) Geol. Soc. London Spec. Publ.

STEEL, R. & RYSETH, A. (1990) The Triassic-Early Jurassic succession in the northern North Sea: megasequence stratigraphy and intra-Triassic tectonics. In: *Tectonic Events Responsible for Britain's Oil and Gas Resources* (Eds Hardman, R.F.P. & Brooks, J.) Geol. Soc. London Spec. Publ. **55**, 139–168.

SURLYK, F. (1978) Submarine fan sedimentation along fault scarps on tilted-fault blocks (Jurassic–Cretaceous boundary, East Greenland). *Bull. Gron. Geol. Unders.* **123**, 108 pp.

SURLYK, F. (1990) Mid-Mesozoic syn-rift turbidite systems: controls and predictions. In: *Correlation in Hydrocarbon Exploration* (Ed. Collinson, J.D.) Norwegian Petrol. Society, Graham & Trotman, London, pp. 231–241.

THERIAULT, P. (1992) Can a fractal approach discriminate tectonic-generated submarine conglomerates from coarse lowstand deposits? Upper Jurassic Helmsdale Boulder Beds of NE Scotland. *31st Ann. Met. British Sedimentological Research Group*, Abstract, Southampton.

THIRIET, J.P., BUROLLET, P.F., MONTENAT, C. & D'ESTEVON, P.O. (1986) Evolutions tectoniques et sedimentaires néogenes à la transition du Golfe de Suez et de la Mer Rouge: le secteur da Port Safaga (Egypte). *Doc. Trav. IGAL*, Paris.

TURCOTTE, D.L. & EMERMAN, S.H. (1983) Mechanisms of active and passive rifting. *Tectonophysics* **94**, 39–5C.

TURNER, C.C., COHEN, J.M., CONNELL, E.R. & COOPER, D.M. (1987) A depositional model for the South Brae oilfield. In: *Petroleum Geology of Northwest Europe* (Eds Brooks, J. & Glennie, K.) Graham & Trotman, London, pp. 853–864.

UNDERHILL, J.R. (1991) Controls on Late Jurassic seismic sequences, Inner Moray Firth, UK North Sea: a critical test of a key segment of Exxon's original global sea level chart. *Basin Res.* **3**, 79–98.

VINCENS, A. & CASANOVA, J. (1987) Modern background of Natron–Magadi Basin (Tanzania–Kenya); physiography, climate hydrology and vegetation. *Sci. Geol. Bull.* **40**, 1–2, 9–21.

WILSON, J.T. (1966) Did the Atlantic close and then re-open? *Nature* **211**, 676–681.

Spec. Publs Int. Ass. Sediment. (1993) **20**, 129–159

Sea-level changes and extensional tectonics in the Lower Jurassic (northern Helvetic realm, western Switzerland)*

B. LOUP

*Department of Geology and Paleontology, University of Geneva, Rue des Maraîchers 13,
CH-1211 GENEVA 4, Switzerland*

ABSTRACT

Located at the hinge between the stable European continental platform and the northern Alpine Tethys, the Lower Jurassic basin of the northern Helvetic realm (western Switzerland) can be subdivided into several elongate symmetrical WSW–ENE sub-basins. Lateral thickness changes are gradual, and areas of non-deposition act as a major control on facies: Helvetic sandy facies near exposed zones and Dauphiné shaley facies in more distal zones.

The sediments were deposited in shallow marine offshore and foreshore zones. However, due to Alpine tectonics, preservation of sedimentary structures and fossils is poor. The sections studied display systematic events which can be interpreted as transgressive surfaces, sequence boundaries, condensed sections and shallowing-up sequences. The Lower Jurassic sedimentary record can be subdivided into genetically related strata recording relative base-level changes. Sedimentary structures or facies assemblages related to episodic palaeofault activity are absent. At least two orders of cyclicity (2nd and 3rd) have been recognized, and the same 3rd-order sequences are found in different sub-basins and in the two main facies types. The mechanisms controlling sedimentary sequences are therefore either regional (main- basin scale) or global. However, correlation with global events is recognized only in some situations, and eustasy is therefore not the only control on cyclic sedimentation.

In order to investigate the regional processes, the subsidence history has been analysed. The tectonic subsidence curves, with corrections for depositional water depth, compaction, tectonic deformation, erosion, eustasy and Airy compensation, have been compared using three stretching models: uniform extension of the entire lithosphere, crustal and subcrustal extension (both non-uniform or depth-dependent discontinuous stretching models). For the study area, the most suitable model is subcrustal stretching, with more extension in the lower lithosphere than in the crust. As predicted using the model, the ratio of initial fault-controlled to thermal subsidence is low and the studied basin is mainly thermally controlled.

The results from sequence and geohistory analyses are integrated. A possible process controlling sequences, and compatible with a mainly thermally controlled basin, is in-plane stress change. This mechanism can explain all observed features, including the asymmetric sedimentary cycles. The stratigraphical record may therefore result from the superposition of both eustasy and in-plane stress changes.

The northern Helvetic Lower Jurassic basin was an intracratonic sag-basin and not, at that time, a constituent part of the more fault-controlled North Tethyan margin.

INTRODUCTION

In past decades, numerous studies (synopsis in Lemoine & Trümpy, 1987) in the Ligurian Tethys have emphasized extensional tectonics as a major control on basin subsidence, stratigraphy and

* Swiss National Science Foundation, projects 2.107-0.86 and 20-26218.89.

facies. For example, such extensional tectonics have been recorded in the South Tethyan margin, the Briançonnais realm *sensu lato* and the western French Alps of the Grenoble–Briançon transect (see Figs 3 & 4 for locations, and following sections for references). These palaeogeographical realms depict a classical passive margin geometry, with tilted basement blocks and sedimentary wedges. Palaeofaults can be observed or inferred from scarp breccias, olistostromes, turbiditic fans, characteristic sedimentary sequences (e.g. thinning- and fining-upward megacycles) or volcanism. Changes in sediment thickness are rapid and subsidence rates are generally high.

On the other hand, the North Tethyan Helvetic realm generally displays a more 'passive' character with reduced sedimentary thicknesses, low subsidence rates and apparently minor structural activity. This suggests an intracratonic rather than a passive-margin evolution. Thus, from a 'field' point of view, the evolution of the Helvetic realm is very different from the other domains.

Conclusive evidence of palaeofaults is absent in the study area. However, normal extensional faulting with passive-margin geometry has previously been evoked to explain basin subsidence, the stratigraphical record and the location of Alpine thrusts (Trümpy, 1949; Baer, 1959; Dolivo, 1982; Lemoine et al., 1986; Gillcrist et al., 1987; Lemoine & Trümpy, 1987; Burkhard, 1988). These assumptions were mainly based on the following interpretations.

1 Shoreline reworking and storm deposits made of millimetre- to centimetre-sized pebbles of Traissic dolomites and basement rocks as distal facies of scarp breccias.

2 Gradual thickness changes as sedimentary wedges.

3 Some Alpine faults and thrusts as former palaeofaults (for instance, the Rote Kuh–Gampel fault (Fig. 2) was described by Dolivo (1982) as a major Early Jurassic fracture).

4 Shallowing-up sandy sequences as an indirect record of progressively uplifted basement blocks.

Thus, despite clear differences in evolution, scale and overall characteristics between the Helvetic domain and other palaeogeographical realms, a passive margin with tilted basement blocks bounded by normal faults interpretation has been proposed. The approach to solving this problem of consistency is twofold (Loup, 1990, 1991): (1) facies and sequence analysis (direct means), and (2) geohistory analysis and comparison with stretching models (indirect means). The underlying aim in these investigations is to make allowance for more external (e.g. global sea-level changes) or more internal (e.g. regional or local tectonics) controls on sedimentation and basin history.

Although correlation between Jurassic depocentres and Late Palaeozoic basins and structures can

be directly noted in some cases (for instance the 'Arbignon Syncline' resting on top of the Permo-Carboniferous 'Dorénaz Syncline'), the influence of Late Variscan structures on the Mesozoic palaeogeography is difficult to estimate in the study area. For example, domains with a high concentration of Variscan granitoids (Mont Blanc and Aar massifs) may have played a particular role during Mesozoic subsidence (e.g. Trümpy, 1982). The faults bounding the Late Palaeozoic basins lie oblique to the Mesozoic facies zones and are not reactivated (e.g. Trümpy, 1958). In contrast, the Variscan influence during the Alpine inversion and nappe thrusting can be important.

After an overview of the general sedimentary trends of the Helvetic Lower Jurassic, the first aim of this study is to obtain a better understanding of the overall evolution of the sedimentary record, using sequence stratigraphy. Due to Alpine tectonics, preservation is poor in the southern Helvetics (Wildhorn nappe). However, the northern parts are more suitable for stratigraphical and sedimentological investigations. Here, all sections studied, including up to 400 m of shallow-marine siliciclastics and carbonates, display systematic events: flooding surfaces, reworking, condensed horizons and shallowing-upward sequences. The stratigraphical record can be subdivided into genetically related strata recording base-level changes. This control can be local (sub-basin scale), regional (main-basin scale) or global (eustatic). A comparison with two charts of 'supra-regional' events is presented: (1) the chart of Haq et al. (1987, 1988) which is thought to be representative of the Earth, and (2) the transgressive–regressive events of Hallam (1988) mainly representative of Europe.

Recent studies, which used the technique of geohistory analysis described by Van Hinte (1978), emphasized distinctive features of some Alpine and European marginal palaeogeographical realms: Funk (1985) for the central Helvetic Alps of the Glarus transect, Funk et al. (1989) and Wildi et al. (1989a, b) for an area between the Jura platform and the Helvetic realm (see Fig 3 for locations). These papers, with particular emphasis on the application of the technique to tectonized settings, described the total or basement subsidence (without backstripping). They were therefore unsuitable for the investigation of a mechanism responsible for basin initiation and development. A study of the tectonic subsidence and related driving mechanisms was carried out by Rudkiewicz (1988) in the western French Alps on the Grenoble–Briançon transect (Dauphiné to Piemont domains).

A second aim of this paper is to describe quantitatively the subsidence in the northern part of the Helvetic realm of western Switzerland, with some points of comparison in the southern Helvetics. After discussion of the different corrections (bathymetry compaction, tectonic deformation, erosion and eustasy) and of the backstripping technique used, the tectonic curves are compared using three theoretical models (uniform, crustal and subcrustal stretching) in order to determine the likely driving mechanism of basin initiation and evolution. Each of these stretching models is characterized by a specific initial fault-controlled to thermal subsidence ratio. Depending on the most suitable model, a more tectonic or alternatively a more thermal control on basin evolution can be interpreted. Major tectonic control would be more probable in young passive margins or in rift systems, whereas thermal control is more likely to occur in older margins or in intracratonic sags (e.g. review in Allen & Allen, 1990).

Finally, the results from facies/sequential and geohistory analyses are integrated. A mechanism

controlling sequences is discussed which is compatible with the driving subsidence mechanism. In conclusion, a basin model is suggested and some further questions are posed.

GEOLOGICAL AND PALAEOGEOGRAPHICAL SETTINGS

The study area (Fig. 1) is located in the Alps of western Switzerland and is part of the Helvetic units in a broad sense. It is composed of crystalline basement massifs and of their sedimentary cover. The basement massifs (external massifs of Aiguilles Rouges, Mont Blanc, Gastern and Aar) were thrust onto the European foreland during the Miocene. The sedimentary cover occurs in autochthonous and parautochthonous positions (Arbignon syncline, Mont Blanc cover, Raron syncline) or in thrust nappes (Morcles and Doldenhorn 'infrahelvetic' nappes, and Diablerets, Gellihorn and Wildhorn Helvetic nappes). To the northwest, these units are overlain by the allochthonous Prealps of Penninic (mainly) and Austroalpine origins. To the southeast, the study area is hidden under the

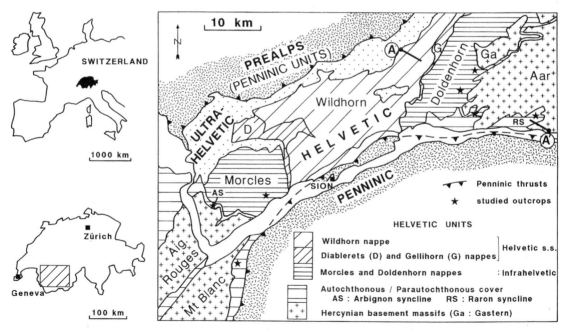

Fig. 1. Simplified tectonic map of the Helvetic units with surrounding Penninic and Prealpine units. The outcrops studied (black stars) are part of the autochthonous and parautochthonous sedimentary cover of the Hercynian basement massifs, and of the infrahelvetic Morcles and Doldenhorn nappes. See Fig. 2 for cross-section A–A'.

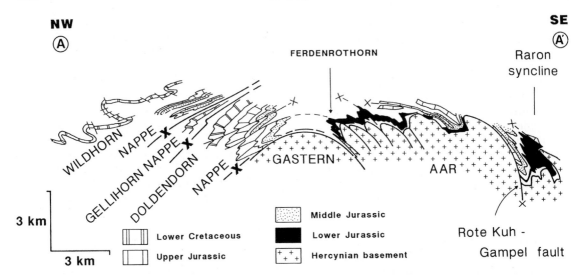

Fig. 2. Simplified cross-section through the southwestern part of the Aar massif (see Fig. 1 for location) with position of the Ferdenrothorn section; only structurally competent formations are shown. (Modified from Masson *et al.*, 1980.) See section B in Fig. 4 for the corresponding palinspastic cross-section.

Penninic thrust front (Brig–Sion–Courmayeur Zone, St-Bernard nappe). For more detail see Trümpy (1980).

A cross-section through the Helvetics (Fig. 2) along the southwestern Aar transect displays a strong decoupling between the different nappes thrust to the north (Wildhorn and Gellihorn Helvetic nappes, and Middle Jurassic to Tertiary parts of the Doldenhorn nappe) and Triassic to Lower Jurassic sediments which, in this transect, remained in autochthonous and parautochthonous positions.

According to numerous studies (synopsis in Burkhard, 1988, figs 12 & 14; Dietrich & Casey, 1989, table 1) based on mineral assemblages and the illite cristallinity index (IC) (Kübler, 1964,

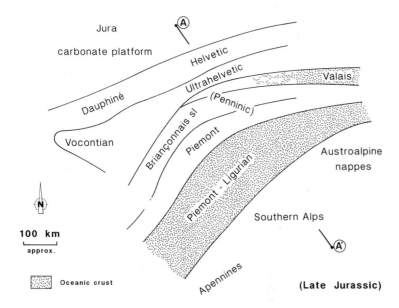

Fig. 3. Palinspastic sketch map of the western and central Alps in Late Jurassic times. (Modified from Boillot *et al.*, 1984.) The Helvetic realm is located at the hinge between the stable European foreland and the northern Alpine Tethys. See Fig. 4 for cross-section A–A'.

1968, 1990), metamorphism parameters for the study area and the Helvetic Alps of western Switzerland generally are typical of the epizonal grade (IC < 0.25), with anchizonal values (0.42 > IC > 0.25) in some northern locations. This thermal metamorphism partly explains the non-preservation of microfossils. Another consequence is the impossibility of reconstructing the thermal history of the Helvetic basins. For example, the rifting and relaxation phases identified from the subsidence history cannot be confirmed by the thermal maturities of the sediments corresponding to these periods of basin evolution.

In a palinspastic framework (Fig. 3) representing the near maximum development of the Ligurian Tethys (Late Jurassic), the Helvetic realm is located at the hinge between the European foreland (Jura carbonate platform) and the North Tethyan margin (northern Valais trough to southern pre-Piemont slope). It is about 60 km wide (from 40 to 80 km depending on the 'unfolding' hypothesis) and approximately 300 km away from the postulated South Penninic mid-ocean ridge. Southwestward,

this Helvetic realm passes into the Dauphiné domain of the Grenoble transect.

In a general palinspastic cross-section (Fig. 4a), the Helvetic realm is defined as part of the European continental platform to the north of the Alpine Tethys and seems structureless in contrast to the following domains, where half-graben geometry is a prominent feature.

1 The South Tethyan margin: e.g. Furrer (1985) and Eberli (1988) for the lower and central Austroalpine units, and Bernoulli (1964), Laubscher and Bernoulli (1977), Bernoulli *et al.* (1979b), Castellarin (1980) and Winterer and Bosellini (1981) for the southern Alpine units and the Apennines.

2 The Briançonnais high of the Swiss transect: e.g. Baud *et al.* (1979), Baud and Septfontaine (1980), Mettraux (1989), Mettraux and Mosar (1989).

3 The western French Alps of the Grenoble–Briançon transect (Dauphiné to Briançonnais domains): e.g. Barféty and Gidon (1984), Boillot *et al.* (1984), Lemoine *et al.* (1986), Gillcrist *et al.* (1987), Dumont (1988), Grand (1987, 1988) and Grand *et al.* (1987).

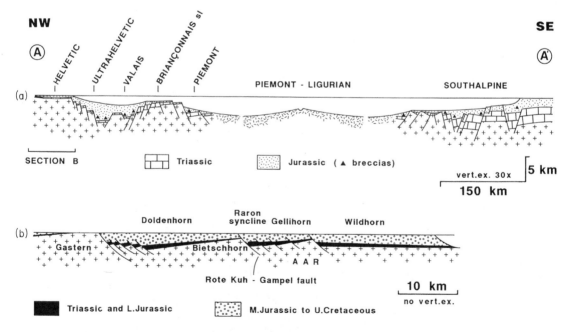

Fig. 4. (a) Palinspastic section of the Alpine Tethys ocean in Late Jurassic times (modified from Laubscher & Bernoulli, 1977). See Fig. 3 for approximate location of the transect A–A'. The narrow Helvetic realm is structureless with respect to the other domains where half-graben geometry is a prominent feature. (b) Detail of section B in the Helvetic realm *sensu lato*, along the southwestern part of the Aar massif (adapted from Burkhard, 1988). This section represents the pre-inversion state of the tectonic cross-section of Fig. 2, according to the traditional interpretation of this platform as a sequence of tilted blocks. However, this paper shows that this interpretation has to be modified.

However, a closer look at the Helvetic realm (Fig. 4b) also suggests numerous tilted basement blocks bounded by normal faults (Burkhard, 1988; with data from Baer (1959) and Dolivo (1982)). A similar scheme, with less pronounced structural activity, is proposed for the Aiguilles Rouges–Mont Blanc transect by Huggenberger (1985). As stated by the several authors listed below, this structural activity is recorded in facies, depositional sequences and subsidence history:

1 Trümpy (1949), Helvetic Alps of the Glarus transect;

2 Baer (1959) and Burkhard (1988), western Aar massif transect (see Fig. 4b);

3 Dolivo (1982), Raron syncline, with Rote Kuh–Gampel fault (Figs 2 & 4b);

4 Lemoine et al. (1986), Aiguilles Rouges–Mont Blanc transect;

5 Gillcrist et al. (1987), Morcles nappe and Chamonix syncline; and

6 Lemoine and Trümpy (1987), Helvetic realm as a whole.

For this study, the Early Jurassic has been chosen because it was a critical period of basin evolution, i.e. the rifting phase, which can be correlated with the Central Atlantic opening history (e.g. Ziegler, 1982, 1988). If normal extensional faulting occurred, Lower Jurassic sediments should contain evidence of this activity.

The first tensional events on the Western Alps transect are known from Carnian age sediments of the Briançonnais realm (Baud & Mégard-Galli, 1975). However, the major rifting phase probably commenced in the Early Jurassic, with the break-up of the large and generally thick Middle to Upper Triassic carbonate platform. In the Helvetic realm however, reduced sedimentation occurred during the Triassic.

GENERAL EVOLUTION OF LOWER JURASSIC SEDIMENTARY BASINS OF THE HELVETIC PLATFORM

The Lower Jurassic sedimentary basins of the west-

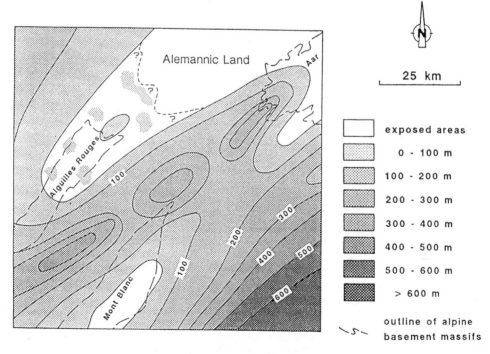

Fig. 5. Isopach map of Lower Jurassic sediments for the study area, in a palinspastic framework (thicknesses not decompacted; contour interval, 100 m). The main basin is subdivided into several sub-basins. See text for discussion of the Aiguilles Rouges area. The Alpine basement massifs are outlined.

ern part of the Helvetic shelf may be described as a simple pattern of thickness changes and of facies distribution.

Sediment thicknesses replaced in a palinspastic map compiled from Trümpy (1971), Ferrazzini and Schuler (1979), Funk (in Trümpy, 1980), Zwahlen (1986), Burkhard (1988) and Wildi *et al.* (1989a) show elongated WSW–ENE oriented sedimentary basins and zones of non-deposition (Fig. 5). The data available indicate regular sediment thickness changes from 0 to 600 m. The shape of the main basin and sub-basins is almost symmetrical. The broader exposed area corresponds to the 'Alemannic Land' (Trümpy, 1949) which is a promontory of the more easterly Bohemian massif.

The basin axes are more or less parallel to Variscan structures such as pinched Permo-Carboniferous synclines or amphibolite dykes. Nevertheless, this parallelism is mainly due to Alpine tectonics and the influence of Late Palaeozoic structures is discreet. The Late Palaeozoic faults were either not reactivated, or moved only slightly with small-scale throws, and in palinspastic

reconstructions the Mesozoic facies zones lie oblique to the former basins. However, the Variscan structures are partly responsible for the heterogeneity of the crust and they probably played a role during the Mesozoic subsidence history by modifying the heat conduction pattern and the crustal densities (for example, the high Variscan granitoid concentration in the Mont Blanc and Aar massifs). This role is, however, difficult to assess precisely (see also Trümpy, 1982).

Two main facies types (Fig. 6) are present. In the shaley Dauphiné facies, shales and limestones predominate, with only a few thin fine-grained siliciclastic intervals. The sandy Helvetic facies, on the contrary, comprises thick coarse sandstones and fine conglomerates with only subordinate shaley horizons. This facies occurs close to areas of non-deposition which partly correspond to present-day basement massifs (outlined in Fig. 6).

These massifs have therefore played a major role in the control on the geometry and stratigraphical record (e.g. Frank, 1930; Trümpy, 1949, 1971).

Concerning the Arbignon Syncline atop of the

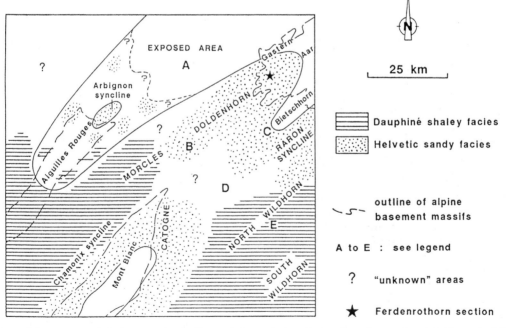

Fig. 6. Main Lower Jurassic facies types in a palinspastic framework, with location of the main sub-basins and of the Ferdenrothorn section. Letters A to E refer to the palaeogeographical realms discussed in the section on geohistory analysis. 'Unknown' areas are zones which cannot be observed, due to erosion or to their deep burial beneath other tectonic units. See Fig. 4(b) for the equivalent palinspastic cross-section along the southwestern Aar transect.

Aiguilles Rouges massif (Figs 5 & 6), two hypotheses have been discussed. The Lower Jurassic series correspond either to a graben bounded by the reactivated faults of the underlying Permo-Carboniferous basin or to a remnant of a more widely deposited cover. Non-deposition on the surrounding areas occurred in the first hypothesis, and pre-Bajocian erosion of Liassic sediments in the second hypothesis. Detailed field studies (Loup, 1991) support the second hypothesis which has been adopted for Figs 5 and 6 (see also Trümpy, 1971).

THE FERDENROTHORN SECTION

The Ferdenrothorn section is one of the 30 sections studied. It has undergone little deformation, is well dated by ammonites, and can be considered as representative of the Helvetic facies zone (Fig. 6).

Four main lithologies are present in this 140-m-thick section (Fig. 7).

1 Thin offshore shaley intervals (not exceeding 10 m) with a few interbedded siliceous limestones (mean bed thickness 8 cm) appear at the base and below the top of the section.

2 Dolomites and dolomitic marls not exceeding 20 m rest on the crystalline pre-Alpine basement after a thin arkosic bed. They represent the first marine sediments and can be interpreted as an early dolomitized tidal-flat, with abundant flat pebbles and tidal channels. They are thought to be of Carnian–Norian age, although there is no biostratigraphical evidence.

3 Bioclastic, crinoidal, siliceous and sandy bedded limestones (mean bed thickness 40 cm) were deposited in the lower shoreface and offshore zones, sometimes storm dominated.

4 Finally, three sand bodies constitute the framework of the section. Sedimentary structures such as bioturbation, parallel lamination, trough cross-bedding, hummocky cross-stratification, herringbone cross-stratification and tidal bundles suggest depositional environments between offshore-transition and lower shoreface zones, grading into upper shoreface and foreshore zones.

Biostratigraphical control based on ammonite faunas has been given by Meister and Loup (1989) (Fig. 7). The recognized biohorizons correlate with the standard northwest European zones and Tethyan affinities were not found. Unfortunately, this control is unevenly distributed: it is very good in shaley horizons and in limestones, but there is no control in the lower, middle and upper more sandy parts of the section.

Characteristic events are observed in all of the studied sections. They are: surfaces across which abrupt changes in water depth are recognized, marker beds appearing repeatedly, and groups of beds with a distinctive vertical evolution. They correspond to transgressive surfaces, reworking, condensed horizons and shallowing-upward sequences (Fig. 7).

Transgressive surfaces

At the base of the section, which corresponds to the Early or Middle Hettangian, a discontinuity in sedimentation is marked by a sharp change from shoreface amalgamated storm sandstones with abundant monospecific pelecypods to offshore shales with ammonites of the *Angulata* Zone (Fig. 7). Such abrupt environmental changes from shallower to deeper environments are interpreted as marine-flooding surfaces (e.g. Van Wagoner *et al.*, 1988).

At the base of the uppermost Sinemurian to Lower Pliensbachian limestones and of Toarcian shales (Fig. 7), this bathymetric change is accompanied by slight erosion of the shoreface during relative sea-level rise and by condensation (phosphatic phases). In such cases, this specific surface is more precisely interpreted as a ravinement surface (e.g. Demarest & Kraft, 1987; Weimer, 1988).

Reworking

Well-rounded lithoclasts, with sizes ranging from 200 μm to 1 cm are often present in the stratigraphical record. They consist of basement, dolomite, sandstone, limestone and shale pebbles; the two former elements dominate. The situation of these pebbly layers in regard to the lower and upper strata and their interpretation is variable.

As discussed above, reworking can be associated with abrupt changes from shallower to much deeper water and they are interpreted as ravinement surfaces.

In an alternative situation, the pebbles can be associated with similar abrupt changes but from deeper to shallower environments, as for example at the base of the Upper Pliensbachian sandstones (Fig. 7). In such cases, reworking denotes an abrupt basinward shift of coastal facies, corresponding to a sequence boundary *sensu* Vail *et al.* (1977; see also Vail *et al.*, 1984; Haq *et al.*, 1987; Posamentier &

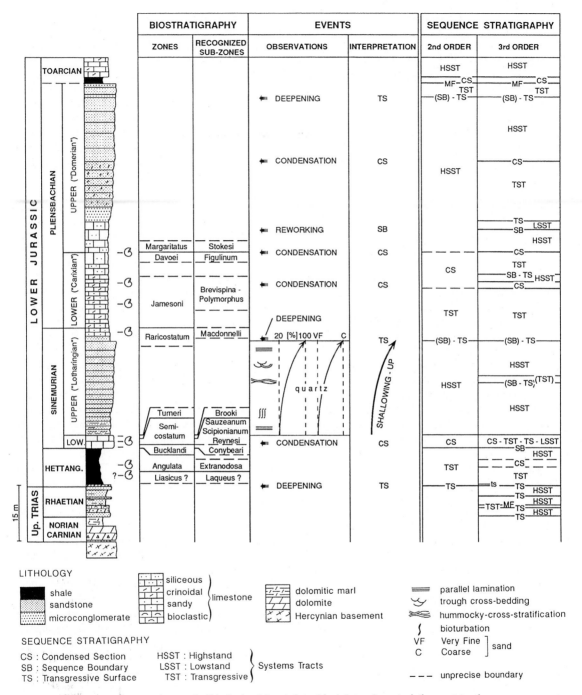

Fig. 7. The Ferdenrothorn section: main lithologies, biostratigraphical data, characteristic events and sequence stratigraphical interpretation. Biostratigraphy is based on ammonite fauna (Meister & Loup, 1989); the zonation used corresponds to the standard northwest European zones and subzones. See text for further discussion.

Vail, 1988). In contrast to the transgressive surfaces, in this case the sequence boundaries are less pronounced.

Reworking may not be related to particular surfaces across which abrupt water depth changes are noted. The pebbles are a constitutive element of a given facies and are related directly to the interpreted depositional environment. They represent continuity in the stratigraphical record and environment rather than a hiatus and environmental change. Here, several features are observed: intraformational conglomerates (in the Upper Triassic dolomites and dolomitic marls); bedload material of migrating megaripples; and storm activity allowing beach-pebbles to be spread into the shoreface and offshore zones.

Condensed horizons

Very low sedimentation rates are indicated by condensed horizons, some of which are mineralized. Condensation alone is obvious in the 3-m-thick Lower Sinemurian limestones (Fig. 7) with abundant *Gryphaea*, nautiloids, foraminifera [*Involutina liasica* (Jones)], microgastropods and ammonites. The *Bucklandi, Semicostatum* and *Turneri* Zones have been recognized. The average sedimentation rate is about 1 mm/1000 years (based on the observed thickness). Due to the long exposure to diagenetic processes, the compaction of these beds may not be significant.

Condensed, mineralized (phosphatic) horizons can be observed in the middle Lower Pliensbachian crinoidal and sandy bedded limestones (Fig. 7), with numerous stacked fossils including brachiopods, belemnites and ammonites of the *Jamesoni* Zone.

These condensed horizons are interpreted as condensed sections (Vail *et al.*, 1984; Loutit *et al.*, 1988). Thus, lithological units can display either very low (condensed sections) or high sedimentation rates (e.g. the sand bodies). This point, as well as biostratigraphical data, demonstrate that the time increments are very unevenly spaced, which suggests episodic sedimentation.

Shallowing-up sequences

The Upper Sinemurian sandstones (Fig. 7) show a systematic trend beginning with laminated muddy sandstones followed by highly bioturbated, fine-grained sandstones, hummocky cross-stratified sandstones, wide trough cross-bedded sandstones, and finally parallel-laminated sandstones with a large heavy mineral content. Detrital quartz shows an upward increase in both percentage and grain-size. This whole sand body can clearly be interpreted as a shallowing-up sequence, beginning in storm-dominated offshore-transition and lower shoreface zones, grading into longshore current-dominated upper shoreface and foreshore zones. It represents a prograding beach-face sequence. However, it is not clear whether it corresponds to a beach, barrier island or delta system setting (e.g. Reineck & Singh, 1980; Reinson, 1984; Walker, 1984; Elliott, 1986).

Interpretation

Based on the previous interpretation of the systematic events as transgressive surfaces (TS), condensed sections (CS) and sequence boundaries (SB), on the seaward- or landward-stepping character of the lithological units and on facies analysis, the section can be subdivided into genetically related strata or sequences (Fig. 7). At least two orders of cycles can be recognized, the main 'field' signal being the superposition of these two cycles. According to the few available ages, the longer cycles have a periodicity of 8–9 Ma (2nd order, e.g. Vail *et al.*, 1991) and the smaller cycles have a periodicity of 2.5–4.5 Ma (3rd order). Even smaller cycles, called here 'elementary sequences', can be distinguished. These are single beds or groups of beds bounded by marine-flooding surfaces. Their duration cannot be calculated precisely, but is probably within the Milankovitch frequency band. The discussion of these elementary sequences is beyond the scope of this paper.

The two main sequential orders are related to base-level changes due either to eustasy or to regional or even local topographical changes of the basement. A superimposition of both mechanisms is also possible.

The 3rd-order sequences identified in the Ferden-rothorn section are also recognized elsewhere, for example in other sub-basins, as such the southern cover of the Mont Blanc massif ('Catogne' area of Fig. 6), about 100 km from the section discussed, and in the other main facies (Dauphiné shales), as such the Morcles domain (Fig. 6).

CONTROLS ON
CYCLIC SEDIMENTATION

As stated above, the process required to control the sedimentary sequences has to be cyclic, '3rd-order-periodic', with a high rate of change and effective at least at a regional or basin scale and perhaps at a global scale. In this case, whether related to orbital changes or to 'global tectonics' (plate movements, plate reorganizations, mid-ocean ridge activity, major intraplate stress field reorganizations), correlation with known global events should be possible.

In Fig. 8, the proposed 3rd-order interpretation of the Ferdenrothorn section and correlative ages based on ammonite faunas are compared with two 'supra-regional' charts of sea-level fluctuations: (1) the global chart of Haq *et al.* (1987, 1988) is thought to represent the eustatic sea-level changes, which are recognizable on a global scale, whereas (2) Hallam (1988) identified transgressive-deepening and regressive-shallowing events effective mainly at a European scale (some of these events may also be recognized on other continents). For the Lower Jurassic of the Helvetic realm, a limitation for this comparison is the relatively poor biostratigraphical control, restricted to only parts of the Ferdenrothorn section. It is also very difficult to assign an absolute age to a stratigraphical limit or systems tract from biostratigraphical data. Nevertheless, this transition from biochronostratigraphy to absolute ages is useful when making comparisons with the high-resolution global chart of Haq *et al.* (1987, 1988). The resolution of Hallam's (1988) curve is predominantly defined by the ammonite zone and the discussed transition is not required here. One must also emphasize that the Jurassic time-scale is 'not fixed' and that substantial differences exist between different versions (see for instance Funk, 1985, fig. 3; Morton, 1987).

The correlation with the Haq *et al.* (1987, 1988) global chart is possible in only two cases (Fig. 8). The 193 Ma condensed section and the 203–204 Ma condensed section to highstand systems tract can be recognized in the sections studied and elsewhere. On the other hand, the 201 Ma sequence boundary appears to correspond to a transgressive systems tract or condensed section in the Haq *et al.* (1987, 1988) chart. Similarly, the 198 Ma 'downlap surface', or change from landward- to seaward-stepping units, corresponds to a type-2 sequence boundary (Fig. 8).

Some of Hallam's (1988) transgressive-deepening events correlate with transgressive surfaces observed in the Ferdenrothorn (Lower to Middle Hettangian, Lower Sinemurian and Upper Sinemurian, Fig. 8). The regressive-shallowing event in the mid-Sinemurian ('sub-*Raricostatum*' Zone),

which occurred somewhere between the *Turneri* and *Oxynotum* Zones, could correlate with the shallowing-up sequence of the Upper Sinemurian sandstones. Similarly, the 'end Early Pliensbachian' event may be recognized as a sequence boundary. The other transgressive-deepening and regressive-shallowing events are more problematic.

Concerning the lowest Jurassic sea-level, Hallam's point of view (1978, 1981, 1984, 1988) is in agreement with our observations. This level is located in the Early–Middle Hettangian rather than in the Early Sinemurian, as stated by Vail *et al.* (1977, 1984) and Haq *et al.* (1987, 1988) (Fig. 8).

Thus, correlation with 'supra-regional' sea-level variation charts is only partly possible, and cannot be applied throughout the section. Some events observed in the study area may be of global or of European significance, whereas other events are restricted to the basins studied. The global picture is therefore distorted by regional tectonics (see also Hallam, 1988). A regional control on depositional sequences, effective at the main-basin scale, is superimposed on mechanisms of wider influence. Geohistory analysis is an indirect means of investigating the regional processes acting in the basin.

GEOHISTORY ANALYSIS

The technique of subsidence analysis or 'geohistory analysis' (Van Hinte, 1978) was originally described by Sleep (1971) and extensively discussed by Watts and Ryan (1976), Steckler and Watts (1978) and Van Hinte (1978). It is generally applied to Recent or undeformed ancient sedimentary basins in order to predict the hydrocarbon potential of an area. Recent papers (Funk, 1985; Wildi *et al.*, 1989a) discussed its applicability to the tectonized Alpine cases of the Glarus transect and of the European marginal platform. Some sections from the Grenoble–Briançon transect have been studied by Rudkiewicz (1988) (see Loup, 1991, for other studies in the Alpine context).

The subsidence history has been reconstructed from outcrop data for 35 sections in the study area, using the complete stratigraphical record. Two sections from the Wildhorn realm are also discussed because of their different and more complex subsidence patterns.

The subsidence curves have been computed by the 'Backstripp'-program developed by R. Schegg and B. Loup (Department of Geology, University of Geneva), which is written in MODULA (by

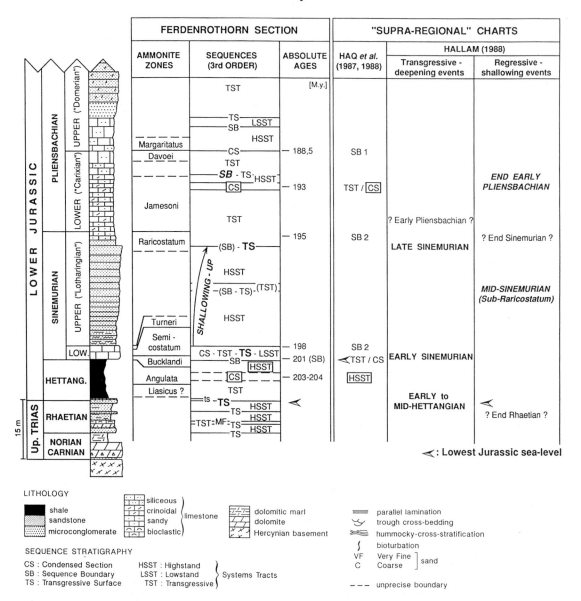

Fig. 8. Comparison of 3rd-order sequences based on the dated part of the Ferdenrothorn section with the global chart of Haq *et al.* (1987, 1988) and with transgressive-deepening and regressive-shallowing periods of Hallam (1988). The interpreted events of the Ferdenrothorn section that can be recognized in the global chart (Haq *et al.*, 1987, 1988) and in Hallam's (1988) 'chart' are indicated by open squares and bold capitals, respectively. The 'Absolute ages' are based on ammonite fauna (Meister & Loup, 1989) within the chronostratigraphical frame of Haq *et al.* (1987, 1988).

R. Schegg) and runs on Apple Macintosh®. This software allows corrections for:

1 depositional water depth — minimum and maximum values at the base and top of each lithological unit;

2 compaction — according to Sclater and Christie

(1980), or Dykstra (1987), or with 'own' porosity/ depth relationships, and each of these three methods with or without early diagenetic processes (for example, cementation in grainstones preventing complete mechanical pore volume reduction);

3 tectonic deformation;

4 erosion;
5 eustasy — long-term curve of Haq *et al.* (1987, 1988).

The calculation of tectonic subsidence is given by the equation developed by Steckler and Watts (1978), Sclater and Christie (1980) and Bond and Kominz (1984):

$$Y = \Phi \left\{ S \left(\frac{\rho_m - \rho_s}{\rho_m - \rho_w} \right) - \Delta SL \left(\frac{\rho_w}{\rho_m - \rho_w} \right) \right\}$$
$$+ (W_d - \Delta SL) \qquad (1)$$

where Y = tectonic subsidence; Φ = basement response function; S = sediment thickness corrected for compaction (and erosion and tectonic deformation); ρ_m = mean mantle density; ρ_s = mean bulk sediment density; ρ_w = water density; W_d = depositional water depth; ΔSL = sea-level change relative to present-day level.

The corrections taken into consideration in subsidence processing are briefly discussed below, with additional comments on the dating of sediments, reconstruction of the complete stratigraphical section and the backstripping method.

Depositional water depth

Bathymetric estimations are mainly made from palaeontological and palaeoecological data, the petrographical composition and the sedimentary structures of the rocks. These estimates are fairly reliable for sediments deposited above the wave base. Errors may be important for sediments below the wave base. The Lower Jurassic part of the reconstructed sections has been studied in detail and water depth values should be reliable. However, for the Middle Jurassic to Tertiary part of the stratigraphical record, data from the literature are sometimes sparse or may be of insufficient reliability to enable sound evaluations of bathymetry. Therefore, somewhat simplistic assumptions have been made in some cases, introducing possible errors into the subsidence computations. However, the sediments discussed here have been deposited in relatively shallow waters throughout the Mesozoic and Cenozoic (a maximum estimation of 300 m for the Upper Cretaceous Seewen limestones is proposed; see Wildi *et al.*, 1989a, for complete discussion).

Despite these uncertainties and correlative errors in subsidence computation, palaeobathymetric correction is very important and should not be disregarded (see also Bertram & Milton, 1988; Célérier, 1988). The shape of the resulting subsidence curve

can be substantially changed with regard to the uncorrected curve (Fig. 9a). Here, minimum and maximum water depths have been assigned to the base and top of each lithological unit.

Compaction

In the Helvetic nappes *sensu lato*, porosities are generally reduced to less than 1 or 2%. As indicated by classical porosity–depth relationships established in offshore wells (e.g. Sclater & Christie, 1980; Bond & Kominz, 1984; Baldwin & Butler, 1985), the maximum overburden is insufficient to explain these very low values by mechanical compaction only. Therefore, cementation (or 'chemical' compaction) must also be considered. In the Alpine case, tectonic thinning is also partly responsible for the porosity reduction by dynamic recrystallization. Unfortunately, we cannot determine the time of cementation, its duration or the origin of the pore water inducing cementation (research on this topic is in progress; see also Bond & Kominz (1984) for a 'delithification procedure'). The question of overpressuring in shales is another difficulty of the decompaction technique (see Addis & Jones (1985) for more detailed discussion on the principles of the decompaction technique).

Just as for bathymetric estimations, data from the literature are insufficient in the Alpine case to allow systematic correction for early cementation, overpressuring or for any other process intervening during sediment diagenesis. Therefore, a pragmatic and somewhat simplistic model of sediment compaction during burial, rather than a data-based model, was chosen. The decompaction procedure used here is from Sclater and Christie (1980), where porosity is assumed to follow an exponential relationship with depth:

$$\Phi = \Phi_o\, e^{-cz}$$

where Φ = porosity at any depth z; Φ_o = porosity at the surface; c = lithological compaction coefficient; z = depth in kilometres.

Five lithological types with parameters given in Table 1 have been considered. We are aware that compaction is a much more complex process than implied here or by Sclater and Christie (1980), with sole consideration being given to mechanical compaction. However, a more suitable correction is not yet possible for the case of Alpine tectonized nappes. Nevertheless, the subsidence curves obtained with 'complete' or 'partial' (as here) corrections are still

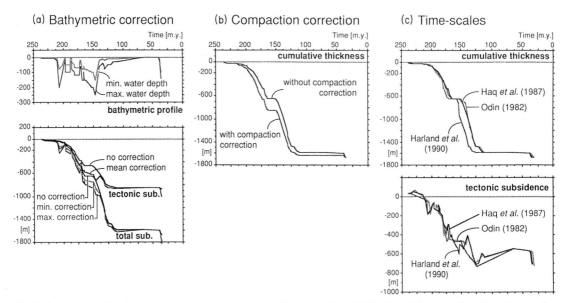

Fig. 9. Importance of some corrections introduced in subsidence analysis. (a) Bathymetric correction: the shape and slope of the corrected and uncorrected curves may be significantly changed (Ferdenrothorn section, without eustatic correction: bathymetric profile, total subsidence without decompaction and tectonic subsidence with decompaction). (b) Compaction correction, using the technique of Sclater and Christie (1980) with five lithologies (Ferdenrothorn section, cumulative thickness without bathymetric and eustatic corrections). (c) Choice of a time-scale: Odin (1982), Haq *et al.* (1987) and Harland *et al.* (1990) (Ferdenrothorn section, cumulative thickness without any correction and tectonic subsidence with decompaction, bathymetric and eustatic corrections).

very similar. The amplitude of the curves may change but their overall shape is preserved (e.g. Bond & Kominz, 1984; see also Morton, 1987).

In our opinion, compaction is the second most important parameter, with bathymetry, which must be considered in subsidence analysis. The shape of the corrected curve is significantly different from the uncorrected subsidence curve (Fig. 9b).

Tectonic deformation

Due to Alpine orogeny, tectonic deformation may strongly modify sediment thicknesses. For example,

33% tectonic thinning was recorded by Ramsay and Huber (1983) in the Lower Liassic shales of the core of the Morcles nappe. In some cases, tectonic restoration using several strain markers was performed through the complete section (e.g. Huggenberger, 1985, for the Morcles nappe). However, information on tectonic thinning is usually very sporadic and only representative of a small part of the section (e.g. Schläppi, 1980; Dolivo, 1982; Burkhard, 1988). Strain analysis methods, as described by Panozzo (1987) and Schmid *et al.* (1987), are in progress and should give better estimations.

Table 1. Compaction variables

Lithology	Initial porosity Φ_0(%)	Lithological coefficient c (km^{-1})	Reference
Limestone	45	0.54	Sawyer *et al.* (1982)
Sandstone	49	0.27	Sclater and Christie (1980)
Shales	63	0.51	Sclater and Christie (1980)
Dolomite	31	0.22	Schmoker and Halley (1982) Heidlauf *et al.* (1986)
Shaley sand	56	0.39	Sclater and Christie (1980)

Erosion

Erosion is generally badly documented. It is difficult to interpret a hiatus as a result of non-deposition or erosion. Even more difficult is the quantification of missing sediment. However, correction is possible in some cases, as for the pre- (or syn-) Eocene erosion in the Alps and in the eastern part of the plateau and Jura Mountains, where sediments of Late Cretaceous age are missing (e.g. Trümpy, 1980).

Eustasy

Although the amplitudes of eustatic sea-level changes are still subject to controversy (e.g. Hays & Pitman, 1973; Vail *et al.*, 1977; Pitman, 1978; Watts & Steckler, 1979; Hardenbol *et al.*, 1981; Hallam, 1984, 1988; Kominz, 1984; Haq *et al.*, 1987, 1988), the overall trend (lst order) is generally confirmed by these authors (see also Morton, 1987). Moreover, Burton *et al.* (1987) and Kendall and Lerche (1988) suggest that if 'relative' changes in sea-level (including a tectonic component) can be determined from several methods, their amplitude cannot be assessed without preliminary assumptions of other variables controlling stratigraphy and basin evolution.

As stated in a previous section, one of the controls on sequences is thought to be of regional influence as correlation with 'global' events is possible in only a few cases. Therefore, eustatic changes were overprinted and partly hidden by a regional process of greater effect. For this reason, and despite the difficulties mentioned above, eustatic corrections in metres have been made, using the long-term curve of Haq *et al.* (1987, 1988).

Dating of sediments

The dating of sediments is based on biostratigraphical evidence. Unfortunately, partly due to Alpine deformation and metamorphism, biostratigraphical control is very sparse: some sections are well dated whereas others lack stratigraphical fossils. When no data are available for one profile, dating has been made by lithological and facies correlation with adjacent dated sections. This can lead to substantial dating errors due to diachronous facies zones (see Wildi *et al.*, 1989a, for discussion).

In order to compare the different subsidence diagrams, it is necessary to use the same time-scale.

Important differences are noted between the available scales (e.g. Morton, 1987). Depending on the choice made, the shape of the subsidence curves obtained may vary significantly, for example from increasing to decreasing subsidence. However, the overall or long-term shape is preserved, as illustrated by Fig. 9c. The time-scale and biostratigraphical zonal schemes used are from Haq *et al.* (1987, 1988).

Reconstructing a complete stratigraphical section

In the Helvetic nappes *sensu lato*, the original pile of lithological units was often disrupted by Alpine tectonics, especially by shearing in the incompetent layers. Subsidence history curves are therefore based on composite sections using data from several different locations. The complete original sedimentary pile can be re-established using precise palinspastic reconstructions. The general map of Wildi *et al.* (1989a) has been completed for western Switzerland, using detailed studies of Schläppi (1980), Dolivo (1982), Bugnon (1986), Burkhard (1988) and Zwahlen (1986).

This palinspastic reconstruction is also necessary for studying the Mesozoic subsidence patterns in space.

Backstripping

The effects of sediment loading have been removed assuming Airy compensation, which means a lithosphere with zero lateral strength or a basement response function Φ set equal to 1 in equation (1).

This isostatic model can only be valid in early stages of basin development. For later stages, a flexural compensation with increasing lateral strength due to cooling of the lithosphere (Watts, 1978) should be used. Lateral heat flow was also ignored. However, Steckler and Watts (1978, 1982) and Bond and Kominz (1984) demonstrated that an Airy compensation during whole-basin development, using a one-dimensional-model (without lateral heat flow) rather than a two-dimensional-model (with flexure and lateral heat flow), introduces only small differences in the amplitude of the curves. It does not affect the shape of the reconstructed tectonic subsidence or modelled curves. These discrepancies are even smaller than those between 'fully' and 'partially' decompacted curves (see section on 'Compaction', and Bond & Kominz, 1984). Also, computation is much easier and does not require assumptions on sediment load geometry and on the increasing flexural rigidity of the lithosphere, which are sources of further possible errors.

Conclusion

The above paragraphs have emphasized the simplifications, problems and possible errors involved with computation of subsidence curves. However, we think that a coherent approach makes comparison between different curves qualitatively possible. Quantitative comparison becomes increasingly more hypothetical.

STRETCHING MODELS

The tectonic subsidence curves have been compared using three stretching models (Fig. 10) (see Karner et al. (1987) for a similar approach in the Wessex Basin): uniform lithospherical stretching (McKenzie, 1978); crustal stretching (references below); and subcrustal stretching (references below).

The crustal (Fig. 10b) and subcrustal (Fig. 10c) models are two variations of the non-uniform or depth-dependent discontinuous stretching model. In contrast to a homogeneous extension of the whole lithosphere as in McKenzie's (1978) model, a detachment in the lithosphere allows different stretching factors in the upper and lower lithospheric layers. The detachment can theoretically run below, in or obliquely through the crust or obliquely through the whole lithosphere. It is often placed at the base of the crust in practical applications. This classical decoupling horizon between the crust and subcrustal lithosphere has been chosen (as in Royden & Keen, 1980; Hellinger & Sclater, 1983). The crust is thinned by a factor δ whereas thinning of the subcrustal lithosphere is given by β. This model was mainly developed by Royden and Keen (1980), Sclater et al. (1980b) and Hellinger and Sclater (1983). It has found numerous applications on passive continental margins as well as in intracratonic settings. Some difficulties are inherent to the depth-dependent discontinuous stretching model. These are that the detachment required is not systematically proven, the different stretching factors imply a space problem (e.g. Karner & Dewey, 1986; Karner et al., 1987) and a process allowing for differential stretching of the lithospheric layers has to be found.

Other models such as continuous depth-dependent stretching (Rowley & Sahagian, 1986), dyke intrusion (Royden et al., 1980), melt segregation (Beaumont et al., 1982) or deep crustal metamorphism (Falvey, 1974; Falvey & Middleton, 1981) were not considered because of the lack of evidence for such processes (for a review see Karner & Dewey, 1986; Allen & Allen, 1990).

Depending on the relative amounts of stretching in the crust (factor δ) and subcrustal lithosphere (factor β, also representing the degree of heating) modifying an initial or pre-rift configuration, the following results can be considered.

1 In the uniform lithospheric stretching model of McKenzie (1978) (Fig. 10a), the crust and subcrustal lithosphere are stretched by the same factor ($\delta = \beta$). Stretching has two consequences: failure of the crust and upward migration of the lithosphere–asthenosphere boundary, which is almost an isotherm, inducing thermal anomaly or disequilibrium. The instantaneous initial fault-controlled subsidence is therefore followed by long-term thermal subsidence due to thermal recovery or cooling of the lithosphere.

2 In the crustal stretching model (Fig. 10b) (e.g. Hellinger & Sclater, 1983; Royden et al., 1983), the crust extends more than the lower lithospheric layers ($\delta > \beta$). The subcrustal lithosphere may remain undisturbed ($\delta > \beta = 1$). The thermal perturbation is small, as shown in the depth–temperature diagram of Fig. 10b, and the thermal subsidence component is reduced. This model therefore predicts high initial subsidence relative to small-amplitude thermal subsidence.

For example, this model can successfully explain the subsidence in the centre of the Pattani Trough (Hellinger & Sclater, 1983), the Vienna Basin (Royden et al., 1983), the Ridge Basin (Karner & Dewey, 1986) and the Wessex Basin (Karner et al., 1987).

3 In the subcrustal stretching model (e.g. Royden & Keen, 1980) (Fig. 10c), the lower lithospheric layers are more stretched than the crust ($\beta > \delta$). This higher stretching of the subcrustal lithosphere induces major thermal disturbance as indicated by the depth–temperature diagram of Fig. 10c, and consequently large long-term thermal subsidence. The initial or fault-controlled subsidence component is much reduced.

For example, this model can successfully explain the subsidence in the Labrador Margin (Royden & Keen, 1980), the central intra-Carpathian Basins (Sclater et al., 1980b), the Pannonian Basin (Royden *et al.*, 1983) and the northern part of the Dutch Central Graben (Kooi *et al.*, 1989).

In summary, the main point concerning the three models chosen is that they predict variable partitioning between initial fault-controlled subsidence and thermal subsidence. The asymptotic value of the total subsidence (initial and thermal components) is given by the stretching factor of the crust. Thinning of the subcrustal lithosphere determines the partitioning of the two components (Fig. 10).

As an illustration and for selected pairs of δ and β

values and other parameters used (see below and Table 2), the predicted ratios of initial to thermal subsidence are respectively (Fig. 10): uniform stretching, 0.86 (δ = β = 1.25); crustal stretching, 3.35 (δ = 1.25/β = 1.0); and subcrustal stretching, 0.48 (δ = 1.25/β = 1.4). The theoretical curves were computed according to the formulae of McKenzie (1978) (corrected by Jarvis & McKenzie, 1980; Sclater & Christie, 1980) for the uniform stretching model, and according to Hellinger and Sclater (1983) for depth-dependent discontinuous stretching.

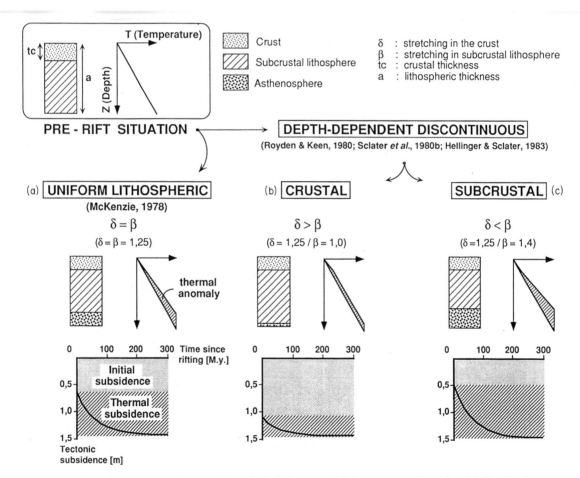

Fig. 10. Various models of stretching modifying the initial or pre-rift lithospheric configuration. (a) The simplest model is uniform lithospheric or depth-independent stretching (McKenzie, 1978), where the crust and subcrustal lithosphere are thinned by the same amount (δ = β); the rapid initial subsidence phase is followed by the long-term cooling of the lithosphere. A variation on this model is provided by the depth-dependent discontinuous stretching model (Royden & Keen, 1980; Sclater *et al.*, 1980b; Hellinger & Sclater, 1983), where thinning of the upper and lower lithospheric layers is variable. (b) In crustal stretching above a detachment (δ > β), thermal perturbation is less prominent and thermal subsidence is therefore very limited; initial subsidence represents the greatest part of the subsidence. (c) In contrast, subcrustal stretching with δ < β implies major thermal disturbance or anomaly and high thermal subsidence; initial subsidence is reduced.

Parameter	Value	Definition
ρ_m	3.33 g/cm³	Mean mantle density at 0°C
ρ_c	2.8 g/cm³	Mean crust density at 0°C
ρ_w	1.0 g/cm³	Water density
Tm	1333°C	Temperature at base of lithosphere
α	$3.28 \times 10^{-5}/°C$	Thermal expansion coefficient
τ	62.8 Ma	Thermal time constant
tc	31.2 km	Initial thickness of the continental crust (e.g. Cochran, 1981)
a	125 km	Initial thickness of the lithosphere (e.g. Cochran, 1981)

Table 2. Parameters for model computations (mostly from Parsons & Sclater, 1977)

The classical uniform stretching model (McKenzie, 1978) and its variations assume instantaneous rifting. In our case, rifting lasted at least 30 Ma, from Late Triassic to late Middle Jurassic (Callovian–Oxfordian) (Boillot *et al.*, 1984), that is, from the first possible rifting event linked to the Tethyan opening to the first formation of oceanic crust on the Western Alps transect. Rather than being a continuous process, rifting must be seen as the addition of several successive tensional phases. Therefore, models have been calculated with finite rifting allowing cooling during rifting (for extended discussion see Jarvis & McKenzie, 1980; Cochran, 1983). In the case discussed here, subsidence has been set equal to 0 at the beginning of rifting, and equal to the tectonic subsidence (initial fault-controlled and thermal components) at the time corresponding to the end of rifting. Between these two points, the subsidence is considered as linear. This is an approximation with respect to the curves established with Cochran's (1983) formulae. However, this simplification has no consequence for finding the possible subsidence driving mechanism.

To calculate theoretical curves, initial or pre-rift crustal and lithospheric thicknesses must be known. These two parameters must be chosen carefully as their influence on the theoretical curves is significant. This demand can satisfactorily be fulfilled in undeformed continental margins or cratonic basins by geophysical prospecting (gravimetry or deep seismic profiling). However, in the Alpine case, access to the pre-extension thicknesses is only possible by comparison with present values of neighbouring unstretched and undeformed areas. In the stable European foreland the crustal and lithospheric thicknesses are respectively 25 to 35 km and 70 to 130 km (Müller *et al.*, 1980; Panza *et al.*,

1984; Freeman & Müller, 1990). These values are consistent with the standard theoretical thicknesses of 31.2 and 125 km respectively (e.g. Sclater *et al.*, 1980a; Cochran, 1981; Dewey, 1982), which have been adopted here for model computations.

The other parameters are taken from Parsons and Sclater (1977) (Table 2). In order to make comparison with models easier, a best-fit curve relative to the computed tectonic subsidence curve has been calculated.

TECTONIC SUBSIDENCE AND STRETCHING MODELS

Due to possible errors in constructing the tectonic geohistory curves (e.g. sparse biostratigraphy, tectonic effects, compaction) and to several assumptions concerning the computed theoretical models (e.g. pre-rift crustal and lithospheric thicknesses), stretching factors obtained should be taken as approximate values rather than absolute stretching amounts. The appropriate δ-value, for stretching in the crust, is given by fitting the asymptotic values of the modelled and of the reconstructed subsidence curves. The amount of stretching in the lower lithosphere (β) is then given by the theoretical model showing the best correlation with the reconstructed curve.

Of the 35 reconstructed sections covering the whole area, only one or two representative curves for each palinspastic realm of Fig. 6 are discussed. All curves from the same realm show similar shapes but different amplitudes (see Fig. 11 for the total and tectonic subsidence curves and Fig. 12 for the 'smoothed' tectonic curves compared with the models). Our own observations and data from the

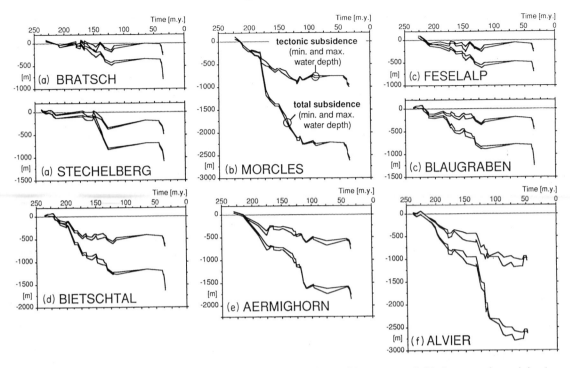

Fig. 11. Selection of representative diagrams with total and tectonic subsidence curves (with decompaction, minimal and maximal bathymetric correction, eustatic correction and Airy compensation). (a) to (e) also refer to the palinspastic realm of Fig. 6. Same horizontal and vertical scales for all diagrams.

literature (references for the presented sections listed in Table 3) have been used for reconstruction of the sections.

The *Morcles* geohistory curve (Fig. 12b) is representative for the central Morcles–Doldenhorn sub-basin (Fig. 6, realm B), located between the Aiguilles Rouges–Gastern and Mont Blanc–

Bietschhorn basement highs. Although of relatively limited extent, this section shows the maximum subsidence in the realm and suggests subcrustal stretching with δ smaller than β (1.15 and 1.3 respectively). Homewood and Lateltin (1988) have suggested a 'heterogeneous' stretching model for this palaeogeographic area (based on total decom-

Table 3. Location (Swiss coordinates) and main references to the discussed sections

Profile	Coordinates	Main references
Aermighorn	621/154.5 and 595/124	Zwahlen, 1986; Moser, 1985, 1987
Alvier	748/219	See Funk, 1985
Bietschtal	630/130	Swiderski, 1919; Schenker, 1946; Baer, 1959; Dolivo, 1982
Blaugraben	633.3/132.6	Swiderski, 1919; Schenker, 1946; Baer, 1959; Masson *et al.*, 1980; Dolivo, 1982
Bratsch	621/130	Lugeon, 1914; Baer, 1959; Bugnon, 1986
Feselalp	622/132	Lugeon, 1914; Baer, 1959; Bugnon, 1986
Morcles	582/116 to 582/124	Bonnard, 1926; Antal, 1971; Badoux, 1972; Masson *et al.*, 1980; Huggenberger, 1985
Stechelberg	636/155 to 633.3/154.6	Bruderer, 1924; Krebs, 1925; Collet & Paréjas, 1931; Masson *et al.*, 1980

pacted cumulative sedimentary thicknesses).

A similar model is valid for the area located to the south of the Mont Blanc–Bietschhorn high and corresponding to the palaeogeographic realm of the Catogne–Raron syncline (Fig. 6, realm D). The *Bietschtal* section (Fig. 12d) suggests a more pronounced decoupling between the crust and subcrustal lithosphere than in the Morcles–Doldenhorn area, with δ and β values of 1.1 and 1.3 respectively.

Areas with low tectonic subsidence display relatively flat curves which are difficult to compare with theoretical models. This occurs in the area surrounding the Bietschhorn high (Fig. 6, realm C), discussed in the *Feselalp* and *Blaugraben* sections (Fig. 12c). The exponential decrease displayed by these curves suggests thermal cooling which was already active in early basin development.

Areas of non-deposition (Fig. 6, realm A) suffered no subsidence until around 150 Ma (Late Jurassic). This is a typical feature of large areas which remained 'positive' during the Early and Middle Jurassic and were first flooded in the early Late Jurassic. This fact is evident from the *Stechelberg* (northern Aar massif) and *Bratsch* (Bietschhorn high) curves (Fig. 12a). As stated above, it is difficult to decide whether episodic deposition and subsequent erosion affected some parts of these areas, or if non-deposition occurred throughout the Early and Middle Jurassic (*pro parte*).

In the southern part of the Helvetic shelf, the northern and central Wildhorn areas (Fig. 6, realm E) are of interest because of the completely different and more complex subsidence pattern exhibiting two main rapid phases: the first in the Early Jurassic, and the second during the early Late Jurassic.

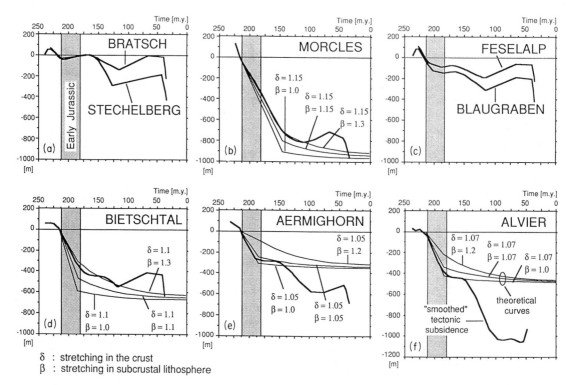

Fig. 12. Tectonic subsidence ('smoothed' curve) and stretching models (see Alvier profile). The tectonic subsidence curves of the Morcles, Bietschtal, Aermighorn and Alvier sections are compared using the three theoretical curves obtained from the stretching models. The Morcles (b) and Bietschtal (d) profiles suggest subcrustal extension (δ < β), whereas the Aermighorn (e) and Alvier sections (f) display uniform (δ = β) and crustal stretching (δ > β). Curves with low subsidence (Feselalp and Blaugraben, c) display exponential decrease in earliest basin evolution. Areas of non-deposition (Bratsch and Stechelberg, a) show a rapid subsidence phase only in Late Jurassic times. (a) to (e) refer to the palinspastic realm of Fig. 6. See text for complete discussion.

For the Early Jurassic phase, the *Aermighorn* curve (Fig. 12e) suggests a process which would lie between uniform stretching with $\delta = \beta = 1.05$ and crustal stretching with $\delta = 1.05$ and $\beta = 1.0$. The strong Late Jurassic to Early Cretaceous phase coincides with a similar subsidence in the former zones of non-deposition.

The last curve discussed was established in the Alvier area (see also Funk, 1985; Wildi *et al.*, 1989a), about 100 km eastward of the western Aar massif transect. This location is laterally equivalent to the southern part of the Wildhorn nappe (Fig. 6). However, the *Alvier* section has suffered less Alpine deformation than the Wildhorn nappe. This geohistory diagram (Fig. 12f) strongly suggests crustal stretching with extension almost confined to the crust ($\delta = 1.07$, $\beta = 1.0$). From the North to South Wildhorn palaeogeographical realms, a slight increase in the difference between the stretching amounts in the crust and in the subcrustal lithosphere can be observed: 0.0 to 0.05 ($\delta = 1.05$, $\beta = 1.05$ to $\delta = 1.05$, $\beta = 1.0$) for Aermighorn (North Wildhorn) and 0.07 ($\delta = 1.07$, $\beta = 1.0$) for Alvier (South Wildhorn). This suggests an increased decoupling between the crust and subcrustal lithosphere, from north to south.

Although beyond the scope of this paper, the characteristic Tertiary high subsidence phase present in most curves must be emphasized. This event is not part of the basin expansion phase but is part of the basin inversion and foreland formation phase (e.g. Homewood *et al.*, 1986).

Geohistory analysis has shown, for the northern part of the Helvetic shelf (realms A to D), that thermal subsidence (effective at a regional scale) is much higher than initial subsidence (of a more local or 'sub-basin-scale' extent). Rudkiewicz (1988) came to similar conclusions for the Grenoble–Briançon transect, with $\delta < \beta$ and a major thermal anomaly under the Briançonnais domain. In the case discussed here and as indicated by the modelled curves, typical initial to thermal subsidence ratios range between 0 and 0.3. The low amplitude curves (Fig. 12c) suggest that thermal cooling was active during early basin evolution. As a consequence, only minor initial or fault-controlled subsidence occurred. Subsidence and basin formation could occur without major implication of the upper brittle layers of the crust. This supports the results from sequence stratigraphy analysis, suggesting regional or 'main-basin scale' processes (see above sections). In our opinion, there is no necessity to invoke normal faulting and to draw such faults on palinspastic cross-sections as proposed by previous workers (e.g. Baer, 1959; Dolivo, 1982; Burkhard, 1988; see Fig. 4b). The fact that faults cannot be directly observed in the field or indirectly deduced from sedimentary facies and geometry does not indicate that they once existed but are now preserved as thrust faults or hidden by thrusting (e.g. fig. 18 of Gillcrist *et al.* (1987) for the Morcles nappe). Their non-existence or very discreet character as concluded from sequential and geohistory analyses provides a much simpler explanation for their present-day absence.

The subcrustal stretching suggested, valid for the northern Helvetic domain, raises the question of strain compatibility, which must be considered for the area as a whole. This can be done either by contemporaneous thrusting or foreland-basin formation in adjacent terranes (Karner & Dewey, 1986), or by crustal stretching in other realms along the 'margin', extension being relayed by crustal or lithospheric detachments (Wernicke, 1985; Gibbs, 1987). From geohistory analysis, subcrustal stretching in the northern areas could be balanced by crustal stretching in the more southern domains.

INTEGRATING THE RESULTS OF SEQUENTIAL AND GEOHISTORY ANALYSES

Before further discussion, the different steps and results from the sequence stratigraphy and geohistory analyses are summarized.

1 Sedimentation is cyclic and periodic. These cycles are related to local (sub-basin), regional (main basin) or global base-level changes.

2 The 3rd-order cycles demonstrated in only a limited area (Ferdenrothorn) can also be recognized elsewhere. Sequences cut through facies zones and 'tectonic' subdomains. The process controlling cyclic sedimentation is of at least regional influence (main basin).

3 A correlation with global charts demonstrates only partial correspondence with eustatic events. Cyclic sedimentation is the result of the superimposition of eustasy (changes of absolute sea-level) and regional tectonic processes (restricted changes of basement topography). Some areas remained positive during the Early Jurassic and therefore sediment supply was probably sufficient to fill the space created by both subsidence and sea-level changes ('accommodation'; Jervey, 1988).

4 Geohistory analysis and comparison of the tectonic subsidence curves using three theoretical models suggest subcrustal stretching. This extension mechanism predicts limited fault-controlled subsidence and high thermal relaxation phases. Thermal processes have a more regional extent and a gradual character compared with more local and episodic influences for the initial faulting. The basin studied is affected mainly by thermally controlled processes.

The sequences discussed are controlled by a process which is capable of forming sedimentary cycles in a similar way to 'glacio-eustasy'. This process must also be compatible with a predominantly thermally driven subsidence. A possibility is in-plane stress changes affecting an already deflected plate, as first described by Cloetingh *et al.* (1985). Karner (1986) also investigated the stratigraphical responses to in-plane stress in intracratonic settings. (See also Cloetingh, 1986, 1988a, b; Cloetingh *et al.*, 1987; Lambeck *et al.*, 1987; Kooi & Cloetingh, 1989.)

For example, intraplate stresses, with values of several kilobars, have been recognized in North America by Zoback and Zoback (1980) and in the Indo-Australian plate by Cloetingh and Wortel (1986), with compressive stress values of 3–5 kbar. Stress variations are related to changing conditions at the plate boundaries and to the various forces acting on the plate.

Several workers (Karner, 1986; Cloetingh, 1988a; Allen & Allen, 1990) have underlined the need to apply huge stresses (several kilobars to a few tens of kilobars) to an undeformed lithosphere to produce deformation (buckling limit). However, these stresses are reduced to much lower values (a few hundreds of bars) if there is pre-existing deformation. Deformation may result from the load of accumulating sediments deflecting the lithosphere. Geohistory analysis indicates that rapid thermal relaxation is active during early basin evolution. The sedimentary load is compensated regionally rather than locally and a deflection of the lithosphere probably existed. Watts (1978) stated that the elastic thickness and related flexural rigidity of the oceanic lithosphere increases with age as it cools away from the oceanic ridge. This can be applied to the continental lithosphere whose rheological and flexural properties are globally similar to those of the oceanic lithosphere, that is, increasing elastic thickness with age after a major heating process (e.g. Watts *et al.*, 1982). Consequently, the deflected area

becomes broader through time as sediments accumulate and sediments progressively onlap the basement ('continuous onlap' of Fig. 13a) (Watts, 1982; Watts *et al.*, 1982; Watts & Thorne, 1984). Provided sediment supply remains unchanged, this process may be recognized in the sections studied by increasing depositional water depth of more distal sand-poor facies with inferred landward shoreline migration.

In-plane stress variations may cause topographical changes of the basement, causing relative sea-level fluctuations of a few tens of metres for a few hundred bars (about 100 m for a few kilobars). These fluctuations may occur within a 1 My period and the resulting rate of change is about 1 cm/ 1000 years (Cloetingh *et al.*, 1985; Cloetingh, 1988b). This rate is in agreement with the value proposed by Pitman and Golovchenko (1983) for glacially induced cycles, with rates greater than 1 cm/1000 years and magnitudes greater than 100 m. In-plane stress change is therefore a suitable process controlling 3rd-order cycles *sensu* Vail (e.g. Cloetingh *et al.*, 1985; Karner, 1986; Cloetingh 1988a, b). For several authors (e.g. Cloetingh, 1986, 1988a; Cloetingh *et al.*, 1987; Lambeck *et al.*, 1987) the sea-level curve is, in fact, a palaeostress curve or a mirror of the tectonic evolution of a region including both rifting and compressional phases. The 1st order trend represents major tectonic events related to major plate reorganizations (see also Bally, 1980, 1982; Ziegler, 1982). The smaller trends (2nd and 3rd orders) would correspond to the regional plate response to global changes, depending also on the sedimentary load and on the inhomogeneities of the crust.

An increase in horizontal tensile stress (Fig. 13c) would result in a rapid relative sea-level rise and enhanced coastal onlap. This is preserved in the field as a rapid change from shallower to deeper environments (transgressive surface). The new space created is filled with landward-stepping sedimentary units (transgressive systems tract), until reduced accommodation forces sediments to prograde in shallowing-up sequences (prograding highstand systems tract). As stated in previous sections (see also Fig. 7), transgressive surfaces are the prominent features, whereas sequence boundaries are more subtle and often coincide with the following transgressive surface: lowstands are absent or very thin, which is also not surprising in a shallow-ramp setting. The resulting coastal onlap curve is therefore in opposition to the classical geometry

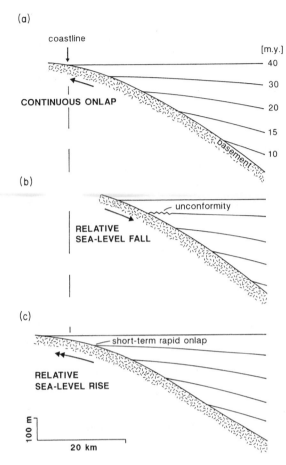

(a)

coastline

↓   [m.y.]

─────────────────────── 40

─────────────────── 30

CONTINUOUS ONLAP

──────────────── 20

─────────── 15

basement ──── 10

(b)

─── unconformity

**RELATIVE
SEA-LEVEL FALL**

(c)

─── short-term rapid onlap

**RELATIVE
SEA-LEVEL RISE**

100 m

20 km

Fig. 13. Effects of in-plane stresses on basin-margin stratigraphy. (Compiled and modified from Karner, 1986; Cloetingh, 1988b.) (a) No intraplate stress: characteristic onlap pattern in the absence of intraplate stress. The progressive and continuous onlap is due to the thermal and mechanical recovery of the lithosphere following rifting (e.g. Watts, 1982). (b) Compressive stress: effects of a 500-bar compressional stress at 30 My. Compression induces uplift of the basin edge, coastline migration towards the basin centre, and truncation of the older sediments resulting in an unconformity. (c) Tensile stress: application of a 500-bar tension at 30 My. Tension produces basin edge subsidence, coastline migration towards the continent and short-term rapid onlap, resulting in a transgressive surface.

('slow' transgression and abrupt shallowing) (see fig. 5 of Karner (1986) for a complete discussion of the onlap patterns predicted by the application of tensile or compressive stresses). However, this trend agrees with several phases of increasing extensional stress from the general tensile regime of the Early

Jurassic extension phase discussed. Extension was probably discontinuous. These phases alternate with periods of relaxed tensional forces.

The prominent sequence boundary at the base of the Upper Pliensbachian sandstones and fine conglomerates (Fig. 7) may be related to the application of horizontal compressive stresses, or sudden relaxation of major tensional stresses inducing a rapid downward shift and correlative unconformity (Fig. 13b). Angular unconformity in the Arbignon Syncline provides evidence for this event (Figs 1 & 5) (Lugeon, 1930; Trümpy, 1945; Loup, 1991). It is dated as post-Early Domerian (Upper Pliensbachian) to pre-Bajocian (Trümpy, 1945). This event is almost time-equivalent to the Mid-Cimmerian tectonic phase. It is well documented in western and central Europe is responsible for uplifts and truncations in some places or renewed subsidence in other areas (Ziegler, 1982). Correlations between tectonic phases and relative regional sea-level changes in different regions have been documented: for the North Sea region by Ziegler (1982, 1988; North Sea tectonics being in turn related to the Atlantic and Tethys Oceans), Cloetingh (1988a) and Kooi *et al.* (1989), and by Karner *et al.* (1987) for the Wessex basin.

The asymmetry of 'field water-depth curves' has been presented by Einsele (1985) and Einsele and Bayer (1991) as a typical feature of shallow epeiric seas or intracontinental basins with low subsidence rates and minor synsedimentary tectonic movements. The type and intensity of asymmetry vary along a beach-to-basin profile and are also a function of the relative rates of subsidence, sea-level and sediment influx. Sedimentary cycles predicted by the models, as well as those observed in the epicontinental Jurassic sea of southern Germany, are interpreted as the product of sea-level oscillations (Einsele, 1985; Einsele & Bayer, 1991). Although the subsidence rate can vary along the profile, it is fixed at a particular location. It is demonstrated here that asymmetric sedimentary cycles can also result from subsidence rates varying with time at a given location within the basin, as a result of in-plane stress changes.

Some problems concerning the effectiveness of in-plane stress changes in intracontinental or marginal settings have been discussed by Karner (1986). The main objection concerns the nonparallelism of the sediment-basement and Moho interfaces in stretched domains. Topographical effects related to in-plane stress variations of these two interfaces would interfere, with very small topographical changes resulting. This is particularly relevant to young basins (Karner, 1986). However, one can argue that in-plane stresses should be mainly effective on young and rapidly loaded margins because of a major deflection due to low flexural rigidity (e.g. Cloetingh et al., 1985; Cloetingh, 1988b; Allen & Allen, 1990).

Although much work on in-plane stresses has been published, little is known about their periodicities and about their relationship with 3rd-order sea-level cycles.

TYPE OF BASIN

The basin model should be characterized by major thermal subsidence, initial subsidence being reduced, by small sediment thicknesses and by low tectonic subsidence rates (maximum value of 1 cm/1000 years for the Lower Jurassic of the study area). Normal faulting is not recorded in the facies and depositional sequences. These features are typical of intracratonic but not of passive-margin settings. A likely possibility is a 'saucer-shaped' basin or 'sag-basin'. Although of totally different scales, type-examples such as the North American Palaeozoic Michigan Basin (Wilson & Burke, 1972; Haxby et al., 1976; Allen & Allen, 1990, and references therein) and the African Neogene Chad Basin (Burke, 1976) are characterized by little structural activity. The limited subsidence and correlative reduced sediment thicknesses are probably due to thermal recovery of the lithosphere after heating. As major thermal disturbance occurs (asthenospheric diapirs in the case of the Michigan Basin; Haxby et al., 1976), uplift of a thermal dome or annulus is important. The uplifted terrains supply sediment into the basin. Water and sediment loading seem to contribute significantly in accelerating subsidence. However, sediment loading cannot be the only driving mechanism.

A scheme redrawn from Gibbs (1987) (Fig. 14A) and based on regional analysis of North Sea data, displays major structural activity concentrated in one narrow area (half-graben), whereas sag- and hanging-wall basins are controlled by thermo-flexural subsidence. Similar interpretations with normal simple shear of the entire lithosphere, and correlative variable initial to thermal subsidence ratios have been proposed by Wernicke (1985) for the Basin and Range Province. These models may

be used to explain the results from our subsidence analysis with different subsidence driving mechanisms along the transect. Subcrustal stretching (small fault-controlled and high thermal subsidence components) in the northern Helvetic shelf may be balanced by uniform to crustal stretching (higher initial subsidence and reduced thermal subsidence) in the Wildhorn area.

A similar contrast between domains with high-amplitude fault-controlled subsidence and others with low and more flexural subsidence has been found in experiments (e.g. Allemand et al., 1989; Fig. 14B).

These two 'models' assume depth-dependent rheology, with detachment either through the crust (Gibbs, 1987) or below the crust (Allemand et al., 1989). The resulting geometry is asymmetric. Although a basal detachment at the crustal–subcrustal boundary has been assumed in model computations for this study, the geometry of the decoupling surface cannot be estimated. Possible models are: cutting through the crust and reaching the Moho (as in Gibbs, 1987); cutting through the entire lithosphere (as in Wernicke, 1985); and paralleling the Moho discontinuity, extension being asymmetric (as in Allemand et al., 1989).

DISCUSSION AND CONCLUSIONS

Normal extensional faulting active since the Early Jurassic has been proposed as a major control on sedimentation on the Helvetic shelf (e.g. Baer, 1959; Dolivo, 1982; Burkhard, 1988). The Helvetic realm in a broad sense was considered as displaying an 'Atlantic' margin geometry from the Early Jurassic to 'Middle'/Late Cretaceous, prior to Tertiary Alpine tectonics. This scheme was mainly based on the interpretation of reworked millimetre- to centimetre-sized pebbles of basement and sedimentary rocks as distal facies of scarp breccias, and on sediment thickness changes as sedimentary wedges. Furthermore, some Alpine reverse faults and thrusts have been interpreted as inverted palaeo-faults (Dolivo, 1982, Rote Kuh–Gampel fault; Gillcrist et al., 1987; Lemoine & Trümpy, 1987).

However, recent field work has confirmed the absence of proximal chaotic scarp breccias corresponding to the more distal fine conglomerates. These are more easily interpreted as storm-deposits or as reworked sediment at sequence boundaries. Thickness changes are progressive and due to differential subsidence. The proposed sedimentary wedges are in fact tectonic wedges due to post-depositional tectonic processes (thickening or thinning). Concerning the Rote Kuh–Gampel fault, often cited as the palaeofault type-example, new investigations have shown that Triassic sediments of the 'footwall block' and basement of the 'hanging-wall block' are not separated by a palaeofault plane but are in original stratigraphical positions. The now tilted depositional surface was reactivated during the Tertiary by Alpine tectonics.

There is no evidence for episodic faulting or instability (slumping, turbidites). On the contrary, the stratigraphical record can be subdivided into genetically related strata or sequences, related to cyclic and periodic base-level changes. Only the two main superimposed orders of cycles have been

presented. Systems tracts and related surfaces extend across the sub-basins and facies zones. Therefore, the process controlling cyclic sedimentation is of at least regional influence. Correlation with global events is only partly possible. Eustasy (glacio- or tectono-eustasy) cannot have been the only control on sedimentation during the Early Jurassic. The stratigraphical record is thought to be the result of the superposition of eustasy and regional tectonic processes. This has been highlighted, but not precisely documented, by Bernoulli *et al.* (1979a).

The most likely endogenous process is in-plane stress variations, inducing topographical changes in the basement and relative sea-level fluctuations. Their physical characteristics are suitable for the control of cyclic sedimentation, and their influence in the studied basins has been demonstrated by

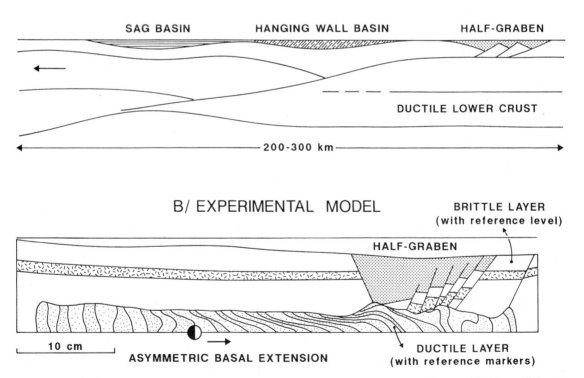

Fig. 14. Comparison of (A) linked extensional basins (data interpretation from the North Sea, modified from Gibbs, 1987) with (B) experimental models (modified from Allemand *et al.*, 1989). In the two cases, major structural activity with formation of a half-graben is confined to one narrow area, whereas wide domains undergo flexure with low subsidence, as sag- and hanging-wall basins. Detachment through (A) or below the crust (B) and depth-dependent rheology result in an asymmetric geometry.

geohistory analysis. Based on this technique, the mechanism of basin formation has also been investigated. For the northern part of the Helvetic shelf, subcrustal stretching with less extension in the crust than in the subcrustal lithosphere ($\delta < \beta$) is the best-fit model for the calculated tectonic subsidence curves. This model suggests that subsidence is mainly thermally driven, with minor fault-controlled subsidence. Uniform and crustal stretching is significant in the southern Helvetic area.

Comparisons with North Sea basins and with an experimental model suggest a 'sag-basin' geometry for the northern Helvetic realm in the Early Jurassic. During this period, this area was not directly connected to the North Tethyan margin where fault-controlled subsidence was more effective.

OPEN QUESTIONS

As stated in previous sections, some errors may slightly alter the results presented. More work is needed to obtain further valuable quantitative results.

Depending on the stretching factors in the crust and in the subcrustal lithosphere, and on the initial crustal thickness, depth-dependent extension may predict initial uplift (e.g. fig. 3 of Royden & Keen, 1980). This may be recorded in the stratigraphy. In this study, the extension of the subcrustal lithosphere is insufficient to imply substantial uplift. However, large areas remain positive until the Middle Jurassic. This may be the result of lateral heat flow coming from the rifted area inducing a peripheral thermal bulge, or the result of a high concentration of Late Variscan granitoids in these positive areas (regional modification of the density and/or of the thermal structure of the basement).

Although numerous papers on in-plane stress tectonics have been published, the applicability of this process, which depends on the geodynamic setting of the basin, for explaining 2nd- or 3rd-order cycles is still subject to controversy. The 2nd-order cycles may be defined as the result of changes in predominant tensional and compressional regimes or as plate boundary readjustments (e.g. Cloetingh, 1988b; Vail & Eisner, 1989). The 3rd-order cycles may represent the irregular local or regional response of the crust to the more regional stress state (Cloetingh *et al.*, 1985; Karner, 1986; Cloetingh, 1988a, b; Vail *et al.*, 1991). Another problem is finding a cause for the periodic changes.

A connection between the northern Helvetics and other domains is 'needed' to balance the extension amounts in the crust and in the subcrustal lithosphere and to find the 'half-graben' area required by the asymmetric 'sag basin' hypothesis. Although the evolution portrayed by the northern Helvetic shelf is close to the development of intracratonic basins such as the northerly located Paris Basin or Jura platform, this link is more easily found with more southern areas, that is in the North Tethyan margin. The southern Helvetic domain displays crustal stretching and could therefore account for the balancing of extension. In this first solution, the detachment required is of a Helvetic scale. A similar geometry with a northward dipping detachment has recently been proposed by Stampfli and Marthaler (1990): the North Tethyan margin is regarded as the 'upper plate' and the Jura platform and the complete Helvetic realm are considered as a 'rim-basin', whereas the pre-Piemontais Breccia nappe corresponds to the fault-controlled basin ('break-away'). This poses difficulties as the required detachment would cut through east–west trending sinistral transform systems. For example, wrench systems postulated by Trümpy (1988) in the Valais–Tauern and north-eastern Piemont belt were active during the Middle Jurassic to mid-Cretaceous, but were already presaged in Early Jurassic rifting. Another east–west oriented major strike-slip zone is proposed by Schmid *et al.* (1990) in the Schams nappes of central Switzerland (Briançonnais palaeogeographical realm) and in the Breccia nappe of western Switzerland (Pre-Piemont domain) with associated transpressional and transtensional basins.

In our opinion, the first solution is the most viable, that is, subcrustal stretching in the north may be balanced by crustal stretching in the southernmost parts of the Helvetic realm and the postulated intralithospheric detachment would be of a limited extent. This asymmetric extensional system may possibly be included in a larger North Tethyan heterogeneous stretching.

ACKNOWLEDGEMENTS

This research work has been supported by the Swiss National Science Foundation, grants 2.107-0.86 and 20-26218.89, which is greatly acknowledged. I thank Walter Wildi (Geneva) for much stimulating discussion during this research. Roland Schegg (Geneva) was invaluable solving computer problems. André Strasser (Fribourg) allowed me to benefit from his experience in sequence stratigraphy. He and Philip Allen (Oxford) provided valuable comments on a previous version of this text. I am indebted to the two anonymous reviewers whose comments and criticisms allowed much improvement of the typescript. Bernhardt Ujetz (Geneva) very kindly reviewed the English for the final version.

REFERENCES

ADDIS, M.A. & JONES, M.E. (1985) Volume changes during diagenesis. *Mar. Petrol. Geol.* **2**, 241–246.

ALLEMAND, P., BRUN, J.P., DAVY, P. & VAN DER DRIESSCHE, J. (1989) Symétrie et asymétrie des rifts et mécanismes d'amincissement de la lithosphère. *Bull. Soc. Géol. France* **8/V/3**, 445–451.

ALLEN, P.A. & ALLEN, J.R.L. (1990) *Basin Analysis: Principles and Applications*. Blackwell Scientific Publications, Oxford, 451 pp.

ANTAL, J.W. (1971) The structure of the core of the Nappe de Morcles. PhD thesis, University of Lausanne, Switzerland, 34 pp.

BADOUX, H. (1972) Tectonique de la Nappe de Morcles entre Rhône et Lizerne. *Matér. Carte Géol. Suisse (n.s.)* **143**, 78 pp.

BAER, A. (1959) L'extrémité occidentale du massif de l'Aar. *Bull. Soc. Neuchât. Sci. Nat.* **82**, 160 pp.

BALDWIN, B. & BUTLER, C.O. (1985) Compaction curves. *Bull. Am. Assoc. Petrol. Geol.* **69/4**, 622–626.

BALLY, A.W. (1980) Basins and subsidence — a summary. In: *Dynamics of Plate Interiors* (Eds Bally, A.W., Bender, P.L., McGetchin, T.R. Walcott, R.I. Am. Geophys. Un. & Geol. Soc. Am. Washington, DC, Geodyn. Ser. **1**, 5–20.

BALLY, A.W. (1982) Musings over sedimentary basin evolution. *Philos Trans. R. Soc. London* **A 305**, 325–338.

BARFÉTY, J.C. & GIDON, M. (1984) Un exemple de sédimentation sur un abrupt de faille fossile: le Lias du versant est du massif du Taillefer (Zone Dauphinoise, Alpes Occidentales). *Rev. Géol. Dyn. Géogr. Phys.* **25/4**, 266–276.

BAUD, A. & MÉGARD-GALLI, J. (1975) Evolution d'un bassin carbonaté du domaine alpin durant la phase préocéanique: cycles et séquences dans le Trias de la zone briançonnaise des Alpes centrales et des Préalpes. *Proc. 9th Int. Sediment. Congr. (IAS)*, Nice 1975, **5**, 45–52.

BAUD, A. & SEPTFONTAINE, M. (1980) Présentation d'un profil palinspastique de la nappe des Préalpes médianes en Suisse occidentale. *Eclogae Geol. Helv.*, **73/2**, 651–660.

BAUD, A., MASSON, H. & SEPTFONTAINE, M. (1979) Karsts et paléotectonique jurassique du domaine Briançonnais des Préalpes. In: *Symp. sur la Sédimentation Jurassique Ouest-européenne*, Paris 1977. Spec. Publ. Assoc. Sédiment. Français **1**, 441–452.

BEAUMONT, C., KEEN, C.E. & BOUTILIER, R. (1982) On the evolution of rifted continental margins: comparison of models and observations for the Nova Scotian margin. *Geophys. J. R. Astron. Soc.* **70**, 667–715.

BERNOULLI, D. (1964) Zur Geologie des Monte Generoso (Lombardische Alpen). *Matér. Carte Géol. Suisse (n.s.)* **118**, 135 pp.

BERNOULLI, D., CARON, C., HOMEWOOD, P., KÄLIN, O. & VAN STUIJVENBERG, J. (1979a) Evolution of continental margins in the Alps. *Schweiz. Mineral. Petrogr. Mitt.* **59**, 165–170.

BERNOULLI, D., KÄLIN, O. & PATACCA, E. (1979b) A sunken continental margin of the Mesozoic Tethys: the northern and central Apennines. In: *Symp. sur la Sédimentation Jurassique Ouest-européenne*, Paris 1977. Spec. Publ. Ass. Sédiment. Français **1**, 197–210.

BERTRAM, G.T. & MILTON, N.J. (1988) Reconstructing basin evolution from sedimentary thickness: the importance of paleobathymetric control, with reference to the North Sea. *Basin Res.* **1/4**, 247–257.

BOILLOT, G., MONTADERT, L., LEMOINE, M. & BIJU-DUVAL, B. (1984) *Les marges continentales actuelles et fossiles autour de la France*. Masson, Paris, New York, 342 pp.

BOND, G.C. & KOMINZ, M.A. (1984) Construction of tectonic subsidence curves for the Early Paleozoic miogeocline, southern Canadian Rocky Mountains: implications for subsidence mechanisms, age of breakup and crustal thinning. *Bull. Geol. Soc. Am.* **95/2**, 155–173.

BONNARD, E.G. (1926) Monographic géologique du Haut de Cry. *Matér. Carte Géol. Suisse (n.s.)* **57/4**, 58 pp.

BRUDERER, W. (1924) Les sédiments du bord septentrional du massif de l'Aar du Trias à l'Argovien. *Bull. Lab. Géol. Géogr. Phys. Minér. Paléont.* **37**, 86 pp.

BUGNON, P.C. (1986) Géologie de l'Helvétique à l'extrémité sud-ouest du Massif de l'Aar. PhD thesis, University of Lausanne, Switzerland, 106 pp.

BURKE, K. (1976) The Chad Basin: an active intracontinental basin. *Tectonophysics* **36**, 197–206.

BURKHARD, M. (1988) L'Helvétique de la bordure occidentale du massif de l'Aar (évolution tectonique et métamorphique). *Eclogae Geol. Helv.* **81/1**, 63–114.

BURTON, R.C., KENDALL, C.G. ST. C., & LERCHE, I. (1987) Out of our depth: on the impossibility of fathoming eustasy from the stratigraphic record. *Earth Sci. Rev.* **24**, 237–277.

CASTELLARIN, A. (1980) Italie, introduction à la géologie régionale: Jurassique et Crétacé. *Oceanol. Acta, 26th Int. Geol. Cong.*, Paris 1980, *Colloq. G13*, Excursion Guidebook, Trip **122**, 65–70.

CÉLÉRIER, B. (1988) Paleobathymetry and geodynamic models for subsidence. *Palaios* **3**, 454–463.

CLOETINGH, S. (1986) Intraplate stresses: a new tectonic mechanism for fluctuations of relative sea-level. *Geology* **14**, 617–620.

CLOETINGH, S. (1988a) Intraplate stress: a new element in basin analysis. In: *New Perspectives in Basin Analysis* (Eds Kleinspehn, K.L. & Paola, C.), *Coll. "Frontiers in Sedimentary Geology"*, Springer, New York, pp. 205–230.

CLOETINGH, S. (1988b) Intraplate stresses: a tectonic cause for third-order cycles in apparent sea level? In: *Sea-level Changes: an Integrated Approach* (Eds Wilgus, C.K., Hastings, B.S., Kendall, C.G.St.C., Posamentier, H.W., Ross C.A. & Van Wagoner, J.C.) Spec. Publ. Soc. Econ. Paleont. Miner. **42**, 19–29.

CLOETINGH, S. & WORTEL, R. (1986) Stress in the Indo-Australian Plate. *Tectonophysics* **132**, 49–67.

CLOETINGH, S., MCQUEEN, H. & LAMBECK, K. (1985) On a tectonic mechanism for regional sea level variations. *Earth Planet. Sci. Lett.* **75**, 157–166.

CLOETINGH, S., LAMBECK, K. & MCQUEEN H. (1987) Apparent sea-level fluctuations and a palaeostress field for the North Sea region. In: *Petroleum Geology of North West Europe* (Eds Brooks, J. & Glennie, K.) Graham and Trotman, London, Dordrecht, Boston, pp. 49–57.

COCHRAN, J.R. (1981) Simple models of diffuse extension and the pre-seafloor spreading development of the continental margin of the northeastern Gulf of Aden. In: *Geology of Continental Margins*, (Eds Blanchet, R. & Montadert, L.) pp. 155–165. Oceanol. Acta, 26th Int. Geol. Cong., Paris 1980, Colloq. C3.

COCHRAN, J.R. (1983) Effects of finite rifting times on the development of sedimentary basins. *Earth Planet. Sci. Lett.* **66**, 289–302.

COLLET, L.W. & PARÉJAS, E. (1931) Géologie de la chaîne de la Jungfrau. *Matér. Carte Géol. Suisse (n.s.)* **63**, 64 pp.

DEMAREST, J.M. & KRAFT, J.C. (1987) Stratigraphic record of Quaternary sea-levels: implications for more ancient strata. In: *Sea-level Fluctuations and Coastal Evolution.* (Eds Nummedal, D., Pilkey, O.H. & Howard, J.D.) Spec. Publ. Soc. Econ. Paleont. Miner. **41**, 223–239.

DEWEY, J. (1982) Plate tectonics and the evolution of the British Isles. *J. Geol. Soc. London* **139**, 371–412.

DIETRICH, D. & CASEY, M. (1989) A new tectonic model for the Helvetic nappes. In: *Alpine Tectonics* (Eds Coward, M.P., Dietrich, D. & Park, R.G.) Spec. Publ. Geol. Soc. London **45**, 47–63.

DOLIVO, E. (1982) Nouvelles observations structurales au SW du massif de l'Aar entre Visp et Gampel. *Matér. Carte Géol. Suisse (n.s.)* **157**, 82 pp.

DUMONT, T. (1988) Late Triassic–early Jurassic evolution of the western Alps and their European foreland: initiation of the Tethyan rifting. *Bull. Soc. Géol. France* **8/IV/4**, 601–612.

DYKSTRA, J. (1987) Compaction correction for burial history curves: application to Lopatin's method for source rock maturation determination. *Geobyte* **November 1987**, 16–23.

EBERLI, G.P. (1988) The evolution of the southern continental margin of the Jurassic Tethys Ocean as recorded in the Allgäu Formation of the Austroalpine Nappes of Graubünden (Switzerland). *Eclogae Geol. Helv.* **81/1, 175–214.**

EINSELE, G. (1985) Response of sediments to sea-level changes in differing subsiding storm-dominated marginal and epeiric basins. In: *Sedimentary and Evolutionary Cycles* (Eds Bayer, U. & Seilacher, A.), pp. 68–97. Lecture Notes Earth Sci. **1**. Springer, Berlin, Heidelberg, New York.

EINSELE, G. & BAYER, U. (1991) Asymmetry in transgressive–regressive cycles in shallow seas and passive continental margin settings. In: *Cycles and Events in Stratigraphy* (Eds Einsele, G., Ricken, W. & Seilacher, A.) Springer, Berlin, Heidelberg, pp. 660–681.

ELLIOTT, T. (1986) Siliciclastic shorelines. In: *Sedimentary Environments and Facies,* 2nd edn (Ed. Reading, H.G.), Blackwell Scientific Publications, Oxford, pp. 155–188.

FALVEY, D.A. (1974) The development of continental margins in plate tectonic theory. *J. Austr. Petrol. Expl. Ass.* **14**, 95–106.

FALVEY, D.A. & MIDDLETON, M.F. (1981) Passive continental margins: evidence for a pre-breakup deep crustal metamorphic subsidence mechanism. In: *Geology of Continental Margins*, (Eds Blanchet, R. & Montadert, L.) pp. 103–114. Dceanol. Acta, 26th Int. Geol. Cong. Paris 1980, Colloq. C3.

FERRAZZINI, B. & SCHULER, P. (1979) Eine Abwicklungskarte des Helvetikums zwischen Rhône und Reuss. *Eclogae geol. Helv.* **72/2**, 439–454.

FRANK, M. (1930) Beiträge zur vergleichenden Stratigraphie und Bildungsgeschichte der Trias-Lias-Sedimente im Alpin-germanischen Grenzgebiet der Schweiz. *Neues Jhb. Min. Geol. Pal., Beil.-Bd* **64**, 325–426.

FREEMAN, R. & MÜLLER, ST. (Ed.) (1990) The European Geotraverse, Part 6. *Tectonophysics* **176/1–2** 1–244.

FUNK, H. (1985) Mesozoische Subsidenzgeschichte im helvetischen Schelf der Ostschweiz. *Eclogae Geol. Helv.* **78/2**, 249–272.

FUNK, H., WILDI, W., AMATO, E., LOUP, B. & HUGGENBERGER, P. (1989) The pattern of Mesozoic subsidence on the proximal European margin of Tethys (Western part). *Abstr. Europ. Union Geosci.* Strasbourg 1989, Terra, **1/1**, 79.

FURRER, H. (Ed.) (1985) Field workshop on Triassic and Jurassic sediments in the Eastern Alps of Switzerland (25th–29th August 1985). *Mitt. Geol. Inst. ETH u. Univ. Zürich*, **248**, 82 pp.

GIBBS, A.D. (1987) Development of extension and mixed-mode sedimentary basins. In: *Continental Extensional Tectonics* (Eds Coward, M.P., Dewey, J.F. & Hancock, P.L.) Spec. Publ. Geol. Soc. London **28**, 19–33.

GILLCRIST, R., COWARD, M. & MUGNIER, J.L. (1987) Structural inversion and its controls: examples from the Alpine foreland and the French Alps. *Geodinamica Acta* **1/1**, 5–34.

GRAND, T. (1987) *Structures en extension et leur influence sur les déformations postérieures dans le domaine téthysien (Bourg d'Oisans, Alpes Occidentales fran*çaises, et Troodos, Chypre). PhD thesis, University of Grenoble, France, 244 pp.

GRAND, T. (1988) Mesozoic extensional inherited structures on the European margin of the Ligurian Tethys: the example of the Bourg d'Oisans half-graben, western Alps. *Bull. Soc. Géol. France* **8/IV/4**, 613–621.

GRAND, T. DUMONT, T. & PINTO-BULL, F. (1987) Distensions liées au rifting téthysien et paléochamps de contrainte associés dans le bassin liasique de Bourg d'Oisans (Alpes occidentales). *Bull. Soc. Géol. France* **8/III/4**, 699–704.

HALLAM, A. (1978) Eustatic cycles in the Jurassic. *Palaeogeogr. Palaeoclim. Palaeoecol.* **23**, 1–32.

HALLAM, A. (1981) A revised sea-level curve for the Early Jurassic. *J. Geol. Soc. London* **138**, 735–743.

HALLAM, A. (1984) Pre-Quaternary sea-level changes. *Ann. Rev. Earth Planet. Sci.* **12**, 205–243.

HALLAM, A. (1988) A reevaluation of Jurassic eustasy in the light of new data and the revised EXXON curve. In: *Sea-level Changes: an Integrated Approach* (Eds Wilgus, C.K., Hastings, B.S., Kendall, C.G.St.C., Posamentier, H.W., Ross, C.A. & Van Wagoner, J.C.) Spec. Publ. Soc. Econ. Paleont. Miner. **42**, 261–273.

HAQ, B.U., HARDENBOL, J. & VAIL, P.R. (1987) Chronology of fluctuating sea levels since the Triassic. *Science* **235**, 1156–1167.

HAQ, B.U., HARDENBOL, J. & VAIL, P.R. (1988) Mesozoic and Cenozoic chronostratigraphy and cycles of sea-level change. In: *Sea-level Changes: an Integrated Approach* (Eds Wilgus, C.K., Hastings, B.S., Kendall, C.G.St.C., Posamentier, H.W., Ross, C.A. & Van Wagoner, J.C.) Spec. Publ. Soc. Econ. Paleont. Miner. **42**, 71–108.

HARDENBOL, J., VAIL, P.R. & FERRER, J. (1981) Interpreting paleoenvironments, subsidence history and sea-level changes of passive margins from seismic stratigraphy. In: *Geology of Continental Margins*, (Eds Blanchet, R. & Montadert, L.) pp. 33–44. Oceanol. Acta, 26th Int. Geol. Cong., Paris 1980, Colloq. C3.

HARLAND, W.B., AMSTRONG, R.L., COX, A.V., CRAIG, L.E., SMITH, A.G. & SMITH, D.G. (1990) *A Geologic Time Scale — 1989*. Cambridge University Press, Cambridge, 263 pp.

HAXBY, W.F., TURCOTTE, D.L. & BIRD, J.M. (1976) Thermal and mechanical evolution of the Michigan Basin. In: *Sedimentary Basins of Continental Margins and Cratons* (Ed. Bott, M.H.P.) Tectonophysics **36**, 57–75.

HAYS, J.D. & PITMAN, III, W.C. (1973) Lithospheric plate motion, sea level changes and climatic and ecological consequences. *Nature* **246**, 18–22.

HEIDLAUF, D.T., HSUI, A.T. & KLEIN, G. (1986) Tectonic subsidence analysis of the Illinois Basin. *J. Geol.* **94/6**, 779–794.

HELLINGER, S.J. & SCLATER, J.G. (1983) Some comments on two-layer extensional models for the evolution of sedimentary basins. *J. Geophys. Res.* **88**, 8251–8259.

HOMEWOOD, P. & LATELTIN, O. (1988) Classic Swiss clastics (flysch and molasse). *Geodinamica Acta* **2/1**, 1–11.

HOMEWOOD, P., ALLEN, P.A. & WILLIAMS, G.D. (1986) Dynamics of the Molasse Basin of western Switzerland. In: *Foreland Basins* (Eds Allen, P.A. & Homewood, P.) Int. Assoc. Sediment. Spec. Publ. **8**, 199–217.

HUGGENBERGER, P. (1985) *Faltenmodelle und Verformungsverteilung in Deckenstrukturen am Beispiel der Morcles-Decke (Helvetikum der Westschweiz)*. PhD thesis, University of Zürich, Switzerland, **78–61**, 193 pp.

JERVEY, M.T. (1988) Quantitative geologic modelling of siliciclastic rock sequences and their seismic expression. In: *Sea-level Changes: an Integrated Approach* (Eds Wilgus, C.K., Hastings, B.S., Kendall, C.G.St.C., Posamentier, H.W., ross, C.A. & Van Wagoner, J.C.) Spec. Publ. Soc. Econ. Paleont. Miner. **42**, 47–69.

JARVIS, G.T. & MCKENZIE, D.P. (1980) Sedimentary basin formation with finite extension rates. *Earth Planet. Sci. Lett.* **48**, 42–52.

KARNER, G.D. (1986) Effects of lithospheric in-plane stress on sedimentary basin stratigraphy. *Tectonics* **5**, 573–588.

KARNER, G.D. & DEWEY, J.F. (1986) Rifting: lithospheric versus crustal extension as applied to the Ridge Basin of southern California. In: *Future Petroleum Provinces of the World* (Ed. Halbouty, M.T.) Am. Assoc. Petrol. Geol. Mem. **40**, 317–337.

KARNER, G.D., LAKE, S.D. & DEWEY, J.F. (1987) The thermal and mechanical development of the Wessex Basin, southern England. In: *Continental Extensional Tectonics* (Eds Coward, M.P., Dewey, J.F. & Hancock, P.L.)

Spec. Publ. Geol. Soc. London **28**, 517–536.

KENDALL, C.G.St.C. & LERCHE, I. (1988) The rise and fall of eustasy. In: *Sea-level Changes: an Integrated Approach* (Eds Wilgus, C.K., Hastings, B.S., Kendall, C.G.St.C., Posamentier, H.W., Ross, C.A. & Van Wagoner, J.C.) Spec. Publ. Soc. Econ. Paleont. Miner. **42**, 3–17.

KOMINZ, M.A. (1984) Oceanic ridge volumes and sea-level changes — an error analysis. In: *Interregional Unconformities and Hydrocarbon Accumulation* (Ed. Schlee, J.S.). Am. Assoc. Petrol. Geol. Mem. **36**, 109–127.

KOOI, H. & CLOETINGH, S. (1989) Some consequences of compressional tectonics for extensional models of basin subsidence. *Geol. Rundsch.* **78**, 183–195.

KOOI, H. CLOETINGH, S. & REMMELTS, G. (1989) Intraplate stresses and the stratigraphic evolution of the North Sea Central Graben. *Geol. Mijnbouw* **68**, 49–72.

KREBS, J. (1925) Geologische Beschreibung der Blümlisalpgruppe. *Matér. Carte Géol. Suisse (n.s.)* **54/3**, 75 pp.

KÜBLER, B. (1964) Les argiles, indicateurs du métamorphisme. *Rev. Inst. Françc Pétrole* **19**, 1093–1112.

KÜBLER, B. (1968) Evaluation quantitative du métamorphisme par la cristallinité de l'illite. *Bull. Cent. Rech. Expl.-Prod. Elf-Aquitaine* **2/2**, 385–397.

KÜBLER, B. (1990) Cristallinité de l'illite et mixed-layers: brève révision. *Schweiz. Mineral. Petrogr. Mitt.* **70**, 89–93.

LAMBECK, K., CLOETINGH, S. & MCQUEEN, H. (1987) Intraplate stresses and apparent changes in sea-level: the basins of northwestern Europe. In: *Sedimentary Basins and Basin-forming Mechanisms* (Eds Beaumont, C. & Tankard, A.J.) Can. Soc. Petrol. Geol. Mem. **12**, 259–268.

LAUBSCHER, H. & BERNOULLI, D. (1977) Mediterranean and Tethys. In: *The Ocean Basins and Margins, Vol. 4A: The Eastern Mediterranean* (Eds Nairn, A.E.M., Kanes, W.H. & Stehli, F.G.) Plenum Press, London, pp. 1–28.

LEMOINE, M. & TRÜMPY, R. (1987) Pre-oceanic rifting in the Alps. *Tectonophysics* **133**, 305–320.

LEMOINE, M., BAS, T., ARNAUD-VANNEAU, A. *et al.* (1986) The continental margin of the Mesozoic Tethys in the Western Alps. *Mar. Petrol. Geol.* **3/3**, 179–200.

LOUP, B. (1990) Sea-level changes and extensional tectonics in the Lower Jurassic of the Helvetic realm (western Switzerland). *Abstr. 13th Int. Sediment. Congr. (IAS)*, Nottingham 1990, 318–319.

LOUP, B. (1991) *Evolution de la partie septentrionale du domaine helvétique en Suisse occidentale au Trias et au Lias: contrôle par subsidence thermique et variations du niveau marin*. PhD thesis, University of Geneva, Switzerland, **2508**, 247 pp.

LOUTIT, T.S., HARDENBOL, J., VAIL, P.R. & BAUM, G.R. (1988) Condensed sections: the key to age determination and correlation of continental margin sequences. In: *Sea-level Changes: an Integrated Approach* (Eds Wilgus, C.K., Hastings, B.S., Kendall, C.G.St.C., Posamentier, H.W., Ross, C. A. & Van)Wagoner, J.C.) Spec. Publ. Soc. Econ. Paleont. Miner. **42**, 183–213.

LUGEON, M. (1914) Les Hautes Alpes calcaires entre la Lizerne et la Kander. *Matér. Carte Géol. Suisse (n.s.)* **30**, 360 pp.

LUGEON, M. (1930) Trois tempêtes orogéniques: la Dent de Morcles. *Soc. Géol. France, Livre Jubilaire 1830–1930*, pp. 499–512.

MASSON, H., HERB, R. & STECK, A. (1980) Helvetic Alps of Western Switzerland. In: *Geology of Switzerland, a Guide-book, part B: Geological Excursions* (Ed. Trümpy, R.) Schweiz. Geol. Komm., Wepf, Basel, pp. 109–153.

McKENZIE, D. (1978) Some remarks on the development of sedimentary basins. *Earth Planet. Sci. Lett.* **40**, 25–32.

MEISTER, CH. & LOUP. B. (1989) Les gisements d'ammonites liasiques (Hettangien à Pliensbachien) du Ferdenrothorn (Valais, Suisse): analyses paléontologique, biostratigraphique et aspects lithostratigraphiques. *Eclogae Geol. Helv.* **82/3**, 1003–1041.

METTRAUX, M. (1989) *Sédimentologie, paléotectonique et paléocéanographie des Préalpes médianes (Suisse Romande) du Rhétien au Toarcien.* PhD thesis, University of Fribourg, Switzerland, **947**, 135 pp.

METTRAUX, M. & MOSAR, J. (1989) Tectonique alpine et paléotectonique liasique dans les Préalpes Médianes en rive droite du Rhône. *Ecolgae Eeol. Helv.* **82/2**, 517–540.

MORTON, N. (1987) Jurassic subsidence history in the Hebrides, NW Scotland. *Mar. Petrol. Geol.* **4**, 226–242.

MOSER, H.J. (1985) *Strukturgeologische Untersuchungen in der Rawil-Depression.* PhD thesis, University of Bern, Switzerland, 147 pp.

MOSER, H.J. (1987) Die Zone von Ayent als triasischliasisches Substrat der helvetischen Decken der Westschweiz. *Eclogae Geol. Helv.* **80/1**, 1–15.

MÜLLER, ST., ANSORGE, J., EGLOFF, R. & KISSLING, E. (1980) A crustal cross section along the Swiss geotraverse from the Rhinegraben to the Po Plain. *Eclogae Geol. Helv.* **73/2**, 463–483.

ODIN, G.S. (1982) The Phanerozoic time scale — revisited. *Episodes* **1982/3**, 3–9.

PANOZZO, R. (1987) Two-dimensional train determination by the inverse SURFOR wheel. *J. Struct. Geol.* **9/1**, 115–119.

PANZA, G.F., CALCAGNILE, G., SCANDONE, P. & MÜLLER, S. (1984) Die geologische Tiefenstruktur des Mittelmeerraumes. In: *Ozeane und Kontinente: ihre Herkunft, Geschichte und Struktur* (Ed. Giesc, P.). Spektrum der Wissenschaft, Heidelberg, pp. 132–142.

PARSONS, B. & SCLATER, J.G. (1977) An analysis of the variation of ocean floor bathymetry and heat flow with age. *J. Geophys. Res.* **32**, 803–827.

PITMAN, III, W.C. (1978) The relationship between eustasy and stratigraphic sequences of passive margins. *Bull. Geol. Soc. Am.* **89**, 1389–1403.

PITMAN, III, W.C. & GOLOVCHENKO, X. (1983) The effect of sea-level change on the shelfedge and slope of passive margins. In: *The Shelfbreak: Critical Interface on Continental Margins* (Eds Stanley, D.J. & Moore, G.T.) Spec. Publ. Soc. Econ. Paleont. Miner. **33**, 41–58.

POSAMENTIER, H.W. & VAIL, P.R. (1988) Eustatic controls on clastic deposition II — sequence and systems tract models. In: *Sea-level Changes: an Integrated Approach* (Eds Wilgus, C.K., Hastings, B.S., Kendall, C.G.St.C., Posamentier, H.W., Ross, C.A. & Van Wagoner, J.C.) Spec. Publ. Soc. Econ. Paleont. Miner **42**, 125–154.

RAMSAY, J.G. & HUBER, M.I. (1983) *The Techniques of Modern Structural Geology, Vol. 1: Strain Analysis.* Academic Press, London, 307 pp.

REINECK, H.E. & SINGH I.B. (1980) *Depositional Sedimentary Environments (with Reference to Terrigenous Clastics),* 2nd edn. Springer, Berlin, 549 pp.

REINSON, G.E. (1984) Barrier island and associated strandplain systems. In: *Facies Models*, 2nd edn (Ed. Walker, R.G.) *Geosci. Can. Repr. Ser.* **1**, 119–140.

ROWLEY, D.B. & SAHAGIAN, D. (1986) Depth-dependent stretching: a different approach. *Geology* **14**, 32–35.

ROYDEN, L. & KEEN, C.E. (1980) Rifting processes and thermal evolution of the continental margin of eastern Canada determined from subsidence curves. *Earth Planet. Sci. Lett.* **51**, 343–361.

ROYDEN, L., SCLATER, J.G. & VON HERZEN, R.P. (1980) Continental margin subsidence and heat flow: important parameters in formation of petroleum hydrocarbon. *Bull. Am. Assoc. Petrol. Geol.* **64**, 173–187.

ROYDEN, L., HORVATH, F., NAGYMAROSY, A. & STEGENA, L. (1983) Evolution of the Pannonian Basin system 2. Subsidence and thermal history. *Tectonics* **2**, 91–137.

RUDKIEWICZ, J.L. (1988) Quantitative subsidence and thermal structure of the European continental margin of the Tethys during Early and Middle Jurassic times in the western Alps (Grenoble–Briançon transect). *Bull. Soc. Géol. France* **8/IV/4**, 623–632.

SAWYER, D.S., TOKSÖZ, M.N., SCLATER, J.G., & SWIFT, B.A. (1982) Thermal evolution of the Baltimore Canyon trough and Georges Bank basin. In: *Studies in Continental Margin Geology* (Eds by Watkins, J.S. & Drake, C.L.) Am. Assoc. Petrol. Geol. Mem. **34**, 743–764.

SCHENKER, M. (1946) Geologische Untersuchung der mesozoischen Sedimentkeile am Südrand des Aarmassivs zwischen Lonza und Baltschiedertal. *Matér. Carte Géol. Suisse (n.s.)* **86**, 60 pp.

SCHLÄPPI, E. (1980) *Geologische und tektonische Entwicklung der Doldenhorndecke und zugehöriger Elemente.* PhD thesis, University of Bern, Switzerland, 154 pp.

SCHMID, S.M., PANOZZO, R. & BAUER, S. (1987) Simple shear experiments on calcite rocks: rheology and microfabric. *J. Struct. Geol.* **9/5–6**, 747–778.

SCHMID, S.M., RÜCK, P. & SCHREUS, G. (1990) The significance of the Schams nappe for the reconstruction of the paleotectonic and orogenic evolution of the Penninic Zone along the NFP-20 East traverse (Grisons, eastern Switzerland). *Soc. Géol. Suisse Mém.* **1**, 263–287.

SCHMOKER, J.W. & HALLEY, R.B. (1982) Carbonate porosity versus depth: a predictable relation for South Florida. *Bull. Am. Assoc. Petrol. Geol.* **66/12**, 2561–2570.

SCLATER, J.G. & CHRISTIE, P.A.F. (1980) Continental stretching: an explanation of the post-Mid-Cretaceous subsidence of the Central North Sea basin. *J. Geophys. Res.* **85/B7**, 3711–3739.

SCLATER, J.G., JAUPART, C. & GALSON, D. (1980a) The heat flow through oceanic and continental crust and the heat loss of the Earth. *Rev. Geophys. Space Phys.* **18**, 269–311.

SCLATER, J.G. ROYDEN, L. HORVATH, F., BURCHFIEL, B.C., SEMKEN, S. & STEGENA, L. (1980b) The formation of the Intra-Carpathian basins as determined from subsidence data. *Earth Planet. Sci. Lett.* **51**, 139–162.

SLEEP, N.H. (1971) Thermal effects of the formation of Atlantic continental margins by continental breakup. *Geophys. J. R. Astron. Soc.* **24/4**, 325–350.

STAMPFLI, G.M. & MATHALER, M. (1990) Divergent and convergent margins in the north-western Alps — confrontation to actualistic models. *Geodinamica Acta* **4/3**, 159–184.

STECKLER, M.S. & WATTS, A.B. (1978) Subsidence of the Atlantic-type continental margin off New York. *Earth Planet. Sci. Lett.* **41**, 1–13.

STECKLER, M.S. & WATTS, A.B. (1982) Subsidence history and tectonic evolution of Atlantic-type continental margins. In: *Dynamics of Passive Margins* (Ed. Scrutton, R.A.) Am Geophys. Un. & Geol. Soc. Am., Washington DC, Geodyn. Ser. **6**, 184–196.

SWIDERSKI, B. (1919) La partie occidentale du Massif de l'Aar entre la Lonza et la Massa. *Matér. Carte Géol. Suisse (n.s.)* **47/1**, 68 pp.

TRÜMPY, R. (1945) Le Lias autochtone d'Arbignon, Groupe de la Dent de Morcles. *Eclogae Geol. Helv.* **38/2**, 421–429.

TRÜMPY, R. (1949) Der Lias der Glarner Alpen. *Denkschr. Schweiz. Natf. Ges.* **79/1**, 192 pp.

TRÜMPY, R. (1958) Remarks on the pre-orogenic history of the Alps. *Geol. Mijnbouw* **20**, 340–352.

TRÜMPY, R. (1971) Sur le Jurassique de la zone Helvétique en Suisse. *Ann. Inst. geol. publ. hung.* **54/2**, 369–382.

TRÜMPY, R. (1980) *Geology of Switzerland — a Guide-book, Part A: an Outline of the Geology of Switzerland; Part B: Geological Excursions.* Schweiz. Geol. Komm., Wepf, Basel, 334 pp.

TRÜMPY, R. (1982) Alpine paleogeography: a reappraisal. In: *Mountain Building Processes* (Ed. Hsü, K.J.) Academic Press, London, pp. 149–156.

TRÜMPY, R. (1988) A possible Jurassic–Cretaceous transform system in the Alps and the Carpathians, Spec. Pap. Geol. Soc. Am. **218**, 93–109.

VAIL, P.R. & EISNER, P.N. (1989) Stratigraphic signatures separating tectonic, eustatic and sedimentologic effects on sedimentary sections. In: *Mesozoic Eustasy Record on Western Tethyan Margins, 2nd Congr. Franç. Sédiment., Lyon 1989.* Spec. Publ. Ass. Sédiment. Françcis, **11**, 62–64.

VAIL, P.R., MITCHUM, R.M., TODD, R.G. *et al.* (1977) Seismic stratigraphy and global changes of sea level. In: *Seismic Stratigraphy — Application to Hydrocartion Explo ration* (Ed. Payton, C.E.) Am. Assoc. Petrol. Geol. Mem. **26**, 49–212.

VAIL, P.R., HARDENBOL, J. & TODD, R.G. (1984) Jurassic unconformities, chronostratigraphy and sea level changes from seismic stratigraphy and biostratigraphy. In: *Interregional Unconformities and Hydrocarbon Accumulation* (Ed. Schlee, J.S.) Am. Assoc. Petrol. Geol. Mem. **36**, 129–144.

VAIL, P.R., AUDEMARD, F., BOWMAN, S.A., EISNER, P.N. & PEREZ-CRUZ, C. (1991) The stratigraphic signatures of tectonics, eustasy and sedimentology — an overview. In: *Cycles and Events in Stratigraphy* (Eds Einsele, G., Ricken, W. & Seilacher, A.) Springer, Berlin, Heidelberg, pp. 617–659.

VAN HINTE, J.E. (1978) Geohistory analysis — application of micropaleontology in exploration geology. *Bull. Am. Assoc. Petrol. Geol.* **62/2**, 201–222.

VAN WAGONER, J.C., POSAMENTIER, H.W., MITCHUM, R.M., VAIL, P.R., SARG, J.F., LOUTIT, T.S. *et al.*, (1988) An overview of the fundamentals of sequence stratigraphy and key definitions. In: *Sea-level Changes: an Integrated Approach* (Eds Wilgus, C.K., Hastings, B.S., Kendall, C.G.St. C., Posamentier, H.W., Ross, C.A. & Van Wagoner, J.C.) Spec. Publ. Soc. Econ. Paleont. Miner. **42**, 39–45.

WALKER, R.G. (1984) Shelf and shallow marine sands. In: *Facies Models*, 2nd edn (Ed.Walker, R.G.) *Geosci. Can. Repr. Ser.* **1**, 141–170.

WATTS, A.B. (1978) An analysis of isostasy in the world's oceans, 1, Hawaiian-Emperor seamount chain, *J. Geophys. Res.* **83/B12**, 5989–6004.

WATTS, A.B. (1982) Tectonic subsidence, flexure and global changes of sea level. *Nature* **297**, 469–474.

WATTS, A.B. & RYAN, W.B.F. (1976) Flexure of the lithosphere and continental margin basins. In: *Sedimentary Basins of Continental Margins and Cratons* (Ed. Bott, M.H.P.) *Tectonophysics* **36**, 25–44.

WATTS, A.B. & STECKLER, M.S. (1979) Subsidence and eustasy at the continental margin of eastern North America. In: *Deep Drilling Results in the Atlantic Ocean: Continental Margins and Paleoenviroments* (Eds Talwani, M., Hay, W.F. & Ryan, W.B.F.). *Maurice Ewing Ser.* **3**, 218–234.

WATTS, A.B. & THORNE, J.A. (1984) Tectonics, global changes in sea-level and their relationship to stratigraphic sequences at the U.S. Atlantic continental margin. *Mar. Petrol. Geol.* **1**, 319–339.

WATTS, A.B., KARNER, G.D. & STECKLER, M.S. (1982) Lithospheric flexure and the evolution of sedimentary basins. *Philos. Trans. R. Soc. London* **A305**, 249–281.

WEIMER, R.J. (1988) Record of relative sea-level changes, Cretaceous of Western Interior, USA. In: *Sea-level Changes: an Integrated Approach* (Eds Wilgus, C.K., Hastings, B.S., Kendall, C.G.St.C., Posamentier, H.W., Ross, C.A. & Van Wagoner, J.C.) Spec. Publ. Soc. Econ. Paleont. Miner. **42**, 285–288.

WERNICKE, B. (1985) Uniform-sense normal simple shear of the continental lithosphere. *Can. J. Earth Sci.* **22**, 108–125.

WILDI, W., FUNK, H., LOUP, B., AMATO, E. & HUGGENBERGER, P. (1989a) Mesozoic subsidence history of the European marginal shelves of the Alpine Tethys (Helvetic realm, Swiss Plateau and Jura). *Eclogae Geol. Helv.*, **82/3**, 817–840.

WILDI, W., LOUP, B., AMATO, E. & FUNK, H. (1989b) Subsidence pattern of the European marginal shelves and basin of the Tethys. *Abstr. 28th Int. Geol. Congr.*, Washington 1989, *Symposium B8* **3**, 359.

WILSON, J.T. & BURKE, K. (1972) Two types of mountain building. *Nature* **239**, 448–449.

WINTERER, E.L. & BOSSELINI, A. (1981) Subsidence and sedimentation on Jurassic passive continental margin, Southern Alps, Italy. *Bull. Am. Assoc. Petrol. Geol.* **65/3**, 394–421.

ZIEGLER, P.A. (1982) *Geological Atlas of Western and Central Europe.* Shell Int. Petrol. Maat. B.V. Elsevier, Amsterdam, 130 pp.

ZIEGLER, P.A. (1988) Evolution of the Arctic–North Atlantic and the Western Tethys. *Am. Assoc. Petrol. Geol. Mem.* **43**, 198 pp.

ZOBACK, M.L. & ZOBACK, M.D. (1980) State of stress in the conterminous United States. *J. Geophys. Res.* **85**, 6113–6156.

ZWAHLEN, P. (1986) *Die Kandertalstörung, eine transversale Diskontinuität im Bau der Helvetischen Decken.* PhD thesis, University of Bern, Switzerland, 105 pp.

Spec. Publs Int. Ass. Sediment. (1993) **20**, 161–169

The Tertiary Queen Charlotte Basin: a strike-slip basin on the western Canadian continental margin

J.R. DIETRICH,*
R. HIGGS,† K.M. ROHR‡ *and* J.M. WHITE*

** Institute of Sedimentary and Petroleum Geology, Geological Survey of Canada,
3303–33 St. NW., Calgary, Alberta, Canada; T2L 2A7;
† Maraven S.A., Apartado 829, Caracas 1010A, Venezuala; and
‡ Pacific Geoscience Centre, Geological Survey of Canada,
P.O. Box 6000, Sidney, British Columbia V8L 4B2, Canada*

ABSTRACT

The Queen Charlotte Basin is a Tertiary strike-slip basin located adjacent to a 500-km-long transform segment of the Pacific–North America plate boundary, offshore Canada. The basin contains Eocene to Pliocene deposits in a number of sub-basins in excess of 3000 m deep, oriented either parallel or at moderately oblique angles to the plate margin. The basin-fill consists of syn-extension volcanics and clastics in grabens and half-grabens, overlain by a regional blanket of post-extension clastics. Clastic depositional settings include a range of non-marine and marine environments. Pliocene compression and uplift resulted in local inversion, deformation and erosion of parts of the basin fill. The basin's evolution is interpreted in terms of a tectonic cycle of transtension, passive subsidence, and transpression, a typical development history for strike-slip basins.

INTRODUCTION

The Queen Charlotte Basin is a 60 000 km² Tertiary basin located along the Pacific continental margin of Canada (Fig. 1). The basin underlies Graham Island, parts of Moresby Island and large areas of offshore Dixon Entrance, Hecate Strait and Queen Charlotte Sound (Fig. 2). Exploration by petroleum companies in the 1960s and early 1970s included the acquisition of several thousand kilometres of seismic reflection data and the drilling of seven onshore and eight offshore wells, all of which were dry and therefore abandoned (Fig. 2). The first publication summarizing the regional geology of the offshore basin was based on this early seismic and well data (Shouldice, 1971). In 1987, an extensive basin analysis programme was initiated by the Geological Survey of Canada to provide new geological and geophysical data and interpretations for the Queen Charlotte region. Individual components of the basin analysis have been previously presented, including descriptions of the sedimentology and biostratigraphy of the basin fill from studies of outcrops and well samples and core (Higgs, 1989, 1991; Patterson, 1989; White, 1990, 1991) and interpretations of deep seismic reflection data acquired in 1988 (Rohr & Dietrich, 1991).

The purpose of this paper is to present a general overview of the Tertiary geology of the basin, summarized from our studies of seismic, sedimentological and biostratigraphical data. We describe the plate tectonic setting and regional structure and stratigraphy of the basin and outline our interpretation of the basin's tectonic evolution. A regional isopach and fault map derived from seismic and well data (Fig. 2), two geological cross-sections (Figs 3 & 5) and segments of two seismic reflection profiles (Figs 4 & 6) are included to illustrate some aspects of the structure and lithology of the basin fill. A few comments on depositional environments are included in the text, but for detailed descriptions of the basin-fill sedimentology the reader is referred to Higgs (1991).

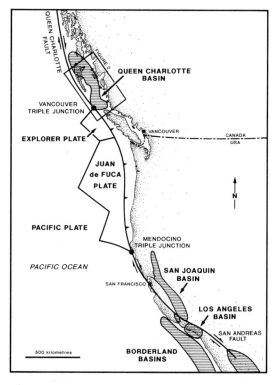

Fig. 1. Location of the Queen Charlotte Basin and present-day plate tectonic setting of the Pacific continental margin of western Canada and the United States.

PLATE TECTONIC SETTING

The regional trend of the Queen Charlotte Basin is northwest, parallel to the present-day plate boundary between oceanic crust of the Pacific and Explorer plates and continental crust of the North America plate (Fig. 1). The plate boundary offshore Canada consists of the southern portion of the Queen Charlotte fault, a right-lateral transform fault, and the northern portion of the Explorer plate subduction zone. The Vancouver triple junction adjacent to the southern end of the Queen Charlotte Basin marks the contact between the Pacific, Explorer and North America plates. Along the United States continental margin, the Pacific, Juan de Fuca and North America plates are in contact at the Mendocino triple junction at the northern end of the San Andreas right-lateral transform fault (Fig. 1). Numerous Neogene strike-slip basins (including the Los Angeles, San Joaquin and Borderland ba-

sins) occur south of the Mendocino triple junction, across the broad and structurally complex transform margin in the coastal California region (Crowell, 1974; Blake *et al.*, 1978; Howell *et al.*, 1980).

The Tertiary Queen Charlotte Basin developed within the North America plate, adjacent to a transform segment of the plate boundary offshore Canada. The Cenozoic plate tectonic history of the Queen Charlotte region has been described by Engebretson *et al.* (1985), Engebretson (1989) and Hyndman and Hamilton (1991). Major Pacific plate motion changes in the Eocene (at about 40 Ma) resulted in Pacific–North America plate interactions in the region changing from near-orthogonal convergence to right-lateral strike-slip motion. The predominant plate interactions from Late Eocene to Early Pliocene were parallel or divergent strike-slip. A significant clockwise shift in the direction of Pacific plate motion in the Pliocene (at about 5 Ma) led to convergent strike-slip across the plate margin, which has continued to the present day. The Vancouver triple junction was probably located south of the Queen Charlotte Basin area until the Pleistocene (at about 1 Ma) when it shifted (northwestward) to its present position adjacent to Queen Charlotte Sound (Riddihough *et al.*, 1980).

BASIN STRUCTURE

The Queen Charlotte Basin contains up to 7000 m of Eocene to Pliocene clastics and volcanics, unconformably overlying a complexly deformed 'basement' of Mesozoic sedimentary, volcanic and plutonic rocks. The Tertiary isopach and fault map (Fig. 2) indicates that the offshore portion of the basin consists of a number of sub-basins in excess of 3000 m deep, most of which are extensional fault-bounded half-grabens or grabens (Fig. 2). The extensional faults display cross-sectional geometries which vary from planar and steep dipping to listric (Figs 3–6). Contractional structures, including reverse faults and folds, are common in the north-western half of the basin (Fig. 2). Many of the reverse faults are reactivated (inverted) extensional faults (Fig. 5). The inverted faults commonly bound or form the core of inversion folds (Figs 2, 5 & 6). Some of the contractional structures in the basin have been described from seismic profiles as positive flower structures and interpreted as strike-slip features (Rohr & Dietrich, 1991).

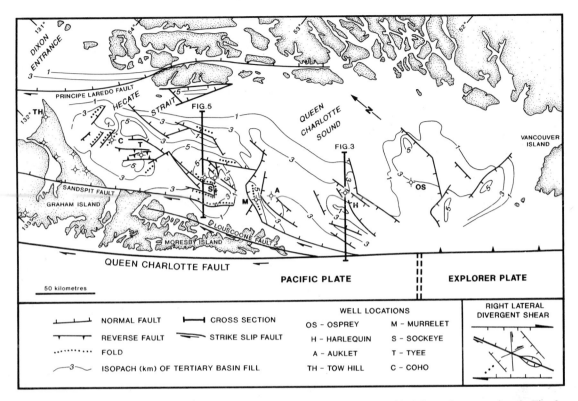

Fig. 2. Regional Tertiary isopach and fault map of the Queen Charlotte Basin and locations of cross-sections in Figs 3 and 5. Many of the basin structure trends match idealized structural patterns for a system of right-lateral shear aligned in a northwest direction, parallel to the plate boundary. (Model adapted from Harding *et al.*, 1985.)

The faults and isopach contours are most commonly north- or northwest-trending and, less commonly, east- or northeast-trending (Fig. 2). Regionally extensive faults in the basin include the northwest-trending Louscoone, Sandspit and Principe Laredo faults. The northwest-trending structural fabric is parallel or subparallel to the present-day Queen Charlotte transform fault. Local structural patterns are complicated by east- or northeast-trending faults and folds which intersect or cut northwest-trending structures (e.g. areas of the Sockeye and Murrelet wells; Fig. 2). East-trending faults occur in a right-stepping, en echelon pattern in the area of the Coho well and east-trending half-grabens occur at the extreme southeastern end of the basin (north of Vancouver Island).

The regional structural trends observed in the basin are considered evidence of strike-slip tectonism. Structural trends predicted from shear strain ellipse models closely match the observed trends in the basin. Specifically, an idealized strain ellipse for right-lateral divergent shear (Harding *et al.*, 1985) aligned in a northwest direction (parallel to the transform plate boundary) outlines the main northwest and north structural trends and secondary east and northeast trends (Fig. 2). The contemporaneous development of extensional faults (and associated depocentres) both parallel and oblique to the plate margin is an important structural relationship and one that is typical of divergent strike-slip or oblique rift systems (Harding *et al.*, 1985; Tron & Brun, 1991). Consistent with model-predicted structural patterns, some of the faults in the basin are interpreted as strike-slip or oblique-slip faults (Fig. 2). Onshore studies have indicated that the northwest-trending Louscoone fault is a strike-slip fault with up to 20 or 30 km of right-lateral displacement (Sutherland Brown, 1968; Lewis, 1991). Other similarly oriented faults, including the Sandspit and Principe Laredo faults, are interpreted to be oblique-slip faults with components of northeast-side-down, dip-slip and right-lateral, strike-slip offsets (Fig. 2). Strike-slip offsets are inferred from the common occurrence of transverse geometries which are characteristic of strike-slip faults (Christie-Blick & Biddle, 1985), including near-vertical fault planes and upward branching fault patterns (Figs 5 & 6; Rohr & Dietrich, 1991). The east- and northeast-trending faults in the basin are interpreted to be conjugate strike-slip faults, as predicted by the shear model (Fig. 2). Northeast-trending, left-lateral faults have been identified in outcrop studies on Graham Island (Lewis, 1990). The east-trending faults with south-side-down offsets at the southern end of the basin match the model-predicted trend and geometry for (right-lateral) oblique-slip faults.

BASIN STRATIGRAPHY

The Tertiary basin fill can be divided into two main

depositional units; a lower unit of siliciclastics, volcaniclastics and volcanics in half-grabens and grabens and an upper unit of siliciclastics overstepping the half-grabens and grabens without obvious fault boundaries (Figs 3 & 5). In the Queen Charlotte Sound area, where the basin is open to the Pacific Ocean, the upper unit forms a south-westward-thickening, continental margin wedge (Fig. 3). We refer to the two depositional units as syn-extension and post-extension sections, recognizing the fundamental control of extensional faulting during deposition of the lower basin-fill unit. The syn-extension sections consist of up to 5000 m of Eocene to Middle or Late Miocene deposits, with the bulk of the strata of Miocene age (Figs 3–6). The fill of each half-graben or graben shows considerable variability in lithology, depositional facies, thickness and structural attitude. The older parts of the syn-extension sections commonly contain volcanics and interbedded clastics (Figs 3 & 5). On seismic profiles, high amplitude reflections in syn-extension

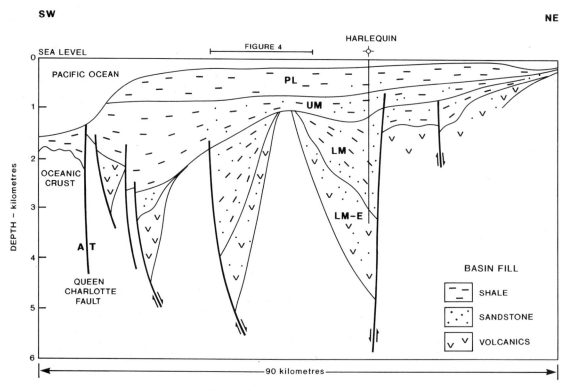

Fig. 3. Geological cross-section (derived from seismic and well data) across the southern Queen Charlotte Basin and adjacent Queen Charlotte fault (location on Fig. 2). The Tertiary basin fill consists of Pliocene (PL) and Upper Miocene (UM) marine clastics and Eocene (E) to Lower Miocene (LM) volcanics and marine and non-marine clastics.

sections are interpreted to be generated by interbedded volcanic flows and clastics (Figs 4 & 6). The volcanics were probably erupted via active extensional faults, an interpretation supported by chemical and physical volcanology from onshore areas (Hyndman & Hamilton, 1991) and the seismic imaging of wedge-shaped, high amplitude reflection packages thickening towards half-graben-bounding faults (Fig. 4).

Syn-extension alluvial fan and fan-delta conglomerates and sandstones have been identified in outcrops and onshore well cores (Higgs, 1991) and may be common in many of the offshore half-grabens/grabens. Seismic indications of fan or fan-delta deposition within syn-extension sections include offlapping wedge-shaped reflection packages (Fig. 4) and low angle seismic clinoforms (Fig. 6). Both footwall- and hanging-wall-sourced fans are indicated by sedimentary wedges thickening into and away from fault scarps (Figs 4 & 6, respectively). In addition to volcanic flows and coarse-grained fan deposits, the syn-extension sections contain fluvial and floodplain sediments consisting of interbedded sandstone, siltstone, shale and coal (Higgs, 1991). A thick section of fine-grained, non-marine deposits occurs in the half-graben penetrated by the Sockeye well (Fig. 5). The close proximity of fine-grained, coal-bearing strata to a major fault scarp indicates that deposition in this continental half-graben was dominated by axial-sediment transport.

In the southern (Queen Charlotte Sound) part of the basin, interbedded volcanics and non-marine clastics in the lower part of the syn-extension fill are overlain by marine shelf and slope deposits (Shouldice, 1971; Patterson, 1989). A half-graben containing shallow-marine deposits is penetrated by the Harlequin well (Fig. 3). The marine sediments consist of interbedded shale and bioturbated, shelf sandstones (Higgs, 1991).

The syn-extension deposits are unconformably overlain by up to 2000 m of post-extension deposits of Pliocene and (locally) Late Miocene age. The

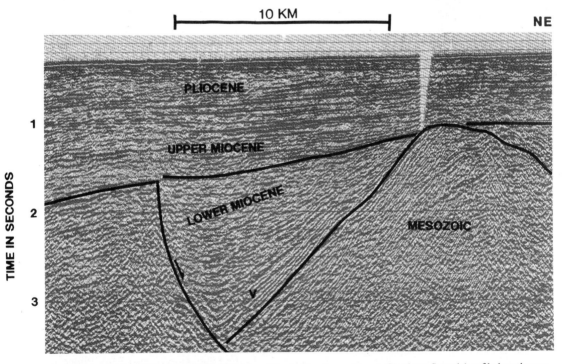

Fig. 4. Interpreted reflection seismic profile across one of the half-grabens, Queen Charlotte Sound (profile location on Fig. 3). A Middle Miocene unconformity separates syn-extension and post-extension sections. A wedge-shaped package of high amplitude reflections (V) at the base of the syn-extension fill may be generated from interbedded volcanic flows and clastics. Reflections are upturned against the half-graben-bounding fault, possibly indicating fault drag.

J.R. Dietrich et al.

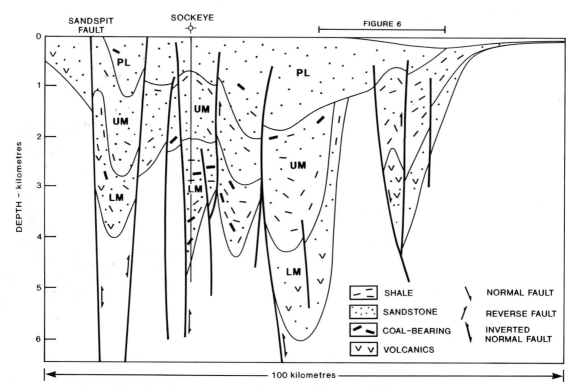

Fig. 5. Geological cross-section (derived from seismic and well data) across the north-central Queen Charlotte Basin (location on Fig. 2). The Tertiary basin fill consists of Pliocene (PL) non-marine and marine clastics and Upper Miocene (UM) and Lower Miocene (LM) to Eocene (E) non-marine clastics and volcanics. The basin fill is cut by numerous normal and reverse faults and is folded and partially eroded in basin-margin areas. The Sandspit fault is an oblique-slip, half-graben-bounding fault, reactivated as a reverse fault.

post-extension deposits are more areally extensive than underlying strata. The base of the post-extension section is locally an erosional unconformity, truncating underlying rocks (Fig. 6). The age of the syn-extension/post-extension unconformity varies from Middle Miocene (Figs 3 & 4) to Late Miocene (Figs 5 & 6). The post-extension deposits consist largely of coarsening-up sandstone and shale successions deposited in delta-plain, tidal shelf and storm-dominated shelf environments (Higgs, 1991). Delta-plain deposits are abundant in the northern parts of the basin, while marine deposits are more prevalent to the south.

BASIN EVOLUTION

The development of the Queen Charlotte Basin

occurred in three main phases. The first phase involved Eocene to Miocene extensional faulting and deposition of syn-extension clastics and (locally) volcanics within half-grabens and grabens. The north- and northwest-oriented half-grabens and grabens are interpreted to have developed in response to transtension within a margin-parallel zone of distributed right-lateral shear. The transtension was initiated in the Middle Tertiary by a major shift in Pacific–North America plate interactions from orthogonal convergence to strike-slip. The peak of extensional faulting and deposition occurred during the Miocene, probably a time of maximum divergence in relative plate motions.

In the Late Miocene and Early Pliocene, blanket deposition of clastics occurred during a post-extension phase characterized by regional (thermal?) subsidence. The timing of the syn-extension

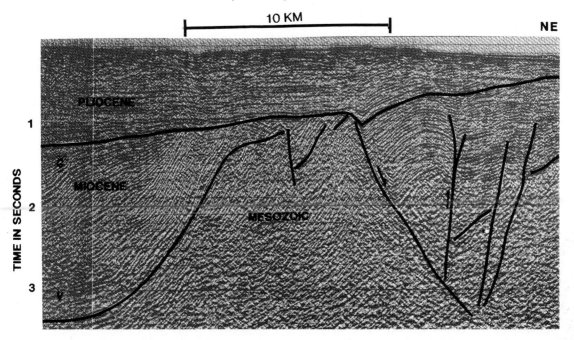

Fig. 6. Interpreted seismic reflection profile across portions of two half-grabens and intervening basement horst (profile located on Fig. 5). An erosional unconformity of Late Miocene age separates syn-extension and post-extension sections. High amplitude reflections (V) within the syn-extension section may be generated from interbedded volcanics and clastics. Slope clinoforms (C) may indicate prograding fan deposits. Syn- and post-extension strata have been locally inverted and folded.

to post-extension transition varies from Middle to Late Miocene in different parts of the basin.

The third major phase of Queen Charlotte Basin development involved Pliocene to Recent transpressional tectonism, associated with a shift from divergent (or parallel) strike-slip to convergent strike-slip plate interactions. Sub-basins were locally uplifted, extensional faults were reactivated with reverse displacements, and inversion folds and positive flower structures developed in both syn- and post-extension sediments. The northwestern basin flank (encompassing the present day area of Graham and Moresby Island; Fig. 2) was uplifted and eroded during the transpression phase.

CONCLUSIONS

The Queen Charlotte Basin is a 'strike-slip basin' following the general definition used by Mann *et al.* (1983) of a basin in which sedimentation is accompanied by significant strike-slip faulting. Evidence for strike-slip tectonism includes the similarity be-

tween mapped and shear-model-predicted structural trends, seismically-imaged geometries of offshore structures and onshore identification of strike-slip offset on the Louscoone and other fault zones. In a regional perspective, the basin consists of a complex mosaic of small and partly coalescing sub-basins, distributed across a 150- to 200-km-wide zone. The pattern of sub-basins is comparable to the Borderland region of California, where a network of small strike-slip basins occurs within a broad transform zone (Howell *et al.*, 1980). The tectonic evolution of the Queen Charlotte Basin is interpreted in terms of successive phases of transtension, subsidence and transpression, a typical development cycle for strike-slip basins (Reading, 1980).

Sedimentation in the basin occurred in a wide range of marine and non-marine environments. The increased abundance of marine deposits in the southern half of the basin in both syn- and post-extension sections indicates regional asymmetry in subsidence rates, possibly associated with longitudinal basin tilting (a common characteristic of strike-slip basins; Christie-Blick & Biddle, 1985). Typical of extensional tilt-block basins (Leeder & Gawthorpe, 1987), deposition in the Queen Charlotte Basin half-grabens and grabens included axial and lateral sediment-transport systems. The half-graben fills consist of variable proportions of footwall- and hanging-wall-derived deposits. Although not specifically diagnostic of a strike-slip basin, the diversity of depositional environments and abrupt lateral and vertical facies transitions do reflect the significant tectonic controls on local sedimentation patterns.

ACKNOWLEDGEMENTS

Access to industry seismic data and permission to publish the basin isopach map (derived, in part, from industry data) was kindly provided by Chevron Canada Resources Ltd and Shell Canada Ltd. Reviews of an initial draft of this paper by D. Cant, H. Olsen and E. Johannessen helped in our preparation of a revised text. Geological Survey of Canada contribution number 44490.

REFERENCES

BLAKE, M.C., CAMPBELL, R.H., DIBBLEE, T.W. *et al.* (1978) Neogene basin formation in relation to plate tectonic evolution of San Andreas Fault System, California. *Am. Assoc. Petrol. Geol. Bull.* **62/3**, 344–372.

CHRISTIE-BLICK, N. & BIDDLE, K.T. (1985) Deformation and basin formation along strike-slip faults. In: *Strike-slip Deformation, Basin Formation and Sedimentation* (Eds Biddle, K. & Christie-Blick, N.) Soc. Econ. Paleontol. Mineral. Spec. Publ. **37.**, 1–34.

CROWELL, J.C. (1974) Origin of Late Cenozoic basins in southern California. In: *Tectonics and Sedimentation* (Ed. Dickinson, W.R.), Soc. Econ.Paleontol. Mineral. Spec. Publ. **22**, 190–204.

ENGEBRETSON, D.C. (1989) Northeast Pacific–North America plate kinematics since 70 Ma. In: *Programme and Abstracts, Symposium on northeast Pacific-North America plate interactions throughout the Cenozoic*, Geological Association of Canada, pp.9–11.

ENGEBRETSON, D.C., COX, A. & GORDON, R.G. (1985) Relative motions between oceanic and continental plates in the Pacific basin. *Geol. Soc. Am. Spec. Pap.* **206**, 55 pp.

HARDING, T.P., VIERBUCHEN, R.C. & CHRISTIE-BLICK, N. (1985) Structural styles, plate-tectonic settings and hydrocarbon traps of divergent (transtensional) wrench faults. In: *Strike-slip Deformation, Basin Formation and Sedimentation* (Eds Biddle, K. & Christie-Blick, N.) Soc. Econ. Paleontol. Mineral. Spec. Publ. **37**, 51–71.

HIGGS, R. (1989) Sedimentological aspects of the Skonun Formation, Queen Charlotte Islands, British Columbia.

Geol. Sur. Can. Curr. Res. Pap. **89-1H**, 87–94.

HIGGS, R. (1991) Sedimentology, basin-fill architecture and petroleum geology of the Tertiary Queen Charlotte Basin, British Columbia. In: *Evolution and Hydrocarbon Potential of the Queen Charlotte Basin, British Columbia* (Ed. Woodsworth, G.) Geol. Sur. Can. Pap. **90-10**, 337–371.

HOWELL, D.G., CROUCH, J.K., GREENE, H.G., McCULLOCH, D.S. & VEDDER, J.G. (1980) Basin development along the Late Mesozoic and Cainozoic, California margin: a plate tectonic margin of subduction, oblique subduction and transform tectonics. In: *Sedimentation in Oblique-slip Mobile Zones* (Eds Ballance, P. and Reading, H.) *Int. Assoc. Sediment. Spec. Publ.* **4**, 43–62.

HYNDMAN, R.D. & HAMILTON, T.S. (1991) Cenozoic relative plate motions along the northeastern Pacific margin and their association with Queen Charlotte area tectonics and volcanism. In: *Evolution and Hydrocarbon Potential of the Queen Charlotte Basin, British Columbia* (Ed. Woodsworth, G.) Geo. Sur. Can. Pap. 90-10, 107–126.

LEEDER, M.R. & GAWTHORPE, R.L. (1987) Sedimentary models for extensional tilt-block/half-graben basins. In: *Continental Extensional Tectonics* (Eds Coward, M., Dewey, J. & Hancock, P.) Geol. Soc. London Spec. Publ. **28**, 139–152.

LEWIS, P.D. (1990) New timing constraints on Cenozoic deformation in the Queen Charlotte Islands, British Columbia. *Geol. Sur. Can. Curr. Re. Pap.* **90-1F**, 19–22.

LEWIS, P.D. (1991) Dextral strike-slip faulting and associated extension along the southern portion of the Louscoone Inlet fault system, southern Queen Charlotte Islands, British Columbia. Geol. Sur. Can. Curr. Res. Pap. **91-1A**, 383–391.

MANN, P., HEMPTON, M.R., BRADLEY, D.C. & BURKE, K. (1983) Development of pull-apart basins. *J. Geol.* **91**, 529–554.

PATTERSON, R.T. (1989) Neogene foraminiferal biostratigraphy of the southern Queen Charlotte Basin. In: *Contributions to Canadian Paleontology.* (Eds Norford, B.S. & Ollernshaw, N.C.) Geol. Surv. Can. Bull. **396**, 229–265.

READING, H.G. (1980) Characteristics and recognition of strike-slip systems. In: *Sedimentation in Oblique-slip Mobile Zones* (Eds Ballance, P. & Reading, H.) Int. Assoc. Sediment. **4**, 7–26.

RIDDIHOUGH, R.P., CURRIE, R.G. & HYNDMAN, R.D. (1980) The Delwood Knolls and their role in triple junction tectonics off northern Vancouver Island. *Can. J. Earth Sci.* **17**, 577–593.

ROHR, K.M. & DIETRICH, J.R. (1991) Deep seismic reflection survey of the Queen Charlotte Basin. In: *Evolution and Hydrocarbon Potential of the Queen Charlotte Basin, British Columbia* (Ed. Woodsworth, G.) Geol. Surv. Can. Pap. **90-10**, 127–133.

SHOULDICE, D.H. (1971) Geology of the western Canadian continental shelf. *Can. Soc. Petrol. Geol. Bull.* **19, no. 2**, 405–436.

SUTHERLAND BROWN, A. (1968) Geology of the Queen Charlotte Islands. *Br. Columbia Dep. Mines Petrol. Resour. Bull.* **54**, 226 pp.

TRON, V. & BRUN, J. (1991) Experiments on oblique rifting in brittle–ductile systems. *Tectonophysics* **188**, 71–84.

WHITE, J.M. (1990) Evidence of Paleogene sedimentation on Graham Island, Queen Charlotte Islands, west coast Canada. *Can. J. Earth Sci.* **27**, 533–538.

WHITE, J.M. (1991) Palynostratigraphy of the Tow Hill Well in the Skonun Formation, Queen Charlotte Basin. In: *Evolution and Hydrocarbon Potential of the Queen Charlotte Basin, British Columbia* (Ed. Woodsworth, G.), *Geol. Surv. Can. Pap.* **90-10**, 373–380.

Spec. Publs Int. Ass. Sediment. (1993) **20**, 171–181

Middle Devonian to Dinantian sedimentation in the Campine Basin (northern Belgium): its relation to Variscan tectonism

P. MUCHEZ*† *and* V. LANGENAEKER‡

* *Fysico-chemische geologie, K.U. Leuven, Celestijnenlaan 200C, B-3001 Heverlee, Belgium;*
†*Senior research assistant National Fund for Scientific Research; and*
‡*Historische geologie, K.U. Leuven, Redingenstraat 16bis, B-3000 Leuven, Belgium*

ABSTRACT

The relationship between synsedimentary faulting and sedimentation has been investigated in the Devonian–Dinantian strata of the Campine Basin (northern Belgium). The basin was formed during the Middle Devonian and was enhanced by widespread extensional faulting in the Chadian/Early Arundian and in the Asbian. A phase of inversion occurred at the end of the Dinantian. The Devonian is characterized by the deposition of siliciclastic sediments in one or more half-graben structures. Facies patterns developed in a S–N direction, with Chadian/Early Arundian siliciclastic carbonates deposited near the Brabant–Wales Massif, and open marine, shallow subtidal limestones bordered to the north by cryptalgal boundstones. During the Asbian, microbial buildups developed on the shelf and at the shelf margin. An approximately E–W trending growth fault possibly formed the boundary between this shelf and the basinal environment. On the shelf, synsedimentary faults are oriented subparallel to the northern margin of the Brabant–Wales Massif and are generally NW–SE trending in the western part and E–W trending in the eastern part. The Brabant–Wales Massif appears to have exerted a major influence on the fault directions.

The S–N trending facies patterns, the *c.* E–W oriented growth fault and the E–W direction of the extensional faults in the eastern part of the Campine Basin are compatible with N–S extension, active during the Middle Devonian to Dinantian interval.

Major synsedimentary tectonism during the Visean was broadly synchronous in northern England, North Wales, Ireland (Dublin area) and northern Belgium. Basin initiation began later in Britain and Ireland (Late Devonian). According to Leeder (1982) the N–S extension, which was active north of the Brabant–Wales Massif during part of the Devonian and during the Dinantian, occurred as a result of back-arc stretching to the north of an active plate collision zone, situated in the South Brittany–Massif Central–Vosges area. Synsedimentary tectonism in the Campine Basin is synchronous with the onset of compressional phases in the Rhenohercynian Basin and the closure of this back-arc basin.

INTRODUCTION

The role of extensional tectonics in Dinantian sedimentation has been well documented in northern England and North Wales (Bott, 1976, 1987; Leeder, 1976, 1982; Gawthorpe, 1987; Gutteridge, 1987; Gawthorpe *et al.*, 1989). Leeder (1982, 1987) suggested that this extensional tectonism occurred due to back-arc stretching in a zone north of active plate collision. However, a fixist model for the Variscides has been proposed by Paproth (1987).

Evidence for similar Dinantian tectono-sedimentary interactions in continental Europe is scarce (Muchez *et al.*, 1987a; Poty, 1989) and has neither been compared with evidence in Britain nor placed in a regional context. Recent correlations between the Rhenohercynian Variscides of Germany and southwest England indicate sedimentary and tectonic continuity between both parts of the Variscan belt (Franke & Engel, 1982; Holder & Leveridge,

1986) and make extrapolations of tectonic models over larger parts of Europe plausible. Earlier studies had suggested the absence of continuity between the two areas (Matthews, 1977, 1978).

During the Late Dinantian (Visean), an important difference existed in tectonic setting between the areas south and north of the Brabant–Wales Massif (Muchez, 1989). In the area north of this Massif, i.e. the Campine Basin, sedimentation took place in a block-faulted basin. The Visean strata of the Campine Basin are known only from boreholes, except at its eastern extension around Visé (Fig. 1), where several outcrops are present. Extensive exploration by seismic methods, however, has allowed the major fault trends in the Campine Basin to be determined (Vandenberghe, 1984; Dreesen et al., 1987). South of the Brabant–Wales Massif in Belgium, the Early Devonian (Gedinnian to Emsian) was characterized by extensional faulting (Meilliez, 1989) which evolved to compressional folding and

thrusting in Namurian and Westphalian times, with inversion of earlier extensional faults (Bless et al., 1989). A similar inversion model has been applied to the Irish and British Variscides (Price & Todd, 1988; Powell, 1989).

The aim of this paper is to:

1 document the relationship between synsedimentary faulting and sedimentation in the Campine Basin, north of the Brabant–Wales Massif in Belgium;

2 compare the periods of synsedimentary tectonism in Britain, Ireland and northern Belgium and to discuss the relation with large-scale extensional tectonism;

3 discuss the influence of pre-existing Caledonian structures on the fault directions;

4 integrate the data from sequences north and south of the Brabant–Wales Massif within the plate-tectonic evolution of western Europe.

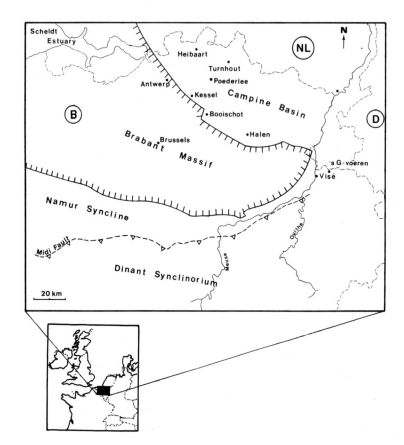

Fig. 1. Location of the studied boreholes: Visé, the Brabant Massif, the Namur syncline and the Dinant synclinorium of southern Belgium. The barbed lines mark the northern and southern border of the Dinantian strata.

EVOLUTION OF THE CAMPINE BASIN DURING THE DEVONIAN AND DINANTIAN

Reconstruction of the basin evolution is based on seismic reflection data, thickness variations and facies analysis. Seismic data (Belgian Geological Survey) show a WNW–ESE to WSW–ENE trending growth fault (called the Hoogstraten fault) in the Upper Carboniferous (Fig. 2), which separates two areas with different geological structures (Vandenberghe, 1984). Based on a comparison with the basins north of the Brabant–Wales Massif in England (Anderton *et al.*, 1979), Vandenberghe (1984) suggested that basin deepening, north of the Hoogstraten fault, may have started in the Dinantian with the deposition of Culm-type sediments. The area south of the Hoogstraten fault is characterized by NW–SE and NNW–SSE faults which cross-cut the top of the Visean strata. The Dinan-

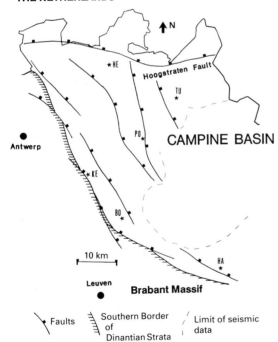

THE NETHERLANDS

10 km

Antwerp

CAMPINE BASIN

Leuven

Brabant Massif

Faults | Southern Border of Dinantian Strata | Limit of seismic data

Fig. 2. Diagram indicating location of major faults cross-cutting the top of the Dinantian in the Campine Basin, as identified in seismic profiles (partly after Dreesen *et al.*, 1987). Boreholes referred to are: Booischot (Bo), Halen (Ha), Heibaart (He), Kessel (Ke), Poederlee (Po) and Turnhout (Tu).

tian itself is characterized by the almost complete absence of good reflectors due to the high acoustic impedance at the contact of Namurian shales and Dinantian carbonates, and consequent loss of seismic wave energy (Dreesen *et al.*, 1987).

Although synsedimentary faulting during the Dinantian cannot be identified from the seismic profiles, borehole data indicating important thickness and/or facies variations across faults are taken as evidence favouring synsedimentary movement. In this text the stages of the geological time scale of Harland *et al.* (1989) are used.

The oldest sediments above the Caledonian basement are Givetian (Middle Devonian) in age in the Booischot borehole, towards the south, and Frasnian (Late Devonian) in age in the Heibaart borehole to the north (Legrand, 1964; Bless *et al.*,1976; Streel & Loboziak, 1987). The Middle Devonian, comprising 203 m of red continental conglomerates in the Booischot borehole (Table 1; Fig. 3), is absent in the more northerly situated Heibaart borehole (Bless *et al.*, 1976). This sedimentation pattern may reflect either the influence of a half-graben structure (Fig. 4a) or several grabens.

During the Late Devonian 175 m of Frasnian marine shales and Famennian sandstones were deposited in the Heibaart area (Table 1). This contrasts with the 385 m of Frasnian conglomerates, sandstones and shales and Famennian shales and sandstones in the Booischot borehole (Legrand, 1964). Although the tectonic configuration remained unchanged from the Middle Devonian (Fig. 4B), comparable thicknesses of Famennian shales and sandstones in both areas suggest that tilting or block-faulting became inactive by the end of the Late Devonian. Strata of Hasterian and Ivorian age have not been recognized in the Campine Basin. Whether this is due to non-deposition is uncertain, since the dolostones above the Famennian have not yielded biostratigraphical dates.

The Visean strata occur in several boreholes (Fig. 1) and can be assigned to five lithostratigraphical megasequences: Chadian–Early Arundian, Late Arundian, Holkerian, Asbian and Brigantian. However, the Brigantian is absent from the Booischot, Heibaart, Kessel and Poederlee boreholes and is intensively modified by dolomitization, neomorphism and karstification in the Turnhout borehole.

Close to the Brabant–Wales Massif, the Chadian–Early Arundian sequence consists of shallow, subtidal open marine bioclastic wackestones and packstones (Fig. 5A) overlain by a restricted la-

SEQUENCE \ BOREHOLE	Booischot	Halen	Heibaart	Kessel	Poederlee	Turnhout
BRIGANTIAN		201 m				34 m
ASBIAN		155 m	>142 m		>145 m	53 m
HOLKERIAN OR REWORKED SEQUENCE		20 m	42 m			23 m
LATE ARUNDIAN	>131 m	94 m	125 m	> 29 m		>10 m
CHADIAN / EARLY ARUNDIAN	64 m	110 m	0-32 m	> 82 m		>417 m
DOLOSTONES OF UNKNOWN AGE	66 m	181 m	n.p.			
FRASNIAN / FAMENNIAN	385 m		175 m			
GIVETIAN	203 m		0 m			

Table 1. Variation in the thickness of the sequences or stages of the Devonian and Dinantian in the different boreholes. (After Legrand, 1964; Bless *et al.*, 1976; Muchez *et al.*, 1987a.)

—·—·— erosion surface; ———— end of borehole; n.p., not present;
············ biostratigraphical dating problems.

goonal assemblage of often sandy, pelitic wackestones and packstones with a low biotic diversity (Fig. 5B). Periods of sedimentation in a restricted environment alternated with the development of dolocretes in supratidal marshes (Muchez & Viaene, 1987). North of this area, at Heibaart, there was little or no sedimentation (Table 1). A synsedimentary NW–SE trending fault, parallel to the Brabant-Wales Massif, appears to have separated these areas (Fig. 6a). The limited sedimentation upon the Heibaart High contrasts with the sedimentation at Turnhout to the east. Here, several hundred metres (>400 m) of intertidal or very shallow subtidal, planar laminated cryptalgal boundstones (Fig. 5C) and shallow subtidal peloidal packstones and grainstones with an open-marine biota (Fig. 5D) were deposited. At Halen, southeast of Turnhout, between 100 m and perhaps up to 290 m of bioclastic packstones and grainstones were deposited above the fairweather wave base. A NNW–SSE fault, recognized on seismic reflection profiles, probably separated the Heibaart High from the Turnhout and Halen areas during the Chadian and Early Arundian. The position of the intertidal to very shallow subtidal boundstones, approximately 25 km north of a coastal area, can be compared with the position of the loferites in the Northern Limestone Alps found on tidal flats near the margin of a shelf (Zankl, 1971). This view is consistent with the presence of the Hoogstraten fault, 8 km north of Turnhout (Fig. 2), which could have formed the boundary between the shelf and deep water of the Campine Basin (Vandenberghe, 1984).

The Late Arundian is characterized by sedimen

tation of peloidal, bioclastic packstones and grainstones on an open marine shelf (Fig. 6b). These are of uniform thickness except in the Turnhout area where erosion occurred during the next stage. Periods of deposition of open-marine shelf limestones alternated with storm deposits formed in a more restricted environment near the Brabant–Wales Massif.

During the Holkerian (Fig. 6c), the Heibaart High became an area where open-marine bioclastic wackestones (Fig. 5E) were deposited below significant wave activity (Muchez *et al.*, 1987b). Limestone fragments of Chadian or Arundian age surrounded by a Holkerian matrix (Fig. 5F; Muchez, 1988) and storm deposits (Fig. 7A) occur in the reworked sequence of the Halen borehole. Lithoclasts have also been found in the Heibaart area. In the Turnhout borehole, the reworked sequence, characterized by foraminifera of Chadian/ Arundian to Early Asbian age (Bless *et al.*, 1976), and the absence of unequivocally Holkerian strata indicate that there was little sedimentation in this area during this stage. The presence of reworked fragments and the establishment of a deeper depositional environment at Heibaart compared with Turnhout are both consistent with reactivation of the faults during the Holkerian.

The contrast in sedimentation between the Turnhout and the Heibaart–Poederlee areas continued in the Asbian. In the Turnhout area, 53 m of cryptalgal boundstones and bioclastic packstones, grainstones and rudstones were deposited. They are interpreted as reef-associated sediments by Mamet *et al.* (1978), although no real buildup has been

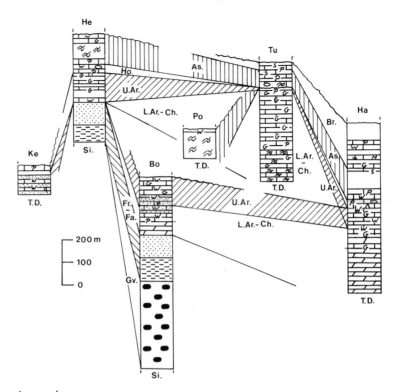

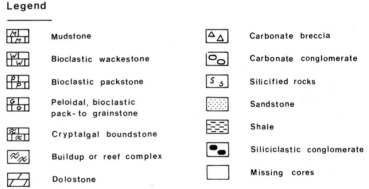

Fig. 3. Lithological logs of the boreholes Booischot (Bo), Halen (Ha), Heibaart (He), Kessel (Ke), Poederlee (Po) and Turnhout (Tu). Data from Legrand (1964), Bless *et al.* (1976) and Muchez *et al.* (1987a). T.D. Terminal depth; Si. Silurian; Gv. Givetian; Fr. Frasnian; Fa. Famennian; Ch. Chadian; L.Ar. Lower Arundian; U.Ar. Upper Arundian; Ho. Holkerian; As. Asbian: Br. Brigantian.

Legend

Mudstone	Carbonate breccia
Bioclastic wackestone	Carbonate conglomerate
Bioclastic packstone	Silicified rocks
Peloidal, bioclastic pack- to grainstone	Sandstone
Cryptalgal boundstone	Shale
Buildup or reef complex	Siliciclastic conglomerate
Dolostone	Missing cores

recognized. As in the Visé area (Muchez & Peeters, 1986), a high energy environment and relatively low subsidence rates probably limited its growth. In the Heibaart–Poederlee area more than 145 m of microbial reef complex were constructed on the shelf (Fig. 7B) and more than 140 m of microbial buildup at the shelf margin (Fig. 7C, D), as previously described by Muchez *et al.* (1987a, b, 1990). This area of strong subsidence was separated from the Turnhout area by the NNW–SSE trending fault (Fig. 6d). Southwest of Turnhout, in the Halen area,

a thick sequence (155 m) of Asbian limestones was cyclically deposited in open marine and restricted environments. The difference between the Turnhout and Halen areas could be the result of southwards tilting of a block containing both areas, or could be explained by several synsedimentary faults parallel to the Brabant–Wales Massif (Fig. 6d). The Hoogstraten fault just north of the Heibaart buildup most likely formed the boundary between the shelf and the basin. Additional evidence for a 'basinal' environment north of this fault comes from the

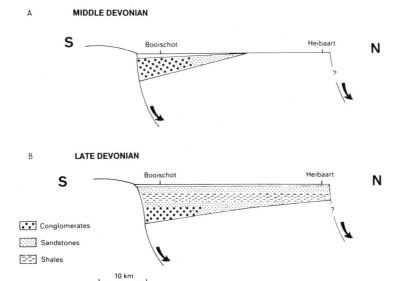

Fig. 4. Half-graben model for the Devonian of the Campine Basin: (A) Middle Devonian; (B) Late Devonian.

Gravenvoeren borehole at the eastern extremity of the Brabant–Wales Massif (Fig. 1). Here, the Asbian is characterized by mass flow deposits (Muchez, 1988). These deposits are situated just north of a shallow shelf around Visé.

The Brigantian is absent in several boreholes (e.g. Heibaart, Poederlee). This could be due to erosion or non-deposition during a widespread regression coinciding with the tectonic movements of the Sudetic Phase at the end of the Visean (Graulich, 1963). This regression and tectonic phase is also responsible for the stratigraphical gap encompassing the Pendleian to Arnsbergian and locally even to the Kinderscoutian along the northern flank of the Brabant Massif (Bouckaert, 1967). The thick Brigantian in the Halen borehole (>201 m, Table 1) must have been deposited in a strongly subsiding part within the shelf. In Derbyshire (England) a comparably thick Brigantian sequence developed within an intrashelf basin (Gutteridge, 1989).

We conclude that in the Campine Basin, tilted-block tectonism started in the Middle Devonian and that widespread synsedimentary extensional faulting occurred during the Chadian/Early Arundian and during the Holkerian/Asbian.

SYNSEDIMENTARY FAULT ACTIVITY AND LARGE-SCALE EXTENSIONAL TECTONISM

The periods of synsedimentary fault activity in

northern England–North Wales and Ireland (Dublin area) have been investigated by Gawthorpe *et al.* (1989) and Nolan (1989) respectively. Basin initiation in the Late Devonian and Tournaisian was followed by further widespread extensional reactivation in the Late Chadian/Early Arundian and in the Asbian. Comparison with the data of the Campine Basin (Fig. 8) shows that basin initiation started earlier in northern Belgium. However, major extension occurred during the same stages of the Visean.

Three conflicting tectonic models have been proposed for the Late Devonian and Dinantian basin development:

1 large-scale dextral shear (e.g. Badham, 1982);
2 E–W extension (e.g. Russell, 1971, 1976; Hazeldine, 1984, 1988, 1989);
3 N–S extension (e.g. Leeder, 1976, 1982; Bott, 1987).

Indeed, strike-slip faulting within a right-lateral shear zone extending between the Appalachians and the Urals is postulated for the Late Carboniferous–Early Permian (Arthaud & Matte, 1977). However, this does not justify an extrapolation to Dinantian times (see also Bott, 1987).

In the Campine Basin, facies patterns developed along S–N trends but the faults on the shelf have a NW to NNW direction, subparallel to the Brabant–Wales Massif (Fig. 2). In the eastern part and at the eastern end of the Campine Basin (around Visé), the faults still run parallel to the northern margin of the Massif but have an approximately E–W direc-

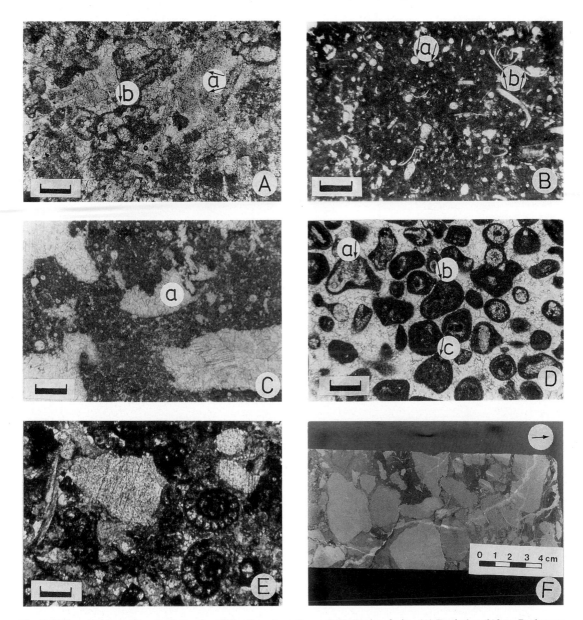

Fig. 5. Slab and thin-section photographs of Chadian, Arundian and Holkerian facies. (A) Booischot 610 m. Packstone with crinoids (a) and foraminifera (b). Scale bar = 360 μm. (B) Booischot 594.8 m. Wackestone with calcispheres (a) and ostracods (b). Scale bar = 180 μm. (C) Turnhout 2647 m. Cryptalgal boundstone with fenestrae (a). Scale bar = 360 μm. (D) Turnhout 2324 m. Grainstone with micritized and coated crinoids (a), grapestones (b) and peloids (c). Scale bar = 360 μm. (E) Heibaart 1277 m. Wacke- to packstone with an open-marine biota. Scale bar = 240 μm. (F) Halen 980 m. Conglomerate with fragments of Chadian or Arundian age in a Holkerian matrix.

tion (Poty, 1989). The change of the fault directions along, and parallel to, the northern margin of the Brabant–Wales Massif indicates that this Caledonian structure exerted an important control on the orientation of the faults. In northern England and North Wales, the regional structural style is also related to the arcuate trend of the Caledonian basement (Leeder, 1982).

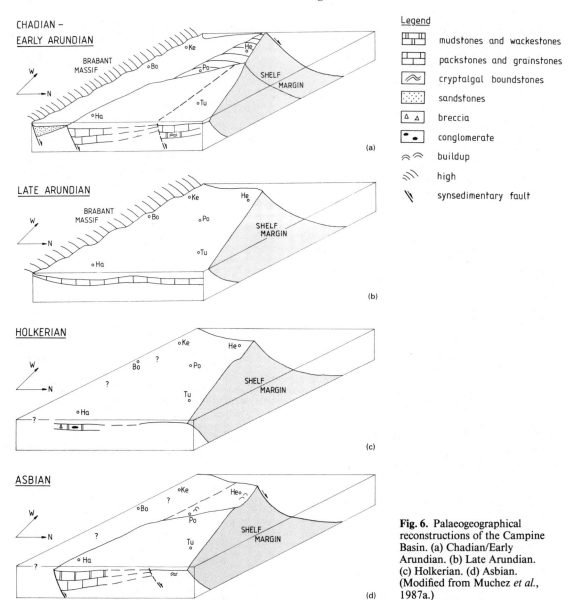

Fig. 6. Palaeogeographical reconstructions of the Campine Basin. (a) Chadian/Early Arundian. (b) Late Arundian. (c) Holkerian. (d) Asbian. (Modified from Muchez *et al.*, 1987a.)

The major structures, such as the approximately E–W trending Hoogstraten fault, the E–W orientation of the faults in the eastern part of the Campine Basin and the N–S facies changes are consistent with a N–S extension. Although an oblique slip displacement along the NW–SE and NNW–SSE faults would be predicted by this model, this cannot be tested with the available data.

CONCLUSIONS

Synsedimentary faults were active in the Campine Basin (northern Belgium) during the Devonian and Dinantian. Basin formation began in the Middle Devonian and was followed by extensional reactivation in the Chadian/Early Arundian and in the Holkerian/Asbian. An inversion phase occurred at

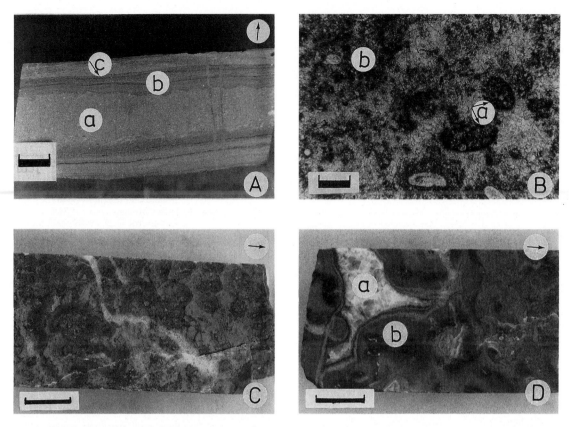

Fig. 7. Slab and thin-section photographs of Holkerian and Asbian facies. (A) Halen 976 m. Storm deposit. The lower part of this deposit consists of a bioclastic packstone (a) and the upper part is a clayey bioclastic wackestone (b) and a mudstone (c). Scale bar = 1 cm. (B) Poederlee 1645.5 m. Peloidal (a) and clotted (b) textures typical of the core of a reef mound. Scale bar = 360 µm. (C) Heibaart 1174 m. Globular framework in the core of a buildup. Scale bar = 2 cm. (D) Heibaart 1164 m. Large stromatactoid cavity (a) surrounded by lime mud (b). The irregular mottled pattern of this mud suggests an organic origin. Scale bar = 2 cm.

the end of the Dinantian. A major E–W fault may have formed the boundary between the shelf and the basinal environment. On the shelf, synsedimentary faults have a NW–SE and an E–W trend, subparallel to the northern margin of the Brabant–Wales Massif. Indeed, the Brabant–Wales Massif appears to have exerted a major influence on fault orientations within the basin.

The phases of major synsedimentary tectonism in northern England, North Wales and Ireland (Dublin area) were broadly coeval with those in northern Belgium. However, basin initiation began earlier in the Campine Basin. Facies patterns and major fault directions are consistent with N–S extension during the Middle and Late Devonian and the Dinantian. According to Leeder (1982, 1987) this type of

tectonism occurred due to back-arc stretching to the north of an active plate collision zone, situated in the South Brittany–Massif Central–Vosges area.

In the Rhenohercynian Basin in the southern part of Belgium, extensional faulting was important during the Early Devonian (Meilliez, 1989). An interplay between back-arc extension and back-arc compression started during the Middle Devonian and an important compressional phase straddled the Devonian–Carboniferous boundary in the Rhenohercynian Basin (Ziegler, 1984, 1989). Extensional faulting within the Campine Basin is of Middle Devonian to Dinantian age and is therefore synchronous with the onset of compressional phases in the Rhenohercynian Basin and the closure of this back-arc basin.

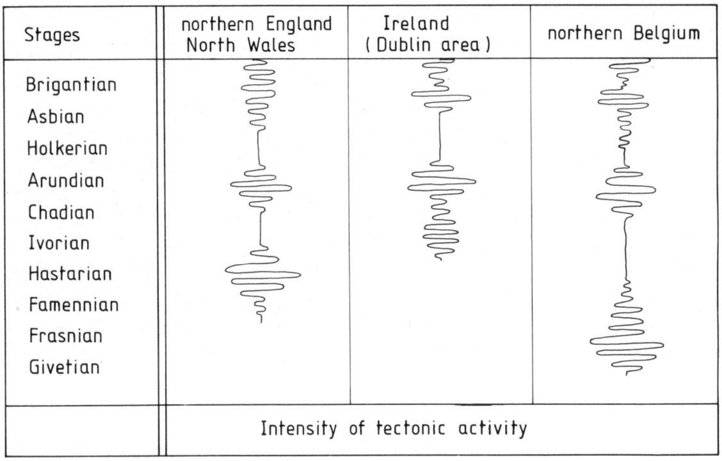

Fig. 8. Comparison of the phases of tectonic activity in the Middle Devonian to Visean basin evolution in northern England, North Wales, Ireland and northern Belgium. (After Muchez *et al.*, 1987a; Gawthorpe *et al.*, 1989; Nolan, 1989.)

ACKNOWLEDGEMENTS

We are grateful to M. Dusar, M. Wilkinson, R. Swennen, N. Vandenberghe, W. Viaene and J. Walsh for their constructive comments on earlier versions of this manuscript. We thank J. Bouckaert, General Inspector of the Geological Survey of Belgium for permission to study the cores and to use the seismic date. Mr C. Moldenaers kindly prepared the thin sections and Ms G. Willems drew the figures.

We also thank Kluwer Academic Publishers for permission to reprint with modification Fig. 6 and to use published data in Table 1 and Figs 3 and 8.

Philippe Muchez gratefully acknowledges the grant of a NATO fellowship which enabled this text to be prepared.

REFERENCES

ANDERTON, R., BRIDGES, P.H., LEEDER, M.R. & SELLWOOD, B.W. (1979) *A Dynamic Stratigraphy of the British Isles.* George Allen & Unwin, London, 301 pp.

ARTHAUD, F. & MATTE, PH. (1977) Late Paleozoic strike-slip faulting in southern Europe and northern Africa: result of a right-lateral shear zone between the Appalachians and the Urals. *Geol. Soc. Am. Bull.* **88**, 1305–1320.

BADHAM, J.P.N. (1982) Strike-slip orogens — an explanation for the Hercynides. *J. Geol. Soc. London* **139**, 493–504.

BLESS, M.J.M., BOUCKAERT, J. & PAPROTH, E. (1989) The Dinant Nappes: a model of tensional listric faulting inverted into compressional folding and thrusting. *Bull. Soc. Belge Géol.* **95**, 89–99.

BLESS, M.J.M., BOUCKAERT, J., BOUZET, PH. *et al.* (1976) Dinantian rocks in the subsurface north of the Brabant and Ardenno–Rhenish massifs in Belgium, the Netherlands and the Federal Republic of Germany. *Meded. Rijks Geol. Dienst* **27**, 81–195.

BOTT, M.H.P. (1976) Formation of sedimentary basins of graben type by extension of the continental crust. *Tectonophysics* **36**, 77–86.

BOTT, M.H.P. (1987) Subsidence mechanisms of Carboniferous basins in northern England. In: *European Dinantian Environments* (Eds Miller, J., Adams, A.E. & Wright, V.P.) John Wiley, New York, pp. 21–32.

BOUCKAERT, J. (1967) Namurian transgression in Belgium. *Ann. Soc. Géol. Pologne* **37**, 145–150.

DREESEN, R., BOUCKAERT, J., DUSAR, M., SOILLE, J. & VANDENBERGHE, N. (1987) Subsurface structural analysis of the Late-Dinantian carbonate shelf at the northern flank of the Brabant Massif (Campine Basin, N-Belgium). *Toelicht. Verhand. Geologische en Mijnkaarten van België* **21**, 37 pp.

FRANKE, W. & ENGEL, W. (1982) Variscan sedimentary basins on the continent and relations with south-west England. *Proc. Ussher Soc.* **5**, 259–269.

GAWTHORPE, R.L. (1987) Tectono-sedimentary evolution of the Bowland Basin, N. England, during the Dinantian. *J. Geol. Soc. London* **144**, 59–71.

GAWTHORPE, R.L., GUTTERIDGE, P. & LEEDER, M.R. (1989) Late Devonian and Dinantian basin evolution in northern England and North Wales. In: *The Role of Tectonics in Devonian and Carboniferous Sedimentation in the British Isles* (Eds Arthurton, R.S., Gutteridge, P. & Nolan, S.C.) *Yorks. Geol. Soc. Occas. Publ.* **6**, 1–23.

GRAULICH, J.-M. (1963) La phase sudète de l'orogène varisque dans le synclinorium de Namur à l'est du Samson. *Bull. Soc. Belge Géol.* **71**, 181–199.

GUTTERIDGE, P. (1987) Dinantian sedimentation and the basement structure of the Derbyshire Dome. *Geol. J.* **22**, 25–41.

GUTTERIDGE, P. (1989) Controls on carbonate sedimentation in a Brigantian intrashelf basin (Derbyshire). In: *The Role of Tectonics in Devonian and Carboniferous Sedimentation in the British Isles* (Eds Arthurton, R.S., Gutteridge, P. & Nolan, S.C.) *Yorks. Geol. Soc. Occas. Publ.* **6**, 171–187.

HARLAND, W.B., ARMSTRONG, R.L., COX, A.V., CRAIG, L.E., SMITH, A.G. & SMITH, D.G. (1989) *A Geological Time Scale 1989*. Cambridge University Press, Cambridge.

HASZELDINE, R.S. (1984) Carboniferous North Atlantic palaeogeography: stratigraphic evidence for rifting, not megashear or subduction. *Geol. Mag.* **121**, 443–463.

HASZELDINE, R.S. (1988) Crustal lineaments in the British Isles: their relationship to Carboniferous basins. In: *Sedimentation in a Synorogenic Basin Complex* (Eds Besly, B.M. & Kelling, G.) Blackie, Glasgow, pp. 53–68.

HASZELDINE, R.S. (1989) Evidence against crustal stretching, north–south tension and Hercynian collision, forming British Carboniferous basins. In: *The Role of Tectonics in Devonian and Carboniferous Sedimentation in the British Isles* (Eds Arthurton, R.S., Gutteridge, P. & Nolan, S.C.) *Yorks. Geol. Soc. Occas. Publ.* **6**, 25–33.

HOLDER, M.T. & LEVERIDGE, B.E. (1986) Correlation of the Rhenohercynian Variscides. *J. Geol. Soc. London* **143**, 141–147.

LEEDER, M.R. (1976) Sedimentary facies and the origins of basin subsidence along the northern margin of the supposed Hercynian Ocean. *Tectonophysics* **36**, 167–179.

LEEDER, M.R. (1982) Upper Palaeozoic basins of the British Isles — Caledonide inheritance versus Hercynian plate margin processes. *J. Geol. Soc. London* **139**, 479–491.

LEEDER, M.R. (1987) Tectonic and palaeogeographic models for Lower Carboniferous Europe. In: *European Dinantian Environments* (Eds Miller, J., Adams, A.E. & Wright, V.P.) Wiley, New York, pp. 1–20.

LEGRAND, R. (1964) Coupe résumée du forage de Booischot (province d'Anvers). *Bull. Soc. Belge Géol. Paleont. Hydr.* **72**, 407–409.

MAMET, B., MORTELMANS, G. & ROUX, A. (1978) Algues viséennes du sondage de Turnhout (Campine, Belgique). *Ann. Soc. Géol. Belge.* **101**, 351–383.

MATTHEWS, S.C. (1977) The Variscan foldbelt in southwest England. *N. Jb. Geol. Paläont. Abh.* **154**, 94–127.

MATTHEWS, S.C. (1978) Caledonian connexions of Variscan tectonism. *Z. Dtsch. Geol. Ges.* **129**, 423–428.

MEILLIEZ, F. (1989) Tectonique distensive et sédimentation à la base du Dévonien, en bordure NE du Massif de Rocroi (Ardenne). *Ann. Soc. Géol. Nord* **57**, 281–295.

MUCHEZ, PH. (1988) *Sedimentologische, diagenetische en geochemische studie van de Dinantiaan strata ten noorden van het Brabant Massief (Bekken van de Kempen)*. PhD thesis, Katholieke Universiteit Leuven, Belgium, 311 pp.

MUCHEZ, PH. (1989) Comparison of the Visean strata north and south of the London–Brabant Massif. *Bull. Soc. Belge Géol.* **98**, 155–162.

MUCHEZ, PH. & PEETERS, C. (1986) The occurrence of a cryptalgal reef structure in the Upper Visean of the Visé area (The Richelle quarries). *Ann. Soc. Géol. Belge.* **109**, 573–577.

MUCHEZ, PH. & VIAENE, W. (1987) Dolocretes from the Lower Carboniferous of the Campine–Brabant Basin, Belgium. *Pedologie* **37**, 187–202.

MUCHEZ, PH., VIAENE, W., WOLF, M. & BOUCKAERT, J. (1987a) Sedimentology, coalification pattern and paleogeography of the Campine–Brabant Basin during the Visean. *Geol. Mijnbouw* **66**, 313–326.

MUCHEZ, PH., CONIL, R., VIAENE, W., BOUCKAERT, J. & POTY, E. (1987b) Sedimentology and biostratigraphy of the Visean carbonates of the Heibaart (DzH1) borehole (northern Belgium). *Ann. Soc. Géol. Belge.* **110**, 199–208.

MUCHEZ, PH., VIAENE, W., BOUCKAERT, J. *et al.* (1990) The occurrence of a microbial buildup at Poederlee (Campine Basin, Belgium): biostratigraphy, sedimentology, early diagenesis and significance for Early Warnantian paleogeography. *Ann. Soc. Géol. Belge.* **113**, 329–339.

NOLAN, S.C. (1989) The style and timing of Dinantian synsedimentary tectonics in the eastern part of the Dublin Basin, Ireland. In: *The Role of Tectonics in Devonian and Carboniferous Sedimentation in the British Isles* (Eds Arthurton, R.S., Gutteridge, P. & Nolan, S.C.) *Yorks Geol. Soc. Occas. Publ.* **6**, 83–97.

PAPROTH, E. (1987) The Variscan front north of the Ardenne–Rhenish Massifs. *Ann. Soc. Géol. Belge.* **110**, 279–296.

POTY, E. (1989) Similar tectono-sedimentary evolutions and important lateral changes in a block-faulting system. *Ann. Soc. Géol. Belge.* **112**, 250–251.

POWELL, C.M. (1989) Structural controls on Palaeozoic basin evolution and inversion in southwest Wales. *J. Geol. Soc. London* **146**, 439–446.

PRICE, C.A. & TODD, S.P. (1988) A model for the development of the Irish Variscides. *J. Geol. Soc. London* **145**, 935–939.

RUSSELL, M.J. (1971) North–south geofractures in Scotland and Ireland. *Scott. J. Geol.* **8**, 75–84.

RUSSELL, M.J. (1976) A possible Lower Permian age for the onset of ocean floor spreading in the northern North Atlantic. *Scott. J. Geol.* **13**, 315–323.

STREEL, M. & LOBOZIAK, S. (1987) Nouvelle datation par miospores du Givetien-Frasnien des sédiments non marins du sondage de Booischot (Bassin de Campine, Belgique). *Bull. Soc. Belge Géol.* **96**, 99–106.

VANDENBERGHE, N. (1984) The subsurface geology of the Meer area in North Belgium, and its significance for the occurrence of hydrocarbons. *J. Petrol. Geol.* **7**, 55–66.

ZANKL, H. (1971) Upper Triassic carbonate facies in the northern Limestone Alps. In: *Sedimentology of Parts of Central Europe* (Ed. Müller, G.) pp. 147–185. Guidebook VIII Int. Sed. Congr.

ZIEGLER, P.A. (1984) Caledonian and Hercynian crustal consolidation of Western and Central Europe — a working hypothesis. *Geol. Mijnbouw* **63**, 93–108.

ZIEGLER, P.A. (1989) *Evolution of Laurussia*. Kluwer Academic Publishers, Dordrecht, 102 pp.

Spec. Publs Int. Ass. Sediment. (1993) **20**, 183–202

A basin reappraisal of the Proterozoic Torridon Group, northwest Scotland

P.G. NICHOLSON*

*Department of Geology and Applied Geology,
University of Glasgow, Glasgow, Scotland G12 8QQ, UK*

ABSTRACT

Existing models of Torridon Group (Applecross Formation) genesis suggest deposition by at least two large alluvial fans within a 75-100-km-wide rift graben cutting Archaean crust. Results presented here reveal the presence of large-scale bar structures (up to 9 m thick) within the alluvium, indicating the occurrence of major rivers within the fluvial system. Palaeocurrent analyses show a greater degree of variability in sediment transport direction, at outcrop level, than previously recognized, whereas regional palaeoflow vectors do not exhibit radial patterns nor do they suggest that any active basin-bounding faults influenced sedimentation.

An alluvial braidplain origin for the Applecross Formation is preferred to the existing alluvial fan model, based on the evidence for rivers over 500 km long (determined by a palaeohydrological reconstruction), together with the consistently E–SE regional palaeoflow. These scale considerations, combined with the lack of evidence for both syndepositional tectonism within the basin and for active basin-bounding fault margins, collectively favour deposition within a significantly larger basin than the relatively self-contained rift graben previously proposed. A larger scale, later stage, extensional basin formed by post-rifting thermal relaxation processes, accompanied by a component of passive subsidence-driven normal faulting along the Outer Isles Fault, is proposed for deposition of the Torridon Group.

INTRODUCTION

'Torridonian' is an informal term used to refer to the entire Middle to Upper Proterozoic succession of predominantly fluvial clastics (red beds) situated along the northwest coast of Scotland. These essentially undeformed and unmetamorphosed sediments rest unconformably on Archaean to Lower Proterozoic Lewisian Gneiss, and crop out in a belt roughly 20–30 km wide by 200 km long stretching from Cape Wrath in the north-northeast to the Isle of Rhum in the south-southwest (Fig. 1). The aim of this paper is to re-evaluate the nature of the basin in which the Torridon Group, which constitutes the majority of this alluvium, was deposited.

Stratigraphical setting

Stratigraphically, 'Torridonian' deposits are subdivided into the Stoer, Sleat and Torridon groups (Fig. 2). The Stoer and Torridon groups are separated by an angular unconformity (Lawson, 1965; Stewart, 1969) and have previously been dated at 968 and 777 Ma respectively (Moorbath, 1969; Stewart, 1982), although some authors note that these Rb–Sr whole-rock ages may be anywhere from 100 to 250 Ma too young (Smith *et al.*, 1983; Stewart, 1988a, p.98). Recent provenance work by Rogers *et al.* (1989 and personal communication, discussed further below) also suggests that the Torridon Group age could be as much as 200 My too young. A sedimentation date of *c.* 900 Ma for the Torridon Group is, therefore, likely to be more realistic, although it is accepted that the original

*Present address: Thai Shell Exploration and Production Company, P.O. Box 345, Bangkok 10501, Thailand.

Rb–Sr ages (Moorbath, 1969; Stewart, 1982) could equally be too old. The Sleat Group is found only within the Kishorn Nappe of Skye, having been transported westwards from its original position by Caledonian compression. This group is undated (it has suffered greenschist metamorphism), but conformably underlies the Torridon Group (Stewart, 1988b). Reviews of the Stoer and Sleat Groups are provided by Stewart (1988a, 1988b) and Nicholson (1992a).

Dominating the 'Torridonian' outcrop belt are the coarse, red, fluvial arkoses of the Torridon Group. This group reaches a maximum stratigraphical thickness of between 5 and 6 km, and is subdivided into Diabaig (0.6 km maximum thickness), Applecross (c. 3.0 km thick), Aultbea (c. 1.5 km thick) and Cailleach Head (0.8 km thick) formations (Stewart, 1988b, fig. 9.3; Fig. 2). At the base of this succession lie the grey shales and siltstones (and local breccias and sandstones) of the Diabaig Formation, which formed in topographical lows on the Lewisian Gneiss basement. Diabaig deposits are overstepped by the coarse sandstones

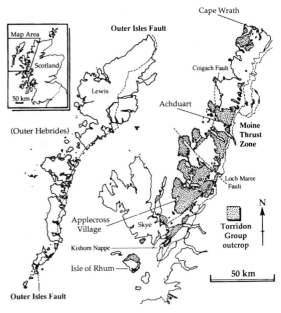

Fig. 1. Location map showing distribution of Torridon Group outcrops.

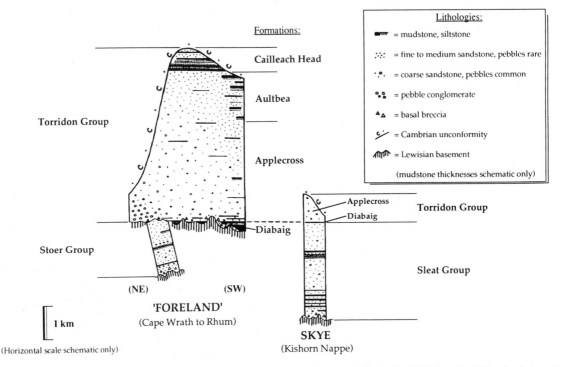

Fig. 2. Schematic stratigraphy of the 'Torridonian'. (Modified after Stewart 1969, fig. 2; 1975, fig. 10; 1988a, fig. 8.4; 1988b, fig. 9.3.) Torridon Group lithologies based on Nicholson (1992b).

and pebble conglomerates of the Applecross Formation. Applecross sediments gradually pass upwards into more homogeneous, medium-grained, Aultbea sandstones. The overlying grey shale-to-sandstone cyclothems of the Cailleach Head Formation are thought to represent freshwater deltaic deposits (Stewart, 1988b).

The vast majority of Torridon Group outcrops belong to the Applecross Formation, and hence the basin reappraisal presented here focuses on the characteristics of these deposits.

Existing models of Torridon Group deposition

Existing interpretations of Torridon Group deposition are based on two models published by Williams (1966, 1969a) and Stewart (1982).

Based on his analysis of textural and palaeocurrent variations within Applecross alluvium of the Cape Wrath region, Williams (1966, 1969a) presented a palaeogeographical reconstruction for the Applecross Formation which suggested that it was deposited by at least two large coalescing alluvial fans formed at the base of a northwestwardly-retreating upland source area. By combining other workers' regional palaeoflow vectors with the results of his own analysis in the Cape Wrath district, Williams obtained radial sediment transport patterns within the Applecross alluvium and also concluded that vertical and lateral (downslope) textural fining within the Cape Wrath succession further suggested deposition by retreating fans.

Stewart (1982) proposed that both the Stoer and Torridon groups were deposited in fault-bounded rift basins created by extension within the Lewisian basement. In his model, the Torridon Group basin consisted of a 75–100-km-wide graben bounded to the west-northwest by the Minch Fault, and to the east-southeast by a normal fault that was subsequently reactivated (during Caledonian compression) to become the Moine Thrust. Stewart indicated that sediment of Laxfordian age (*c.* 1500–1700 Ma) was supplied to this basin from the flanking west-northwest horst block, which served as the upland source for the large alluvial fans proposed by Williams (1966, 1969a).

Study approach

Much of the Applecross Formation was re-examined during the parent study (Nicholson, 1992b) in order to evaluate the validity of existing depositional models. The parent study broadly consists of two levels: detailed analyses of individual sandbodies and stratigraphical logs in specific localities, and a regional analysis of lithosome characteristics (facies, palaeoflow, etc.) along the entire outcrop belt length. For the purposes of this Torridon Group basin reappraisal, the following aspects of Applecross alluvium are considered: possible evidence for syndepositional tectonism, local and regional palaeoflow variability, and scale of the preserved bar structures.

DESCRIPTION OF APPLECROSS ALLUVIUM

A 20 m stratigraphical log from the Applecross Formation is shown in Fig. 3. This log, taken from a locality at the northern end of the Applecross peninsula (type area), is representative of the majority of the formation, and is situated approximately 650 m stratigraphically above the local Lewisian unconformity.

Textures and structures

The Applecross lithosome predominantly consists of coarse-grained heterogeneous sandstone which is often very coarse to granular. It contains sporadic well-rounded pebbles, typically 10–30 mm in diameter, of 'exotic' (non-Lewisian) lithologies (see Williams (1969b) for a description of pebble types). Subordinate amounts of homogeneous medium- to fine-grained sandstone, siltstone and mudstone are also typical. In the extreme northern (Cape Wrath) region of the outcrop belt, pebble and cobble conglomerates are common, and it is here that the exotic clasts are largest. Petrographically, Applecross sandstones are classified as arkoses (McBride, 1963), and are immature in that they contain an abundance of subangular, relatively unaltered feldspar grains.

Applecross alluvium is abundantly cross-stratified (Fig. 3). Three types of cross-strata sets are recognized in this study: 'trough', 'compound', and 'planar-tabular'. Trough cross-stratification, usually in cosets, is the most commonly occurring type of cross-strata within the Applecross succession, with individual sets typically 5–30 cm in thickness and 0.5–3.0 m in width. These cross-sets are widely reported from other ancient fluvial deposits (Cant & Walker, 1976; Allen, 1983; Rust & Gibling, 1990).

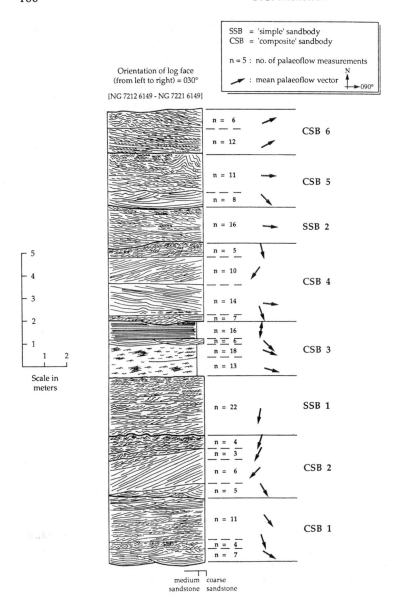

Orientation of log face
(from left to right) = 030°

[NG 7212 6149 - NG 7221 6149]

SSB = 'simple' sandbody
CSB = 'composite' sandbody

n = 5 : no. of palaeoflow measurements

↗ : mean palaeoflow vector

n = 6   CSB 6
n = 12

n = 11   CSB 5
n = 8

n = 16   SSB 2

n = 5
n = 10   CSB 4
n = 14
n = 7
n = 16
n = 6   CSB 3
n = 18
n = 13

n = 22   SSB 1

n = 4
n = 3   CSB 2
n = 6
n = 5

n = 11   CSB 1
n = 4
n = 7

5
4
3
2
1
1   2

Scale in
meters

medium   coarse
sandstone sandstone

Fig. 3. Applecross Formation
20-m stratigraphical log [Rubha na
Fearn, NG 7212 6149–
NG 7221 6149]. All structures
other than ripples drawn to scale.

Larger trough sets 30–50 cm thick and 3–5 m wide (shown in Fig. 4) occur less frequently, and occasionally single troughs reach a maximum of 50–70 cm in thickness and over 6 m in width. Compound cross-sets are downcurrent-dipping, and hence this term is used *sensu* Banks (1973) and Allen (1984, vol. 1, p.356, fig. 9-7). A wide variety of compound set and coset geometries occur in the Applecross lithosome, including tabular, wedge-shaped and lenticular forms, with foresets usually planar to gently concave-up. Compound set thicknesses vary from 5 to 50 cm, whereas compound cosets are typically 1–3 m in thickness but may reach 5–9 m in height in the largest bar structures. Planar-tabular sets occur least frequently within the lithosome. These sets are usually isolated, typically 0.75–1.5 m in thickness, and often extend laterally downcurrent for several tens of metres.

Fig. 4. Large-scale (4 m wide) trough cross-stratified set, Applecross Formation [Bealach na Ba, NG 777 416].

Applecross sandbodies: description and terminology

The Applecross lithosome consists of sandbodies which are typically 1–3 m in thickness and over 50–100 m in width and/or length, and hence the sandbodies exhibit sheet geometries as defined by Friend *et al.* (1979). Their lateral extent in both flow-transverse and flow-parallel sections is usually limited by available exposure.

Applecross sandbodies are composed of 'sedimentation units', defined in this study as genetically related groups of one or more facies which form distinctive depositional packages. For example, individual sedimentation units commonly consist of a group of facies sharing a similar palaeoflow direction relative to adjacent units, or (where a range of grain sizes are present), are often identified by an upward-coarsening textural pattern between different facies (Nicholson, 1992b). These sedimentation units are analogous to 'storeys' of Friend *et al.* (1979), the 'complexes' of Allen (1983), and the 'cosets' and 'complex-cosets' of Haszeldine (1983); however, they do not necessarily follow an implied temporal sequence as do the 'storeys', nor do they always consist of cosets and/or amalgamations of more than one facies type (a sedimentation unit may, for example, consist of an individual tabular set of planar cross-strata). Consequently, the term 'sedimentation unit' is preferred.

Two types of Applecross sandbody are recognized based on their internal structure: 'simple' and 'composite'. A 'simple' sandbody consists of a single sedimentation unit (such as 'SSB 1' in Fig. 3), whereas a 'composite' sandbody is composed of two or more sedimentation units (for example, 'CSB 2' in Fig. 3).

Palaeocurrent variability in Applecross sandbodies

Composite sandbodies display a higher degree of structural complexity and a higher degree of palaeocurrent variability than simple sandbodies. This is evident from the stratigraphical log presented in Fig. 3, where individual sedimentation units within composite sandbodies show markedly different mean palaeoflows (e.g. 'CSB 2' and 'CSB 4', Fig. 3). Moreover, palaeocurrent variability within sandbodies increases as the scale of bar structures increases, as has been demonstrated in studies of other ancient alluvium (Bluck, 1980, 1981). One of the larger Applecross bar structures, a 6-m-thick composite sandbody which consists of down-current-dipping cosets of compound cross-strata, is shown in Fig. 5a. The dip directions of the coset surfaces are interpreted as the accretion directions of the original bar, following other studies of similar large-scale ancient fluvial bars (Bluck, 1980; Haszeldine, 1983). The palaeocurrent directions of the smaller scale cross-strata (which include trough sets) within each coset represent the migration directions of smaller bedforms on the larger parent bar structure (Bluck, 1980; Haszeldine, 1983). Palaeocurrent rose diagrams corresponding to these two different scales of structure are shown in Fig. 5b.

(a)

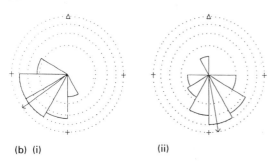

(b) (i)   (ii)

Fig. 5. (a) Composite sandbody consisting of large-scale (6 m thick) compound cross-stratification, Applecross Formation [Shieldaig, NG 816 522]. Individual sedimentation units are bounded by major downcurrent-dipping accretion surfaces as indicated by arrow. Profile face = 190° ←. (b) Equal area rose diagrams (methodology of Nemec, 1988) for the composite sandbody in (a) [Shieldaig, NG 816 522]. Dotted circles are drawn at intervals of 10% of the total number of readings. (i) Accretion directions: total number = 17, maximum radius = 5, mean = 234, standard deviation = 26°. (ii) Cross-strata: total number = 51, maximum radius = 15, mean = 170°, standard deviation = 42°.

Collectively, these data suggest a large-scale, probably medial bar that was building downstream, with smaller bedforms migrating obliquely across and down the inclined bar face. Overall, it is necessary to consider the potentially high degree of palaeocurrent variability (within composite sandbodies) at outcrop level when acquiring data for regional palaeoflow analyses.

SYNDEPOSITIONAL TECTONIC ACTIVITY

Hinxman (in Peach *et al.*, 1907, p. 273) stated that the 'Torridonian's' thickness and relative petrographical homogeneity indicated 'a long-continued uniformity in the conditions of deposit'. Nonetheless, certain structural and textural variations within Applecross alluvium have been used by previous authors as evidence for syndepositional

tectonism. Two such lines of evidence are reinvestigated here.

'Fining-upward fan cycles'

Stewart (1982) indicated that fining-upward cycles are stacked upon each other at the base of the Applecross Formation between the Inverpolly and Dundonnell forests. These cycles were interpreted as alluvial fan deposits generated by periodic uplift along the basin-bounding Minch Fault to the west-northwest, followed by subsequent scarp degradation and fan retreat.

An 80-m stratigraphical log through these cycles in one locality (Stewart, 1982 and personal communication) is presented in Fig. 6. At the base of this log are the intercalated grey shales, siltstones and fine sandstones of the Diabaig Formation, interpreted as lacustrine deposits (Stewart, 1988b). The overlying arkoses belonging to the Applecross

Formation are subdivided into sections 'A', 'B' and 'C' (for convenience) in Fig. 6. Sections 'A' and 'C' consist of typical coarse, heterogeneous, cross-stratified Applecross sandstone containing sporadic pebbles. Trough and a variety of compound types of cross-bedding predominate and sandbodies range in thickness from 1 to 9 m, with composite types 3–5 m thick most common. Section 'B' consists of medium- to coarse-grained, also heterogeneous, horizontally-laminated to ripple cross-laminated sandstone which only rarely contains pebbles; 5–10-cm-thick sets of trough cross-strata are also common; current lineations occur on some horizontal surfaces. According to Stewart (1982 and personal communication), sections 'A' and 'B' together represent a single fining-upward cycle, while section 'C' forms the lower part of an overlying second cycle.

A more detailed 40-m stratigraphical log is also shown in Fig. 6. Of particular significance is the 9 m-thick composite sandbody at the base of section 'C' (labelled 'CSB 1' in Fig. 6). A photomosaic (Fig. 7a) and line drawing (Fig. 7b) of this composite sandbody indicate that it consists of an amalgamation of several individual compound cross-stratified sedimentation units, collectively indicative of a large bar structure. As with the previous large bar structure in Fig. 5a, orientation data were measured both for the accretion directions of the major descending surfaces of the sedimentation units, and for the smaller scale cross-strata within the units. The resulting rose diagrams (Fig. 7c) display an oblique relationship between their mean palaeoflow vectors, and once again the structure is interpreted as a large medial bar building approximately downstream, with smaller bedforms migrating obliquely across and down the bar face. A possible reconstruction of the original bar form is shown in Fig. 7d. Similar bar structures (composite sandbodies) of 3–7 m thickness also occur in section 'A' of the 80-m log (Fig. 6).

The 40-m log in Fig. 6 also shows the interdigitating relationship between the 9-m bar structure at the base of section 'C', and the finer horizontally-stratified to rippled sandstones at the top of section 'B'. These interdigitations, occurring over a roughly 2 m vertical thickness, are indicative of the lateral equivalence of the two sections ('B' and 'C') during sedimentation. In addition, mean palaeoflow

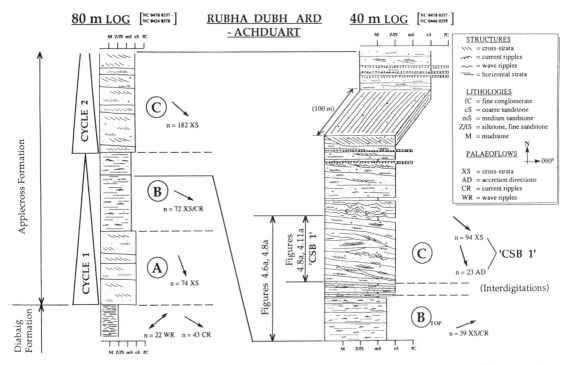

Fig. 6. Schematic stratigraphical logs through 'fining-upward cycles' (full explanation provided in text) [Rubha Dubh Ard–Achduart, NC 0424 0378–NC 0478 0357]. (After Stewart, 1982.)

(a)

(b)

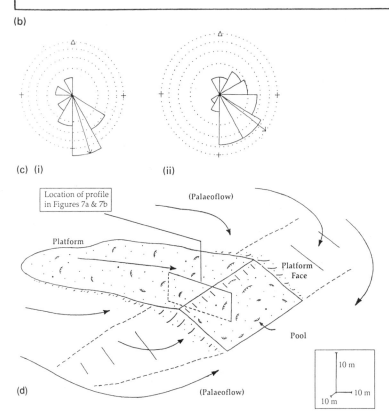

(c) (i)　　　　　　　(ii)

(d)

Fig. 7. (a) Composite sandbody ('CSB 1' in Fig. 6) consisting of large-scale compound cross-stratified sedimentation units, overlying slightly finer-grained horizontally-laminated deposits (again refer to Fig. 6). Scale provided by 1.5-m-staff in centre of photograph. Total thickness of bar structure is 9 m. Profile face = 165° →. [Achduart, NC 0448 0358]. Fig. 7 (b) Line drawing of composite sandbody in (a) Scale is again indicated by 1.5-m-staff. Profile face = 165° →. Fig. 7 (c) Equal area rose diagrams (methodology of Nemec, 1988) for the composite sandbody in (a) and (b) [Achduart, NC 0448 0358]. Dotted circles are drawn at intervals of 10% of the total number of readings. (i) Accretion directions: total number = 23, maximum radius = 12, mean = 162°, standard deviation = 23°. (ii) Cross-strata: total number = 94, maximum radius = 26, mean = 126°, standard deviation = 43°. Fig. 7 (d) Medial bar interpretation of the large-scale composite sandbody at Achduart [NC 0448 0358]. The relative position of (a) and (b) is also indicated.

vectors for the large bar of 'C' and the upper part of 'B' are approximately orthogonal as shown (Fig. 6). Collectively, given the interdigitations and the orthogonal palaeoflow relationship between these coarser and finer sections, and the existence of large bar structures within the coarser zones, these cycles are not thought to represent the deposits of alluvial fans formed in response to periodic tectonism. Instead are interpreted as the products of a large river system. The finer horizontally stratified and rippled sediments are suggestive of sheetfloods and waning-stage flows on a floodplain, whereas the coarser bar structures indicate the presence of channels commonly 3–7 m deep with a maximum depth of 9 m. A fluvial (autocyclic) origin for these cycles is thus favoured over the previously proposed tectonic (allocyclic) process.

Soft-sediment deformation structures

Soft-sediment deformation structures ('water escape' or 'quicksand' structures) are ubiquitous in the coarse- to medium-grained sandstones of both the Applecross and Aultbea formations. Several types of structure are visible, including vertical intrusions of sediment reaching up to several metres in height (analogous to diapirs and termed 'cusps' by Selley (1969)), a variety of folds ranging from complex convolutions to gentle synclines often linked with sharp anticlines (cusps), and overturned cross-strata (recumbent foresets). Most structures show that the predominant direction of sediment motion was vertical, and that deformation occurred by the upward expulsion of water (Selley, 1969; Nicholson, 1992b). The majority of structures also display only one phase of deformation (Selley, 1969; Nicholson, 1992b), and are commonly confined to the upper part of individual sandbodies (Nicholson, 1992b) (see 'SSB 1', 'CSB 2' and 'CSB 5', Fig. 3).

Selley (1969) listed three possible mechanisms which could have generated Torridon Group soft-sediment deformation structures. The first, favoured by Williams (1970), involves the direct fluidization of sediment by rising groundwater springs associated with a hydrostatic pressure gradient (high water table) on the distal slopes of an alluvial fan. However, this mechanism is discounted, since these springs (once established in a given locality) would have probably deformed several beds many times. Moreover, vertical fluidiza-

tion by rising springs would almost certainly not produce structures that are typically confined to the upper zones of individual sandbodies (the deformation would be expected to occur throughout entire beds). The two alternative deformation mechanisms (Selley, 1969) both involve initial liquefaction of the sediment, followed by fluidization caused by the expulsion of porewater. In one case, liquefaction is triggered by earthquake activity, whereas in the other, it is triggered by microseisms from turbulent current activity. Although these two mechanisms have each been observed in Recent sediments and have each been simulated experimentally (Selley, 1969), certain features of Torridon Group alluvium suggest that a 'fluvial' trigger was more likely than a 'tectonic' alternative. Deformation structures are often truncated by the base of the overlying sandbody (Stewart, 1988b, p. 108, fig. 9.9; also see 'CSB 5' in Fig. 3); Stewart (1988b) stated that it was unlikely that sufficient earthquake activity would have occurred in order to coincide with the deposition of every second bed, and suggested that a flood-related process was a more probable trigger. Although either an earthquake or a fluvial trigger could produce deformation structures confined to the upper portion of sandbodies (Owen, 1987 and personal communication), indirect evidence again tends to favour a fluvial trigger. Palaeocurrents occasionally deformed the Torridon Group sediments directly (overturned foresets) and were potentially vigorous given the inferred scale of some of the Applecross rivers and their flood discharges (see discussions below). McKee *et al.* (1967) recorded deformation structures in flash flood deposits that were confined to the upper portions of the trenches they dug. Selley's (1969) simulation of Torridon Group deformation structures demonstrated that a fluvial liquefaction-triggering mechanism was plausible, whereas Allen (1984, vol. 2, p. 300) included river flood waves among other phenomena which can cause liquefaction (such as explosions, tsunamis and earthquakes). Furthermore, Torridon Group field reconnaissances (Nicholson, 1992b) reveal no preferential regional distribution of soft-sediment deformation structures that could suggest a genetic link with synsedimentary faulting, as, for example, demonstrated in the Northumberland basin by Leeder (1987). Consequently, a 'fluvial' as opposed to 'tectonic' origin is favoured for the Torridon Group deformation structures.

PALAEOCURRENT ANALYSES

Discussion of previous palaeocurrent analysis

Certain aspects of Williams' (1966, 1969a) methodology for both measuring and analysing palaeocurrent data may have led to an incorrect interpretation of regional sediment transport patterns within the Applecross Formation.

1 Figures 3, 5b, 6 and 7c illustrate that there is a significant amount of palaeoflow variability at outcrop level within the Applecross succession, as has been documented in numerous other modern and ancient braided river deposits (for example, see Bluck, 1974, 1980). In his analysis, Williams (1969a) subdivided the Cape Wrath region into 500-m by 500-m grids, and randomly measured (where possible) a maximum of two trough cross-set axes and a maximum of two palaeocurrents from other cross-strata for each grid area. Williams' four palaeocurrent measurements per 500-m by 500-m area of outcrop seem unlikely to be able to give an accurate representation of local palaeoflow direction.

2 Before analysing his data, Williams subdivided the Cape Wrath district into 110°-trending zones (1969a, fig. 10) into which he grouped all measurements. The basis for his choice of 110°-trending zones was that, if the sediments were deposited by alluvial fans with fan medians oriented roughly 110°, then within these zones each facies would have a relatively uniform mean palaeoflow direction. Such an approach involves a degree of circular reasoning which could have biased his data and ultimately produced an incorrect final interpretation.

3 Williams (1966, 1969a) directly equated his different facies in the Cape Wrath area with relative stratigraphical and geographical positions on a retreating alluvial fan (coarsest deposits at the base of the succession and in the west, finer at the top and in the east). However, the Cape Wrath area is traversed by numerous faults, there are no visible stratigraphical markers within the Applecross Formation, and the entire Torridon Group succession is strongly diachronous (Smith *et al.*, 1983). Consequently, geographically different outcrops are not necessarily similar in age, and their palaeocurrents cannot be combined with certainty to give the mean palaeoflow for a particular facies. Furthermore, several different facies types (each with their own different mean palaeoflow) commonly occur together within individual bar structures at Cape Wrath (Nicholson, 1992b), which further weakens Williams' attempt to equate stratigraphically specific facies with relative positions on a fan.

4 Regionally, the radial palaeoflow pattern for Williams' second fan to the south (1969a, fig. 12) was only obtained by using an upcurrent extrapolation of the mean palaeocurrent vector of Maycock (1962). During this earlier study, Maycock was not concerned with the relative stratigraphical position of his measurements, and recorded data over the total thickness (roughly 3000 m) of the Applecross Formation. Using a similar approach to that of Williams, Lawson (1970) combined his own palaeocurrent data from the Gairloch area with all available regional data from other workers, and by extrapolating these mean palaeoflow vectors upstream, found that a least five point sources for alluvial fans could be proposed. Lawson concluded that the validity of Williams' method of extrapolating local palaeocurrent measurements over large distances (thereby producing radial transport patterns) was 'open to serious question' (Lawson, 1970, p. 149).

Palaeocurrent analysis and results of present study

A primary objective of this study was to obtain a regional palaeocurrent database for the Applecross Formation, from which the existing palaeoflow model could be evaluated and, if necessary, revised. The sampling procedure adopted for measuring palaeocurrents was dictated by the potentially high degree of palaeocurrent variability both within and between sandbodies at outcrop level, and by the absence of stratigraphical markers within the succession.

Palaeocurrent data were collected from two 'sampling levels' within the Applecross Formation, in order to provide a relative stratigraphical framework for the measurements, to obtain representative sediment transport directions in both the lower and upper parts of the formation, and to test for any time-related changes in palaeoflow across the outcrop belt. These two levels were defined on the basis of their stratigraphical height above the underlying Lewisian Gneiss (the lack of chronostratigraphy within the Torridon Group does not permit a more rigorous definition of data sampling levels). If reference is made to Stewart (1988b, fig 9.3), the two levels correspond to approximate stratigraphical heights of '500–1000 m' ('lower Applecross') and

Table 1. Regional palaeoflow data for the Applecross Formation (see also Fig. 8)

Locality (see Fig. 8)		Palaeoflow data			
		n	*N*	mean	σ
Lower Applecross Formation					
'a'	Sheigra (Cape Wrath) [NC 184614–NC 188634]	159	31	91°	25°
'b'	Handa Island [NC 126475–NC 146486]	325	58	110°	26°
'c'	Quinag [NC 210264–NC 207270]	123	30	129°	21°
'd'	Coigach [NC 074028–NC 085020]	136	30	127°	24°
'e'	Camusnagaul [NH 077891–NH 077874]	134	31	106°	30°
'f'	Carn Dearg (Gairloch) [NG 765775–NG 741791]	147	30	134°	29°
'g'	Fearnmore (N Applecr.) [NG 725610–NG 708605]	246	49	140°	37°
'h'	Raasay [NG 546437–NG 561471]	180	38	141°	37°
'i'	Rhum (E): Scresort [NM 409998–NG 423002]	203	40	118°	33°
Upper Applecross Formation					
'j'	Culkein (Stoer) [NC 041338–NC 025341]	172	34	95°	31°
'k'	Reiff [NB 963142–NB 970173]	158	35	121°	36°
'l'	Ardmair [NH 112983–NH 131988]	143	32	126°	30°
'm'	Slaggan [NG 839943–NG 841962]	189	37	113°	33°
'n'	Liathach (Torridon) [NG 920560–NG 919566]	144	30	142°	29°
'o'	Bealach na Ba (S Applecr.) [NG 775424–NG 779410]	137	30	146°	35°
'p'	Rhum (N): Sham. Insir [NG 366045–NG 389037]	206	44	132°	36°

n, total number of palaeocurrent measurements at given locality; *N*, number of sandbodies from which measurements were taken; mean, mean of all mean vectors from '*N*' sandbodies; σ, Batschelet's standard deviation of all mean vectors from '*N*' sandbodies.

'2000–3000 m' ('upper Applecross') above the basement unconformity. Measurement localities within the 'lower Applecross' were chosen in order to avoid the initial effects of basement relief on local palaeoflow directions.

A number of localities were selected within each sampling level (as listed in Table 1), either on the basis of favourable exposure for measurements, and/or on their geographical location relative to other localities. At each locality, a mean palaeocurrent vector was determined using the following procedure. For individual sandbodies with an essentially unidirectional palaeoflow (simple individual sandbodies or outcrops consisting of 10–30 cm thick sets of trough cross-strata), a minimum of three palaeocurrents were measured and a mean vector for the sandbody was obtained. For sandbodies which displayed a more complex internal palaeoflow (primarily composite sandbodies), a minimum of six measurements were taken in order to obtain a mean palaeocurrent direction. For each locality, a minimum of 30 sandbodies were measured randomly over a minimum stratigraphical thickness of 100 m, resulting in a minimum of 120 palaeocurrent readings. A mean palaeoflow vector for each locality was then determined from the collective palaeocurrent means of the sandbodies observed. Trough cross-set axes constitute the majority of readings at each locality, as these structures provide a more accurate indication of regional palaeoflow as compared to other cross-set types (e.g. High & Picard, 1974).

Regional palaeoflow data and the resulting mean vectors for each locality within the two sampling levels are presented in Table 1, and Figs 8a and 8b. For each individual locality, palaeoflows between

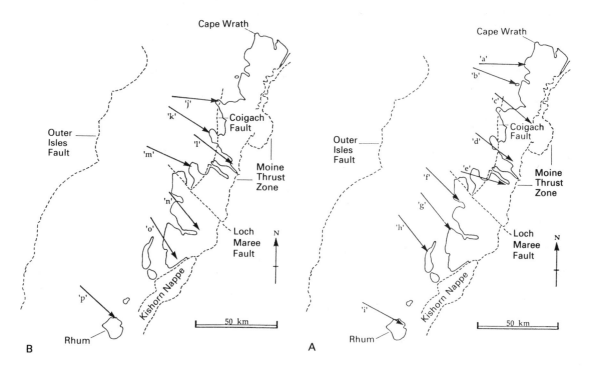

Fig. 8. Mean palaeoflow vectors in (a) the Lower Applecross Formation and (b) the Upper Applecross Formation (data presented in Table 1).

different sandbodies displayed a substantial degree of variability, typically over 90° (recall Fig. 3). However, when the mean vectors from individual localities are considered collectively, a much lower degree of regional palaeoflow variability is apparent (Figs 8a & 8b). The data confirm an overall E–SE regional palaeoflow for the Applecross Formation, but also indicate that the various mean palaeocurrent vectors for the different localities do not exhibit radial patterns when extrapolated upcurrent. These results fail to confirm deposition by scarp-generated alluvial fans to the west. In addition, the vectors show no evidence of deflection along the eastern margin of the outcrop belt (near the Moine Thrust zone), as might be expected had there been a basin-bounding normal fault in the vicinity. Consequently, the results suggest that regional sediment transport patterns were not controlled by basin-bounding normal faults lying along or near the present-day Outer Isles Fault and Moine Thrust zone.

PALAEOHYDROLOGICAL RECONSTRUCTION

Reasonable estimates of the palaeohydrology of Applecross rivers can be obtained through comparisons with empirical results from modern rivers, and with similar analyses performed by other workers on ancient braided stream alluvium. Although these calculations are inherently vulnerable to large degrees of error due to the assumptions involved, nevertheless they provide a useful impression of the scale of the fluvial system under investigation.

From the outset, one must consider the prospective differences in pre-Devonian fluvial systems when compared to their modern counterparts. The absence of vegetation in pre-Devonian systems, probably the most significant difference, likely resulted in greater sediment loads and discharge variabilities, as well as much-reduced bank stability in bedload-dominated rivers (Schumm, 1968; Cotter, 1978; Fuller, 1985), the latter factor favouring

Table 2. Results of palaeohydrological reconstructions for Applecross rivers

	'Shallow' Applecross rivers			'Deep' Applecross rivers		
Mean bankfull depth, D (m)	2			7		
Mean bankfull width, W (m)	200	400	800	350	700	1400
Form ratio, F (W/D)	100:1	200:1	400:1	50:1	100:1	200:1
Bankfull discharge, Q_b (@1.5 m/s mean flood velocity) (m³/s)	300	600	1200	1840	3675	7350
Av. annual discharge, Q_{av} (m³/s)	75	150	300	460	920	1840
Drainage basin area, A_d (km²)	10 000–20 000			100 000+		
River length, L (km)	200+			500+		

channels with high width-to-depth ratios. Other aspects, such as super-continental masses and the corresponding possibility of larger drainage systems (Potter, 1978), also merit consideration.

All results from the Applecross palaeohydrological reconstruction are presented in Table 2, and are discussed under separate headings below. The hydrology and deposits of three modern sand-dominated fluvial systems, the Platte, Tana and South Saskatchewan rivers, are referred to for comparison.

River depths

The abundant cross-stratification which occurs within the vast majority of Applecross sandbodies, can be grouped broadly into one of two categories. Small- to medium-scale cross-strata are most common, such as 1–3-m-thick cosets of trough cross-stratification with individual sets each 10–30 cm in thickness, or compound-tabular sets 1–2 m thick (Fig. 3). Large-scale compound cross-strata 5–9 m in height (Figs 5a & 7a) occur less frequently. Regarding the scale of Applecross palaeochannels, it is accepted that cross-strata thicknesses represent minimum values of formative flow depth. For the purposes of this reconstruction, conservative estimates of bankfull depth are made based on observed cross-strata thicknesses. Consequently, the Applecross palaeorivers in which the majority of sediment was deposited fall into two broad categories:

1 commonly occurring 'shallow rivers', roughly 1–3 m deep, with a mean value of bankfull depth of approximately 2 m;

2 occasionally occurring 'deep rivers', roughly 5–9 m deep, with a mean value of bankfull depth of

approximately 7 m. These bankfull depth estimates again represent conservative values, since they do not account for the fact that the cross-strata were formed by flows somewhat greater in height than the original bedforms, and that the original bedforms were greater in thickness than the cross-strata that have been ultimately preserved.

An alternative method of estimating the depth of Applecross rivers is provided by the relationships between dune height and water depth plotted by Allen (1984, vol. 1, p. 333, fig. 8-20). For 'shallow' Applecross rivers, an average (preserved) trough cross-strata thickness of 20 cm (representing the most commonly occurring 10–30-cm-thick Applecross trough sets) is increased by 50% to obtain a conservative estimate of original mean dune height (a method used by Rust & Gibling, 1990). This 30-cm value corresponds to bankfull flow depths of 2–3 m (using the 'mean' relationships in Allen's fig. 8-20, namely $H/h = 0.100$–0.167). Within the Platte River, three-dimensional dunes of 30–50 cm in height are common at bankfull depths of 1.0–2.5 m (Crowley, 1983). Collectively, and given that 1–2 m thick compound-tabular cross-stratified sets also occur within the succession (Fig. 3), a mean bankfull depth estimate of 2 m for the 'shallow' Applecross rivers is reasonable. The largest sets of trough cross-strata within the Applecross succession are typically between 30 and 50 cm thick with widths of 3–5 m, and can often be traced downcurrent on plan surfaces for over 20 m (Fig. 4). Again referring to Allen (1984, fig. 8-20), an estimated original mean dune height of 60 cm for these trough sets corresponds to flow depths of approximately 3–6 m. Comparable dunes up to 1.5 m in height have been recorded from 'major channels' (3–5 m deep) of the South Saskatchewan River (Cant &

Walker, 1978). Deposition within the 'deep' Apple-cross rivers is suggested by the largest trough sets of 50–70 cm thickness and 5–7 m width that occasionally occur within the lithosome: an original dune height of 90 cm in Allen's fig. 8-20 indicates bankfull depths of 6–9 m.

River widths

Although formulae for determining widths have been proposed for meandering stream deposits (see Ethridge & Schumm, 1978), the diversity of braided alluvium has to date prevented the recognition of similar relationships in these fluvial systems. In the absence of flow-transverse exposures where channel margins are preserved, river widths must be estimated.

Individual Applecross sandbodies (typically 1–3 m thick) can often be traced laterally for over 100 m in both flow-parallel and flow-transverse directions. Maximum observed lateral extent (limited by available exposure) exceeds 300 m. The sheet-like sandbody geometry, together with the absence of recognizable channel margins within the succession, suggests that channel widths were large and that channel perimeter slopes were probably only a few degrees, as might be expected in channels with substantial bank instability. Fuller (1985) indicates that form ratios (sandbody width/depth) of up to 1000:1 are to be anticipated in pre-Devonian sandy, bedload-dominated fluvial channels. Although the maximum observed form ratio is roughly 200:1 within the Applecross Formation, higher values (given adequate exposure) are likely. Consequently (for calculation purposes), three estimates of channel width are made for each of the two Applecross palaeoriver categories (see Table 2). For 'shallow' (2 m deep) rivers, estimated widths of 200, 400 and 800 m (form ratios of 100, 200, and 400:1) are roughly comparable to the 450–600 m Platte River width (Smith, 1971), while the width estimates of 350, 700 and 1400 m for 'deep' (7 m depth) Applecross channels (form ratios of 50, 100 and 200:1) are comparable to both the South Saskatchewan River (a confined valley width of 1000 m: Cant & Walker, 1978) and the Tana River (widths between 600 and 2000 m: Collinson, 1970).

River discharges

A simple method of estimating bankfull discharge

(Q_b), used by Allen (1983) in his study of the Devonian 'Brownstones', is adopted here. The cross-sectional channel area is reduced by an arbitrary factor of one-half to account for sediment bodies obstructing flow, and is then multiplied by a mean flood velocity of 1.5 m/s to obtain bankfull discharge. A flood velocity of 1.5 m/s seems to provide a realistic estimate: South Saskatchewan River flood velocities of 1.75 m/s in major channels and 1.5 m/s in minor channels and on sand flats were reported by Cant and Walker (1978) and Cant (1978), respectively. Moreover, the resulting Applecross bankfull discharge results compare favourably with those from modern sandy braided rivers. For example, a 300-m³/s discharge in a 200-m wide, 2-m deep Applecross channel is comparable to the 1971 flood of 271 m³/s in a 275–460-m wide, 1.25-m deep reach of the Platte River (Blodgett & Stanley, 1980). A maximum recorded flood discharge of 3200 m³/s for the Platte River (Smith, 1971) also indicates that the 600–1200-m³/s discharges obtained for 400–800-m wide 'shallow' Applecross rivers are not unrealistic. For 'deep' Applecross rivers 350–700 m wide, the bankfull discharge range of 1840–3675 m³/s is similar to flood discharges of between 1500 to 3800 m³/s for the South Saskatchewan River (Cant & Walker, 1978), and of 1800–3800 m³/s for the Tana River (Collinson, 1970). Discharges of 7100–7400 m³/s for the 1400-m wide, 7-m deep Applecross rivers are larger but roughly equivalent to 5200 m³/s maximum discharge recorded by Doeglas (1962) for the 6.6-m deep, 700-m wide Durance River.

Data presented by Leopold *et al.* (1964, table 7-13) show that modern streams which have relatively similar bankfull discharges (Q_b) may have grossly dissimilar average (mean) annual discharges (Q_{av}). By averaging the ratios of Q_{av}/Q_b for rivers with the six largest bankfull discharges (300–900 m³/s) given by Leopold *et al.* (1964), a result of 0.25 is obtained. Average annual discharges (Q_{av}) for Applecross palaeorivers (Table 2) are thus assumed to be approximately one-quarter of their bankfull flood discharges (Q_b). This factor again may also be conservative, as Miall (1976) obtained Q_{av}/Q_b ratios of one-quarter to one-third in his Cretaceous braided alluvium study.

Drainage basin areas

Estimates of drainage basin area (A_d) can be made by referring to data given by Leopold *et al.* (1964, table 7-13) and Gregory and Walling (1973, fig. 5.9) for rivers from varied climatic and topographical settings. These data show that rivers with bankfull discharges ranging between 300 and 1200 m^3/s have typical drainage areas of 10 000–20 000 km^2, a result which can be applied directly to 'shallow' Applecross streams. For 'deep' Applecross palaeo-orivers, reference is made to a data table from Czaya (1983, p. 54) for the 25 largest rivers in Europe (which again include a variety of climatic and topographical settings). European rivers with Q_{av} values ranging between 510-2200 m^3/s have drainage basin areas typically averaging between 50 000 and 200 000 km^2. Given the average annual discharges of 460–1840 m^3/s obtained for 'deep' Applecross rivers, a drainage basin exceeding 100 000 km^2 is not unrealistic.

River lengths

A useful table relating drainage basin area to approximate river length is provided by Czaya (1983, p. 48), and is reproduced here as Table 3. 'Shallow' Applecross channels with drainage areas of 10 000–20 000 km^2 correspond to river lengths exceeding 200 km and may be termed 'large minor rivers', whereas 'deep' Applecross rivers with A_d values of roughly 100 000 km^2 were 'small major rivers' likely in the vicinity of 500 km long. Alternatively, the relationship provided by Leopold *et al.* for estimating river length (1964, p. 145 and fig. 5-6): $L = 1.4(A_d)^{0.6}$ (where L = length in miles, and A_d = area in square miles), gives results of 320 km for 'shallow' Applecross channels and (probably somewhat over-optimistically) 1275 km for 'deep' Applecross rivers, using A_d values of 10 000 km^2 and 100 000 km^2 respectively. By comparison, the Platte River (including either one of its two major forks) is roughly 750 km long, with a basin covering over 200 000 km^2 (Crowley, 1983). Therefore, it is quite conceivable that a river system similar to the

Table 3. Relationship between drainage basin area and river length for the world's rivers (After Czaya, 1983)

River type (terminology of Czaya, 1983)	Drainage basin area (km^2)	River length (km)
Minor river (small)	1000–10 000	100–200
Minor river (large)	10 000–100 000	200–500
Major river (small)	100 000–1 000 000	500–2500
Major river (large)	> 1 000 000	> 2500

Platte in scale, and/or channels similar in size and discharge to the South Saskatchewan and Tana rivers (with corresponding drainage basin areas of the order of 10^6 km^2), were responsible for deposition of the Applecross Formation.

TORRIDON GROUP BASIN REAPPRAISAL

Depositional environment and basin scale

Two features must be considered when evaluating the depositional environment of Applecross alluvium: the evidence for major rivers within the fluvial system, and the uniformly E–SE palaeoflow within the succession combined with the absence of radial sediment transport patterns regionally. These features suggest that the Applecross Formation was deposited on a major alluvial braidplain, as opposed to the large alluvial fans previously proposed. Although it is accepted that the difference between alluvial fan and alluvial braidplain environments is often of little consequence (particularly in distal fan settings), a clear distinction is deemed necessary here. Whereas alluvial fans form at the base of a mountain front or other upland area, regional palaeoflow data from this study fail to indicate the existence of the previously proposed scarp-bounded upland source area to the immediate W–NW over the Outer Hebrides (Williams, 1966, 1969a). Moreover, the implied depths, discharges and lengths of the larger Applecross rivers favour an alluvial braidplain setting.

The concept of depositional system scale is crucial to the Torridon Group basin reappraisal. The existing model envisages the Torridon Group being deposited in a 75–100-km-wide rift graben, with the flanking horst areas acting as sediment source in the west (Minch-Fault-scarp-sourced alluvial fans) and basin margin in the east (basin-bounding normal fault lying near the present Moine Thrust zone) (Stewart, 1982). However, certain characteristics of this alluvium cannot be reconciled to such a self-contained graben model. A 200-km-wide, *c.* 5-km-thick fluvial succession with recurring evidence for major rivers and a uniformly E–SE regional palaeoflow collectively indicates that the Torridon Group basin was probably larger in scale than the previously proposed rift graben. Furthermore, the Applecross and Aultbea formations' relative homogeneity of petrography and facies (a 4–5-km thickness of coarse-to medium-grained cross-stratified sandstone with a variable pebble content) also supports a depositional system scale greater than that generated from an immediately flanking western horst block and deposited by an inter-rift drainage network. Such overall uniformity contrasts clearly with the typical deposits of smaller fault-bounded extensional basins as, for example, described from the East African Rift System (Frostick *et al.*, 1986) or modelled by Leeder and Gawthorpe (1987), which are characterized by a much more varied suite of lithologies and facies, together with more variable palaeoflow patterns.

Although regional palaeoflow data fail to indicate the existence of a basin-bounding fault in the east, the absence of converging transport directions when extrapolated upstream does not preclude the possibility that normal movement (easterly downthrow) on the Outer Isles Fault bounded the Torridon Group basin to the west. Assuming the reconstructed lengths of 200–500 + km for Applecross rivers are correct, then it could have been possible that such rivers were antecedent, and that they entered a half-graben from the W–NW. However, when these river lengths, the aforementioned alluvium characteristics and the depositional system scale considerations are all taken collectively, it seems more probable (in the absence of any direct evidence for such a boundary fault) that the Outer Isles Fault was unlikely to have been a controlling (basin-bounding) feature of Torridon Group deposition (despite marking the present-day western limit of the 'Torridonian' (Binns *et al.*, 1974; Stein, 1988)), and that the Torridon Group basin was substantially larger in scale.

The palaeogeographical reconstruction for the Applecross Formation presented by Nicholson (1992a), shown here in Fig. 9, assumes that the Torridon Group was originally deposited over the Outer Hebrides to the west of the Outer Isles Fault, and was subsequently eroded. The fact that Torridon Group sedimentation also occurred to the east of the present-day Moine Thrust zone is confirmed by the Applecross klippen within the thrust belt, which have been transported westwards by Caledonian compression.

The source of Torridon Group detritus was formerly thought to be Laxfordian in age (*c.* 1500–1700 Ma), originating from a distance not more that 100 km west of the Outer Isles Fault (Stewart, 1982). However, Rogers *et al.* (1989) have identified Grenvillian (*c.* 1100 Ma) and Archaean (*c.* 2700 Ma) detrital zircons within the Applecross

lithosome, together with *c.* 1650 Ma detritus previously interpreted as Laxfordian and thought to represent the sole Applecross sediment source (Stewart, 1982, 1988b). These detrital zircon results also support the likelihood of a larger drainage area, a more distal provenance (further onto the Laurentian Shield) and, accordingly, a major depositional system with large-scale rivers. Similarly, much of the Applecross 'exotic' pebble suite has been matched with crustal blocks in Greenland and Labrador (Williams, 1969b; Allen *et al.*, 1974), further supporting the prospect that a large-scale fluvial system was responsible for Torridon Group deposition. Although Cliff and Rex (1989) have found Grenville cooling ages in biotites from the Lewisian Complex on Harris and Lewis, it is uncertain whether this area was actually involved in a major Grenvillian detritus-forming event. Moreover, given the length of Applecross rivers and the inferred basin scale, it is most likely that the source hinterland for Torridon Group detritus was substantially more distant than the neighbouring Outer Hebrides.

Revised basin model

Extensional basin models typically describe a two-stage process of basin evolution, whereby an initial phase of active rifting is followed by a secondary phase of thermal subsidence caused by lithospheric cooling and thickening (McKenzie, 1978; Wernicke, 1985; Coward, 1986). The thermal basin is significantly greater in lateral extent than the initially formed fault-bounded rift(s). Moreover, the two different stages of basin development can often be recognized by their characteristic deposits, with rift sediments typically recording the effects of active extension and syndepositional tectonism, and thermal basins frequently exhibiting more 'tectonically quiet' transgressive fills.

It is proposed here that the Torridon Group basin was a second stage extensional basin formed by thermal relaxation processes, as opposed to an initial stage, smaller-scale rift undergoing active extension. A schematic cross-section of the Torridon Group basin is shown in Fig. 9. This interpretation is supported by the absence of unequivocal features supporting syndepositional tectonism, the absence of evidence for basin-bounding faults affecting sedimentation, the alluvium's relative petrographical and facies uniformity over its *c.* 5 km thickness (including a complete absence of volcanic

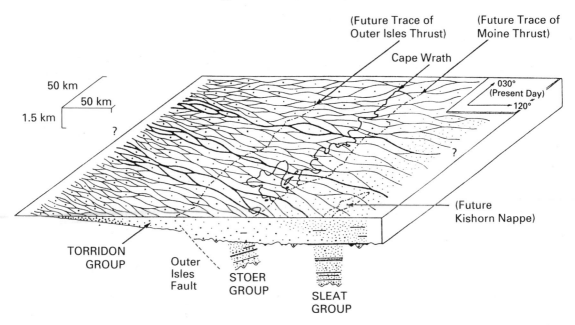

Fig. 9. Palaeogeographical reconstruction and schematic cross-section of the Torridon Group thermal relaxation basin during deposition of the Applecross Formation (approximate stratigraphical level 2000 m above Torridon Group base). Passive normal movement on the Outer Isles Fault (which was not a basin-bounding structure) is thought to have accommodated thermal subsidence. Palaeogeography modified after Nicholson (1992a). Latitude of cross-section is not intended to be specific. Same key to lithologies as in Fig. 2 (Stoer and Sleat group thicknesses also from Fig. 2). Braided river pattern diagrammatic only.

deposits), and the aforementioned aspects of fluvial system scale. The consistently E–SE palaeoflow within the succession indicates that the basin floor slope was apparently not disrupted during sedimentation, and that the pattern of fluvial deposition was allowed to continue essentially undisturbed from the base to the top of the succession, producing what is effectively a single fining-upward mega-cycle. Deposition concluded with the return of lacustrine/shallow marine conditions at the top of the Torridon Group (the Cailleach Head Formation). Collectively, this entire tectonically quiescent, transgressive scenario seemingly suits a thermal relaxation origin for the Torridon Group basin.

Recent studies of the mineralization in microcracks within the Lewisian Gneiss basement and the Stoer Group (Hay, 1988; Hay *et al.*, 1988) appear to support a thermal basin model for the Torridon Group. Two major crack-sealing phases, 'prehnite + albite + calcite' and 'pumpellyite + quartz + calcite', were identified as chronologically post-dating Stoer Group deposition but clearly pre-dating Tor-

ridon Group sedimentation (as the mineralized cracks do not continue above the unconformable contact with the Applecross Formation). Hay *et al.* (1988) suggested that these two mineralization phases were associated with a hydrothermal system developed during early 'Torridonian' rifting. Moreover, their results indicate that such a hydrothermal system had already begun to decay prior to the onset of Torridon Group deposition, supporting the thermal relaxation model.

Given the absence of volcanics within the Torridon Group, and considering that the underlying crust has an essentially uniform thickness of 26–30 km (Brewer *et al.*, 1983) indicating that significant crustal attenuation has not occurred, then the extension factor (β) for the Torridon basin would not be expected to be greater than 2 (Dewey, 1982, pp. 392–393, fig. 21). This value is similar to the maximum β suggested for the northern North Sea (Allen & Allen, 1990, pp. 45–46). However, in order to generate *c.* 5 km of thermal subsidence for a maximum β of 2, a time span of roughly 120 My is required (McKenzie, 1978, p. 29, fig. 4); if (for example) a more conservative maximum duration of 50–80 Ma is assumed for sedimentation and subsidence, then clearly unacceptable β values of between 3 and 4 are necessary.

Allen and Allen (1990, p. 58) note that the homogeneous lithospheric stretching model of McKenzie (1978) commonly underestimates the amount of thermal subsidence that has occurred in actual basins over known time periods. In the North Sea Viking Graben, Badley *et al.* (1988) account for this discrepancy by demonstrating that thermal subsidence was accelerated by subsidence-driven planar-normal faulting. These authors indicate that such faulting, which accommodates asymmetric subsidence, is of a 'passive' nature and does not constitute an active phase of rifting. This passive faulting is typically located along major crustal fractures that were previously active during initial rifting.

It is suggested here that Torridon Group thermal subsidence was accelerated and accommodated by subsidence-driven planar-normal movement along the Outer Isles Fault (as shown in Fig. 9). This faulting was of a passive, non-rotational nature, as regional palaeoflow data for the Torridon Group show no evidence for active basin-floor tilting (which, by contrast, is apparent in the Stoer Group). Moreover, any such subsidence-driven faulting could not have had a disruptive effect on the overall pattern of alluvial sedimentation within the basin, given the aforementioned uniformity (in texture and structure) of the Torridon Group.

Accepting a thermal relaxation origin for this basin, two possible relationships between the Torridon Group and other Proterozoic successions of the Northwest Highlands merit consideration. Firstly, it is conceivable that the Sleat Group represents the initial rifting stage of extensional basin development. This hypothesis is supported by the Sleat Group's much more diverse suite of facies (indicating a repetitive, at times cyclical, interplay between alluvial fan, braided stream, deltaic and lacustrine environments), its petrological differences relative to the Torridon Group (Sleat Group's deposits are less mature and also contain large amounts of volcanic detritus), and its apparently conformable relationship underlying the Torridon Group (Stewart, 1988b). Secondly (but seemingly less probable based on existing provenance and age data) is the possibility that part of the Moine assemblage is a distal lateral equivalent of the Torridon Group. The scale of Applecross rivers, together with their appar-

ently unconstrained easterly palaeoflow, favours such a correlation. However, any such 'Sleat–Torridon' and 'Torridon–Moine' relationships remain largely speculative given the present uncertainties in age, palaeogeography and provenance of these successions.

ACKNOWLEDGEMENTS

This paper constitutes part of the author's PhD thesis. Financial support provided by a University of Glasgow Postgraduate Scholarship, together with an Overseas Research Students Award from the Committee of Vice-Chancellors and Principals (UK), is gratefully acknowledged. My sincere thanks to Brian Bluck and Peter Haughton for their assistance throughout this project. Signe-Line Røe and Stuart Haszeldine are also thanked for their helpful comments on the manuscript. Sandy Stewart kindly provided grid references for the Achduart exposures.

REFERENCES

ALLEN, J.R.L. (1983) Studies in fluviatile sedimentation: bars, bar-complexes and sandstone sheets (low-sinuosity braided streams) in the Brownstones (L. Devonian), Welsh Borders. *Sediment. Geol.* **33**, 237–293.

ALLEN, J.R.L. (1984) *Sedimentary Structures: Their Character and Physical Basis. Developments in Sedimentology* **30**, Elsevier, Amsterdam (2 vols), 1256 pp.

ALLEN, P.A. & ALLEN, J.R. (1990) *Basin Analysis: Principles and Applications.* Blackwell Scientific Publications, Oxford, 451 pp.

ALLEN, P., SUTTON, J. & WATSON, J.V. (1974) 'Torridonian' tourmaline–quartz pebbles and the Precambrian crust northwest of Britain. *J. Geol. Soc. London* **130**, 85–91.

BADLEY, M.E., PRICE, J.D. RAMBECH DAHL, C. & AGDESTEIN, T. (1988) The structural evolution of the northern Viking Graben and its bearing upon extensional modes of basin formation. *J. Geol. Soc. London* **145**, 455–472.

BANKS, N.L. (1973) The origin and significance of some downcurrent-dipping cross-stratified sets. *J. Sediment. Petrol.* **43**, 423–427.

BINNS, P.E., MCQUILLIN, R. & KENOLTY, N. (1974) *The Geology of the Sea of Hebrides.* NERC Institute of Geological Sciences, Report **73/14**, 34 pp.

BLODGETT, R.H. & STANLEY, K.O. (1980) Stratification, bedforms, and discharge relations of the Platte braided river system, Nebraska. *J. Sediment. Petrol.* **50**, 139–148.

BLUCK, B.J. (1974) Structure and directional properties of some valley sandur deposits in southern Iceland. *Sedimentology* **21**, 533–554.

BLUCK, B.J. (1980) Structure, generation and preservation

of upward fining, braided stream cycles in the Old Red Sandstone of Scotland. *Trans. R. Soc. Edin Earth Sci.* **71**, 29–46.

BLUCK, B.J. (1981) Upper Old Red Sandstone, Firth of Clyde. In: *Field Guides to Modern and Ancient Fluvial Systems in Britain and Spain* (Ed. Elliot, T.) International Association of Sedimentologists, Second International Conference on Fluvial Sedimentology, Keele, England, pp. 5.14–5.20.

BREWER, J.A. et al. (1983) BIRPS deep seismic reflection studies of the British Caledonides. *Nature* **305**, 206–210.

CANT, D.J. (1978) Bedforms and bar types in the South Saskatchewan River. *J. Sediment. Petrol.* **48**, 1321–1330.

CANT, D.J. & WALKER, R.G. (1976) Development of a braided-fluvial facies model for the Devonian Battery Point Sandstone, Quebec. *Can. J. Earth Sci.* **13**, 102–119.

CANT, D.J. & WALKER, R.G. (1978) Fluvial processes and facies sequences in the sandy braided South Saskatchewan River, Canada. *Sedimentology* **25**, 625–648.

CLIFF, R.A. & REX, D.C. (1989) Evidence for a 'Grenville' event in the Lewisian of the northern Outer Hebrides. *J. Geol. Soc. London.* **146**, 921–924.

COLLINSON, J.D. (1970) Bedforms of the Tana River, Norway. *Geogr. Ann.* **52A**, 31–56.

COTTER, E. (1978) The evolution of fluvial style, with special reference to the Central Appalachian Palaeozoic. In: *Fluvial Sedimentology* (Ed. Miall, A.D.) Can. Soc. Petrol. Geol. Mem. **5**, 361–383.

COWARD, M.P. (1986) Heterogeneous stretching, simple shear and basin development. *Earth Planet. Sci. Lett.* **80**, 325–336.

CROWLEY, K.D. (1983) Large-scale bed configurations (macroforms), Platte River Basin, Colorado and Nebraska: primary structures and formative processes. *Geol. Soc. Am. Bull.* **94**, 117–133.

CZAYA, E. (1983) *Rivers of the World.* Cambridge University Press, Cambridge, 246 pp.

DEWEY, J.F. (1982) Plate tectonics and the evolution of the British Isles. *J. Geol. Soc. London* **139**, 371–412.

DOEGLAS, D.J. (1962) The structure of sedimentary deposits of braided rivers. *Sedimentology* **1**, 167–190.

ETHRIDGE, F.G. & SCHUMM, S.A. (1978) Reconstructing paleochannel morphologic and flow characteristics: methodology, limitations, and assessment. In: *Fluvial Sedimentology* (Ed. Miall, A.D.) Can. Soc. Petrol. Geol. Mem. **5**, 703–721.

FRIEND, P.F., SLATER, M. J. & WILLIAMS, R.C. (1979) Vertical and lateral building of river sandstone bodies, Ebro Basin, Spain. *J. Geol. Soc. London* **136**, 39–46.

FROSTICK, L.E., Renaut, R.W., Reid, I. & Tiercelin, J.-J. (Eds) (1986) *Sedimentation in the African Rifts.* Geol. Soc. London Spec. Publ. **25**, 382 pp.

FULLER, A.O. (1985) A contribution to the conceptual modelling of pre-Devonian fluvial systems. *Trans. Geol. Soc. S. Afr.* **88**, 189–194.

GREGORY, K.J. & WALLING, D.E. (1973) *Drainage Basin Form and Process.* Edward Arnold, London, 456 pp.

HASZELDINE, R.S. (1983) Fluvial bars reconstructed from a deep, straight channel, Upper Carboniferous coalfield of northeast England. *J. Sediment. Petrol.* **53**, 1233–1247.

HAY, S.J. (1988) *Permeability — Past and Present — in Continental Crustal Basement.* PhD thesis, University of Glasgow (unpublished).

HAY, S.J. et al. (1988) Sealed microcracks in the Lewisian of NW Scotland: a record of 2 billion years of fluid circulation. *J. Geol. Soc. London* **145**, 819–830.

HIGH, L. & PICARD, M.D. (1974) Reliability of cross-stratification types as paleocurrent indicators in fluvial rocks. *J. Sediment. Petrol.* **44**, 158–168.

LAWSON D.E. (1965) Lithofacies and correlation within the lower 'Torridonian'. *Nature* **207**, 706–708.

LAWSON, D.E. (1970) *The 'Torridonian' rocks of the Gairloch area and parts of Wester Ross Sutherland, NW Scotland.* PhD thesis, University of Reading (unpublished).

LEEDER, M.R. (1987) Sediment deformation structures and the palaeotectonic analysis of sedimentary basins, with a case study from the Carboniferous of northern England. In: *Deformation of Sediments and Sedimentary Rocks* (Eds Jones, M.E. & Preston, R.M.F.) Geol. Soc. London Spec. Publ. **29**, 137–146.

LEEDER, M.R. & GAWTHORPE, R.L. (1987) Sedimentary models for extensional tilt-block/half-graben basins. In: *Continental Extensional Tectonics* (Eds Coward, M.P., Dewey, J. & Hancock, P.L.). Geol. Soc. London Spec. Publ. **28**, 139–152.

LEOPOLD, L.B., WOLMAN, M.G. & MILLER, J.P. (1964) *Fluvial Processes in Geomorphology.* W.H. Freeman, San Francisco, 522 pp.

MAYCOCK, I.D. (1962) *The Torridonian sandstone round Loch Torridon, Wester Ross.* PhD thesis, University of Reading (unpublished).

MCBRIDE, E.F. (1963) A classification of common sandstones. *J. Sediment. Petrol.* **33**, 664–669.

MCKEE, E.D., CROSBY, E.J. & BERRYHILL, J.R. (1967) Flood deposits, Bijou Creek, Colorado, June 1965. *J. Sediment. Petrol.* **37**, 829–851.

MCKENZIE, D. (1978) Some remarks on the development of sedimentary basins. *Earth Planet. Sci. Lett.* **40**, 25–32.

MIALL, A.D. (1976) Palaeocurrent and palaeohydrologic analysis of some vertical profiles through a Cretaceous braided stream deposit, Banks Island, Arctic Canada. *Sedimentology* **23**, 459–483.

MOORBATH, A. (1969) Evidence for the age of deposition of the Torridonian sediments of north-west Scotland. *Scott. J. Geol.* **5**, 154–170.

NEMEC, W. (1988) The shape of the rose. *Sediment. Geol.* **59**, 149–152.

NICHOLSON, P.G. (1992a) Northwestern British Isles: Upper Proterozoic Torridonian. In: *Atlas of Palaeogeography and Lithofacies* (Eds Cope, J.C.W., Ingham, J.K. & Rawson, P.F.) Geol. Soc. London Mem. **13**, 5–7.

NICHOLSON, P.G. (1992b). *Sedimentation of the fluvial 'Torridonian' Applecross Formation, NW Scotland.* PhD Thesis, University of Glasgow.

OWEN, G. (1987) Deformation processes in unconsolidated sands. In: *Deformation of Sediments and Sedimentary Rocks* (Eds Jones, M.E. & Preston, R.M.F.) Geol. Soc. London Spec. Publ. **29**, 11–24.

PEACH, B.N. et al. (1907) *The Geological Structure of the Northwest Highlands of Scotland.* Geological Survey of Great Britain, Memoir, 668 pp.

POTTER, P.E. (1978) Significance and origin of big rivers. *J. Geol.* **86**, 13–33.

ROGERS, G., KROGH, T.E., BLUCK, B.J. & KWOK, Y.Y. (1989) Sedimentary provenance ages of the 'Torridonian' sandstone using single grain U–Pb zircon analysis. *Abstracts, Developments in Sedimentary Provenance Studies,* BSRG & Petroleum Group Joint Meeting, Geological Society, London.

RUST, B.R. & GIBLING, M.R. (1990) Braidplain evolution in the Pennsylvanian South Bar Formation, Sydney Basin, Nova Scotia, Canada. *J. Sediment. Petrol.* **60**, 59–72.

SCHUMM, S.A. (1968) Speculations concerning paleohydrologic controls of terrestrial sedimentation. *Geol. Soc. Am. Bull.* **79**, 1573–1588.

SELLEY, R.C. (1969) Torridonian alluvium and quicksands. *Scott. J. Geol.* **5**, 328–346.

SMITH, N.D. (1971) Transverse bars and braiding in the Lower Platte River, Nebraska. *Geol. Soc. Am. Bull.* **82**, 3407–3420.

SMITH, R.L., STEARN, J.E.F. & PIPER, J.D.A. (1983) Palaeomagnetic studies of the Torridonian sediments, NW Scotland. *Scott. J. Geol.* **19**, 29–45.

STEIN, A.M. (1988) Basement controls upon Hebridean basin development. *Basin Res.* **1**, 107–119.

STEWART, A.D. (1969) Torridonian rocks of Scotland reviewed. In: *North Atlantic: Geology and Continental Drift* (Ed. Kay, M.) Am. Assoc. Petrol. Geol. Mem. **12**, 595–608.

STEWART, A.D. (1975) 'Torridonian' rocks of western Scotland. In: *A Correlation of Precambrian Rocks in the British Isles* (Eds Harris, A.L. *et al.*) Geol. Soc. London Spec. Rep. **6**, 43–51.

STEWART, A.D. (1982) Late Proterozoic rifting in NW Scotland: the genesis of the 'Torridonian'. *J. Geol. Soc. London* **139**, 413–420.

STEWART, A.D. (1988a) The Stoer Group, Scotland. In: *Later Proterozoic Stratigraphy of the Northern Atlantic Regions* (Ed. Winchester, J.A.) Blackie, Glasgow, pp. 97–103.

STEWART, A.D. (1988b) The Sleat and Torridon Groups. In: *Later Proterozoic Stratigraphy of the Northern Atlantic Regions* (Ed. Winchester, J.A.) Blackie, Glasgow, pp. 104–112.

WERNICKE, B. (1985) Uniform-sense normal simple shear of the continental lithosphere. *Can. J. Earth Sci.* **22**, 108–125.

WILLIAMS, G.E. (1966) Palaeogeography of the 'Torridonian' Applecross Group. *Nature* **209**, 1303–1306.

WILLIAMS, G.E. (1969a) Characteristics and origin of a Precambrian pediment. *J. Geol.* **77**, 183–207.

WILLIAMS, G.E. (1969b) Petrography and origin of pebbles from 'Torridonian' strata (Late Precambrian), NW Scotland. In: *North Atlantic: Geology and Continental Drift* (Ed. Kay, M.) Am. Assoc. Petrol. Geol. Mem. **12**, 609–629.

WILLIAMS, G.E. (1970) Origin of disturbed bedding in the Torridon Group sandstones. *Scott. J. Geol.* **6**, 409–411.

Spec. Publs Int. Ass. Sediment. (1993) **20**, 203–221

Stratigraphical evolution of a Proterozoic syn-rift to post-rift basin: constraints on the nature of lithospheric extension in the Mount Isa Inlier, Australia

K.A. ERIKSSON*, E.L. SIMPSON† *and* M.J. JACKSON‡

*Department of Geological Sciences, Virginia Polytechnic Institute and State University, Blacksburg, Virginia 24061, USA;
†Department of Physical Sciences, Kutztown University, Kutztown, Pennsylvania 19530, USA; and
‡Australian Geological Survey Organization, Canberra 2061, Australia

ABSTRACT

Three cover sequences are recognized in the Mount Isa Inlier, Australia; the second consists of the *c.* 17-km-thick Haslingden Group and the overlying *c.* 1.8-km-thick Quilalar Formation and correlatives. This cover sequence accumulated at 1800–1720 Ma over a time period of between 30 and 80 My. Stratigraphical evolution of the Haslingden Group is compatible with a two-stage rift model. The first syn-rift fill consists of felsic volcanics and coarse alluvial sediments. An overlying 2-km-thick interval of quartz arenites of tidal origin accumulated in a post-rift, thermal-relaxation basin. The second syn-rift fill commences with 6 km of continental basalts, the upper half of which contain interflow siliciclastic sediments. An overlying sedimentary interval was derived from a basement high in the centre of the syn-rift basin. The Quilalar Formation and correlatives conformably overlie the syn-rift deposits and also non-conformably overlie basement. Depositional systems were restricted to either north–south trending marginal *platform* or central *trough* palaeogeographical settings in an elongate, intracratonic basin with minimum dimensions of 300 by 100 km. The close temporal relationship of the Quilalar Formation and correlatives to the syn-rift deposits favours a post-rift, thermal-relaxation origin for the basin. This contention is supported by the elongate basin shape and the symmetrical distribution of depositional systems within the basin. The axis of the post-rift basin is located above a pre-existing basement high. The homogeneous-stretching model of McKenzie thus is inappropriate for the second syn-rift to post-rift basin. A model proposed by Coward is more appropriate; this model involves heterogeneous stretching and includes a wide zone of moderate extension in the upper crust and a narrow zone of pronounced extension in the lower crust and lithospheric mantle. Thermally-driven uplift above the zone of lower crustal and mantle thinning results in erosion of syn-rift volcanics and sediments followed by thermal subsidence. This model accounts for the location of the axis of the post-rift basin above a basement high and the occurrence of thick volcanics in the preserved syn-rift basin.

INTRODUCTION

The nature of lithospheric extension remains an ongoing debate with the two most widely cited models being those involving pure shear versus simple shear. In the pure shear model of McKenzie (1978), both the crust and lithospheric mantle extend uniformly over the entire width of the basin. The upper crust behaves in a brittle manner, whereas in the lower crust and lithospheric mantle, pure shear is accommodated by ductile flow. In the simple shear model of Wernicke (1981, 1985), the lithosphere extends along a low-angle detachment fault that penetrates the entire lithosphere. Both models predict that subsidence in response to lithospheric extension takes place in two stages, an initial stage resulting from thinning and associated rifting of the upper crust (syn-rift) and a later stage related to cooling and thickening of the lithosphere (post-rift). Different approaches have been employed in an attempt to distinguish between the two models. In the North Sea, patterns of initial and thermal subsidence, geohistory analysis, as well as the location, volume and composition of magmatism have been used to argue in favour of pure shear models (e.g. Sclater & Christie, 1980; Giltner, 1987; Klemperer & White, 1989; White, 1989; Latin & White, 1990) whereas deep seismic reflection data have been the basis for simple shear models (e.g. Gibbs, 1984; Beach *et al.*, 1987).

The Early Proterozoic Mount Isa Inlier in Australia (Fig. 1) contains two rift-to-sag (syn-rift to post-rift) sequences of intracratonic basin evolution

(cover sequences II and III, Fig. 2; Blake, 1987; Wyborn *et al.*, 1988). The earliest of these phases is the best preserved and consists of up to 17 km of syn-rift sedimentary and volcanic rocks (Haslingden Group) overlain by a 1.8-km-thick sedimentary interval (Quilalar Formation) considered to represent a post-rift, thermotectonic response to rifting (Jackson *et al.*, 1990). Age control for this sequence is limited and the basin fill has been disrupted by post-depositional faulting and folding. Nevertheless it is possible to model the stratigraphical evolution of the Haslingden Group and Quilalar Formation based on: (1) maximum decompacted thicknesses of the syn-rift and post-rift deposits; (2) palaeogeographical reconstructions based on facies analysis; (3) estimates of the duration of syn-rift and post-rift subsidence; (4) the configuration of the post-rift or thermotectonic basin and its spatial relationship with respect to preserved rift deposits; and (5) the presence of thick rhyolitic and basaltic volcanics within the rift sequence. We recognize two stretching phases, both of which were succeeded by thermal relaxation in response to declining heat flow, and conclude that neither pure shear nor simple shear lithospheric models are appropriate for this Early Proterozoic sequence. We suggest that the data are more compatible with the stretching model of Coward (1986) that combines heterogeneous lithospheric stretching and associated simple shear. The evolution of a comparable age foreland-basin succession in Canada has been modelled by Grotzinger and McCormick (1988) and Grotzinger and Royden (1990) whereas this is the first attempt at modelling the stratigraphical evolution of a Precambrian syn-rift to post-rift basin.

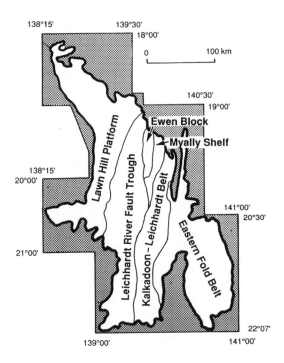

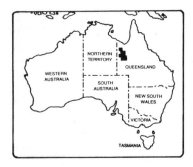

Fig. 1. Location map of the Mount Isa Inlier showing the dominant tectonic elements. (From Blake, 1987.)

GEOLOGY OF THE MOUNT ISA INLIER

The Mount Isa Inlier (Fig. 1) is located in northwest Queensland on the North Australian Craton. The craton formed between 2000 and 1900 Ma (Page *et al.*, 1984); development of the Mount Isa orogenic belt commenced around 1800 Ma and was terminated with deformation and greenschist- to amphibolite-facies regional metamorphism between 1610 and 1510 Ma (Fig. 2; Page & Bell, 1986). The Mount Isa Inlier consists of four north–south trending tectonic or structural elements (Fig. 1). The Kalkadoon–Leichhardt Belt separates an extremely thick western succession occupying the

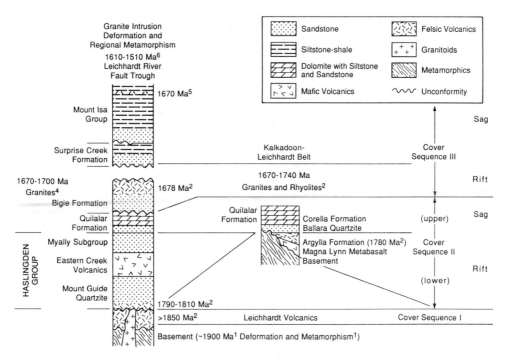

Fig. 2. Generalized lithostratigraphical columns for cover sequences in the Leichhardt River Fault Trough and Kalkadoon–Leichhardt Belt of the Mount Isa Inlier. (Adapted from Derrick *et al.*, 1980; Plumb *et al.*, 1981; Blake, 1987.) Geochronological data from: (1) Blake (1987); (2) Page (1983a); (3) Page (1983b); (4) Page *et al.* (1981); (5) Page (1981); (6) Page and Bell (1986).

Leichhardt River Fault Trough, from a moderately thick and more intensely deformed and metamorphosed succession in the Eastern Fold Belt (Blake, 1987). A relatively thin and flat-lying succession makes up the Lawn Hill Platform.

Primary stratigraphical subdivision in the Mount Isa Inlier is between basement and cover rocks. Basement is exposed mainly in the Kalkadoon–Leichhardt Belt and comprises an assemblage of paragneiss, schist, quartzite and migmatite intruded by felsic plutonics. These rocks were involved in amphibolite-facies regional metamorphism and deformation at around 1900 Ma (Blake, 1987). Where attaining their maximum thickness in the Leichhardt River Fault Trough, cover rocks are subdivisable into three sequences. Cover sequence I consists of felsic volcanic rocks (Leichhardt Volcanics) dated at *c.* 1870 Ma (U–Pb zircon; Page, 1983a) that are intruded by comagmatic granitoid batholiths, of *c.* 1850 Ma age (U–Pb zircon; Page, 1978; Fig. 2).

Lower cover sequence II is represented by the Haslingden Group in the Leichhardt River Fault Trough and by the Magna Lynn Metabasalt and the Argylla Formation (Fig. 2) on the eastern margin of the Kalkadoon–Leichhardt Belt and extending into the Eastern Fold Belt. Upper cover sequence II is represented by the Quilalar Formation to the west and the Ballara Quartzite and overlying Corella Formation east of the Kalkadoon–Leichhardt Belt (Fig. 2; Derrick *et al.*, 1980). Sedimentation of cover sequence II was terminated by uplift and gentle folding of the Haslingden Group and Quilalar Formation and equivalents. The Corella Formation is intruded by granites and rhyolite dykes.

Cover sequence III is developed only west of the Kalkadoon–Leichhardt Belt where it unconformably overlies cover sequence II. Red bed conglomerates and sandstones, and subordinate siltstone and carbonate beds (Bigie Formation) make up lower cover sequence III. An overlying volcanic interval consists of up to 700 m of potassic rhyolite and trachyte, tuff, agglomerate, and minor basalt (Fig. 2; Derrick, 1982). Felsic volcanics are dated at 1678 ± 1 Ma (U–Pb zircon; Page, 1983a).

A major period of granitoid intrusion accompanied or preceded deposition of upper cover sequence III. Granitoids west of the Kalkadoon–Leichhardt Belt are coeval with the felsic volcanics, ranging in age from 1700–1670 Ma (Fig. 2; U–Pb zircon; Page *et al.*, 1981). East of the Kalkadoon–Leichhardt Belt, granitoids and rhyolites have a similar minimum age to those to the west but range down to 1737 ± 15 Ma (Fig. 2; U–Pb zircon; Page, 1978, 1983b).

Upper cover sequence III is represented by the Surprise Creek Formation that accumulated in local linear basins and the overlying Pb–Zn hosting Mount Isa Group that has a more widespread distribution (Blake, 1987). The Surprise Creek Formation is an upward-fining sequence of interbedded conglomerate, sandstone, and mudstone whereas the Mount Isa Group consists of 7 km of lacustrine siliciclastics, carbonates, and evaporites (Neudert & Russell, 1983). A tuff bed in the Mount Isa Group is dated at 1670 ± 20 Ma (U–Pb zircon; Page, 1981).

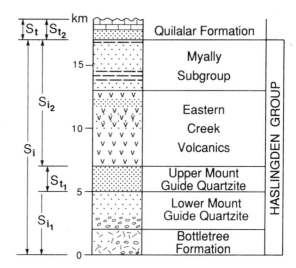

Fig. 3. Stratigraphy of the Haslingden Group and Quilalar Formation in the Leichhardt River Fault Trough. For a single-stage rift: S_i, initial (syn-rift) subsidence; S_t, thermal (post-rift) subsidence. For a two-stage rift, subscripts 1 and 2 refer to first and second rifts, respectively. (Adapted from Blake, 1987.)

STRATIGRAPHICAL EVOLUTION AND PALAEOGEOGRAPHY OF COVER SEQUENCE II

Cover sequence II was studied mainly in that part of the Mount Isa Inlier which is at greenschist metamorphism. Despite the metamorphism, faulting and folding to which the sediments and volcanics have been subjected, primary textures and structures are well preserved (e.g. Jackson *et al.*, 1990; Simpson & Eriksson, 1991). Within structural blocks coherent stratigraphical successions are recognizable and these can be correlated between blocks. No evidence exists for volume loss associated with metamorphism and deformation; pressure solution is rare. Thus, measured thicknesses of stratigraphical units represent true compacted thicknesses.

The Haslingden Group has a cumulative thickness of up to 17 km (Fig. 3). Petrography of sandstones in this group is illustrated on Fig. 4; Table 1 lists facies and interpreted environments of deposition. The basal Bottletree Formation is *c.* 2 km thick and consists of dacitic to rhyolitic and subordinate basaltic volcanics, and alluvial conglomerates and lithic to feldspathic sandstones. Conglomerates contain clasts of basement metamorphics, felsic volcanics of cover sequence I, and granites. The *c.* 3-km-thick, overlying Lower Mount Guide Quartzite consists of conglomerates and lithic to feldspathic sandstones (Fig. 4) derived from the east and predominantly of braided-alluvial origin. Thin intervals of turbidites and stromatolitic dolomites probably represent lacustrine deposits (Table 1). The Upper Mount Guide Quartzite is a *c.* 2-km-thick unit of supermature quartz arenites (Fig. 4) that consists of stacked, metre-scale cycles or parasequences (Table 1; Simpson & Eriksson, 1991). Stacked parasequences define thinning- and thickening-upward packages tens of metres thick. Cycles are attributed to drowning related to high-frequency relative sea-level changes, followed by shoaling from shallow–subtidal to intertidal–supratidal environments. Packages, in contrast, reflect longer-term changes in accommodation space related to lower-frequency, relative sea-level fluctuations (Eriksson & Simpson, 1990).

The overlying Eastern Creek Volcanics are up to 6 km thick and consist of two intervals of subaerial basalts separated by a siliciclastic unit up to 750 m thick (Fig. 3). The basalts are continental tholeiites of subalkaline affinity (Bultitude & Wyborn, 1985; Blake, 1987). Intercalated within the upper half of the Eastern Creek Volcanics are tabular, siliciclastic sedimentary horizons from 1 to 40 m thick. Sandstones predominate and consist mainly of quartz, and felsic volcanic, quartz arenite and locally

Table 1. Facies and depositional environments in cover sequence II

Stratigraphical unit	Facies	Depositional environment
Bottletree Formation	*Continental* — alternating felsic volcanic and subordinate sedimentary intervals. Sediments consist of massive conglomerates, massive and horizontal- and cross-stratified sandstones and 1- to 2-cm-thick pelites arranged in 2- to 5-m-thick fining-upward sequences	Ignimbrites and alluvial sheet-flood deposits
Lower Mount Guide Quartzite	*Continental* — facies include massive conglomerates, pebble lags, and sandstones containing medium-scale trough cross-bedding passing vertically into medium- to small-scale trough cross-bedding. Subordinate Bouma-type beds and stromatolitic dolomites *Palaeocurrents* — westerly	Braided-alluvial with marginal lacustrine systems
Upper Mount Guide Quartzite	*Tide-dominated shelf* — facies consist of cross-bedded and thin-bedded arenites. Cross-bedded arenites consist of tabular cross-bed sets and cosets with sigmoidal or concave-up pause planes, compound cross-bed cosets with a hierarchy of bounding surfaces, and large-scale trough cross-bed cosets. Thin-bedded arenites contain: unidirectional, oscillatory, combined-flow, flat-topped, and ladderback ripples, washed-out ripple patches, desiccation cracks, adhesion ripples and warts, swash laminations and low-angle wind-ripple stratification. Facies are associated in stacked parasequences *Palaeocurrents* — cross-beds: west-southwesterly; unidirectional ripples: southeasterly flow; combined-flow and wave ripples: northwest–southeast flow.	Subaqueous subtidal dunefield complexes that shoaled upwards into intertidal-to-supratidal sandflat (interpretations from Simpson & Eriksson, 1991). Parasequences are separated by flooding surfaces. Well-developed cyclicity related to high-frequency fluctuations of relative sea-level (Eriksson & Simpson, 1990)
Eastern Creek Volcanics	*Continental* — subaerial flood basalts with intercalated siliciclastic sedimentary units. Thicker siliciclastic units have areal extent of 100s of square km; thinner units persist along strike for 10s to 100s of m. Facies include: (1) massive clast-supported conglomerates overlain by horizontally stratified sandstones, (2) cross-stratified, massive and horizontally stratified conglomerates overlain by planar, horizontally- and trough cross-stratified sandstones, (3) fining-upward sequences consisting of conglomerate lags overlain by trough cross-bedded sandstones capped by either horizontal stratification or small-scale trough cross-laminations, (4) sandstones containing tabular-planar to tabular-tangential cross-bed cosets capped by erosional trough cross-beds, (5) trough cross-laminated and horizontally stratified sandstones, and (6) conglomerates and cross-bedded sandstones overlain by low-angle subcritically climbing translatent stratification *Palaeocurrents* — northerly	Braided-fluvial and subordinate aeolian systems (interpretations from Eriksson & Simpson, 1993). (1) Longitudinal-bar and bar-top deposition, (2) downstream accretion of gravel sheets that were gradually abandoned, (3) vertical aggradation of mid-channel bars, (4) migration of transverse or linguoid bars that were incised during falling-water stage, (5) sheet-flood processes in ephemeral rivers, (6) aeolian reworking of fluvial deposits. Facies association 1 deposited by transverse higher-gradient streams that supplied sediment from the east to a lower-gradient, longitudinal system

(Continued on p. 208)

Table 1. (*continued*)

Stratigraphical unit	Facies	Depositional environment
Myally Subgroup (Leichhardt River Fault Trough) Alsace Formation	*Tide-dominated shoreline* — facies consist of tabular sets of trough and planar cross-bedded arenites that display a vertical decrease in scale of structures, and thin-bedded arenites containing oscillatory, combined-flow, unidirectional, flat-topped and ladderback ripples, adhesion ripples and warts, and swash laminations. Cross-bedded facies dominate lower half of formation whereas upper half contains both facies randomly interbedded *Palaeocurrents* — highly variable with dominant mode to southeast	Structures present are consistent with a tide-dominated setting in which lower interval is dominantly subtidal shelf and upper interval subtidal shelf to intertidal–supratidal flat. Poorly developed cyclicity may be related to episodic subsidence
Bortalla Formation	*Storm-dominated shelf* — facies consist of black mudstone, Bouma-type beds, massive, horizontally stratified, hummocky cross-stratified sandstones passing upwards into trough cross-bedded and thin-bedded sandstones with oscillatory, combined-flow, unidirectional, flat-topped and ladderback ripples, desiccation cracks, and adhesion warts and ripples. Facies are arranged in three thickening- and coarsening-upward parasequences; the uppermost is gradational into the basal Whitworth Formation (Fig. 5) *Palaeocurrents* — cross-beds: northerly; combined-flow ripples give northwest–southeast flow	Vertical succession of facies records shoaling from outer shelf through shoreface into intertidal–supratidal conditions on a storm-dominated shelf-coast. Progradation from east
Whitworth Formation	*Continental* — facies include wind-rippled arenites, making up cosets of cross-strata that contain rare evaporitic casts, (2) sets of horizontal to low-angle stratification, and (3) isolated sets of cross-strata. Additional arenite facies include: horizontal stratification, horizontal stratification overlain by trough cross-strata, and cosets of medium-scale trough cross-beds *Palaeocurrents* — southwesterly	Wind-ripple facies interpreted as: (1) deposits of small dunes, (2) interdune, sand-sheet or plinth deposits, and (3) thin preservation of dune deposits. Additional facies are interpreted as ephemeral-stream deposits (Simpson & Eriksson, 1993). Progradation from east over Bortalla Formation
	Tide-dominated shelf — facies consist of cross-bedded and thin-bedded arenites. Cross-bedded arenites are dominated by tabular sets and cosets with concave-up and sigmoidal pause planes. Thin-bedded arenites contain oscillatory, combined-flow, unidirectional, flat-topped, ladderback and interference ripples, washed-out ripple patches, desiccation cracks, and adhesion ripples and warts. Facies are randomly interbedded *Palaeocurrents* — cross-beds: southwesterly	Internal structures within cross-beds record tidal currents. Structures within thin-bedded arenites formed in intertidal to supratidal settings (Simpson & Eriksson, 1993). Poorly developed cyclicity may be related to episodic subsidence

(*Continued on p. 209*)

Table 1. (*continued*)

Stratigraphical unit	Facies	Depositional environment
Lochness Formation	*Continental* — facies consist of interbedded sandstones and mudstones and include: (1) horizontal stratification and tabular sets of planar and trough cross-beds; (2) rippled arenites containing oscillatory, chevron, combined-flow unidirectional, double-crested, flat-topped, ladderback and interference ripples, desiccation cracks, and adhesion ripples and warts; and (3) mottled horizons. Facies are associated in couplets 1–3 or 2–3 *Palaeocurrents* — cross-beds: southwesterly	Ephemeral-river and lacustrine deposits with associated soils
Myally Subgroup (western flank Kalkadoon–Leichhardt Belt)	Facies consist of (1) boulder conglomerates up to 250 m thick. Conglomerates are clast-supported. Clasts are well-rounded and consist of felsic volcanics and massive quartz arenites up to 40 cm diameter. (2) Cross-bedded and thin-bedded quartz arenites. Crossbeds are tabular-tangential with pause planes. Thin-bedded arenites are wave rippled. Facies 1 and 2 developed separately or facies 1 is overlain by facies 2. Sediment derived from east	Facies 1: alluvial deposits of hyperconcentrated flows Facies 2: tide-dominated shelf comparable to lower Whitworth Formation
Myally Subgroup (Basal Quilalar and Basal Ballara on eastern flank Kalkadoon–Leichhardt Belt)	Facies include conglomerates and feldspathic arenites in stratigraphical successions up to 700 m thick and lenticular along strike over 2 to 10 km. Conglomerates are mainly clast-supported. Clasts are well-rounded and consist of felsic volcanics, massive quartz arenites, stratified feldspathic arenites, basalts and vein quartz up to 25 cm in diameter. Beds are single clast to 10 m thick. Arenites are trough cross-bedded and horizontally stratified. The two facies are developed either separately or define fining-upward sequences. Sediment derived from west	Thick conglomerates are alluvial, hyperconcentrated flow deposits. Thinner conglomerates are channel-bar and channel-lag deposits of braided rivers. Arenites represent channel-bar deposits
Lower Quilalar Formation	*Continental* — facies include: (1) clast-supported massive conglomerates overlain by horizontally stratified and trough cross-bedded sandstones; and (2) sandstones with large-scale trough cross-beds composed of inversely graded stratification *Palaeocurrents* — limited to facies 2 — southerly *Petrology* — feldspathic arenites; *clast types* — quartz porphyry and granite with subordinate vein quartz and quartz arenite	Ephemeral-river (1) and aeolian deposits (2) (see Jackson *et al.*, 1990)
	Tide-dominated shelf — facies include cross-bedded and thin-bedded arenites. Cross-beds consist of tabular-planar sets with curved and sigmoidal-shaped pause planes and compound cosets with a hierarchy of bounding surfaces. Thin-bedded arenites contain oscillatory, combined-flow, unidirectional, flat-topped and ladderback ripples, washed-out ripple patches, desiccation cracks, and adhesion ripples and warts. Facies associated in stacked parasequences that define thickening- and thinning-upward parasequence sets *Palaeocurrents* — cross-beds: southerly to southwesterly; oscillatory; combined-flow and unidirectional ripples give north–south flow *Petrology* — quartz arenites	Structures within cross-stratified arenites are characteristic of tide-dominated shelves. Thin-bedded arenites accumulated in intertidal–supratidal settings (see Jackson *et al.*, 1990). Well-developed cyclicity reflects high-frequency fluctuations of relative sea-level. Stacking patterns of parasequences related to different rates of accommodation space creation

(*Continued on p. 210*)

Table 1. (*continued*)

Stratigraphical unit	Facies	Depositional environment
	Storm-dominated shelf — facies consist of black mudstone, Bouma-type beds, massive, horizontally stratified, and hummocky cross-stratified sandstones passing upwards into trough cross-stratified sandstones. Facies arranged in thickening- and coarsening-upward parasequences and parasequence sets and subordinate thinning- and fining-upward parasequence sets	Vertical sequences record progradation of a storm-dominated shelf (Jackson *et al.*, 1990). Recognizable subenvironments are outer and inner shelf, and shoreface. Parasequences are separated by flooding surfaces. Control on cyclicity same as above
	Palaeocurrents — southeasterly	
	Petrology — feldspathic arenites	
	Wave-dominated shoreline — facies include: (1) megarippled arenite; (2) interbedded wave-rippled arenite and mudstone; (3) cosets of trough cross-stratified arenite; and (4) cosets of low-angle, cross-stratified arenite. Facies 1–4 are associated in stacked parasequences	Vertical sequence of structures records progradation of wave-dominated shoreline from lower to upper shoreface to foreshore (Jackson *et al.*, 1990). Megaripples define ravinement or flooding surfaces at top of parasequences. Well-developed cyclicity related to high-frequency fluctuations of relative sea-level
	Paleocurrents — cross-strata: northerly; wave ripples: northwest–southeast flow	
	Petrology — quartz arenite	
Upper Quilalar and Corella Formations	*Continental* — facies consist of (1) horizontal- and trough cross-bedded coarse ferruginous sandstone; and (2) current rippled and desiccated fine dolomitic sandstone	(1) Channelized ephemeral river, and (2) alluvial flat
	Carbonate shelf — facies include: (1) matrix-supported, intraclast breccias overlain by trough cross-bedded coarse, sandy, oolitic dolostones; (2) desiccated dolomitic mudstones with evaporite pseudomorphs; and (3) columnar stromatolite biostromes and low-relief domal bioherms. Facies 1 and 2 associated in stacked parasequences. Facies 3 randomly interbedded	Intertidal (1), supratidal (2), and shallow subtidal (3). Cyclicity related to high-frequency relative sea-level fluctuations
	Storm-dominated carbonate shelf — facies include: (1) matrix-supported, intraclast conglomerates overlain by massive and wavy laminated dolomitic siltstone, and dolomitic mudstone, in decimetre-thick Bouma-type beds; and (2) parallel, thin-bedded and laminated, graded dolomitic siltstone-mudstone	Subtidal

derived basalt rock fragments (Fig. 4). Local conglomerates contain felsic volcanic, granite and quartz arenite clasts. Felsic volcanic fragments were derived by uplift and erosion of cover sequence I whereas quartz arenite clasts represent recycled Upper Mount Guide Quartzite. Sedimentation took place during prolonged hiatuses in volcanism in braided-alluvial and subordinate aeolian environments (Table 1). Facies, (including palaeocurrent) analysis, have identified two interacting alluvial systems: a relatively high-gradient and coarse-grained, transverse system that supplied siliciclastic

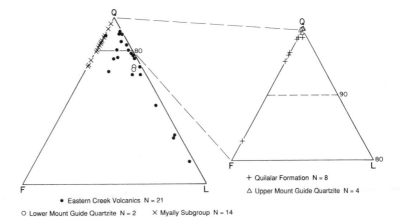

Fig. 4. Petrography of the Haslingden Group and Quilalar Formation. Q, monocrystalline quartz; F, feldspar; L, total lithics including polycrystalline quartz.

+ Quilalar Formation N = 8
△ Upper Mount Guide Quartzite N = 4

● Eastern Creek Volcanics N = 21
○ Lower Mount Guide Quartzite N = 2   × Myally Subgroup N = 14

debris from eastern highlands and a lower-gradient, sandy, longitudinal or trunk system that reworked sediment down a south-to-north palaeoslope parallel to the axis of the Leichhardt River Fault Trough (Fig. 5). The Magna Lynn Metabasalts on the eastern margin of the Kalkadoon–Leichhardt Belt are correlated with the Eastern Creek Volcanics (Blake, 1987).

The Myally Subgroup is the uppermost subdivision of the Haslingden Group; in the Leichhardt River Fault Trough it attains a thickness of 4 km and consists of the Alsace, Bortalla, Whitworth and Lochness Formations (Fig. 6). Sandstones in the Myally Subgroup are quartzose to feldspathic (Fig. 4) and contain scattered pebbles that include felsic and mafic volcanics. Discontinuous pebbly sandstones of local derivation in the Alsace Formation have been related to synsedimentary transpressional faulting (W. Nijman, personal communication, 1988). Rare-earth and trace-element data from mudstones in the Myally Subgroup provide additional information on provenance composition. Higher Eu/Eu* values, lower light:heavy rare-earth element, La/Sc and Th/Sc ratios, and greater concentrations of Cr, Ni, V and Co than in sediments derived exclusively from granitic crust reflect dilution of differentiated felsic provenances by mafic volcanic components (Eriksson *et al.*, 1992). Sedimentary environments represented in the arenaceous Alsace and Whitworth formations include braided-alluvial, aeolian and shallow-marine, whereas the more argillaceous Bortalla Formation is exclusively of storm-shelf origin. Alluvial and lacustrine conditions prevailed during deposition of the Lochness Formation and well-developed soils are present in the Lochness Formation (Table 1). The

Fig. 5. Palaeogeographical model for siliciclastic units intercalated within the Eastern Creek Volcanics illustrating interacting transverse and longitudinal alluvial systems. Longitudinal rivers flowed from south to north. Siliciclastic debris was derived exclusively from an eastern provenance consisting of the Upper Mount Guide Quartzite and cover sequence I composed of felsic volcanics and related granites, and metaquartzites. Inset shows location of palaeogeographical reconstruction. LHP, Lawn Hill Platform; LRFT, Leichhardt River Fault Trough; K–LB, Kalkadoon–Leichhardt Belt; EFB, Eastern Fold Belt.

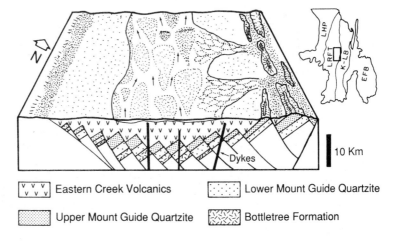

Dykes

10 Km

| v v v Eastern Creek Volcanics

:::::: Lower Mount Guide Quartzite

Upper Mount Guide Quartzite

Bottletree Formation

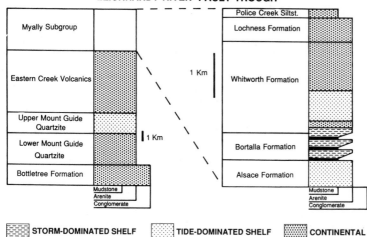

Fig. 6. Genetic stratigraphy of the Haslingden Group. Volcanics in the Bottletree Formation and Eastern Creek Volcanics are entirely subaerial.

vertical transition from Bortalla to Whitworth to Lochness indicates progressive shoaling and infilling of the basin. Progradation took place from east to west. On the eastern margin of the Leichhardt River Fault Trough, the Myally Subgroup is less than 500 m thick. Facies consist of quartz arenites, and boulder conglomerates up to 250 m thick that contain quartz arenite and felsic volcanic clasts (Table 1). The Argylla Formation east of the Kalkadoon–Leichhardt Belt is up to 1200 m thick and is correlated with the Myally Subgroup (Blake, 1987). This formation consists mainly of dacitic to rhyolitic volcanics and also contains sandstones. Locally the Argylla Formation is capped by alluvial boulder conglomerates and feldspathic arenites with subordinate basalts and felsic tuffs (Table 1). Stratigraphical thicknesses are up to 700 m. Felsic volcanics, quartz arenites and locally derived sandstones are the dominant clast types. Facies patterns and palaeocurrent data indicate that the Kalkadoon–Leichhardt Belt was a topographical high during deposition of the Myally Subgroup and correlatives to the east. Conglomeratic, alluvial sediments were deposited adjacent to both the western and eastern margins of the Kalkadoon–Leichhardt Belt (Fig. 7). Early in the depositional history of the Whitworth Formation, these coarse sediments graded westwards in the Leichhardt River Fault Trough into braided-alluvial and storm-shelf mudstones and arenites (Fig. 7). Associated with transgression in later Whitworth time, a tide-dominated sandy shelf became established in the Leichhardt River Fault Trough and onto the west-

ern flank of the Kalkadoon–Leichhardt Belt. Available data suggest that subaerial conditions prevailed throughout deposition of Myally Subgroup equivalents east of the Kalkadoon–Leichhardt Belt (Fig. 7). Clast types, sandstone petrography and mudstone geochemistry indicate that Myally Subgroup and correlatives were derived from a source terrane consisting of quartz arenites of the upper Mount Guide Quartzite, felsic igneous rocks comparable to cover sequence I, and the mafic Eastern Creek–Magna Lynn volcanics (Eriksson et al., 1992).

Upper cover sequence II accumulated across all structural elements in the Mount Isa Inlier (Fig. 1). The Quilalar Formation is up to 1800 m thick and consists of lower siliciclastic and upper dolomite–sandstone members. Sandstones range in composition from quartz arenites to subarkoses (Fig. 4). It comformably overlies the Haslingden Group in the Leichhardt River Fault Trough and non-conformably onlaps basement and cover sequence I in the Kalkadoon–Leichhardt Belt (Fig. 2). On the eastern flank of the Kalkadoon–Leichhardt Belt and in the Eastern Fold Belt, the Ballara Quartzite overlies the Argylla Formation both conformably and unconformably or conformably overlies the coarse sediments correlated with the Myally Subgroup (Blake, 1987). Detailed facies analysis of the Lower Quilalar Formations and Ballara Quartzite has permitted discrimination of four depositional systems (Fig. 8; Table 1). Continental, tide-dominated shelf and wave-dominated shoreline depositional systems were restricted to north–south

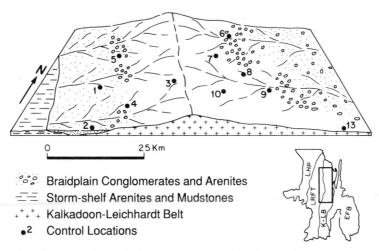

Fig. 7. Palaeographical model for the Myally Subgroup west of the Kalkadoon–Leichhardt Belt, and sedimentary correlatives of the Myally Subgroup east of the Kalkadoon–Leichhardt Belt at time of progradation, on a palinspastic base from Jackson *et al.* (1990). Note that the Kalkadoon–Leichhardt Belt was a topographical high supplying coarse detritus to both the west and east. The Kalkadoon–Leichhardt Belt consisted of felsic volcanics of cover sequence I overlain by the Upper Mount Guide Quartzite and Eastern Creek Volcanics. Control locations are the same as for upper half of Fig. 9. Inset shows location of the palaeogeographical reconstruction prior to palinspastic restoration. LHP, Lawn Hill Platform; LRFT, Leichhardt River Fault Trough; K-LB, Kalkadoon–Leichhardt Belt; EFB, Eastern Fold Belt.

trending, eastern and western marginal platform palaeogeographical settings whereas a central trough was dominated by the storm-dominated depositional system (Fig. 8; Jackson *et al.*, 1990). Metre-scale cyclicity characterizes the shallow-marine depositional systems. On a basin scale, depositional systems are associated in systems tracts that make up stacked sequences. The two scales of cyclicity are attributed to relative sea-level changes of different frequency (Jackson *et al.*, 1990). Symmetrical distribution of the depositional systems favours an intracratonic setting for the basin. Palinspastic reconstruction indicates that the basin was elongate (Fig. 9) with minimum dimensions of 300 by 100 km (Jackson *et al.*, 1990). The axis of the basin, represented by the trough facies, is located over a pre-existing basement high flanked on either side by preserved rift deposits (cf. Figs 7 & 9). Similar platformal (continental and carbonate shelf) and trough (storm-dominated shelf) associations are recognized in the carbonate-rich Upper Quilalar and Corella Formations (Table 1).

The age of cover sequence II is constrained by U–Pb zircon dating of felsic volcanics in the Bottletree Formation (1808^{+22}_{-17} and 1790^{+10}_{-8} Ma; Page, 1983a) and rhyolite dykes that cross-cut the Corella Formation (1737 ± 15 Ma; Page, 1983b). Thus, cover sequence II developed over a maximum time

period of 80 Ma and may represent as short a period of time as 30 Ma.

TECTONIC SETTING

The Mount Isa Inlier is considered variously to have developed in continental-margin and intracratonic settings (Glikson, 1980; Plumb *et al.*, 1981; Condie, 1982; Wyborn & Blake, 1982; Wyborn *et al.*, 1988). Volcanics in lower cover sequence II, as throughout the inlier, are distinctly bimodal, consisting predominantly of continental tholeiites and ignimbrites. Bimodal volcanics typify continental rifts (Burke & Kidd, 1980) and particularly wet rifts characterized by high strain rates (Barberi *et al.*, 1982). McKenzie and Bickle (1988) show that in the presence of mantle plumes, significant thicknesses (5–12 km) of tholeiitic basalts can be generated in intracratonic extensional settings by partial melting of the asthenosphere with β values less than 2.0, whereas in the absence of plumes, β values greater than 3.0 are required to produce tholeiitic basalts. Mudstones in the Haslingden Group also have a bimodal signature; their trace element and rare earth element geochemistry records mixing in different proportions of mafic and felsic provenance components (Eriksson *et al.*, 1992). Other evidence

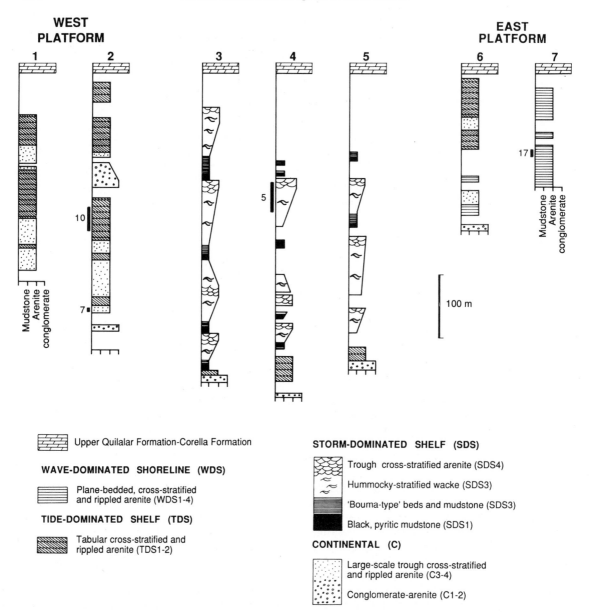

Fig. 8. Genetic stratigraphy of the Lower Quilalar Formation–Ballara Quartzite illustrating lithofacies that make up the WDS, TDS, SDS and C depositional systems. The base of the Upper Quilalar Formation and the Corella Formation is used as a datum. (Adapted from Jackson *et al.*, 1990.)

in the Haslingden Group for rifting includes boulder conglomerates and abundant feldspathic sandstones in the Bottletree Formation, Lower Mount Guide Quartzite and Myally Subgroup. The elongate, symmetrical shape of the basin in which the Lower Quilalar Formation–Ballara Quartzite accumulated and the symmetrical distribution of depositional systems within the basin are the best lines of evidence for an intracratonic setting. Stratigraphical evolution is consistent with the Haslingden Group representing an extensional, probable synrift sequence; thick quartz arenites of the Upper

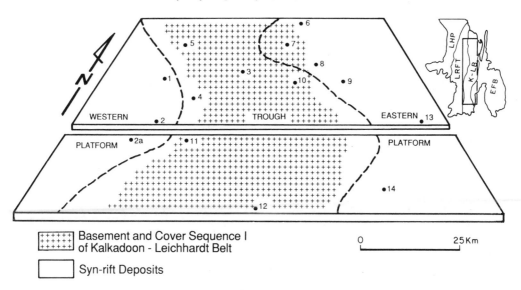

Fig. 9. Distribution of palaeogeographical settings recognized in the post-rift Lower Quilalar Formation–Ballara Quartzite on a palinspastic base (from Jackson *et al.*, 1990). Localities 1–7 are the same as on Fig. 7. Localities 8–14 were assigned to the respective palaeographical setting on the basis of reconnaissance observations. Note that basin axis represented by the *trough* is located above a pre-existing basement high consisting of basement rocks and cover sequence I, and *not* above preserved syn-rift deposits. The syn-rift Haslingden Group is located beneath the *western platform*, and syn-rift correlatives of the Haslingden Group are located beneath the eastern platform. Insert shows location of the palaeographical reconstruction prior to palinspastic restoration. LHP, Lawn Hill Platform; LRFT, Leichhardt River Fault Trough; K-LB, Kalkadoon–Leichhardt Belt; EFB, Eastern Fold Belt.

Mount Guide Quartzite are less compatible with a rift setting. The more widespread distribution of the Eastern Creek Volcanics and Myally Subgroup and their correlatives than the Bottletree Formation and Mount Guide Quartzite suggests that the extensional basin expanded through time. The Quilalar Formation and correlative Ballara Quartzite and Corella Formation are interpreted as a post-rift deposit that developed in response to cooling and thickening of the lithosphere.

The limited evidence in the Haslingden Group for transpression is compatible with the inferred rift setting. Strike-slip faulting takes place concurrently with extension along transfer faults (Gibbs, 1984) and reverse faults are typically associated with oblique-slip extension in rifts (Gibbs, 1987).

NATURE OF LITHOSPHERIC EXTENSION

The Haslingden Group contains a distinctive tenfold stratigraphy (Fig. 3). In addition, a number of the formations contain an internal stratigraphy,

notably the thick Bottletree Formation and Eastern Creek Volcanics. The complete Haslingden Group stratigraphy is not preserved at any one locality within the Leichhardt River Fault Trough, and thus the 17-km thickness (Fig. 3) is a composite from a number of localities. Outcrops of the Bottletree Formation are confined to the southern half of the Leichhardt River Fault Trough whereas the Myally Subgroup thins from *c.* 4 km in the north to less than 2 km south of Mount Isa. The maximum thickness of the Haslingden Group in any part of the Leichhardt River Fault Trough can be estimated at *c.* 15 km. Such a thickness is not unrealistic in rift systems. The little-deformed Midcontinent Rift contains up to 32 km of volcanic and sedimentary rock (Behrendt *et al.*, 1988). Also, seismic data indicate that the extensional, intracratonic, Late Precambrian–Early Palaeozoic Amadeus and Palaeozoic Canning basins in Australia contain upwards of 10 km and 14 km of sediments, respectively (Lindsay & Korsch, 1989; Drummond *et al.*, 1991).

For purposes of subsidence modelling, *maximum* thicknesses are used; these are 15 km for the Haslingden Group and 1.8 km for the Quilalar

Formation. Stratigraphical units vary in thickness indicating variable amounts of subsidence but in this analysis we are concerned only with estimating the maximum amount of stretching (β) in any part of the Mount Isa Inlier.

We first model cover sequence II as a single-stage, syn-rift to post-rift sequence that developed in response to uniform extension (McKenzie, 1978) or non-uniform extension (Royden, 1990). Both models invoke instantaneous stretching and thus are valid only if cover sequence II developed in c. 30 Ma (the lower age limit) rather than c. 80 Ma (the upper age limit). Because the axis of the thermotectonic (post-rift) basin is located above basement and cover sequence I (Fig. 9) rather than above the syn-rift Haslingden Group, only the syn-rift phase can be modelled. To determine the amount of stretching necessary to accommodate the syn-rift deposits it is necessary to identify the amount of subsidence related to initial stretching as distinct from that resulting from sediment, volcanic and water loading. This is achieved by backstripping the Haslingden Group using the method of Royden (1990) and assuming an average compaction factor for the Haslingden Group as that of quartz sandstones (0.23, Bond et al., 1983). Calculations show that initial subsidence of 3.2 km will accommodate the 15-km-thick Haslingden Group. This amount of initial subsidence requires crustal extension values (β) for uniform extension, and upper crustal extension values (β) for non-uniform extension of greater than 3.0 (Fig. 10), clearly showing that the Haslingden Group could not have developed in response to a single stretching event.

The implication of the above is that cover sequence II must reflect more than one period of extension within the 30 to 80 Ma time period available. The 2-km-thick quartz arenite Upper Mount Guide Quartzite is a likely candidate as an early post-rift, thermotectonic response. For realistic β values of 2.0, insignificant thicknesses of sediment develop in response to thermal relaxations in less than 20 Ma (McKenzie, 1978). It is thus reasonable to assume that the post-rift basins encompass 20 to 30 Ma each with a maximum of 30 to 40 Ma available for accumulation of the two syn-rift sequences. Mafic and felsic volcanics in the Midcontinent Rift system accumulated at rates of $1.3 ^{+1.8}_{-0.6}$ mm/year (Davis & Paces, 1990). Based on these subsidence rates, the Bottletree Formation and the Eastern Creek Volcanics would represent maximum time spans of 5 Ma and 10 Ma respec-

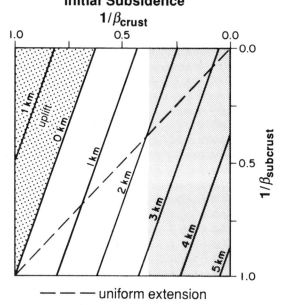

Fig. 10. Model of the Haslingden Group as a single rift showing that a 3.2-km initial subsidence (shaded area), derived by backstripping the Haslingden Group, requires crustal extension values for uniform extension, and upper crustal extension values for non-uniform extension of greater than 3.0. (Graph is from Royden, 1990.)

tively, with 20 to 30 Ma represented by the Lower Mount Guide Quartzite and Myally Subgroup.

The upper Mount Guide Quartzite is located above the Bottletree Formation and lower Mount Guide Quartzite, and thus both the initial and thermal subsidence phases of the first syn-rift to post-rift sequence (Fig. 3) can be modelled. Initial subsidence to accommodate the decompacted 5.0-km-thick Bottletree Formation and lower Mount Guide Quartzite is 1.35 km (Fig. 11), whereas that required for the 7-km-thick Bottletree Formation, Lower Mount Guide Quartzite and Upper Mount Guide Quartzite is 1.85 km. Subsidence related to thermal contraction is thus c. 0.5 km and is estimated to have taken place over 20 to 30 Ma (Fig. 12; Eriksson & Simpson, 1990). From Fig. 11, β values can be estimated as c. 2.0 for uniform extension. For non-uniform extension, the amount of upper crustal extension can be estimated as c. 1.5, whereas subcrustal β values are c. 2.0 (Fig. 12).

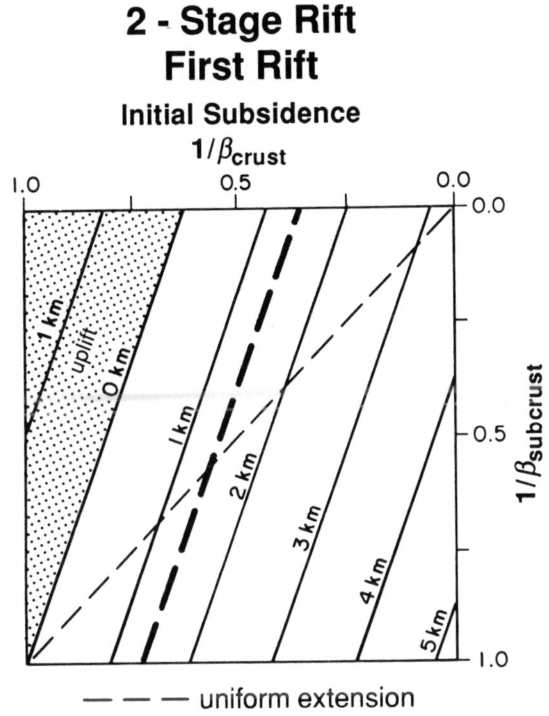

2 - Stage Rift
First Rift

Initial Subsidence

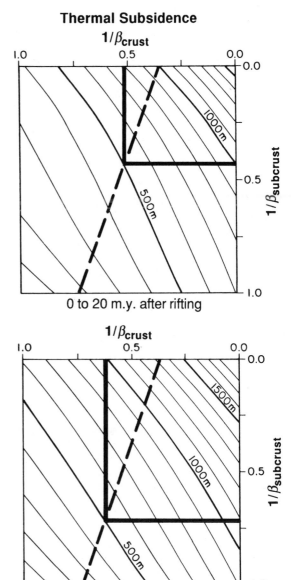

— — — uniform extension

Fig. 11. Model for the first rift (treating the Haslingden Group as a two-stage rift) in which the post-rift basin is located above the syn-rift deposits. Initial subsidence to accommodate the Bottletree Formation and Lower Mount Guide Quartzite is *c.* 1.35 km; this curve is superimposed on Fig. 12. From this figure crustal extension values can be estimated as *c.* 2.0 for uniform extension. (Graph is from Royden, 1990.)

All values are compatible with an intracratonic basin.

Whereas the lower syn-rift to post-rift sequence can be explained using an extension model involving deformation distributed throughout the lithosphere (e.g. McKenzie, 1978; Royden *et al.*, 1980), the upper sequence (Fig. 3) is more difficult to accommodate in such a model. In these models the thermal basin is superimposed symmetrically over the earlier extensional basin. The location of the axis of the second (younger) post-rift basin above basement and cover sequence I (Fig. 9) rather than above syn-rift deposits must be accounted for in any extension model proposed for the younger syn- and post-rift sequence. An alternative lithospheric extension model is that of Wernicke (1981, 1985), which involves displacement on a large-scale, gently-dipping detachment fault that penetrates the

Fig. 12. Two-stage rift (first rift). Thermal subsidence to accommodate the Upper Mount Guide Quartzite is *c.* 0.5 km. Available age data suggest that the Upper Mount Guide Quartzite basin evolved over 20 to 30 Ma. Thus, upper crustal and subcrustal extension values can be estimated as *c.* 1.5 and 2.0, respectively, for non-uniform extension. (Graphs are from Royden, 1990.)

entire lithosphere. In this model, the zone of maximum stretching and asthenospheric upwelling is located in the lower crust and lithospheric mantle rather than beneath the zone of upper crustal extension where stretching factors are small. A thermal-relaxation basin develops above the zone of maximum stretching if erosion, associated with heating and uplift, precedes cooling. The resulting thermal basin may (Coward, 1986) or may not (White, 1989) be offset with respect to the preserved rift deposits. This model could account for the location of the thermal basin at the top of cover sequence II but is not compatible with the location of the 6-km-thick Eastern Creek Volcanics west of the zone of maximum stretching. Latin and White (1990) have shown that small volumes, if any, of basaltic melt, are produced by partial melting of the asthenosphere in the Wernicke simple shear model (Wernicke, 1985).

Coward (1986) proposed a heterogeneous or non-uniform extension model that predicts lateral offset of the axis of the thermal basin with respect to preserved syn-rift deposits. This model includes a wide zone of moderate extension in the upper crust and a narrow zone of more pronounced stretching in the lower crust and lithosphere mantle. Thermally driven uplift above the zone of lower crustal and lithospheric mantle thinning, resulting in local

erosion of syn-rift deposits, is followed by thermal subsidence (Fig. 13). In this model the lower half of the Eastern Creek Volcanics lacking siliciclastic sediments can be related to initial stretching (Fig. 13A). Subsequent thermal uplift resulted in unroofing of the Upper Mount Guide Quartzite and cover sequence I, including cogenetic granites, and their erosion and deposition in the upper half of the Eastern Creek Volcanics (Fig. 13B). Evidence for reworking of mafic (Eastern Creek–Magna Lynn) and felsic (cover sequence I) volcanics and Upper Mount Guide Quartzites into the Myally Subgroup and correlatives to the west and east of the Kalkadoon–Leichhardt Belt is compatible with this stage of basin development. The Coward model also accounts for the location of the post-rift basin with respect to preserved syn-rift deposits, and pre-rift basement and cover sequence I (Fig. 13C) and also accommodates evidence for expansion of the syn-rift basin with time. In addition, thick mafic volcanic intervals are compatible with this model (Coward, 1986; Latin & White, 1990). Based on an estimated time span of 20 to 30 My and using backstripping techniques of Royden (1990), at least 0.6 km of thermal subsidence is required to accom-

Faults propagate away
from rift

A

Decoupling on previous fabric   **LOWER EASTERN CREEK - MAGNA LYNN**

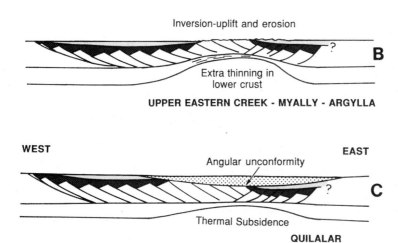

Inversion-uplift and erosion

B

Extra thinning in
lower crust

UPPER EASTERN CREEK - MYALLY - ARGYLLA

WEST   **EAST**

Angular unconformity

C

Thermal Subsidence

QUILALAR

Fig. 3. Evolution of the second syn-rift to post-rift basin. (Modified from Coward, 1986.) (A) Where preserved, Lower Eastern Creek Volcanics and Magna Lynn Metabasalt lack siliciclastic sediments. (B) Upper Eastern Creek Volcanics contain numerous siliciclastic sedimentary horizons; sediments were supplied from an uplifted provenance to the east by transverse rivers and reworked by a longitudinal trunk system. Overlying Myally Group sediments were derived from a central basement high. (C) The axis of the symmetrical thermal-relaxation basin is located above the pre-existing basement high.

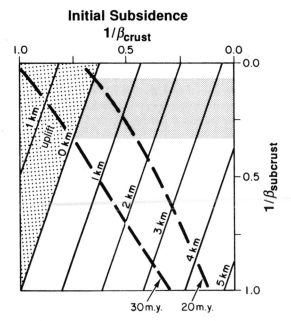

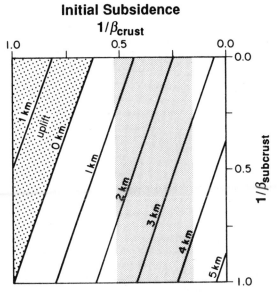

Fig. 14. Model for the second stage of rifting (treating the Haslingden Group as a two-stage rift) in which the axis of the post-rift basin is located above a pre-existing basement high. Model is based on non-uniform extension that approximates the model of Coward (1986) in requiring decoupling of mid-crustal depths. Thermal subsidence to accommodate the Quilalar Formation and correlatives is *c.* 0.6 km. Curves for 20 and 30 My for this amount of thermal subsidence are shown. Because uplift preceded thermal subsidence, subcrustal extension can be estimated as greater than 3.0. (Graph is from Royden, 1990.)

Fig. 15. Two-stage rift (second rift). Initial subsidence to accommodate the Eastern Creek Volcanics and Myally Subgroup is *c.* 2.5 km (shaded area) requiring upper crustal extension of greater than 2.0. Maximum amount of upper crustal extension and amount of subcrustal extension cannot be estimated. (Graph is from Royden, 1990.)

modate the Quilalar Formation and correlative Ballara Quartzite and Corella Formation. The location of the basin axis above basement and cover sequence I indicates that uplift preceded subsidence. Amounts of uplift are not known, but subcrustal extension factors of greater than 3.0 can be estimated (Fig. 14) if the Coward (1986) model is visualized as approximating a non-uniform extension model with decoupling at mid-crustal depths. Similarly, by backstripping the Eastern Creek Volcanics and Myally Subgroup, initial subsidence can be calculated as 2.5 km, requiring upper crustal extension factors of greater than 2.0 (Fig. 15).

CONCLUSIONS

The syn-rift Haslingden Group and post-rift Quila-

lar Formation and correlatives attain maximum thicknesses of 15 km and 1.8 km respectively, permitting maximum stretching factors (β) to be estimated. Treating the Haslingden Group as a single-stage, syn-rift deposit requires unrealistic β values for an intracratonic basin, suggesting that extension took place in two stages. A 2-km-thick interval of quartz arenites of tidal origin in the Haslingden Group is considered to represent the first post-rift response to thermotectonic subsidence. Reasonable β estimates are obtained for the lowermost syn-rift to post-rift sequence for both uniform and non-uniform extension. The location of the axis of the uppermost post-rift basin above a pre-existing basement high implies uplift prior to thermal subsidence. A model comparable to that of Coward (1986) is proposed to explain the location of the post-rift basin and evidence for recycling of older syn-rift volcanics into younger syn-rift deposits. Utilizing a non-uniform extension model for the younger syn-rift to post-rift sequence gives β estimates that are compatible with an intracratonic basin.

ACKNOWLEDGEMENTS

This study was supported jointly by the National Science Foundation through Grant EAR-87–07357 to K.A.E. and by the Bureau of Mineral Resources, Geology and Geophysics. We thank D.H. Blake and G.S. Lister for discussions and guidance with the palinspastic reconstruction. We benefited from comments by M.P. Coward and an anonymous reviewer. S. Chiang and D. Thomson are thanked for assistance with manuscript preparation. M.J.J. publishes with permission of the Director, Australian Geological Survey Organization.

REFERENCES

BARBERI, F., SANTACROSE, R. & VARET, J. (1982) Chemical aspects of rift magmatism. In: *Continental and Oceanic Rifts* (Ed. Palmason, G.), Geodynamic Series, vol. 8, Geol. Soc. Am./Am. Geophys. Union, Washington, DC, pp. 223–258.

BEACH, A. BIRD, T. & GIBBS, A. (1987) Extensional tectonics and crustal structure: deep seismic reflection data from the northern North Sea Viking graben. In: *Continental Extensional Tectonics* (Eds Coward, M.P., Dewey, J.F. & Hancock, P.L.) Spec. Publ. Geol. Soc. London **28**, 467–476.

BEHRENDT, J.C., GREEN, A.G., CANNON, W.F. *et al.* (1988) Crustal structure of the Midcontinent rift system: results from GLIMPCE deep seismic reflection profiles. *Geology* **16**, 81–85.

BLAKE, D.H. (1987) Geology of the Mount Isa Inlier and environs, Queensland and Northern Territory. *Bur. Mineral Resour. Bull.* **225**, 83 pp.

BOND, G.C., KOMINZ, M.A. & DEVLIN, W.J. (1983) Thermal subsidence and eustasy in the Lower Paleozoic miogeocline of western North America. *Nature* **306**, 775–779.

BULTITUDE, R.J. & WYBORN, L.A.I. (1985) Distribution and geochemistry of volcanic rocks in the Duchess–Urandangi Region, Queensland. Bureau Mineral Resources, *J. Austr. Geol. Geophys.* **7**, 99–112.

BURKE, K. & KIDD, W.S.F. (1980) Volcanism on Earth through time. In: *The Continental Crust and Its Mineral Deposits* (Ed. Strangeway, D.W.), Geol. Assoc. Can. Spec. Pap. **20**, 503–522.

CONDIE, K.C. (1982) Early and Middle Proterozoic supercrustal successions and their tectonic settings. *Am. J. Sci.* **282**, 341–357.

COWARD, M.P. (1986) Heterogeneous stretching, simple shear and basin development. *Earth Planet. Sci. Lett.* **80**, 325–336.

DAVIS D.W. & PACES, J.B. (1990) Time resolution of geologic events on the Keweenaw Peninsula and implications for development of the Midcontinent Rift system. *Earth Planet. Sci. Lett.* **97**, 54–64.

DERRICK, G.M. (1982) A Proterozoic rift zone at Mount Isa, Queensland, and implications for mineralization.

Bereau Mineral Resources, *J. Austr. Geol. Geophys.* **7**, 81–92.

DERRICK. G.M., WILSON, I. H. & SWEET, I.P. (1980) The Quilalar and Surprise Creek Formations — new Proterozoic units from the Mount Isa Inlier: their regional sedimentology and application to regional correlation. Bureau Mineral Resources, *J. Aust. Geol. Geophys.* **5**, 215–223.

DRUMMOND, B.J., SEXTON, M.J., BARTON, T.J. & SHAW, R.D. (1991) The nature of faulting along the margins of the Fitzroy Trough, Canning Basin, and implications for the tectonic development of the trough. *Expl. Geophys.* **22**, 111–116.

ERIKSSON, K.A. & SIMPSON, E.L. (1990) Recognition of high-frequency sea-level fluctuations in Proterozoic siliciclastic tidal deposits, Mount Isa, Australia. *Geology* **18**, 474–477.

ERIKSSON, K.A. & SIMPSON, E.L. (1993) Braided-alluvial sedimentation in a mafic volcanic setting: tectonic implications. Int. Assoc. Sediment. Spec. Publ.

ERIKSSON, K.A. TAYLOR, S.R. & KORSCH, R.W. (1992) Geochemistry of 1.8–1.67 Ga mudstones and siltstones from the Mount Isa Inlier, Queensland, Australia: provenance and tectonic implications. *Geochim. Cosmochim. Acta.* **56**, 899–909.

GIBBS, A. (1984) Structural evolution of extensional basin margins. *J. Geol. Soc. London* **141**, 609–620.

GIBBS, A. (1987) Development of extension and mixed-mode sedimentary basins. In: *Continental Extensional Tectonics* (Eds Coward, M.P., Dewey, J.F. & Hancock, P.L.) Spec. Publ. Geol. Soc. London **28**, 19–33.

GILTNER, J.P. (1987) Application of extension models to the Northern Viking Graben. *Norsk Geol. Tids.* **67**, 339–352.

GLIKSON, A.Y. (1980) Precambrian sial–sima relations: evidence for earth expansion. *Tectonophysics* **63**, 193–234.

GROTZINGER, J. & MCCORMICK, D.S. (1988) Flexure of the Early Proterozoic lithosphere and the evolution of Kilohigok Basin (1.9 Ga), northwest Canadian Shield. In: *New Perspectives in Basin Analysis* (Eds Kleinspehn, K. & Paola, C.), Springer-Verlag, New York, pp. 405–430.

GROTZINGER, J. & ROYDEN, L. (1990) Elastic strength of the Slave Craton at 1.9 Gyr and implications for the thermal evolution of the continents. *Nature* **347**, 64–66.

JACKSON, J.A. (1987) Active normal faulting and crustal extension. In: *Continental Extensional Tectonics* (Eds Coward, M.P., Dewey, J.F. & Hancock, P.L.), Spec. Publ. Geol. Soc. London **28**, 3–18.

JACKSON, M.J., SIMPSON, E.L. & ERIKSSON, K.A. (1990) Facies and sequence stratigraphic analysis in an intracratonic thermal-relaxation basin: the Early Proterozoic, lower Quilalar Formation and Ballara Quartzite, Mount Isa Inlier, Australia. *Sedimentology* **37**. 1053–1078.

KLEMPERER, S.L. & WHITE, N.J. (1989) Coaxial stretching or lithospheric simple shear in the North Sea? Evidence from deep seismic profiling and subsidence. In: *Extensional Tectonics and Stratigraphy of the North Atlantic Margins* (Eds Tankard, A.J. & Balkwill, H.R.) Mem. Am. Assoc. Petrol. Geol. **46**, 511–522.

LATIN, D. & WHITE, N. (1990) Generating melt during lithospheric extension: pure shear vs. simple shear. *Geology* **18**, 327–331.

LINDSAY, J.F. & KORSCH, R.J. (1989) Interplay of tectonics and sea-level changes in basin evolution: an example from the intracratonic Amadeus Basin, central Australia. *Basin Res.* **2**, 3–25.

MCKENZIE, D.P. (1978) Some remarks on the development of sedimentary basins. *Earth Planet. Sci. Lett.* **40**, 25–32.

MCKENZIE, D. & BICKLE, M.J. (1988) The volume and composition of melt generated by extension of the lithosphere. *J. Petrol.* **29**, 625–679.

NEUDERT, M.K. & RUSSELL, R.E. (1983) Shallow water and hypersaline features from the middle Proterozoic Mt Isa Sequence. *Nature* **293**, 284–286.

PAGE, R.W. (1978) Response of U–Pb and Rb–Sr total-rock and mineral systems to low-grade regional metamorphism in Proterozoic igneous rocks, Mount Isa, Australia. *J. Geol. Soc. Austr.* **25**, 141–164.

PAGE, R.W. (1981) Depositional ages of the stratiform base metal deposits at Mount Isa and McArthur River, Australia, based on U–Pb dating of concordant tuff horizons. *Econ. Geol.* **76**, 648–658.

PAGE, R.W. (1983a) Timing of superposed volcanism in the Proterozoic Mount Isa Inlier, Australia. *Precam. Res.* **21**, 223–245.

PAGE, R.W. (1983b) Chronology of magmatism, skarn formation and uranium mineralization, Mary Kathleen, Queensland, Australia. *Econ. Geol.* **85**, 838–853.

PAGE, R.W. & BELL, T.H. (1986) Isotopic and structural responses of granite to successive deformation and metamorphism. *J. Geol.* **94**, 365–379.

PAGE, R.W., BOWER, M.G., ZAPASNIK, T.K., GUY, D.B. & HYETT, N.C. (1981) Mount Isa Project (Geochronology Laboratory). *Bureau Mineral Resources, Record* 1981-46, pp. 139–142.

PAGE, R.W., MCCULLOCH, M.T. & BLACK, L.P. (1984) Isotopic record of major Precambrian events in Australia. *Proc. 27th Geol. Congr.,* vol. 5, *Precambrian Geology,* pp. 25–72.

PLUMB, K.A., DERRICK, G.M., NEEDHAM, R.S. & SHAW, R.D. (1981) The Proterozoic of northern Australia. In: *The Precambrian of the Southern Hemisphere* (Ed. Hunter, D.R.) Elsevier, Amsterdam, pp. 205–307.

ROYDEN, L. (1990) Evolution of extensional basins (thermal and subsidence histories). *Short Course Notes, Geol. Soc. Am.,* 53 pp.

ROYDEN, L., SCLATER, J.G. & VON HERZEN, R.P. (1980) Continental margin subsidence and heat flow: important parameters in formation of petroleum hydrocarbons. *Bull. Am. Assoc. Petrol. Geol.* **64**, 173–187.

SCLATTER, J.G. & CHRISTIE, P.A.F. (1980) Continental stretching: an explanation of the post-mid-Cretaceous subsidence of the Central North Sea basin. *J. Geophys. Res.* **85**, 3711–3739.

SIMPSON, E.L. & ERIKSSON, K.A. (1991) Depositional facies and controls on parasequence development in siliciclastic tidal deposits from the Early Proterozoic, upper Mount Guide Quartzite, Mount Isa Inlier, Australia. *Mem. Can. Soc. Petrol. Geol.* **16**, 371–387.

SIMPSON, E.L. & ERIKSSON, K.A. (1993) This eolianites interbedded within a flurial and marine succession: early Proterozoic Whitworth Formation Mount Isa Inlier, Australia. *Sediment. Geol.* **86**.

WERNICKE, B. (1981) Low-angle normal faults in the Basin and Range Province: nappe tectonics in an extending orogen. *Nature* **291**, 645–648.

WERNICKE, B. (1985) Uniform-sense normal simple shear of the continental lithosphere. *Can. J. Earth Sci.* **22**, 108–125.

WHITE, N. (1989) Nature of lithospheric extension in the North Sea. *Geology* **17**, 111–114.

WYBORN, L.A.I. & BLAKE, D.H. (1982) Reassessment of the tectonic setting of the Mount Isa Inlier, Queensland; source, chemistry, age and metamorphism. Bureau Mineral Resources, *J. Austr. Geol. Geophys.* **8**, 51–69.

WYBORN, L.A.I., Page, R.W. & McCulloch, M.T. (1988) Petrology, geochronology, and isotope geochemistry of the post-1820 Ma granites of the Mount Isa Inlier: Mechanisms for the generation of Proterozoic anorogenic granites. *Precam. Res.* **40/41**, 509–541.

Convergent
Plate-Margin Basins

Spec. Publs Int. Ass. Sediment. (1993) **20**, 225–257

Controls on sedimentation at convergent plate margins

D.I.M. MACDONALD*

*British Antarctic Survey, NERC, High Cross, Madingley Road,
Cambridge CB3 0ET, UK*

ABSTRACT

Convergent margins are areas of geological complexity. Basins which occur in the same position
relative to a volcanic arc may owe their origin to more than one process of formation. Sedimentation
in convergent-margin basins is controlled by a wide variety of tectonic processes, which may also
influence and bias the preserved sedimentary record. As a result, there are no unique sedimentological
indicators of basin position or type. However, the sedimentary rocks record many of the important
events, which can be synthesized to reconstruct the history of a convergent margin. Autocyclic
sedimentological models are not always appropriately applied to convergent- margin sedimentary
systems. This is particularly true of fan models of deep-marine turbidite sedimentation. Any
sedimentological model developed for a convergent-margin basin should include allocyclic controls
relevant to the tectonic setting. Such models should be built within a rational plate tectonic framework.

INTRODUCTION

Since the days of Hall and Dana, geologists have
been interested in the links between sedimentation
and tectonics (Dott, 1978). Plate tectonic theory
provided the crucial link between conceptual geo-
syncline models and the reality of modern plate
margins (Dewey & Bird, 1970). Early efforts were
aimed mainly at classifying the variety of basin
types relative to position on different types of
margin (Dickinson, 1974). In recent years, work has
been more concerned with the processes by which
basins form and are filled.

This text is a brief review of the current state of
knowledge of sedimentation in convergent-margin
basins, and of changing research patterns over the
20 years of plate tectonic theory. There are four
main aims.

1 To provide a general introduction to the principles and problems of basin classification in relation
to tectonic setting.

2 To consider the major aims and approaches of
sedimentologists working on such basins.

3 To review briefly the major features of
convergent-margin basins. Each major basin type is
illustrated by discussion of two examples from the
ancient record. In each case, one example is drawn
from the Scotia arc–Antarctic Peninsula region,
which represents one of the longest lived arc-basin
systems known, with activity from Permian until
Neogene times. There are a number of late Mesozoic arc-related basins in the area which illustrate
well the controls on sedimentation in convergent-
margin basins. The stratigraphy of these has been
summarized by Macdonald and Butterworth (1991).

4 To examine the applicability of autocyclic sedimentological models to convergent-margin basins.

The perspective is that of a field-based sedimen-
tologist, considering factors which I consider important as controls, either on sedimentation, or on the
post-depositional history of the rock. My concern is
that sedimentary and tectonic models derived from
studies of modern environments are not always
applied appropriately to ancient examples. Inevita-
bly, much is left out, due to space constraints and
personal bias. However, there are a number of
excellent modern summaries of current research in
basin analysis (see for example Allen & Allen,

*Present address: Cambridge Arctic Shelf Programme,
Department of Earth Sciences, University of Cambridge,
Downing Street, Cambridge CB2 3EQ, UK.

1990), which cover the full range of topics much more thoroughly. Specific basin types are also covered by relatively recent special publications on trench and fore-arc areas (Leggett, 1982), marginal basins (Kokelaar & Howells, 1984), foreland basins (Allen & Homewood, 1986), and strike-slip basins (Ballance & Reading, 1980).

CLASSIFICATION

Basin classification systems

The seminal work on the plate tectonic setting of sedimentary basins was by Dickinson (1974). His scheme was descriptive, relying on position relative to the plate margin. He recognized nine basin types, in five major settings, but did not include strike-slip basins. Subsequent schemes have expanded the scope of the criteria used, and the number of basin types. Bally and Snelson (1980) defined 19 basin types based on the nature of the underlying litho-

sphere, and the position of the basin with respect to large-scale crustal sutures. The seemingly inexorable rise in the number of basin types continues: both Ingersoll (1988) and Miall (1990) recognize 23 types. Unfortunately they are not the same 23. Ingersoll has further subdivisions of some types. For instance, he distinguishes nine different types of fore-arc, six of which have fore-arc basins. The situation can be further complicated if basins change type with time; see, for example, the polyhistory basins described by Klemme (1980). There is an obvious danger that if too many criteria are applied, then no two basins will be classified in the same group. If the process goes too far, basin analysis degenerates into philately.

Allen and Allen (1990) pointed out that most classification systems ignore the underlying mechanisms of basin formation, and proposed a genetic classification based on the three mechanisms by which a basin can form: rifting, crustal loading and strike-slip (Fig. 1). Genetic systems have the twin advantage of cutting across the active/passive mar-

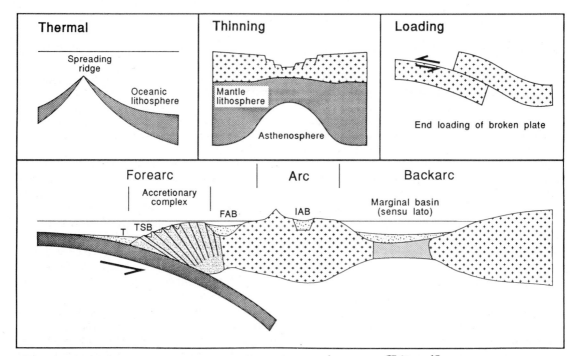

Fig. 1. The three fundamental mechanisms of basin formation (after Allen & Allen, 1990), and the major sites of sediment accumulation at convergent margins discussed in this paper. T, trench; TSB, trench slope basin; FAB, fore-arc basin; IAB, intra-arc basin; note that an ensialic setting is shown here, but the terminology applies equally to intra-oceanic convergent margins.

gin division, and simplicity. By definition, they cannot spawn the large number of types and sub-types which arise from 'geographically' based classification systems. They are also to be encouraged, as discovery of the underlying mechanism of basin formation should be one of the ultimate goals of basin analysis. They do, however, have the disadvantage that deciding the mechanism of basin formation is an *interpretation*. This is not particularly helpful to the field geologist, who deals with incomplete data, and cannot always make a reliable judgement. Even in extensively explored areas, basins can have such a complex history that defining the relative importance of several causative mechanisms is impossible (see for example the discussion of the Los Angeles Basin in Biddle, 1991). For these reasons, I intend to use a modification of Dickinson's (1974) original, positional classification scheme (Fig. 1).

Types of plate margin

Plate margins can be divergent (extensional), convergent (consuming), conservative (transform or strike-slip), or collisional (suturing). Since preservability of sedimentary basins on ocean crust is low, due to their ultimate subduction, field sedimentologists mostly study basins formed on or adjacent to continental margins. This paper is concerned with convergent margins, which involve ocean/continent and ocean/ocean consuming plate boundaries.

In the 'simple' case of orthogonal subduction of ocean crust below a continental margin, the overall setting is compressive, but there is also significant extension (Vine & Smith, 1981), a process which creates most of the major basins. Basin types can alter radically with time; extensional basins can close, leading to crustal loading and development of foreland basins. If subduction is oblique, the situation becomes even more complex, with strike-slip motion in discrete zones in the margin. As a result, convergent-margin basins can form by several processes, and a wide variety of basin associations can be present, and hence the distinction of these basins from those formed on other types of margin is not always straightforward.

The processes by which basins associated with other margins form, or are preserved, may lead to similar deposits to those of convergent margins. For instance, collision and suturing margins form where two areas of continental crust collide. However, as most of the former distance between the continents must have been removed by subduction of ocean crust, the collision margin may preserve convergent margin units. In the early stages of opening of a continental divergent margin, there is much volcanism and active faulting which heavily influences sedimentation. In this respect their deposits can be hard to distinguish from convergent-margin sediments. Almost any plate boundary or continental margin can have areas of transform faulting due to either oblique subduction or oblique extension. Irregularities in the boundary lead to transtensional basins, which are closed by transpression. Such basins have a short life and can occur as parts of any type of margin. Major transform margins can succeed earlier margin types without a major change in plate geometry (e.g. Engebretson *et al.*, 1985). As a result, strike-slip basins can be created which dismember an older margin history. The most notable example of this is in the Neogene of California, where a convergent margin has become a strike-slip one (Crowell, 1974).

Discussion

Basin classification is arbitrary, and depends to a large extent on the viewpoint and aims of the classifier; as a result, there is no one 'correct' classification scheme. In addition, the *processes* which form a basin are independent of the type of plate margin. The limited number of basin-forming processes can combine in many different ways on any type of margin, both in space and time. As a result there are no *unequivocal* indicators of type of continental margin in the sedimentary record of any basin.

AIMS AND APPROACHES

Sedimentary geologists have had two main aims in the study of convergent-margin basins. The first was to develop methods of distinguishing basin types in the geological record using sedimentological criteria (see for example the notable collection of papers on strike-slip basins edited by Ballance & Reading, 1980). The second was to assess the extent of allocyclic (external, tectonic) control on the development of sedimentary systems (see review in Macdonald, 1986). These two aims have not always been explicitly stated, but one or other has been implicit in most sedimentological studies of convergent-margin basins.

Distinguishing basin types

Facies analysis

The study of sedimentary facies in convergent-margin basins was an extension of the work being done on the relationship of facies to depositional environments throughout the 1970s (Reading, 1986). The approach was uniformitarian, relating knowledge of modern basins to ancient examples. For the most part, facies were interpreted in terms of standard models, although departures from standard models have been examined (Macdonald, 1986). In very few cases were new models developed specific to the basin setting (but see, for example, Whitham, 1989). As no sedimentary environment is unique to a particular basin type (Reading, 1980), and as the number of basin types recognized exceeds the number of sedimentary

environments resolvable in the rock record, the hope of being able to distinguish basin type purely using facies analysis has proved to be a chimera.

Petrographic analysis

Sandstone composition has long been used as an indicator of provenance, and hence as a guide to tectonic setting. Systematic methods to overcome the effects of grain-size variation on the detrital mode of rock fragments were developed by Gazzi (1966) and Dickinson (1970). During the 1970s, research by Dickinson and his co-workers was aimed at relating sandstone composition to plate tectonic setting, using discriminant fields (e.g. Ingersoll & Suczek, 1979). These studies have been distilled into a general provenance model (Fig. 2; Dickinson, 1984). Geochemical methods have also been used to try to discriminate sands from different settings (van de Kamp & Leake, 1985).

The major drawback of such techniques is that they show where the sands come from, rather than where they are deposited; this point was discussed by Maynard *et al.* (1982). While there are certainly broad associations between source and basin, it is by no means fixed. Potter (1986) pointed out how the physiography of South America influences the distribution of sands of different types. The entire Argentine coast (largely an extensional margin) is currently receiving sands derived from the Andean (convergent) margin. The datasets used in construction of the various ideal petrographic plots (see for example Fig. 2) may also be biased towards too few examples.

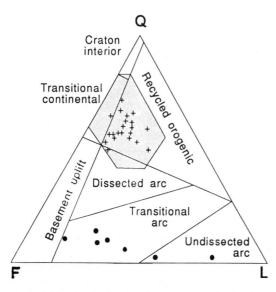

Fig. 2. QFL triangle showing the provenance fields of Dickinson (1984) together with data points from known continental margin settings. Circles: deep-marine sands from fore-arc areas of modern continental margin arcs; shaded area: field of modern trailing-edge sands (both from Maynard *et al.*, 1982); crosses: Neogene sands from the Bengal and Nicobar fans (Ingersoll & Suczek, 1979). All of the individual samples plotted here could become part of an accretionary prism; this diagram illustrates the potential dangers of placing undue reliance on petrographic evidence in determining plate tectonic setting.

Subsidence curves

Basins formed by different tectonic mechanisms may have different subsidence histories, which can be determined using the geohistory techniques of van Hinte (1978). Allen *et al.* (1986) compared foreland and extensional basin subsidence histories. They pointed out that some foreland basins may be confused, in terms of their subsidence histories, with the initial extensional subsidence associated with highly stretched continental lithosphere. Geohistory techniques and subsidence analysis cannot be used alone as a determinant of tectonic setting. They can, however, help to determine the mode of basin formation. Geohistory analysis can also be used as a guide to uplift (e.g. Fortuin & de Smet, 1991).

Basin analysis

This term has been used in a variety of different ways; however, I take it to mean the study of sedimentary basins as geodynamic entities. This involves the use of techniques from any earth science discipline, including those discussed in the previous three sections, to help elucidate a basin's history, and hence its tectonic setting. It is quite clear that no single technique is sufficient to characterize a sedimentary basin. Sedimentologists must step outside the traditional boundaries of their subject, and indeed, outside the boundaries of sedimentary geology if they are to place their work in a tectonic context. To return to the first of the aims that I outlined above, it is only by integrating data from as many sources as possible that basin type can be determined.

Allocyclic controls

The aim of assessing the extent of external control on sedimentary systems has also become more difficult to achieve, as the number of recognized external controls on sedimentary systems has increased. Most sedimentological models are expressed in autocylic terms, with all control on sedimentation patterns coming from natural switching of environments (e.g. the fan model of deep-marine coarse clastic sedimentation, as outlined by Walker, 1978). Most real examples in the rock record contain at least some parts of the sequence which do not fit 'ideal' models, and these have traditionally been explained as tectonically controlled at source (e.g. the turbidite systems discussed by Pickering, 1981). In addition, there is now a realization that the influences of eustatic sea-level change (Macdonald, 1991) and changing climate (Frostick & Reid, 1989) can have a profound influence on sedimentation patterns even in tectonically active basins. With the large number of possible allocyclic controls, the degrees of freedom in the problem commonly outweigh our ability to resolve them.

Current aims

In the light of the problem of too few discriminant features for too many basin types, sedimentologists are increasingly working on the less complex problems associated with extensional settings. For instance, a recent thematic set of papers on tectonics and sedimentation edited by Pickering (1991) contained a high proportion of papers either on divergent-margin basins (e.g. Eyers, 1991) or concerned with the detail of sedimentary response to normal fault movement in other types of margin (Leeder *et al.*, 1991). Mathematical modelling is also easier in divergent- and passive-margin settings, where there are fewer degrees of freedom. Marine sediments accumulate where there is space (accommodation) available to them (Jervey, 1988). The controlling factors are expressed in the *accommodation model*:

$$A_t = D_{t-1} + \Delta E_t + S_t$$

where A = accommodation; ΔE = eustatic sea-level change; S = subsidence; and D = water depth at $t-1$ (A, ΔE and S all at time t).

This model has been applied particularly to extensional basins which follow thermal subsidence laws. In this case the subsidence can be predicted and the model can be used to explain shifting stratigraphical patterns in terms of eustatic sea-level change. This is the basis of the sequence stratigraphical model (van Wagoner *et al.*, 1988), which uses a linear subsidence pattern and constant sedimentation rate (Fig. 3). It should be stressed, however, that the model is applicable to basins on any type of margin, even one undergoing uplift.

Recent research on extensional basins shows that the approach of using linear subsidence is oversimplified. White and McKenzie (1988) used a two-layer model, in which stretching in the crust and lithospheric mantle are differently distributed, to predict stratigraphical onlap without sea-level change. The simple accommodation model as stated above also ignores the effects of sediment loading and compaction on basin development. Despite these reservations, passive margins are probably easier to model, and most of the detailed, quantitative work will come from such areas, at least in the near future.

CONVERGENT-MARGIN BASINS

The behaviour of convergent-margin basins is ultimately controlled by the subducting slab and the stress state of the overriding plate. As a result the tectonic (and hence sedimentary) responses in basins in different parts of the margin are linked, and difficult to separate.

Time (Ma)

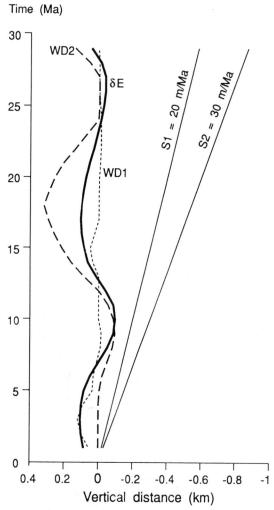

Fig. 3. Sequence stratigraphical model, showing variation in water depth (WD1 + 2) near the edge (S1) and centre (S2) of a linearly subsiding basin with constant sedimentation rate and fluctuating sea-level (δE). The space made available for new sediment is a function of these parameters. (Redrawn after Esso (UK) workshop exercise, 1988.)

Dynamics of subduction

There are a number of syn- and post-depositional tectonic processes which control sedimentation, deformation, uplift, and preservation of the sedimentary record in convergent-margin basins.

Topography of the ocean plate

As newly created ocean floor moves away from

mid-ocean ridges it cools and sinks; age and depth are directly related by a thermal decay function. Old sea floor is also smoother due to thicker sediment cover. These are the primary topographical controls. There are secondary topographical features such as seamounts, aseismic ridges, transform faults and microcontinental blocks which also have an effect.

When convergence rates are very low, or when old, cold ocean floor is being subducted, the ocean plate has a tendency to sink (rollback). This process can cause extension in the overriding plate, leading to frontal collapse of the accretionary prism and back-arc extension (Barker, 1982). In contrast, as progressively younger oceanic crust is subducted, the ridge crest approaches and enters the trench. De Long *et al.* (1979) proposed general models for the subduction of ridges which involved uplift, heating and erosion of the overriding plate. They suggested that uplift occurs due to the arc moving up the approaching ridge flank. However, young ocean lithosphere is very thin and has little flexural strength; it is more likely that uplift is due to the heating pulse as the ridge is subducted (Larter & Barker, 1991). Uplift can be substantial and can lead to considerable subaerial erosion. In some settings where convergence rates are low, or where the upper plate and the seaward limb of the spreading ridge are part of the same plate (Barker, 1982), the ridge will not be subducted. In other cases either a new trench forms seaward of the old ridge crest, or the ridge is subducted and is thought to control arc magmatism (Cande & Leslie, 1986). There is still debate on the topic, but the thermal models of De Long *et al.* (1979), which suggest that the effect is restricted to within 40 km of the trench with the main effect lasting for only 10–20 My (Fig. 4), are probably minimum estimates. The effects of this heating on metamorphism of the accretionary complex are not fully understood.

Subduction of younger, rougher crust can increase tectonic erosion. In Neogene and Recent trench systems this effect is best seen on the inner trench wall, which becomes steepened (Barker, 1982; Cande & Leslie, 1986). In ancient rocks, the effects of subduction erosion may not be obvious at the erosion site. However, removal of the toe of the accretionary prism can lead to extension in the fore-arc (Leggett *et al.*, 1987). Such fore-arc extension can affect both deep and shallow levels of the prism (Nell, 1990).

Secondary topography on the ocean plate can also

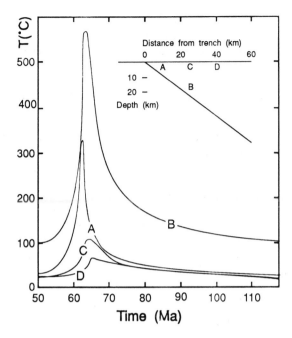

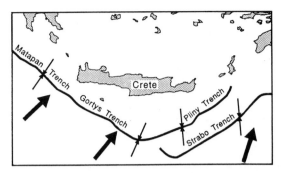

Fig. 5. Geometry of plate convergence at the Hellenic Arc, showing near-orthogonal subduction at the western end and oblique subduction or strike-slip at the eastern end (after Le Pichon *et al.*, 1982). Small arrows show direction of maximum shortening, based on sea-floor structures.

Fig. 4. Temperature changes with time as a result of ridge-crest–trench collision at four locations in the accretionary wedge predicted by the thermal models of De Long *et al.* (1979).

have an effect on deposition and preservation of sediments. Aseismic ridges can dam and divert sediment in the trench, as where the Ninety-east Ridge acts as a barrier between the Bengal and Nicobar fans at the Sunda Trench (Moore *et al.*, 1982). Subduction of such ridges may also uplift the whole fore-arc and produce local unconformities (Flint *et al.*, 1991). Westbrook (1982) suggested that oblique ridge subduction can cause depositional asymmetry and structural heterogeneity in the accretionary complex. Seamount subduction has received more attention, with two main models. The seamount can be subducted as a rigid block causing compression and uplift in front and extension and subsidence behind; passage of a seamount can cause melange formation. Alternatively the trenchward side can be extended by normal faulting, leading to much more subdued topography before subduction. Fryer and Smoot (1985) proposed that the distinction between the two types lay in seamount size: small features, less than 40 km in diameter, behave as part of the ocean plate, whereas larger seamounts are less 'digestible'. Fisher *et al.* (1991) suggest a model whereby obstacles are 'streamlined' by re-

moval of projections and thrusting during subduction. Whatever the process, it is likely that the sedimentary record in the accretionary complex will be considerably disrupted by subduction of large secondary features. In an extreme example, it has been proposed that seamount subduction can lead to fore-arc basin development (Collot & Fisher, 1989).

Convergence angle

The pull exerted by the down-dip slab means that in most modern subduction zones convergence direction is within 25° of the perpendicular (Jarrard, 1986). As most arc–trench systems are portions of small circles, convex with respect to the downgoing plate, there must be oblique subduction along at least part of every margin. The present-day Hellenic Arc is a good example (Fig. 5; Le Pichon *et al.*, 1982). Oblique convergence can be accommodated by oblique deformation in the fore-arc (e.g. Kemp, 1987a), or by partitioning into trench-normal compression and trench-parallel strike-slip (Lundberg & Reed, 1991). Any strike-slip faulting parallel to the trench is commonly concentrated in a narrow zone separating the fore-arc from the rest of the overriding plate, although pre-existing basement structures can influence its position (Jarrard, 1986). Such strike-slip faulting may play a part in the formation of fore-arc basins (Storey & Nell, 1988). In addition, it could be that the presence of old crystalline rocks close to the trench (e.g. at the Middle America Trench; McMillen *et al.*, 1982), commonly held to be due to subduction erosion, may be the result of strike-slip faulting.

Terrane docking

The term terrane (or suspect terrane) has been used to refer to '... fault-bounded geologic entities of regional extent that differ in significant ways from their neighbours' (Jones, 1990). The concept recognizes the fact that many parts of mountain belts cannot be genetically linked to areas with which they are juxtaposed. There has been great debate about the usefulness of the concept, with claims that (at its extreme) terrane analysis has no regard for the mode of origin or lithotectonic relationships of terranes and takes a backward step from plate tectonics (Sengör & Dewey, 1990). In reply, Jones (1990) claimed that establishing the genetic significance of lithic assemblages and discovering genetic linkage between terranes is the ultimate goal of terrane analysis.

The extreme case of dispersal of terrane fragments along major transform margins is outside the scope of this paper. However, terranes are important even on margins displaying orthogonal subduction. Exotic blocks, such as microcontinents, can exert the same topographical effects as outlined for seamounts and aseismic ridges, with the exception that they are unlikely to be subducted. Structural and palaeontological studies of accreted blocks and/or seamounts yield a great deal of information about the dynamics and palaeogeography of the subducted plate (see for example Matsuda &

Isozaki, 1991; Holdsworth & Nell, 1992). Accretion of larger blocks causes relocation of subduction zones, as in the case of the elongate continental fragments rifted from the northern Gondwana margin from Permian until Jurassic times and accreted to the southern margin of Laurasia (Audley-Charles, 1988). The last, and largest of these events was the northward passage of India, resulting in full-scale continental collision. Thus the scale of the block is the key element in determining whether the 'receiving' margin continues to be convergent or turns into a collision margin. Terrane analysis also focuses attention on the role of strike-slip faulting. This can considerably confuse one's view of the geological history of a margin. With care, however, and a sufficiently large view, the origins of blocks can be traced, and the history of even complex polygenetic margins unravelled (e.g. Pigram & Davies, 1987; Bluck & Dempster, 1991).

The role of fluids

The discovery of mud volcanoes in trench sediments at the toe of the Barbados Ridge Complex was the first direct evidence for very high pore-fluid pressures associated with subduction (Westbrook & Smith, 1983). In the Barbados example it was thought that fluid in the sediments of the ocean plate was being overpressured by the weight of the advancing prism. This overpressure zone controlled the position of the décollement horizon (Fig. 6), and allowed lower layers of sediment to be subducted undeformed. Mud volcanoes are also

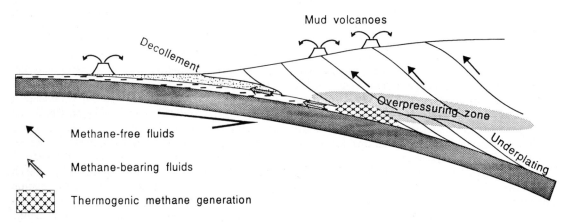

Fig. 6. Sources and migration pathways of fluids in accretionary prisms. (After Moore, 1989; Westbrook & Smith, 1983.)

found on top of the Barbados Ridge Complex, indicating upward flow of fluid along thrusts rooted on the décollement (Brown & Westbrook, 1988). Recent work has been stimulated by the discovery that the porosity structure of the prism can be imaged by seismic methods (Karig, 1985; Vrolijk, 1991). It is now realized that fluids are crucial in sediment accretion, deformation, diagenesis and metamorphism (Langseth & Moore, 1990). Some fluids are also released with magmas of the volcanic arc (Lundberg & Reed, 1991).

Most fluids are probably pore waters, expelled by burial compaction. However, some waters originate from dehydration reactions such as the smectite–illite transition (Vrolijk, 1991); thermogenic methane also contributes to fluid flow along the basalt décollement zone (Moore, 1989). In addition to lubricating thrust movements, fluids are important in bulk deformation. Westbrook (1982) suggested that the chaotic deformation of the northern part of the Barbados Ridge Complex was a result of fine grain size of the sediment and high fluid pressure. This suggestion is supported by theoretical studies of pore-fluid dilation and failure (Moore, 1989) which predict web-like structures similar to those found in real examples (Langseth & Moore, 1990).

Processes of accretion

The process by which accretion or erosion occurs is the sum of all the controls discussed in this section (von Huene, 1986). Karig and Sharman (1975) suggested that the accreted volume in the upper plate of six accretionary prisms in the western Pacific significantly exceeds the volume of trench sediment, hence there must be accretion of ocean crust material. In contrast, subduction erosion requires the entire sediment layer of the ocean plate to be subducted. This latter view is supported by geochemistry of subduction-related basalts, which shows that ocean floor sediment is one component in their genesis (Ellam & Hawkesworth, 1988). Thus there must be two end-member states: either all or none of the sediment on the ocean plate is subducted. In accreting systems, the pathway by which sediment enters the complex will determine the sedimentary record. There are two main processes: frontal accretion and underplating. Both are controlled by the position of the décollement below the accretionary prism, which in turn is controlled by the porosity structure of the trench fill (West-

brook & Smith, 1983). Thornburg and Kulm (1987) suggested a general model of sediment accretion, based on the idea that the two plates will decouple along the zone of weakest, most overpressured sediment (von Huene, 1986), usually the oceanic pelagic layer. As a result, most frontally accreted slices will be a series of coarsening-upward cycles.

Material carried farther down the subduction zone can be added to the base of the complex by underplating (Fig. 6). Most rocks added by underplating will be hemipelagic and pelagic deposits or ocean-plate lavas; they will probably be metamorphosed to greenschist or blueschist facies (Moore, 1989). Little or no sedimentological evidence will survive this process.

Accretion lengthens the complex, while underplating and back-thrusting thicken, broaden and raise it. Most of the internal deformation serves to maintain a stable wedge profile (Platt, 1986). Close to the plate interface, rocks are deformed by simple shear, but as they are uplifted, they pass into a zone dominated by pure shear (Fig. 7; Karig, 1985). If the complex then undergoes frontal collapse, high-level rocks will be in an extensional regime. As a result, the preserved record can be of highly deformed and dismembered strata, and distinction between the various depositional settings could be very difficult.

Discussion

My major thesis is that external controls override autocyclic processes, and hence determine depositional geometry. Post-depositional processes bias the preserved sedimentary record. These ideas are examined below with reference to the various types of convergent-margin basin.

THE FORE-ARC

Trenches

The fore-arc includes the accretionary complex plus trench-slope basins, and the fore-arc basin. While the trench is not strictly part of the fore-arc, it is considered here because it is the main depositional area where sediment accumulates before incorporation into the accretionary complex. Trenches are linear, commonly arcuate, depressions with a steep inner (landward) wall and a more gentle outer (seaward) slope. Thrusts propagating through the

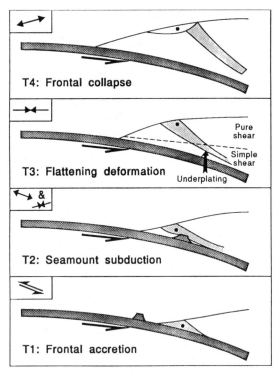

Fig. 7. Four-stage cartoon illustrating the effects of accretionary processes on a single slice (shaded), and the stresses (small arrows) within that slice. Boundary between dominantly simple and pure shear regimes from Karig (1985).

ocean crust can create a trench with complex bottom topography (e.g. Kulm *et al.*, 1982), and sedimentation is coeval with deformation at the inner trench wall. Sediment arrives at the trench from four sources: carried on the incoming plate; *in situ* hemipelagic sedimentation; clastic supply from the continental side of the complex; and clastic supply from one end of the trench. The volumetric balance between these sources may vary greatly from trench to trench.

The thickness of sediment on the incoming plate depends on its age and whether it was open to coarse, terrigenous, clastic sedimentation. Pelagic and hemipelagic sedimentation is dominant on the deep ocean floor: black shales, red clays, cherts, and siliceous and calcareous oozes are all found (Jenkyns, 1986). The character of these sediments depends on the depth of the plate with respect to the calcite compensation depth (CCD) (Davies & Gorsline, 1976). Coarse clastic sediments may be present and their deposition is greatly facilitated by

low sea-level (Moore *et al.*, 1982; Haq, 1991). Background sedimentation in the trench is hemipelagic, and may resemble that on the incoming plate.

In many fore-arcs, inboard trapping mechanisms are very efficient, and there is little direct lateral supply to the trench (e.g. the northern Chile Trench: Thornburg & Kulm, 1987). Lateral supply of arc material may be important if there is no outer arc high, or if the fore-arc is cut by a major fault, as in the area of the Sunda Strait, where the Sumatran Fault provides a pathway to the Sunda Trench (Hamilton, 1979). In many trenches, lateral supply is confined to locally derived material emanating from small canyons on the lower trench slope (von Huene, 1974). Slumping on the inner wall is an important mechanism for transport of material into some trenches. Moore *et al.* (1976) identified a slide which has transported more than 900 km^3 of sediment into the Sunda Trench off Burma. In contrast, the same trench contains no slumps off Sumatra and Java (Moore *et al.*, 1982). Piper *et al.* (1973) described slump masses up to 10 km across with relief of 300–400 m on the floor of the Aleutian Trench, which have deflected axial flow along the trench. Laterally supplied deposits are commonly called fans (either lateral fans: Pickering *et al.*, 1989; or trench fans: Thornburg & Kulm, 1987), and discussed in terms of sedimentological fan models. However, this terminology is misleading, as it implies autocyclic control of facies distribution, whereas the main control is external to the sedimentary system. Lateral 'fans' tend to be distorted along the trench (e.g. in the Chile Trench: Thornburg & Kulm, 1987). This is partly as a result of topographical confinement and partly due to a longitudinal slope along the trench axis. Schweller and Kulm (1978) proposed a general model relating clastic sediment supply to plate convergence rate (Fig. 8). They distinguished between elongate trench wedges and large radial fans overriding the trench to rest on the ocean plate.

Clastic supply from one end is the largest volumetric source of sediment in many modern trenches. Trench–fore-arc systems can end against transform faults, continental collision zones or at a cusp with another trench system (Fig. 9). This provides a potential pathway for sediment to enter the system. As the sediment is derived from another plate, which can comprise crust of any type or age, there is no relation between the sediment *source* and the trench within which it is *deposited*. This is one of the main drawbacks of petrography as a tool for

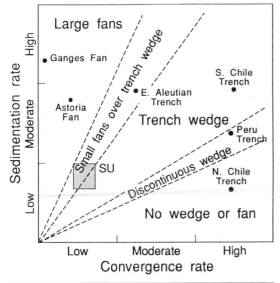

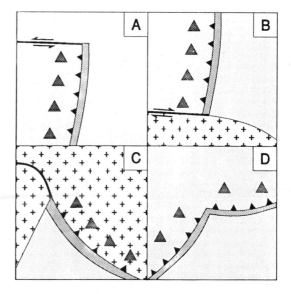

Fig. 9. Possible geometries for termination of an arc–trench system. (A) Ocean–ocean transform fault (e.g. north end of the Tonga–Kermadec arc), (B) ocean–continent transform (e.g. south end of Lesser Antilles arc), (C) continental collision zone (e.g. Burma–Sunda system), (D) arc–arc cusp (e.g. Japan–Kurile cusp).

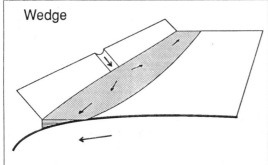

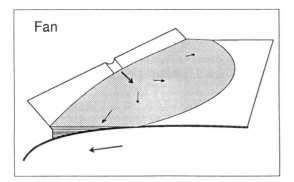

Fig. 8. Trench sedimentation model of Schweller and Kulm (1978) relating sediment supply and convergence rate for some modern trench settings. The field for the Palaeozoic example of the Southern Uplands (SU) is taken from Leggett (1980). The lower diagrams illustrate the distinction between fans and trench wedges.

distinguishing tectonic setting. Sediment is carried along the trench by turbidity currents, but again fan terminology is inappropriate, as the system is confined by trench topography. Facies of distorted lateral fans and of end-fed systems will be very similar, with a main axial flow and marginal or levee areas. The Aleutian Trench fill (fed from one end) is composed of a channel–overbank system more than 350 km long (Piper *et al.*, 1973). This channel is diverted by slump masses on the inner trench wall. A similar channelized system has developed from distorted lateral fans in the Chile Trench (Fig. 10; Thornburg & Kulm, 1987). While the facies are similar to those used in many fan models (e.g. Walker, 1978), it is clear that external factors are the dominant control on facies architecture.

The sedimentology of trench-fill deposits bears more resemblance to the linear flysch troughs of the Caledonian and Alpine–Carpathian chains (e.g. Lovell, 1970) than to the radial fans described from

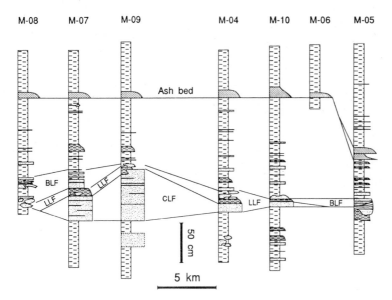

Fig. 10. Facies of the fill of the Chile Trench from Thornburg and Kulm (1987), showing the geometry of a channelized sand in an otherwise sheeted basin (letter-number code gives core number).

CLF — Channel facies: non-graded, laminated, and graded & laminated sand

LLF — Levee facies: graded sand, graded silt

BLF — Basin 2 facies: graded & laminated silt

the California borderland (Normark, 1974), which are the basis of most sedimentological fan models. Ancient linear turbidite troughs tend to be sand-rich and are characterized by axial palaeocurrent patterns and proximal to distal sedimentological variation (see Macdonald, 1986, table 4). The sedimentological changes include a downcurrent decrease in any or all of: grain size, sandstone/shale ratio, sandstone bed thickness, and percentage of sandstone bed amalgamation, plus an overall decline in the thickness of the unit. It might be possible to distinguish between a series of deformed lateral fans and a single end-fed system on the basis of proximal to distal changes. If there are many sites of lateral input, changes will not be consistent. For example, the Sunda Trench (Fig. 11) is end-fed, and its fill thins dramatically from more than 5 km off the Nicobar Islands to less than 100 m south of Java (Moore *et al.*, 1982). In contrast, the fill of the Chile Trench (Thornburg & Kulm, 1987) is much thinner, there are many sites of input, and although it thins northward, the thinning is much less marked. The northward thinning in the Chile Trench is accompanied by an overall decline in the percentage of sandstone-rich facies.

The ancient example of the Cretaceous Chugach

Terrane of southern Alaska (Fig. 11) has been interpreted as a trench-fill by Nilsen and Zuffa (1982). It is intermediate in size between the two modern examples quoted. Palaeocurrents show that the system was fed from the eastern end, but there are also sites of lateral input at the western end. There are two parallel facies belts, extending along the entire 2200 km of the terrane; both are deformed into south-facing and south-verging folds. The northern (inboard) belt comprises thick units of hemipelagic mudstone, with thin-bedded siltstone and mudstone turbidites, plus olistostromes and slide blocks. In the western central portion of the terrane there are also conglomeratic channel deposits. This facies association is interpreted as slope deposits of the inner trench wall, and seems relatively consistent along the length of the belt. In contrast, the seaward belt varies markedly along the length of the terrane from 'inner fan' to 'basin plain'. Although Nilsen and Zuffa (1982) interpret the facies in terms of a fan model, the Chugach Terrane is in no real sense a fan. Facies distribution is controlled by basin shape (the trench) and by distance from source. Without more detailed information, further interpretation is difficult, but it would appear that the system grew by aggradation

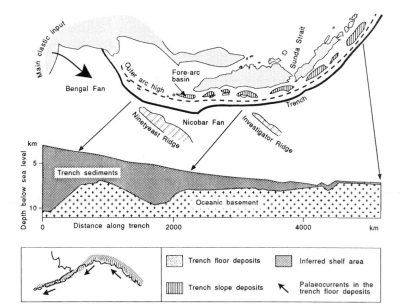

Fig. 11. Geometry and along-axis sediment thickness variation of the modern Sunda Trench (Moore *et al.*, 1982) compared with the Cretaceous Chugach Terrane of south Alaska (Nilsen & Zuffa, 1982).

rather than progradation, as the facies belts appear to be fixed in position through time, leading to a system where the main sedimentological variation is lateral rather than vertical. The Chugach terrane illustrates three problems of general relevance to the study of all ancient convergent-margin basins:

1 The major orocline in the central area is almost certainly a secondary feature. Later deformation such as this is common along convergent margins, and limits the value of palaeogeographical data from pre-deformation units.

2 None of the surface features used in the interpretation of modern deposits are preserved. This is particularly important in the distinction of the northern and southern facies belts, interpreted as inner trench wall and trench floor respectively. In modern basins, this distinction is easily made on topographical grounds. In ancient examples it is a subjective decision based on facies interpretation, and hampered by imprecise time resolution.

3 Facies models developed from studies of modern environments and ancient sequences not affected by contemporaneous tectonics may not be appropriate for convergent-margin basins.

Trench-slope basins

Trench sediments are incorporated into the accretionary prism as discrete slices separated by oceanward-propagating and verging thrust faults

with the geometry of an emergent imbricate fan. The surface of the accretionary prism is not smooth, and small basins develop, unconformable on the accreted material, and bounded on the landward side by the sole thrust of the previously accreted slice. Trench-slope basins are elongate parallel to the accretionary complex and have a high length-to-width ratio. Basin width and depth of fill tend to increase upslope (Fig. 12), as some thrusts become inactive. The increase in width is accompanied by a change from having their fill actively deformed, to a more passive situation (Moore & Karig, 1976; Aalto, 1982).

The fill of trench-slope basins is highly variable. In the deepest basins, the sediment is commonly hemipelagic clay and biogenic oozes. These are often redeposited (Moore & Karig, 1976), as is shown by the sapropel turbidites of the Hellenic trench-slope basins (Anastasakis & Piper, 1991). Terrigenous deposits occur in higher basins; these are commonly arc-derived (Moore *et al.*, 1980). Pelagic carbonates are also present in higher basins, reflecting deposition above the CCD (Moore *et al.*, 1980). The composition of coarse material delivered to the slope basins will depend on whether the trench-slope break is emergent, and whether supply from the arc is trapped in the fore-arc basin.

Underwood and Bachman (1982) proposed a general model for sedimentation in slope basins (Fig. 12), with background hemipelagic mudstone

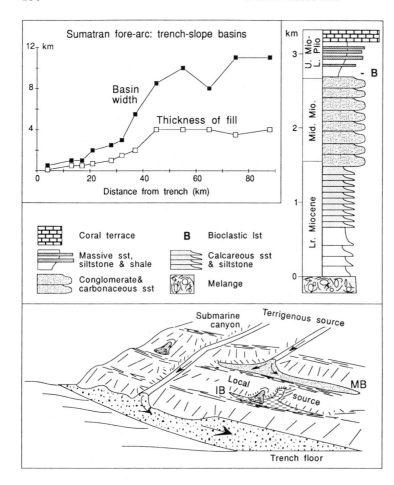

Fig. 12. Graph showing the progressive upslope increase in size of trench-slope basins on the Sumatran fore-arc (data from Moore & Karig, 1976), with the stratigraphy of Nias Island, a mature trench-slope basin fill (Moore *et al.*, 1980). The lower diagram is a model for the distinction of immature (IB) from mature (MB) trench-slope basins (Underwood & Bachman, 1982).

and locally-derived slumps present in most basins. *Immature* slope basins are starved of any but locally derived and hemipelagic sediments. *Mature* trench-slope basins are open to terrigenous sediment and are filled by massive coarse sandstones, classical sandstone turbidites and mudstone turbidites, with subordinate pebbly sandstone and conglomerate. These facies represent submarine fan associations, although the 'fans' are elongate down the length of the basin. The work of Moore *et al.* (1980) on the Neogene slope-basin deposits of Nias would appear to support this model. The succession is 3 km thick (Fig. 12) and comprises a coarsening-upward cycle with terrigenous, arc-derived sandstones and conglomerates overlying marls and thin turbidites. Limestone becomes more common upsection. They interpret this succession as a submarine fan prograding over a basin plain and controlled by increased clastic supply, with concomitant uplift

through the CCD. In contrast, Aalto (1982) interpreted an olistostrome unit overlain by a retrograding fan sequence in the Franciscan Complex of northern California as reflecting uplift, filling, isolation and bypassing of a trench-slope basin. These two field examples indicate that *external* controls, such as sediment supply and basin shape, are more important than sedimentological controls in determining facies patterns.

Oceanographic factors other than CCD are also likely to control fine-grained sedimentation in trench-slope basins. Gorsline (1987) has outlined some of the controls in the California borderland basins (Fig. 13). These are strike-slip basins, but are analogous to trench-slope basins for oceanographic purposes, in lying at varying depths, being tectonically active, and having intermittent physical barriers. Since trench-slope basin deposits are mainly turbidites and are commonly deformed,

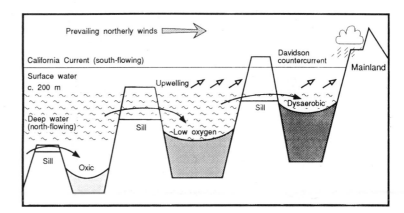

Fig. 13. Structural and oceanographic control of oxygen distribution in the California Borderland basins. This model could apply to any setting with tectonically active basins at increasing depths. (After Gorsline, 1987.)

they can be difficult to distinguish from accreted trench deposits.

Ancient accretionary complexes

The evidence reviewed above suggests that the best preservation potential is in sand-rich systems, deposited in a trench above relatively old ocean crust. The wide range of external controls means that accretionary complexes will vary enormously. The range of this variation is well illustrated by the examples of the Southern Uplands of Scotland (Ordovician–Silurian) and the LeMay Group of Alexander Island, Antarctica (Jurassic–?Tertiary).

Southern Uplands

The Southern Uplands were the first ancient example interpreted as an accretionary complex (Leggett *et al.*, 1979). This interpretation was aided by the remarkable stratigraphical control given by graptolites. The complex is a series of thrust-bound slices, each consisting of a lower unit of dark fissile shale (± basalt and chert), overlain by thin–medium-bedded sandstone turbidites (Fig. 14; Leggett, 1980). There are subordinate slump deposits, and channellised sandstones and conglomerates. Slices get progressively younger southwards from Arenig to Wenlock, as does the onset of sand-rich deposition. This overall stratigraphical pattern also holds good at a local scale within the complex (see Kemp, 1987b.)

Within each slice, mudstones represent the bulk of the time, and are thought to have been pelagic deposits on the ocean plate. Sandstones show both strike-parallel and strike-normal palaeocurrents

(Leggett, 1980; Kemp, 1987b) and are interpreted as trench deposits. Leggett (1980) suggested that on the model of Schweller and Kulm (1978; see Fig. 8) the Southern Uplands represented small fans over a trench wedge. Both he and Kemp (1987b) used a model of sand supplied to lateral fans turning to flow along a topographically confined trench. In Llandovery–Wenlock times, subduction rate slowed and the ocean plate was blanketed by turbidite sands. Material is thought to have been added to the complex largely by frontal accretion (Leggett, 1980, 1987; Kemp, 1987a; Leggett *et al.*, 1979, 1982), although as Leggett (1987) has pointed out, the poor nature of the exposure undoubtedly masks many of the structural complexities. There are no deposits identified as trench-slope basins.

Within the overall picture of diachronous southward onset of sand-rich sedimentation, there are inconsistencies, which illustrate the importance of external controls on sedimentation. Leggett (1980) ascribed the late onset of sandstone deposition in his Slices 6 and 7 to control by an outer trench high on the ocean plate, with sand bypassing the high to rest on ocean crust later being accreted on Slice 8. The isolated Ashgill sandstones of Slice 8 have been interpreted by D.M. Casey (oral communication, 1981) as a result of low sea-levels during Late Ordovician glaciation.

Alexander Island

The LeMay Group of Alexander Island, Antarctica (Fig. 15) is in complete contrast to the Southern Uplands in being penetratively deformed, lithologically heterogeneous, and with little stratigraphical control. The complex consists of polyphase de-

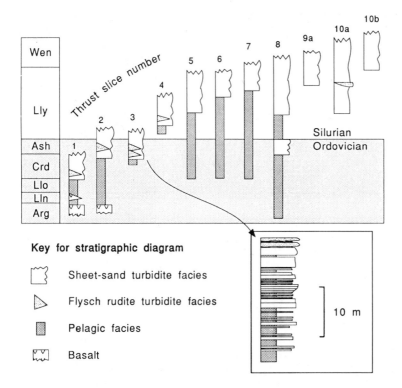

Key for stratigraphic diagram

⬡ Sheet-sand turbidite facies

▷ Flysch rudite turbidite facies

▨ Pelagic facies

〰 Basalt

Fig. 14. Stratigraphical correlations in the Southern Uplands. Inset shows detail of the pelagic/turbidite transition in Tweeddale. (Redrawn after Leggett, 1980.)

formed greywackes, basalts, cherts and mélange. The unit was first interpreted as an accretionary complex by Suárez (1976) on the basis of general lithology and its position relative to the magmatic arc. It is almost completely unfossiliferous, so workers had to rely on structural and lithostratigraphical mapping to demonstrate the large-scale geometry (Tranter, 1987; Nell, 1990). A synthesis of available radiometric dates coupled with the discovery of stratigraphically useful radiolaria (Holdsworth & Nell, 1992) has allowed demonstration of ocean-ward younging of both sedimentation and deformation (Nell, 1990).

Tranter (1987) recognized four major facies associations. He interpreted an association of conglomerate, sandstone and mudstone turbidites, with subordinate pebbly mudstone and slumps as trench-slope basin deposits. Thin–medium-bedded sandstone turbidites were thought to represent frontally accreted trench deposits. An association of pillowed and massive basalts, radiolarian chert, siltstone and mudstone were interpreted as ocean floor deposits, while vesicular basalts with interbedded hyaloclastite breccia and tuffs represented an accreted seamount. In parts of the complex there are major mélange belts, and strata added by underplating are exposed (Tranter, 1987; Nell 1990; Doubleday & Tranter, 1992). The original accretionary structure appears to have been masked by a complex series of processes including shallow-level extension and compression caused by frontal collapse, strike-slip transpression, and late-stage compressive shortening (Tranter, 1987; Nell, 1990; Nell & Storey, 1991).

Conclusions

The LeMay Group probably represents deeper levels of a more complexly deformed system than the Southern Uplands. It probably had lower rates of sediment supply or higher rates of convergence, and may be a more 'typical' exposure of an accretionary complex in these respects. From a sedimentological viewpoint, however, new insights into depositional processes will come from areas such as the Southern Uplands, where there is good biostratigraphical control, and deformation is not wholly disruptive.

Fore-arc basins

Fore-arc basins lie oceanward of the volcanic arc, bounded on their seaward side by the trench-slope

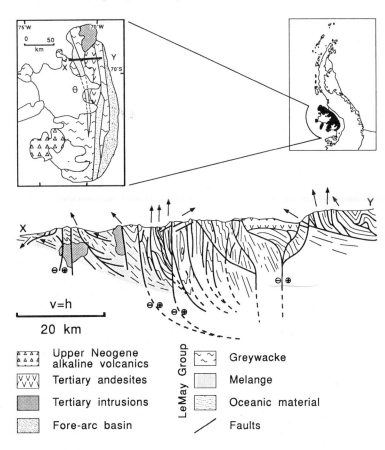

Fig. 15. Geological map of Alexander Island, Antarctica, showing the three dominant associations of the LeMay Group accretionary complex in relation to the (partly) coeval fore-arc basin, and the later Tertiary igneous rocks. The cross-section X–Y is after Nell (1990).

break. They are discontinuous along the length of the margin, and are highly variable in size, depth and sediment thickness. They vary even along one margin such as the Sunda Arc, where Moore *et al.* (1982) recognized a series of basins ranging from 150–500 km long by 50–100 km wide, lying in water depths of 500–4000 m, with sedimentary fill up to 6 km thick. Sediment enters these basins via submarine canyons from the arc margin, with 2–9 canyons per basin.

Processes of formation

Fore-arc basins are created by relative uplift of the accretionary complex barring oceanward dispersal of sediment (Seely, 1979). Not all arcs have a fore-arc basin, especially in younger ocean–ocean convergent margins. Seely (1979) explained the genesis of fore-arc basins in terms of crustal types and position of the initial rupture, and suggested that their evolution is controlled by the processes of

seaward accretion and underplating. The behaviour of the subducting slab, particularly the subduction of aseismic ridges, can influence directly the sedimentary and tectonic evolution of fore-arc basins (Collot & Fisher, 1989; Daly, 1989; Flint *et al.*, 1991). Daly suggested that the degree of coupling between the two plates, and hence of ocean plate influence on basin development in the overriding plate, is a function of the convergence rate.

It appears that fore-arc basins exist merely because of the presence of a physical barrier to seaward. This suggests that they can be created without the action of any of the fundamental mechanisms of basin formation (Fig. 1; Allen & Allen, 1990). However, there are at least three other processes which may operate.

1 If subduction is oblique, and resolves into margin-normal and parallel components, any strike-slip movement could be concentrated at the natural zone of weakness between the accretionary complex and arc (see for example Howell *et al.*,

1980). Some fore-arc basins may have formed by strike-slip movement, and some are certainly influenced by it in their later stages (Nell & Storey, 1991).

2 Where arc material and old ocean crust extend seaward below an accretionary complex (as in Barbados; Westbrook, 1982), there may be an element of crustal loading in the inception and development of the fore-arc basin.

3 As has already been mentioned, there can be significant extension at convergent plate margins (Vine & Smith, 1981). Whether this creates the fore-arc basin or merely enhances a pre-existing feature, is uncertain, but major normal faults commonly occur at the margins (e.g. Herzer & Exon, 1985).

Sedimentology

Coulbourn and Moberly (1977) demonstrated that the depocentres of three fore-arc basins off northern Chile and Peru migrated landward in response to uplift and back-rotation of the trench-slope break. This pattern is not, however, universal. Ingersoll (1982) showed how the depocentre of the Great Valley Group migrated seaward, probably as a result of the supply of very large volumes of arc-derived material. In both examples the later stages of sedimentation are coeval with uplift and deformation of the seaward basin margin. Sedimentation is also coeval with the later stages of trench and trench-slope basin sedimentation, so distinction of fore-arc basin from accretionary complex in ancient examples is not always straightforward on either stratigraphical or structural grounds. However, as a general rule, fore-arc basin sequences are younger and less deformed than the adjacent parts of the accretionary complex.

There is comparatively little detailed information on modern fore-arc basins, as they are relatively shallow and close to shore, and generally beyond the scope of the Deep Sea Drilling Programme/Ocean Drilling Programme (DSDP/ODP). Coulbourn and Moberly (1977) provided a good overview of the structure of the fore-arc basins of northern Chile, based on seismic work. Moore *et al.* (1982) pointed out that in the Sumatran fore-arc, almost all the arc-derived detritus is trapped in the fore-arc basins. They also interpreted the flat floors and flat internal reflectors as indicating deposition from laterally extensive ('basin plain') turbidity currents. Shallow-marine fore-arc basins are less well-known, although Saito (1991) has shown how latest

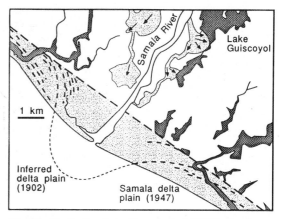

Fig. 16. Illustration from Kuenzi *et al.* (1979) showing some of the effects of the 1902 eruption of Santa Maria on the Samalá River in the eastern part of the Guatemalan fore-arc basin. The river bed was raised 10–15 m, forming an elongate alluvial fan, damming marginal lakes, and causing seaward progradation of a considerable delta.

Pleistocene–Holocene sea-level changes have influenced sedimentation in the shallowest part of the fore-arc of northeast Japan, with formation of a regionally extensive coastal plain sequence, overlain by a transgressive sheet sand.

The terrestrial portions of fore-arc basins are relatively well-known in comparison, with the coastal plain of Guatemala being particularly well-studied (Vessel & Davies, 1981). Terrestrial deposits tend to be restricted to the arcward portion of fore-arc basins, remote from the syndepositional tectonism of the seaward basin margin. Sedimentation is, however, still influenced by tectonics, with arc volcanism as the major control (Davies *et al.*, 1979; Kuenzi *et al.*, 1979). Large-volume eruptions, in particular, have a catastrophic effect on river systems, raising their beds and leading to rapid vertical aggradation, development of an elongate alluvial fan system, and production of a significant coastal delta (Fig. 16; Kuenzi *et al.*, 1979).

Ancient examples

Ancient fore-arc basins are dominated by deep-marine turbidite sedimentation (Ingersoll, 1978, 1982; Butterworth *et al.*, 1988). Shallowing-up-ward trends have been identified in a significant number of basins (Ingersoll, 1982; Butterworth & Macdonald, 1991). Smaller-scale deepening up-ward has also been reported (Morris & Busby-Spera, 1988).

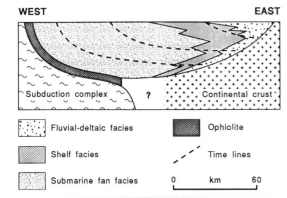

Fig. 17. Cross-section of the Great Valley Group basin of California, showing the predominance of submarine fan deposits. The Jurassic–Cretaceous, Albian–Cenomanian and Cretaceous–tertiary boundaries are shown. (After Howell *et al.*, 1980.)

The Great Valley Group of California is probably the best-known example of an ancient fore-arc basin. The basin is Late Jurassic–Early Tertiary in age, rooted partly on continental crust and partly on accretionary complex, and grew from about 30–40 km wide at 125 Ma to 200 km wide at 25 Ma (Ingersoll, 1978; Howell *et al.*, 1980). More than half of the basin fill is formed of submarine fan facies (Fig. 17); shallower facies form a transgressive–regressive cycle restricted to the eastern (continental) portion of the basin. The length of the cycle attests to *relative* stability throughout deposi-

tion (Howell *et al.*, 1980). Palaeocurrents define an orthogonal pattern, with supply from the eastern (arc) margin and north–south longitudinal transport (Ingersoll, 1982). Ingersoll explained major changes to sedimentation pattern in terms of changes in Pacific plate kinematics. As the basin broadened in response to westward migration of the subduction zone and eastward migration of the arc, it was also filling. By latest Cretaceous time, most deposition was in shallow marine or terrestrial environments, and volcanic detritus was able to pass over the basin into the trench. By the end of deposition in Palaeogene times, the active arc lay far to the east, and detritus was being derived from the eroded roots of the Cretaceous arc. It would appear that external tectonics shaped the large-scale sedimentary events (Ingersoll, 1982). However, the depositional sequence of the turbidite facies is explained in terms of autocyclic submarine fan models (Ingersoll, 1978).

The Jurassic–Cretaceous fore-arc basin of Alexander Island, Antarctica (Fig. 18) also displays a combination of auto- and allocyclic controls. The overall fill shallows upward from deep marine to terrestrial (Butterworth & Macdonald, 1991; Moncrieff & Kelly, 1993). Most detritus is arc-derived, with a marked evolution from volcanic to plutonic material, reflecting erosion of the source area (Butterworth, 1991). Palaeocurrents indicate long-lived westerly transport, normal to the basin axis. The basal unit represents disruption of a pre-existing submarine fan by tectonically-induced slumping.

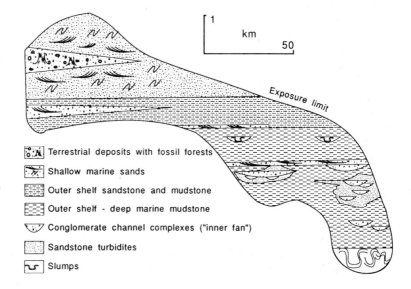

Fig. 18. North–south section along the eastern exposures of the Fossil Bluff Group fore-arc basin (for location, see Fig. 15), illustrating the large-scale facies architecture, and the overall shallowing upward. (After Butterworth *et al.*, 1988; Moncrieff & Kelly, 1993.)

The overlying submarine fan unit is variable along strike (Butterworth *et al.*, 1988), but the greatest thickness of conglomerate-rich, inner- fan channel facies appears to be controlled by the site of the main slumping. Shifting of sediment bodies within this unit was controlled both by autocyclic channel avulsion processes and by external control of the source (Butterworth, 1988). The submarine fan unit is abruptly overlain by a widespread, thin shallow-marine unit, probably reflecting a eustatic sea-level fall (Butterworth, 1991). The rest of the basin-fill comprises outer-shelf mudstones with interbedded, sharp-margined sand bodies, which seem to be tectonically controlled, capped by a regressive–transgressive cycle involving a considerable terrestrial incursion into the southern part of the basin. The lack of a basal erosion surface to the terrestrial unit, plus a reversion to more volcanic-rich detritus in the terrestrial interval suggests that this regression was a response to a tectonic event (A.C.M. Moncrieff, personal communication).

Discussion

These two examples indicate that although fore-arc basins are relatively stable through time, deposition is strongly influenced by tectonics. Overall shallowing upward seems to be a characteristic of fore-arc basins. This is easily explained if the basin exists because of a physical barrier in the accretionary complex, which is being raised by underplating. It could well be that departures from this pattern (e.g. Morris & Busby-Spera, 1988) are a sign of other factors such as strike-slip faulting or renewed extension. The preservation potential of fore-arc basins is high. They are unlikely to be subject to major penetrative deformation unless there is a radical change in margin type. They can, however, be dismembered by later tectonics (Howell *et al.*, 1980).

THE ARC

Sedimentary rocks are deposited within arc areas either in fault-bounded intra-arc basins created by extensional or strike-slip processes, or in the distal apron of volcanic edifices. Almost all of the sediment involved is arc-derived volcaniclastic material. As Suthren (1985) has pointed out, there are a number of problems in the study and interpretation of volcaniclastic sequences, including lack of

modern analogues, the complex interaction of volcanic and sedimentary processes, and extensive diagenetic alteration. In addition, since most arcs are physically elevated and thus prone to erosion, the preservation potential of intra-arc basins is low, and there are few interpreted from the geological record (relative to other convergent-margin basin types).

Volcanic and sedimentary processes

Volcanic and primary volcaniclastic rocks can be analysed in facies terms in exactly the same way as sedimentary rocks. The problem is that the degree of secondary reworking by sedimentary processes can be very hard to quantify (for example distinguishing between a graded bed due to primary airfall and one due to reworking, transport and deposition in a turbidity current). Indeed, there are many parallels between deposition from a pyroclastic flow and deposition from a turbidity current (Sparks *et al.*, 1973; Lowe, 1982). Any intra-arc basin can be entirely filled with either primary or secondary volcaniclastic detritus, but it is more likely that successions comprise interbeds of both, and hence require careful analysis. This problem is discussed further in the excellent accounts of volcanic rocks and volcaniclastic facies given by Fisher and Schmincke (1984) and Cas and Wright (1987).

Ancient examples

Intra-arc basins have been little-studied in pre-Neogene deposits, possibly as a result of the difficulties outlined above. The work of Busby-Spera (1984, 1985) provides an honourable exception. The area she described from Mineral King, California, is preserved as a roof pendant in the Sierra Nevada batholith. It reveals a complex series of volcanic and sedimentary rocks, pre-, syn- and post-formation of a large submarine caldera (Fig. 19). While the tectonics are clearly related to development of the caldera, and the most spectacular deposits are associated with collapse of the caldera wall during eruption (Busby-Spera, 1984), there are significant reworked deposits, attesting to the continuation of normal marine processes (Busby-Spera, 1985). The influence of volcanism on sedimentation in an ancient terrestrial intra-arc basin has been stressed by Smith (1987). He pointed out that fluvial facies models do not allow for the very large amounts of sediment being deliv-

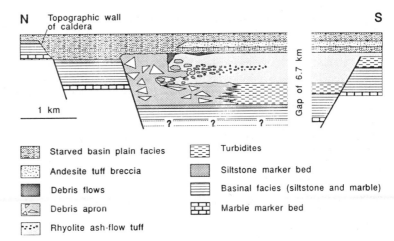

Fig. 19. Geometry of Early Mesozoic rocks of the Vandever Mountain caldera, Mineral King, California, before disruption by Cretaceous plutonism. (After Busby-Spera, 1984.)

ered to the drainage basin. This is very similar to the fore-arc basin examples discussed earlier.

The intra-arc basins of the northern Antarctic Peninsula (Fig. 20; Farquharson, 1984) formed under rather more stable conditions with much less volcanic influence. A Jurassic–Tertiary magmatic arc was established on a Late Palaeozoic–Early

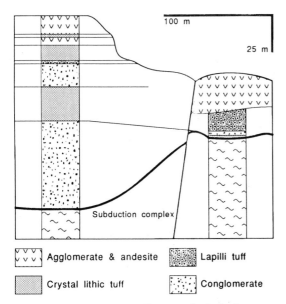

Fig. 20. Geometry of a small intra-arc basin in the northern Antarctic Peninsula, illustrating a change from fluvial facies, derived from the underlying basement and deposited in a fault-bounded basin, to volcanic facies as the arc developed; note syndepositional basin-margin fault. (After Farquharson, 1984.)

Mesozoic subduction complex, uplifted in Jurassic times with formation of a series of small fault-bounded alluvial basins, which received dominantly quartzose detritus. Sedimentation was in alluvial fan, fluvial and lacustrine delta settings, influenced by contemporaneous faulting. The upper part of the succession is formed almost entirely of tuffs in most basins, overlain in turn by the extensive arc volcanic suite (Farquharson, 1984), which probably records the onset of major arc volcanism in Late Jurassic times (Miller *et al.*, 1990).

In the California example, sedimentation was controlled almost entirely by volcanic processes, including shaping the depositional basin by caldera collapse. Normal sedimentary processes dominated only when volcanism became quiescent or was located in distal localities. The Antarctic example provides a complete contrast. The basins were probably created by uplift and extension associated with development of the arc, but non-volcanic sedimentary processes dominated. Even the volcanic input was largely from airfall. Most intra-arc basins are likely to lie between these two extremes.

THE BACK-ARC AREA

Origin of basins behind arcs

The terminology used for basins 'behind' (i.e. on the side away from the subduction zone) arcs is confused, reflecting both confusion on the part of some researchers, and the variety of origins of basins referred to as marginal, back-arc, or retro-arc. The definition of marginal basins proposed by Kokelaar and Howells (1984, p. 1) is 'Relatively small semi-isolated oceanic basins, spatially associated with active or inactive volcanic-arc and trench systems...'. This is only applicable to intra-oceanic basins, although the authors go on to include ensialic basins in an expansion of the definition. Uyeda and Kanamori (1979) classified back-arc basins into five categories, depending on the crustal setting and mode of opening. Recently the terminology for such basins has been improved in a new scheme by B.C. Storey (Table 1; personal communication, October 1991).

All of the four basin types in Storey's scheme are created by extension, but only in true back-arc basins (either oceanic or ensialic) are subduction, arc volcanism and back-arc extension contemporaneous. This is an important point for the distinction of boundary basins, such as the Gulf of California, where the arc is inactive during formation of the basin. The amount of contemporaneous spreading is also highly variable, with none in trapped ocean basins, and rates varying from 2 to 10 cm/year in other types (Uyeda & Kanamori, 1979). As a result there is a greater variability in the sedimentary fill of marginal basins than in any other setting on convergent margins.

Sedimentation in modern marginal basins

Much of the knowledge of modern back-arc basins (*sensu stricto*) comes from DSDP work in the western Pacific (e.g. Klein, 1975; Hussong & Uyeda, 1981). Early models of sedimentation stressed the isolation from terrigenous deposits, and identified the arc as the primary sediment source, with minor contributions from the remnant arc and spreading centre (Karig, 1975). Carey and Sigurdsson (1984) pointed out that volcanological factors can also be important, and used the example of Site 450 in the Parece Vela Basin (Fig. 21), where extensional volcanism is followed by arc volcanism, which then gives way to pelagic sedimentation. At the time of writing there have been a number of recent ODP cruises to the western Pacific: Leg 124 to the Sulu Sea, Legs 127 and 128 to the Japan Sea, and Leg 135 to the Lau Basin. The preservation potential of basins of this sort is probably low unless the basin is involved in accretion or collision tectonics and then they are likely to be present in a fragmented state.

Ensialic back-arc basins can also have sediment derived from the continental side of the basin, creating a bimodal fill, which has been suggested as a characteristic of such basins (Dalziel *et al.*, 1975).

Table 1. Classification scheme for marginal basins proposed by B.C. Storey at the meeting on 'Magmatism and causes of continental breakup' held at the Geological Society of London, October 1990. Note that the shaded areas are non-back-arc marginal basins.

Nomenclature		Cause	Example
MARGINAL BASINS	Back-arc basin	Subduction-related extension	Bransfield Strait
	Small ocean	Branch of mid-ocean ridge	Tasman Sea
	Trapped ocean	Plate re-organisation	Aleutian Sea
	Boundary basin	Plate boundary re-adjustments	Gulf of California

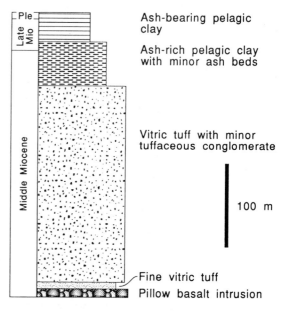

Fig. 21. Lithological log of DSDP Site 450 in the Parece Vela Basin illustrating the declining importance of arc volcanism. (After Kroenke *et al.*, 1981.)

The Bransfield Strait, off the northern Antarctic Peninsula, is one of the best-known modern examples. It is probably less than 4 My old (Barker & Dalziel, 1983), with a maximum of 500–1000 m of sediment fill. The young volcanic axial ridge is a prominent, sediment-free feature of the basin and sedimentation on either side of this barrier is dominated by draped hemipelagic units deposited during sea-level highstands, and coarse glacial sediments deposited during lowstands (Jeffers & Anderson, 1991).

Some marginal basins can have a very thick fill, with high sedimentation rates, up to 600 m/My (see Macdonald & Tanner, 1983, for further discussion). The amount of volcanic detritus entering a marginal basin is a function of the productivity of the arc. Sigurdsson *et al.* (1980) have presented an elegant model for sedimentation in the Grenada Basin (Fig. 22) which relates the partitioning of sediment between the fore- and back-arc areas to topography and direction of the prevailing winds and currents. The bulk of the recent basin fill has come from the northern island of Dominia as major subaqueous pyroclastic-turbidity flows which feed longitudinally down the basin, interdigitating with a volcaniclastic apron. The apron is formed of direct input from other volcanic islands and small fans of reworked material produced by west-flowing currents between the islands (Sigurdsson *et al.*, 1980; Carey & Sigurdsson, 1984; Whitham, 1989). Preservation of pyroclastic flow deposits is enhanced by steep slopes into the marine marginal basin. In contrast, air-fall ashes dominate on the fore-arc side, reflecting the influence of the tropospheric westerly winds, and the gentle slopes which cause pyroclastic flows to lose their integrity (Sigurdsson *et al.*, 1980). This study remains one of the best demonstrations of direct climatic effect on sedimentation.

Ancient examples

The literature on ancient back-arc basins is complicated by the nomenclature problems outlined above, and by the need to use detailed geochemistry to isolate the basin type (e.g. Storey & Alabaster, 1991). Two examples illustrate the range of marginal basins interpreted from the ancient record: the Lower Palaeozoic Welsh Basin has preserved a very good record of extensive explosive intrabasinal volcanism, while the Mesozoic Rocas Verdes–Magallanes system of southern South America and South Georgia is a good example of a polyhistory basin.

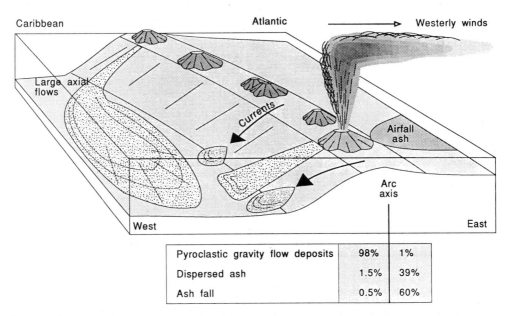

	West	East
Pyroclastic gravity flow deposits	98%	1%
Dispersed ash	1.5%	39%
Ash fall	0.5%	60%

Fig. 22. Model for sedimentation in the Grenada Trough illustrating the importance of axial and marginal supply, plus sediment partitioning due to climatic factors. (After Sigurdsson *et al.*, 1980; Carey & Sigurdsson, 1984.)

The Ordovician volcanic rocks of Wales record a transition from Tremadocian arc volcanism to back-arc extension in Arenig–Caradoc times (Kokelaar *et al.*, 1984), although recent interpretations have cast doubt on the relationship of the Welsh volcanic rocks to a specific arc terrane (Kokelaar, 1988). Whatever the gross tectonic setting, however, Snowdonia is a classic area for the study of the deposits of a highly active marginal basin. A recent publication by Howells *et al.* (1991) is a magnificent compilation of many years of detailed field work, recording in detail the geometry of volcanic and associated sedimentary facies in the Caradocian rocks. Sedimentation was controlled by active tectonism related to caldera evolution and by primary volcanic processes (see Fig. 23 for one example of the complexity of the resulting interbedded lava, pyroclastic and reworked deposits).

The northern limits of the Scotia arc contain remnants of a large arc–basin system of Late Jurassic–Early Cretaceous age (Fig. 24), dismembered by Tertiary faulting during opening of the Scotia Sea. The basin was originally at least 1000 km long, but probably not more than 100–200 km wide. It was interpreted as an extensional back-arc basin by Dalziel *et al.* (1974, 1975). Recent geochemical work from South Georgia, however, suggests the possibility of formation as a boundary basin, analogous to the Gulf of California (Alabaster & Story, 1990). The most detailed sedimentology comes from South Georgia, where a moderately deformed, arc-derived turbidite unit has been thrust over an intensely deformed, continent-derived siliciclastic turbidite unit (Macdonald *et al.*, 1987). The volcaniclastic unit is 6–8 km thick. Localized deposits of high-density turbidity flows were derived from the arc margin and interbedded with turbidites deposited by currents flowing down the basin axis (Macdonald & Tanner, 1983). The range of facies is restricted, with no conglomerates, no channels, and no thick, mud-rich sections. There is no systematic vertical variation, but the axial turbidites display a strong proximal-to-distal variation. Macdonald (1986) argued that a fan interpretation was inappropriate and presented a model of a marginal slope apron interbedded with a basin axis system (Fig. 24). The main controls on sedimentation were the shape of the basin and contemporaneous faulting on the margin, which kept the whole depositional system anchored, producing an aggrading pattern with systematic lateral variation, rather than a prograding one with vertical variation. The main sedimentological problem with the back-arc basin model is the low percentage of primary volcaniclastic material. Almost all of the sand-grade material is reworked, with most sandstones containing seven or eight distinct types of volcanic rock fragments. The boundary basin model of Storey and

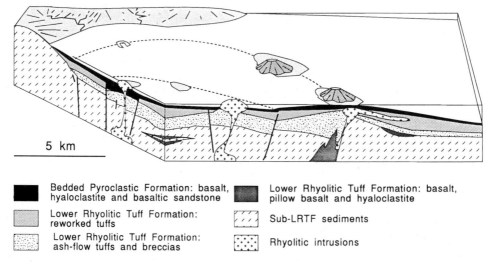

5 km

■ Bedded Pyroclastic Formation: basalt, hyaloclastite and basaltic sandstone

▨ Lower Rhyolitic Tuff Formation: basalt, pillow basalt and hyaloclastite

▥ Lower Rhyolitic Tuff Formation: reworked tuffs

▨ Sub-LRTF sediments

⣿ Lower Rhyolitic Tuff Formation: ash-flow tuffs and breccias

⣿ Rhyolitic intrusions

Fig. 23. Reconstruction of the Snowdon volcanic centre at the end of Lower Rhyolitic Tuff Formation times (After Howells *et al.*, 1991). Note the control of sediment distribution by the caldera and the disruptive effects of high-level intrusions.

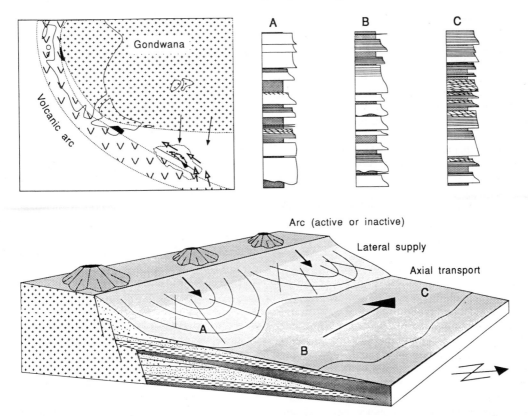

Fig. 24. Palaeogeographical sketch showing the position of South Georgia within the Rocas Verdes marginal-basin system in Early Cretaceous times (After Macdonald & Tanner, 1983); major ophiolites are shown in black. The depositional model for the volcaniclastic turbidites of South Georgia shows lateral variation in thickness and structure of typical turbidite beds in relation to the slope apron/axial system. Logs A–C all show 10 m of typical sandstone facies.

Alabaster (1991), with an inactive arc being eroded on the seaward side of the marginal basin, fits the sedimentology much better, and it could be that much of the Rocas Verdes basin could have developed in a manner analogous to the Gulf of California.

The volcaniclastic basin-fill unit is less well-known from South America, but what is known appears to be consistent with the model presented for South Georgia, although there are significant conglomerate units in addition to the sandstones (Winn & Dott, 1978). The continental side of the basin, absent from South Georgia, is preserved, and appears to have behaved as a passive margin throughout the development of the basin (Wilson, 1991). The whole basin closed in Mid-Cretaceous times with strong deformation and uplift, leading to formation of a foreland basin with a fold-and-thrust belt (Wilson, 1991).

DISCUSSION

Common problems

Although each convergent-margin basin type has individual characteristics, they have some problems in common. These include the interpretation of deep-marine turbidite systems, provenance characterization, the origins of cyclicity, and the effects of later deformation.

Deep-marine systems

Many convergent-margin basins are deep marine, which makes direct observations of recent examples difficult. Interpretation is also hampered because the worldwide Holocene sea-level rise has led to many submarine fans becoming inactive. As a result there is debate over the true significance of features on modern fans and their applicability to ancient examples (see discussion by Nilsen, 1980). By extension, the environmental significance of features of ancient deep-marine, convergent-margin basins is unclear.

Fan models usually show radial sediment bodies, emanating from a point source at a basin margin. All are expressed in autocyclic terms, with changing facies distribution coming from migration of environments, and the principal control being within the sedimentary system. Events external to the system, such as contemporaneous tectonism or sea-level change, are regarded as extraordinary in that they disrupt 'ordinary' sedimentary processes. As such they are not regarded as integral to the model. In contrast, the examples discussed have suggested that turbidite systems in convergent-margin basins tend to be sand-rich and characterized by axial palaeocurrent patterns, cyclic vertical growth and proximal-to-distal sedimentological variation. Any model of turbidite sedimentation applied in convergent-margin settings must be allocyclic, recognizing extrabasinal tectonic activity, sea-level change, topographic confinement, and basinal tectonism as the primary controls on sedimentation. Differences between fan systems in different basin types arise from pre-existing basin shape and size, differing provenance pathways, and the length of time that the basin is stable.

Volcanism and provenance

The most obvious and volumetrically important volcanism at convergent margins is in the volcanic arc. As has been shown earlier, much of the detritus from the arc is trapped in the fore-arc basin, with little volcanic material reaching the trench. There is a tendency to treat this as a general rule; this has been enhanced by the predominance of modern studies of end-fed systems, which are rich in continental detritus (e.g. Ingersoll & Suczek, 1979). Arc-derived rocks, however, do reach some trenches. This can either be through a dissected fore-arc, or from end-feed in an arc–arc cusp. There may also be active volcanoes on the outer trench slope, either hotspot-related or part of a separate arc system (Leggett, 1987, figs 2 & 3). The extent of volcaniclastic sedimentation in trenches was given prominence by the reinterpretation of the Southern Uplands as a back-arc by Stone *et al.* (1987) on the basis of petrographic evidence. The rebuttal of Stone *et al.*'s argument by Leggett (1987, see especially p. 747) is a very good example of the complexity of provenance at a convergent margin, and a timely reminder of the dangers of undue reliance on petrography in the interpretation of ancient sequences.

The origins of cyclicity

The debate on the origins of sedimentary cycles is as old as the science of stratigraphy. Various mechanisms have been proposed to account for cycles at a number of different scales. These forcing factors can be broadly grouped into two categories: (1) *internal* (or tectonic), where the primary mechanism depends on a process occurring within the Earth (generally in the lithosphere); and (2) *external* where the forcing factor is climatic or extraterrestrial. These factors, however, are not independent of each other. First, there are primary links between factors both within each major group, such as plate configuration and intraplate stress, and also between groups, such as plate configuration and major climatic change. Second, each primary cause can operate at a number of scales. For instance, Milankovitch forcing can cause climatic cycles with periodicities from 20 000 to 100 000 years (Fischer *et al.*, 1990). Third, and most important, all of the primary mechanisms operate on the sedimentary record through a proxy, usually change in relative sea-level.

The sequence stratigraphical model (Fig. 3; Posamentier *et al.*, 1988) is an elegant synthesis, combining an *external* factor ('eustatic' sea-level variation) with an *internal* one (tectonic subsidence or uplift) to model relative sea-level change, and hence sediment distribution. Although it is commonly thought of as a passive-margin tool, there is no reason why it should not be applied to active margin basins, nor does the model require sea-level changes to be either global or synchronous. The primary causes of the two vertical changes used in the model are irrelevant in the first instance, but problems do arise when one tries to work back to an original mechanism for the relative sea-level change inferred. For instance, Cathles and Hallam (1991) have shown how intraplate stress can mimic the effects of global eustatic sea-level change. There are several factors which can individually change base level and hence cause cyclicity; since these factors can also interact in many different combinations, Burton *et al.* (1987) questioned the possibility of ever isolating a single factor such as eustatic change.

These arguments apply to basins in any setting. Convergent-margin basins, however, are different from passive margins in having generally steeper slopes, and being mostly deep marine. Hence relatively little of the basin is within the zone of any short-term sea-level fluctuations that could be caused by either changing ice volume or intraplate stress. In addition, horizontal plate movements can be translated into large vertical displacements at active margins. Fortuin and de Smet (1991, fig. 5) used geohistory techniques to demonstrate that vertical movements in the Banda Arc occurred at rates comparable to those inferred in the 'global' sea-level curve of Haq *et al.* (1987), but were several orders of magnitude larger. Data of this sort has led to the feeling that cyclicity in convergent-margin basins must always be tectonically induced.

However, small-scale sea-level changes do not need to impinge directly on the sediment–water interface in order to influence the sedimentary record. Shanmugam and Moiola (1988) pointed out that submarine 'fans' develop primarily during periods of low sea-level in both active- and passive-margin settings. The narrow shelves characteristic of convergent margins may enhance this effect, and conformable contacts below sharp-based turbidite sandstone packages could be the expression of sequence boundaries in active-margin basins.

Effects of later tectonism

All convergent margins evolve, either due to a natural progression, such as decreasing age of subducted crust, or by a catastrophic event, such as collision or major plate reorganization. The products of the earlier tectonic phase will tend to be obscured by the later ones. In some cases, such as closure of an ensialic marginal basin and establishment of a foreland basin, the change can be relatively easy to deduce (e.g. Wilson, 1991). Catastrophic transformations, in contrast, can remove or obscure the older record almost completely.

Major continental collision can lead to significant subduction of one or other of the opposed plates and dismemberment of pre-existing active margins (e.g. Klootwijk *et al.*, 1985). Continent–arc (Audley-Charles, 1988) and arc–arc (Ito, 1987) collisions also produce complex collisional mosaics, which make it difficult to trace the earlier history of the margin. Large-scale strike-slip faulting can disperse fragments of active margins. Jones (1990) documented the history of western North America, where terranes were accreted to a convergent margin from Late Palaeozoic until Cenozoic times. This margin was then dismembered by Cenozoic strike-slip faulting which has moved fragments northwards. Although Jones has said quite forcefully that the discovery of genetic linkage is the ultimate goal of terrane analysis, the temptation to describe a dismembered system purely in terms of separate terranes is strong. Dewey (1982) discussed this problem in terms of tectonic *weather* and tectonic *climate*: if the local history cannot be synthesized, then a larger region must be considered. This is particularly true when trying to interpret small palaeotectonic studies in terms of large-scale neotectonic systems (see Dewey, 1982).

SUMMARY

Convergent margins are areas of great complexity. The preserved sedimentary record of such margins is always incomplete and can be hard to interpret. However, all convergent-margin basins are part of a dynamic system, and hence display strong external control of the sedimentary systems, generally due to syndepositional tectonism. Sedimentological models for such basins must incorporate 'external' dynamic factors, or be incapable of describing the system fully. Figure 25 is an attempt to show some of the dynamic factors which must be built into sedimentological models. These are generalized, but should hold for any type of convergent-margin basin. Plate tectonics gives a conceptual framework which can provide the external dynamics for sedimentological models. To quote Dewey (1982, p. 376): 'Plate tectonics does not make geology any easier; on the contrary, it makes it a great deal more complicated in that we are constrained to build and test models within a rational plate tectonic framework rather than to erect ad hoc constructions . . .'.

ACKNOWLEDGEMENTS

I am grateful to my colleagues Paul Doubleday, Andrew Moncrieff, Bryan Storey and Mike Thomson for reading early drafts of the manuscript, to Lynne Frostick and Ron Steel for their constant encouragement, and to Gill McDonnell for cheerfully typing all the versions of this manuscript.

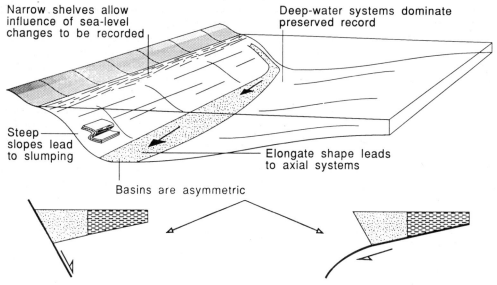

Narrow shelves allow influence of sea-level changes to be recorded

Deep-water systems dominate preserved record

Steep slopes lead to slumping

Elongate shape leads to axial systems

Basins are asymmetric

Syndepositional tectonism anchors sedimentary systems, prevents free migration of environments, and leads to aggradation rather than progradation. Frequency of tectonic activity prevents stabilisation of the system.

Fig. 25. General model of the allocyclic factors which should be built into sedimentary models of convergent-margin basins.

REFERENCES

AALTO, K.R. (1982) The Franciscan Complex of northernmost California: sedimentation and tectonics. In: *Trench–Forearc Geology* (Ed. Leggett, J.K.) Spec. Publ. Geol. Soc. London **10**, 419–432.

ALABASTER, T. & STOREY, B.C. (1990) A modified Gulf of California model for South Georgia, north Scotia Ridge, and implications for the Rocas Verdes back-arc basin, southern Andes. *Geology* **18**, 497–500.

ALLEN, P.A. & ALLEN, J.R. (1990) *Basin Analysis: Principles and Applications.* Blackwell Scientific Publications, Oxford, 451 pp.

ALLEN, P.A. & HOMEWOOD, P. (Eds) (1986) Foreland basins. Int. Ass. Sediment. Spec. Publ. **8**, 453 pp.

ALLEN, P.A., HOMEWOOD, P. & WILLIAMS, G.D. (1986) Foreland basins: an introduction. In: *Foreland Basins* (Eds Allen, P.A. & Homewood, P.). Int. Ass. Sediment. Spec. Publ. **8**, 3–12.

ANASTASAKIS, G.C. & PIPER, D.J.W. (1991) The character of seismo-turbidites in the S-1 sapropel, Zakinthos and Strofadhes basins, Greece. *Sedimentology,* **38**, 717–733.

AUDLEY-CHARLES, M.G. (1988) Evolution of the southern margin of Tethys (North Australian region) from Early Permian to Late Cretaceous. In: *Gondwana and Tethys* (Eds Audley-Charles, M.G. & Hallam, A.). Spec. Publ. Geol. Soc. London **37**, 79–100.

BALLANCE, P.F. & READING, H.G. (Eds) (1980) Sedimentation in oblique-slip mobile zones. Int. Ass. Sediment. Spec. Publ. **4**, 265 pp.

BALLY, A.W. & SNELSON, S. (1980) Realms of subsidence. In: *Facts and Principles of World Petroleum Occurrence* (Ed. Miall, A.D.) Can. Soc. Petrol. Geol. Mem. **6**, 9–75.

BARKER, P.F. (1982) The Cenozoic subduction history of the Pacific margin of the Antarctic Peninsula; ridge crest–trench interactions. *J. Geol. Soc. London* **139**, 787–801.

BARKER, P.F. & DALZIEL, I.W.D. (1983) Progress in geodynamics in the Scotia Arc region. *Am. Geophys. Union, Geodynamics Series* **9**, 137–170.

BIDDLE, K.T. (Ed.) (1991) Active margin basins. Mem. Am. Ass. Petrol. Geol. **52**, 324 pp.

BLUCK, B.J. & DEMPSTER, T.J. (1991) Exotic metamorphic terranes in the Caledonides: Tectonic history of the Dalradian block, Scotland. *Geology* **19**, 1133–1136.

BROWN, K.M. & WESTBROOK, G.K. (1988) Mud diapirism and subcretion in the Barbados Ridge Complex. *Tectonics* **7**, 613–640.

BURTON, R., KENDALL, C.G.St.C. & LERCHE, I. (1987) Out of our depth: on the impossibility of fathoming eustasy from the stratigraphic record. *Earth Sci. Rev.* **24**, 237–277.

BUSBY-SPERA, C.J. (1984) Large-volume rhyolite ash flow eruptions and submarine caldera collapse in the lower Mesozoic Sierra Nevada, California. *J. Geophys. Res.* **89**, B10, 8417–8427.

BUSBY-SPERA, C.J. (1985) A sand-rich submarine fan in the lower Mesozoic Mineral King caldera complex, Sierra Nevada, California. *J. Sediment. Petrol.* **55**, 376–391.

BUTTERWORTH, P.J. (1988) Sedimentology and stratigraphy

of part of the Mesozoic Fossil Bluff Group, Alexander Island, Antarctica. Unpublished PhD thesis. Council for National Academic Awards, London, 351 pp.

BUTTERWORTH, P.J. (1991) The role of eustasy in the development of a regional shallowing event in a tectonically active basin: Fossil Bluff Group (Jurassic–Cretaceous), Alexander Island, Antarctica. In: *Sedimentation, Tectonics and Eustasy* (Ed. Macdonald, D.I.M.) Int. Ass. Sediment. Spec. Publ. **12**, 307–329.

BUTTERWORTH, P.J. & MACDONALD, D.I.M. (1991) Basin shallowing from the Mesozoic Fossil Bluff Group of Alexander Island and its regional tectonic significance. In: *Geological Evolution of Antarctica* (Eds Thomson, M.R.A., Crame, J.A. & Thomson, J.W.). Cambridge University Press, Cambridge, pp. 449–453.

BUTTERWORTH, P.J., CRAME, J.A., HOWLETT, P.J. & MACDONALD, D.I.M. (1988) Lithostratigraphy of Upper Jurassic–Lower Cretaceous strata of eastern Alexander Island, Antarctica. *Cretaceous Res.* **9**, 249–264.

CANDE, S.C. & LESLIE, R.B. (1986) Late Cenozoic tectonics of the Chile trench. *J. Geophys. Res.* **91**, B1, 471–496.

CAREY, S. & SIGURDSSON, H. (1984) A model of volcanigenic sedimentation in marginal basins. In: *Marginal Basin Geology* (Eds Kokelaar, B.P. & Howells, M.J.) Spec. Publ. Geol. Soc. London **16**, 37–58.

CAS, R.A.F. & WRIGHT, J.V. (1987) *Volcanic Successions: Modern and Ancient.* Allen & Unwin, London, 528 pp.

CATHLES, L.M. & HALLAM, A. (1991) Stress-induced changes in plate density, Vail sequences, epeirogeny, and short-lived global sea level fluctuations. *Tectonics* **10**, 659–671.

COLLOT, J.-Y. & FISHER, M.A. (1989) Formation of forearc basins by collision between seamounts and accretionary wedges: an example from the New Hebrides subduction zone. *Geology* **17**, 930–933.

COULBOURN, W.T. & MOBERLY, R. (1977) Structural evidence of the evolution of fore-arc basins off South America. *Can. J. Earth Sci.* **14**, 102–116.

CROWELL, J.C. (1974) Origin of Late Cenozoic basins in Southern California. In: *Tectonics and Sedimentation* (Ed. Dickinson, W.R.) Spec. Publ. Soc. Econ. Paleont. Miner., Tulsa **22**, 190–204.

DALY, M.C. (1989) Correlations between Nazca-Farallon plate kinematics and forearc basin evolution in Ecuador. *Tectonics* **8**, 769–790.

DALZIEL, I.W.D., DE WIT, M.J. & PALMER, K.F. (1974) Fossil marginal basin in the southern Andes. *Nature, London* **250**, 291–294.

DALZIEL, I.W.D., DOTT, R.H., WINN, R.D. & BRUHN, R.L. (1975) Tectonic relations of South Georgia Island to the southernmost Andes. *Bull. Geol. Soc. Am.* **86**, 1034–1040.

DAVIES, T.A. & GORSLINE, D.S. (1976) Oceanic sediments and sedimentary processes. In: *Chemical Oceanography* (Eds Riley, J.P. & Chester, R.) 2nd edn. Academic Press, London, pp. 1–80.

DAVIES, D.K., VESSELL, R.K., MILES, R.C., FOLY, M.G. & BONIS, S.B. (1979) Fluvial transport and downstream sediment modifications in an active volcanic region. In: *Fluvial Sedimentology* (Ed. Miall, A.D.) Mem. Can. Soc. Petrol. Geol. **5**, 61–84.

DE LONG, S.E., SCHWARZ, W.M. & ANDERSON, R.N. (1979) Thermal effects of ridge subduction. *Earth Planet. Sci. Lett.* **44**, 239–246.

DEWEY, J.F. (1982) Plate tectonics and the evolution of the British Isles. *J. Geol. Soc. London* **139**, 371–412.

DEWEY, J.F. & BIRD, J.M. (1970) Mountain belts and the new global tectonics. *J. Geophys. Res.* **75**, 2625–2647.

DICKINSON, W.R. (1970) Interpreting detrital modes of greywacke and arkose. *J. Sediment. Petrol.* **40**, 695–707.

DICKINSON, W.R. (1974) Plate tectonics and sedimentation. In: *Tectonics and Sedimentation* (Ed. Dickinson, W.R.) Spec. Publ. Soc. Econ. Paleont. Miner., Tulsa **22**, 1–27.

DICKINSON, W.R. (1984) Interpreting provenance relations from detrital modes of sandstones. In: *Provenance of Arenites* (Ed. Zuffa, G.G.) NATO ASI series. **C148**, D. Reidel, Dordrecht, pp. 333–361.

DOTT, R.H. JR. (1978) Tectonics and sedimentation a century later. *Earth Sci. Rev.* **14**, 1–34.

DOUBLEDAY, P.A. & TRANTER, T.H. (1992) Modes of formation and accretion of oceanic material in the Mesozoic fore-arc of central and southern Alexander Island, Antarctica: a summary. In: *Recent Progress in Antarctic Earth Scieces* (Eds Yoshida, Y., Kaminuma, K. & Shiraisi, K.). Terrupub, Tokyo, pp. 377–382.

ELLAM, R.M. & HAWKESWORTH, C.J. (1988) Elemental and isotopic variations in subduction related basalts: evidence for a three component model. *Contrib. Mineral. Petrol.* **98**, 72–80.

ENGEBRETSON, D.C., COX, A. & GORDON, R.G. (1985) Relative motions between oceanic and continental plates in the Pacific basins. *Spec. Pap. Geol. Soc. Am.* **206**, 59 pp.

EYERS, J. (1991) The influence of tectonics on early Cretaceous sedimentation in Bedfordshire, England. *J. Geol. Soc. London* **148**, 405–414.

FARQUHARSON, G.W. (1984) Late Mesozoic, non-marine conglomeratic sequences of northern Antarctic Peninsula (the Botany Bay Group). *Br. Antarct. Surv. Bull.* **65**, 1–33.

FISCHER, A.G., DE BOER, P.L. & PREMOLI SILVA, I. (1990). Cyclostratigraphy. In: *Cretaceous Resources, Events and Rhythms* (Eds Ginsburg, R.N. & Beaudoin, B.) NATO ASI Series **C304**, D. Reidel, Dordrecht, pp. 139–172.

FISHER, M.A., COLLOT, J.-Y. & GEIST, E.L. (1991) Structure of the collision zone between Bougainville Guyot and the accretionary wedge of the New Hebrides island arc, southwest Pacific. *Tectonics* **10**, 887–903.

FISHER, R.V. & SCHMINCKE, H.-V. (1984) *Pyroclastic Rocks.* Springer-Verlag, Berlin, 472 pp.

FLINT, S., TURNER, P. & JOLLEY, E.J. (1991) Depositional architecture of Quarternary fan-delta deposits of the Andean fore-arc: relative sea-level changes as a response to aseismic ridge subduction. In: *Sedimentation, Tectonics and Eustasy* (Ed. Macdonald, D.I.M.) Int. Ass. Sediment. Spec. Pub. **12**, 91–103.

FORTUIN, A.R. & DE SMET, M.E.M. (1991) Rates and magnitudes of late Cenozoic vertical movements in the Indonesian Band Arc and the distinction of eustatic effects. In: *Sedimentation, Tectonics and Eustasy* (Ed. Macdonald, D.I.M.) Int. Ass. Sediment. Spec. Publ. **12**, 79–89.

FROSTICK, L. & REID, I. (1989) Is structure the main control of river drainage and sedimentation in rifts? *J. Afr. Earth Sci.* **8**, 165–182.

FRYER, P. & SMOOT, N.C. (1985) Processes of seamount subduction in the Mariana and Izu-Bonin trenches. *Mar. Geol.* **64**, 77–90.

GAZZI, P. (1966) Le arenarie del flysch sopracretaceo dell'Appennino modenese; correlazioni con il flysch di Monghidoro. *Mineral. Petrogr. Acta* **12**, 69–97.

GORSLINE, D.S. (1987) Deposition in active margin basins: with examples from the California borderland. In: *Deposition in Active Margin Basins* (Ed. Gorsline, D.S.). Soc. Econ. Paleont. Miner. Pacific Section **54**, 33–51.

HAMILTON, W. (1979) Tectonics of the Indonesian region. *Prof. Pap. U.S. Geol. Surv.* **1078**, 345 pp.

HAQ, B.U. (1991) Sequence stratigraphy, sea-level change, and significance for the deep sea. In: *Sedimentation, Tectonics and Eustasy* (Ed. Macdonald, D.I.M.) Int. Ass. Sediment. Spec. Publ. **12**, 3–39.

HAQ, B.U., HARDENBOL, J. & VAIL, P.R. (1987) Chronology of fluctuating sea-levels since the Triassic. *Science* **235**, 1156–1167.

HERZER, R.H. & EXON, N.F. (1985) Structure and basin analysis of the southern Tonga forearc. In: *Geology and Offshore Resources of Pacific Island Arcs — Tonga Region* (Eds Scholl, D.W. & Vallier, T.L.), Circum-Pacific Council for Energy and Mineral Resources, Houston, pp. 55–73.

HOLDSWORTH, B.K. & NELL, P.A.R. (1992) Mesozoic radiolarian faunas from the Antarctic Peninsula: age, tectonic and palaeoceanographic significance. *J. Geol. Soc. London* **149**, 1003–1020.

HOWELL, D.G., CROUCH, J.K., GREENE, H.G., McCULLOCH, D.S. & VEDDER, J.G. (1980) Basin development along the late Mesozoic and Cainozoic California margin: a plate tectonic margin of subduction, oblique subduction and transform tectonics. In: *Sedimentation in Oblique-slip Mobile Zones* (Eds Ballance, P.F., & Reading, H.G.) Int. Ass. Sediment. Spec. Publ. **4**, 43–62.

HOWELLS, M.F., REEDMAN, A.J. & CAMPBELL, S.D.G. (1991) *Ordovician (Caradoc) Marginal Basin Volcanism in Snowdonia (North-west Wales).* HMSO, London, 191 pp.

HUSSONG, D.M. & UYEDA, S. (1981) Tectonic processes and the history of the Mariana Arc: a synthesis of the results of Deep Sea Drilling Project Leg 60. *Initial Reports of the Deep Sea Drilling Project* **60**, 909–929. US Government Printing Office, Washington, DC.

INGERSOLL, R.V. (1978) Petrofacies and petrologic evolution of the Late Cretaceous fore-arc basin, northern and central California. *J. Geol.* **86**, 335–353.

INGERSOLL, R.V. (1982) Initiation and evolution of the Great Valley forearc basin of northern and Central California, USA. In: *Trench–Forearc Geology* (Ed. Leggett, J.K.) Spec. Publ. Geol. Soc. London **10**, 459–467.

INGERSOLL, R.V. (1988) Tectonics of sedimentary basins. *Bull. Geol. Soc. Am.* **100**, 1704–1719.

INGERSOLL, R.V. and SUCZEK, C.A. (1979) Petrology and provenance of Neogene sand from Nicobar and Bengal fans, DSDP 211 and 218. *J. Sediment. Petrol.* **49**, 1217–1228.

ITO, M. (1987) Middle to Late Miocene foredeep basin successions in an arc–arc collision zone, northern Tanzawa Mountains, central Honshu, Japan. *Sediment. Geol.* **54**, 61–91.

JARRARD, R.D. (1986) Relations among subduction parameters. *Rev. Geophys.* **24**, 217–284.

JEFFERS, J.D. & ANDERSON, J.B. (1991) Sequence stratigraphy of the Bransfield Basin, Antarctica: implications for tectonic history and hydrocarbon potential. In: *Antarctica as an Exploration Frontier* (Ed. St. John, B.) Am. Ass. Petrol. Geol., Stud. Geol. **31**, 13–29.

JENKYNS, H.C. (1986) Pelagic environments. In: *Sedimentary Environments and Facies*, 2nd edn (Ed. Reading, H.G.) Blackwell Scientific Publications, Oxford, pp. 343–397.

JERVEY, M.T. (1988) Quantitative geological modeling of siliciclastic rock sequences and their seismic expression. In: *Sea-level Changes: an Integrated Approach* (Eds Wilgus, C.K., Hastings, B.S., Kendall, C.G.St.C., Posamentier, H., Ross, C.A. & van Wagoner, J.) Soc. Econ. Paleontol. Mineral., Spec. Publ. **42**, 47–70.

JONES, D.L. (1990) Synopsis of late Palaeozoic and Mesozoic terrane accretion within the Cordillera of western North America. *Philos. Trans. R. Soc. London.* **A331**, 479–486.

KARIG, D.E. (1975) Basin genesis in the Philippine Sea. *Initial Reports of the Deep Sea Drilling Project*, US Government Printing Office, Washington, DC, **31**, 857–879.

KARIG, D.E. (1985) Kinematics and mechanics of deformation across some accreting forearcs. In: *Formation of Active Ocean Margins* (Ed. Nasu, N.) Terra Scientific Publishing Company, Tokyo, pp. 155–177.

KARIG, D.E & SHARMAN, G.F. (1975) Subduction and accretion in trenches. *Bull. Geol. Soc. Am.* **86**, 377–389.

KEMP, A.E.S. (1987a) Tectonic development of the Southern Belt of the Southern Uplands accretionary complex. *J. Geol. Soc. London* **144**, 827–838.

KEMP, A.E.S. (1987b) Evolution of Silurian depositional systems in the Southern Uplands, Scotland. In: *Marine Clastic Sedimentology* (Eds Leggett, J.K. & Zuffa, G.G.) Graham & Trotman, London, pp. 124–155.

KLEIN, G. DE V. (1975) Sedimentary tectonics in Southwest Pacific marginal basins based on Leg 30 Deep Sea Drilling Project cores from the South Fiji, Hebrides and Coral Sea Basins. *Bull. Geol. Soc. Am.* **86**, 1012–1018.

KLEMME, H.D. (1980) Petroleum basins — classification and characteristics. *J. Petrol. Geol.* **3**, 187–207.

KLOOTWIJK, C.T., CONAGHAN, P.J. & POWELL, C.M. (1985) The Himalayan Arc: large-scale continental subduction, oroclinal bending and back-arc spreading. *Earth Planet. Sci. Lett.* **75**, 167–183.

KOKELAAR, B.P. (1988) Tectonic controls of Ordovician arc and marginal basin volcanism in Wales. *J. Geol. Soc. London* **145**, 759–775.

KOKELAAR, B.P. & HOWELLS, M.F. (Eds) (1984) Marginal basin geology. *Spec. Publ. Geol. Soc. London* **16**, 322 pp.

KOKELAAR, B.P. HOWELLS, M.F., BEVINS, R.E., ROACH, R.A. & DUNKLEY, P.N. (1984). The Ordovician marginal basin of Wales. In: *Marginal Basin Geology* (Eds Kokelaar, B.P. & Howells, M.F.) Spec. Publ. Geol. Soc. London **16**, 245–269.

KROENKE, L., SCOTT, R., *et al.* (1981) Site 450: east side of the Parece Vela Basin. *Initial Reports of the Deep Sea Drilling Project.* US Government Printing Office, Washington, DC, **59**, 355–403.

KUENZI, W.D., HORST, O.H. & McGEHEE, R.V. (1979). Effect of volcanic activity on fluvial–deltaic sedimentation in a modern arc–trench gap, southwestern Guatemala. *Bull. Geol. Soc. Am.* **90**, 827–838.

KULM, L.D., RESIG, J.M. THORNBURG, T.M. & SCHRADER, H.J.

(1982) Cenozoic structure, stratigraphy and tectonics of the central Peru forearc. In: *Trench–Forearc Geology* (Ed. Leggett, J.K.). Spec. Publ. Geol. Soc. London **10**, 151–169.

LANGSETH, M.G. & MOORE, J.C. (1990) Introduction to special section on the role of fluids in sediment accretion, deformation, diagenesis, and metamorphism in subduction zones. *J. Geophys. Res.* **95**, B6, 8737–8741.

LARTER, R.D. & BARKER, P.F. (1991) Neogene interaction of tectonic and glacial processes at the Pacific margin of the Antarctic Peninsula. In: *Sedimentation, Tectonics and Eustasy* (Ed. Macdonald, D.I.M.) Int. Ass. Sediment. Spec. Publ. **12**, 165–186.

LEEDER, M.R., SEGER, M.J. & STARK, C.P. (1991) Sedimentation and tectonic geomorphology adjacent to major active and inactive normal faults, southern Greece. *J. Geol. Soc. London* **148**, 331–343.

LEGGETT, J.K. (1980) The sedimentological evolution of a Lower Palaeozoic accretionary fore-arc in the Southern Uplands of Scotland. *Sedimentology* **27**, 401–417.

LEGGETT, J.K. (Ed.) (1982) Trench–forearc geology. *Spec. Publ. Geol. Soc. London* **10**, 576 pp.

LEGGETT, J.K. (1987) The Southern Uplands as an accretionary prism: the importance of analogues in reconstructing palaeogeography. *J. Geol. Soc. London* **144**, 737–752.

LEGGETT, J.K., McKERROW, W.S. & EALES, M.H. (1979) The Southern Uplands of Scotland; a Lower Palaeozoic accretionary prism. *J. Geol. Soc. London* **136**, 755–770.

LEGGETT, J.K., McKERROW, W.S. & CASEY, D.M. (1982). The anatomy of a Lower Palaeozoic accretionary forearc: The Southern Uplands of Scotland. In: *Trench–Forearc Geology* (Ed. Leggett, J.K.) Spec. Publ. Geol. Soc. London **10**, 494–520.

LEGGETT, J.K. LUNDBERG, N., BRAY, C.J. *et al.* (1987) Extensional tectonics in the Honshu fore-arc, Japan: integrated results of DSDP Legs 57, 87 and reprocessed multichannel seismic reflection profiles. In: *Continental Extensional Tectonics* (Eds Coward, M.P., Dewey, J.F. & Hancock, P.L.) Spec. Publ. Geol. Soc. London **28**, 593–609.

LE PICHON, X., HUCHON, P., ANGELIER, J. *et al.* (1982) Subduction in the Hellenic Trench: probable role of a thick evaporitic layer based on sea beam and submersible studies. In: *Trench–Forearc Geology* (Ed. Leggett, J.K.). Spec. Publ. Geol. Soc. London **10**, 319–333.

LOVELL, J.P.B. (1970) The palaeogeographical significance of lateral variations in the ratio of sandstone to shale and other features of the Aberystwyth Grits. *Geol. Mag.* **107**, 147–158.

LOWE, D.R. (1982) Sediment gravity flows: II. Depositional models with special reference to the deposits of high-density turbidity currents. *J. Sediment Petrol.* **52**, 279–297.

LUNDBERG, N. & REED, D.L. (1991) Continental margin tectonics: forearc processes. *Rev. Geophys., Supplement*, 794–806.

MACDONALD, D.I.M. (1986) Proximal to distal sedimentological variation in a linear turbidite trough: implications for the fan model. *Sedimentology* **33**, 243–259.

MACDONALD, D.I.M. (Ed.) (1991) Sedimentation, tectonics and eustasy: sea-level changes at active margins. *Int. Ass. Sediment. Spec. Publ.* **12**, 518 pp.

MACDONALD, D.I.M. & BUTTERWORTH, P.J. (1991) The stratigraphy, setting and hydrocarbon potential of the Mesozoic sedimentary basins of the Antarctic Peninsula. In: *Antarctica as an Exploration Frontier* (Ed. St. John, B.) Am. Ass. Petrol. Geol., Stud. Geol. **31**, 101–125.

MACDONALD, D.I.M. & TANNER, P.W.G. (1983) Sediment dispersal patterns in part of a deformed Mesozoic back-arc basin on South Georgia, South Atlantic. *J. Sediment. Petrol.* **53**, 83–104.

MACDONALD, D.I.M., STOREY, B.C. & THOMSON, J.W. (1987) *South Georgia. BAS Geomap Series, Sheet 1, 1: 250,000*, Geological Map and supplementary text, British Antarctic Survey, Cambridge, 63 pp.

MATSUDA, T. & ISOZAKI, Y. (1991). Well-documented travel history of Mesozoic pelagic chert in Japan: from remote ocean to subduction zone. *Tectonics* **10**, 475–499.

MAYNARD, J.B., VALLONI, R. & YU, H.-S. (1982) Composition of modern deep-sea sands from arc-related basins. In: *Trench–Forearc Geology* (Ed. Leggett, J.K.) Spec. Publ. Geol. Soc. London **10**, 551–561.

McMILLEN, K.J., ENHEBOLL, R.H., MOORE, J.C., SHIPLEY, T.H. & LADD, J.W. (1982) Sedimentation in different tectonic environments of the Middle America Trench, southern Mexico and Guatemala. In: *Trench–Forearc Geology* (Ed. Leggett, J.K.) Spec. Publ. Geol. Soc. London **10**, 107–119.

MIALL, A.D. (1990) The plate-tectonics framework of sedimentary basins. *J. Sedim. Soc. Japan* **33**, 3–13.

MILLAR, I.L., MILNE, A.J. & WHITHAM, A.G. (1990) Implications of Sm–Nd garnet ages for the stratigraphy of northern Graham Land, Antarctic Peninsula. *Zbl. Geol. Paläont. Teil I, Heft 1/2*, 97–104.

MONCRIEFF, A.C.M. & KELLY, S.R.A. (1993) Lithostratigraphy of the uppermost Fossil Bluff Group (Early Cretaceous) of Alexander Island, Antarctica: history of an Albian regression. *Cret. Res.* **14**, 1–5.

MOORE, D.G., CURRAY, J.R. & EMMEL, F.J. (1976) Large submarine slide (olistostrome) associated with Sunda Arc subduction zone, northeast Indian Ocean. *Mar. Geol.* **21**, 211–226.

MOORE, G.F. & KARIG, D.E. (1976) Development of sedimentary basins on the lower trench slope. *Geology* **4**, 693–697.

MOORE, G.F., BILLMAN, H.G., HEHANUSSA, P.E. & KARIG, D.E. (1980). Sedimentology and paleobathymetry of Neogene trench-slope deposits, Nias Island, Indonesia. *J. Geol.* **88**, 161–180.

MOORE, G.F., CURRAY, J.R. & EMMEL, F.J. (1982) Sedimentation in the Sunda Trench and forearc region. In: *Trench–Forearc Geology* (Ed. Leggett, J.K.) Spec. Publ. Geol. Soc. London **10**, 245–258.

MOORE, J.C. (1989) Tectonics and hydrogeology of accretionary prisms: role of décollement zone. *J. Struct. Geol.* **11**, 95–106.

MORRIS, W.R. & BUSBY-SPERA, C.J. (1988) Sedimentologic evidence of a submarine canyon in a forearc basin, Upper Cretaceous Rosario Formation, San Carlos, Mexico. *Bull. Am. Ass. Petrol. Geol.* **72**, 717–737.

NELL, P.A.R. (1990) Deformation in an accretionary mélange, Alexander Island, Antarctica. In: *Deformation Mechanisms, Rheology and Tectonics* (Eds Knipe, R.J. & Rutter, E.H.) Spec. Publ. Geol. Soc. London **54**, 405–416.

NELL, P.A.R. & STOREY, B.C. (1991) Strike-slip tectonics within the Antarctic Peninsula forearc. In: *Geological Evolution of Antarctica* (Eds Thomson, M.R.A., Crame, J.A. & Thomson, J.W.) Cambridge University Press, Cambridge, pp. 443–448.

NILSEN, T.H. (1980) Modern and ancient submarine fans: discussion of papers by R.G. Walker and W.R. Normark. *Bull. Am. Ass. Petrol. Geol.* **64**, 1094–1101.

NILSEN, T.H. & ZUFFA, G.G. (1982) The Chugach Terrane, a Cretaceous trench-fill deposit, southern Alaska. In: *Trench–Forearc Geology* (Ed. Leggett, J.K.) Spec. Publ. Geol. Soc. London **10**, 213–227.

NORMARK, W.R. (1974) Submarine canyons and fan valleys: factors affecting growth patterns of deep-sea fans. In: *Modern and Ancient Geosynclinal Sedimentation* (Eds Dott, R.H. & Shaver, R.H.) Spec. Publ. Soc. Econ. Paleont. Miner., Tulsa **19**, 56–68.

PICKERING, K.T. (1981) Two types of outer-fan lobe sequence, from the late Precambrian Kongsfjord Formation submarine fan, Finnmark, North Norway. *J. Sediment. Petrol.* **51**, 1277–1286.

PICKERING, K.T. (1991) Tectonics and sedimentation. *J. Geol. Soc. London* **148**, 315–316.

PICKERING, K.T., HISCOTT, R.N. & HEIN, F.J. (1989) *Deep Marine Environments.* Unwin Hyman, London, 416 pp.

PIGRAM, C.J. & DAVIES, H.L. (1987) Terranes and the accretion history of the New Guinea orogen. *B.M.R.J. Aust. Geol. Geophys.* **10**, 193–211.

PIPER, D.J.W., VON HUENE, R.E. & DUNCAN, J.R. (1973) Late Quaternary sedimentation in the active eastern Aleutian trench. *Geology* **1**, 19–22.

PLATT J.P. (1986) Dynamics of orogenic wedges and the uplift of metamorphic rocks. *Bull. Geol. Soc. Am.* **97**, 1037–1053.

POSAMENTIER, H.W., JERVEY, M.T. & VAIL, P.R. (1988) Eustatic controls on clastic deposition. I, conceptual framework. In: *Sea Level Changes: An Integrated Approach* (Eds Wilgus, C.K., Hastings, B.S., Kendall, C.-G.St.C., Posamentier, H.W., Ross, C.A. and Van Wagoner, J.C.) Spec. Publ. Soc. Econ. Paleontol. Miner. **42**, 109–124.

POTTER, P.E. (1986) South America and a few grains of sand: part 1 — beach sands. *J. Geol.* **94**, 301–319.

READING, H.G. (1980) Characteristics and recognition of strike-slip fault systems. In: *Sedimentation in Oblique-Slip Mobile Zones* (Eds Ballance, P.F. & Reading, H.G.) Int. Ass. Sediment. Spec. Publ. **4**, 7–26.

READING, H.G. (Ed.) (1986) *Sedimentary Environments and Facies*, 2nd edn. Blackwell Scientific Publications, Oxford, 615 pp.

SAITO, Y. (1991) Sequence stratigraphy in the shelf and upper slope in response to the latest Pleistocene–Holocene sea-level changes off Sendai, northeast Japan. In: *Sedimentation, Tectonics and Eustasy* (Ed. Macdonald, D.I.M.) Int. Ass. Sediment. Spec. Publ. **12**, 133–150.

SCHWELLER, W.J. & KULM, L.D. (1978) Depositional patterns and channelized sedimentation in active Eastern Pacific trenches. In: *Sedimentation in Submarine Canyons, Fans and Trenches* (Eds Stanley D.J. & Kelling, G.) Dowden, Hutchinson and Ross, Stroudsburg, PA, pp. 311–324.

SEELY, D.R. (1979) The evolution of structural highs bordering major forearc basins. In: *Geological and Geophysical Investigations of Continental Margins* (Eds Watkins, J.S., Montadent, L. & Dickerson, P.W.) Mem. Am. Ass. Petrol. Geol. **29**, 245–260.

SENGÖR, A.M.C. & DEWEY, J.F. (1990) Terranology: vice or virtue? *Philos. Trans. R. Soc. London* **A331**, 457–477.

SHANMUGAM, G. & MOIOLA, R.J. (1988) Submarine fans: characteristics, models, classification and reservoir potential. *Earth Sci. Rev.* **24**, 383–428.

SIGURDSSON, H., SPARKS, R.S.J., CAREY, S.N. & HUANG, T.C. (1980) Volcanogenic sedimentation in the Lesser Antilles arc. *J. Geol.* **88**, 523–540.

SMITH, G.A. (1987) The influence of explosive volcanism on fluvial sedimentation: the Deschutes Formation (Neogene) in central Oregon. *J. Sediment. Petrol.* **57**, 613–629.

SPARKS, R.S.J., SELF, S. & WALKER, G.P.L. (1973) Products of ignimbrite eruption. *Geology,* **1**, 115–118.

STONE, P., FLOYD, J.D., BARNES, R.P. & LINTON, B.C. (1987) A sequential back-arc and foreland basin thrust duplex model for the Southern Uplands of Scotland *J. Geol. Soc. London* **144**, 753–764.

STOREY, B.C. & ALABASTER, T. (1991) Tectonomagmatic controls on Gondwana break-up models: evidence from the proto-Pacific margin of Antarctica. *Tectonics* **10**, 1274–1288.

STOREY, B.C. & NELL, P.A.R. (1988) Role of strike-slip faulting in the tectonic evolution of the Antarctic Peninsula. *J. Geol. Soc. London* **145**, 333–337.

SUÁREZ, M. (1976) Plate-tectonic model for southern Antarctic Peninsula and its relation to the southern Andes. *Geology* **4**, 211–214.

SUTHREN, R.J. (1985) Facies analysis of volcaniclastic sediments: a review. I: *Sedimentology: Recent Developments and Applied Aspects* (Eds Brenchley, P.J. & Williams, B.P.J.) Spec. Publ. Geol. Soc. London **18**, 123–146.

THORNBURG, T.M. & KULM, L.D. (1987) Sedimentation in the Chile Trench: Depositional morphologies, lithofacies, and stratigraphy. *Bull. Geol. Soc. Am.* **98**, 33–52.

TRANTER, T.H. (1987) The structural history of the LeMay Group of central Alexander Island, Antarctic Peninsula. *Br. Antarct. Surv. Bull.* **77**, 61–80.

UNDERWOOD, M.B. & BACHMAN, S.R. (1982) Sedimentary facies associations within subduction complexes. In: *Trench–Forearc Geology* (Ed. Leggett, J.K.) Spec. Publ. Geol. Soc. London **10**, 537–550.

UYEDA, S. and KANAMORI, H. (1979) Back-arc opening and the mode of subduction. *J. Geophys. Res.* **84**, 1049–1061.

VAN WAGONER, J.C., POSAMENTIER, H.W. MITCHUM, R.M. *et al.* (1988) An overview of the fundamentals of sequence stratigraphy and key definitions. In: *Sea-level Changes: an Integrated Approach* (Eds Wilgus, C.K., Hastings, B.S., Kendall, C.G.St.C., Posamentier, H.W., Ross, C.A. & Van Wagoner, J.) Spec. Publ. Soc. Econ. Paleont. Miner., Tulsa **42**, 39–46.

VAN DE KAMP, P.C. & LEAKE, B.E. (1985) Petrography and geochemistry of feldspathic and mafic sediments of the northeastern Pacific margin. *Trans. R. Soc. Edinburgh* **76**, 411–449.

VAN HINTE, J.E. (1978) Geohistory analysis — application of micropalaeontology in exploration geology. *Bull. Am. Ass. Petrol. Geol.* **62**, 201–222.

VESSELL, R.K. & DAVIES, D.K. (1981) Non-marine sedimentation in an active forearc basin. In: *Recent and Ancient Non-marine Depositional Environments: Models for Exploration* (Eds Etheridge, F.G. & Flores, R.M.) Spec. Publ. Soc. Econ. Paleont. Miner., Tulsa **31**, 31–45.

VINE, F.J. & SMITH, A.G. (Eds) (1981) Extensional tectonics associated with convergent plate boundaries. *Philos. Trans. R. Soc. London* **A300**, 219–442.

VON HUENE, R. (1974) Modern trench sediments. In: *The Geology of Continental Margins* (Ed. Burk, C.A. & Drake, C.L.) Springer-Verlag, New York, pp. 93–103.

VON HUENE, R. (1986) To accrete or not accrete, that is the question. *Geol. Rundsch.* **75**, 1–15.

VROLIJK, P. (1991) On the mechanical role of smectite in subduction zones. *Geology* **18**, 703–707.

WALKER, R.G. (1978) Deep-water sandstone facies and ancient submarine fans; models for exploration and stratigraphic traps. *Bull. Am. Ass. Petrol. Geol.* **62**, 932–966.

WESTBROOK, G.K. (1982) The Barbados Ridge Complex: tectonics of a mature forearc system. In: *Trench–Forearc Geology* (Ed. Leggett, J.K.) Spec. Publ. Geol. Soc. London **10**, 275–290.

WESTBROOK, G.K. & SMITH, M.J. (1983) Long décollements and mud volcanoes: Evidence from the Barbados Ridge Complex for the role of high pore-fluid pressure in the development of an accretionary complex. *Geology* **11**, 279–283.

WHITE N. & MCKENZIE, D. (1988) Formation of the "Steer's head" geometry of sedimentary basins by differential stretching of the crust and mantle. *Geology* **16**, 250–253.

WHITHAM, A.G. (1989) The behaviour of subaerially produced pyroclastic flows in a subaqueous environment: evidence from the Roseau eruption, Dominica, West Indies. *Mar. Geol.* **86**, 27–40.

WILSON, T.J. (1991) Transition from back-arc to foreland basin development in the southernmost Andes: Stratigraphic record from the Ultima Esperanza District, Chile. *Bull. Geol. Soc. Am.* **103**, 98–111.

WINN, R.D. & DOTT, R.H. (1978) Submarine fan turbidites and resedimented conglomerates in a Mesozoic arc-rear marginal basin in southern South America. In: *Sedimentation in Submarine Canyons, Fans & Trenches* (Eds Stanley, D.J. & Kelling, G.) Dowden, Hutchinson & Ross, Stroudsburg, PA, pp. 362–373.

Spec. Publs Int. Ass. Sediment. (1993) **20**, 259–276

Thrust-generated, back-fill stacking of alluvial fan sequences, south-central Pyrenees, Spain (La Pobla de Segur Conglomerates)

D. MELLERE*

*Dipartimento di Geologia, Geofisica e Paleontologia, Via Giotto 1,
Universita' di Padova, Italy*

ABSTRACT

La Pobla de Segur Conglomerates (Collegats Formation) represent a Late Eocene and Oligocene alluvial fan complex developed within an intermontane basin, in the south-central part of the Pyrenean Chain, to the south of the Axial Zone antiformal stack (Nogueres). The conglomerates onlap a previously deformed substratum of Mesozoic and Palaeozoic rocks. They are organized into a series of stacked wedge-shaped bodies reaching a cumulative thickness of about 3500 m. In a vertical succession, five main allogroups (Pessonada, Ermita, Pallaresa, Senterada and Antist), comprising more than 20 alluvial fan lobes have been recognized. The allogroups are distinguished on the basis of underlying major unconformity surfaces, clast composition and palaeocurrent patterns. The structural relationship between successive alluvial fan lobes and their substratum has allowed the establishment of a chronology of deformation in this thrust-controlled basin. Each of the five allogroups can be tied to a specific stage of structuring.

The basal Pessonada allogroup is composed of small, local alluvial fans deposited at the southern active margin of the basin. Relationships between tectonic activity and the contemporaneous sedimentation have demonstrated a break-back thrusting sequence. The Ermita allogroup consists of units representing small fan-deltas which prograded into a shallow lake. The widespread fine-grained materials (floodbasin and lacustrine) suggest a period of initial subsidence linked to the onset of deep thrusting along the northern margin of the basin. The basinwards progradation of the polymict alluvial fan conglomerates of the Pallaresa allogroup define the third stage. Tectonic activity had shifted towards the northern margin of the basin and was characterized by the emergence of backthrusts related to the paroxysmal growth of the adjacent Nogueres structure. The imbricated system of backthrusts led to the development of a minor basin to the north, whose infill (Senterada allogroup) defines the fourth stage: alluvial units, bounded by tectonic structures, are deformed in progressive unconformities demonstrating their contemporaneous deposition with emplacement of the back-thrusts. Only the youngest alluvial deposits of the Antist allogroup are not deformed and record the cessation of activity in the Nogueres Zone (fifth stage). The general north–south geometry of the Collegats Formation shows a stepwise, backfilling pattern of deposition for the allogroups. There was a northward migration of the sedimentary depocentres as a response to deep seated thrust emplacement along the northern margin. The Collegats Formation, rather than being merely post-tectonic infill, has been shown to be Late Eocene–Oligocene syntectonic deposits.

INTRODUCTION

Syntectonic sediments are infrequently preserved in the stratigraphical record as they accumulate in unstable areas where rates of erosion often exceed those of accumulation (Heller *et al.*, 1986; Burbank

*Present address: Geologisk Institutt, Allegaten 41, Universitet i Bergen, 5007 Bergen, Norway.

et al., 1992; Jordan *et al.*, 1988). The high subsidence rates associated with the continental intramontane basins of the south-central Pyrenees, however, is such that preservation was excellent and allowed a detailed study of the relationships between thrust movement and contemporaneous clastic sedimentation. Conglomeratic packages are deformed in progressive angular unconformities (Riba, 1976) during their deposition by the emplacement of adjacent thrusts (Riba, 1976; Anadón *et al.*, 1986; Martinez *et al.*, 1988; Verges & Muñoz, 1990). Clastic deposition is not only synchronous with thrust emplacement, but often the same conglomeratic packages can be shown to be bounded by the thrust sheets themselves. The occurrence of synorogenic deposits has been critical to the establishment of the relative ages of thrust emplacement and the kinematics of deformation.

In the south-central Pyrenees, accretion of deep-seated duplexes below the upper structural units of the Axial Zone of the Chain (Nogueres) created at shallower levels the out-of-sequence thrusting as well as backthrusting, which cross-cut previous structures (Fig. 1). Strong subsidence rates affected the intramontane basins located in the front of the vertical stacking units.

The purpose of this paper is to describe the clastic infill of one of these basins (La Pobla de Segur Basin) and determine its relationships with the tectonic structures. Both out-of-sequence thrusting and backthrusting are here recorded within a thick conglomeratic succession (Figs 3 and 5). Base-level fluctuations induced by tectonic loading controlled the location of depocentres. In the classical model of a piggy-back basin succession (Ori & Friend, 1984), crustal loading produced by the emplacement of thrust sheets (Beaumont, 1981; Jordan, 1981) induces the systematic displacement of the sedimentary depocentres towards the stable foreland. In the particular physiography of an intramontane basin located between an antiformal stack

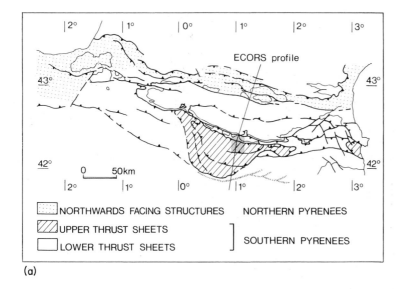

(a)

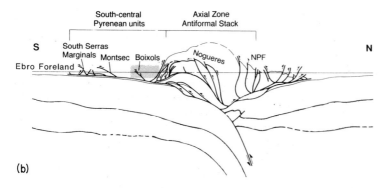

(b)

Fig. 1. Structural sketch of the Pyrenees also showing location of study area (a) and the ECORS seismic profile (b). NPF, North Pyrenean foreland.

and an old and inactive thrust that becomes tilted by later emplacement of deep thrust units, the depocentres migrate progressively away from the thrust front toward the vertically stacking wedge (see also Roure *et al.*, 1990). Detailed structural and stratigraphical analysis, backed up by palaeocurrent data and compositional studies on the continental deposits of La Pobla de Segur Basin demonstrates the northward shifting of depocentres towards the core of the Pyrenean fold-belt with time.

GEOLOGICAL SETTING

The Pyrenean chain represents an Alpine collision belt located between the Iberian and European plates. Convergence occurred during Late Cretaceous (Campanian) to Early Miocene time as a consequence of the sinistral displacement of the Iberian plate and it culminated with the partial closure of the Bay of Biscay (Boillot & Capdevila, 1977) and with the inversion of the Lower Cretaceous extensional structures. The original distance between the two plates was calculated to be in the order of 100–150 km (Grimaud *et al.*, 1982; Olivet *et al.*, 1984; Boillot, 1986) and this is consistent with the total shortening established in the central Pyrenees from the construction of crustal balanced cross-sections (Roure *et al.*, 1989; Muñoz, 1992). The publication of the ECORS seismic profile (Fig. 1), a deep seismic reflection traverse extending across the entire chain, from the Aquitaine Basin in France to the Ebro foreland in Spain (ECORS Pyrenean team, 1988; Choukroune & ECORS team, 1989), has allowed a better understanding of the evolution of the Pyrenees. The structural configuration of the chain (Fig. 1) is characterized by a basement antiformal stack, which consists primarily of Cambrian to Triassic rocks, and is bounded at both sides by imbricated thrust systems (Muñoz, 1992) involving cover deposits of Mesozoic and Early Palaeogene age, displaced respectively northwards toward the Aquitaine Basin and southwards toward the Ebro foreland. In the south-central Pyrenees, the system of thrusted cover deposits is referred to as the South-central Pyrenean Units (Seguret, 1972), and consists of three main thrust sheets called, from south to north, the Serras Marginals, the Montsec and the Boixols thrust sheets (Fig. 1). The characteristics of syntectonic sediments indicate that these thrust-sheets were carried in a piggy-back fashion from Late Cretaceous to Miocene times. Similarly, the nature of the infill of the Southern Pyrenean Foreland Basins indicates that their evolution was strongly linked to the pattern of thrusting (Puigdefàbregas & Souquet, 1986; Puigdefàbregas *et al.*, 1986, 1989, 1992; Mutti & Sgavetti, 1987; Mutti *et al.*, 1988; Burbank *et al.*, 1992). During the Late Eocene and Oligocene the tectonic physiography of the thrust belt evolved in response to two ongoing tectonic processes: (1) the progression of the thrust system toward the Ebro foreland, recorded by the emplacement of the Serra Marginals thrust sheet; and (2) the incorporation of Hercynian basement units within the Axial Zone antiformal stack in the hinterland and below the South–central Pyrenean Units (Muñoz, 1992). The thrust-generated basement wedging resulted in a strong increase in topographical relief in the Axial Zone of the chain, and was accompanied by widespread deposition of continental sediments throughout the South-central Pyrenean Units. La Pobla de Segur Conglomerates (Rosell & Riba, 1966), known also as the Collegats Formation (Mey *et al.*, 1968) was deposited during this stage of the Pyrenean chain evolution, within an intramontane basin (La Pobla de Segur Basin) localized over the Boixols thrust sheet and the southern extremity of the Axial Zone antiformal stack (Nogueres) (Fig. 2).

The Boixols thrust sheet was emplaced southwards during the Late Cretaceous, as demonstrated by its relationships with the syntectonic deposits of the Aren Sandstones (Nagtegaal *et al.*, 1983; Simó, 1985; Simó & Puigdefàbregas, 1985; Mutti & Sgavetti, 1987). It is bounded to the south by the Boixols thrust, and to the north by the Morreres backthrust (Late Eocene–Oligocene?). This last structure represents a passive roof thrust (Banks & Warburton, 1986) related to the stacking of basement thrust sheets below (Muñoz, 1992). It truncates previously developed folds, and displaces the Lower Cretaceous succession over the Triassic sequences. The Morreres backthrust partitioned the intramontane basin into two sub-basins: La Pobla de Segur Basin in its hanging-wall, and Senterada Basin in its footwall (Figs 2, 3 and 4). The internal structure of the Boixols thrust sheet, the 'relative basement' of La Pobla de Segur Basin, is characterized by an asymmetric syncline between the south-dipping Morreres backthrust and the north-dipping Boixols thrust (Figs 4 and 9). The frontal structure consists of the Sant Corneli hanging-wall anticline, a concentric fold which borders the southern flank of the study region. Other minor folds and thrust

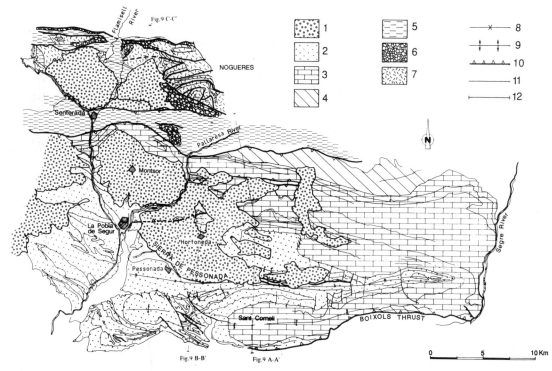

Fig. 2. Geological map of study area with the major structural features. Locations of geological sections (Fig. 9) are also shown. Symbols are as follows: 1, Upper Eocene–Oligocene La Pobla de Segur Conglomerates; 2, Upper Cretaceous; 3, Lower Cretaceous; 4, Jurassic; 5, Upper and Middle Triassic, 6, Lower Triassic; 7, Palaeozoic; 8, syncline; 9, anticline; 10, thrust; 11, trace of bedding; 12, line of cross-section. The map is based in part on the data of Rosell (1967), Mey *et al.*, (1968), Simó (1985) and Garcia (1989).

sheets also occur. Most of these structures are Late Cretaceous in age, although tilting of the northern limb of the synform continued during the Eocene as the Nogueres units developed.

La Pobla de Segur Conglomerates cover an area of 170 km^2 and reach a cumulative thickness of 3500 m or more. They represent an alluvial fan complex comprising more than 20 interfingering alluvial fan lobes that prograded into floodbasin and shallow lacustrine environments. The age of the Formation is poorly constrained. The only bio-stratigraphical data consists of mammal remains within the lowermost lacustrine interval and suggests a Late Eocene age (Casanovas, 1975). Below this marker layer, the 1200-m-thick conglomeratic succession is Middle to Late Eocene in age as it unconformably overlies a Middle Eocene continental and marine succession exposed to the west of the study area. The age of the upper 2500 m of conglomerates is conjectural, in the absence of bio-stratigraphical data, but it is possibly Oligocene.

The Formation has for long been considered as a post-orogenic infill of an intramontane basin which had not suffered any tectonic activity since Late Cretaceous time. The unconformity at its base was considered to be associated with the major compressional Pyrenean phase (de Sitter, 1961: Mey *et al.*, 1968). Rosell and Riba (1966) provided a new contribution regarding the stratigraphy and the geometric disposition of La Pobla de Segur Conglomerates. They pointed out that the succession was tilted to the north due to progressive uplift of the southern flank of the basin, and therby suggested the syntectonic character of the deposit. This tilting of the flank would have also caused the displacement of sedimentary depocentres in the same direction. Robles (1984) and Robles and Ardevol (1984) gave accounts of the stratigraphy and the sedimentology of the alluvial fan–lacustrine environment transition along both sides of Pallaresa River. They suggested that the cyclicity within the major alluvial-fan succession was linked to tectonic activity.

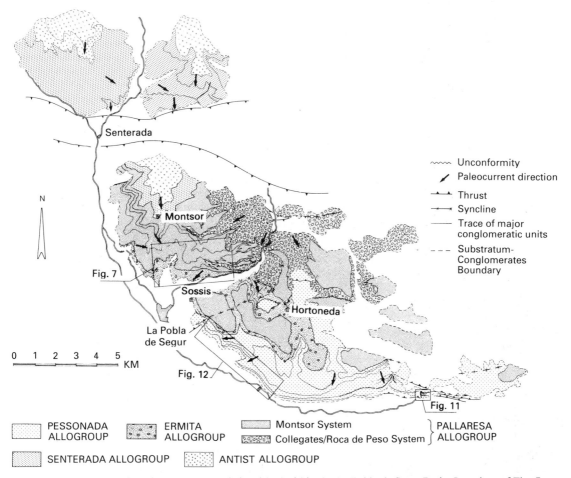

Fig. 3. General outline of the four allogroups distinguished within the La Pobla de Segur Basin. Locations of Figs 7, 11 & 12 are shown. (Mellere, 1992)

Detailed field studies in the region, in particular along the contact between the conglomerates and the substratum, have shown that the deposition of the Collegats Formation was strongly affected and controlled by tectonic structures. In the present work, a new north–south subdivision of the stratigraphical record of the La Pobla de Segur and Senterada Basins is proposed on the basis of the location and distribution of major unconformities related to the main tectonic structures.

STRATIGRAPHY

Detailed stratigraphical and sedimentological field studies in the continental succession of La Pobla de Segur Basin indicate that the establishment of traditional lithostratigraphical and chronostratigraphical units is extremely difficult, if not impossible. The lack of accurate age-dating prevents the application of chronostratigraphy, while facies relationships (frequent interfingering between alluvial-fan, fan-delta and lacustrine deposits) are spatially too complex for lithostratigraphical units to be recognized and established. The same problems emerge with the application of the concepts derived from seismic stratigraphy (Mitchum *et al.*, 1977; Vail *et al.*, 1977; Posamentier *et al.*, 1988; Van Wagoner *et al.*, 1990). The cyclicity, recognizable within the alluvial–lacustrine succession, occurs at too many scales for depositional sequences to be correlated. Allostratigraphy (NACSN, 1983) has

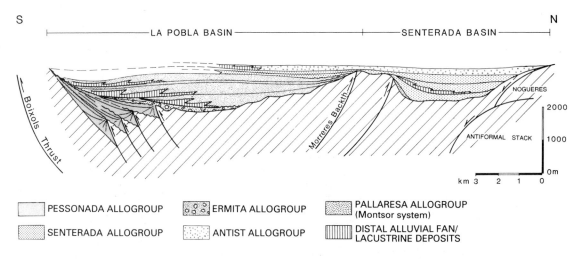

Fig. 4. Schematic cross-section showing the geometric relationships between the conglomerates and the tectonic structures. The cross-section is parallel to the trace of the C–C′ geological section shown in Fig. 9 (see Fig. 2 for location).

been chosen as the most useful way to subdivide the continental succession of La Pobla de Segur Basin. Unlike traditional lithostratigraphical subdivisions and depositional sequences, allo-units are non-interpretative and their application is independent of the agent causing the bounding discontinuity. Allostratigraphical units are herein applied in a more creative way with respect to the original definition, extending the recognized unconformities to their laterally conformable, time-equivalent, horizons. On this basis, the continental succession has been subdivided into five major allogroups (200–1000 m thick) defined and identified by bounding, basin-wide (5–20 km long) erosive to angular unconformities and by their lateral paraconformities. The spectacular continuity of the outcrops has allowed the establishment of a relative chronostratigraphy. From the oldest to the youngest, the allogroups will be referred to as: Pessonada, Ermita, Pallaresa, Senterada and Antist. The allogroups can be constituted by two or more interfingering alluvial-fan and/or fan-delta units, often showing different composition and palaeocurrents. Although referred to different source areas, all the alluvial fans belonging to a particular allogroup, are bounded by the same unconformable surfaces at base and top. Within the allogroups, the major pulses of progradation and retreat of alluvial-fan/fan-delta lobes (30–200 m thick) are referred to as sequences.

The origin of the unconformities is shown to be related to tectonic activity. The tectonosedimentary infill of the basin can be traced through five main stages, corresponding with the deposition of each of the five main allogroups. The relationship between the tectonic structures and the five allogroups is shown in Fig. 4.

DESCRIPTION OF THE ALLOGROUPS

The Pessonada allogroup

The Pessonada allogroup (Robles & Ardevol, 1984) represents the oldest and southernmost group of the succession. It is localized immediately to the north of the Sant Corneli anticline (Fig. 3) where it reaches a thickness of about 1000 m. It unconformably overlies a 'basement' of Cretaceous units. Its upper bounding surface is defined by an erosional surface over which conglomeratic packages of Ermita allogroup onlap. The Pessonada allogroup consists of a proximal–medial alluvial-fan assemblage, generally prograding south-southwestwards. It is composed almost exclusively of cobble to boulder, clast-supported conglomerates with subordinate sandy and lutitic facies. The conglomerates are mostly monomict. Clast composition, mainly Middle and Upper Cretaceous limestones with only subordinate Upper Mesozoic sandstones, reflects a local source area. At the contact with the palaeorelief, the

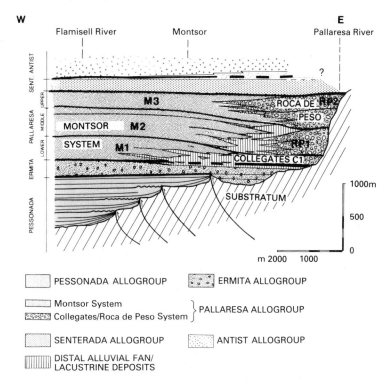

Fig. 5. West–east cross-section showing the stratigraphical relationships between the main units of La Pobla de Segur Conglomerates. This E–W scheme is based on 15 stratigraphical sections measured along Pallaresa River. The lowermost Pessonada clastic units are, in this section, deformed by out-of-sequence thrusting.

conglomerates are coarser with subangular large blocks (up to 2 m in diameter) within a cobble–boulder matrix.

The Group is vertically arranged into four clastic units (100–400 m thick) (Figs 4 and 5), or sequences, identified on the basis of their bounding discontinuities. In a southeast direction, the bounding surfaces constitute progressive angular unconformities (Fig. 3) which evolve northwestwards into paraconformities. The sequences result from the successive progradation and retrogradation of small alluvial fans. Along an axial section their lateral extent usually attains 2–3 km, while along a transverse section it ranges 4–6 km. The sequence boundary is traced at the base of single progradational units (Fig. 6). The fan retreat that took place during the final evolutionary stage of each unit resulted in the widespread accumulation of fine-grained, alluvial fan-fringe deposits (Fig. 6). Often, these fine-grained sediments are eroded by the subsequent unconformity. No significant grain-size variation characterized the vertical arrangement of the stacked alluvial-fan sequences, except for an indistinct retrogradational tendency in the youngest unit, expressed by a relative decrease in thickness of

the conglomeratic packages and by the development of more extensive fine-grained, alluvial fan-fringe and floodbasin deposits.

The Ermita allogroup

The Ermita allogroup (Robles, 1984; Robles & Ardevol, 1984) represents a fan-delta system which emerged from the south-west margin of the basin and prograded into a shallow lacustrine environment. Towards the south it unconformably overlies the lower Pessonada allogroup, while towards the north it onlaps the Mesozoic palaeorelief (Fig. 3). The upper boundary is represented by an erosional surface, above which the polymict units of Pallaresa allogroup lie (Figs 3 and 4). The Ermita allogroup crops out mainly along the right margin of Pallaresa River (Fig. 7) where it reaches a thickness of 200–250 m. It consists of monomict 'white' conglomerates and subordinately of fine-grained floodbasin and lacustrine deposits with occasional coal layers. Palaeocurrents are generally directed towards the west–south-west (Fig. 3). Conglomeratic composition reflects a local source area located within the Mesozoic carbonate succession of the basin margin.

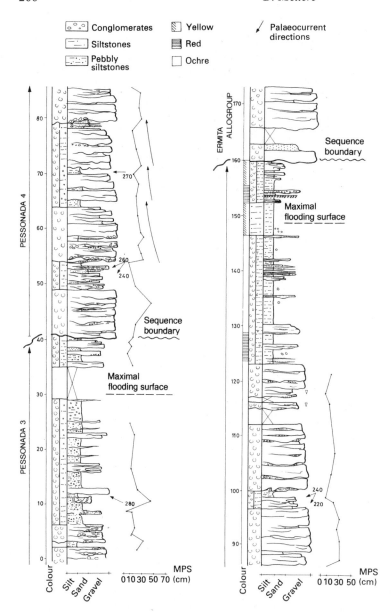

Fig. 6. Representative stratigraphical section of the last sequence of the Pessonada allogroup, measured on the conglomeratic cliff that borders the western margin of the basin (see Figs 3 and 12). Sequence boundary is traced at the base of the prograding conglomeratic packages. The sequences (Pessonada 3 and 4) are defined at their tops by the development of alluvial fan-fringe deposits. The general vertical stacking of the conglomeratic units is aggradational–retrogradational followed by a slight pleogradational trend. A surface of maximum flooding has been traced, corresponding with the maximum expansion of lacustrine/floodbasin deposits.

The allogroup is vertically arranged into three small sequences (each 50–100 m thick) of fan-delta progradation separated by floodbasin and lacustrine intervals (20–40 m thick in their medium–proximal area, 40–70 m thick in their distal reaches). Their bounding surfaces are defined by erosional and angular unconformities of minor rank with respect to the allogroup boundaries, and recognizable only in the most easterly areas, where the coarsest alluvial deposits occur. Basinwards (southwestwards),

the unconformities evolve into paraconformities. The sequences show a progressive retrogradational stacking pattern with an upward decrease in thickness of the conglomeratic fan-delta wedges and a progressive increase of the floodbasin and lacustrine intervals (Fig. 8). The Ermita allogroup was deposited during the latest Eocene as indicated by the mammal remains found in its lacustine intervals (Casanovas, 1975).

Fig. 7. Spectacular exposure along the Pallaresa River (location outlined in Fig. 3) showing the stratigraphical relationships between the lower monomict Ermita and the upper Pallaresa (Montsor system) allogroups. El, E2 and E3 indicate respectively the sequences of the Ermita allogroup; M1, M2 and M3 the sequences of the Montsor System prograding southeastwards; C1 and RP1 the alluvial fan units of the Collegats–Roca de Peso system, prograding south-eastwards.

The Pallaresa allogroup

The Pallaresa allogroup crops out mainly to the north of Pallaresa River (Figs 3 and 7) where it reaches a thickness of about 1000 m. The spectacular outcrop along the Pallaresa and Flamisell Rivers has allowed the allogroup to be examined in three dimensions. Northwards and eastwards it onlaps the Mesozoic margin of the basin. Along Flamisell River the allogroup overlies, across an erosional unconformity, the monomict Ermita allogroup. Toward the basin, along Pallaresa River, the unconformity evolves to a paraconformity. In the southern part of the basin (Sierra de Pessonada) the allogroup onlaps an erosional and angular unconformity surface developed over Ermita and Pessonada units. Onlap angle ranges between 10 and 15°. The upper bounding surface of the allogroup is an unconformity, below the Senterada allogroup.

The Pallaresa allogroup is composed of two convergent alluvial fan systems (Figs 3 and 5): Montsor system, with palaeocurrents directed to the east, and Collegats–Roca de Peso system, with palaeocurrent directed to the west (Rosell & Riba, 1966; Robles, 1984).

The Montsor system represents the major alluvial fan complex of the basin, covering a surface of around 50 km². It is exposed in spectacular red cliffs just to the north of the town of La Pobla de Segur and along the Flamisell River. It is dominated by reddish polymict, clast-supported, alluvial fan conglomerates, cobble- to boulder-grade, and by coarse to fine sandstones. Between the clasts, matrix is typically of coarse-grained sandstones. The framework clasts are composed of Lower Triassic sandstones and conglomerates, Carboniferous volcano-clastic sandstone, Devonian limestone and only subordinately by lower Mesozoic limestones.

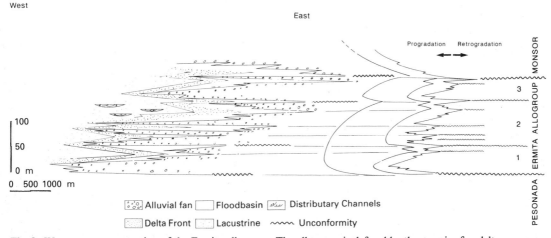

Fig. 8. West–east representation of the Ermita allogroup. The allogroup is defined by three major fan-delta sequences, bounded eastwards by erosive unconformities. The allogroup shows a distinctive retrogradational stacking pattern. Within each sequence minor unconformities bound the prograding fan-delta lobes.

This composition reflects a source area localized within the Axial Zone of the Pyrenean chain. In a vertical succession (Fig. 5) the Montsor system can be subdivided into three main prograding alluvial-fan units (each one 200–400 m thick and referred to as M1, M2 and M3) bounded by major lacustrine intervals (50–70 m thick in the downfan reaches). Each alluvial fan unit is composed of a coarsening-and-thickening-upward stack of progradational wedges (30–50 m at their proximal ends) separated by minor lacustrine intervals (Robles, 1984). Compositional data and palaeocurrent directions indicate that the alluvial units of the Montsor system represent major fans, with a radius of at least 20–30 km.

Towards the east the Montsor system interfingers with the minor, partially monomictic system of Collegats–Roca de Peso. The Collegats–Roca de Peso system crops out along the eastern side of Pallaresa Valley where it reaches a thickness of about 800 m. Eastwards and northwards it onlaps the Mesozoic palaeorelief. It represents an alluvial-fan/fan-delta complex developed at the same time as the Montsor system and has the same bounding unconformities at base and top. It is dominated by 'white' conglomerates and subordinately by floodbasin and lacustrine siltstones and fine sandstones. Clast composition, mainly Cretaceous and Jurassic limestones and dolostones with minor Triassic red sandstones, reflects a provenance area localized within the surrounding Mesozoic succession. The contrast in composition between the two convergent systems of Collegats-Roca de Peso and Montsor is one of the most spectacular features of the Pallaresa allogroup. Vertically, the Collegats–Roca de Peso system is composed by a stack of 3 fan-delta units, three C1, RP1 and RP2 (Fig. 5). The stacking pattern of these sequences is similar to that recognized in the Montsor allogroup. The progradation and the retreat of the fan-delta wedges are contemporaneous with the main pulses of progradation and retrogradation of the Montsor system. Although they represent minor wedges, the fans in the collegats–Roca de Peso system show a three-dimensional geometry easily recognizable on both sides of Pallaresa Valley. The fans are about 3 km long and 2 km wide with a thickness rapidly decreasing from 200 m in their proximal part to pinch-out downfan. Their base is sharp and erosive over floodbasin and lacustrine sediments. The relationship between the two convergent systems is spectacularly exposed along the Pallaresa Valley (Fig. 7).

A detailed study shows that sequence boundaries within the Pallaresa allogroup can be traced from one system to the other (Fig. 5). The first two sequences (M1 and C1) of both Montsor and Collegats–Roca de Peso systems are defined, in a north–south direction, by the same erosional unconformities and, moving towards the centre of the basin, along the Pallaresa River, by their relative paraconformities. Northwards, these sequences simply onlap the Mezosoic 'basement'. The upper sequences of the Pallaresa allogroup (M2–RP1 and M3–RP2) display a slightly different configuration. Sequence boundaries can always be followed physically at the base of the converging conglomeratic units. However, their northwards relationship with the Mesozoic basement is not defined by a simple onlap, but by a progressive angular unconformity related to thrust emplacement.

Unlike the Ermita allogroup, Pallaresa sequences are vertically arranged in a progradational stacking pattern. Initially progradation occurs within a relatively well-developed alluvial plain and lacustrine setting: later the floodbasin and lacustrine intervals become subordinate with respect to the proximal and distal alluvial-fan conglomerates until the coarse-grained alluvial-fan conglomerates dominate.

The Senterada allogroup

The Senterada allogroup (Figs 3, 4 and 5) crops out at the top of the La Pobla de Segur Basin succession, where it unconformably overlies the Montsor allogroup, and, in the Senterada Basin, along both sides of the Flamisell River to the north of the village of Senterada. It reaches a thickness of about 1000 m and unconformably covers a substratum above the Nogueres unit, which consists of a Triassic succession to the south and Carboniferous and Devonian sandstones and limestones to the north. Its upper bounding surface is represented by an unconformity with the youngest Antist allogroup. The Senterada allogroup consists of polymict alluvial-fan deposits, with conglomerates, pebble to boulder grade, and with subordinate alluvial-fringe and floodbasin sandstones and siltstones. Conglomerates range from clast- to matrix-supported. Clast composition indicates a source area within the Nogueres unit and the most internal part of the Axial Zone of the chain. Matrix composition is typically red to violet shale and fine sandstone of Upper Triassic (Keuper) deposits.

Although conglomerates are composed of Triassic and Palaeozoic lithologies, the relative percentage of the lithotypes involved varies throughout the succession. Three major conglomeratic units (100–300 m thick) have been distinguished on the basis of their bounding unconformities. Each unit shows a general retrogradational arrangement with a stacking of fining-upward, minor cycles. Palaeocurrents are directed towards the south-southwest.

The Antist allogroup

The Antist allogroup is the youngest part of the conglomeratic succession and overlies the Senterada allogroup in both the Senterada and the La Pobla basins (Fig. 3) where it reaches a thickness of around 300 m. Its lower bounding surface is an erosional and angular unconformity: its upper boundary is the present topographical surface. The Antist allogroup is characterized by very coarse-grained, massive, alluvial-fan conglomerates (cobbles to block) clast- to matrix-supported, which prograde towards the south. Matrix consists of coarse to very coarse sandstones. Clast composition is dominated by Carboniferous sandstones and by Devonian limestones with a minor percentage of Triassic lithotypes.

RELATION WITH TECTONICS

The structural relationships between the conglomerates and their substratum have been defined through a detailed field mapping and the construction of serial N–S geological sections. The sections were traced though erosional 'windows' which allowed delineation of the structure of the Mezosoic substratum in an E–W direction. Sketches of the sections are shown in Fig. 9. The structures recognized in one section were then extrapolated to the following sections. The relationships between tectonics and sedimentation have been used to establish the kinematic history and the relative age of the thrust emplacement.

The structures in the southern part of the basin (Sierra de Pessonada) and their relationships with sedimentation

The structures in the Sierra de Pessonada are characterized by an imbricated thrust system which controlled and affected the deposition of the Pes-

sonada allogroup. To the northeast of Sant Corneli anticline, the four clastic units of the allogroup are bounded by progressive angular unconformities related to thrusting. Laterally, towards the west, the unconformities pass to paraconformities (Figs 9 and 12). The first unit unconformably overlies the previously folded Upper Cretaceous substratum (Fig. 10). Towards the north, this unit is deformed, showing a progressive angular unconformity (Fig. 11) related to the emplacement of the first thrust of the Sierra de Pessonada imbricated system. At the contact with the thrust, the conglomerates display an offlap–onlap disposition, as in the model proposed by Riba (1976), with a fan-like geometry open towards the southwest (Fig. 11). Based on these observations, thrusting and alluvial-fan sedimentation was coeval. The most proximal part of the unconformity, where the beds are subvertical, is dominated by scree facies with blocks (2–3 m in diameter) derived from the folded substratum. The top is eroded and covered, in angular unconformity, by the second clastic unit of the allogroup.

The second conglomeratic package is deformed by a new thrust emplaced further inwards on the hanging-wall of the first thrust it is thrusted inwards by a thrust developed in the hanging-wall of the previous one. It shows the same geometric relationship (progressive unconformity and erosion at the top) as the first conglomeratic unit (Fig. 9). Similar relationships occur for the third and fourth units of the allogroup (Fig. 9). The thrusts in Sierra de Pessonada represent subordinate structures, with a displacement in the order of a few metres. Towards the west they usually evolve into blind thrusts expressed by kink folds. The northernmost and youngest thrust, near the village of Hortoneda, shows an exceptional displacement of 200 m in the Cretaceous succession. This major structure resulted in the subdivision of the La Pobla de Segur Basin into two sub-basins: the first with the depocentre localized in the footwall of this youngest thrust (Sierra de Pessonada), and infilled by the Pessonada allogroup; the second with the depocentre in the hanging-wall of the thrust (Pallaresa River), and infilled by the Ermita and Pallaresa allogroups.

Kinematic reconstruction of thrust sequence

The thick conglomeratic cover generally hinders recognition of the deep structures and the true

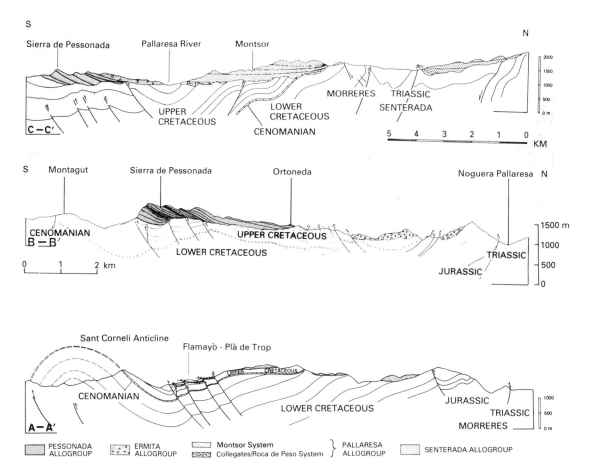

Fig. 9. Geological sections across La Pobla de Segur and Senterada basins (traces in Fig. 2).

Fig. 10. Photograph of the basal unconformity between the folded upper Mesozoic substratum and the first unit of the Pessonada allogroup at the northern flank of Sant Corneli anticline. To the north this conglomeratic unit displays a spectacular progressive angular unconformity (Fig. 11). P1, P2 and P3 indicate the sequences of the Pessonada allogroup cropping out in this sector of the basin.

Fig. 11. Progressive angular unconformity in the first clastic unit of the Pessonada allogroup related to the emplacement of the first out-of-sequence thrust of the Sierra de Pessonada imbricated system (location in Fig. 3). Note the offlap–onlap disposition of the beds (indicated by arrows).

geometry of the Sierra de Pessonada thrust system. Graphic and geometric methodologies(branch line maps — Hossak, 1985; Diegle, 1986) cannot be applied. However, the geometric relationships shown by the sequences in the Pessonada allogroup at the contact with the tectonic structures allowed a reliable kinematic and chronological reconstruction of thrust emplacement. Each unit of the Pessonada allogroup overlaps a thrust located in the footwall of a younger thrust that cuts it. The thrust which has affected the deposition of the first alluvial fan is the oldest, while the uppermost thrust, responsible for the deformation of the last alluvial fan of the allogroup is the youngest (Fig. 4). This geometry clearly shows an overstep (Elliott & Johnson, 1980) or a break-back thrusting sequence (Butler, 1987), with a progressive shift of deformation towards the north. Similar structural relationships have been found further to the east of the study area, along the

Segre River valley, in the Upper Eocene strata of Serres Marginals by Martinez *et al.* (1988) and by Vergés and Muñoz (1990).

The structural disposition of the clastic units

Synsedimentary tectonic activity in the Sierra de Pessonada is evidenced also by the sigmoidal stratal geometry displayed by each clastic unit. As can be inferred from the cross-sections (Fig. 9), conglomerate beds are tilted northwards, with progressively increasing (15 to 40°) and successively decreasing (40 to 10°) dip angles. This geometry defines an apparent 'downlap' of the clastic units. Each sequence, moreover, onlaps the lower one. Geometric relationships are easily recognizable both in the cross-sections (Fig. 9) and in the conglomeratic cliff that borders the Sierra de Pessonada on the west flank of the basin (Fig. 12).

Fig. 12. General view of the conglomeratic units of Sierra de Pessonada (see Fig. 3 for location). The conglomerates dip 30° to the north due to the tilting of the southern margin of the basin. They are unconformably overlain by the Ermita allogroup. Note the sigmoidal shape of the Pessonada sequences (1, 2, 3 and 4).

The structures in the northern margin of the basin

Tectonic structures in the northern margin of the basin are represented by the backthrust system, which is testimony to the activity of the Axial Zone antiformal stack. The backthrusts can be considered as an 'in-sequence' imbricated system of passive roof thrusts (Banks & Warburton, 1986) vergents towards the north. Like the structures recognized in Sierra de Pessonada, the backthrust emplacement was contemporaneous with conglomeratic sedimentation. The backthrusts deform the units of the Pallaresa and Senterada allogroups. Synsedimentary deformation is proved by the progressive angular unconformities displayed by the upper units of the Pallaresa and by the sequences of Senterada allogroups. Structural relationships are recognizable in cross-sections B and C (Fig. 9) respectively along the Pallaresa and Flamisell rivers. Along the Pallaresa River, the sequences RP1 and RP2 of the Collegats–Roca de Peso system are folded by small backthrusts, a minor expression of the most important Morreres structure which defines the northern border of La Pobla de Segur Basin (Figs 3 and 4). The fans are progressively reduced in thickness toward the palaeorelief and they are vertically bounded by angular unconformities (Figs 3 and 9). Farther to the west, along the Flamisell River, the M2 and M3 sequences of the Montsor system show the same relationships (Fig. 9). The reddish conglomerates are folded into a progressive angular unconformity which demonstrates the synsedimentary activity of the Morreres backthrust. The same geometric relationships are displayed by the Senterada allogroup. The clastic units of the allogroup are deformed in a large progressive angular unconformity open towards the north (Fig. 9). Bed dips progressively decrease from 50° to horizontal along the Flamisell River and then slightly increase to the north.

DISCUSSION

Structural data suggest a strong tectonic control on sedimentation, and demonstrate that deposition of La Pobla de Segur Conglomerates was contemporaneous with the later, Late Eocene–Oligocene, compressional history of the Pyrenean chain.

The first stage of basin infill is characterized by the progressive tilting of the basin towards the north during the sedimentation of the Pessonada allogroup. This hypothesis is confirmed by the out-of-sequence thrusting in Sierra de Pessonada, the 'onlap–downlap' disposition of the alluvial fan sequences towards the north and by palaeocurrent analysis. Palaeocurrents are directed south-southwestwards, in a direction opposite to that of the northward downlap geometry of the conglomeratic units. The southern margin of the basin, represented by the Sant Corneli anticline (Late Cretaceous age), should therefore have been active during the Late Eocene, at least during the deposition of the youngest allogroup.

There are two possible interpretations:
1 an out-of-sequence reactivation of the Boixols thrust during the conglomeratic sedimentation;
2 the presence of deep structures which induced a ramp in the sole thrust and therefore the growth of the Sant Corneli anticline with the subsequent tilting of the basin.
Deep-seated structures are responsible for the tectonic wedging in the Axial Zone of the chain, but with the data available both hypotheses can be accepted.

The deposition of the Ermita allogroup was not apparently affected by any tectonic structure: no major progressive unconformities related to thrust emplacement have been seen. However, towards the eastern basin margin, the floodbasin and lacustrine intervals are progressively reduced and the three alluvial fans are vertically stacked and bounded by unconformities (Fig. 8), suggesting a synsedimentary tectonic activity. There should also have been a tilting of the southern margin of the basin during the deposition of the Ermita sequences, and this as seems to be indcated by the general retrogradational stacking pattern of the conglomeratic units (Fig. 8).

The main feature during this stage of the basin infill is the development of an extensive floodbasin and a lacustrine environment. Two main depositional systems, one alluvial and the other lacustrine become intimately related. Tectonic loading produced by the stacking of the Axial Zone units is thought to be responsible for the subsidence of the basin, creating the conditions for a lacustrine environment.

The deposition of Pallaresa allogroup marks a change in the palaeogeography of the basin. The basin infill was no longer dominated only by small alluvial fans emerging from the surrounding relief: a major alluvial fan related to an axial drainage system now prograded into the basin. The most visible expression is a sharp change in the composition of the clasts and in the palaeocurrent directions. In this context Collegats–Roca de Peso represents a minor system, more localized and strongly linked to the dynamics of the surrounding relief. Tectonic activity, during this stage of basin infill, is concentrated along the northern margin of the basin, as evidenced by the synsedimentary deformation. The growth of the antiformal stack within the Axial Zone of the chain would have instigated the erosional processes which produced the build-up and the strong southward progradation of the polymict fans. The southern margin of the basin had only subordinate activity with respect to the growth of the Axial Zone antiformal stack.

The imbricated system of backthrusts caused the formation of a small depocentre (Senterada Basin) towards the north, close to the Nogueres unit. Sedimentation of the Senterada allogroup occurred contemporaneously with the emplacement of the backthrusts, and consequently with the growth of the Nogueres antiformal stack. Only the conglomerates of the youngest Antist allogroup are not affected by tectonic activity. Beds are horizontal and unconformably cover the Senterada sequences in both the Senterada and La Pobla de Segur basins, overlying the previously formed structures. They are slightly folded towards the north where they onlap the Palaeozoic succession of the Nogueres. Deposition of the Antist allogroup reflects the cessation of tectonic activity in the Nogueres units and probably constitutes the only true, post-tectonic infill of the basin. The different stages of basin evolution are shown in Fig. 13.

CONCLUSIONS

La Pobla de Segur Conglomerates (Late Eocene–Oligocene age) represent the syntectonic infill of an intramontane basin that originated during the Late Cretaceous and reactivated by Late Eocene–Oligocene compression. Deposition was controlled constantly by compressive structures induced by the deformation of the southern (Sant Corneli anticline) and northern (Nogueres) margin of the basin. Major tectonic unconformities allow five allogroups to be defined (Pessonada, Ermita, Pallaresa, Senterada and Antist). The lower Pessonada and Ermita Allogroups are characterized by a complex of monomict, small alluvial fans and fan deltas. Their sedimentation was controlled by the dynamics of the Mesozoic succession along the southwestern margin of the basin. The upper Pallaresa and Senterada allogroups are polymict alluvial fans whose deposition was controlled by the dynamics of the northern margin made up of Mesozoic and Palaeozoic rocks.

Two main tectonic systems have been recognized:
1 local out-of-sequence thrusts, in the southern margin of La Pobla de Segur Basin;
2 backthrusts in the northern margin of the basin. Each structure was emplaced contemporaneously with clastic deposition. The relationship between tectonics and sedimentation was defined by progressive angular unconformities and by the progressive onlap of conglomeratic packages.

Initially the tectonic deformation is concentrated in the southern part of the basin. Structural geometries (out-of-sequence thrusting and progressive onlap–downlap of the Pessonada sequences towards the north) and synsedimentary deformation (progressive angular unconformities and northwards direction of sedimentary depocentres) suggest a basin tilting toward the north during the deposition of the Pessonada allogroup. Basin tilting is thought to have been induced by out-of-sequence reactivation of the Boixols thrust and/or by a ramp in the sole thrust originating from deep-seated thrusting.

Later, tectonic activity was concentrated in the northern margin of the basin and was coeval with the deposition of the Pallaresa and Senterada allogroups. Synsedimentary deformation is evidenced by progressive angular unconformities induced by the emplacement of backthrusts (passive roof structures of the Axial Zone antiformal stack) and by a southward migration of a depocentre containing the polymict units of the Pallaresa allogroup. The lacustrine environment which developed during the deposition of the Ermita allogroup, before the sedimentation of the Pallaresa units, is thought to be linked to a stage of maximum subsidence in the basin induced by the thrust wedging in the Axial Zone of the chain (antiformal stack).

The youngest conglomerates of the Hau allogroup unconformably cover the earlier ones and the tectonic structures. They record the cessation of tectonic activity in the Nogueres units.

ACKNOWLEDGEMENTS

This paper is part of a PhD thesis the author wrote

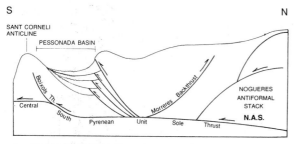

S   N

1 - OUT OF SEQUENCE THRUSTING SUTURATED BY SYNTECTONIC
    MONOGENIC CONGLOMERATES (PESSONADA ALLOGROUP)

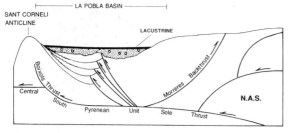

2 - DEPOSITION OF ERMITA ALLOGROUP
  - LACUSTRINE EVENT

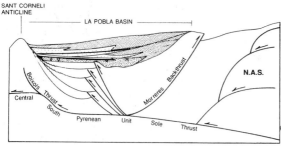

3 - STRONG PROGRADATION OF POLIMICTIC FANS
  - PROGRESSIVE UNCONFORMITY RELATED TO THE EMERGENCE OF BACKTHRUST

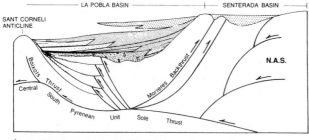

4 - SYNTECTONIC INFILL OF THE NORTHERNMOST BASIN

Fig. 13. (*Above and opposite.*) Idealized reconstruction of the tectonosedimentary evolution of La Pobla de Segur Conglomerates. No vertical and horizontal scale implied.

at the University of Barcelona under the supervision of F. Massari (University of Padova), M. Marzo (University of Barcelona) and C. Puigdefàbregas (Geological Service of Catalunya). I am most indebted to my advisors for their constructive comments and suggestions. I would particularly like to thank J. Verges, J.A. Muñoz and L. Cabrera for their critical appraisal of an earlier version of this

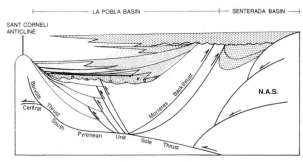

5 - PROGRADATION OF THE YOUNGEST CONGLOMERATE UNIT COVERING THE TECTONIC STRUCTURES

Fig. 13. (*Continued*)

manuscript, an anonymous reviewer and Ron Steel for their suggestions and improvements to the final text. The work was funded by DGICYT PB91-0805 project.

REFERENCES

ANADÓN, P., CABRERA, L., COLOMBO, F., MARZO, M. & RIBA, O. (1986) Syntectonic intraformational unconformities in alluvial fan deposits, eastern Ebro Basin margins (NE Spain). In: *Foreland Basins* (Eds Allen, P.A. & Homewood, P.), Int. Assoc. Sediment. Spec. Publ. **8**, 259–271.

BANKS, C.J. & WARBURTON, J. (1986) Passive-roof duplex geometry in the frontal structures of the Kirthar and Sulaiman mountain belts, Pakistan. *J. Struct. Geol.* **8(3/4)**, 229–237.

BEAUMONT, C. (1981) Foreland basins. *Geophys. J. R. Astron. Soc.* **65**, 291–329.

BOILLOT. G. (1986) Comparison between the Galicia and Aquitaine margins. *Tectonophysics* **129**, 243–255.

BOILLOT, G. & CAPDEVILA, R. (1977) The Pyrenees, subduction and collision? *Earth Planet. Sci. Lett.* **35**, 251–260.

BURBANK, D.W., VERGES, J., MUÑOZ, J.A. & BENTHAM, P. (1992) Coeval hindward and forward-imbrication thrusting in the Central Southern Pyrenees, Spain Timing and rates of shortening and deposition. *Geol. Soc. Am. Bull.* **104**, 3–17.

BUTLER, R.W.H. (1987) Thrust sequences. *J. Geol. Soc. London* **144**, 619–634.

CASANOVAS, M.L. (1975) Estratigrafia y Paleontologia del yacimiento ludiense de Roc de Santa (Area del Noguera Pallaresa). Unpublished PhD thesis, University of Barcelona.

CHOUKROUNE, P. & ECORS TEAM (1989) The ECORS Pyrenean deep seismic profile reflection data and the overall structure of an orogenic belt. *Tectonics* **8(1)**, 23–39.

DE SITTER, U. (1961) La phase tectogénique des Pyrenees. *Bull. Carte Geol. Fr.* **t. 35**, Paris.

DIEGLE, F.A. (1986) Topological constraints on imbricate thrust networks, examples from the Mountain City Window, Tennessee, U.S.A. *J. Struct. Geol.* **8**, 269–279.

ECORS PYRENEAN TEAM (1988) The ECORS deep seismic survey across the Pyrenees. *Nature* **331**, 508–511.

ELLIOTT, D. & JOHNSON, M.R.W. (1980) Structural evolution in the northern part of the Moine Thrust Belt, NW Scotland. *Trans. R. Soc. Edinburgh, Earth Sci.* **71**, 69–96.

GARCIA J. (1989) El Cretacico inferior entre los Rios Segre y Ribagorzana. *Servei Geologic de Catalunya*, internal report.

GRIMAUD, S., BOILLOT, G., COLETTE, B.J., MAUFFRET, A., MILES, P.R. & ROBERTS, D.B. (1982) Western extension of the Iberian–European plate boundary during the Early Cenozoic (Pyrenean) convergence, a new model. *Mar. Geol.* **45**, 63–77.

HELLER, P.L., BOWDLER, S.S., CHAMBERS, H.P. *et al.* (1986) Time of initial thrusting in the Sevier Orogenic belt, Idaho–Wyoming and Utah. *Geology* **14**, 388–391.

HOSSAK, J.R. (1985) The role of thrusting in the Scandinavian Caledonides. In: *The Tectonic Evolution of the Caledonide–Appalachian Orogen* (Ed. Gayer, R.W.) Fried. Vieweg & Sohn, Braunschweig, Weisbaden, pp. 98–116.

ISSC (International Subcommission on Stratigraphic Classification) (1987) Unconformity-bounded stratigraphic units. *Geol. Soc. Am. Bull.* **98**, 232–237.

JORDAN, T.E. (1981) Thrust loads and foreland basin evolution, Cretaceous, Western United States. *Am. Assoc. Petrol. Geol. Bull.* **65**, 2506–2520.

JORDAN, T.E., FLEMING, P.B. & BEER, J.A. (1988) Dating thrust fault activity by use of Foreland Basin Strata. In: *New Perspectives in Basin Analysis* (Eds Kleinspehn, K. & Paola, C.). Springer-Verlag, New York, pp. 307–330.

MARTINEZ, A., VERGES, J. & MUÑOZ, J.A. (1988) Secuencias de propagación del sistema de cabalgamientos de la terminación oriental del manto del Pedraforca y relación con los conglomerados sinorogenicos. *Acta Geol. Hisp.* **23**, 119–128.

MELLERC, D. (1992) I conglomerati di La Pobla de Segur: stratigrafia fisica e relazioni Tettonica–Sedimentazione. Unpublished PhD thesis, 203 pp. University of Padora, Italy.

MEY, P.H.W., NAAGTEGAAL, P.J.C., ROBERTI, K.J. & HARTEVELT, J.J.A. (1968) Lithostratigraphic subdivision of post-Hercynian deposits in the south-central Pyrenees, Spain. *Leidse Geol. Mededel.* **41**, 221–228.

MITCHUM, R.M., VAIL, P.R. & THOMSON, S. (1977) Seismic stratigraphy and global changes of sea level, part 2: the depositional sequence as a basic unit for stratigraphic analysis. In: *Seismic Stratigraphy – Applications to Hydrocarbon Exploration* (Ed. Payton, C.E.) Am. Assoc. Petrol. Geol. Mem. 26, 53–62.

MUÑOZ, J.A (1992) Evolution of a Continental Collision Belt: ECORS Pyrenees Crustal Balanced Cross-section. In: *Thrust Tectonics* (Ed. McClay, K.). Unwin Hyman, London, pp. 235–246.

MUÑOZ, J.A., MARTINEZ, A. & VERGES, J. (1986) Thrust sequences in the eastern Spanish Pyrenees. *J. Struct. Geol.* 8(3/4), 399–405.

MUTTI, E. & SGAVETTI, M. (1987) Sequence stratigraphy of the Upper Cretaceous Aren strata in the Orcau–Aren region, south-central Pyrenees, Spain: distinction between eustatically and tectonically controlled depositional sequences. *Ann. Univ. Ferrara. Sez. Sc. della Terra* 1, 22 pp.

MUTTI, E., SEGURET, M. & SGAVETTI, M. (1988) Sedimentation and deformation in the Tertiary sequences of the southern Pyrenees. *A.A.P.G. Mediterranean Basins Conference*, Field Trip 7, 153 pp.

NAAGTEGAAL, P.J.C., VAN VLIET, A. & BOUWER, J. (1983) Syntectonic coastal offlap and concurrent turbidite deposition: the Upper Cretaceous Aren Sandstone in the south-central Pyrenees, Spain. *Sediment. Geol.* 34, 185–218.

NACSN (1983) North American Stratigraphic Code. *Am. Assoc. Petrol. Geol. Bull.* 65/15, 841–875.

OLIVET, J.L., BONNIN, J., BEUZART, P. & AUZENDE J.M. (1984) Cinematique de L'Atlantique nord et central. Centre National pour l'exploration des oceans. Rapports scientifiques et techniques 54, 108 pp.

ORI, G.G. & FRIEND, P.F. (1984) Sedimentary basins formed and carried piggyback on active thrust sheets. *Geology* 12, 475–478.

POSAMENTIER, H.W. & VAIL, P.R. (1988) Eustatic controls on clastic deposition II — sequence and systems tract models. In: *Sea-level Change – an Integrated Approach* (Eds Wilgus, C.K., Hastings, B.S., Posamentier, H., Van Wagoner, J., Ross, C.A. & Kendall, C.G.St.C.) Soc. Econ. Paleontol. Mineral. Spec. Publ. 42, 125–154.

POSAMENTIER, H.W., JERVEY, M.T. & VAIL, P.R. (1988) Eustatic control on clastic deposition I — conceptual framework. In: *Sea-level Change – an Integrated Approach* (Eds Wilgus, C.K., Hastings, B.S., Posamentier, H., Van Wagoner, J., Ross, C.A. & Kendall, C.G.St.C.) Soc. Econ. Paleontol. Mineral. Spec. Publ. 42, 109–123.

PUIGDEFÀBREGAS, C. & SOUQUET, P. (1986) Tectonosedimentary cycles and depositional sequences of the Mesozoic and Tertiary from the Pyrenees. *Tectonophysics* 129, 173–203.

PUIGDEFÀBREGAS, C., MUÑOZ, J.A. & MARZO, M. (1986) Thrust belt development in the Eastern Pyrenees and related depositional sequences in the southern foreland basin. In: *Foreland Basins* (Eds Allen, P.A. & Homewood, P.). Int. Assoc. Sediment. Spec. Publ. 8, 229–246.

PUIGDEFÀBREGAS, C., COLLINSON, J., CUEVAS, J.L. *et al.* (1989) Alluvial deposits of the successive foreland basin stages and their relation to the Pyrenean thrust sequence. *4th Int. Conf. Fluvial Sedim., Excursion Guidebook* (Eds Marzo, M. & Puigdefàbregas, C.), 175 pp.

PUIGDEFÀBREGAS, C., MUÑOZ, J.A. & VERGES, J. (1992) Thrusting and foreland basin evolution in the Southern Pyrenees. In: *Thrust Tectonics* (Ed. McClay, K.). Unwin Hyman, London, pp. 247–254.

RIBA, O. (1976) Syntectonic unconformities of the Alto Cardener Pyrenees: a genetic interpretation. *Sediment. Geol.* 15, 213–233.

ROBLES, S.O. (1984) El complejo sedimentario aluvial y lacustre de edad paleogena de La Pobla de Segur, entre los rios Noguera Pallaresa y Flamisell (Prepirineo de Lerida). *Ilerda* 45, 119–143. Lleida.

ROBLES, S.O. & ARDÈVOL, L. (1984) Evolución paleogeográfica y sedimentológica de la cuence lacustre de Sossis (Eoceno Superior, Prepirineo de Lerida): ejemplo de la influencia de la actividad de abánicos aluviales en el desarrollo de una cuenca lacustre asociada. Publ. Depart. Estratigrafia U.A.B., Barcelona, pp. 233–267.

ROSELL, J. (1967) Estudio geologico del sector del Prepirineo comprendido entre los rios Segre y Noguera Ribagorzana (Prov. de Lerida). *Pirineos* 21, 9–214.

ROSELL, J. & RIBA, O. (1966) Nota sobre la disposicion sedimentaria de los Conglomerados de Pobla de Segur (Provincia de Lerida). Instituto Estudios Pirenaicos, Zaragoza, pp. 1–16.

ROURE, F., CHOUKROUNE, P., BERASTEGUI, X. *et al.* (1989) ECORS deep seismic data and balanced cross-sections: geometric constraints to tracing the evolution of the Pyrenees. *Tectonics*, 8, 41–50.

ROURE, F., HOWELL, D.G., GUELLEC, S. & CASERO, P. (1990) Shallow structures induced by deep-seated thrusting. In: *Petroleum and Tectonics in Mobile Belts* (Ed. Letouzey, J.), pp. 15–30.

SEGURET, M. (1972) Etude tectonique des nappes, et séries décolles de la partie centrale du versant sud des Pyrenées. Pub. USTELA, Ser. Geol. Struct. 2, Montpellier.

SIMÓ, A. (1985) Secuencias deposicionales del Cretacio superior de la Unidad del Montsec (Pirineo Central). Unpublished PhD Thesis, Universidad de Barcelona, 325 pp.

SIMÓ, A. & PUIGDEFÀBREGAS, C. (1985) Transition from shelf to basin on an active slope, Upper Cretaceous, Tremp area, Southern Pyrenees. *Exc. Guide-book 6th European Regional Meeting*, Lerida, Spain, pp. 63–108.

VAIL, P.R., MITCHUM, R.M., JR., Todd, R.G. *et al.* (1977) Seismic stratigraphy and global changes of sea level. In: *Seismic Stratigraphy – Applications to Hydrocarbon Exploration* (Ed. Payton, C.E.) Am. Assoc. Petrol. Geol. 26, Mem., 49–205.

VAN WAGONER, J.C., MITCHUM, R.M., CAMPION, K.M. & RAHMANIAN, V.D. (1990) Siliciclastic sequence stratigraphy in well logs, cores, and outcrops: concepts for high-resolution correlation of time and facies. *Am. Assoc. Petrol. Geol. Methods Expl. Ser.* 7, 55 pp.

VERGES, J. & MUÑOZ, J.A. (1990) Thrust sequences in the Southern Central Pyrenees. *Bull. Soc. Geol. France* 8, VI(2), 265–271.

Spec. Publs Int. Ass. Sediment. (1993) **20** 277–290

Evolution of a Pliocene fan-delta:
links between the Sorbas and Carboneras Basins, southeast Spain

A.E. MATHER*

*Department of Earth Sciences, University of Liverpool, Brownlow St.,
PO Box 147, Liverpool L69 3BX, UK*

ABSTRACT

The connection between a partially enclosed, fluvially dominated, tectonically elevated sedimentary basin (Sorbas Basin) and a less elevated basin with marine connections (Carboneras Basin) is examined. The two basins are separated by faulted, and differentially uplifted, basement across which a connection was maintained throughout the basin evolution but which became increasingly restricted as a result of continued tectonism. In the Pliocene the connection between the two basins had been reduced to a shallow valley through which the fluvial system of the Sorbas basin exited, to form part of a small fan-delta complex in the northern part of the Carboneras Basin. This complex was also fed by a more proximal source area located in the uplifted basement of the Sierra Alhamilla. As a result of its proximity to the fan-delta complex, sediment from this latter source area reached the fan delta fairly rapidly and dominated its early history. Maximum sediment discharge from the more distant Sorbas Basin source area only occurred after progradation of the developing fluvial system in the Sorbas Basin enabled coarse clastics to be delivered to the fan-delta front. Maximum sediment supply from the larger, Sorbas-Basin-sourced portion of the fan-delta therefore dominated the later history of the fan-delta complex. Initially high sediment yield from the Sorbas Basin was able to keep pace with contemporaneous deformation between the two basins. With continued uplift and regression, however, incision of the fan-delta complex and overlying fluvial deposits ensued.

INTRODUCTION

Work on the fill of the Neogene sedimentary basins of southeast Spain has largely concentrated on the palaeoenvironments of the marine deposits of the Messinian and older basin fill (Dronkert, 1976; Roep *et al.*, 1979; Ott d'Estevou, 1980; Dabrio *et al.*, 1981; Van de Poel *et al.*, 1984; among others) or more recently on the Pleistocene dissectional history (Harvey, 1987; Harvey & Wells, 1987; Mather *et al.*, 1991). Little research has been directed towards the evolution of the basins during the Plio-Quaternary. It is the aim of this paper to examine how a small Plio-Quaternary fan-delta complex (nomenclature after Nemec & Steel, 1988),

which forms the connection (previously outlined by Harvey, 1987; Harvey & Wells, 1987) between two of the Neogene Basins (Sorbas and Carboneras Basins), reflects changing rates of structural deformation and sediment supply.

REGIONAL TECTONICS

The Carboneras Basin is one of a series of east–west orientated basins developed in the internal zone of the Betic Codillera of southeast Spain (Fig. 1). This belt originated from the relative movements between the African and Iberian plates, which began in the Early Mesozoic (Smith & Woodcock, 1982) and climaxed with continental collision of the plates in the Burdigalian (Dewey *et al.*, 1973; Bourrouilh

*Present address: Department of Geographical Sciences, University of Plymouth, Drake Circus, Plymouth, Devon PL4 8AA, UK.

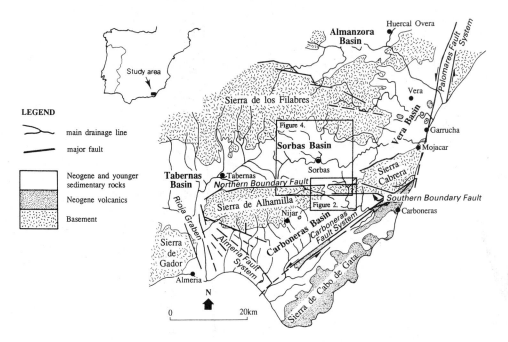

Fig. 1. Simplified sketch map of the Almería region, southeast Spain, showing study area, principal lines of drainage and major tectonic units. (After Harvey, 1990.)

& Gorsline, 1979). The collision was marked by the emplacement of nappes onto the external zones of the Betic–Rif Arc, a continuous orogenic arc located in the western Mediterranean. The continuity of the arc, which stretches from Minorca to Calabria via Gibralter, Morocco and Tunisia, indicates a lack of significant movement between the African and Iberian plates since the Oligocene (Smith & Woodcock, 1982)

Since the emplacement of the nappes a general north–south compressive regime has existed (Ott D'Estevou, 1980; Hall, 1983; Montenat *et al.*, 1987; Coppier *et al.*, 1989; Sanz de Galdeano, 1990). Much of this movement is taken up along major left lateral, strike-slip faults (Bousquet, 1979; Weijermars, 1987) which form a wide, left lateral shear zone (Montenat *et al.*, 1987). This faulting helped define the Neogene sedimentary basins (Fig. 1), generating localized areas of compression and extension (Hall, 1983).

The basement is currently undergoing uplift as a result of the Oligocene nappe emplacement (Weijermars *et al.*, 1985) and the compressional regional tectonics.

PALAEOGEOGRAPHY

The study area occupies the northern margin of the Carboneras Basin, southeast Spain (Figs 1 & 2). The Carboneras Basin lies to the south of the Sorbas Basin (Fig. 1) with which it maintained marine connections until the Late Messinian (Ott D'Estevou, 1980). The Carboneras Basin was the only remaining external connection between the Sorbas and adjacent basins after the end of the Late Messinian (Mather, 1991). Late Messinian and Pliocene deposits within the Sorbas Basin (Cariatiz and Gochar Formation, Fig. 3) are dominantly subaerial in origin. They comprise a suite of alluvial fans which developed along the basin margin and prograded over coastal plain deposits which existed in the southern and more central parts of the basin (Mather, 1991). The fans amalgamated to form a braided fluvial system. This fluvial system consisted of two distinctive feeder systems from the north and western margins of the basin, and one from the south (Fig. 4). These supplied a longitudinal drainage area, which exited from the Sorbas Basin into the Carboneras Basin via a low between the

the Sierras Alhamilla and Cabrera (Fig. 4). At the same time the northern area of the Carboneras Basin was still occupied by the Plio-Mediterranean sea (Addicot *et al.*, 1978; Van de Poel *et al.*, 1984), into which coarse detritus was carried to form a hitherto unrecognized fan-delta. It is this fan-delta which forms the focus of this study.

STRATIGRAPHY OF STUDY AREA

In this account of the Pliocene and Pleistocene deposits of the study area, two formations have been recognized. The lower is a sequence of bioturbated sandstone, the Cuevas Viejas Formation of Addicot *et al.* (1978) and the upper a sequence of conglomerates (the fan-delta deposits of this paper). The latter cannot be clearly related to any units described elsewhere in the Almeria/Carboneras Basin, forming part of the localized stratigraphy of the northern margin of the basin. Because the stratigraphy of the study area is different from that described by other workers (Addicot *et al.*, 1978; Roep & Van Harten, 1979; Van de Poel *et al.*, 1984)

the conglomerates have been assigned to a new formation — the Polopos Formation.

The Polopos Formation is defined as mainly well-bedded cobble and boulder conglomerates up to about 90 m in thickness which are set erosively into/intercalated with the top of the Cuevas Viejas Formation. The type locality for the Polopos Formation is near the village of Polopos in the Rambla de Lucainena (Fig. 2). The formation can be divided into three principal stratigraphical units (Fig. 3). These are a Lower and Upper Conglomerate Unit, separated by a laterally persistent red sandstone, which is here referred to as the Ibañez Sandstone Member, after its type locality of Cortijo Ibañez (Figs 2 & 3). This member acts as a useful marker horizon. The base of the Polopos Formation is generally marked by a sharply defined erosive surface into the underlying Cuevas Viejas Formation, and a switch to conglomerate-dominated deposition. The top is capped by gently unconformable Quaternary terrace gravels.

The conglomerate units can further be divided, on the basis of clast assemblage, into two systems. One is dominated by dark coloured low-grade metamor-

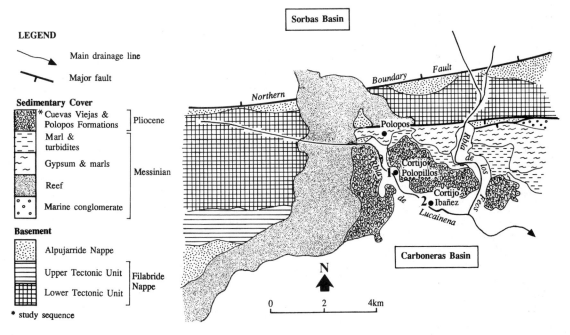

Fig. 2. Simplified geological sketch map of study area showing localities referred to in the text. (Modified from Weijermars *et al.*, 1985.) Type localities: 1, Polopos Formation (GR 825968); 2, Ibañez Sandstone Member (GR 844956).

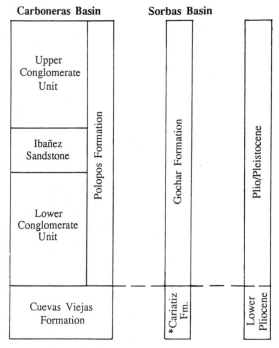

Fig. 3. Stratigraphical nomenclature used in this study and correlation between the Pliocene deposits in the Carboneras and Sorbas Basins. Stratigraphical nomenclature for the Sorbas Basin is modified from Dronkert (1976). The Cariatiz Formation* (from Mather, 1991) incorporates the Zorreras Member of previous work.

phic mica schists and phyllitic material derived from the Sierra Alhamilla (Figs 1 & 2) (Lucainena System), the other contains abundant amphibole–mica-schists derived from the Sierra de los Filabres

of the Sorbas Basin (Feos System) (Figs 1 & 4).

The Cuevas Viejas Formation is dated from microfauna (see below). The age of the Polopos Formation is inferred from its stratigraphical relations.

CUEVAS VIEJAS FORMATION

This formation is present as marine, thoroughly bioturbated, yellow to green sands, silts and muds which contain locally abundant macro- and microfauna. These fauna have been used to allocate a Zanclian (Early Pliocene) age to the deposits (Addicot *et al.*, 1978). This makes it the temporal equivalent of the top Zorreras Member (Fig. 3) (Ruegg, 1964; Dronkert, 1976; Roep *et al.*, 1979) of the Sorbas Basin which has been dated, from microfauna, as approximately Early Pliocene in age (Montenat & Ott D'Estevou, 1977; Ott D'Estevou, 1980). The Cuevas Viejas Formation is the dominant Pliocene lithology in the Polopos area (Carboneras Basin). At the top of the sequence the sands are intercalated with sheets of clast-supported conglomerates belonging to the basal Polopos Formation.

THE POLOPOS FORMATION

The Polopos Formation conglomerates (Figs 5 & 6) can be divided into two drainage systems, according to provenance, established from the lithology of the clasts. The Lucainena system of the Polopos For-

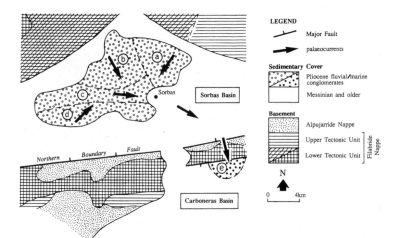

Fig. 4. Simplified sketch map of the Pliocene fluvial conglomerates of the Sorbas Basin showing drainage paths and systems (a,b,c, and d) delimited by Mather (1991) and position of the Pliocene marine conglomerates of the fan-delta deposits (e) of the Carboneras Basin.

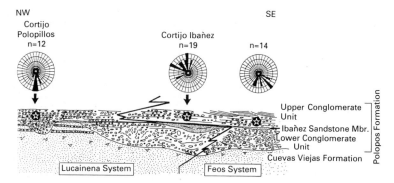

Fig. 5. Schematic reconstruction of the facies relationships within the Polopos Formation. Sketch is approximately normal to main palaeoflow direction and covers an area approximately 1 km across (see Fig. 2 for localities). Palaeocurrents are derived from imbrication. Stars indicate location of sampling points. The Polopos Formation reaches a maximum thickness of approximately 90 m. Section is not to scale. Thick dividing line in centre of section indicates the approximate contact between the sediments of the Lucainena system (left of sketch) and Feos system (right of sketch).

mation comprises a clast assemblage dominated by Messinian carbonate, black mica-schists (Veleta-Montenegro Unit, lower part of the lower Filabride Nappe; Platt *et al.*, 1983; Weijermars *et al.*, 1985; Frizon de Lamotte *et al.*, 1989) and Triassic limestone from the Alpujarride Nappe (Figs 2 & 6) all of which outcrop in the Sierra Alhamilla (Figs 1 & 2). It is typically lacking in clasts from the higher grade amphibole-mica-schists of the Tahal Unit, upper

part of the lower Filabride Nappe, Sierra de los Filabres of the Sorbas Basin (Frizon de Lamotte *et al.*, 1989) which characterize the clast assemblage of the Feos system (Fig. 8). The main sections through the Lucainena system crop out in the Rambla de Lucainena (Fig. 2). Main sections through the Feos system are located in the Rambla de los Feos (Fig. 2).

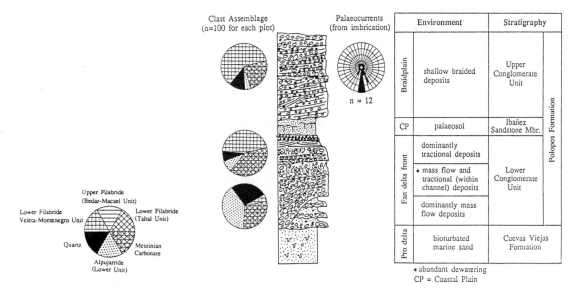

Fig. 6. Schematic log through the Lucainena system showing characteristic clast assemblage, palaeocurrent and sedimentological features. Log covers approximately 50 m of vertical section.

POLOPOS FORMATION:
LOWER CONGLOMERATE UNIT

Facies of the Lucainena system

Mass flow deposits

The Lower Conglomerate Unit is dominated by mass flow deposits in the lower part of the Lucainena system (Fig. 6). These mass flow deposits range from cohesive (poorly sorted, matrix-supported conglomerates) to cohesionless (matrix free, randomly packed conglomerates) in nature (nomenclature is taken from Postma, 1986). The more cohesive flows tend to display a matrix-supported framework, are poorly graded and contain outsized, subvertical-standing clasts. In some flows the percentage of matrix increases upwards, with better packing of the conglomerates in the lower parts of the bed. In the lower parts of the sections these conglomerates may be intercalated with the underlying, marine Cuevas Viejas Formation. Some of the depositing flows were probably fully turbulent in nature, the sediments displaying strong vertical and lateral (down palaeoslope) sorting from conglomerate to sandstone.

Some of the debris flow deposits show characteristics of bipartite flows (Nemec & Steel, 1984), with the development of associated lower sandstone beds and overlying conglomerate beds. The sands are texturally the same as the matrix of the overlying conglomerates, and appear to be gradational with them over a short distance. This apparent bipartite nature to the beds suggests a mechanism of flow separation of genetically-related clast-rich and clast-poor flows, the genesis of which has been discussed by many workers (Lewis *et al.*, 1984; Nemec & Steel, 1984; Postma, 1986). The conglomerate beds may also show lateral changes from a poorly sorted and organized fabric in more central parts of the bed to a better sorted, better organized fabric at the margins, suggesting that the sediment mixture was becoming increasingly intermixed with water at its margins.

Dewatering structures

Early deformation of the conglomerates is common only in the lower parts of the Lower Conglomerate Unit (Fig. 6). At times it completely obliterates any original bedding.

The deformation takes the form of vertically elongate structures infilled with sandstone matrix containing scattered conglomerate clasts. Commonly the clasts are aligned vertically, particularly along the margins of the structure. The coarsest material is generally found at the bottom of the structures. The structures have sharp vertical margins and do not affect the surrounding sediment.

The sharply defined margins of these structures makes them unlike load structures (cf. Lowe, 1975). Similarly their geometry, size (up to 1 m in width) and lack of internal sedimentary structures rules out scour, neptunean dyke or biogenic origin. The features appear to have most in common with the pillars described by Postma (1983). These are formed by paths of fluidization (see also Lowe, 1975; Middleton & Southard, 1978). Typically the coarser fraction sinks downwards through the finer material which loses its frictional strength as a result of liquefaction. The finer material than resediments on top. Vertically aligned clasts tend to be discoidal in shape, and probably align with weakest resistance to the upward moving, escaping pore water. The larger rounder clasts tend to remain 'suspended' in the more central part of the structure, perhaps as a result of the upward pressure exerted on their surface area by the escaping pore fluid.

Traction deposits

At some localities traction deposits laterally fill channels (Fig. 7). Other deposits are sub-horizontally bedded and contain imbricated clasts. Rare, broad shallow troughs and cross-beds up to 70 cm high may be present. These are locally associated with patches of slumped conglomerate and sandstone.

At the top of the Lower Conglomerate Unit deposits atypical of the sequence are developed. In contrast to the underlying conglomerates (described above) which form most of the sequence, these uppermost gravels and conglomerates are well-bedded, well-rounded, quartz-rich and display well-developed internal organization in the form of good size and shape segregation of clasts. This together with the roundness of the clasts and abundance of quartz is typical of wave-reworked gravels (see reviews by Clifton, 1973; Bourgeois & Leithold, 1984; Nemec & Steel, 1984).

Facies of the Feos system

The Lower Conglomerate Unit of the Feos system is not as well exposed as its equivalent in the Lucainena system. The deposits are dominantly mass flow (Fig. 8) and comprise cohesive and weakly cohesive debris flows similar to those described from the Lower Conglomerate Unit of the Lucainena System, displaying rapid lateral and vertical changes in the degree of textural organization. Incorporated in the thicker, cohesive debris flow deposits are large (average 1 m) well-rounded blocks of conglomerate. These conglomerate boulders are typically composed of well-rounded clasts which are well-sorted and quartz-rich, suggesting a shallow-marine origin, probably from nearshore deposits of the fan-delta complex.

Depositional environment

The depositional environment of the Lower Conglomerate Unit appears to have been submarine from the following lines of evidence.

1 The basal conglomerates are intercalated with marine sandstone of the Cuevas Viejas Formation.

2 Wave reworked gravels are developed at the top of the Lower Conglomerate Unit.

3 Many features of the mass flow deposits require a water-rich environment for their genesis. For example, the flows show rapid variations in their fluidal properties, often becoming more dilute towards their edge, suggesting intermixing with water at their margins. Similarly the beds locally display an upward increase in matrix content, a feature considered by Nemec and Steel (1984) to be typical of subaqueous debris flows.

4 Dewatering structures are abundant in the lower and middle parts of the sequence, together with rare slumps. It is generally difficult to fluidize very coarse-grained deposits (Lowe, 1975; Postma, 1983). This suggests a water-rich sediment, at least at the time of deformation, in order to generate sufficient pore pressures. This deformation occurred soon after deposition, sometimes affecting only one sedimentary bed.

These deposits probably represent a fan-delta front. The restricted geographical distribution of the geological sections through the conglomerates suggests a fan-delta of limited geographical extent, the sediments being rapidly dumped adjacent to the margins of the basin. The delta front was dominated by mass flows. The fan-delta deposits of the Lower Conglomerate Unit show a clear vertical transition from mass-flow dominated fan-delta slope to tractional dominated delta top. Locally, beach deposits were preserved at the top of the subaqueous portion of the fan-delta deposits.

This period was terminated by the deposition of the Ibañez Sandstone Member which is described below.

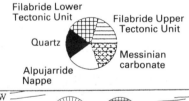

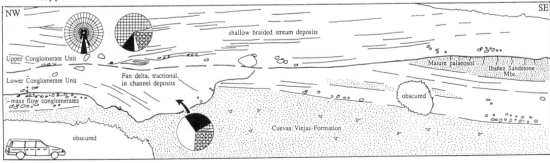

Fig. 7. Sketch taken from photo mosaic of the stratigraphical relations in the western part of the Lucainena system (GR 825 969). Note gentle unconformity between the Upper and Lower Conglomerate Units. For clast assemblage data, *n* = 100. Palaeocurrent data is taken from imbrication, *n* = 12.

POLOPOS FORMATION: IBAÑEZ SANDSTONE MEMBER

The Ibañez Sandstone Member comprises a red sandstone which is continuous across the area of the fan-delta (Fig. 5) and contains linear horizons of amalgamated carbonate glaebules, together with a colour banding. This colour banding ranges in intensity and definition, reaching a maximum colour of 5YR 5/6. The sandstone varies in thickness across the area (Fig. 5) with a maximum development of 1–15 m at Cortijo Ibañez (Figs 2 & 5). The carbonate nodular development evident in the Ibañez Sandstone Member, together with the strong red colouration is typical of pedogenesis (Wright & Allen, 1989). The original sediments were probably predominantly parallel laminated silts (a weak lamination is preserved in places, and may account for the planar developments of the carbonate at some localities). The presence of grey mottles with red haloes suggests hydromorphic modification of the soil as a result of periodic waterlogging (Buurman, 1975, 1980; Fitzpatrick, 1980), probably related to water-table fluctuations. Intensity of pedogenesis varied across the area both temporally and spatially, as is demonstrated by the varied maturity (colour intensity and degree of carbonate development) in both a vertical and lateral sense.

Deposition of the Ibañez Sandstone was terminated by a second influx of conglomerates, the Upper Conglomerate Unit, described below.

POLOPOS FORMATION: THE UPPER CONGLOMERATE UNIT

As in the case of the Lower Conglomerate Unit, the conglomerates of the Upper Conglomerate Unit can be divided into two systems on the basis of clast assemblage; the Lucainena system (low-grade metamorphic mica-schists and phyllite from the Sierra Alhamilla) and the Feos system (amphibole–mica-schists from the Sierra de los Filabres).

Facies of the Lucainena system

The Upper Conglomerate Unit of the Lucainena system erosively, and with mild unconformity, overlies the Ibañez Sandstone Member.

The Upper Conglomerate Unit comprises a basal conglomerate containing large (in excess of 1 m) Messinian carbonate clasts. The conglomerates of

this unit are typically thinly bedded. Bedding is dominantly horizontal and contains trough and planar cross-stratification. This is typically small scale (less than 0.5 m in height). The conglomerates are well imbricated with a north-to-south palaeocurrent direction (Figs 5, 6 & 7). The conglomerates are interbedded with rare haematitic sands, which may contain carbonate nodules. Small erosive channels filled with poorly sorted, chaotic conglomerate clasts may be cut into the better sorted gravels.

These deposits are the product of tractive flow, which generated small bedforms. The lack of height to any of the bedforms probably reflects fairly shallow flow conditions. The characteristics of the haematitic sands suggest pedogenesis.

Facies of the Feos system

Mass flow deposits

These are rare within the Upper Conglomerate Unit facies association and confined to the lower part of the westernmost exposure of the Feos system within the study area. The deposits comprise cohesive and cohesionless flows similar to those described for the Lower Conglomerate Unit of the Lucainena system. Occasionally present within this system are surging debris flow deposits (Nemec & Steel, 1984). These display a normal to inverse grading which becomes more pronounced when traced laterally. The more cohesive nature of the deposit at the margin of the flow seems to infer water loss from the margins of the flow lobe.

Traction deposits

Parallel-bedded deposits dominate the Upper Conglomerate Unit of the Feos system (Fig. 8). Palaeoflow is dominantly from north to south. Conglomerates are typically moderately sorted, well-imbricated and well-bedded. Bedding is typically sub-horizontal and may be weakly channelized. Some of the beds are affected by broad, shallow, lenticular scours, erosive into one another and infilled with laminated sandstones and gravels. Post-depositional modification of the bedding is restricted to minor reorientation of clasts adjacent to minor fluid escape pathways.

In the basal part of the upper conglomerate the sequence is dominated by laminated sands and cross-bedded gravels which infill erosional scours, with cross-beds up to 65 cm high, but more com-

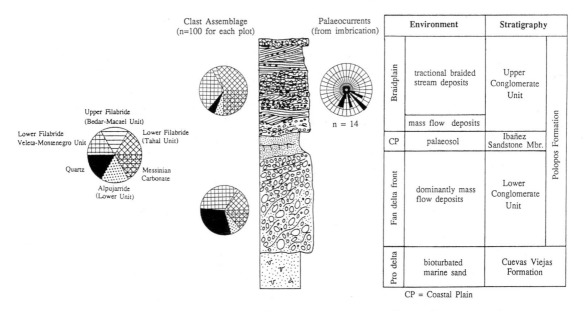

Fig. 8. Schematic log through the Feos system showing characteristic clast assemblage, palaeocurrent and sedimentological features. Log covers approximately 50 m of vertical section.

monly smaller. Cross-beds tend to be planar when developed in the coarser gravel and trough when developed in the sandstones. Beds are typically normally graded. The sequence is capped by a strong development of thinly laminated, broad, sandy trough cross-beds (Figs 5 & 8).

Locally, in the west of the system, decimetre-scale planar cross-beds are developed in granular material. They appear to infill shallow scours. Flow was highly fluctuating with haematitic mudstone drapes developed over some of the erosional hollows. The size of the cross-beds is typically small, which implies that flow depths were dominantly shallow.

Depositional environment

The Upper Conglomerate Unit characteristically shows a lack of any mass flow deposits, and instead is dominated by deposits generated by unidirectional tractive flow. The association with weakly developed palaeosols indicates a subaerial environment. The typically small size of the bedforms indicates that flow conditions were fairly shallow, and dominantly of an unconfined nature, with only small channels developed, whilst the presence of sandstone drapes, together with rapidly varying sediment types, indicates an environment with rapidly varying flow conditions. This is indicative

of a braidplain environment, perhaps in the distal area of a small alluvial fan.

Mass flow deposits are locally present in the basal sections of the Upper Conglomerate Unit in the west of the study area. The flows have characteristics of both subaerial (possible water loss at the margins of the flows in the surging debris flow deposits) and subaqueous (increasing matrix percent at the top of the flow) origin. This implies an environment which is probably close to a standing water deposit. The combination of a relatively high water table and possible periodic inundations by the adjacent water body may have been responsible for the deposition of these beds.

PALAEOENVIRONMENTAL SYNTHESIS

The deposits of the Cuevas Viejas and Polopos Formations record a vertical passage in depositional environment from marine, pro-delta (Cuevas Viejas Formation), delta front (the Polopos Formation, Lower Conglomerate Unit), through coastal plain (Polopos Formation; Ibañez Sandstone Member) to fluvial (Polopos Formation, Upper Conglomerate Unit). The clast assemblages of the conglomerate units indicate two dominant source areas for this fan-delta. The Lucainena system is sourced in the Sierra Alhamilla whereas the clast assemblage and palaeocurrents of the Feos system indicate derivation from the fluvial Gochar-Formation drainage of the Sorbas Basin (Figs 4 & 8). The deposits of the fan-delta complex are of restricted geographical distribution and suggest a fan-delta of limited geographical extent, the sediments having been rapidly dumped adjacent to the margins of the basin. The delta front (the lower portion of the Lower Conglomerate Unit) was dominated by mass flows.

In the case of the Lucainena system, dewatering structures were fairly abundant in the delta-front conglomerate deposits. Postma (1983) considers these structures to be the product of steep depositional slopes. This, however, does not appear to be the case for the deposits of the Lucainena system as available outcrop does not imply an original steep depositional dip. It appears more likely that in this particular case the dewatering of the sediments was stimulated by rapid loading and local tectonism. The fan-delta deposits of the Lower Conglomerate Unit show a clear vertical transition from mass flow dominated fan-delta slope to tractional-dominated delta top. Locally, beach deposits were preserved at the top of the subaqueous portion of the fan-delta deposits.

In the Feos system, the fan-delta system prograded out from a steep depositional slope into the Pliocene Mediterranean Sea in the Carboneras Basin. The steep slope is inferred from the steeply dipping, thickly bedded mass flow conglomerates and steep, inclined bedding. These were probably generated from large-scale sediment slides. Unfortunately there is little exposure of these delta-front deposits.

In both the Lucainena and Feos Systems the transition from this subaqueous environment to delta top (Lower Conglomerate Unit) and overlying subaerial environments (Ibañez Sandstone Member and Upper Conglomerate Unit) is fairly rapid.

The deposits of the Lower Conglomerate Unit were succeeded by the deposition of sandstone, modified by pedogenesis—the Ibañez Sandstone Member. This was probably deposited in a coastal plain setting and records a period of reduced gravelly input to the area. This period was followed by gentle folding, creating a low angle unconformity.

The conglomerates which overlie the Ibañez Sandstone (the Upper Conglomerate Unit) record the progradation of the fluvial element of the fan delta into the Carboneras Basin. Palaeocurrents (derived from imbrication), combined with provenance (derived from the lithology of the clast assemblage) indicate that although the two drainage systems (Lucainena and Feos systems) existed, the Feos system was now showing a marked expansion relative to the Lucainena system. To the west the Feos system encroached on the areas formerly dominated by the Lucainena portion of the fan-delta (Fig. 5). Local palaeocurrent anomalies (Fig. 5) may reflect the channel switching of a distal braided river or perhaps palaeocurrent deflection as a result of fan morphology.

The deposits of the fan-delta system were finally capped by the fluvial terrace deposits of the palaeo-Feos river. Continued uplift stimulated incision of the Quaternary river systems into the former fan surface (Mather, 1991). Fluvial terrace deposits related to this incision can be traced from the Sorbas Basin into the Carboneras Basin (Harvey & Wells, 1987).

CONTROLS ON FAN-DELTA DEVELOPMENT

Tectonic controls

Tectonics were important during the development of the Polopos System and through to the Quaternary, when fluvial terrace gravels were disturbed (Dumas *et al.*, 1978; Harvey & Wells, 1987). During the deposition of the Polopos Formation tectonism generated a gentle unconformity between the Upper Conglomerate Unit and Ibañez Sandstone Member. In addition there is considerable post-depositional deformation in the form of faulting and gentle folding. The faulting is fairly early, before the main cementation of the conglomerates. Sediments may show dewatering adjacent to the faults, developing pillar and pocket dewatering structures (Postma, 1983) and realignment of clasts. The deformation reaches a maximum in the area of thickest sediment accumulation of the Ibañez Sandstone Member (Figs 2 & 5, Cortijo Ibañez). This implies that synsedimentary subsidence may have been important during the deposition of the Pliocene fan-delta system. Thick Quaternary fluvial deposits also appear to develop preferentially in this area, perhaps indicating continued subsidence during the Quaternary. The deformation probably relates to differential uplift of the Sierra Alhamilla. Uplift rates calculated from the present elevation of Pliocene marine deposits indicate uplift rates of approximately 0.16 mm/year on the Sorbas Basin side of the Sierra Alhamilla and 0.13 mm/year along the Carboneras side (Mather, 1991) implying differential uplift of the sierras. This uplift of the Sierras Alhamilla/Cabrera is probably responsible in part for the diminishing connection between the Sorbas and Carboneras Basins. During the Late Messinian the connection was fairly broad. During deposition of the fan-delta this access was restricted to the Feos area and by the Pleistocene was restricted to incisional valley-constrained deposits of the palaeo-Feos river (Harvey & Wells, 1987).

Relative sea-level controls

As a result of uplift of the basins, and eustatic fall in sea-level since the Pliocene (Vail *et al.*, 1977), regressional sea-level fall was important in the development of the fan-delta system. An increase in sediment supply combined with decline in sea-level ensured a rapid progradation of the fluvial system, and a rapid transition from the delta to fluvial environments.

Source area controls

Source area controls on sediment supply were of great importance in the development of the fan-delta. The Lucainena system was largely supplied by material from the Sierra Alhamilla. The maximum sediment discharge from this relatively proximal source area reached the fan-delta complex at a fairly early stage in its development. In the case of the Feos system, however, the period of maximum sediment discharge to the Carboneras Basin was only possible after a time lag which enabled the developing fluvial system in the source area (Sorbas Basin) to prograde into the Carboneras Basin and supply course detritus to the fan-delta (Mather, 1991). Hence, the period of maximum sediment input from the larger, Sorbas-sourced, Feos system occurred in the later stages of fan-delta development, leading to a domination of the fan-delta complex by the Feos system in the later stages of its existence.

wide gap in the Sierra Cabrera. With uplift of the sierras this connection became more restricted.

The northern area of the Carboneras Basin, in the vicinity of Polopos, was dominated in the Early Pliocene by the development of a fan-delta complex. This complex was sourced by two main feeders, the Lucainena and Feos systems (Fig. 9a).

The Lucainena system was initially the most important feeder of the fan-delta complex. It comprised a fairly gentle delta front, and was dominated by tractional processes. It was probably similar to the shallow, shelf-type fan-deltas referred to by Massari and Colella (1988).

From the available data it is apparent that the Plio-Pleistocene drainage from the Sorbas Basin exited via a gap in the Sierra Cabrera into the Carboneras Basin (Figs 4 & 9), which at this time was occupied by the Pliocene Mediterranean Sea. The fluvial conglomerates exited from the constraints of the mountain front into the Carboneras Basin, to form the 'Feos system' of the Polopos fan-delta complex (Fig. 9). This portion of the delta had a fairly steep front, and may well have been of Gilbert type (Gilbert, 1885). With the development of the river systems of the Gochar Formation in the Sorbas Basin the Feos system grew in importance and eventually dominated the fluvial part of the delta complex (Fig. 9b).

The main controls on the development of the fan-delta complex were sediment supply in the source areas. The Sorbas-sourced fan-delta (Feos system) was able to keep pace with the contemporaneous high uplift rates along the Alhamilla–Cabrera axis as a result of high sediment and water discharge. The Lucainena system, however, sourced from the Sierra Alhamilla, was smaller in terms of sediment yield and was gradually defeated by the contemporaneous deformation.

Continued uplift of the Sierra Alhamilla and declining sediment yield, as a result of sediment exhaustion and possibly climatic change (less effective runoff to generate sediment supply) eventually led to increased isolation of the sedimentary source areas from which the fan-delta was fed. Uplift eventually stimulated incision by the palaeo-drainage.

CONCLUSIONS

During the Early Pliocene, connection between the Carboneras Basin and the Sorbas Basin was via a

ACKNOWLEDGEMENTS

Field research was funded by NERC grant GT4/87/ GS/53. I would like to thank Adrian Harvey and

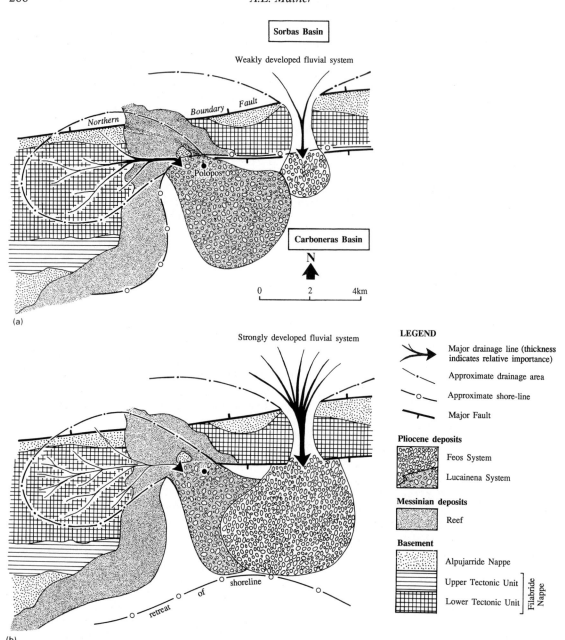

Fig. 9. Palaeogeographical reconstruction of the Polopos Formation fan-delta complex for the Lower Conglomerate Unit (a) and the Upper Conglomerate Unit (b).

Pat Brenchley for their critical discussions and for reading through various drafts of the manuscript, and the critical comments of Th. B. Roep. Thanks are also extended to my field assistants, Freya and Merlin, and the hospitality of Bill and Lindy at Cortijo Urra Field Centre, Sorbas.

REFERENCES

ADDICOT, W.O., PARKE, D.S. JR, BUKRY, D. & POORE, R.Z. (1978) Neogene stratigraphy of Southern Almeria Province, Spain: an overview. *Geol. Surv. Am. Bull.* **1454**, 59 pp.

BOURGEOIS, J. & LEITHOLD, E.L. (1984) Wave-reworked conglomerates-depositional processes and criteria for recognition. In: *Sedimentology of Gravels and Conglomerates* (Eds Koster, E.M. & Steel, R.J.) Can. Soc. Petrol. Geol. Mem. **10**, 1–32.

BOURROUILH, R. & GORSLINE, D.S. (1979) Pre-Triassic fit and Alpine tectonics of continental blocks in the western Mediterranean. *Geol. Soc. Am. Bull.* **90**, 1074–1083.

BOUSQUET, J.C. (1979) Quaternary strike-slip faults in southeastern Spain. *Tectonophysics* **52**, 277–286.

BUURMAN, P. (1975) Possibilities of palaeopedology. *Sedimentology* **22**, 289–298.

BUURMAN, P. (1980) Palaeosols in the Reading Beds (Paleocene) of Alum Bay, Isle of Wight, U.K. *Sedimentology* **27**, 593–606.

CLIFTON, H.E. (1973) Pebble segregation and bed lenticularity in wave-worked versus alluvial gravel. *Sedimentology* **20**, 173–187.

COPPIER, G., GRIVEAUD, P., LAROUZIERE, F.D. DE, MONTENAT, C. & OTT D'ESTEVOU, PH. (1989) Example of Neogene tectonic indentation in the Eastern Betic Cordilleras: the Arc of Aguilas (Southeastern Spain). *Geodinamica Acta* **3**, 37–51.

DABRIO, J., ESTEBAN, M. & MARTIN, J.M. (1981) The coral reef of Nijar, Messinian (uppermóst Miocene), Almería Province, SE Spain. *J. Sediment. Petrol. (Paris)* **51**, 521–539.

DEWEY, J.F., PITMAN III, W.C., RYAN, W.B.F. & BONNIN, J. (1973) Plate tectonics and the evolution of the Alpine system. *Geol. Soc. Am. Bull.* **84**, 3137–3180.

DRONKERT, H. (1976) Late Miocene evaporites in the Sorbas Basin and adjoining areas. *Mem. Soc. Geol. It.* **16**, 341–361.

DUMAS, B., GUEREMY, P., LHENAFF, R. & RAFFY, J. (1978) Géomorphologie et néotectonique dans la région d'Almeria (Espagne du sudest). In: *Relief et Néotectonique de pays Méditerranèens*. RCP (CNRS, Paris) Publication, **461**, 123–170.

FITZPATRICK, E.A. (1980) *Soils: Their Formation, Classification and Distribution.* Longman Group, London, 353 pp.

FRIZON DE LAMOTTE, D., GUÉZOU, J.C. & ALBERTINI, M.A. (1989) Deformation related to Miocene westward translation in the core of the Betic Zone. Implications on the tectonic interpretation of the Betic orogen (Spain). *Geodinamica Acta* (Paris) **3**, 267–281.

GILBERT, G.K. (1885) The topographic features of lake shores. *Ann. Rep. U.S. Geol. Surv.* **5**, 69–123.

HALL, S.H. (1983) *Post Alpine tectonic evolution of SE Spain and the structure of fault gouge.* PhD thesis, University of London.

HARVEY, A.M. (1987) Patterns of Quaternary aggradational and dissectional landform development in the Almeria region, southeast Spain: a dry-region tectonically active landscape. *Die Erde* **118**, 193–215.

HARVEY, A.M. (1990) Factors influencing Quaternary alluvial fan development in southeast Spain. In: *Alluvial Fans: a Field Approach* (Eds Rachocki, A.H. & Church, M.) John Wiley, Chichester, pp. 247–269.

HARVEY, A.M. & WELLS, S.G. (1987) Response of Quaternary fluvial systems to differential epeirogenic uplift: Aguas and Feos River systems, south-east Spain. *Geology* **15**, 689–693.

LEWIS, D.W., LAIRD, M.G. & POWELL, R.D. (1984) Debris flow deposits of Early Miocene age, Deadman Stream, Marlborough, New Zealand. *Sediment. Geol.* **27**, 83–118.

LOWE, D.R. (1975) Water escape structures in coarse grained sediments. *Sedimentology* **22**, 157–204.

MASSARI, F. & COLELLA, A. (1988) Evolution and types of fan-delta systems in some major tectonic settings. In: *Fan Deltas: Sedimentology and Tectonic Settings* (Eds Nemec, W. & Steel, R.J.), Blackie & Son, London, pp. 103–122.

MATHER, A.E. (1991) *Late Cainozoic drainage evolution of the Sorbas Basin, SE Spain.* PhD thesis, University of Liverpool.

MATHER, A.E., HARVEY, A.M. & BRENCHLEY, P.J. (1991) Halokinetic deformation of Quaternary river terraces in the Sorbas Basin, southeast Spain. *Z. Geomorphol.* **82**, 87–97.

MIDDLETON, G.V. & SOUTHARD, J.B. (1978) Mechanisms of sediment movement. *Soc. Econ. Paleontol. Mineral. Short Course,* **3**.

MONTENAT, C. & OTT D'ESTEVOU, PH. (1977) Présence du Pliocene marin dans la bassin de Sorbas (Espagne méridionale). Conséquences paleogeographiques et tectoniques. *C.R. Somm. Soc. Geol. Fr.* **4**, 209–211.

MONTENAT, C., OTT D'ESTEVOU, PH., LAROUZIÈRE, F.D. DE, & BEDU, P. (1987) Originalité géodynamique des bassins néogénes du domaine bétique oriental. Total Compagnie Francaise des Pétroles, Paris. *Notes et Memoirs* **21**, 11–41.

NEMEC, W. & STEEL, R.J. (1984) Alluvial and coastal conglomerates: their significant features and some comments on gravelly mass-flow deposits. In: *Sedimentology of Gravels and Conglomerates* (Eds Koster, E.M. & Steel, R.J.) Can. Soc. Petrol. Geol. Mem. **10**, 1–32.

NEMEC, W. & STEEL, R.J. (1988) What is a fan delta and how do we recognize it? In: *Fan Deltas: Sedimentology and Tectonic Settings* (Eds Nemec, W. & Steel, R.J.), Blackie & Son, London, pp. 3–13.

NEMEC, W., STEEL, R.J., PORĘBSKI, S.J. & SPINNANGR (1984) Domba conglomerate, Devonian, Norway: Process and lateral variability in a mass flow-dominated lacustrine fan-delta. In: *Sedimentology of Gravels and Conglomerates* (Eds Koster, E.M. & Steel, R.J.) Can. Soc. Petrol. Geol. Mem. **10**, 295–320.

OTT D'ESTEVOU, PH. (1980) *Evolution dynamique du bassin Néogene de Sorbas (Cordileres Betiques Orientales, Espagne).* PhD thesis, University of Paris VII, Paris, 264 pp.

PLATT, J.P. VAN DEN EECKHOUT, B., JANZEN, E., KONERT, G., SIMON, O.J. & WEIJERMARS, R. (1983) The structure and tectonic evolution of the Aguilon fold-nappe, Sierra Alhamilla, Betic Cordilleras, SE Spain. *J. Struct. Geol.* **2**, 397–410.

POSTMA, G. (1983) Water escape structures in the context of a depositional model of a mass flow dominated conglomeratic fan-delta (Abrioja Formation, Pliocene), Almeria Basin, SE Spain. *Sedimentology* **30**, 91–103.

POSTMA, G. (1986) Classification for sediment gravity flow deposits based on flow conditions during sedimentation *Geology* **14**, 291–294.

ROEP, TH. B & VAN HARTEN, D. (1979) Sedimentological and ostracodological observations on Messinian post-evaporite deposits of some southeastern Spanish Basins. *Ann. Geol. Pays Hellen. VIIth International Congress on Mediterranean Neogene*, Athens, pp. 1037–1044.

ROEP, TH.B., BEETS, D.J., DRONKERT, H. & PAGNIER, H. (1979) A prograding coastal sequence of wave-built structures of Messinian age, Almeria, Spain. *Sediment. Geol.* **22**, 135–163.

RUEGG, G.J.H. (1964) Geologische onderzoekingen in het bekken van Sorbas, S.E. Spanje. Internal Report Geol. Inst. Univ. Amsterdam, 67 pp.

SANZ DE GALDEANO, C. (1990) Geologic evolution of the Betic Cordilleras in the Western Mediterranean, Miocene to the present. *Tectonophysics* **172**, 107–119.

SMITH, A.G. & WOODCOCK, N.H. (1982) Tectonic syntheiss of the Alpine–Mediterranean region: a review. In: *Alpine–Mediterranean Geodynamics* (Eds Berckhemer, H. & Hsu, K). Am. Geophys. Union Geodynam. Ser. 7, 15–38.

VAIL, P.R., MITCHUM, JR, R.M. & THOMPSON, S. (1977) Seismic stratigraphy and global changes of sea level, Part 4: Global cycles of relative changes of sea level. In: *Seismic Stratigraphy—Applications to Hydrocarbon Exploration* (Ed. Payton, C.E.) Am. Assoc. Petrol. Geol. Mem. **26**, 83–97.

VAN DE POEL, H.M., ROEP, TH. B. & PEPPING, N. (1984) A remarkable limestone breccia and other features of the Mio-Pliocene transition in the Agua Armarga basin (SE Spain). *Geol. Méditerranéene* **XI**, 265–276.

WEIJERMARS, R. (1987) The Palomares brittle–ductile shear zone of southern Spain. *J. Struct. Geol.* **9**, 139–157.

WEIJERMARS, R., ROEP, TH. B., C. VAN DEN EECKOUT, B., POSTMA, G. & KLEVERLAAN, K. (1985) Uplift history of a Betic fold nappe inferred from Neogene–Quaternary sedimentation and tectonics (in the Sierra Alhamilla and Almería, Sorbas and Tabernas Basins of the Betic Cordilleras, SE Spain. *Geol. Mijnbouw* **64**, 397–411.

WRIGHT, V.P. & ALLEN, J.R.L. (1989) *Palaeosols in Siliclastic Sequences*. Postgraduate Research Institute for Sedimentology Short Course Notes 001, Reading University, 98 pp.

Spec. Publs Int. Ass. Sediment. (1993) **20**, 291–299

Tectonic generation of lowstand turbidite units in a foreland basin, Marchean Apennines, Italy

N. CAPUANO

Istituto di Geologia Applicata, Universita di Urbino, Via Muzio Oddi, 14, 61029 Urbino, Italy

ABSTRACT

Eight distinctive and cyclically arranged coarse-clastic bodies have been recognized within the Early Pliocene succession of the foreland basin of Montecalvo-in-Foglia, in the northern Marchean Apennines. Each clastic body is the product of a series of high-energy sediment gravity flows which originated by catastrophic events provoked by active tectonics but subject also to the significant effect of sea-level oscillation. Despite their widespread distribution, the clastic bodies were deposited rapidly, thus forming well-defined lithological markers, permitting the definition of the geometry and palaeogeography of the basin.

INTRODUCTION AND GENERAL MODEL

Several Pliocene marine foreland basins outcrop along the Apennine Chain (Fig. 1). The sedimentary basin of Montecalvo-in-Foglia (MCF) is located on the Pedeapennine margin, shows a synclinal structure (Fig. 1) and was formed on an active thrust (*piggyback thrust basin*, Ori & Friend, 1984; Capuano *et al.*, 1989 Capuano, 1990 or *overstep thrust-sequence basin*, Boyer & Elliot, 1982). This basin is one of the better preserved cases in the Apennine chain, which originated and evolved during the later compressive tectonic phases (syn-orogenic molasse stage, according to Ricci Lucchi, 1986) from Late Miocene to Early Pliocene. It was completely filled by terrigenous sediments within the biozone range distribution of *Globorotalia puncticulata* and *Globorotalia bononiensis* (from 4.2 Ma to 2.9 Ma after Rio & Sprovieri, 1986). A stratigraphical succession characterized by two turbiditic depositional systems was recognized in the basin (Fig. 2). The lower stratigraphical sequence, 300 m thick, is composed mainly of fine-grained turbidites (facies D of Mutti & Ricci Lucchi, 1975) which originated from deep-water, dilute turbidity currents during periods of highstand of sea-level (Mutti, 1985). The upper stratigraphical sequence, 142 m thick, is characterized by eight coarse-grained turbiditic units, i.e. sandstones, pebbly

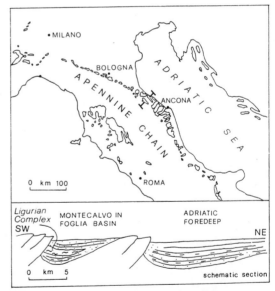

Fig. 1. Distribution of some of the Italian marine Pliocene deposits and location of the investigated area with tectonic detail in a section across the Montecalvo-in-Foglia basin.

sandstones and conglomerates, interbedded within hemipelagic mudstones. The vertically repeated

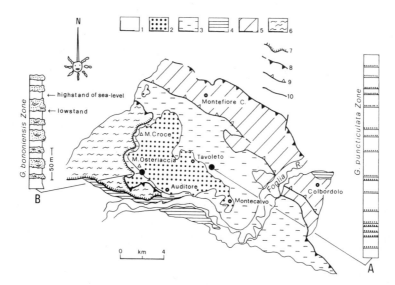

Fig. 2. Geological sketch map of the Montecalvo-in-Foglia basin showing the Pliocene turbidite stratigraphy sequence accumulated within an interval of about 1.3 My. 1, Terraced alluvial sediments (Pleistocene); 2, conglomerates and sandstones (Early Pliocene, *G. puncticulata* zone); 3, pelites and sandstones (Early Pliocene, *G. puncticulata* zone); 4, marls (Early Pliocene, *Sphaeroidinellopsis* sp.); 5, Messinian deposits; 6, allochthonous Marecchia Valley sheet; 7, contact between Umbro–Marchigiano–Romagnola succession and allochthonous Marecchia Valley sheet; 8, thrust; 9, backthrust; 10, faults. Representative vertical sections of (A) fine-grained turbidites (lower sequence) and (B) coarse-grained turbidite units (upper sequence) are also shown.

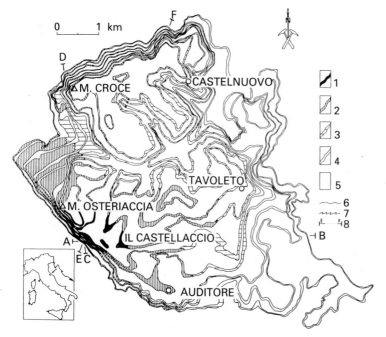

Fig. 3. Map showing lateral extension of the Pliocene of Montecalvo-in-Foglia basin. 1, Proximal zone: sandstones and sandstone–conglomeratic facies (S/G ratio ≫ 1). *Rapid sedimentation from high-density turbidity currents.* 2, Proximal–intermediate zone: sandstones and sandstone–mudstones facies (S/M > 5). 3, Distal–intermediate zone: sandstone–mudstone facies (1.2 < S/M < 5). *Deposition from high-density turbidity currents, initially from suspension followed by traction.* 4, Distal zone: mudstone–sandstone facies (M/S < 1.2). *Sand deposited from low-energy turbidity currents; mud deposited from suspension and turbidity currents.* 5, Pelitic epibathyal facies interbedded between the coarse-grained bodies. 6, Stratigraphical limit. 7, Facies boundaries. 8, Cross-section trace in Fig. 4.

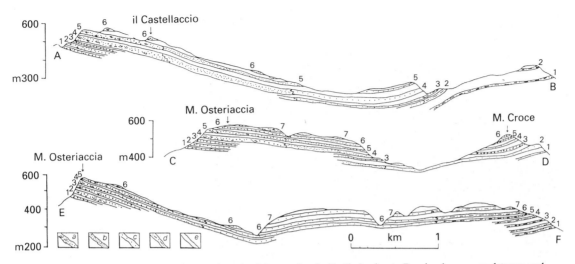

Fig. 4. Cross-sections through the Pliocene deposits, Montecalvo-in-Foglia basin: A, Proximal zone: sandstones and sandstone–conglomeratic facies. B, Proximal–intermediate zone: sandstones and sandstone–mudstone facies. C, Distal–intermediate zone: sandstone–mudstone facies. D, Distal zone: mudstone–sandstone facies. E, Epibathyal pelites. Numbers 1–7 refer to the coarse-grained turbidite bodies.

turbidite units (Figs 3, 4, 5 & 6) show a variable thickness from proximal to distal areas.

The turbidite units were derived directly by resedimentation from older, structurally higher situated fan-delta deposits. The turbidite sands and gravels have a strong deltaic–littoral textural imprint; hence, derivation from stored nearshore coastal sands and fan-delta complexes can be inferred. During periods of tectonic uplift, large amounts of sand and gravel were introduced into

the basin within the interval of resultant relative lowstand of sea-level. In the subsiding basinal area, coarse-grained sedimentation episodes do not represent shallowing (regressive) events, but reflect the falling base-level (sea-level) events. During these periods of lowstand, rivers bypass the shelf and deposit their sediment load directly onto the slope, so that sand and gravel would be funnelled through submarine canyons onto deep-sea fans (Vail *et al.*, 1984). The 'catastrophic' model for deposition of

Fig. 5. Proximal zone showing coarse-grained turbidite bodies (numbers 1, 2, 3 & 4) interbedded within pelitic epibathyal facies. Vertical view of the southwestern basin margin.

Fig. 6. Cross-section (SW–NE) showing onlap of a coarse-grained turbidite body (number 6) on pelitic–sandy facies deposited on a submarine slope probably originated by tectonic uplift.

such turbidite units is discussed by Coleman *et al.* (1983). The resedimentation process is outlined in Figs 7 and 8.

TURBIDITE SYSTEMS AND DEPOSITIONAL SEQUENCES

The Pliocene sediments of the MCF basin are of turbiditic nature and they show a notable tectonic influence on their deposition. Such influence is seen within two depositional sequences which are characterized by geometries that pass from parallel and continuous (lower sequence, Fig. 2) to discontinuous and cyclic. In the upper sequence, eight informal units (composite clastic bodies or depositional lobes) were identified and mapped, constituted by one or more turbiditic elements (*sensu* Mutti & Normark, 1987) characterized by different associations of facies. The proximal zone showed

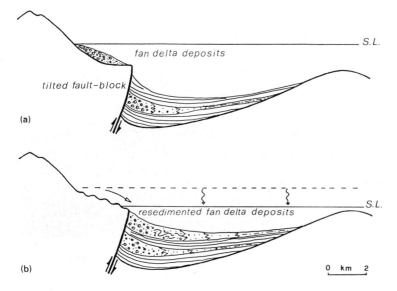

Fig. 7. Depositional model: relative fluctuations of the sea-level and their potential influence both on the construction of a fan-delta (a) and on its demolition and consequent sediment supply for developing a submarine fan at the base-slope apron (b).

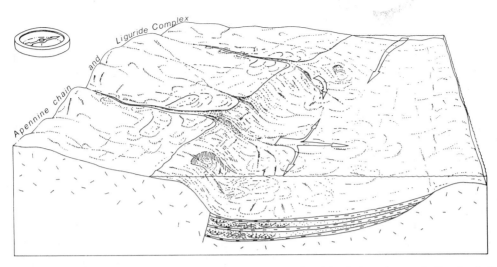

Fig. 8. Palaeogeographical reconstruction: pictorial view of a terraced fan-delta during a lowstand of sea-level when it undergoes remobilization into deep water.

coarse-grained sediments, organized in massive and amalgamated lenticular bodies with erosive bases (lobe deposits). Distally, the bodies present better internal organization and become generally thinner (lobe-fringe deposits, Fig. 9 & Table 1). The geo-

metrical features of the clastic bodies suggest an origin either from mass transportation (debris flow, slumping) or from confined turbiditic currents.

In this context, it seems likely that the variation of facies, internal geometry and position of deposi-

turbidite bodies number	max thickness of proximal area	min thickness of distal area
1	11.7 m	6.5 m
2	41.8	7.1
3	8.5	5.5
4	22.0	7.0
5	24.5	6.0
6	18.0	10.4
7	22.0	13.0
8	9.5	2.5

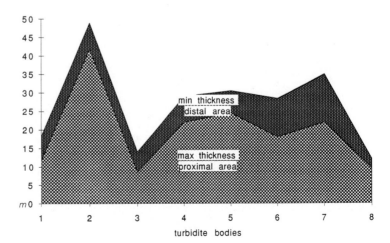

Fig. 9. Histogram of the thickness of the coarse-grained turbidite bodies in MCF basin.

Table 1. Proximal and distal sedimentological features of the coarse-grained turbidite bodies of MCF basin.

	Proximal area	Distal area
1 Bed thickness	Thick	Thin
2 Grain size:		
gravel	Coarse–medium	Absent
sand	Medium	Fine–medium
3 Arenite/rudite ratio	High	Absent
4 Arenite/pelite ratio	High	Low
5 Bed amalgamation	Moderate–high	Absent
6 Depositional intervals	Disorganized	Organized
7 Bed internal structures	Massive	Parallel lamination
	Chaotic levels (slump, slurried bed, debris flow, slump-scar wedging)	Rare cross-lamination
	Channel	
	Lag	
	Armoured mud balls	
	Clay chips	Horizontal rip-up clast orientation
	Gastropods, oysters, bivalves	Diffuse organic fragments
	Moderate bioturbation	
		Vegetal debris

tional system is closely related to the combined effects of the continuous tectonic uplift of the northwest part of the basin and eustatic fluctuations of the sea-level.

The present distribution and volume of turbidite bodies is shown in Fig. 10. The lithological composition of the turbidite bodies is expressed graphically in Fig. 11. Four lithofacies associations were recognized on the basis of the sand/gravel and sand/mud ratios (Capuano *et al.*, 1991). Facies and facies associations are documented and interpreted in Figs 3 and 4 (see geological map and profiles, Table 1). The geometrical reconstruction is shown in Fig. 12.

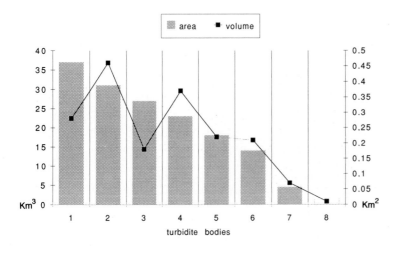

Fig. 10. Diagram of the present area and volume of the coarse-grained turbidite bodies in MCF basin.

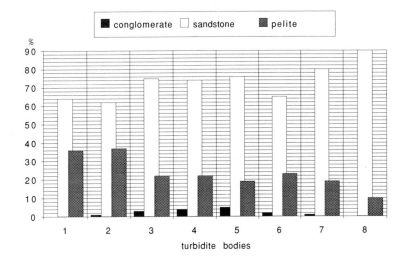

Fig. 11. Histogram of the lithological composition of the coarse-grained turbidite bodies of MCF basin.

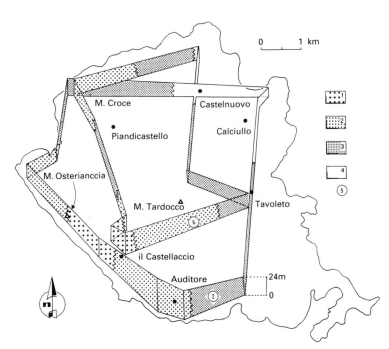

Fig. 12. Geometrical reconstruction of the coarse-grained turbidite bodies numbered 2 and 6. 1, Proximal zone; 2, proximal–intermediate zone; 3, distal–intermediate zone; 4, distal zone; 5, a number indentifying coarse grained turbidite bodies of MCF basin.

LATERAL VARIABILITY AND SOURCE AREAS

The lateral continuity of the facies suggests that the coarse-grained bodies were deposited by catastrophic gravitative events, induced and controlled by the thrust activity of the active west-northwest margin of the basin (Figs 1 & 2).

The vertical repetition of the facies association can be interpreted following the models of sequence stratigraphy (Vail *et al.*, 1984; Mitchum, 1985; Posamentier & Vail, 1988). The coarse-grained clastic sediments could reflect periods of relative lowstand of sea-level (Posamentier & Vail, 1988) whereas the draping, interbedded pelitic sediments reflect the less frequent and low volume turbidity currents during a relative sea-level highstand (Posamentier & Vail, 1988).

The MCF basin provides an example of a tectonically-controlled turbidite basin fed by a deltaic system located along its western margin (Fig. 8).

The correlation between the sections, provenance, clastic body geometry and granulometric data, suggests that the main source area (southwest flank) was closely related with the emergent Apennines chain. Only at a later stage, with the beginning of the deposition of turbidite clastic body number 3, did the basin start to be supplied also by source areas located in the northwestern sector (Figs 3, 4 & 8).

tilting of the depositional surfaces, and eustatic sea-level fall. The rapid sea-level fluctuations in the Pliocene probably contributed to the changes in sedimentation style, and to a significant increase in the volume of turbidity currents.

It is concluded that the causes of cyclicity in the coarse-grained turbidite sequence are complex. They include factors both internal and external to the sedimentary environment, such as subsidence and tectonic uplift, relative sea-level variation, climate changes during the Pliocene (Valleri et al., 1986) and global astronomical cycles.

CONCLUSIONS

The foreland basin of Montecalvo-in-Foglia, confined by two adjacent thrusts systems, is characterized by two turbidite systems: one deposited during periods of highstand of sea-level, with low sedimentation rates and muddy deposits: the other, very coarse-grained with high sedimentation rates, was deposited during periods of lowstand of sea-level. High frequency tectonic uplift 'events' within a short interval of time, about 1.3 My are thought to be responsible for relative sea-level lowstand. Turbidite depocentres migrated northeastwards in response to the thrust-tectonics that have characterized the evolution of the Umbria–Marchean Apennines during the Early Pilocene.

The coarseness and volume of the turbidite units appear to be related both to eustatic sea-level and/or to regional tectonism. Both factors combine to create the basic framework within which turbidite bodies were deposited. Turbidite systems reflect an alternation of thrust activity and quiescence. The regional distribution of facies and facies associations, and the orogenic trends, shows that the foreland basin was narrow, elongated (NW–SE) and confined within the Adriatic foredeep. It was characterized by an internal margin with continuous syn-sedimentary tectonic activity (slumps, cannibalistic input, allochthonous Ligurian sheets have been identified). Mineralogical composition and dispersal patterns suggest a hinterland characterized by an emergent Apennines chain, partly covered by allochthonous chaotic units (Liguride Complex of the Montecalvo area; Fig. 8).

Nevertheless, the coarse-grained turbidite facies are seen to be the result of the collapse of the shelf as well as the interaction between tectonic uplift,

REFERENCES

Boyer, S. & Elliot, D. (1982) Thrust system. *Am. Assoc. Petrol. Geologists Bull.* **66**, 1196–1230.

Capuano, N. (1990) Sedimentazione e tettonica del Pliocene nord-marchingiano. *Atti Soc. Tosc. Sci. Nat. Mem., Ser. A* **97**, 65–84.

Capuano, N., Pappafico, G. & Pera, M. (1991) Geological map of the intra-Pliocene sequence of Montefeltro area: geometry and facies distribution of the coarse-grained bodies. *Neue Descrive Carta Geol. d'Italia* **46**, 475–480.

Capuano, N., Tonelli, G., Veneri, F., Ricci Lucchi, F. & Barbieri, R (1989) Cannibalized fan-delta deposits of Montecalvo in Foglia. Marchean Apennines, Italy: a problem of identification. *Giorn. Geol., Ser. 3* **51/1**, 45–60.

Coleman, J. M., Prior, D. B. & Lindsay, J. F. (1983) Deltaic influences of shelf-edge instability processes. *The Shelfbreak: Critical Interface on Contimental Margins* (Eds Stanley, D.J. & Moore, G.T.) Soc. Econ. Paleontol. Mineral. Spec. Publ. **33**, 121–137.

Mitchum, Jr. R.M. (1985) Seismic stratigraphic expression of submarine fan. In: *Seismic Stratigraphy II: an Integrated Approach to Hydrocarbon Exploration* (Eds Berg, O.R. & Woolverton, D.G.) Am. Assoc. Petrol. Geol. Mem. **39**, 117–138.

Mutti, E. (1985) Turbidite systems and their relations to depositional sequences. In: *Provenance of Arenites* (Ed. Zuffa, G.G.) NATO ASI Series C**148**. Reidel, Dordrecht, pp. 65–93.

Mutti, E. & Normark, W.R. (1987) Comparing examples of modern and ancient turbidite systems: problems and concepts. In: *Marine Clastic Sedimentology* (Eds Legget, J.K. & Zuffa, G.G.) 1–38.

Mutti, E. & Ricci Lucchi, F. (1975) Turbidite facies and facies associations, *9th Int. Congr. Sedim.*, Nice, Guidebook Field Trip **11**, pp. 21–36.

Ori, G.G. & Friend, P.F. (1984) Sedimentary basins formed and carried piggyback on active thrust sheets. *Geology* **12**, 475–478.

Posamentier, H.W. & Vail P.R. (1988) Eustatic controls on clastic deposition ll — sequence and system-tract models. In: *Sea level Change — an Integrated Approach* (Eds Wilgus, C.K., Hastings, B.S., Kendall, C.G.St.C., Posamentier, H.W., Ross, C.A. & Van Wagoner, J.C.) Soc. Econ. Paleontol. Mineral. Spec. Publ. **42**, 125–154.

RICCI LUCCHI, F. (1986) The Oligocene to Recent foreland basins of the northern Apennines. In: *Foreland Basins* (Eds Allen, P.A. & Homewood, P.) Int. Ass. Sedimentol. Spec. Publ. **8**, 105–139.

RIO, D. & SPROVIERI, R. (1986) Biostratigrafia integrata del Pliocene–Pleistocene inferiore mediterraneo in un' ottica di Stratigrafia Sistematica. *Boll. Soc. Pal. It.* **25** (1), 65–85.

VAIL, P.R., HARDENBOL, J. & TODD, R.G. (1984) Jurassic unconformities, chronostratigraphy and sea level changes from seismic stratigraphy and biostratigraphy. *Interregional Unconformities and Hydrocarbon Accumulation* (Ed. Schlee, J.S) Am. Assoc. Petrol. Geol. Mem. **36**, 1129–1144.

VALLERI, G., MONECHI, S. & PIRINI RADRIZZANI, L. (1986) Pliocene and Pleistocene marine stratigraphy. *Mem. Soc. Geol. It.* **31**, 153–165.

Spec. Publs Int. Ass. Sediment. (1993) **20**, 301–333

Tectonic controls on non-marine sedimentation in a Cretaceous fore-arc basin, Baja California, Mexico

M.M. FULFORD *and* C.J. BUSBY

Department of Geological Sciences, University of California, Santa Barbara, CA 93106, USA

ABSTRACT

The El Gallo Formation is a 1300-m-thick sequence of Upper Campanian fluvial sedimentary rocks which are exposed along the west coast of Baja California, Mexico. The formation was deposited along the eastern margin of the Rosario embayment of the Late Cretaceous Peninsular Ranges fore-arc basin. Palaeocurrent directions measured from imbrication and trough cross-bedding, combined with evidence of compositional and textural immaturity and lack of alteration of sandstones throughout the formation, indicate that deposition occurred in a transverse fluvial system. Sedimentation along the basin margin was controlled by the interplay between high and gradually increasing basin subsidence rates and variable sediment influx and to a lesser degree a rising eustatic sea-level.

Single-crystal laser probe $^{40}Ar/^{39}Ar$ dates on individual sanidine grains from four rhyolitic tuff horizons within the El Gallo Formation permit the calculation of depositional rates that indicate that the depocentre was undergoing rapid subsidence (approximately 600 m/My). The effect of high and increasing subsidence rates on deposition can be recognized in the overall thickness and 'retrogradational' character of the deposit, as depositional environments changed from a proximal alluvial plain to braidplain through to floodplain and finally to a tidally-influenced coastal plain.

Rapid vertical movements of both the basin floor and the volcanoplutonic arc source terrane largely controlled sedimentation in the El Gallo Formation. Coarse-grained non-marine deposits at the base of the formation overlie Lower Campanian bathyal marine deposits (Punta Baja Fm) along a 15° angular unconformity. The magnitude of this relative sea-level change (000–2000 m) requires tectonic uplift of the basin, although erosion along the unconformity may have been enhanced during two possible eustatic low stands proposed for this time frame. This unconformity thus records tectonic uplift of the basin. Subsequent accumulation of non-marine deposits of the El Gallo Formation resulted from a combination of rapid tectonic subsidence of the basin and a contemporaneous 3rd-order sea-level rise. An abrupt lithological change from gravelly to sandy deposits between the La Escarpa and El Disecado members of the El Gallo Formation is inferred to be the result of downfaulting and sediment trapping along the fore-arc basin margin. Higher in the section, sandstone compositions record an increase in plutonic relative to volcanic and sedimentary fragments coincident with an apparent increase in sediment supply to the basin. This reflects an uplift event in the volcanoplutonic arc source terrane.

INTRODUCTION

Recent applications of facies analysis to the reconstruction of fluvial systems have resulted in the recognition of the importance of the interaction of basin movement (tectonic uplift and subsidence), eustasy and sediment supply in controlling non-marine sedimentation patterns (Ethridge *et al.*, 1987; Galloway, 1989). Many recent studies of ancient fluvial systems have concentrated on deposits in tectonically active areas such as intermontane or foreland basins (McLean & Jerzykiewicz, 1978; Puigdefàbregas & Van Vliet, 1978; Kraus & Middleton, 1987; Lawrence & Williams, 1987; Shuster & Steidtmann, 1987) and rift or extensional terranes (Steel & Aasheim, 1978; Graham, 1983; Turner & Whateley, 1983). Ancient fluvial deposits in fore-arc basins, however, have been subject to few studies, possibly due to the predominance of marine sedimentation in the fore-arc basin setting. Although a few studies of modern non-marine fore-arc basin deposits do exist, they focus primarily on volcanoclastic deposition and the effects of volcanism on sedimentation in the fore-arc basin (Kuenzi *et al.*, 1979; Vessel & Davies, 1981). As a result, the current literature yields little information on fluvial facies architecture in this tectonically active setting. The aim of this paper is to document such a fluvial system (the El Gallo Formation), deposited along the Late Cretaceous Peninsular Ranges fore-arc basin margin. The fluvial deposits comprise a 1300-m-thick retrogradational sequence of alluvial plain through to flood plain deposits capped by tidal deposits. Accumulation rates determined for the fluvial sequence demonstrate the great accommodation potential (Posamentier *et al.*, 1988) created along the margin largely by basin subsidence.

The effects of tectonics, eustasy and sediment influx on sedimentation patterns along the fore-arc margin are evaluated through the integration of facies analyses, palaeocurrent measurements, and petrographic data, combined with an analysis of subsidence rates calculated from statistically distinct mean ^{40}Ar/^{39}Ar dates of several laterally continuous tuffs occurring throughout the formation. The sizes of the rivers and the types of fluvial palaeoenvironments are also evaluated on the basis of the inferred relief and size of the drainage area and its proximity to the basin. The results indicate that although deposition of the El Gallo Formation coincided with an inferred global sea-level rise (3rd-order curve of Haq *et al.*, 1987), tectonically induced subsidence occurred at a much faster rate and, along with possible downfaulting along the basin margin and uplift in the source area, was the dominant control of sedimentation along the fore-arc basin margin in Late Campanian time.

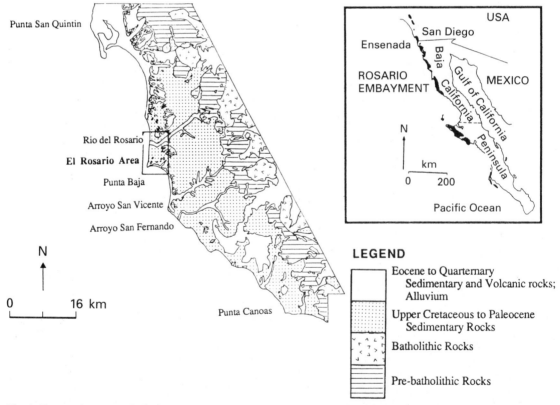

Fig. 1. Reconnaissance geological map of the Rosario embayment. (Modified from Gastil *et al.*, 1975.) Location of Upper Cretaceous fore-arc basin sedimentary rocks along the Pacific margin (shown in black on inset map). Detail of study area (outlined: El Rosario area) shown in Fig. 3.

REGIONAL GEOLOGICAL SETTING

Upper Cretaceous sedimentary rocks of the Peninsular Ranges fore-arc basin complex occur discontinuously, for a distance of approximately 500 km, along the Pacific coast of southern California and Baja California (Fig. 1; Beal, 1948; Gastil, 1975). The Peninsular Ranges fore-arc basin complex is an arc-massif-type fore-arc basin complex (classification scheme of Dickinson & Seely, 1979) that formed atop the western, trenchward margin of a Cretaceous volcanoplutonic arc (Bottjer & Link, 1984). The strata of this complex thus provide a record of fore-arc subsidence patterns as well as uplift and erosion of the arc massif source region. Basal strata of the Peninsular Ranges fore-arc basin complex are non-fossiliferous, non-marine deposits that may range in age from Cenomanian to Early Campanian; these are unconformably overlain by Early Campanian to Palaeocene marine and non-marine strata (Beal, 1948; Kilmer, 1963). Numerous authors have suggested that syndepositional faults provided controls on basin shape and bathymetry, distribution of depositional systems and cyclic sedimentation in the Peninsular Ranges fore-arc basin complex (Gastil & Allison, 1966; Boehlke & Abbott, 1986; Cunningham & Abbott, 1986; Morris *et al.*, 1987; Morris & Busby-Spera, 1988; Morris, 1992).

The Rosario embayment (Fig. 1) is the most areally extensive and by far the best exposed segment of the Peninsular Ranges fore-arc basin complex. Its fill consists of alternating non-marine and deep-marine deposits of Late Cretaceous to Palaeocene age (Fig. 2). Post-depositional faults in the Rosario embayment are few, generally with minor offsets, and strata in the study area dip gently to the northeast or east (Fig. 3).

THE EL GALLO FORMATION

The El Gallo Formation comprises approximately 1300 m of Late Campanian non-marine sedimentary rocks in the El Rosario area (Fig. 2). In the

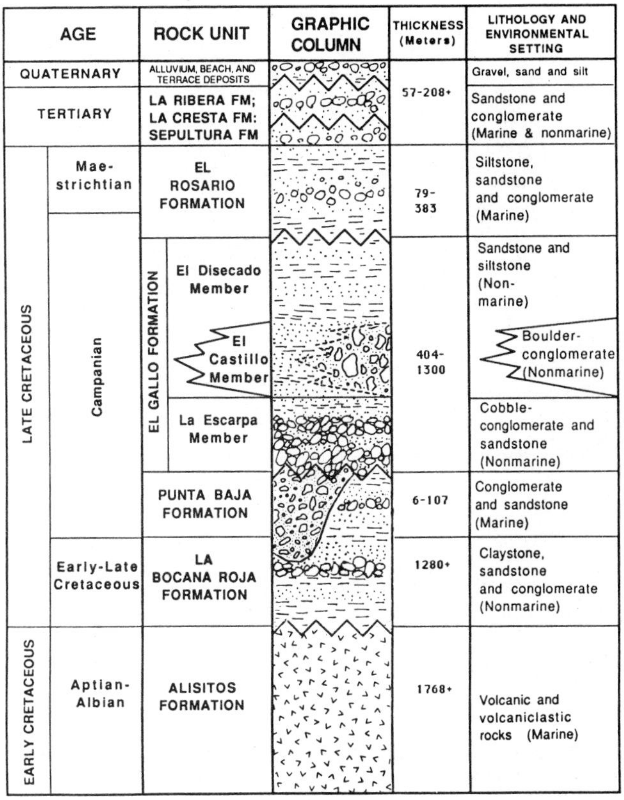

AGE		ROCK UNIT	GRAPHIC COLUMN	THICKNESS (Meters)	LITHOLOGY AND ENVIRONMENTAL SETTING
QUATERNARY		ALLUVIUM, BEACH, AND TERRACE DEPOSITS		57-208+	Gravel, sand and silt
TERTIARY		LA RIBERA FM; LA CRESTA FM: SEPULTURA FM			Sandstone and conglomerate (Marine & nonmarine)
LATE CRETACEOUS	Maestrichtian	EL ROSARIO FORMATION		79-383	Siltstone, sandstone and conglomerate (Marine)
	Campanian	EL GALLO FORMATION — El Disecado Member		404-1300	Sandstone and siltstone (Non-marine)
		El Castillo Member			Boulder-conglomerate (Nonmarine)
		La Escarpa Member			Cobble-conglomerate and sandstone (Nonmarine)
		PUNTA BAJA FORMATION		6-107	Conglomerate and sandstone (Marine)
	Early-Late Cretaceous	LA BOCANA ROJA FORMATION		1280+	Claystone, sandstone and conglomerate (Nonmarine)
EARLY CRETACEOUS	Aptian-Albian	ALISITOS FORMATION		1768+	Volcanic and volcaniclastic rocks (Marine)

Fig. 2. Generalized stratigraphical column for the El Rosario area, Baja California. (Modified from Kilmer, 1963.)

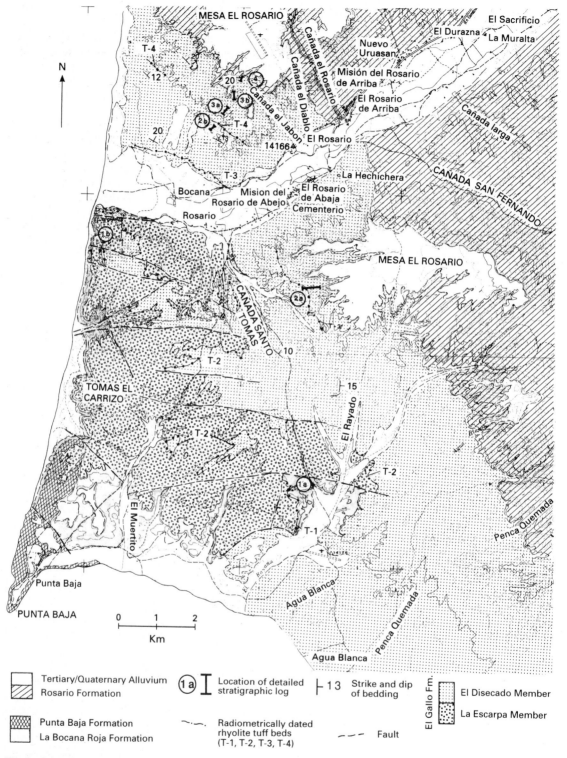

Fig. 3. Geological map of the El Rosario study area, Baja California. (Modified from Kilmer, 1963.) Circled numbers refer to stratigraphical logs in Figs 8, 9, 12 & 19.

western part of the study area (Fig. 3) it overlies submarine canyon deposits (Punta Baja Fm), which yield middle bathyal marine foraminifera, with an angular unconformity of up to 15° (Boehlke & Abbott, 1986). The unconformity represents a hiatus of approximately 5 My because: (1) the upper part of the Punta Baja Formation yields Early Campanian microfossils and a reversed magnetic isochron (Boehlke & Abbott, 1986; Filmore & Kirschvink, 1989), and (2) isotopic ages from a tuff at the base of the El Gallo Formation (described below) correspond to a Late Campanian age. In the southern and eastern part of the study area the El Gallo Formation overlies early Late Cretaceous fluvial deposits of the La Bocana Roja Formation (Figs 2 & 3). To the east of the study area, the El Gallo Formation abuts against and onlaps the Early Cretaceous volcanoplutonic arc basement (Alisitos Formation of Fig. 2; Kilmer, 1963; Morris, 1992).

The El Gallo Formation was originally defined by Kilmer (1963) who divided the formation into three members (Fig. 2). The conglomeratic La Escarpa Member, which lies at the base of the formation near the present-day coast southwest of El Rosario, and the conformably overlying sand- and silt-rich El Disecado Member are the focus of this study, due to the accessibility of excellent exposures and the abundance of datable rhyolitic tuffs which served as marker beds (Fig. 3). The contact between the La Escarpa and El Disecado Members is covered within the study area, but appears to young southward, based on correlation of a laterally continuous tuff marker bed. The third member, the El Castillo Member, comprises boulder conglomerates, and is restricted to the east side of the basin where it interfingers eastward with fault talus breccias shed from north–south trending basin-bounding faults (Morris, 1992). Minor approximately east–west trending intrabasinal faults that offset beds of the La Escarpa and lower El Disecado Members in the study area appear to die out within the upper part of the El Disecado Member (Fig. 3; Kilmer, 1963).

The El Gallo Formation was originally interpreted as a non-marine deposit originating in a flat basin or lacustrine environment (Kilmer, 1963). Evidence cited for the non-marine origin of the deposit includes the presence of terrestrial vertebrate fossils, abundant and large pieces of fairly unabraded fossilized wood, roots *in situ*, and palaeosol development throughout the El Disecado Member and minor palaeosols in the siltstones within the La Escarpa Member. Additionally we find that conglomerates in the La Escarpa Member show an abundance of possible non-marine characteristics, as outlined by Nemec and Steel (1984), including a dominantly clast-supported fabric, dominant transverse-to-flow a-axis orientation of the larger clasts, scoured bases and uncommon normal and inverse grading. We interpret the upper 15 m of the El Gallo Formation as tidal deposits on the basis of sedimentary structures, as described below. This, in turn, is conformably overlain by the marine Rosario Formation.

SEDIMENTOLOGY AND STRATIGRAPHY

We recognize four lithological successions within the El Gallo Formation based on upsection changes in lithology and bed thicknesses (Table 1). Succession 1 occurs at the base of the formation and comprises the La Escarpa Member of Kilmer (1963). It is approximately 150 m thick and forms an overall upward-fining sequence of conglomerates and lesser sandstones. Succession 2 comprises the base of the El Disecado Member and is an approximately 650-m-thick fining- and thinning-upward succession of sandstone and siltstone. Succession 3 is a 400-m-thick coarsening-upward sequence of sandstone and siltstone and succession 4 consists of approximately 100 m of sandstone and siltstone units showing no apparent vertical trends in bed thickness or grain size. Succession 3 and succession 4 comprise the middle and upper parts of the El Disecado Member, respectively.

Badland topography and sparse vegetation in the study area have resulted in excellent outcrop exposure (Fig. 4). The lateral extent of the exposures, however, is limited by the 10 to 20° northeastward dip of the beds which is perpendicular to the average palaeocurrent direction. Also, the outcrops are dissected by closely-spaced gullies and arroyos. Thus, the possible identification and classification of the sandstone body architecture, which is often an important factor in the qualitative determination of the sinuosity of ancient fluvial systems (Friend *et al.*, 1979), is somewhat restricted.

The following sections describe the lithostratigraphy of each succession. The facies schemes of Miall (1977, 1978, 1985) are followed. A detailed description of the common lithofacies we recognize in successions 2, 3 and 4 (El Disecado Member) is presented in Table 2. Palaeocurrent data were col-

Table 1. General description of significant characteristics of successions 1, 2, 3 and 4. Facies codes refer to Miall's lithofacies code (Miall, 1977, 1978) for succession 1 and facies code presented in Table 2 for successions 2, 3 and 4.

	Succession 1	Succession 2	Succession 3	Succession 4
Lithology	Pebble–cobble conglomerate, coarse to pebbly sandstone and lesser sandy siltstone	Coarse to fine-grained sandstone and sandy siltstone to silty clay. Cobbles, pebbles, wood (<50 cm length)	Same as succession 2	Same as succession 2. No cobbles or wood
Vertical trend	Apparent upward fining and thickening	Upward fining and thinning	Upward coarsening and thickening	None
Thicknesses of lithological units	Lower 50 m: Conglomerates: 30 cm–1 m Sandstones: <40 cm Upper 50 m: Conglomerates: 1–3 m Sandstones: 50 cm–2 m	Sandstones: 2–12 m Siltstone: 1–15 m	Sandstones: 2–30 m Siltstones: 1–20 m	Sandstones: 2–8 m Siltstones: 1–5 m
Conglomerate: sandstone: siltstone	80 : 20 : 0 (lower 50 m) 50 : 40 : 10 (upper 50 m)	1 : 59 : 40 (lower 400 m) 0 : 40 : 60 (upper 150 m)	0 : 40 : 60 (lower 200 m) 2 : 59 : 39 (upper 200 m)	0 : 50 : 50
Most common facies (Miall's (1977, 1978) lithofacies code used for succession 1, see text. See Table 2 for facies code for successions 2, 3 & 4)	Lower 50 m: Gm, Se, Ss with minor Gp, Gt, Sp Upper 50 m: Gm, Sp, Sh, Sl with minor Gp, Gt, St (upper 50 m)	Thicker sandstone bodies (≤5 m thick): Facies: 2, 3, 5, 7 & 8 Thinner sandstone bodies (<5 m thick): Facies: 2, 3 & 5	Thicker sandstone bodies (≤5 m thick): Facies: 2, 3, 4, 5, 6, 7 & 8 Thinner sandstone bodies (<5 m thick): Facies: 2 & 3	Facies: 2, 3 & 6 Also, large-scale planar cross-bed sets with bimodal cross-bedding and sigmoidal reactivation surfaces Ostrea sp. lag at top
Fluvial architectural elements (after Miall, 1985; see Table 2)	GB, gravelly bars and bedforms; SB, sandy bedforms	CH, channels; SB, sandy bedforms; LS, laminated sand sheets; OF, overbank fines	CH, channels; SB, sandy bedforms; LS, laminated sand sheets; OF, overbank fines; possible FM, foreset macroforms	CH, channels; SB, sandy bedforms; OF, overbank fines
Inferred depositional environment	Alluvial plain to braided stream	Low sinuosity sandy fluvial system with shallow (3 m) channels	'Prograding' low sinuosity sandy fluvial system with shallow (3 m) isolated channels in lower 200 m and stable fixed channels (possibly sinuous) in upper 200 m	Fluvial channels of low sinuosity to tidal flat and tidal channel subenvironments. Marine transgressive lag at top
Inferred tectonic controls	High sediment accumulation rate, due in part to subsidence of the margin	High subsidence rates. Abrupt lithological change from gravelly braided stream deposits to sandy braided stream deposits coincident with downdropping of eastern basin margin along normal fault	High subsidence rate. 'Prograding' fluvial system due to influx of sediment coincident with uplift and partial unroofing of volcanoplutonic arc source terrane	Decrease in depositional gradient and reduction of sediment influx

Fig. 4. Overview of a dominantly fine-grained section of the El Gallo Formation (upper part of succession 2), showing extensive exposures parallel to the palaeocurrent direction (to the left), 10–20° east-northeastward dip and badland topography.

lected from clast imbrication within succession 1 (La Escarpa Member) and from trough and planar cross-stratification in successions 2, 3 and 4 (El Disecado Member) (Fig. 5). Palaeocurrent directions were structurally corrected for those measurements taken from beds dipping more than 10°.

Fig. 5. Palaeocurrent measurements. All rose diagrams plotted on map (inset) represent measurements of imbrication from succession 1 (La Escarpa Member): 30 clasts per site. A rose diagram that combines all data for succession 1 is given. Note low variability in directions for succession 1. Palaeocurrent data for successions 2 to 4 (straight arrows) are sparse and represented on map (inset) by the statistical mean for several closely spaced measurements of large-scale cross-lamination only. Number of measurements per site given (N). Sites in succession 2 noted by an open circle, succession 3 by an open square and succession 4 by an open triangle. Rose diagrams on left side of figure display total measurements for each succession with mean for entire succession given by arrow.

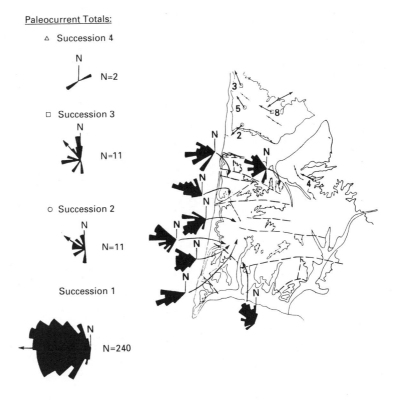

Paleocurrent Totals:

△ Succession 4
N=2

□ Succession 3
N=11

○ Succession 2
N=11

Succession 1
N=240

Table 2. Facies code, description and interpretation for common facies in successions 2, 3 and 4 (El Disecado Member). Referable to Figs 9, 12, 19, Table 1 and all photographs. Bedding thicknesses and set heights after Ingram (1954); thinly bedded and small-scale cross-lamination is 3–10 cm; medium bedded and medium-scale cross-lamination is 10–30 cm; thick bedded and large-scale cross-lamination is 30–100 cm. Set heights approximately > 1 m were not recognized.

Interval thickness	Lithology	Position within sandstone bodies	Description	Interpretation
Bounding surfaces				
1a Scour and lag deposit	Pebbles, cobbles, wood, intraclasts	Base and lower-middle parts of sandstone bodies	Scoured surface (< 1 m) lined with pebbles, cobbles, wood. May also contain intraclasts. Commonly overlain by large-scale trough cross-bedded sand (3)	Channel lag. Scour and transport during high-energy flow conditions. Deposition in waning conditions
1b Scour and lag deposit	Intraclasts only	Base and lower-middle parts of sandstone bodies	Scoured surface (< 1m) lined with intraclasts. Undercuts and intraclast slumps and breccias may occur with this surface. Commonly overlain by large- and medium-scale planar cross-bedded sandstone (2)	Undercut scour and lag. Scour and deposition may have occurred in the same event. May form from moderate fluctuations in flow energy or lateral migration of channel
1c Sharp base, no lag	—	Base of sandstone bodies. May also separate bedding sets	Sharp, flat unscoured surface, no lag. Commonly overlain by overlying large-scale sets of tabular and wedge-shaped planar cross-bedded sandstone (2)	Erosional surface. Created by high flow regime planar bed formation and migration
1d Scour, no lag	—	Base of sandstone bodies. May also separate bedding sets	Uncommon surface, similar to facies 1a, but lacks lag deposit. Commonly overlain by large-scale sets of tabular and wedge-shaped planar cross-bedded sandstone (2)	Erosional surface. Possibly formed by migration of channel
1e Lateral accretion	—	Upper and upper-middle parts of thick (> 5 m) sandstone bodies	Erosional surfaces dipping perpendicular to palaeocurrents, showing inclination of up to 10°. Commonly occurs in large-scale planar and/or trough cross-bedded sandstone (2, 3) and overlain by medium-scale sets of trough to tabular cross-bedded sandstone (5)	Lateral accretion surface, upper point bar. Channel migration results in lateral migration of dunes and plane beds into channel

Common facies

Facies	Thickness	Lithology	Description	Position	Interpretation
2 Large- to medium-scale sets of alternating tabular to wedge-shaped planar cross-bedded sandstone (lithofacies Sp)	1–3 m	Medium- to coarse-grained sandstone. Few small intraclasts	Low angle and horizontal stratification are common. This facies is most commonly overlain by alternating large-scale sets of trough and tabular cross-bedded sandstone (3)	Lower part of sandstone bodies, commonly comprises the basal bed	Basal channel fill. Alternating beds of upper flow regime plane beds and lower flow regime megaripples produced by rapid (vertical) channel filling
3 Large-scale sets of alternating trough and tabular cross-bedded sandstone (lithofacies St)	1–3 m	Coarse (trough) to fine (tabular) sandstone. Cobbles, pebbles, wood and intraclasts may line bases of troughs	Trough cross-beds may be large to medium scale and alternate laterally (within 3 m) and vertically (in 50 cm sets) with low angle tabular cross-beds. Most commonly overlain by small to medium sets of trough and tabular cross-bedded sandstone (5)	Lower part of sandstone bodies	Channel fill. Same as (2) with less high flow regime structures and more lower flow regime dune formation and migration
4 Convolute bedded sandstone (lithofacies Sp)	< 50 cm	Medium- to fine-grained sandstone	Commonly comprises oversteepened forsets of large- to medium-scale tabular cross-bedding (2). Commonly overlain by medium- to small-scale tabular and trough cross-bedded sandstone (5)	Middle parts of thick (> 5 m) sandstone bodies	Uppermost channel fill. Deformation caused by migration of sand bar over unconsolidated channel fill deposits
5 Medium- to small-scale sets of alternating tabular and trough cross-bedded sandstone (lithofacies Sp & St)	1–10 m	Medium to fine-grained sandstone. May contain large pieces of fossil wood	Generally overlies erosional surfaces (1b or 1e) or large-scale sets of trough and tabular cross-bedded sandstone (3). In the thicker sandstone bodies (> 10 m), this facies may comprise up to 10 m of cosets, and may contain roots *in situ* and large wood fragments	Upper parts of sandstone bodies. May comprise entire internal structures of thinner (< 5 m) sandstone bodies	Sand bar deposit. Formed by dune (megaripples) migration in lower flow regime conditions
6 Horizontal stratification (lithofacies Sh)	< 50 cm	Fine-grained sandstone	This facies is rare, but where present the laminae appear to be defined by heavy-mineral concentrations	Tops of sandstone bodies	Bar top deposit. Low flow regime plane bed
7 Structureless fining-upward sequence	< 50 cm	Fine-grained sand, silty sandstone, sandy siltstone, siltstone	Very common upward-fining gradational sequence occurs over 30 cm. Lack of structure possibly due to post-depositional homogenization	Tops of sandstone bodies	Bar top deposit

Continued on page 310

Table 2. (*continued*)

Interval thickness	Lithology	Position within sandstone bodies	Description	Interpretation
8 Thin-bedded silty clay and fine-grained sandstone (lithofacies Fl or Fsc)	<1 m	Silty clay and fine-grained sandstone	Tops of sandstone bodies	Fairly rare facies. Consists of thin beds of laminated fine sandstone and silty claystone	Levee or proximal flood-plain. Upper point bar
9 Alternating sets of microscale and wavy cross-bedded siltstone (lithofacies Fl)	<2 m	Very fine-grained sandstone and claystone	Tops of sandstone bodies. Thin (<2 m) intervals in thick multistoried sandstone bodies	Poorly preserved facies. Common bioturbation. In some places, occurs as discontinuous wavy bedding	Sand bar top
10 Soft-sediment deformation and climbing ripples (lithofacies Fl)	<50 cm	Laminated sandy siltstone	Occurs immediately below a sandstone body	Generally laminated and may be carbonaceous. Small fossil bone fragments, teeth and plant fragments are commonly found in this facies. Soft-sediment deformational features include distorted bedding and convolute laminae. Small (<1 m wide) mass flows (debris-filled channels) also included in this facies	Overbank pond, sand bar top
11 Pedogenic horizons (lithofacies P)	50 cm–20 m	Sandy silt and silty clay	Occurs between sandstone bodies	Palaeosol development ranges from stage 1 to stage 3 (Bown & Kraus, 1987). Complex palaeosols, one palaeosol stacked on top of another, commonly comprise the thicker pedogenic horizons (2–20 m) and commonly contain thin interbeds of laminated or rippled fine- to very fine-grained sandstone that separate sequential palaeosol horizons. A palaeosol is generally characterized by a thick (50 cm–1.5 m) reddish sandy silt horizon overlain by a thin (<30 cm) green sandy siltstone. Rarely, a thin (<20 cm) purplish horizon lies at the top of the reddish zone. Calcrete is very common in the reddish horizons and may form long stringers of large (10 cm dia.) flat rounded calcrete nodules interpreted to be remnant calcrete pans. In the La Escarpa Member, calcrete occurs in small (<5 cm) rounded nodules that are randomly dispersed in bands (50 cm) within the reddish horizons. No original sedimentary structures are preserved	Abandoned overbank areas

Fig. 6. Characteristic fabric and grain size of imbricated clasts in a massive to crudely-cross-bedded conglomerate bed (Gm lithofacies: Miall, 1977, 1978) within succession 1.

Succession 1

Description

Much of the lower part of the El Gallo Formation is covered or inaccessible thus limiting examination to several small outcrops, less than 25 m wide and 15 m high. The east–west trending faults described above have disrupted the La Escarpa Member into fault blocks of varying thicknesses and thus compli-cate the identification of lateral and stratigraphical relationships between isolated outcrops. We estimate a maximum thickness of 150 m for succession 1 on the basis of composite sections and general map relations.

The conglomerate beds throughout succession 1 are dominantly clast-supported and poorly sorted, and consist of well-imbricated, subrounded to rounded pebbles to large cobbles (Fig. 6). The compositions of the clasts are dominantly metasedimen-

Fig. 7. Overview of upper-middle part of succession 1 showing lateral continuity of sandstone and conglomeratic lithological units. Also shows fairly high proportion of sandstone and siltstone relative to conglomerate. Some thinner sandstone units (< 2 m) pinch out laterally over a distance of 20–30 m. Palaeocurrent direction to the left and into the page.

tary and metavolcanic with lesser plutonic and metamorphic clasts. Composition varies with clast size, such that 70% of the pebbles and small cobbles are clastic and volcanic in composition while plutonic rock-types make up approximately one-half of the large-cobble clast size (based on 800 clasts measured from 8 locations, discussed below). Clast shapes are mostly blade-like although the largest clasts tend to have more equant, brick-like shapes. Clast size does not appear to change significantly upsection or laterally, and the common large clast size for these conglomerates is in the large cobble size range (average of 7–8 cm in diameter). The largest clast measured was 45 cm in diameter and was found in an anomalously coarse bed at the base of the succession. Clast imbrication directions from 240 conglomerate clasts throughout succession 1 record a mean of 270° with low variability (Fig. 5).

Although clast size is generally constant, succession 1 becomes more sandstone-rich upsection and,

therefore, comprises an overall 'upward-fining' sequence (Figs 7 & 8). Outcrops at the base of the succession are characterized by a high proportion of conglomerate to sandstone (70–80% conglomerate), with sandstone limited to impersistent lenses within amalgamated conglomerate bodies. Outcrops of the uppermost part of the succession have a lower proportion of conglomerate relative to sandstone (50–60% conglomerate), and commonly contain siltstone. The lower and upper parts of succession 1 are described in turn below.

Outcrops containing the basal contact or stratigraphically equivalent to the lower 50 m of succession 1 are characterized by medium- to thick-bedded (10–60 cm) conglomerates with massive or crude horizontal bedding (Table 1; Gm lithofacies: Miall, 1977, 1978) and are rarely trough cross-bedded or planar bedded (Gt or Gp; Table 1; Fig. 8a). The thickness of the conglomerate bodies ranges from 50 cm to 1 m, rarely exceeding 1 m in

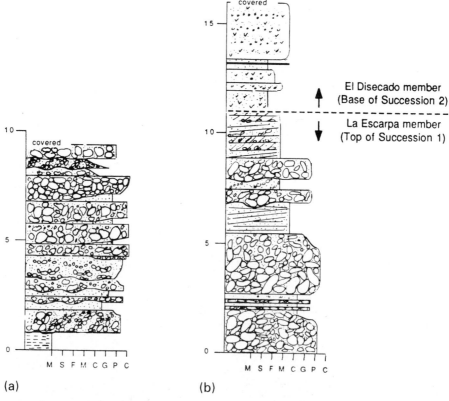

Fig. 8. Representative measured sections from the (a) lower part and (b) upper part of succession 1. Stratigraphical position plotted on Fig. 3. Bases of sections covered. Symbols are also used in stratigraphical logs, Figs 8b, 10, 12 and 20.

thickness. The conglomerate bodies are generally laterally continuous for the extent of the outcrops (20 m). Matrix-supported conglomerates (Gms) are rare, occurring in only two outcrops just above the basal contact. These units comprise poorly sorted 10–30-cm-thick beds of dominantly large-pebble- to small-cobble-sized clasts. In one location, rounded, cobble-sized mudclasts were identified within these matrix-supported beds.

Sandstones in the lower part of succession 1 are generally less than 40 cm thick and occur as laterally discontinuous (<8 m) concave-up lenses of coarse sand containing pebbles and granules. Commonly, these sands have medium-scale, crude pebbly cross-bedding (Se and Ss; Fig. 8a) and appear to be somewhat gradational with the underlying conglomerate unit.

As in the lower part of succession 1, the conglomerates in the upper part of succession 1 are dominantly massive to crudely bedded (Gm) and occasionally cross-bedded (Gt and Gp). Bedding surfaces appear to be concave-up (as a result of scouring processes) although they are difficult to follow due to the apparent amalgamation of beds. Conglomerate bodies are thicker than those lower in the succession, and commonly reach 2–3 m in thickness (Fig. 8b).

The sandstone beds in the upper part of the succession are much thicker and more laterally persistent and commonly comprise 2–3 m of 40–50-cm-thick sets of low-angle cross-bedding (Sl). Some of the smaller sandstone bodies show medium-scale sets of pebbly cross-bedding (Sp). Like sandstone beds in the basal outcrops, the sandstones in the upper part of succession 1 appear to have gradational bases and sharp, erosional tops (Fig. 8b).

Siltstone, which is rarely present in the lower part

LITHOLOGY AND SEDIMENTARY STRUCTURES:

Conglomerate

Sandstone

Siltstone

Trough cross lamination

Planar cross lamination

Tabular and wedge-shaped planar crossbedding

Planar lamination

Paleosol development

Rhyolitic tuff

Granules, pebbles and cobbles

Siltstone-mudstone intraclasts

Flame structures

Calcrete

Ripple cross lamination

Wavy, discontinuous lamination

Convolute lamination

Fossil bone fragments

Fossil wood fragments

Fossilized oyster shell fragments

Bioturbation

BEDDING CONTACTS:

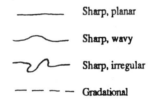

——————   Sharp, planar

——⌣——   Sharp, wavy

⟿⌣——   Sharp, irregular

– – – –   Gradational

GRAIN SIZE:

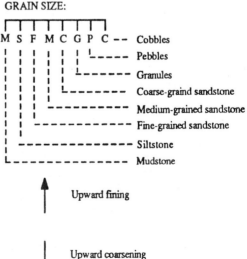

M S F M C G P C – –   Cobbles

Pebbles

Granules

Coarse-graind sandstone

Medium-grained sandstone

Fine-grained sandstone

Siltstone

Mudstone

↑   Upward fining

↓   Upward coarsening

Fig. 8. (*continued from facing page*) Key.

of succession 1, makes up 10–20% of the lithology in the upper part of the succession, where it comprises 1–2-m-thick units that commonly contain scattered granules and coarse sand. Thin (<30 cm) interbeds of sand are rarely present within the siltstone units. The siltstone units commonly show diffuse, approximately 50 cm to 1 m thick, bands of faint red and green colouration, as well as poorly developed horizons of scattered, small rounded calcrete nodules (P). These features are interpreted to be pedogenic in origin (stage 1: Bown & Kraus, 1987) and give this part of the succession a characteristic pinkish colour.

Interpretation

Succession 1 represents a gravelly fluvial deposit characterized by Miall's (1985) architectural elements (Table 1) representing gravelly bars and bedforms (GB) and sandy bedforms (SB). The predominant lithofacies in the lower part of the succession are gravelly bars (Gm) and scour fills (Se). These were probably deposited primarily during flood events, possibly in an alluvial-fan stream or alluvial-plain environment, based on the laterally extensive amalgamated conglomerate units, as well as widespread imbrication, and lack of well-defined bedding or internal stratification (Nemec & Steel, 1984). The high proportion of conglomerate to sandstone in the basal part of succession 1 and the predominance of massive to crudely-bedded gravels (Gm) are similar to deposits in the proximal reaches of the modern Donjek River, Yukon, Canada (Rust, 1978) as well as to the Scott-type braided stream model of Miall (1977) and Miall's 'Model 2' (1985) which may be typical of proximal braided rivers (including alluvial fans with stream flows).

In the upper part of succession 1, an overall upward-fining trend is accompanied by increased development of overbank deposits with incipient palaeosol development, and thicker, multistorey (after Friend *et al.*, 1979), conglomerate units. These features indicate that flow was more confined than during deposition of the lower part of succession 1. The abundance of lower flow-regime sandy bedforms (Sp, Sh, Sl, Fl) in the upper part suggests that the flow was also more consistent and less intermittent. Channelization and sustained flow are characteristics of many braided stream models (McPherson *et al.*, 1987). The middle reaches of the Donjek River (Williams & Rust, 1969; Miall, 1978) record depositional features similar to those de-

scribed here for the upper part of succession 1.

We interpret succession 1 to record dominant flood-stage deposition, initially in bedload streams on an alluvial plain which may have evolved over time into a system of more confined channels in a more distal braidplain setting.

Unidirectional palaeocurrents support the interpretation for deposition of succession 1 on an alluvial plain where sedimentation occurred primarily during high flood-stage events. High flood-stage deposition may result in relatively low variability in imbrication directions due to the minimal effect of bars on downslope flow directions during high water (Rust, 1978). Furthermore, in an alluvial plain, the proximity of neighbouring fans would confine the sweep of the channels and limit the range of preserved palaeocurrent directions. A good example of channel confinement on alluvial plains compared to the channel spread in alluvial fans can be seen in the southern part of Death Valley (Dorn, 1988; fig. 1). Finally, the lack of any significant variability between directions measured at the base of succession 1 and the top of succession 1 may preclude the possibility of a major lateral shift in the locus of deposition over time for this succession.

Succession 2

Description

A laterally continuous rhyolite tuff bed occurs at the lithological change from the conglomeratic La Escarpa Member to the sandstone- and siltstone-rich El Disecado Member (Kilmer, 1963). The contact is poorly exposed in the study area but appears to be gradational over several tens of metres. Succession 2 is a 650-m-thick, overall thinning-upward sequence of alternating coarse- to medium-grained sandstone and siltstone, with subordinate conglomerate that shows an upward decrease in the sandstone to siltstone ratio (Fig. 9 & Table 1). The sandstone to siltstone ratio ranges from approximately 60:40 in the lower part of the succession to 40:60 in the upper part. As with succession 1, the fining- and thinning-upward is an apparent trend, based on 125 m of section exposed at the top, 150 m exposed near the middle, and 50 m near the base of the succession. The remaining intervals of the succession are covered by alluvium and vegetation. The location of the upper boundary of succession 2 is somewhat gradational over 50 m and is

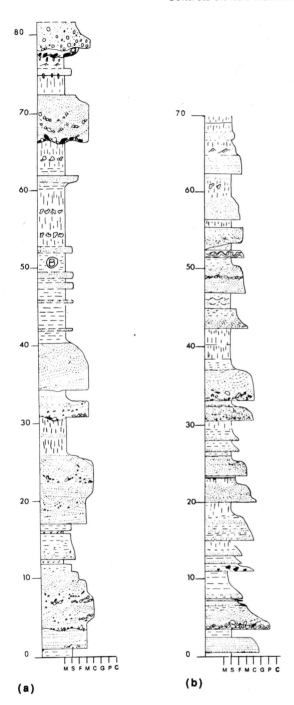

Fig. 9. Representative measured sections from the (a) lower part and (b) upper part of succession 2. Stratigraphical position plotted on Fig. 3. Legend for symbols in Fig. 8.

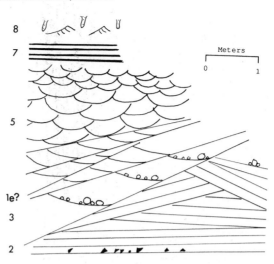

Fig. 10. Outcrop sketch of upward-fining sequence in sandstone body, middle part of succession 2. Facies indicated by numbers and referable to Table 2.

placed where the apparent upward-fining and -thinning trend is replaced by the thickening- and coarsening-upward trend of succession 3.

In vertical profile, sandstone bodies in succession 2 become less multistorey and more isolated upsection. The sandstone bodies in succession 2 range from 2 to 12 m thick, with the thickest occurring nearer the base of the succession (Table 1 and Fig. 9a). The internal arrangement of sedimentary structures within a sandstone body is shown in a detailed stratigraphical column (Fig. 9) and in an outcrop sketch (Fig. 10). The bases of the sandstone bodies are generally scoured but may be sharp and flat (Surface 1c: Table 2). The basal scours have a maximum relief of 1 m and are commonly lined with cobbles, pebbles, intraclasts and fossilized palm and conifer fragments (Morris, 1974); these also occur along scour surfaces within the medium- to large-scale trough cross-bedded sandstone sections (Facies 2) in the lower parts of the sandstone bodies (Fig. 11). Trough cross-bedding (St) alternates with medium- to large-scale tabular to wedge-shaped planar cross-bedding (Sp) in the lower parts of these sandstone bodies (Facies 3). The middle parts of the sandstone bodies commonly comprise medium- to small-scale trough and tabular cross-bedding (Facies 5). Convolute bedding is rare and tends to lie below the medium-scale bedding in the middle parts of the thicker sandstone bodies (Facies 4). A lack of induration of the deposit and possible

Fig. 11. Large intraclasts along internal scoured erosional surface within sandstone body in middle part of succession 3. Note deformed cross-bedding, medium- to small-scale cross-bed sets and ripples. Fossil wood fragments in upper right corner.

destruction by palaeorootlets and/or bioturbation of the upper parts of the sandstone bodies has resulted in a homogeneous appearance to much of the upper portions of many of the sandstone bodies.

Most of the thinner sandstone bodies (<5 m) tend to be single storey, that is, they contain no internal scour surfaces, and show an upward decrease in grain size. Also, vertical organization of sedimentary structures record a gradual decrease in flow regime. In these units, medium- to small-scale trough and tabular cross-bedding (Facies 5: Table 3) tends to be the dominant structure, and coarse

material is virtually absent within the sandstone body, except for a possible lag at the base or small pebbles and intraclasts along the trough cross-beds. Less commonly, the thinner sandstone bodies do not fine upward and comprise only one type of sedimentary structure, particularly low-angle to horizontal cross-bedding (Sh; Facies 6), above a flat rather than scoured base. These are similar to Miall's architectural element LS which he defines as laminated sand sheets (Miall, 1985).

Siltstone units range from one to 15 m, with the thicker siltstone units occurring more commonly

Table 3. Petrographic data. Samples arranged in stratigraphically ascending order. Qm, monocrystalline quartz; Qp, polycrystalline quartz and chert; K, potassic feldspar; P, plagioclase feldspar; Ls, sedimentary and metasedimentary rock fragments; Lv, volcanic and metavolcanic rock fragments; Lp, aplite; Lm, foliated rock fragments.

	Succession																			
	1						2							3					4
Qp	11	6	8	3	6	9	7	4	6	5	6	5	4	1	1	2	4	1	3
Qm	18	20	16	20	21	19	27	26	29	29	36	30	24	40	42	42	33	34	36
K	7	12	13	11	8	6	8	11	9	12	7	6	13	10	11	13	9	12	8
P	28	36	36	41	43	41	33	39	35	29	32	33	37	33	40	29	33	40	25
Lv	15	11	11	6	8	8	11	6	8	9	8	10	8	3	1	3	7	3	14
Ls	13	8	10	5	8	10	7	3	6	4	5	7	6	1	1	2	3	2	5
Lm	1	0	1	0	0	1	2	2	1	0	1	1	1	1	0	0	0	0	2
Lp	4	1	0	2	1	1	2	0	1	1	0	1	1	0	0	1	1	1	4
Biotite	3	4	4	11	4	4	7	8	5	10	3	5	7	7	2	7	9	4	1
Muscovite	1	0	0	0	0	0	0	0	0	0	2	0	0	1	0	0	0	1	0
Hornblende	0	2	1	1	2	1	0	0	1	1	1	1	1	3	1	2	0	2	2

toward the top of the succession. Siltstone commonly grades upward from the underlying sandstone body and tends to show some degree of pedogenesis (Facies 11). In the lower part of succession 2, incipient to moderate pedogenesis (Stage 2 of Bown & Kraus, 1987) can be recognized by definite red and green colour bands and the development of thin beds of large, flat (5–10 cm diameter), rounded, calcrete nodules (P) that may represent remnant calcrete pans. Also, sporadic, thin, carbonaceous layers can be found in the upper parts of these siltstone units and at one location small plant fragments were identified. This pedogenic alteration of the siltstone units, however, becomes less common in the upper part of the succession where siltstone units are more commonly interrupted by thin sandstone interbeds. In many cases, the upper part of the siltstone bodies in the upper part of succession 2 is rippled, bioturbated, has soft-sediment deformational features, and may be carbonaceous. Teeth and scattered bone fragments of amphibians, lizards, archosaurs (crocodilians, ornithischians and sarischians), turtles, birds and mammals, as well as two groups of dinosaurs (hadrosaurs and carnosaurs), have been recovered from localities largely within the siltstone units in the uppermost part of succession 2 (Morris, 1974; Los Angeles Natural History Museum, archives). An ammonite fossil was also recovered from the lower part of the El Disecado Member (Morris, 1974) correlating with succession 2.

A mean palaeocurrent direction of 294° for succession 2 was calculated from eleven trough cross-bed measurements (Fig. 5). Although the data are sparse, the palaeocurrent distribution is vaguely unimodal with fair variability.

Interpretation

Succession 2 marks an abrupt change from the gravelly bed-load stream deposits of succession 1 to mixed-load stream deposits similar to the CH and OF architectural elements of Miall (1985) which represent channels and overbank fines. This change in facies architecture and type of sediment load may record deposition in a more fixed channel system resulting from a decrease in gradient and an increase in stream stability (Jackson, 1978; Galloway, 1985).

The characteristics of the thin, isolated sandstone bodies that are most common in the upper 150 m of succession 2, such as sharp erosional bases, normal grading and preservation of an apparent upward

decrease in flow-regime structures (Fig. 9), are similar to the descriptions of sandy braided stream deposits (Battery Point Formation) by Cant (1978), and crevasse-splay (Flores & Pillmore, 1987) and short-lived channel deposits of meandering streams (Puigdefàbregas & Van Vliet, 1978). The laminated sand sheets (architectural element LS) record flood deposition in interchannel areas similar to deposits in Bijou Creek (McKee *et al.*, 1967) and, therefore, may further record deposition in a low sinuosity sandy fluvial system. The palaeocurrent pattern is also consistent with deposition in a low-sinuosity fluvial system in a floodplain setting and precludes the possibility of a major avulsion being the cause of the upsection decrease in amalgamated channel facies.

The preservation of complete single-storey upward-fining sequences and thin single-storey sandstone bodies within thick, fine-grained non-channel-facies in the upper part of succession 2 was accomplished by rapid overbank sedimentation and fine-grained sheet flooding. High rates of flood-plain aggradation resulted in the burial of these deposits before a channel could be reactivated or migrate laterally (Puigdefàbregas & Van Vliet, 1978). High floodplain aggradation can further be inferred for the upper part of succession 2 on the basis of (1) ubiquitous soft-sediment deformation in both sandstones and siltstone units, (2) the preservation of vertebrate fossils and unlithified volcanic tuffs, and (3) evidence of only poor palaeosol development in the overbank facies throughout the upper 150 m of succession 2.

Although the presence of an ammonite fossil locality within succession 2 indicates a marine incursion, sedimentary structures throughout the succession are indicative of a non-marine depositional environment. The presence of this fossil indicates that at least part of the succession was deposited at or near base level (sea-level).

Succession 3

Description

Succession 3 is an overall thickening- and coarsening-upward sequence of sandstone and siltstone (Fig. 12 and Table 1). The lower 200 m of succession 3 is similar in thickness and lithology to the underlying upper part of succession 2, with a sandstone to siltstone ratio of approximately 40:60. Sandstone bodies range from 2 to 8 m in thickness,

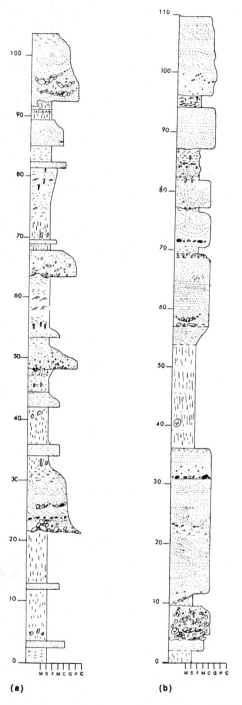

Fig. 12. Representative measured sections from the (a) lower part and (b) upper part of Succession 3. Stratigrapical position plotted on Fig. 3. Sections continue down and up. Legend for symbols in Fig. 8.

and siltstone bodies range from 1 to 15 m in thickness (Table 1). Sandstone bodies in the lower part appear to be laterally discontinuous over several tens of metres (Fig. 13). A gradual increase in the relative proportion of sandstone to siltstone occurs upward through the succession as both the sandstone and siltstone bodies become thicker and pebble- to cobble-sized clasts increase in abundance. The upper 200 m has a sandstone to siltstone ratio of 65:35; sandstone bodies average between 10 and 30 m in thickness and siltstone units average between 3 and 20 m. Sedimentary structures within the thinner sandstone bodies (> 2 m) are similar to those described for succession 2, and consist dominantly of large- to medium-scale trough and tabular cross-bedding (Facies 3: Table 2 and Fig. 14) and large-scale tabular to wedge-shaped planar cross-bedding (Facies 2: Table 2). Palaeocurrents in succession 3 are similar to those of succession 2 and record a mean palaeocurrent direction of 309° for eleven trough cross-bed measurements (Fig. 5).

A notable difference between succession 2 and succession 3 occurs in the very thick sandstone bodies at the top of succession 3 (Fig. 15). These sandstone bodies are generally multistorey with several erosional surfaces bounding coarse-grained intervals near the base and then several more complete fining-upward sequences stacked one onto another in the middle and upper parts. Medium- to small-scale trough and tabular cross-bed sets (Facies 5) are much more common within these thick fining-upward sequences than elsewhere in the El Gallo Formation. At some outcrops, 40- to 50-cm-thick sets of planar cross-bedding are stacked vertically for up to 10 m above the coarser grained and scoured basal parts of the sandstone bodies. Large tree trunks, up to 7 m in length (Fig. 16), are associated with the medium-scale trough cross-beds within these thick sandstone bodies. In one location, large (10–20 cm diameter) roots are preserved in an apparent upright life position within medium-scale trough cross-bedded sands. Convolute laminae, deformed cross-bedding and dewatering structures (Facies 4: Table 2 & Fig. 17) are also much more common in the large multistorey sandstone units at the top of succession 3 than elsewhere in the El Gallo Formation, occurring generally in the middle of the well-developed upward-fining sequences.

Lateral accretion surfaces (Surface 1e: Table 2) are present in the upper parts of the thick sandstone bodies. These surfaces are subtle and very rare, and

Fig. 13. Overview of base of succession 3. Note isolated sandstone bodies in thick, laterally continuous, fine-grained horizons. Palaeocurrent direction to the left.

have only been recognized in the thick sandstone bodies in the upper part of succession 3. Laterally discontinuous siltstone intervals, averaging less than 2 m in thickness, are also common within the upper to middle parts of these thick sandstone bodies. These consist of alternating, medium- and small-scale sets of bioturbated microcross-stratified to wavy-bedded sandstone and siltstone (Facies 9; Table 2).

In the lower 200 m of succession 3, siltstone bodies are generally equal in thickness to the sandstone bodies or thicker, reaching a maximum of 30 m (Table 1). In the upper 100 m of succession 3, siltstone bodies are thinner, ranging from thin (<2 m), discontinuous intervals within the thick sandstone bodies to 10 m in thickness between the sandstone bodies. Throughout succession 3, siltstone commonly grades up from the tops of the underlying sandstones over a distance of 10 to 20 cm (Facies 7: Table 2).

Fig. 14. Sharp erosional surface (1c surface: Table 2) overlain by intraclast-lined large-scale trough cross-beds (Facies 3: Table 2). Note homogenization (due to bioturbation) and indistinct upward fining of upper part of underlying sandstone body.

Fig. 15. Thick, multistoried sandstone body (approximately 12.5 m thick) in lower-upper part of succession 3. Note sharp, flat erosional surface at base (Surface 1c; Table 2), and thin sandstone interbeds in thick underlying siltstone horizon (Facies 8, 11).

Fig. 16. Large piece of fossilized wood and wood fragments. Occurs within medium-scale trough cross-bedded interval (Facies 5: Table 2).

The thicker siltstone bodies (> 5 m) never show original sedimentary structures, possibly due in part to pedogenesis. The thick siltstone bodies commonly have thick (1–2 m) palaeosol profiles with abundant yellow, blue and red mottling (Table 1) indicating long periods of overbank saturation (Bown & Kraus, 1987). The thinner siltstones (< 2 m), however, do retain sedimentary structures and commonly contain climbing ripples, soft-sediment deformational structures and burrows (Facies 9 and 10: Table 2 and Fig. 18), particularly in the lower part of the succession.

Interpretation

The laterally impersistent and upward-fining nature of the sandstone bodies at the base of succession 3 is similar to that at the top of succession 2 (architectural element CH of Miall, 1985) and suggests deposition from short-lived mixed-load

channels in which a gradual waning of flow within the channels allowed bedload deposition to the point of channel abandonment (Galloway, 1985). As in the upper part of succession 2, high floodplain aggradation rates and significant transport of suspended load in these channels allowed the sand bodies to become isolated in fine-grained deposits (Galloway, 1985).

We interpret the thicker sandstone bodies in the upper part of the succession as stacked channel-fill and possibly point bar sequences. The large-scale sets of tabular to wedge-shaped planar cross-beds (Facies 2: Table 2) and trough to tabular cross-beds (Facies 3: Table 2) are interpreted to be channel-fill deposits. The thick intervals of medium-scale trough and tabular cross-bedding (Facies 5: Table 2) are inferred to be the deposits of sandy bedforms, possibly dunes on point bars. The small-scale cross-bed sets and rippled units (Facies 9: Table 2) at the top are typical of point bar top deposits while the fossil-laden, often convoluted, siltstone may be more typical of overbank ponds.

In the upper part of succession 3, the thickness of the larger sandstone bodies, the possible presence of lateral accretion surfaces and point bar deposits within the thick sandstone bodies, and the development of fairly mature palaeosols in overbank deposits interstratified with thick, multistorey sandstone bodies suggest a reduction in floodplain aggradation rate and development of more stable channels,

Fig. 17. Convolute bedding (Facies 4: Table 2). Overturned foresets of large-scale planar cross-beds in the middle of a sandstone body in the upper part of succession 3.

Fig. 18 Medium-scale sets of climbing ripples (Facies 10: Table 2). Note bioturbation throughout and soft-sediment deformation of underlying unit. Lower part of succession 3.

which could aggrade through lateral migration rather than avulsion and switching.

Succession 4

Description

Succession 4 is similar in lithology, bed thickness, sandstone to siltstone ratio (60:40) and sedimentary structures to the lower part of succession 3 (Fig. 19 & Table 1) and the contact between the two successions is gradational. No vertical trend in thickness of sandstone and siltstone bodies, however, can be recognized and the sandstone bodies are more variable in thickness, ranging from 1 to 10 m. Another notable difference is the common presence of a thin layer of mud draped at the top of the sandstone bodies. Definite 1–2-m-thick bands of bright red, green and even purple pedogenic alteration occurs in the overbank sediments, and calcrete horizons are particularly well developed, containing large globular calcareous nodules, possibly replacing roots or filling bioturbated hollows in the soils. One particularly well-developed palaeosol horizon (stage 3 of Bown & Kraus, 1987) extends laterally for approximately 5 km (Fig. 20).

In the upper 5 m of the succession, a 50-cm-thick bed of organic-rich, convoluted, laminated siltstone containing abundant large teeth, small bones, and soft-sediment deformational features occurs below an unusual 30-cm-thick bed that appears to consist of a mass of carbonate-cemented sand-filled burrows. Overlying these beds is a carbonate-cemented, single-pebble-thick layer that occurs at the base of two 1-m-thick sets of tabular, cross-bedded sandstone containing abundant mud-draped sigmoidal reactivation surfaces (Fig. 21). Bimodal palaeocurrent directions of 65° and 235° are recorded in these sets (Figs 5 & 21). A sharp flat erosional surface overlain by a shelly (*Ostrea* sp.) lag marks the contact of the top of the El Gallo Formation with the overlying marine Rosario Formation.

Interpretation

We interpret all but the upper 5–15 m of succession 4 as the deposit of a mixed-load stream, much like the lower part of succession 3, probably representing stream channel deposits of low sinuosity. The increase in pedogenesis in succession 4 compared with successions 1, 2 and 3 suggests that the over-

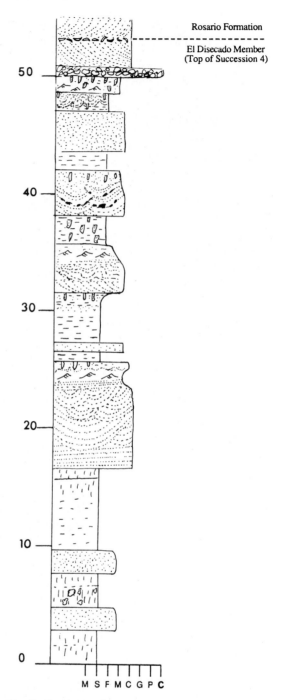

Fig. 19. Representative measured section from the upper part of succession 4. Stratigraphical position plotted on Fig. 3. Section continues down and up. Legend for symbols in Fig. 8.

Fig. 20. Laterally extensive bright red palaeosol horizon (Stage 3: Bown & Kraus, 1987) near upper contact of El Gallo Formation with marine Rosario Formation. Photograph covers approximately 112 m of lateral exposure.

Fig. 21. Inferred tidal depositional features at top of El Gallo Formation (top of succession 4). Large-scale planar cross-beds with bimodal cross-bedding and mud-draped sigmoidal reactivation surfaces.

bank areas were exposed to weathering processes for longer periods of time and must have undergone lower rates of aggradation. Although Bown and Kraus (1987) note that an increase in pedogenesis may reflect distance from an active channel, there is no evidence to suggest that these deposits were located significantly farther from active channels than elsewhere lower in the formation. Thus, the increase in pedogenesis is inferred to reflect a temporal change in environmental setting rather than a lateral change. A decrease in sediment load

combined with a possible decrease in depositional gradient due to a relative rising sea-level should have resulted in straighter channels (Schumm & Khan, 1983; Morton & Donaldson, 1989).

The large-scale bimodal cross-bedded sandstones with mud-draped sigmoidal reactivation surfaces (Fig. 21) are similar to descriptions of tidal sand-wave deposits from the Early Cretaceous seaway of southern England (Allen, 1982). Similar features, including a basal lag, are described as tidal channel deposits by Elliot (1986). The local occurrence of these sets also supports a tidal channel rather than shelfal origin. The unusual organic-rich laminated siltstone and heavily bioturbated sandy bed that occur below these tidal channel deposits are unlike any features recognized elsewhere in the fluvial part of the El Gallo Formation and are also interpreted to be littoral deposits, possibly originating in tidal flat, estuarine or interdistributary bay environments. The close association of these organic-rich and bioturbated beds with inferred tidal channel deposits, however, would suggest a tidal flat origin.

The presence of littoral deposits and a shelly lag at the top of the formation at the contact with overlying marine deposits of the Rosario Formation indicate that a marine transgression terminated fluvial deposition of the El Gallo Formation in the study area.

PETROGRAPHY AND PROVENANCE

Sandstone and conglomerate compositions were analysed in order to ascertain the burial and weathering history of the sandstones, possible compositional variations through time, and provenance of the El Gallo Formation. This information, combined with facies architecture, allows a more complete understanding of the possible tectonic controls on sedimentation. Seventeen medium-grained sandstone samples were analysed (300 points each)

and plotted on the ternary sandstone classification diagrams of Folk (1968; Fig. 22) and on the ternary provenance diagrams of Dickinson *et al.* (1983) and Dickinson and Suczek (1979) (Fig. 23). The Gazzi–Dickinson point-counting method was used with grain categories as defined by Dickinson *et al.* (1983).

Sandstones in the El Gallo Formation range from lithic arkoses at the base (successions 1 and 2) to arkoses at the top (successions 3 and 4). The sandstones are dominantly uncompacted and friable. Sparse carbonate-cemented concretions and horizons, however, occur throughout the formation and record an average of 15% matrix, 10% porosity and 13% cement for five samples from the base of the formation (successions 1 and 2), and 0% matrix, 4% porosity and 33% cement for two samples from the upper part of the formation (successions 3 and 4). In the carbonate-cemented samples, biotite flakes are remarkably undeformed, indicating that cementation must have occurred prior to or in the early stages of burial. Furthermore, clay rinds, which occur on some grain boundaries in the uncemented samples, are not apparent in the cemented samples, suggesting that the samples were cemented not only prior to or during early burial, but also prior to the diagenetic alteration of lithic volcanic and feldspar grains to clays. The limited compaction and cementation of the El Gallo Formation indicates that these rocks were never buried

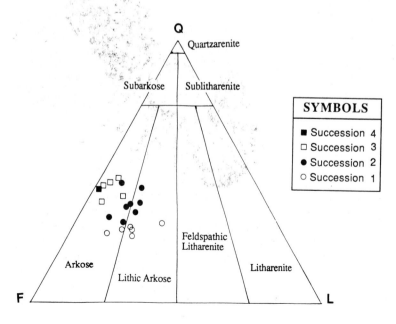

SYMBOLS

- ■ Succession 4
- □ Succession 3
- ● Succession 2
- ○ Succession 1

Fig. 22. Sandstone petrography of the El Gallo Formation, using the classification of Folk (1968). Q, monocrystalline and polycrystalline quartz; F, plagioclase and potassic feldspar, granitic rock fragments (aplite) and foliated rock fragments (gneiss and schist). Note overall compositional immaturity and apparent upsection increase in plutonic fragments. See Table 3 for petrographic data.

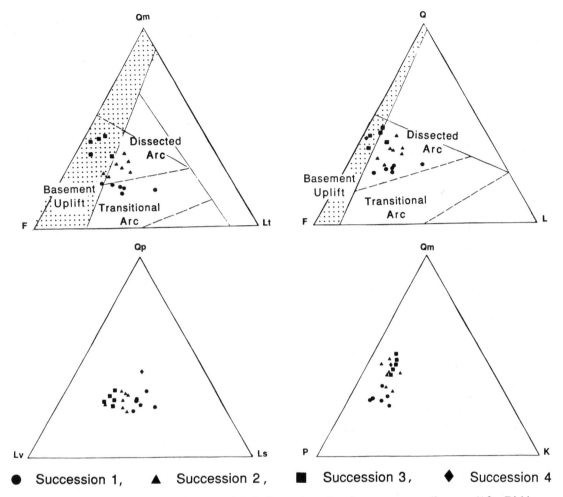

Fig. 23. Sandstone petrographic data from the El Gallo Formation, plotted on provenance diagrams. (After Dickinson and Suzcek, 1979; Dickinson *et al.*, 1983.) Q = Qm + Qp (inc. chert); F = k + p; Lt = Lv + Ls; Qp = Lp(aplite) + Lm(foliated). Stippled area denotes continental block provenance (from Dickinson *et al.*, 1983). Note upsection trends in composition and apparent derivation of succession 3 and 4 sandstones from a continental block provenance. See Table 3 for petrographic data.

deeper than 2.5–3 km (Bjørlykke *et al.*, 1989). There is also very little evidence of leaching of feldspars and other unstable minerals and no precipitation of diagenetic kaolinite. This suggests that these sediments underwent rapid burial, allowing little time for leaching by meteoric water (Bjørlykke, 1988; Bjørlykke *et al*, 1989).

Compositions of conglomerate clasts were studied in detail from eight locations within succession 1 and from one location in each of successions 2 and 3. As in the lithic fragments of the sandstone samples, metasediments and metavolcanics are the

most common clast type, together representing more than three-quarters of the rock types displayed. The volcanics are dominantly mafic, although silicified tuffs were identified at all locations. Of the plutonic clasts, silicic plutonics are the most common although gabbros and diorites are present. At some sites in the La Escarpa Member (succession 1), muscovite–biotite granites were identified.

On the 'QFL' and 'QmFLt' diagrams, the sandstones appear to plot primarily in the dissected magmatic arc field for successions 1 and 2, and appear to plot primarily in the continental block provenance field for successions 3 and 4. Thus, sandstone compositions changed significantly during deposition of succession 3.

Sandstone compositions appear to increase in quartz content upsection, whereas the proportion of sodic to potassic feldspars remains constant. Dickinson and Suczek (1979) interpret this kind of compositional change as an upsection increase in the proportion of plutonically derived grains, thereby recording unroofing of the source area.

Lithic grains in the El Gallo Formation also record an upsection increase in plutonic and metamorphic grains relative to the dominating proportion of metasedimentary and metavolcanic grains. The metasedimentary lithic fragments consist dominantly of silicified or greenschist-facies altered shales and mudstones. The metavolcanic lithic fragments consist dominantly of silicified and greenschist-grade tuffs and ignimbrites, although mafic and intermediate metavolcanic lithic fragments were also recognized. Except for diabases and volcanic fragments with microporphyritic textures, lithic volcanic fragments, particularly mafic to intermediate metavolcanic fragments, are difficult to recognize as they tended to be altered and are often hard to distinguish from matrix or lithic sedimentary fragments. Toward the top of the formation, metamorphic lithic fragments include biotite–muscovite schist. Sphene, although rare, occurs in many of the sandstone samples towards the top of the formation. Chert and quartzite make up a significant proportion of the polycrystalline quartz fraction of the sediments.

Three important points regarding syndepositional processes can be summarized based on the petrography of the El Gallo Formation.

1 The mineralogical immaturity and lack of diagenetic alteration indicate that the sandstones of the El Gallo Formation were derived from a rapidly eroding source area and were buried rapidly.

2 The petrographic composition of the sandstones and conglomerates suggests that the metavolcanic and metasedimentary cover of the arc massif was being eroded simultaneously with the plutonic core. An overall unroofing trend is recognized in the compositional changes from the base of the formation to the top resulting in the youngest sediment having an apparent continental block source.

3 An abrupt compositional change in sandstone compositions records an increase in plutonic and high-grade metamorphic fragments during deposition of succession 3. This compositional change

occurred contemporaneously with the inferred increase in sediment influx to the basin during deposition of succession 3 and may reflect a period of accelerated uplift of the arc massif.

PALAEOENVIRONMENTAL SYNTHESIS

The El Gallo Formation comprises four lithological successions that record an overall retrogradational sequence of deposition from braided to sinuous fluvial systems and finally to a tidal subenvironment followed by a marine transgression across the basin. Preserved upward-fining sequences suggest that channel depths were typically no greater than 5 m and commonly averaged 3 m. Abundant fossilized palm and conifer fragments in the El Gallo Formation, as well as the occasional preservation of roots *in situ,* and the common homogenization of the sandstones, suggest that large trees may have been a major feature on or near the floodplain. Abundant fossil evidence of aquatic species suggests a constant availability of water in the fluvial setting and indicates that flow conditions were probably not ephemeral, and that small lakes may have been present. The presence of calcrete in the pedogenic horizons, therefore, may be less a result of climatic conditions and more a result of the compositional immaturity of the sediments and consequent abundance of available calcium. Although constant water supply is implied, several features suggest that the climate was probably not humid. These features include: (1) the paucity of evidence of dense vegetation, such as lignite, peat deposits and root casts; (2) the presence of calcrete; and (3) the unaltered nature of the sandstones.

Consistent west to northwest palaeocurrent directions throughout alluvial plain to floodplain deposition are perpendicular to the trend of the arc massif, indicating that the fluvial system that deposited the El Gallo Formation must have been part of a transverse drainage system. The transverse character of the El Gallo depositional environment is also supported by: (1) the unaltered nature of unstable minerals and lithic fragments that indicate high relief, short-lived transport and rapid burial; and (2) the apparent lack of deep channels such as would characterize wider rivers associated with longitudinal drainage. The drainage into fore-arc basins is controlled by the topographically uplifted arc. The short distance from the source area to the basin limits the degree of confluence of the small rivers into larger ones. The presence of channel deposits identified in the El Gallo Formation, which suggests that the rivers were only 3–5 m deep, is therefore consistent with transverse drainage across this tectonically active setting.

The predominant depositional controls on each succession within the El Gallo Formation can be evaluated through the synthesis of facies analyses and petrographic data. The following discussion presents a summary of the depositional history of the formation.

Succession 1 occurs at the base of the formation and overlies bathyal marine deposits along a low-angle (15°) unconformity. Uplift of bathyal deposits to non-marine elevations resulting in a 15° discordance between deposits resulted largely from tectonic movements in the basin. Erosion along the angular unconformity, however, may have been accentuated during two inferred eustatic sea-level low stands (Haq *et al.*, 1987; Fig. 24).

Succession 1 represents the initial non-marine deposition resulting from tectonic activity dramatically affecting the fore-arc basin floor. The succession is an upward-fining conglomeratic sequence that records high subsidence rates. The unimodal nature of the palaeocurrents precludes a possible major lateral shift of the depositional system over time to explain the apparent fining-upward trend. Unidirectional palaeocurrent measurements combined with facies interpretations suggest that deposition of this succession was ephemeral and occurred in a retrograding alluvial plain to braid-plain environment.

Succession 2 consists of sandstone and siltstone units that fine and thin upwards. This succession is inferred to have been deposited by a fluvial system of fairly low sinuosity. Increased sediment accumulation and floodplain aggradation due to a change in base level or sediment influx is indicated by an

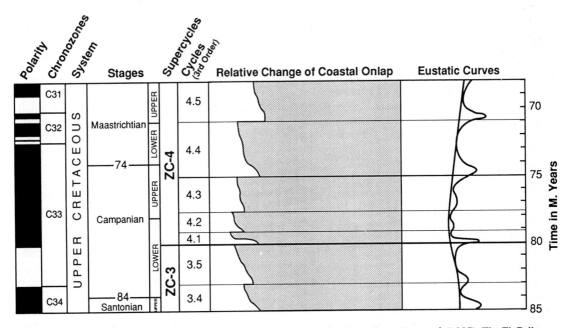

Fig. 24. Eustatic sea-level curves for Campanian to early Maastrichtian time, from Haq *et al.* (1987). The El Gallo Formation ranges in age from 75 to 73 Ma as determined by isotopic ages on tuffs (Fig. 25 and text). Erosion along the angular unconformity between the El Gallo Formation and the underlying Punta Baja Formation could correspond to the type 1 unconformity shown here at 80 Ma and/or the type 2 unconformity shown at 75 Ma, since the Punta Baja Formation yields Early Campanian microfossils and a reversed magnetic isochron (see text). The El Gallo Formation is overlain by the marine Rosario Formation, which contains the Campanian–Maastrichtian boundary near its base; the boundary is thus about 1 My younger than that indicated by the time-scale used in this figure (as discussed by Renne *et al.*, 1991). Due to this problem in correlation of biozones, it is not certain whether eustatic sea-level rose throughout deposition of the El Gallo Formation, or if a medium fall could have occurred early in its deposition. If the latter occurred, tectonic subsidence apparently masked its effects (see text).

overall fining- and thinning-upward trend and by preservation of complete upward-fining sandstone bodies (channels, point bars) within thick overbank deposits towards the top of succession 2. The abrupt lithological change between successions 1 and 2 probably resulted from downfaulting along a basin-margin normal fault. Such an event would trap sediment in the downdropped area next to the fault and would result in a sudden reduction in the coarse fraction of sediment being brought through the fluvial system without a significant change in palaeocurrent direction. This is supported by the geochronology of interbedded tuffs that records an increase in sediment accumulation rates in succession 2, relative to succession 1, as discussed below. East–west trending faults that die out in the lower part of succession 2 indicate that some structural disruption of the basin margin must have occurred during and possibly prior to deposition of succession 2 (Kilmer, 1963).

Succession 3 comprises a coarsening- and thickening-upward sedimentary sequence that is inferred to record a reduction in the rate of floodplain aggradation and increase in influx of coarser material indicating that more fine-grained sediment was bypassing the system. Sediment bypass and reduced floodplain aggradation rates could result from either a decrease in the subsidence rate and/or an increase in the rate of sediment supply to the basin (Beerbower, 1964) resulting in an apparent progradation of the fluvial system. Eustatic curves for this time period show no apparent regression or fluctuations in sea-level (Haq *et al.*, 1987; Fig. 24) that would increase the depositional gradient of the non-marine systems and cause an apparent progradation of the fluvial system. Progradation of the fluvial system could have resulted from an increase in sediment influx. Sandstone compositions indicate that succession 3 marks a change in provenance from dominantly magmatic arc to continental block, which suggests that the source area had been uplifted or rejuvenated at this time, likely increasing the influx of coarse sediment to the fluvial system. Additionally, infilling of the initial downfaulted area (which trapped the coarser sediments during deposition of succession 2 and the lower part of succession 3) may have occurred. The fluvial system would then have begun to prograde away from the mountain front resulting in an upsection increase in grain size and an increase in sediment influx without a substantial change in fluvial style or palaeocurrent direction.

Succession 4 records a marine transgression in the El Rosario area. Preservation of littoral deposits, which are recorded at the top of succession 4, would be unusual during a strictly eustatic sea-level rise since the sediments would be subjected to wave base and storm activity for a prolonged period of time, probably causing them to be eroded. However, if the basin floor was actively subsiding during a eustatic sea-level rise, the accumulation potential would be increased, and littoral deposits would be more likely to be preserved. The preservation of littoral deposits of succession 4 may, therefore, be the result of continued active subsidence of the basin floor during a eustatic sea-level rise. Increased pedogenesis and decreased channel sinuosity, recorded lower in the succession, are inferred to be a result of a decrease in the depositional gradient and relative rate of sediment influx prior to the final marine transgression.

GEOCHRONOLOGY AND SUBSIDENCE ANALYSIS

High-resolution dates of four rhyolite tuffs in the El Gallo Formation were obtained from $^{40}Ar/^{39}Ar$ laser probe analyses of single sanidine crystals, in order to calculate apparent sediment accumulation rates. The methods and detailed analytical results are discussed elsewhere (Renne *et al.*, 1991) and a summary of the results are presented in Fig. 25. The tuffs generally occur in fine-grained facies (Facies 7 & 8: Table 2) and were rarely identified within the channel deposits, suggesting that they were nearly always washed out of the channels by subsequent channel activity. Where tuffs were present in channel facies, bedding is almost completely unrecognizable due to very large-scale soft-sediment deformation that appears to be the result of density contrasts between tuff and sandstone.

The results of the geochronological analyses of the tuffs (Fig. 25) indicate that succession 1 (the La Escarpa Member) accumulated at a rate of 366 ± 92 m/Ma (Renne *et al.*, 1991). Although comparable data are sparse, this rate appears to be about three times the typical rate for alluvial fan sediment accumulation, which rarely exceeds 100 m/Ma (Blatt *et al.*, 1980). This high sediment accumulation rate suggests rapid tectonic subsidence of the basin.

The average sediment accumulation rate of succession 2 is calculated as ±600 m/Ma (Renne *et*

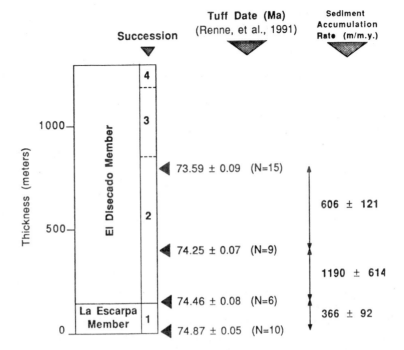

Fig. 25. Stratigraphical positions of rhyolitic tuffs and calculated sediment accumulation rates, El Gallo Formation. Stratigraphical positions of successions 1 to 4 also given. Tuff dates calculated by single-crystal laser probe $^{40}Ar/^{39}Ar$ analyses of large populations (N) of sanidine crystals for each tuff (Renne *et al.*, 1991). Sediment accumulation rates calculated using error in tuff dates and 10% error in stratigraphical thicknesses.

al., 1991). An ammonite fossil found within the lower part of the El Disecado Member (Morris, 1974) and inferred floodplain setting suggest that deposition of succession 2 occurred at or near base level (sea-level). Therefore, the rate of sediment accumulation for succession 2 may be roughly equivalent to the subsidence rate of the basin-margin floor.

A total subsidence rate of approximately 600 m per My falls between typical subsidence rates for strike-slip basins, which may be up to 1–2 km per My, and subsidence rates for rift and foreland basins, which are in the order of 300 m per My (Blatt *et al.*, 1980). The high subsidence rates recorded for these settings are primarily a result of tectonic activity. The similarity of such high sediment accumulation rates recognized in the El Gallo Formation to these other active settings, therefore, suggests that some of the subsidence during El Gallo deposition must also have been due to tectonic activity along the margin of the fore-arc basin.

To evaluate the tectonic component of subsidence versus the cumulative effect of isostasy, eustasy and compaction for the El Gallo Formation, a geohistory analysis was performed on the complete stratigraphical section of the El Gallo Formation, following the methods of Steckler and Watts

(1978) assuming non-flexural (Airy) isostasy (Pinter & Fulford, 1991). The cumulative effects of sedimentation, compaction, and isostatic loading in the basin over the period of deposition were quantified and evaluated. The effects of sediment compaction were evaluated on the basis of unit thicknesses and lithologies, utilizing empirical relationships between porosity and burial depth. Eustatic changes were estimated using the sea-level curves of Haq *et al.* (1987) (Fig. 24). Tight age control was provided for the lower 800 m of the El Gallo Formation by the high precision, single-crystal $^{40}Ar/^{39}Ar$ dates from the interbedded tuffs (Renne *et al.*, 1991) and a final age was extrapolated for the top of the El Gallo Formation assuming the average sedimentation rate calculated for succession 2.

The results of the calculations are displayed on a geohistory diagram (Fig. 26) and show the relative components of eustasy and tectonics versus total subsidence during deposition of the El Gallo Formation. The average rate of tectonic subsidence is between 85.7 and 160.7 m per My. In comparison, the rate of tectonic subsidence along the Gulf of Mexico passive margin during a 35 My period in the Late Cretaceous was approximately 25.7 m per My (Allen & Allen, 1990) and the average rate of thrust-induced tectonic subsidence of the northern

M.M. Fulford & C.J. Busby

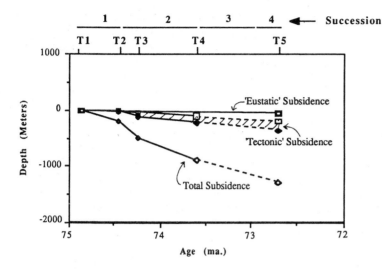

Fig. 26. Geohistory diagram showing relative eustatic and tectonic components of subsidence. Note the importance of tectonic subsidence (85–160 m per My) relative to eustatic (17 m per My) in controlling very high total subsidence rate (600 m per My) which represents the cumulative effects of tectonic and eustatic activity and compaction and isostasy. Tuff dates (T1, T2, T3 & T4) from geochronological analysis of rhyolitic tuffs (Fig. 25; Renne *et al.*, 1991). T5 is an extrapolated date, using average subsidence rate of 600 m per My.

Green River basin during a 20 Ma period in the Late Cretaceous was approximately 35 m/Ma (Shuster & Steidtmann, 1987). The eustatic component driving subsidence during deposition of the El Gallo Formation is equivalent to the rate of sea-level rise over this time period which was approximately 17 m/Ma based on the inferred 3rd-order sea-level curves of Haq *et al.* (1987). This eustatic component of subsidence is less than one-fifth of the tectonic component, and virtually negligible when compared to the total subsidence rate. As with all basins, the total subsidence rate (600 m/Ma) is predominantly a result of compaction and isostatic compensation, which are indirectly controlled by eustasy and tectonics. The overwhelming rate of tectonically controlled subsidence relative to eustatically controlled subsidence during deposition of the El Gallo Formation, therefore, suggests that tectonic activity along the fore-arc basin margin was the primary driving force of the high subsidence rates recognized from the deposits of the El Gallo Formation.

DISCUSSION

Sedimentary facies, petrography and sediment accumulation rates suggest that the El Gallo Formation was deposited along a rapidly subsiding basin margin. Sediment accumulation rates were very high (600 m/Ma) during deposition of the El Gallo Formation and the calculated tectonic component (86–161 m/Ma) of this subsidence is over five times

the rate of subsidence due to a rise in sea-level (17 m per My). The high sediment accumulation rates calculated for the El Gallo Formation are, therefore, primarily a result of tectonically induced subsidence along the fore-arc basin margin in Late Campanian time.

Although high subsidence rates and, to a much lesser extent, a eustatic rise in sea-level can account for the overall retrogradational pattern of deposition, several features of the El Gallo Formation can only be reasonably explained by other syndepositional tectonic events along the basin margin and in the source area. These features include:
1 high accumulation rates of coarse sediment in succession 1 relative to accumulation rates of comparable deposits reported in the literature;
2 abrupt lithological change between succession 1 and succession 2;
3 progradation of the fluvial system in succession 3 concomitant with an increase in plutonic and high-grade metamorphic fragments.

The high accumulation rate of succession 1 reflects a large accumulation potential created by tectonic subsidence. The abrupt lithological change between successions 1 and 2 is inferred to be the result of sediment trapping by downfaulting of the eastern basin margin along normal faults. Progadation and a change in sandstone composition recorded in succession 3 reflect an increase in sediment supply during accelerated uplift of the source area, which resulted in infilling of the half-graben. The sediment trapping and subsequent progradation are similar to that modelled for alluvial sedimentation in grabens by Alexander and Leeder (1987).

In summary, tectonic activity appears to have been the main control on deposition of the El Gallo Formation. Deposition of the El Gallo Formation was preceded by tectonic uplift of the fore-arc basin by approximately 400–2000 m and development of an angular unconformity, between about 80 and 75 Ma. By 75 Ma, subsidence of the basin had begun (succession 1). Subsequent downfaulting along the margin at 74.5 Ma resulted in coarse sediment being trapped closer to the source (succession 2) and structural disruption of the basin floor (Kilmer, 1963). Continued filling of the down-dropped area, combined with uplift in the source area, resulted in an influx of coarse sediment to the fluvial system (succession 3) by 73.6 Ma. Finally, a reduction in sediment influx during ongoing rapid subsidence resulted in tidal deposition (at the top of succession 4) and marine transgression by 73 Ma.

CONCLUSIONS

Stratigraphical, sedimentological and petrological analyses combined with $^{40}Ar/^{39}Ar$ single-crystal laser probe dating of interbedded tuffs throughout the El Gallo Formation in the El Rosario area of Baja California, Mexico, indicate that syndepositional tectonic activity had a significant effect on the sedimentation patterns along the Peninsular Ranges fore-arc basin complex in Late Campanian time. In particular, very high sediment accumulation rates (approximately 600 m/Ma) are recorded and appear to be primarily controlled by tectonic subsidence of the fore-arc basin. The compositional immaturity of sandstones in the El Gallo Formation indicate rapid uplift of the source area. Very limited leaching of the feldspar and unstable minerals by meteoric water is also indicative of rapid uplift and is consistent with a rapid sedimentation rate. Rapid uplift of the source region and rapid subsidence of the basin, shallow (3–5 m) fluvial channels, and palaeocurrent directions oriented perpendicular to the inferred plate boundary suggest deposition from a transverse fluvial system, which is characteristic of an active fore-arc setting. Four distinct lithological successions recognized in the El Gallo Formation appear to record initial alluvial plain sedimentation, recording subsidence that followed major uplift of the fore-arc basin, and subsequent basin-margin downfaulting, followed by a progradation of the fluvial depositional system (uplift in the source area) and, finally, a marine transgression.

Palaeomagnetic data and plate reconstruction models indicate that the plate boundary along the Peninsular Ranges fore-arc was likely undergoing highly oblique subduction in Late Campanian time (Engebretson *et al.*, 1985; Hagstrum *et al.*, 1985). Evidence from many modern fore-arc basins indicates that oblique convergence or regional scale extension may result in syndepositional faulting within fore-arc basins (Scholl *et al.*, 1980; Moberly *et al.*, 1982; Kanamori, 1986). Oblique subduction in Late Campanian time, therefore, was probably responsible for the tectonic uplift of the arc massif and high subsidence rates in the fore-arc basin recorded in the fluvial sedimentary rocks of the El Gallo Formation.

ACKNOWLEDGEMENTS

Acknowledgement is made to the donors of the Petroleum Research Fund, administered by the American Chemical Society and to British Petroleum Exploration Company Ltd, and to the Geological Society of America for support of this research. We are grateful to Knut Bjørlykke and James Boles for numerous discussions and thorough review of this manuscript. We thank Paul Renne for geochronological results and discussions in the field. Field discussions with James Boles, William Morris and Douglas Imperato were also most helpful. We thank Ray Torres and Steven Sutter for assistance in the field. Thorough reviews by J.P. Nyustem and A.C. MacDonald helped us to improve the manuscript substantially.

REFERENCES

Alexander, J. & Leeder, M.R. (1987). Active tectonic control on alluvial architecture. In: *Recent Developments in Fluvial Sedimentology. Contributions from the Third International Fluvial Sedimentology Conference* (Eds Ethridge, F. G. & Flores, R. M.) Spec. Publ. Soc. Econ. Paleont. Miner., Tulsa **39**, 243–252.

Allen, J.R.L. (1982). Mud drapes in sand-wave deposits: a physical model with application to the Folkestone Beds (Early Cretaceous, southeast England). *Proc. R. Soc. London, Ser. A* **306**, 291–345.

Allen, P.A. & Allen, J.R.L. (1990) *Basin Analysis: Principles & Applications*. Blackwell Scientific Publications, Oxford, 451 pp.

Beal, C.H. (1948). Reconnaissance of the geology and oil

possibilities of Baja California, Mexico. *Geol. Soc. Am., Mem.* **31**, 138 pp.

BEERBOWER, J.R. (1964). Cyclothems and cyclic depositional mechanisms in alluvial plain sedimentation. In: *Symposium on Cyclic Sedimentation* (Ed. Merriam, D. F.) Bull. State Geol. Surv. Kansas **169**, 31–42.

BJØRLYKKE, K. (1988). Sandstone diagenesis in relation to preservation, destruction and creation of porosity. In: *Diagenesis, I* (Eds Chlingarian, E.V. & Wolf, K.H.) Developments in Sedimentology **41**, 555–588.

BJØRLYKKE, K., RAMM, M. & SAIGAL, S.G. (1989). Sandstone diagenesis and porosity modification during basin evolution. *Geol. Rundsch.* **78**, 243–268.

BLATT, H., MIDDLETON, G. & MURRAY, R. (1980). *Origin of Sedimentary Rocks.* Prentice-Hall, Inc., New Jersey, 782 pp.

BOEHLKE, J.E. & ABBOTT, P.L. (1986). Punta Baja Formation, a Campanian submarine canyon fill, Baja California, Mexico. In: *Cretaceous Stratigraphy of Western North America* (Ed. Abbott, P.L.). Spec. Publ. Soc. Econ. Paleont. Miner. Tulsa **46**, 91–102.

BOTTJER, D.J. & LINK, M.H. (1984). A synthesis of Late Cretaceous southern California and northern Baja California paleogeography. In: *Tectonics and Sedimentation along the California Margin* (Eds Crouch, J.K. & Bachman, S.B.) Spec. Publ. Soc. Econ. Paleont. Miner. Tulsa **46**, 79–90.

BOWN, T.M. & KRAUS, M.J. (1987). Integration of channel and floodplain suites, I. Developmental sequence and lateral relations of alluvial paleosols. *J. Sedim. Petrol.* **57**, 587–601.

CUNNINGHAM, A.B. & ABBOTT, P.L. (1986). Sedimentology and provenance of the upper Cretaceous Rosario Formation south of Ensenada, Baja California, Mexico. In: *Cretaceous Stratigraphy of Western North America* (Ed. Abbott, P.L.) Spec. Publ. Soc. Econ. Paleont. Miner., Tulsa **46**, 103–118.

CANT, D.J. (1978). Development of a facies model for sandy braided river sedimentation: comparison of the South Saskatchewan River and the Battery Point Formation. In: *Fluvial Sedimentology* (Ed. Mial, A.D.) Can. Soc. Petrol. Geol. Mem. **5**, 627–639.

DICKINSON, W.R. & SEELY, D.R. (1979). Structure and stratigraphy of forearc regions. *Bull. Am. Ass. Petrol. Geol.* **63**, 2–31.

DICKINSON, W.R. & SUCZEK, C.A. (1979). Plate tectonics and sandstone compositions. *Bull. Am. Ass. Petrol. Geol.* **63**, 2164–2182.

DICKINSON, W.R., BEARD, L.S., BRAKENRIDGE *et al.* (1983). Provenance of North American Phanerozoic sandstones in relation to tectonic setting. *Bull. Geol. Soc. Am.* **94**, 222–235.

DORN, R.I. (1988). A rock varnish interpretation of alluvial-fan development in Death Valley, California. *Natl Geogr. Res.* **4**, 56–73.

ELLIOT, T. (1986). Siliclastic shorelines. In: *Sedimentary Environments and Facies* (Ed. Reading, H.G.) Blackwell Scientific Publications, Oxford, 615 pp.

ENGEBRETSON, D., COX, A. & GORDON, R.G. (1985). Relative motions between oceanic and continental plates in the Pacific Basin. *Geol. Soc. Am., Spec. Pap.* **206**, 59 pp.

ETHRIDGE, F.G., FLORES, R.M. & HARVEY, M.D. (1987). *Recent Developments in Fluvial Sedimentology. Contributions from the Third International Fluvial Sedimentology Conference* (Eds Ethridge, F.G., Flores, R.M. & Harvey, M.D.) Spec. Publ. Soc. Econ. Paleont. Miner., Tulsa **39**, (preface), 389 pp.

FILMORE & KIRSCHVINK, (1989). A paleomagnetic constraint on the Late Cretaceous paleoposition of northwestern Baja California, Mexico. *J. Geophys. Res.* **94**, 7332–7342.

FLORES, R.M. & PILLMORE, C.L. (1987). Tectonic control on alluvial paleoarchitecture of the Cretaceous and Tertiary Raton Basin, Colorado and New Mexico. In: *Recent Developments in Fluvial Sedimentology. Contributions from the Third International Fluvial Sedimentology Conference* (Eds Ethridge, F.G., Flores, R.M. & Harvey M.D.) Spec. Publ. Soc. Econ. Paleont. Miner., Tulsa **39**, 311–320.

FOLK, R.L. (1968). *Petrology of Sedimentary Rocks.* Hemphilil's Book Store, Austin, TX, 170 pp.

FRIEND, P.F., SLATER, M.J. & WILLIAMS, R.C. (1979). Vertical and lateral building of river sandstone bodies, Ebro Basin, Spain. *J. Geol. Soc. London* **136**, 39–46.

GALLOWAY, W.E. (1985). Meandering streams modern and ancient. In: *Recognition of Fluvial Depositional Systems and Their Resource Potential* (Eds Flores, R.M., Ethridge, F.G., Miall, A.D. & Galloway, W.E.) Soc. Econ. Paleont. Miner., Tulsa, Short Course **19**, 145–166.

GALLOWAY, W.E. (1989). Genetic stratigraphic sequences in basin analysis, I: architecture and genesis of flooding-surface bounded depositional units. *Bull. Am. Ass. Petrol. Geol.* **73**, 125–142.

GASTIL, R.G. & ALLISON, E.C. (1966). An upper Cretaceous fault-line coast. (abstr) *Bull. Am. Ass. Petrol. Geol.* **50**, 647–648.

GASTIL, R.G., PHILLIPS, R.P. & ALLISON, E.C. (1975). Reconnaissance Geology of the State of Baja California. *Mem. Geol. Soc. Am.* **140**, 170 pp.

GRAHAM, J.R. (1983). Analysis of the upper Devonian Munster Basin, and example of a fluvial distributary system. In: *Modern and Ancient Fluvial Systems* (Eds Collinson, J.D. & Lewin, J.) Int. Assoc. Sediment., Spec. Publ. **6**, 473–484.

HAGSTRUM, J.T., MCWILLIAMS, M., HOWELL, D.G. & GROMME, S. (1985). Mesozoic paleomagnetism and northward translation of the Baja Peninsular. *Bull Geol. Soc. Am.* **96**, 1077–1090.

HAQ, B.U., HARDENBOL, J. & VAIL, P.R (1987). Chronology of fluctuating sea levels since the Triassic. *Science* **235**, 1156–1167.

JACKSON II, R.G. (1978). Preliminary evaluation of lithofacies models for meandering alluvial streams. In: *Fluvial Sedimentology* (Ed. Miall, A.D.) Can. Soc. Petrol. Geol. Mem. **5**, 543–576.

KANAMORI (1986). Rupture process of subduction-zone earthquakes. *Annu. Rev. Earth Sci.* **14**, 293–322.

KILMER, F.H. (1963). Cretaceous and Cenozoic Stratigraphy and Paleontology, El Rosario Area. Unpublished PhD Dissertation, University of California, Berkeley, 149 pp.

KRAUS, M.J. & MIDDLETON, L.T. (1987). Fluvial response to Eocene tectonism, the Bridger Formation, southern Wind River Range, Wyoming. In: *Recent Developments in Fluvial Sedimentology. Contributions from the Third International Fluvial Sedimentology Conference* (Eds

Ethridge, F.G., Flores, R.M. & Harvey M.D.) Spec. Publ. Soc. Econ. Paleont. Miner., Tulsa **39**, 263–368.

Kuenzi, W.D., Horst, O.H. & McGehee, R.V. (1979). Effect of volcanic activity on fluvial–deltaic sedimentation in a modern arc-trench gap, southwestern Guatemala. *Bull. Geol. Soc. Am.* **90**, 827–838.

Lawrence, D.A. & Williams, B.P.J. (1987). Evolution of drainage systems in response to Arcadian deformation: the Devonian Battery Point Formation, Eastern Canada. In: *Recent Developments in Fluvial Sedimentology. Contributions from the Third International Fluvial Sedimentology Conference* (Eds Ethridge, F.G., Flores, R.M. & Harvey, M.D.) Spec. Publ. Soc. Econ. Paleont. Miner., Tulsa **39**, 280–287.

McKee, E.D., Crosby, E.J. & Berryhill, Jr., H.L. (1967). Flood deposits, Bijou Creek, Colorado, June 1965. *J. Sediment. Petrol.* **37**, 829–851.

McLean, J.R. & Jerzykiewicz, T. (1978). Cyclicity, tectonics and coal: some aspects of fluvial sedimentology in the Brazeau–Paskapoo Formations, Coal Valley area, Alberta, Canada. In: *Fluvial Sedimentology* (Ed. Miall A.D.) Soc. Petrol. Geol. Can. Mem. **5**, 441–468.

McPherson, J.G., Shanmugam, G. & Moiola, R.J. (1987). Fan-deltas and braid deltas: Varieties of coarse-grained deltas: *Bull. Geol. Soc. Am.* **99**, 331–340.

Miall, A.D. (1977). A review of the braided-river depositional environment. *Earth Sci. Rev.* **13**, 1–62.

Miall, A.D. (1978). Lithofacies types and vertical profile models in braided river deposits: a summary. In: *Fluvial Sedimentology* (Ed. Miall A.D.) Can. Soc. Petrol. Geol. Mem. **5**, 597–603.

Miall, A.D. (1985) Architectural-element analysis: a new method of facies analysis applied to fluvial deposits. In: *Recognition of Fluvial Depositional Systems and their Resource Potential* (Eds Flores, R.M., Ethridge, F.G., Miall, A.D. & Galloway, W.E.) Soc. Econ. Paleont. Miner., Tulsa, Short Course **19**, 33–81.

Moberly, R., Sheperd, G.L. & Coulbourn, W.T. (1982). Forearc and other basins, continental margin of northern southern Peru and adjacent Equador and Chile. In: *Sedimentation and Tectonics on Modern and Ancient Active Plate Margins Trench, Forearc Geology* (Ed. Leggett, J.K.) Geol. Soc. London Spec. Publ. **10**, 171–189.

Morris, W.J. (1974). Upper Cretaceous "El Gallo" Formation and its vertebrate fauna: In: *Geology of Peninsular California* (Eds Gastil, G. & Lillegraven, J.) 49th Ann. Mtng. Soc. Econ. Paleont. Miner., Pacific Sec. 60–66.

Morris, W.R., Busby-Spera, C.J. & Berry, K. (1987). Tectonic and eustatic controls on forearc sedimentation, upper Cretaceous Rosario Group, Baja California, Mexico. *Geol. Soc. Am., Abs.* **7**, 779.

Morris, W.R. & Busby-Spera, C.J. (1988). Sedimentological evolution of a submarine canyon in a forearc basin, upper Cretaceous Rosario Formation, San Carlos, Mexico. *Bull Am. Assoc. Petrol. Geol.* **72**, 717–737.

Morris, W.R. (1992). Sedimentology and stratigraphy of the Rosario embayment of the Peninsular Ranges forearc basin complex, Baja California, Mexico. Unpublished PhD Dissertation, University of California, Santa Barbara.

Morton, R.A. & Donaldson, A.C. (1989). Hydrology, morphology, and sedimentology of the Guadalupe fluvial–deltaic system. *Geol. Soc. Am. Bull.* **89**, 1030–1036.

Nemec, W. & Steel, R.J. (1984) Alluvial and coastal conglomerates: their significant features and some comments on gravelly mass-flow deposits. In: *Sedimentology of Gravels and Conglomerates* (Eds Koster, E.H. & Steel, R.J.) Can. Soc. Petrol. Geol. Mem. **10**, 1–31.

Pinter, N. & Fulford, M.M. (1991). Late Cretaceous basement foundering of the Rosario embayment, Peninsular Ranges forearc basin: backstripping of the El Gallo Formation, Baja California Norte, Mexico. *Basin Res.* **3**, 215–222.

Posamentier, H.W., Jervey, M.T. & Vail, P.R. (1988). Eustatic controls on clastic deposition I–conceptual framework. In: *Sea-level Changes: an Integrated Approach* (Eds Wilgus, C.K., Hastings, B.S., Posamentier, H., Van Wagoner, J., Ross, C.A. & Kendall, C.G.St.C.) Spec. Publ. Soc. Econ. Petrol. Miner., Tulsa, **42**, 109–124.

Puigdefàbregas, C. & Van Vliet, A. (1978). Meandering stream deposits from the Tertiary of the southern Pyrenees. In: *Fluvial Sedimentology* (Ed. Miall, A.D.) Can. Soc. Petrol. Geol. Mem. **5**, 469–486.

Reineck, H.E. & Singh, I.B. (1973). *Depositional Sedimentary Environments with Reference to Terrigenous Clastics.* Springer-Verlag, New York, 439 pp.

Renne, P.R., Fulford, M.M. & Busby-Spera, C.J. (1991). High resolution ^{40}Ar/^{39}Ar chronostratigraphy of the Late Cretaceous El Gallo Formation, Baja California Norte, Mexico. *J. Geophys. Res. Lett.* **18**, 459–462.

Rust, B.R. (1978). Depositional models for braided alluvium: In: *Fluvial Sedimentology* (Ed. Miall, A.D.). Can. Soc. Petrol. Geol. Mem. **5**, 605–626.

Scholl, D.W., Von Huene, R., Vallier, T.L. & Howell, D.G. (1980). Sedimentary masses and concepts about tectonic processes at underthrust ocean margins. *Geology* **8**, 564–568.

Schumm, S.A. & Khan, H.R. (1983). Experimental study of channel patterns. *Geol. Soc. Am. Bull.* **83**, 1755–1770.

Shuster, M.W. & Steidtmann, J.R. (1987). Fluvial-sandstone architecture and thrust-induced subsidence, northern Green River basin, Wyoming. In: *Recent Developments in Fluvial Sedimentology. Contributions from the Third International Fluvial Sedimentology Conference* (Eds Ethridge, F.G., Flores, R.M. & Harvey M.D.) Spec. Publ. Soc. Econ. Paleont. Miner., Tulsa **39**, 279–286.

Steckler, M.S. & Watts, A.B. (1978). Subsidence of the Atlantic-type continental margin off New York. *Earth Planet. Sci. Lett.* **4**, 1–13.

Steel, R., & Aasheim, S.M. (1978). Alluvial sand deposition in a rapidly subsiding basin (Devonian, Norway). In: *Fluvial Sedimentology* (Ed. Miall, A.D.) Can. Soc. Petrol. Geol. Mem. **5**, 385–412.

Turner, B.R. & Whateley, K.G. (1983). Structural and sedimentological controls of coal deposition in the Nongoma graben, northern Sululand, South Africa. In: *Modern and Ancient Fluvial Systems* (Eds Collinson, J.D. & Lewin, J.) Int. Assoc. Sediment. Spec. Publ. **6**, 457–472.

Vessel, R.K. & Davies, D.K. (1981). Nonmarine sedimentation in an active forearc basin. *Spec. Publ. Soc. Econ. Paleont. Miner., Tulsa* **31**, 31–45.

Williams, P.F. & Rust, B.R. (1969). The sedimentology of a braided river. *J. Sediment Petrol.* **39**, 649–679.

Spec. Publs Int. Ass. Sediment. (1993) **20**, 335–362

Basin-fill patterns controlled by tectonics and climate: the Neogene 'fore-arc' basins of eastern Crete as a case history

G. POSTMA,* A.R. FORTUIN† *and* W.A. VAN WAMEL‡

**Comparative Sedimentology Divison, Utrecht University,
Budapestlaan 4, 3508 TA Utrecht, The Netherlands;
†Centre of Marine Earth Sciences, Free University,
De Boelelaan 1085, 1081 HV Amsterdam, The Netherlands; and
‡Structural Geology Group, Utrecht University,
Budapestlaan 4, 3508 TA Utrecht, The Netherlands*

ABSTRACT

Detailed structural, stratigraphical and sedimentological data show that sedimentation in the Late Miocene 'fore-arc' basins of eastern Crete was controlled by two stages of compression separated by a stage of extension. The compressional stages culminated around the Middle/Upper Miocene boundary and around the Miocene/Pliocene boundary, respectively.

The 3rd-order sedimentary cycles in the basin fills appear to be in phase with major tectonic events resulting from plate tectonic processes in the Mediterranean region. Apart from an extra, 3rd-order sedimentary sequence in the Ierapetra Basin spanning the Early Tortonian, all other sedimentary sequences identified from the east Cretan basin fills are grossly correlatable with those found in other basins around the Mediterranean and with the global 3rd-order cycles.

During the culmination of compression at the Middle/Upper Miocene boundary, sediment transport and basin-fill patterns were directly controlled by folding and reverse faulting due to steepening of the drainage basin relief. The relief-steepening increased the slope's instability and caused rapid infill of available accommodation space with numerous landslides, mass-flow deposits, and fault-scarp derived breccias deposited in rapidly prograding alluvial-fan and fan-delta environments.

During the following extensional stage in Tortonian times, slope instability features are rare. Their absence is related to a reduction of the drainage basin relief due to progressive northwards rotation of fault-blocks. The resultant half-graben fills are strongly tabular, with alternating intervals of fine- and coarse-grained clastic sediments indicating alternating periods of deepening and progradation (parasequences). The texture of the half-graben sediments is more mature and grains are better rounded compared with the clastic sediments deposited during the peak of the previous compression.

During the compressional stage at the end of the Miocene, sedimentation is controlled again by progressive steepening of the basin relief. This is shown by basin-subsidence analysis and structural data, and is recorded in the basin fill by numerous synsedimentary subaqueous slide and mass-flow deposits. In contrast with the former compressional stage, the supply of terrigenous material is very limited, which may be related to an increasingly dry climate in Messinian times.

INTRODUCTION

The relationship between sea-level changes and clastic stratal patterns (systems tracts) has been recently discussed by Posamentier *et al.* (1988). Trends in stratal patterns of passive margins have been used to infer global sea-level fluctuations resulting in a Mesozoic–Cenozoic 3rd-order cycle chart which gives an indication of the relative change in global coastal onlap (e.g. Haq *et al.*, 1988; Van Wagoner *et al.*, 1990). Both the sequence stratigraphical concept and the cycle chart, conceived as a valuable predictive tool for sedimentary geologists, initiated a stream of thought-provoking discussions, regarding how to unravel the important independent variables of deposition rate, tectonics and eustasy (e.g. Burton *et al.*, 1987; Miall, 1991). For passive margins, tectonics have been simplified by assuming a simple linear or curvilinear function for tectonic subsidence allowing evaluation of the eustatic sea-level change from regional relative sea-level changes. For active margins, the tectonic component of the relative sea-level change is much more difficult, if not impossible, to estimate.

Nevertheless, sequence stratigraphical methods have been applied in tectonically active regions, e.g. along active plate margins (e.g. Vail *et al.*, 1990; Cloetingh, 1991; MacDonald, 1991). Systems tracts of active margins may deviate in thickness and timing from those of passive margins due to tectonics, as exemplified by studies of the rates and magnitudes of tectonic vertical movements in the late Cenozoic Banda Arc (Indonesia; De Smet *et al.*, 1989; Fortuin & De Smet, 1991). The latter authors show that tectonics may induce relative sea-level changes of the same order of duration as eustatic changes, but which are an order of magnitude larger. In the latter case, eustatic changes become subdued. Tectonically overprinted eustatic changes may become unrecognizable if the relative sea-level curve is out-of-phase with the eustatic curve.

We investigated the sequence stratigraphical and tectonic framework of the Neogene 'fore-arc' basins on eastern Crete (Fig. 1) with three objectives: (1) can sequence stratigraphical patterns of the eastCretan basins be correlated with global onlap–offlap patterns deduced from passive margins (Haq *et al.*, 1988); (2) to what extent are basin-fill patterns in tectonically active basins controlled by tectonics and to what extent by climate; (3) can correlation between tectonically-active basins be improved by using sequence stratigraphical methods?

To answer these questions, we elucidate the complex interplay between tectonics, sequential patterns and basin-fill processes with emphasis on the Late–Middle and Upper Miocene fill of the Ierapetra Basin, where a sequence of about 1700 m of Neogene sediments is preserved (Fortuin, 1977, 1978). The sequence stratigraphy of the Ierapetra Basin is compared with the basin-fill patterns of similar, but smaller basins of the Sitia district. The inferred relative sea-level changes for eastern Crete appear to be largely in phase with the eustatic changes inferred by Haq *et al.* (1988), but are an order of magnitude larger (Table 1).

TECTONIC SETTING OF THE 'FORE-ARC' BASINS OF CRETE

The island of Crete is situated along the southernmost bend of the east Mediterranean Hellenic Arc, a convergent zone extending between Yugoslavia and Turkey (Fig. 1). The Hellenic Arc and the Ionian Trench are associated with the northward subduction of ocean lithosphere of the African plate under the Aegean continental plate (McKenzie, 1978; Angelier *et al.*, 1982). Subduction must have started in the Late Oligocene/Early Miocene (cf. Meulenkamp *et al.*, 1988), or even earlier (De Boer, 1989) and is still continuing. Throughout the Neogene, the position of Crete has been in the fore-arc realm, between the subduction zone and the associated, southward migrating volcanic arc, which is presently 150 km north of Crete. Hence, a fore-arc-type basin evolution can be expected to dominate the Neogene geology of Crete.

However, Hall *et al.* (1984) pointed out many anomalies between the 'classical' fore-arc system of the Banda Arc and that of the Hellenic Arc. In particular, the lack of evidence for an imbricate or accretionary-wedge structure and the numerous steep normal-faults of Late Cenozoic age on Crete (Drooger & Meulenkamp, 1973; Le Pichon & Angelier, 1979; Angelier *et al.*, 1982; Mercier *et al.*, 1989) led to the general assumption that the Cretan 'fore-arc' basins are controlled mainly by extensional tectonics. Moreover, the evolutionary trend of the basin fills from continental to deep-marine deposits is not typical for a fore-arc basin.

Sedimentary and structural evidence for periods of compression during the Neogene is present on Crete, however, and tectonic structures indicating N–S to NE–SW oriented shortening were pointed out by, among others, Kopp and Richter (1983) and Fortuin and Peters (1984). Recently, Meulenkamp *et al.* (1988) again focused attention on structural and sedimentary evidence for brief periods of regional compression during the Late-Middle and Late Miocene on Crete. They postulated that alternating compressional and extensional tectonics may have been induced by the onset of a roll-back process at about the Middle/Late Miocene boundary, related to the final stages of the collision between Africa, Arabia and Turkey. The resultant lithospheric stretching in the Cretan Sea area would have generated a supracrustal slab, which slid southward over low-angle shear zones (Lister *et al.*, 1984). The sliding of the supracrustal slab may have

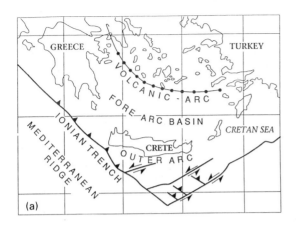

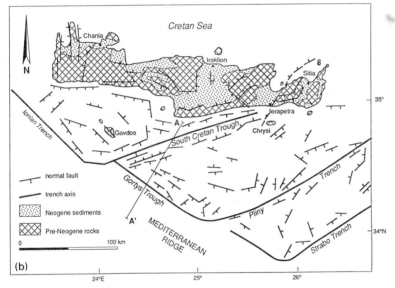

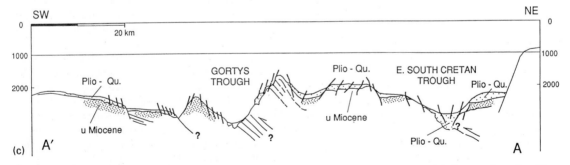

Fig. 1. (a) The Neogene basins of eastern Crete can be considered fore-arc basins because of their tectonic setting. The evolution of the basin fill, however, is not typical for that of a fore-arc basin (Hall *et al.*, 1984). (b) Geological sketch-map of the South Hellenic Arc showing the general distribution of pre-Neogene and Neogene rocks of Crete. The major fault systems have been indicated. (Modified from Fortuin and Peters (1984), Angelier *et al.* (1982) and Mascle *et al.* (1982).) A–A′ seismic profile is shown in (c). (c) Interpretation of a seismic reflection profile across the South Cretan Trough (A–A′). Jongsma *et al.* (1977) and Leité and Mascle (1982) considered the possibility that the offshore structure would be the result of compressional tectonics around the Mio-Pliocene boundary, followed by a period of extensional tectonics (Plio-Quaternary fill). Their tectonic framework matches perfectly with that inferred for the Ierapetra Basin (see Table 1). (Redrawn from Leité & Mascle, 1982.)

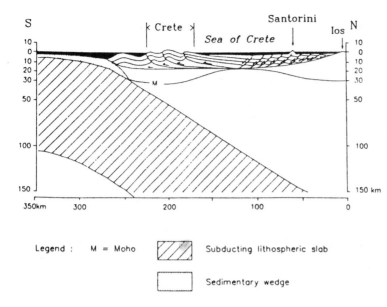

Fig. 2. Schematic N–S transect from Ios over Crete to the Hellenic trench, with a possible model of the tectonics associated with the southward movement (roll-back) of a supracrustal slab, which may have started around the Middle/Upper Miocene boundary. The position of the Moho (M) in the Aegean lithosphere is indicated schematically. (From Meulenkamp et al., 1988.)

generated compressional stresses on its southern edge, alternating with periods of zero transport of the slab, during which almost vertical gravitational stresses dominated and generated almost universally oriented extensional-stress patterns (Fig. 2).

We present further evidence for two periods of compression on eastern Crete: one which culminated at the transition of the Middle to Late Miocene and another at the transition of the Miocene to the Pliocene. The two compressional periods are each succeeded by a relatively long period of extension in the Tortonian and Pliocene/Quaternary, characterized by high subsidence rates in the order of 1–1.5 cm/10 years (below). The age of the most important folds and faults on land, in combination with an estimate of the total vertical movements, has been determined by geohistory analysis techniques (Van Hinte, 1978). Sedimentary and stratigraphical data from selected sections are combined with age and palaeobathymetric data based on the microfauna (plankton/benthos ratios). From graphs of age versus basin-depth and sediment thickness, the vertical movements of the underlying basement are approximated.

PRE-NEOGENE HISTORY OF CRETE

The Cretan pre-Neogene rock succession (Fig. 3) consists of a pile of non-metamorphosed rocks, which overlies metamorphosed rocks of the Phyllite–Quartzite (PQ) Unit (Creutzburg & Seidel, 1975; Baumann et al., 1976; Bonneau, 1982, 1984) and the Plattenkalk Series (Seidel et al., 1982).

The Plattenkalk Series, a well-bedded succession of limestone, is generally considered to represent the parautochtonous basement. The unit was deformed under considerable overburden (high-pressure–low-temperature (HP–LT) metamorphism) during a period of N–S compression at 25–16 Ma (Seidel et al., 1982). In the following Neogene period it underwent deformation due to N–S and NE–SW compression, which resulted in folding and reverse faulting (Meulenkamp et al., 1988). Its less complex deformational history compared with the overlying PQ Unit has already been noted by others (e.g. Hall & Audley-Charles, 1983; Richter & Kopp, 1983; Hall et al., 1984).

The PQ Series is mainly pelitic to phyllitic, with bodies of quartzite, gypsum, volcanics and also slices of metamorphics that must have been derived from a Hercynian basement (Seidel, 1978). The sequence is tectonic rather than stratigraphical, as it contains rocks of different age, origin, and grade of metamorphism (Creutzburg et al., 1977). Hall et al. (1984) consider the possibility that the PQ Unit first underwent E–W deformation in Eocene times and then N–S deformation in Early/Middle Miocene times (see also discussion in Robertson & Dixon, 1984, p. 47). It is noted here, that the N–S deformation in the PQ Unit is likely to be of the same age as the N–S deformation in the Plattenkalk

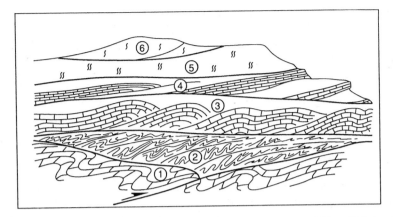

Fig. 3. The pre-Neogene rock succession, with a non-metamorphic succession about 10 km thick, overlying a metamorphic sequence of the Plattenkalk Series and Phyllite–Quartzite (PQ) Unit: 1, parautochtonous Plattenkalk Series (Ionian zone: Ida Sequence) HP–LT metamorphism; 2, Phyllite–Quartzite Series (LP–HT); 3, Tripolitza Limestone Series (dark micritic limestone of the Gavrovo Tripolitza Zone) and Eocene flysch; 4, Pindos–Ethia Series (deep-water limestones, chert and shales); 5, Asteroussia Series, system of ophiolites and limestones, locally intruded by granitic to granodioritic dykes and sills; 6, Ultra-mafic (UM) Series (including flysch with mafic volcanics — diabase, pillow lavas, spilites; ultra-mafic rocks — periodotites mainly serpentinized, metabasalts and andesite. (Redrawn from Bonneau, 1984.)

Series, and thus of late Middle to Upper Miocene age.

The non-metamorphic nappe pile consists of three units and comprises from base to top (Fig. 3; Baumann *et al.*, 1976; Bonneau, 1984; Hall *et al.*, 1984): (1) The Tripolitza Series, representing Triassic to Eocene platform limestones and Eocene flysch; (2) The Pindos Series, containing Triassic to Eocene deep-water radiolarian limestones and turbidites; and (3) the Asteroussia Complex, comprising, among others, recrystallized limestone together with Late Cretaceous igneous intrusions (Bonneau, 1982). We refer to the latter series as the Ultra-mafic Unit (UM Unit).

The Cretan nappes are generally thought to have been emplaced in Eocene–Early Miocene times. For further discussion on this subject, see Creutzburg (1958), Creutzburg and Seidel (1975), Baumann *et al.* (1976), Bonneau (1982, 1984), Hall *et al.* (1984), Robertson and Dixon (1984), Mercier *et al.* (1989), and De Boer (1989).

BIOSTRATIGRAPHY

The distribution of Neogene sediments on Crete is well known and shown on the geological map of Crete (scale 1:200 000) by Creutzberg *et al.* (1977). The map shows a complex pattern of tectonic culminations and depressions (simplified in Fig. 1), with the pre-Neogene basement outcropping as structural highs, and with Neogene and Quaternary preserved in the lows. In eastern Crete, Neogene basin fill starts with terrigenous, dominantly coarse-grained clastic units (? Middle Miocene), followed by open-marine, Late–Middle and Upper Miocene shelf, slope and basin-floor facies in the Ierapetra Basin (Fortuin, 1977, 1978), and by mainly shelf facies in the smaller basins of the Sitia district (Gradstein, 1973; Peters, 1985; Figs 4 & 5). The tectonic deformation, often complex in the Ierapetra Basin, is less complex in the Sitia district.

Due to the impact of tectonics, good basin-wide correlations of rock successions depend heavily on adequate biostratigraphical control. The biostratigraphical subdivision for the Neogene marine succession on Crete has been established by both planktonic foraminifera (Zachariasse, 1975; Zachariasse & Spaak, 1983) and calcareous nanofossils (Theodoridis, 1984; Driever, 1988) in combination with magneto-stratigraphy (Langereis, 1984; Langereis *et al.*, 1984). Although in general biostratigraphical resolution of the Upper Miocene is less well constrained than for the Lower Pliocene (cf. Spaak, 1983), various associations and datum levels enable further subdivision. The faunal changes within the group of keeled globorotaliids and *Neogloboquadrina* are especially useful. As shown by Zachariasse and Spaak (1983), a valuable datum level is the entry of *N. acostaensis*, which Tortonian age. In this paper the N-15 Biozones has been put at the top of the Serravallian stage (Zachariasse & Spaak, 1983). The next datum level is the shift in coiling direction of *N. acostaensis* from sinistral to dextral. Another important level is the dextral to sinistral shift in coiling direction of *N. acostaensis* in mid-Tortonian times, about 8 My ago (Sierro *et al.*, 1990; Table 1).

During the Tortonian–Messinian time span, three intervals with morphological variants of *Globorotalia menardii* can be distinguished in the open-marine facies (so-called *menardii* forms 3, 4 and 5 respectively; see Zachariasse, 1975; Langereis *et al.*, 1984). The last appearance datum levels of the sinistrally coiling *G. menardii* 4 species and the short range of the latest Tortonian dextral form-5 species in the basins of Crete were dated by means of detailed magneto-stratigraphical calibration (Langereis, 1984). The exit of form 4 is calibrated at 6.6 Ma, whereas the entry of the short-range form 5 is at 6 Ma. The beginning of the Messinian stage is characterized by the entry of *G. conomiozea* and is dated at 5.6 Ma (Langereis *et al.*, 1984). If we compare our biozonations with the cycle chart of Haq *et al.* (1988), a good correlation of the *G. menardii* form 3, 4 and 5 occurrences with the successive intervals of sea-level highstands is apparent (Table 1).

According to Zachariasse and Spaak (1983), the latest Miocene was a period of general contraction of the tropical faunal province, during which the Mediterranean planktonic foraminiferal fauna reflect the onset of a progressive cooling trend following the Mediterranean salinity crisis (Hsü *et al.*, 1977).

SEQUENCE STRATIGRAPHICAL AND TECTONIC FRAMEWORK OF THE IERAPETRA BASIN

The geology of eastern Crete, presented in Fig. 4, is based on earlier work of Gradstein (1973), Creutzburg *et al.* (1977), Fortuin (1977), Fortuin & Peters (1984), Peters (1985), and this paper. Offshore, the structure of the Hellenic Arc has been investigated through single-channel seismics and Sea-Beam (Rabinowitz & Ryan, 1970; Jongsma *et al.*, 1977; Angelier *et al.*, 1982; Leité & Mascle, 1982; and others). The gently southward-dipping

(4°) continental slope of the Hellenic Arc consists of northward-dipping crustal blocks bounded by the Hellenic Trench, the South Cretan Trough, and the Pliny and Strabo Trenches (see Fig. 1).

In eastern Crete, the identification and biostratigraphical dating of roughly E–W running sinistral oblique-slip zones and folds are important for the recognition of two compressional events in the Upper Miocene. In addition, we have identified important normal faults which must have accommodated the Tortonian extension. The faults and folds which controlled the basin relief during the Late Miocene have been mapped in detail in the Ierapetra Basin (Figs 5 & 6). Stratigraphical and sedimentological data (Fig. 7) were used to reconstruct the kinematics of these faults (Fig. 8).

The stratigraphy of the Ierapetra Basin is reasonably well known and was investigated in the early 1970s by Fortuin (1977, 1978). More recently, sedimentological aspects of the basin fill were studied in detail by Drinia (1989), Monogiou (1989), Postma (1989) and Postma and Drinia (1993). In the following account of the stratigraphy of the Ierapetra Basin, we investigate the role of tectonics on basin-wide sediment dispersal patterns for the following four periods:

1 the Early–Middle Miocene period (compression);
2 the period around the Middle and Late Miocene boundary (compression);
3 the period during the Tortonian (extension);
4 the period around the Miocene/Pliocene boundary (compression).

The Pliocene period, which is not discussed here, is again dominated by extension with northward tilting of fault blocks (Fortuin, 1978; Peters *et al.*, 1985; Meulenkamp & Hilgen, 1986; Meulenkamp *et al.*, 1988).

Early–Middle Miocene period

Sedimentary facies

Sediments assigned to the ?Early and Middle Miocene period cover the highest allochtonous units of the non-metamorphic nappe pile. They are nonmarine and poorly dated and are overlain by Upper Miocene sediments. Possibly contemporaneous with the continental deposits on Crete are marine Langhian sediments cored in the Strabo trench at the southernmost edge of the Aegean plate (DSDP core 129; Ryan *et al.*, 1973).

Fortuin (1977, 1978) identified two formations in

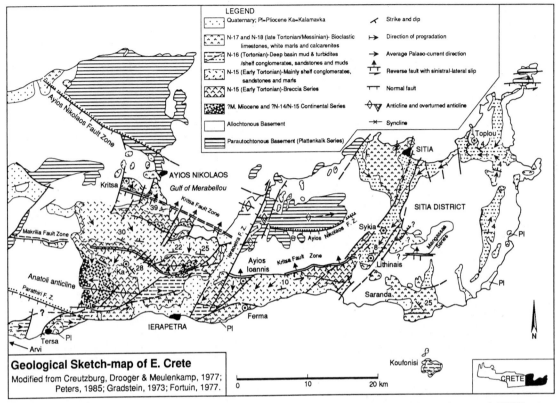

Fig. 4. Geological sketch-map of eastern Crete, based on Fortuin (1977), with data from Creutzburg *et al.* (1977), Fortuin and Peters (1984), Peters (1985), Gradstein (1973) and Meulenkamp *et al.* (1988). The map shows the names of the major faults and folds discussed in the text. Each of the named faults has its own signature. The pre-Neogene has been simplified into parautochtonous (Plattenkalk Series only) and allochtonous basement (see Fig. 3). The Neogene is subdivided according to sedimentary facies/age relationship.

the continental series: the Mithi and the Males formations. The latter overlies the Mithi Formation with an angular unconformity. The Mithi Formation probably represents immature, clastic alluvial-fan sequences, which cover a palaeorelief cut into the highest units of the UM Series. The clast composition, mainly igneous rockfragments stemming from the UM Series, varies laterally.

In contrast with the Mithi Formation, the younger Males Formation comprises mature alluvial sediments of quite different composition, with components derived mainly from the Pindos Series (Fortuin, 1977). The provenance of the characteristic white limestone pebbles is not certain, but Peters (1985) suggested that they may have been derived from the Pindic Kalimini unit from the neighbouring island Karpathos (Aubouin *et al.*, 1976). The westerly palaeocurrent directions measured in the

Ierapetra Basin and elsewhere on Crete throughout the more than 450-m-thick succession (Fortuin, 1977; Peters, 1985; Postma, 1989) are in agreement with this interpretation.

Tectonics

The consistent pattern of westerly palaeocurrent directions in the Males Formation suggests that the rivers were confined in approximately E–W running valleys paralleling the southern margin of the Aegean landmass (Fortuin, 1977). The river valleys, which were connected with the sea farther to the west, may have a width of approximately 20 km. Similar east–west palaeovalleys have been recognized from central Crete south of the Ida Mountains by Creutzburg (1958). The origin of the river valleys may be related to the onset of large-scale

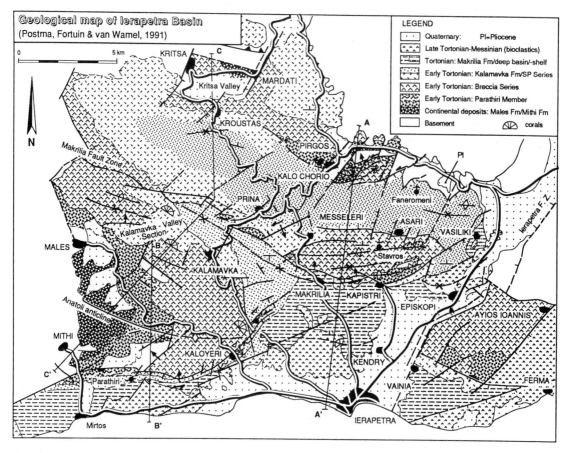

Fig. 5. Geological map of the Ierapetra Basin, modified from Fortuin (1977). The positions of three geological cross-sections (Fig. 6) are indicated on the map. Structural symbols are according to the legend in Fig. 4.

folding of the crystalline basement (Plattenkalk Series) during N–S compression, which started probably after the termination of the HP–LT metamorphism about 16 My ago. During the folding the Plattenkalk must have been at a crustal depth of more than 10 km, according to estimations of the total nappe-pile thickness and evidence for ductile deformation (pressure solution and newly formed platy quartz parallel to the northward dipping fold planes). The same limestones started to deliver detritus in the course of the N-16 Zone (Kalamavka Formation), after significant amounts of Tripolitza material had been supplied to the basin during the compressional period around the Middle/Late Miocene boundary (progressive unroofing of the nappe pile).

The period around the Middle/Late Miocene boundary (compression)

Sedimentary facies

Towards the top of the Males Formation, shallow-marine sandy to gravelly deposits of the Parathiri Member (Fortuin, 1977) onlap over Males flood plain/coastal plain sediments. Deposition occurred in estuarine, coastal and deltaic environments. Stratigraphically higher in the Parathiri section (for locality of the section see Fig. 5), a gradual deepening of the depositional environment becomes apparent by the change from brackish into open-marine fauna. The latter are contained in silty to clayey sediments (Fig. 7).

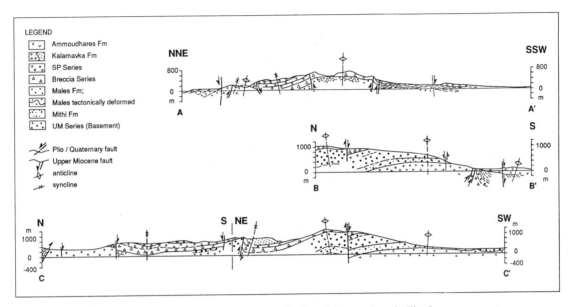

Fig. 6. Geological cross-sections (A–C) of the Ierapetra Basin. Localities are given in Fig. 5.

The brackish-water fauna contain abundant *Terebralia bidentata* and *Ostrea* shells which are not age diagnostic (see also Böger & Willmann, 1979). The marine intercalations at the top of the sequence contain *Globigerina continuosa* (Zachariasse, 1975; Fortuin, 1977; and personal observation), which indicates a Late Serravallian/Early Tortonian (N-15 Zone) age for the end of the Parathiri period. The calcareous nannofossil association from these intervals can be correlated with biozone NN-8 (S. Theodoridis, personal communication), which means that the Parathiri Member coincides with the N-15 Biozone (Haq *et al.*, 1988).

After the onset of the marine transgression, a gradual change in composition occurred through the arrival of immature dark-blue, micritic limestone breccia of the Tripolitza Series. We relate the change in composition to the onset of important tectonic uplift and brittle deformation along the Kritsa Fault Zone (Figs 4 & 8A; see further below). In the south near Parathiri, the change in composition occurred in the shallow-marine realm, whereas in the north it occurred in the continental sediments of the Males Formation. Near the Kritsa Fault Zone in the northern part of the basin (Kritsa Valley and Gulf of Mirabellou) sheared relics of the Mithi Formation are found covered by a thin sequence of brackish-water limestone beds probably equivalent to the top of the Parathiri Member.

Culmination of compressional tectonics — the Breccia Series

The deposits of the Males Formation, including those of the Parathiri Member, are widely covered by the *Breccia Series*, a series of slides, slumps and debris-flow deposits containing exotic large Tripolitza blocks (Figs 9 & 10) and Mithi, Males and brecciated Tripolitza limestone material (Fig. 11). Massive blocks and finer detritus derived from the underlying basement (UM Series) are found near localities where the Mithi and Males sediments had been largely eroded: i.e. in the Kritsa Valley and in the area around Pirgos and Kapistri (NB: the basement east of Pirgos and north of Kapistri consists of crystalline rocks which belong to the Asteroussia Nappe, and not to the Tripolitza Nappe as suggested by the map of Creutzburg *et al.*, 1977). Hence, near the fault zone, the Breccia Series overlie older sediments by an important erosional unconformity, whereas away from the fault zone, in the Parathiri section, the same boundary is a conformity.

Near the Kritsa Fault Zone the Breccia Series reach thicknesses of up to 400–500 m locally (Fig. 7; Pirgos section). The Breccia Series are deposited in alluvial cones and fans which had their apex against the Kritsa Fault Zone (Fig. 8A). Their thickness decreases basinwards (towards the south),

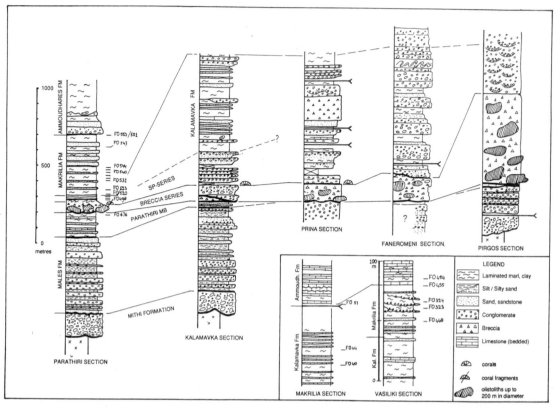

Fig. 7. Stratigraphical columns of the Ierapetra Basin with position of samples quoted in the text. Along a SSW–NNE transect roughly parallel to the axis of main deformation with sections near Parathiri, Kalamavka, Prina, Faneromeni and Kalo Chorio (see Fig. 5). Plankton/benthos ratios for subsequent samples in the Parathiri section: FO498, 64%; FO502, 79%; FO506, 72%; FO512, 78%; FO520, 71%; FO523, 64%; FO531, 84%; FO533, 90%; FO540, 86%; FO542, 94%; FO544, 86%; FO547, 81%; FO551, 58%; FO553, 80%. Inset shows sections near Makrilia and Vasiliki.

although large blocks and breccias occur as far south as in the Parathiri section and near the village of Anatoli, which is about 20 km away from the Kritsa Fault Zone. Near Ayios Ioannis (Fig. 9), there is evidence for south-sloping breccia fans, which interfinger with the west-flowing 'Males' rivers.

Tectonics

The Kritsa Fault Zone (Fig. 4) separates rocks of the highest allochtonous UM Series from the Tripolitza Series. The fault zone most probably continues from northeast of Ierapetra, where it is offset by the younger Ierapetra Fault Zone, along the south coast further down to the east. Near Lithinais, the zone possibly continues where the Tripolitza Series abuts against the allochtonous Mangassa Series (Peters, 1985). Fault movements have been measured to be

reversed, with mainly sinistral-oblique movements along the south coast (near Skinokapsala), and with only a small oblique-slip component in the Kritsa Valley. The Mithi and Males Formations and underlying basement are increasingly folded and otherwise deformed towards the fault zone (Kritsa Valley and east of Kalo Chorio). Sedimentary features of the Breccia Series point to significant slope instability, which resulted in mainly gravitational sediment transport and cannibalism of the Mithi and Males Formations (see also Fortuin & Peters, 1984), in particular in the region near the fault zone. The increase in slope instability must be a direct result of the reverse faulting through steepening of the palaeorelief. The erosional unconformity underlying the Breccia Series is thus directly related to tectonics.

The age of the Breccia Series and its underlying unconformity must correspond with the boundary

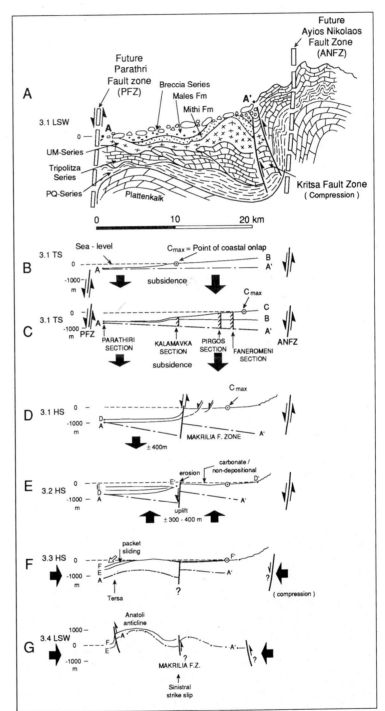

Fig. 8. Reconstruction of fault kinematics based on geohistory analysis techniques. The palaeobathymetry is based on plankton/benthos ratios and sedimentological data (see Fig. 7). The various events, A–G, are correlated with the sequence-chronostratigraphy of Haq *et al.* (1988); see Table 1. The transect runs approximately south–north across the sections near Parathiri, Kalamavka, Prina, and Kritsa Valley (approximately parallel with cross-section 6C). HS, highstand systems tract; LSW, lowstand wedge; TS, transgressive systems tract.

between the N-15 and N-16 Biozones. The Breccia Series are underlain by the Parathiri Member and overlain by a thick series of continental and marine sediments which belong to the early Tortonian (Prina Complex and Kalamavka Formation, Fig. 7 and Table 1).

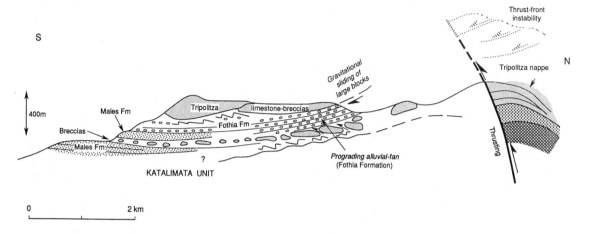

Fig. 9. North–south cross-section through the Late Serravallian–Early Tortonian thrust masses just south of the eastern continuation of the Kritsa Fault Zone near the village of Ayios Ioannis. The Males conglomerates in the south have west-southwest current directions. They interfinger to the north with slides, large blocks (10 m) of crushed and brecciated Tripolitza limestones and breccias. These series are equivalent with the Breccia Series of the Ierapetra Basin and must have originated from the advancing thrust zone, against which they formed south-facing, steep alluvial-fan or scree surfaces. Towards the fault contact, the Males Formation becomes increasingly more deformed, and occurs as large tectonic lenses mixed with pre-Neogene. The Breccia Series are covered by more mature alluvial-fan conglomerates (Fothia Formation, which is equivalent to the Stratified Prina Series and Kalamavka Formation), which accumulated mainly during the Tortonian extension.

In conclusion, we think that the entire nappe-pile underwent progressive deformation in late Middle Miocene times during the Males River period as a result of folding of the underlying crystalline basement (Plattenkalk Series). The east–west folding culminated into intense brittle deformation with reverse faulting in roughly east–west running oblique-slip zones around the Middle/Late Miocene boundary (Fig. 8A). There is clear evidence that the marine transgression of the Parathiri Member pre-

Fig. 10. Detail of an allochtonous 20-m slab of Tripolitza limestone overlying Males conglomerates on the peninsula of Vrionisi (Gulf of Merabellou, Ierapetra Basin). Slickensides indicate sliding of the block from north (15° E) to south (from right to left on the photograph). The block must have been deposited very close to the Kritsa Fault Zone (see maps of Figs 4 & 5).

Fig. 11. The homogeneous, micritic limestone of the Tripolitza Series has an extremely brittle behaviour under stress. The fractures and joints are closely spaced resulting in a completely shattered rock, which easily falls apart into grit and breccia.

ceded the reverse faulting and the related erosional unconformity. The transgression of the Parathiri Member continued during the reverse faulting until the deposition of the Breccia Series.

Following the chronostratigraphical chart of Haq *et al.* (1988), the sea-level rise for the Parathiri Member corresponds to the eustatic sea-level rise 2.6 HS (top Parathiri Member is in both the N-15 and NN-8 zones; Table 1). The relative sea-level rise itself must have been in excess of 150 m (the total thickness of the Parathiri Member without allowing for compaction) and, therefore, is likely to have a tectonic component in addition to the eustatic component. The tectonic component may well originate from the tectonic compression causing synclinal downwarping of the basin. The regression during the Breccia Series reflects both tectonic uplift and rapid progradation due to the increase in sediment supply. The erosional unconformity at the base of the Breccia Series represents a type-1 sequence boundary, whose age is close to the 9.8 Ma type-1 sequence boundary (SB) of Haq *et al.* (1988).

The period during the Tortonian (extension)

Early Tortonian Sedimentary facies

The Breccia Series is covered in the south (near Parathiri) by a *c.* 50-m-thick, condensed marine succession, and in the north between Pirgos and Vasiliki (Fig. 5) by an up to 1000-m-thick series of continental and marine deposits (Fig. 7). Palaeocurrent directions are dominantly towards S–SW throughout the succession. The sedimentary succession is conspicuously tabular bedded (Fig. 12) and contains approximately 10–12 major sedimentary (para-) sequences, each resulting from deepening and successive infilling of the basin (Postma & Drinia, 1993). Near the village of Kalamavka, the succession is approximately 600 m thick, gravelly in the lower part, and sandy in the upper part. The gravelly lower part was referred to as the 'upper structural level of the Prina Complex' by Fortuin and Peters (1984) and as the 'stratified breccias of the Prina Complex' by Fortuin (1977). In this paper, we refer to these series as *Stratified Prina Series* (SP Series). The marly to sandy upper part is called the *Kalamavka Formation* (Fortuin, 1977). The basin-fill architecture and facies enable a detailed reconstruction of the fault kinematics during the Early Tortonian (Postma, 1991). Below, we give a brief account of the sedimentology of the SP Series and Kalamavka Formation along sections that roughly represent a SSW–NNE line, conforming to the direction of maximum tectonic deformation (see Fig. 7).

The condensed sequence of the *Parathiri section* comprises *c.* 20 m of shallow-marine sandy facies, largely reworked by gravitational processes, and 30 m of marls, with microfauna indicating rapid deepening (up to 400 m or more). Planktonic foraminifera at the top of the sequence contain the shift in coiling direction of *N. acostanaesis* (see Table 1) from sinistral to dextral.

In the *Kalamavka section*, the parasequences of the SP Series are up to 32 m thick, retrogradationally stacked and each crudely coarsening upward (Fig. 13). The base of each sequence is generally fine-grained, and contains marly to fine sandy, often

Fig. 12. View of the base of the Stratified Prina Series in the valley north of the village of Kalamavka to show coarsening-upward trends of parasequences up to 32 m thick. Sedimentary logs through this part of the parasequence set have been described by Monogiou (1989). Each parasequence is coarsening upward from marl to coarse gravels and represents a period of delta progradation.

bioturbated beds. The fine-grained base grades upwards into massive and graded sandstone beds with evidence for wave reworking. These beds are overlain by moderately-sorted pebble conglomerates which are occasionally wave-reworked (symmetrical gravel-ripples and swaley cross-stratification). The pebble conglomerates are truncated by poorly-sorted boulder conglomerates of debris-flow origin and covered by stream-dominated alluvial fans and palaeosols. The boulder conglomerates occur as composite, up to 10-m-thick megabeds. The top of each sequence shows a rapid fining-upward tendency, and grades upwards and laterally into ma-

rine marls of the overlying sequence. Coral and stromatolite beds are found intercalated throughout the SP Series.

The above-described sedimentary sequence indicates deepening (marly sequences) and shallowing of the basin by episodic infill of coarse-clastic deltaic deposits. The coarsening-upward sequences resemble progradational sequences of shallow-marine, wave-dominated mouthbar deltas with a shoal-water profile (for delta classification, see Postma, 1990; Fig. 14). Similar deltas are described from ancient fan-deltas of the Lower Carboniferous of Spitsbergen (Kleinspehn *et al.*, 1984) and from

Fig. 13. Detail of a coarsening-upward sequence at the top of the Stratified Prina Series. At the base, distal delta-front sediments (see Fig. 14) with turbiditic sandstones (storm layers) and intercalated gravel layers. Towards the top, proximal delta-front sediments with sandstone and coarse-grained (bouldery) conglomerates.

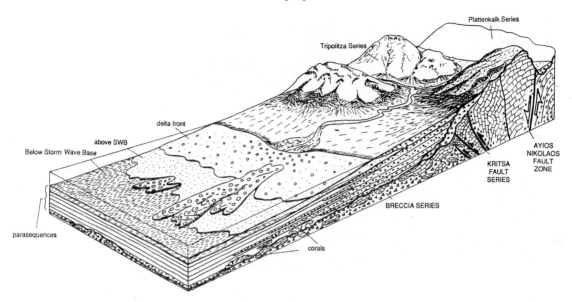

Fig. 14. Block diagram of the stacking of prograding shallow-water type deltas in the SP Series deposited during the N-15 Biozone.

the Montserrat fan-delta (Marzo & Anadon, 1988).

In the Kalamavka section, the SP Series is overlain by a thick marly interval with planktonic foraminifera, including *Globorotalia menardii* (form 3; sample FO 103 from Fortuin, 1977). Benthic foraminifera contained in the marly intervals indicate significant deepening, possibly to depths of 400–900 m (Speelman, 1984; J. A. Manuputty, personal communication and G. J. Van der Zwaan, personal communication). Coral and stromatolites are lacking in the Kalamavka Formation. The sandy intervals consist of rhythmically bedded fine- to coarse-sandy turbiditic layers. Some of the sandstone levels are intensively burrowed and contain *Skolithos, Chondritus, Thalassinoides* and *Zoophycus* trace-fossils (Fig. 15). Towards the footwall (Ayios Nikolaos Fault Zone) and higher up in the Kalamavka succession, the rhythmically bedded sandstones are truncated by channelized boulder conglomerates (e.g. southwest of Vasiliki). The boulder conglomerates are often channelized and embedded in marls (north of Stavros and Asari in the Faneromeni section), and are associated with major submarine sliding (north of the Faneromeni monastery). The sedimentary sequences in the Kalamavka Formation are aggradationally and, at the top of the formation, progradationally stacked.

The deep-water deposits of the Kalamavka Formation, which are marly and sandy basinwards and marly to gravelly towards the footwall, resemble those described from deep-water fan-delta systems (Prior & Bornhold, 1988, 1990; Postma, 1990). The rhythmically-bedded sandstone resembles the sand-splays at the base of the slope, and the channelized boulder conglomerates the active delta slope. The associated shelf was probably very narrow or non-existent near Faneromeni.

In the *Pirgos section*, the Breccia Series are covered by a thick series of stream-dominated alluvial fans. Between Krousta and Prina, an approximately 600-m-thick succession with alluvial fans (belonging to the SP Series) is overlain by a coarse-grained clastic, mainly marine, succession, which has not been studied in detail but is probably time-equivalent to the Kalamavka Formation (Fig. 5). Palaeocurrent directions here vary from northwest over east towards southwest. The highest marl levels of the SP–Kalamavka succession near the Gulf of Merabellou contain *N. falconarae* and primitive, dextral *N. acostaensis* and *G. druryi*, which fall into the early part of the N-16 Biozone (see also next section).

Early Tortonian tectonics

The major normal fault zones (FZ) active during

(a)

(b)

Fig. 15. (a) Fourth-order sequences of the Kalamavka Formation show coarsening- and fining-, and thinning-, upward tendency from the base towards the top (Kalamavka section, along the road near the village of Kalamavka). (b) Rhythmical layering of turbiditic sandstones in 4th-order sequences of the Kalamavka Formation. The turbiditic sandstones probably represent base of slope sand-splays with episodic supply of sands (Kalamavka section, near the village of Kalamavka). Note the vertical *Skolithos burrows*, e.g. just under the hammer (35 cm). (Kalamavka section, along the road near the village of Kalamavka.)

this period of extension are the Ayios Nikolaos FZ and the Parathiri FZ (Fig. 4). The two fault zones delineate a block which comprises the whole of the Ierapetra Basin and includes the Kritsa FZ. The activity of these faults during the early Toronian period is inferred on the basis of (1) the presence of patches of Tripolitza breccia and conglomerates directly south of the Ayios Nikolaos FZ; (2) the significant offset along the Ayios Nikolaos FZ as inferred from the pre-Neogene and Neogene geology (see below; map of Creutzburg *et al.*, 1977); (3) the character of both fault zones, which is normal;

and (4) the absence of evidence for compression in the basin fill during this period. Through relaxation the orientation of the axes of lengthening (between N–NE and S–SW) may well have been similar to the axis of shortening during the preceding period of compression.

Basin subsidence analysis was done for the five logged sections in Fig. 7. A reconstruction of the tilting of the block between the Parathiri and the Ayios Nikolaos FZs has been done using palaeo-bathymetric data inferred from plankton/benthos ratios, and by establishing the palaeoshoreline position. These datum points and the preserved strati-graphical thicknesses were plotted in Fig. 8B–C using the upper surface of the Breccia Series (an excellent marker in the field) as a reference horizon. The obtained subsidence values are minimal, be-cause no correction was made for compaction. The reconstruction shows a significant north- to north-eastward tilting of the fault block during the early Tortonian period. The tilting controlled the archi-tecture and facies of the clastic wedge by a progres-sive reduction of the gradient of the alluvial feeder system.

The reduction in gradient not only leads to a gradual fining of the grain size with time, but also leads to a change in the rate of landward shoreline migration. This rate increases drastically if the gradient of the coastal plain reaches zero, and the timing of the shoreline 'jump' is marked in the stratigraphical column by a sudden fining in grain size and deepening of the basin. Continued subsid-ence then caused further deepening of the basin with sediment aggradation near the footwall. This progressive deepening may have been a reason behind the drowning of the coral and stromatolite beds present in the SP Series.

Near the village of Kalamavka, the effect of the shoreline 'jump' is clearest and occurred at the boundary of the SP Series and the Kalamavka Formation. Towards the footwall, the grain-size effect of the shoreline jump decreases and finally disappears north of the palaeoshoreline. Similarly, basinwards the abrupt change in lithofacies de-creases (cf. Parathiri section). Hence, only around Kalamavka village is there a conspicuous change in lithofacies which makes a distinction between the Prina Complex and the Kalamavka Formation possible (Fortuin, 1977, 1978).

Subsidence along the Ayios Nikolaos FZ must have been considerable. Based on the preserved sediment thicknesses and palaeorelief, a total offset of at least 1500–2000 m can be inferred (Fig. 8A–C). The subsidence must have decreased towards the end of the early Tortonian, probably due to the formation of a new important fault zone, named the Makrilia FZ (Figs 4 & 5).

In conclusion: for the early Tortonian period, we observe an onlap from south to north on top of the Breccia Series. The rapid fining and deepening in the Parathiri section indicates rapid transgression clearly related to extensional tectonics. This exten-sion may heralds the start of the 'roll-back' mecha-nism (Fig. 2), which led to the opening of the Cretan

Sea in the N-16 Biozone (cf. Drooger & Meulen-kamp, 1973; Angelier *et al.*, 1982). Both the north-ward tilting and the overall subsidence of the block in this outer-arc setting may have been caused by the southward sinking of the slab. The tectonically-induced transgression correlates with the 3.1 TR (transgressive systems tract) of Haq *et al.* (1988); see Table 1.

Middle-Tortonian period (N-16 Zone): sedimentary facies

A new basin-fill period is heralded by the gradual change from dominant calcareous clastic sediments (eroded from the Tripolitza and Plattenkalk Series in the north) into brownish, predominantly silici-clastic sands derived from the west. The apparent change in basin configuration is recorded also in the sedimentary development of the Makrilia Forma-tion (Fortuin, 1977, 1978). In the Ierapetra Basin, for example, only 50 m of shelf sediments were deposited north of the Makrilia Fault, and *c.* 400 m of deep-basin mud and turbidites accumulated south of the fault (Figs 7 & 8D). Palaeobathymetry is reconstructed with the help of microfauna and, in addition to sedimentary facies (below), aids in the reconstruction of kinematics of the Makrilia FZ.

Makrilia shelf

Below the chapel on the hill in the village of Prina, greyish homogeneous marls and calcarenites of the early Tortonian are overlain by about 50 m of marl with abundant *Globorotalia menardii*, form 4 (Tjalsma, 1971; Zachariasse, 1979), present in the top of the succession indicating a middle Tortonian age. This rapid transition suggests that much of the lower Tortonian succession, represented in the south by the Makrilia Formation (Figs 5 & 7), is absent. The succession is overlain by bioclastic packstones and marls assigned to the Ammoud-hares Formation (Figs 5 & 7B).

In the section near Vasiliki (Figs 5 & 7B), the Makrilia Formation is dominated by stacked chan-nel fills. Just below the channels is the transition from the early into the middle Tortonian period (see Table 1). The channel fills contain bioturbated marls, fine to coarse-grained sandstones and con-glomerate layers (Fig. 16; Fortuin, 1977). The benthic fauna from these marls indicate inner to outer shelf environment (J.A. Manuputty, personal communication). Just below the base of the overly-

Fig. 16. Stacked channels of the Makrilia Formation near Vasiliki. The channels probably represent 4th-order relative sea-level fluctuations.

ing Ammoudhares Formation, similar faunal assemblages as in the section below the chapel of the village of Prina, including *G. menardii* form 4, occur.

Makrilia turbidite basin

In the Parathiri section, deep-water conditions persist and a continuous record of the N-16 Biozone exists. The Kalamavka Formation grades gradually in the Makrilia Formation by the gradual disappearance of calcarenite detritus stemming from the north and the gradual dominance of the siliciclastic sands from the west. The basal 185 m of the section (see Fig. 7) is characterized predominantly by marls with minor sand influx (ratio 1:4). The middle part of the section is dominated by sandy and gravelly high-density turbidity current deposits, and the upper 50–75 m consists predominantly of marl. The topmost marls of the Makrilia Formation are homogeneous and characterized by somewhat pink colours and intercalated molluscs. This lithology, which also forms the top of the formation farther west and in the northern area (Vasiliki), may represent a period of shallowing, but with little deposition of land-derived materal. Further to the west, in the Arvi area, a similar trend is present, but with a notable increase in conglomeratic beds and outsized

marl clasts. Sediments become finer grained towards the east, where they have been found as far away as the island of Koufonisi (Peters, 1985).

The consistent eastward current directions suggest that the turbidites and marls represent an axial basin-fill, probably related to an eastward prograding delta-system characterized by a delta-fed submarine ramp system (e.g. Heller & Dickinson, 1985; Postma, 1990).

Foraminiferal assemblages of the Makrilia Formation in the Parathiri section are rich in benthic and planktonic fauna and have plankton/benthos ratios (in percentages) between 64 and 94% (Fig. 7). These suggest, in combination with the benthonic assemblages, upper- to middle-bathyal (between 500 and 1000 m) palaeobathymetry (J.A. Manuputty & B.J. van der Zwaan, personal communication; Van Marle *et al.*, 1987). The basal unit coincides with a significant deepening of the basin, as demonstrated by the upward increase of plankton/benthos ratios (from 64–79% in the Kalamavka unit to 78–94% in the Makrilia muds (Fig. 7). The marls in the upper part of the Makrilia Formation indicate shallowing plankton/benthos ratio 58–80%), with benthos indicative of an outer-shelf–upper slope setting. In the type section near Makrilia, just south of the fault zone, plankton/benthos type and ratios are similar to those found in the Parathiri section, but the section is less complete.

The age of the base of the Makrilia Formation in the Parathiri section conforms approximately to the first appearance of *Neogloboquadrina acostaensis* dextral. In the middle part of the Makrilia Formation, where turbiditic sands with easterly current directions are most abundant, sinistral *N. acostaensis* is most common. The top of the section, just below the more calcareous beds of the Ammoudhares Formation, contains abundant *Globorotalia menardii* form 4.

Middle Tortonian: tectonics

Towards the end of the deposition of the SP–Kalamavka succession, which is approximately at the N-15/N-16 boundary (Table 1), an important reorganization of fault-blocks occurred, accompanied by major changes in provenance and palaeo-relief. Subsidence south of the Makrilia FZ persisted throughout the middle Tortonian period, whereas no significant subsidence occurred north of the fault. This resulted in shelf and deep-basin environments during the course of the N-16 Biozone. Reconstructions of the palaeobathymetry show that basin subsidence south of the Makrilia FZ must have been in the order of 400 m.

The Makrilia FZ continues along Vainia and Ferma more to the east (Figs 4 & 5), where the sedimentary succession deposited during the middle Tortonian is incomplete and only 200 m in thickness. The very consistent easterly palaeocurrent directions indicate that the E–W basin axis was tilted towards the east. Also the change in detrital composition between the Kalamavka Formation and the Makrilia Formation (Table 1) to more terrigenous clastic sediments of the UM and PQ Series points to a change in source area due to a tectonically-induced change in basin relief.

In conclusion: both the reorganization of fault-blocks and the gradual change in provenance correspond with the transition from the early into the middle Tortonian. This transition corresponds with the highstand systems tract of the 3rd-order 3.1 cycle of Haq *et al.* (1988); Table 1. In our case, however, there is evidence for an additional sequence boundary, identified on the basis of the change in provenance and the fault-block reorganization. The nature of this boundary in the Ierapetra Basin is probably type 2 (conformable boundary). The turbiditic succession characterizing the middle

of the formation in the Parathiri section (Fig. 7) can be related to a period of renewed uplift (Fig. 8E) on the basis of basin subsidence analysis (shallowing in the Parathiri section) and correlates in time with the SMW (shelf-margin systems tract) of cycle 3.2 (*N. acostaensis* sinistral). The top of the section with *Globorotalia menardii* 4 four correlates with the highstand of cycle 3.2.

The period around the Miocene/Pliocene boundary (compression)

Late Tortonian/Messinian (N-17 Zone): sedimentary facies

Sediments belonging to the N-17 Biozone are characterized by calcareous (bioclastic) and marly sediments with abundant sponge needles. The basal part of the sequence still contains terrigenous clastic sediments, similar to the Makrilia Formation, although mixed with some bioclastic material. Upward in the sequence, a gradual change from predominantly lithoclastic to bioclastic sediments is evident. These sediments are grouped in the Ammoudhares Formation, which is estimated to be 100 m thick. Evaporites (gypsum and white limestones) deposited in the Late Messian belong to the Mirtos Formation (Fortuin, 1977, 1978). The description of the sedimentary facies of this part of the basin fill will show that sedimentation in the course of the N-17 Zone is increasingly controlled by important relief-steepening along the south coast and by aridification of the climate (cf. Chamley & Müller, 1991).

The entire Ammoudhares Formation underwent submarine sliding and slumping (Fortuin, 1977), so that the sections are incomplete (Fig. 17). Essentially, three units can be distinguished: (1) a *c.* 20-m-thick lower unit, rich in gravelly debris-flow deposits and sandy turbidites, overlain by (2) a *c.* 40-m-thick middle unit with thinly-bedded, horizontally-laminated, indurated brownish sands, passing upward into sandy calcarenites alternating with laminated or homogeneous marls, and (3) a *c.* 40-m-thick upper unit with the well-known Messinian facies consisting of whitish homogeneous chalky marls, rich in sponge needles, interbedded laminated sands and algal debris. The top of the sequence is covered by recrystallized, cavernous bioclastic limestones.

Benthic foraminifera present in the marls of the lower and middle units indicate open-marine, prob-

Table 1. Compilation of the litho- and biostratigraphical units of eastern Crete and their main characteristics and correlations with the chronostratigraphical and sequence stratigraphical schemes of Haq *et al.* (1988). The Mediterranean biostratigraphy is based on Zachariasse (1975), Berggren *et al.* (1985) and Sierro *et al.* (1990).

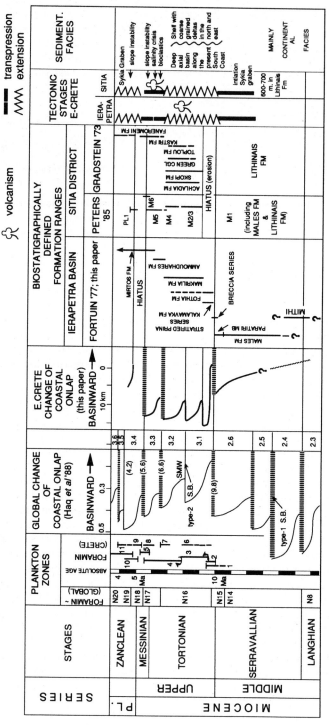

1, *N. falconarae*; 2, *N. acostaensis sinistral*; 3, *N. acostaensis sinistral*; 4, *N. acostaensis dextral*; 5, *G. conomiozea*; 6, *N. menardii* form 3; 7, *N. menardii* form 4; 8, *N. menardii* form 5; 9, *Sphaerodinellopsis*; 10, *G. margaritae*, 11, *G. puncticulata*.

Fig. 17. Slide packets of Ammoudhares sediments, exposed 2.5 km east of Arvi. The calcareous deposits of the Ammoudhares Formation slid over the marly top interval of the Makrilia Formation. The basal layers are strongly deformed. The view is to the west and the picture covers about 100 m.

ably bathyal sedimentation. The age of these marls is Late Tortonian/Early Messinian (N-17), based on the presence of *Globorotalia menardii* form 5 (dated at 6 Ma by Langereis *et al.*, 1984). The entry of the Messinian marker species *Globorotalia conomiozea* (dated at 5.6 Ma by Langereis *et al.*, 1984) can be observed in the upper unit.

The Lower Messinian strata of the middle unit along the south coast are dominated by southward-directed pebbly mudflow, slide and slump deposits, the latter with large (up to 10 m) olistoliths of algal limestones. Near Tersa, channel fills up to 10 m thick occur in muddy slope sediments. These fills contain gravelly to sandy mass-flow deposits with algal olistoliths. The abundance of gravitationally emplaced shelf sediment points to instability of the Ammoudhares shelf-edge in Early Messinian times. Large-scale, southward-directed 'packet' sliding of Ammoudhares sediments over marls of the underlying Makrilia Formation occurred at a later stage in the Early Messinian (Fig. 17; Fortuin, 1977, 1978).

Near Vasiliki and north of the Makrilia FZ slope, instability phenomena are virtually absent in the Ammoudhares succession. The fauna in the marls of the base and middle unit of the Ammoudhares Formation indicate open-marine, outer-shelf to upper-slope environments. The marls of the Ammoudhares Formation can be distinguished from those of the underlying Makrilia Formation by the change from pink into beige coloured marls and by the presence of abundant sponge spicules and algal debris. The rapid environmental change may be related to less restricted conditions of deposition

(Fortuin, 1977, p. 110) due to a relative sea-level rise. Going upward in the section, a distinct shallowing is evident. Thick bioclastic by the dominance of coarse-grained biogenic components in the limestone beds. Similar sections through the Ammoudhares Formation can be found along the Makrilia Fault near Ferma.

An *in situ* stratigraphy of the Late Messinian to Early Pliocene period is lacking. Gypsum deposits up to 25 m thick were found only as allochtonous units in slump, marl-breccia and debris-flow deposits up to 12 m thick along the road east of Mirtos. These chaotic deposits, which are the result of Late Messinian/Early Pliocene sliding (post-gypsum and marl-breccia with Lower Pliocene index species) (Fortuin, 1977, p. 118; see also Peters, 1985) mark the transition to Lower Pliocene marls and sands (Fortuin, 1977). The latter sediments contain *G. margaritae* — *G. puncticulata* of the N-19 Biozone. The unconformable, erosive contacts of the allochtonous units with the underlying displaced strata of the Ammoudhares Formation represent an important depositional hiatus of approximately 1 Ma around the Mio-Pliocene boundary.

Late Tortonian to Early Pliocene (N-17/N-18 Zone): tectonics

In the Late Tortonian to Early Pliocene interval (a period in the order of 1.5–2 Ma; Table 1), the successions along the south coast give ample evidence that sedimentation was progressively more controlled by slope-instability processes and deepening of the basin towards the south. The progressive increase of the instability of the south-facing slope (from small slides to packet sliding) suggests that the palaeorelief was enhanced by tectonic activity and progressive uplift in the north. In contrast, close to the Makrilia FZ and north of it, sedimentation was not controlled by slope-instability processes. After an initial deepening, a gradual shallowing of the basin is apparent, which indicates that movement along the Makrilia FZ may have become negligible at this time (Fig. 8F).

The compressional tectonics during the N-17 and N-18 period must have resulted in the present Anatoli anticline (Figs 4, 5, 6B–C & 8G). Allochthonous gypsum deposits and marl-breccias, the latter containing fauna of the N-18 Zone, indicate that the compression continued until the end of the N-18 Biozone, with the peak in tectonic activity around the Miocene/Pliocene boundary. The Anatoli anticline was submarine at the beginning of the N-17, and must have emerged after deposition of gypsum, therefore probably at the end of the N-18 Zone. The Breccia Series, used as the marker level in Fig. 8, forms a relief of approximately 1000 m (Anatoli anticline), which means that during the latest Messinian/earliest Pliocene period, the uplift was at least 1600 m in the region of Anatoli (Fig. 8F–G). Most of the uplift was probably the result of folding, but part of the compression must have been accommodated by sinistral-reverse oblique-slip faults, according to striations on fault planes of the reactivated Parathiri FZ in the south (Figs 5 & 6B). Simultaneously, the Makrilia FZ has been reactivated to become a sinistral-reverse oblique-slip zone in the Makrilia–Kapistri region, where it is associated with an anticlinal structure, similar to the Parathiri FZ and the Anatoli anticline (Figs 5, 6A & 8G). The first important dextral offset along the Ierapetra Fault also may be of this period.

Although clear evidence for relief steepening during the N-17 period exists (e.g. Anatoli anticline), the tectonic activity did not result in fresh input of land-derived material. This may point to deformation that occurred predominantly below sea-level, without significant emergence and erosion of the hinterland. On the other hand, the absence of land-derived material may also be related to the strong aridification of the climate since latest Tortonian times. The effects of aridification during the N-17 Biozone can be correlated around the Mediterranean and led towards the end of the Messinian to the well-known Mediterranean salinity crisis and the precipitation of gypsum and white and greyish

marls (Hsü *et al.*, 1973, 1977; Weijermars, 1988; Chamley & Müller, 1991).

In conclusion: the coarse clastic deposits and subsequent deepening at the base of the Ammoudhares Formation point to the presence of a sequence boundary (type unknown). The boundary corresponds closely with the onset of tectonic compression in the Ierapetra Basin. In time it corresponds with the type 1 sequence boundary between sedimentary cycle 3.2 and 3.3 of Haq *et al.* (1988). The depositional hiatus at the base of the Pliocene is associated with the important uplift in latest Messinian/Early Pliocene times, and corresponds in time with the global Early Pliocene lowstand (base of 3rd order cycle 3.4).

SEQUENCE STRATIGRAPHY AS A TOOL FOR CORRELATION

The many small, Neogene basins in the Sitia district (eastern Crete) differ from the much larger Ierapetra Basin by more simple tectonic deformation and by the lack of deep-water environments. Depositional environments range from continental to shelf, with condensed stratigraphical succession. Gradstein (1973) studied the basins mainly north of Sykia, and Peters (1985) studied the basins mainly south of that village (Fig. 4). The shallow-marine facies contains few age diagnostic microfauna, so that the biostratigraphical control is poor compared with that of the Ierapetra Basin. The aim of this correlation excercise is to investigate to what extent sequence stratigraphical tools can aid correlation in tectonically-active basins nearby.

Basin-fill history

The history of the basin-fills in the Sitia district is briefly summarized below from the work of Gradstein (1973), Peters (1985) and our own data. The stratigraphy of this area is given in Fig. 18, and a summary of the sequence stratigraphy is shown in Fig. 19. Table 1 shows the range in time of the various formations described below.

In the Ayia Saranda–Pervolakia area, alluvial-fan breccia-conglomerates (M1 in Figs 18 & 19) prograded southwards and interfinger with an axial, westward-draining fluvial system, which is compositionally and texturally similar to the Males Formation. In the Ayia Saranda area the M1 unit is more than 500 m thick, whereas in the Sykia–Ethia

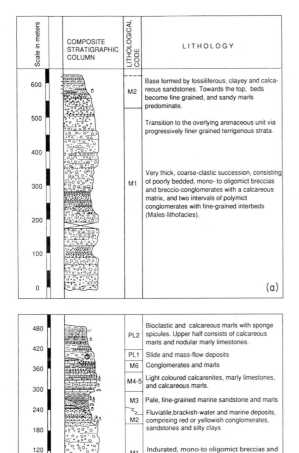

Fig. 18. Stratigraphical and sedimentary logs of the Saranda area and the Sykia–Ethia area. For the columns' positions within the sequence stratigraphical framework see Fig. 19. (Redrawn from Peters 1985.)

area the same breccia-conglomerates are only 200 m thick. The difference in thickness may be ascribed to differential subsidence along the eastern continuation of the Kritsa FZ during the continental? late middle Miocene period.

The *first* transgression (M2) is recorded by shallow-water brackish and later open-marine deposits which onlap towards the north. The approximately 100-m-thick succession overlies an erosional unconformity. It covers both the M1 unit, the supposed Kritsa FZ equivalent near Lithinais, and the pre-Neogene basement farther in the north

and west (Skopi Formation of Gradstein, 1973). The M2 unit itself is poorly dated and diachronous, but it interfingers and is covered by basin marls of the M3 unit (Achladia Formation of Gradstein, 1973), which contains fauna of the N-16 Biozone. It is evident that this first transgression (M2) represents a TR systems tract, which must represent either the sequence boundary at the base of the N-16 Zone or possibly the sequence boundary in the middle of the N-16 Zone (between cycles 3.1 and 3.2). A more precise correlation becomes possible after examination of the whole basin fill (below).

In the southwestern Sitia district and near Sykia, Gradstein (1973) described channel-fill conglomerates and Gilbert-type deltas up to 50 m thick, which truncate the Tortonian marls of the Achladia Formation (M3), and the clastic sediments from the underlying Skopi Formation (M2). The conglomerates have a typical greenish colour and consist of strongly-altered mafic components, quartzites, schists and carbonates, which may have been derived mainly from the PQ Series (Peters, 1985, p. 163). They are truncated by a *second* regional onlap and covered by a thin, coarse-clastic layer belonging to a 2nd TR systems tract. This tract is covered by marine marls and sandstones representing the 'second stage' of the Achladia Formation (Gradstein, 1973), which is also of Tortonian age. Similar Gilbert-type deltas are present in the area around Toplou. These deltas were fed by north-flowing rivers of the Kastri Formation (Gradstein & Van Gelder, 1971; Gradstein, 1973).

The change in composition and grain size recorded by the greenish conglomerates point to rejuvenation of relief, by either base-level lowering or tectonic uplift in the hinterland. Within the N-16 Zone this may correlate with the sandy middle part of the Makrilia Formation, i.e. the SMW systems tract between the 3.1 and 3.2 3rd-order cycles (Table 1). The Gilbert-type deltas may thus represent the lowstand wedge belonging to this sea-level lowstand phase with a type 1 unconformity. This means that the Skopi Formation and correlates with the base of the Makrilia Formation. The sea-level rise and fall during the N-15 Biozone, which is present in the Ierapetra Basin and on Gavdos (de Stigter, 1989), is represented here by a depositional hiatus.

The *third* transgression is recorded by the yellow to white coloured calcarenitic and bioclastic limestones and marls, which overlie the Achladia Formation in the north and centre of the Sitia district.

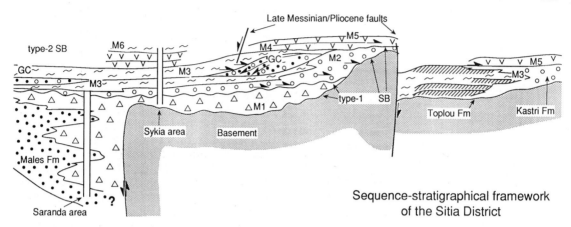

Fig. 19. Sequence stratigraphical interpretation of the Neogene of southeast Crete. Mainly based on data from Peters (1985) and Gradstein (1973).

The limestones are predominantly of shallow marine origin, with evidence for wave reworking and some mass-flow deposition. The thickness of the M4/5 unit (Faneromeni Formation of Gradstein, 1973) increases from Sykia towards the north and east from about 4 m to about 80 m. Syn-sedimentary sliding and mass flows are common in the central part of the formation. The M5 unit of Peters is characterized by predominantly bioclastic limestones. On top of the M6 unit, monomict mass-flow deposits occur near Ethia and on the island of Koufonisi. These are covered by marl-breccias and mass-flow deposits with gypsum clasts of latest Messinian/ earliest Pliocene age which accumulated in the centre of the northeast–southwest graben structure near Sykia and Lithinais (Peters, 1985, p. 166).

The abrupt change from Tortonian marls of the Achladia Formation into bioclastic limestones of the Faneromeni Formation is very similar to the change of the Makrilia Formation into the Ammoudhares Formation, although the striking change must possibly be attributed to an important regional climatic change. The occurrence of mass-flow deposits in the Faneromeni Formation and their progressive increase towards the top of the formation is strikingly similar (see Table 1), and points to regional tectonic uplift at the Mio-Pliocene boundary.

Tectonics during infilling of the Sitia district basins

The onset of the extensional stage in early Torto-

nian times is not shown in the Sitia district, and suggests that the Sitia district formed a palaeohigh causing considerable delay (one 3rd-order sedimentary sequence is missing) in marine onlap. After the onlap, the Sitia district remained a broad shelf throughout the Miocene separated from the deep basin in the south by the Makrilia FZ running east via Ferma south of the present coastline. The shelf situation did not change until the tectonic event around the Mio-Pliocene boundary, which is also in the Sitia district, demarcated by relief steepening and gravitational sediment transport into the Sykia graben.

The regional Mio-Pliocene uplift may be associated with the NNE–SSW oriented axis of shortening we noticed in the Ierapetra Basin. Some anticlinal deformation in the basement east of the Ierapetra Fault seems apparent. At the same time the NE–SW trending faults, which presently delineate the graben system of the Sitia district, may have been active during the tectonics around the Mio-Pliocene boundary. This view is supported by the palaeocurrent directions of the Early Pliocene mass flows, which accumulated in the centre of the graben.

Discussion

Apparently, the 3rd-order sedimentary sequences in the basins of the Sitia district are correlatable with the sequences in the Ierapetra Basin. An important tool for the correlation seems a careful comparison of sequence boundaries and transgressive systems tracts. Even in the shallow basins of east Crete, correlation by maximum-flooding surfaces or by highstand progradation was not feasible, because their characteristics are not unambiguous and the 3rd-order sedimentary sequences are too fragmented by later tectonics.

Another important point in favour of sequence boundary and TR tracts as correlative tools is that in tectonically active areas rates of uplift and rates of subsidence are high. The result is more pronounced in depositional hiatuses (sequence boundaries) and onlap series (transgressive systems tracts). The architecture of a half-graben fill is a typical example of the tectonic control on rate of onlap. In the case of the SP Series and Kalamavka Formation half-graben fill, rotation of the fault-block caused a progressive reduction of the gradient of the feeder system, which led to increased rates of shoreline migration to landward. In such cases, the TR systems tract can become very thick.

CONCLUDING REMARKS

We reconstructed and discussed coastal onlap and offlap patterns for the Ierapetra Basin in relation to tectonics. In the following paper we compared the basin-fill patterns of the Ierapetra Basins with the smaller basins of the Sitia district. If we compare our results of eastern Crete with the 3rd-order global curve of coastal onlap of Haq *et al.* (1988; Table 1), it becomes clear that a good match exists between our local and the global coastal onlap patterns apart from an extra sequence in the early Tortonian period. The in-phase relationship between these two curves is perhaps astonishing, because rates of sea-level change due to the tectonics are four to five times higher than eustatic sea-level changes; yet, a similar pattern emerges.

The tectonic events discussed for eastern Crete can be correlated with other basins around the Mediterranean Sea by comparing sequence stratigraphically. The architecture of basin fills is controlled by changes in base level and thus mainly by sea-level (Posamentier *et al.*, 1988). Hence, a good match of basin-fills around the same active ocean margin means that tectonic events are largely in phase (see also Cloetingh, 1988). For some well-studied basins along the northern margin of the Mediterranean Sea this appears to be the case: e.g.

in southeastern Spain (Weijermars *et al.*, 1985; Kleverlaan, 1989; Sierro *et al.*, 1990), and in southern Italy (Meulenkamp & Hilgen, 1986). The good correlation of the 'Mediterranean sea-level curve' and the global curve of Haq *et al.* (1988) is not surprising, because the Mediterranean had open seaways in the east and west with the Indian and Atlantic Ocean, respectively, until the salinity crisis at the end of the Messinian (Hsü *et al.*, 1973. Weijermars, 1988).

A good match of the sequence stratigraphical frameworks of basins of similar ocean setting can only be achieved if the framework for reconstruction is based on a large enough basin area comprising several fault-blocks. This will avoid misinterpretations of relative sea-level curves which are seemingly out-of-phase due to individual fault-block movements. An example may be the quite different successions preserved at the north and south side of the Makrilia FZ (Fig. 8). Similarly, erosional and non-depositional hiatuses may result in omissions of entire 3rd-order sedimentary sequences, as demonstrated by the rare preservation of the 3rd-order sedimentary sequence of the early Tortonian which is not indicated on the cycle chart of Haq *et al.* (1988).

Along active plate boundaries, the processes controlling sediment diffusion and sedimentary facies define the ultimate thickness and geometry, of parasequences, and depend on many variables of which the most important ones seem to be climate and tectonics. Climate regulates the timing of land-derived sediment transport, the production of carbonate clastics and evaporites, and tectonics controls the basin relief, the preservation potential and the sediment yield.

ACKNOWLEDGEMENTS

We would like to thank J.W. Zachariasse, B.J. van der Zwaan, J.E. Meulenkamp, E. Hogerduyn-Strating, and J.A. Manuputty for discussions. Fieldwork was supported by funds from the University of East Anglia (Norwich, UK), the Rijks University of Utrecht and the Free University (Amsterdam). The text figures were drafted by A. Trappenburg, J.J. van Bergenhenegouwen and I.M. Santoe. N. Pepping and editorial reviewer D.I.M. MacDonald are thanked for their comments on an earlier version of the manuscript.

REFERENCES

ANGELIER, J., LYBERIS, N., LE PICHON, X., BARRIER, E. & HUCHON, P. (1982) The tectonic development of the Hellenic Arc and the sea of Crete: a synthesis. In: *Geodynamics of the Hellenic Arc and Trench* (Eds Le Pichon, X., Augustithis, S.S. & Mascle, J.) *Tectonophysics* **86**, 159–196.

AUBOUIN, J., BONNEAU, M. & DAVIDSON, J. (1976) Contribution à la géologie de l'arc égéen: l'île de Karpathos. *Bull. Soc. Géol. France* (7) **XVIII**, 385–401.

BAUMANN, A., BEST, G., GWOSDZ, W. & WACHENDORF, H. (1976) The nappe pile of eastern Crete. *Tectonophysics* **30**, 33–40.

BERGGREN, W.A., KENT, D. V. & VAN COUVERING, J.A. (1985) The Neogene: Part 2, Neogene geochronology and chronostratigraphy. In: *The Chronology of the Geological Record* (Ed. Snelling, N.J.) Mem. Geol. Soc. London **10**, 211–260.

BÓGER, H. & WILLMANN, R. (1979) Die limnischen Gastropoden aus dem Neogen von Chersonisos (Kreta). *Ann. Géol. Pays Hell.* série (I), 159–162.

BONNEAU, M. (1982) Evolution dynamique de l'Arc Egéen depuis le Jurassique supérieur jusqu'au Miocène. *Bull. Soc. Géol. France* **19**, 87–102.

BONNEAU, M. (1984) Correlation of the Hellenide nappes in the south-east Aegean and their tectonic reconstruction. In: *The Geological Evolution in the eastern Mediterranean* (Eds Dixon, J.E. & Robertson, A.H.F.) Spec. Publ. Geol. Soc. London **17**, 517–525.

BURTON, R., KENDALL, C.G.ST.C. & LERCHE, I. (1987) Out of our depth: on the impossibility of forthcoming eustasy from the stratigraphic record. *Earth Sci. Rev.* **24**, 237–277.

CHAMLEY, H. & MÜLLER, D.W. (1991) Clay mineralogy in southeast Spain during the Late Miocene: climatic, palaeoceanographic and tectonic events in the Eastern Betic seaway. *Geol. and Mijnbouw* **70**, 1–120.

CLOETINGH, S. (1988) Intraplate stresses: a new element in basin analysis. In: *New perspectives in Basin Analysis. Frontiers in Sedimentary Geology* (Eds Kleinspehn, K.L. & Paola, C.) Springer-Verlag, New York, pp. 205–230.

CLOETINGH, S. (1991) Tectonics and sea-level changes: a controversy? In: *Controversies in Modern Geology: a Survey of Recent Developments in Sedimentation and Tectonics* (Eds Müller, D., Weissel, H. & McKenzie, J.) Academic Press, London, pp. 249–277.

CREUTZBURG, N. (1958) Probleme des Gebirgsbaues und der Morphogenese auf der Insel Kreta. *Freiberg. Universität. Neue Folge* **26**, 47 pp.

CREUTZBURG, N. & SEIDEL, E. (1975) Zum Stand der Geologie des Präneogens auf Kreta. *Neues Jahrb. Geol. Palaeont. Abh.* **149**, 363–383.

CREUTZBURG, N., DROOGER, C.W. & MEULENKAMP, J.E. (1977) *Geological map of Crete (scale 1:200,000)*. Inst. Geol. Min. Expl. (IGME), Athens.

DE BOER, J.Z. (1989) The Greek enigma: is development of the Aegean orogeny dominated by forces related to subduction or obduction? *Mar. Geol.* **87**, 31–54.

DE SMET, M.E.M., FORTUIN, A.R., VAN HINTE, J.E. & TJOKROSAPOETRO, S. (1989) Late Cenozoic vertical movements of non volcanic islands in the Banda Arc area. In: *Proceedings of the Snellius II Symposium Jakarta, 1987* (Eds

Van Hint, J.E., Van Weering, Tj. C.E. & Fortuin, A.R.). *Neth. J. Sea Res.* **24**, 263–275.

DODD, J.R. & STANTON JR., R.J. (1991) Cyclic sedimentation in three Neogene basins of California. In: *Sedimentation, Tectonics and Eustasy — Sea-level Changes at Active Margins* (Ed. MacDonald, D.I.M.) Ass. Sedimentol. Spec. Publ. **12**, 201–216.

DRIEVER, B.W.M. (1988) Calcareous nanofossil biostratigraphy and paleo-environmental interpretation of the Mediterranean Pliocene. *Utrecht Micropaleont. Bull.* **36**, 245 pp.

DRINIA, H. (1989) Shallow marine, sandy carbonate clastic sedimentation in a forearc basin (Kalamavka Formation, late Serravallian–early Tortonian, E. Crete). MPhil. thesis, University of East Anglia, Norwich, 123 pp.

DROOGER, C.W. & MEULENKAMP, J.E. (1973) Stratigraphic contributions to geodynamics in the Mediterranean area: Crete as a case history. *Bull. Geol. Soc. Greece* **10**, 193–200.

FORTUIN, A.R. (1977) Stratigraphy and sedimentary history of the Neogene deposits in the Ierapetra region, eastern Crete. *GUA Pap. Geol. Series 1* **8**, 164 pp.

FORTUIN, A.R. (1978) Late Cenozoic history of Eastern Crete and implications for the geology and geodynamics of the southern Aegean sea. *Geol. Mijnb.* **57**, 451–464.

FORTUIN, A.R. & PETERS, J.M. (1984) The Prina Complex in eastern Crete and its relationship to possible Miocene strike-slip tectonics. *J. Struct. Geol.* **6**, 459–476.

FORTUIN, A.R. & DE SMET, M.E.M. (1991) Rates and magnitudes of Late Cenozoic vertical movements in the Indonesian Banda Arc and the distribution of eustatic effects. In: *Sedimentation, Tectonics and Eustasy* (Ed. MacDonald, D.I.M.) In: Ass. Sedimentol. Spec. Publ. **12**, 79–90.

GRADSTEIN, F. (1973) The Neogene and Quaternary deposits in the Sitia district of eastern Crete. *Ann. Géol. Pays Hell.* **24**, 527–572.

GRADSTEIN, F.M. & VAN GELDER, A. (1971) Prograding clastic fans and transition from a fluviatile to a marine environment in Neogene deposits of eastern Crete. *Geol. Mijnb.* **50**, 383–392.

HALL, R. & AUDLEY-CHARLES, M.G. (1983) The structure and regional significance of the Talea Ori Crete. *J. Struct. Geol.* **5**, 167–179.

HALL, R., AUDLEY-CHARLES, M.G. & CARTER, D.J (1984). The significance of Crete for the evolution of the Eastern Mediterranean. In: *The Geological Evolution in the Eastern Mediterranean* (Eds Dixon, J.E. & Robertson, A.H.F.) Spec. Publ. Geol. Soc. London **17**, 499–516.

HAQ, B.U., HARDENBOL, J. & VAIL, P.R. (1988) Mesozoic and Cenozoic chronostratigraphy and cycles of sea-level change. In: *Sea-level changes: an Integrated Approach* (Eds Wilgus, C.K., Hastings, B.S., Posamentier, H., Van Wagoner, J., Ross, C.A. & Kendall, C.G.St.C.) Spec. Publ. Soc. Econ. Paleont. Mineral. **42** 72–109.

HELLER, P.L. & DICKINSON, W.R. (1985) Submarine Ramp facies model for delta fed, sand-rich turbidite systems. *Bull. Am. Ass. Petrol. Geol.* **69**, 960–976.

HSÜ, K.J. CITA, M.B. & RYAN, W.B.F. (1973) Origin of the Mediterranean evaporites. In: *Initial Report D.S.D.P. Vol 2.* (Eds Ryan, W.B.F., Hsü, K.J. & Cita, M.B. *et al.*) US Government Office, Printing Weshington, DC, pp. 1203–1231.

Hsü, K.J., Montadert, L. & Bernoulli, D. *et al.* (1977) History of the Mediterranean salinity crisis. *Nature* 267, 399–403.

Jongsma, D., Wissmann, G., Hinz, K. & Gardè, S. (1977) Seismic studies in the Cretan Sea 2. The southern Aegean Sea: an extensional margin basin without sea-floor spreading? Results of R.V. "Meteor" and R.R.S. "Schackleton" cruises. *"Meteor" Forschungsergeb. Reihe C* 27, 3–30.

Kleinspehn, K.L., Steel, R.J., Johanssen, E. & Netland, A. (1984) Conglomeratic fan-delta sequences, Late Carboniferous–Early Permian, Western Spitsbergen, In: *Sedimentology of Gravels and Conglomerates* (Eds Koster, E.H. & Steel, R.J.) Can. Soc. Petrol. Geol. Mem. 10, 279–294.

Kleverlaan, K. (1989) Neogene history of the Tabernas basin (SE Spain) and its Tortonian submarine fan development. *Geol. Mijnb.* 68, 421–432.

Kopp, K.O. & Richter, D. (1983) Synorogenetische Schuttbildungen und die Eigenstandigkeit der Phyllit-Gruppe auf Kreta. *Neues Jahrb. Geol. Palaeont. Abh.* 165, 228–253.

Langereis, C.G. (1984) Late Miocene magnetrostratigraphy in the Mediterranean. *Geol. Ultoiectina* 34, 178 pp.

Langereis, C.G., Zachariasse, W.J. & Zijderveld, D.J.A. (1984) Late Miocene magnetobiostratigraphy of Crete. *Mar. Micropaleont.* 8, 261–281.

Leité, O. & Mascle, J. (1982) Geological structures on the south Cretan continental margin and Hellenic Trench (eastern Mediterranean). *Mar. Geol.* 49, 199–223.

Le Pichon, X. & Angelier, J. (1979) The Hellenic arc and trench system; a key to the neotectonic evolution of the Eastern Mediterranean area. *Tectonophysics* 60, 1–42.

Listter G., Banga, G. & Feenstra, A. (1984) Metamorphic core complexes of Cordilleran type in the Cyclades, Aegean Sea, Greece. *Geology*, 12, 221–225.

Marzo, M. & Anadon, P. (1988) Anatomy of a conglomeratic fan-delta complex: the Eocene Montserrat Conglomerate, Ebro Basin, northeast Spain. In: *Fan Deltas: Sedimentology and Tectonic Settings* (Eds Nemec, W. & Steel, R.J.) Blackie & Son, London, pp. 319–340.

Mascle, J., Jongsma, D., *et al.* (1982) The Hellenic margin from eastern Crete to Rhodos: preliminary results. In: *Geodynamics of the Hellenic Arc and Trench* (Eds Le Pichon, X., Augustithis, S.S. & Mascle, J.) *Tectonophysics*, 86, 133–147.

MacDonald, D.I.M. (Ed.) (1991) Sedimentation, Tectonics and Eustasy — Sea-level Changes at Active Margins. *Int. Ass. Sedimentol. Spec. Publ.* 12, 518 pp.

McKenzie, D.P. (1978) Active tectonics of the Alpine–Himalayan belt: the Aegean Sea and surrounding regions. *Geophys. J. R. Astron. Soc.* 55, 217–254.

Mercier, J.L., Sorel D., Vergely, P. & Simeakis, K. (1989) Extensional tectonic regimes in the Aegean basins during the Cenozoic. *Basin Res.* 2, 49–71.

Meulenkamp, J.E. & Hilgen, F.J. (1986) Event stratigraphy, basin evolution and tectonics of the Hellenic and Calabro-Sicilian Arcs. In: *The Origin of Arcs* (Ed. Wezel, F.C.) Elsevier, Amsterdam, pp. 327–350.

Meulenkamp, J.E., Jonkers, A. & Spaak, P. (1979) Late Miocene to Early Pliocene development of Crete. *Proc. VI, Coll. Geol. Aegean Region (Athens)*, pp. 137–149.

Meulenkamp, J.E., Wortel, M.J.R., van Wamel, W.A., Spakman, W. & Hoogerduyn Strating, E. (1988). On the Hellenic subduction zone and the geodynamic evolution of Crete since the Late Middle Miocene. *Tectonophysics* 146, 203–215.

Miall, A.D. (1991) Stratigraphic sequences and their chronostratigraphic correlation. *J. Sediment. Petrol.* 61, 497–505.

Monogiou, E. (1989) Shallow marine, gravelly carbonate clastic sedimentation during the compressional tectonics (Prina Complex, Serravallian, E. Crete). MPhil. thesis, University of East Anglia, Norwich, 122 pp.

Peters, J.M. (1985) Neogene and Quaternary vertical tectonics in the south Hellenic Arc and their effect on concurrent sedimentation processes. *GUA Pap. Geol. Series 1* 23, 247 pp.

Peters, J.M., Troelstra, S.R. & van Harten, D. (1985) Late Neogene and Quaternary vertical movements in eastern Crete and their regional significance. *J. Geol. Soc. London* 142, 501–513.

Posamentier, H.W., Jervey, M.T. & Vail, P.R. (1988) Eustatic controls on clastic deposition I — Conceptual framework. In: *Sea-level Changes: an Integrated Approach* (Eds Wilgus, C.K., Hastings, B.S., Posamentier, H., Van Wagoner, J., Ross, C.A. & Kendall, C.G.St. C.) Spec. Publ. Soc. Econ. Paleont. Mineral 42, 109–124.

Postma, F. (1989) *The Males Section, Eastern Crete.* Internal report, 24 pp. Free University, Amsterdam.

Postma, G. (1990) An analysis of the variation in delta architecture. *Terra Nova* 2, 124–130.

Postma, G. (1991) Analysis of tectonic and climatic control on sediment dispersal patterns in a fault-bounded clastic wedge. *British Sedimentological Research Group Meeting, Edinburgh, Book of Abstracts.*

Postma, G. & Drinia, H. (1993) Tectonic and climatic control on architecture and sedimentary facies of a fault-bounded clastic wedge. Basin Res. 5, 103–124.

Prior, D.B. & Bornhold, B.D. (1988) Submarine morphology and processes of fjord fan deltas and related high-gradient systems: modern examples from British Columbia. In: *Fan Deltas: Sedimentology and Tectonic Settings* (Eds Nemec, W. & Steel, R.J.). Blackie, Glasgow, pp. 125–143.

Prior, D.B. & Bornhold, B.D. (1990) The underwater development of Holocene fan deltas. In: *Coarse-Grained Deltas* (Eds Colella, A. & Prior, D.B.) Int. Ass. Sediment. Spec. Publ. 10, 75–90.

Rabinowitz, P.D. & Ryan, W.B.F. (1970) Gravity anomalies and crustal shortening in the eastern Mediterranean. *Tectonophysics* 10, 585–608.

Richter, D. & Kopp, K.O. (1984) Zur Tektonik der untersten geologischen Stockwerke auf Kreta. *Neues Jahrb. Geol. Palaeont. Mh.* 1, 27–46.

Robertson, A.H.F. & Dixon, J.E. (1984) Introduction: aspects of the geological evolution of the Eastern Mediterranean. In: *The Geological Evolution in the Eastern Mediterranean* (Eds Dixon, J.E. & Robertson, A.H.F.) Spec. Publ. Geol. Soc. London 17, 1–74.

Ryan, W.B.F., Hsü, K.L. *et al.* (1973) Init. Rep. of the Deep Sea Drilling Project 13 (site rept. 10), pp. 323–353. US Govern. Print. Office, Washington, DC.

Seidel, E. (1978) *Zur Petrologie der Phyllit-Quarzite Series Kretas.* Habilitationschrift, 145 pp. Universität Braunschweig, Braunschweig.

SEIDEL, E., KREUZER, H. & HARRE, W. (1982) A Late Oligocene/Early Miocene high pressure belt in the External Hellenides. *Geol. Jarhb. E* **23**, 165–206.

SIERRO, F.J., GONZALES, DELGADO, J.A., DABRIO, C.J., FLORES, J.A. & CIVIS, J. (1990). The Neogene of the Gualdalquivir basin (SW Spain). In: *Iberian Neogene Basins* (Eds Agusti, J., Domenech, R., Julia, R. & Martinell, J.) Mem. Esp. **2**, 209–250.

SPAAK, P. (1983) Accuracy in correlation and ecological aspects of the planctonic foraminiferal zonation of the Mediterranean Pliocene. *Utrecht. Micropaleont. Bull.* **28**, 159 pp.

SPEELMAN, A.J. (1984) Benthicforaminifera of the Kalamavka Formation. Crete — a paleobathymetrical investigation. MSc thesis, Free University, Amsterdam, 24 pp.

STIGTER, H. (1989) *Neogene-Recent compressional tectonics in the southern Hellenic Arc.* Internal report, University Utrecht, 34 pp.

THEODORIDIS, S. (1984) Calcareous nannofossil biozonation of the Miocene and revision of the Helicoliths and Discoasters. Utrecht. Micropaleont. Bull. **32**, 272 pp.

THOMAS, E. (1980) Details of *Uvigerina* development in the Cretan Mio-Pliocene. *Utrecht. Micropaleont. Bull.* **23**, 167 pp.

TJALSMA, R.C. (1971) Stratigraphy and foraminifera of the Neogene of the Eastern Gualdalquivir Bain (S. Spain). *Utrecht. Micropaleont. Bull.* **4**, 161 pp.

VAIL, P., AUDERMARD, F., BOWMAN, S.A., EISNER, P.N. & PEREZ-CRUZ, G. (1990) The stratigraphic signatures of tectonics, eustasy and sedimentation, an overview. *Sequence Stratigraphy Workshop, part 4*, May 1991, Free University, Amsterdam, pp. 1–45.

VAN MARLE, L.J., VAN HINTE, J.E. & NEDERBRAGT, A. (1987) Plankton percentage of the foraminifera fauna in seafloor samples from the Australian Irian Jay Continental Margin, eastern Indonesia. *Mar. Geol.* **77**, 151–156.

VAN HINTE, J.E. (1978) Geohistory analysis-application of micro-paleontology in exploration geology. *Am. Ass. Petrol. Geol. Bull.* **62**, 210–222.

VAN WAGONER, J.C., MITCHUM, R.M., CAMPION, K.M. & RAHMANIAN, V.D. (1990) Siliciclastic sequence stratigraphy in well logs, cores, and outcrops: Concepts for high-resolution correlation of time and facies. *Am. Ass. Petrol. Geol. Mem.*, 55 pp.

WEIJERMARS, R. (1988) Neogene tectonics in the western Mediterranean may have caused the Messinian Salinity Crisis and an associated glacial event. *Tectonophysics* **148**, 211–219.

WEIJERMARS, R., ROEP, TH.B., VAN DEN EECKHOUT, B., POSTMA, G. & KLEVERLAAN, K. (1985) Uplift history of a Betic fold nappe inferred from Neogene-Quaternary sedimentation and tectonics in the Sierra Alhamilla and Almeria, Sorbas and Tabernas Basins of the Betic Cordilleras, SE Spain. *Geol. Mijnbow* **64**, 397–411.

ZACHARIASSE, W.J. (1975) Planktonic foraminiferal biostratigraphy of the Late Neogene of Crete (Greece). *Utrecht. Micropaleont. Bull.* **11**, 143 pp.

ZACHARIASSE, W.J. (1979) The origin of *Globorotalia conomiozea* in the Mediterranean and the value of its entry level in biostratigraphic correllations. *Ann. Géol. Pays Hell., Tome hors Série (III)*, 1281–1292.

ZACHARIASSE, W.J. & SPAAK, P. (1983) Middle Miocene to Pliocene palaeoenvironmental reconstruction of the Mediterranean and adjacent Atlantic Ocean: planktonic foraminiferal record of south Italy. In: *Reconstruction of marine paleoenvironments* (Eds Meulenkamp, J.E.) Utrecht. Micropaleont. Bull. **30**, 91–110.

Spec. Publs Int. Ass. Sediment. (1993) **20**, 363–376

Ancient and modern examples of tectonic escape basins: the Archaean Witwatersrand Basin compared with the Cenozoic Maracaibo Basin

I.G. STANISTREET

Department of Geology, University of the Witwatersrand, Private Bag 3, Wits 2050, Johannesburg, South Africa

ABSTRACT

The Archaean Witwatersrand Basin is the one Precambrian basin in which, because of extensive mining and exploration activity, the geological and geophysical data set compares well with that of younger Phanerozoic basins. The later part of the basin history, and that of the succeeding Lichtenburg Basin which resulted from tectonic inversion, record events which occurred at an adjacent plate margin between 2750 and 2690 Ma when the Zimbabwe Craton collided with the Kaapvaal Craton. Prior to this collision the lowermost marine sediments of the Witwatersrand Supergroup were deposited in an epicontinental sea which covered much of the Kaapvaal Craton. In contrast the uppermost sediments of the Witwatersrand Supergroup were deposited in a fault-controlled basin which was far smaller than the earlier depository and in which Laramide style block-faulting interacted with sedimentation dominated by a fluvial apron or bajada which passed proximally into pediment surfaces. The bajada still interacted with a marine water body in the basin centre. Oblique-slip reverse faults defined the basin margin and developed under NE–SW-directed collisional compression which also activated northwest-trending synsedimentary folds. This compression produced source area uplift to the north and west of the basin. Strike-slip components of fault movement are also reflected in shear zones exterior to the basin. The resultant displacement was that of a block tectonically escaping out of the Kaapvaal Craton, a process similar to tectonic escapes described from behind the Himalayan Orogen and in the Aegean area. The tectonic escape was followed by major collisional indentation and the development of an impactogenal rift, the initial form of the succeeding Lichtenburg Basin. An analogous recent basin to the end-phase Witwatersrand Basin, in terms of size, depositional systems and tectonic style is the Maracaibo Basin of northern South America, which is tectonically escaping northward into the Caribbean Sea because of the collision of the Panama Arc with South America.

INTRODUCTION

Society's unremitting thirst for oil and fossil energy supplies has provided a major incentive to understanding the mechanics and basin infill history of younger Phanerozoic basins. Geological and geophysical exploration data in these basins provide a resolution of knowledge which is unsurpassed in older basins, particularly those of Precambrian age. The one Precambrian basin which approaches Phanerozoic basins in terms of our resolution of geological and geophysical knowledge is the Witwatersrand Basin because of society's additional thirst for gold.

Many attempts have been made in the previous five years to provide a single, all-encompassing model for the development of the Witwatersrand Basin during its 300 My history (e.g. Burke *et al.*, 1986; Winter, 1987; Clendenin *et al.*, 1988; Myers *et al.*, 1990). It has become increasingly apparent, however, that the Kaapvaal Craton during the Archaean recorded a history of successive basins showing differing tectono-sedimentary styles during this long period of time, an evolution related to an entire tectonic cycle (Stanistreet & McCarthy, 1991). The present paper focuses on the tectonics and sedimentation of the Witwatersrand Basin near to the end of its evolution at a crucial changeover from compressional to extensional basin development on the Kaapvaal Craton. This changeover is also considered in the context of the tectono-sedimentary changes which led up to and followed this event.

GEOLOGICAL SETTING

Precambrian tectonics on the Kaapvaal Craton

Interest in Precambrian 'intracratonic' basins of southern Africa has grown considerably over the last decade. It has become increasingly clear that the 'intracratonic' label is for most of them inappropriate, and that we see the same basinal responses to tectonism in the Precambrian geological record that we do in the Phanerozoic response to extensional and collisional phases related to Wilson cycles (e.g. Dewey & Burke, 1973; Windley, 1984). Comparable features include: basin inversion (e.g. Clendenin *et al.*, 1990; Myers *et al.*, 1990); impactogenal rifting (Burke *et al.*, 1985); and oblique-slip fault zones on a subcontinental scale controlling sedimentation (Stanistreet *et al.*, 1986). The realization gained from Phanerozoic terranes that collisional deformational stresses can be transferred for over 1000 km into continental interiors (e.g. Tapponier *et al.*, 1982) has totally changed perspectives on the modes of basin formation in the interiors of what were presumably smaller Archaean continental crustal bodies.

Because of the high density of geological and geophysical data available from the Witwatersrand Basin it is not surprising that it and its related basins have been the focus of studies of tectonics and sedimentation and basin modelling over the last five years (see Stanistreet & McCarthy, 1991, for an overview). The Witwatersrand Basin is located in the interior of the Kaapvaal Craton (Fig. 1). Recent papers (Cheney *et al.*, 1988; Cheney, 1990), following the initial suggestions of correlations by Button (1976), have proposed from stratigraphical comparisons that the Pilbara Craton was possibly situated to the (present day) southwest of the Kaapvaal Craton during the late Archaen and that together they made up an Archaean microcontinent (named Vaalbara by Cheney *et al.*, 1988) from at least 3070 Ma (U–Pb zircon dates). Afterwards they were joined by other microcontinents. A most

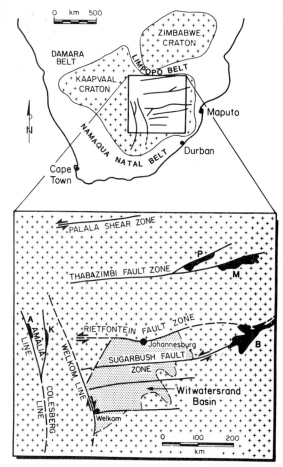

Fig. 1. Location of the Witwatersrand Basin on the Kaapvaal Craton. The position and senses of movement of fault and shear zones thought to be active during end-phase basin development during the collision between the Zimbabwe and Kaapvaal Cratons are shown. Note that they reactivate older lineaments related to greenstone belts. The stippled area indicates portions of the Witwatersrand Basin (*sensu stricto*) in which Central Rand Group as well as West Rand Group rocks have been preserved.

notable example of this was the collision and incorporation of the Zimbabwe Craton which occurred during the Late Archaean (Light, 1982) to form what must have been by Archaean standards a supercontinent, Zimvaalbara. The peak metamorphism associated with collision has been dated at 2700–2600 Ma (Tankard *et al.*, 1982) and much of the related crustal thickening must have taken place prior to this during the end phase of the deposition of the Witwatersrand Supergroup and initiation of the succeeding Lichtenburg Basin (Burke *et al.*,

1985; Stanistreet *et al.*, 1986; Clendenin *et al.*, 1988).

The last event in the Witwatersrand Basin was represented by the outpouring of the Klipriviersberg Group flood basalts which have also been related to the same style of tectonics which controlled the late phase sedimentation of the Witwatersrand Supergroup (McCarthy *et al.*, 1990; Stanistreet & McCarthy, 1991). The tectonic timing of the extrusion of these basalts has been compared with that of the Columbia River basalts of the western USA (Myers *et al.*, 1993). It was Burke *et al.* (1985) who suggested that the extensional rift basin which subsequently received Platberg Group (Fig. 2) and succeeding Pniel Sequence sediments, named the Lichtenburg Basin by Stanistreet and McCarthy (1991), was initially an impactogenal rift also extending orthogonally southwestwards away from the Limpopo collisional belt.

THE WITWATERSRAND BASIN

The stratigraphy of the Witwatersrand Supergroup and related units is shown in Fig. 2 with units deposited or preserved in the Witwatersrand Basin and in the Lichtenburg Basin indicated. The Witwatersrand Supergroup was deposited during the interval 3074 ± 6 Ma (deposition of the underlying Dominion Group) and 2714 ± 8 Ma (deposition of the overlying Klipriviersberg Group established from the single zircon U–Pb dating of Armstrong *et al.* (1986, 1990). The West Rand Group comprises interbedded marine quartz arenites and mudstones (e.g. Eriksson *et al.*, 1981; Mayer & Albat, 1988) with some fluvial interludes (Vos, 1975; Tainton & Meyer, 1990) while the Central Rand Group comprises interbedded fluvial conglomerates and meta-arkoses with many marine transgressive interludes (e.g. Verrezen, 1987; Holland *et al.*, 1990). The rocks have been metamorphosed subsequently up to conditions of 330–350°C at 3 kb pressure (Wallmach & Meyer, 1990). The end of the Witwatersrand Basin is marked by the outpouring of tholeiitic flood basalts, the Klipriviersberg Group. At least 1.7 km thickness of lava flows was extruded over the relatively short period of time of less than 10 Ma (between 2714±6 Ma and 2709±4 Ma (U–Pb zircon dating of Armstrong *et al.*, 1990).

The subsequent sedimentation of the Platberg Group took place in the younger Lichtenburg Basin which represents an extensional tectonic inversion

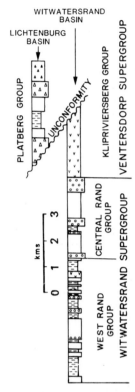

Fig. 2. Stratigraphy of the Witwatersrand Supergroup and related units. The West Rand Group is dominated by marine quartz arenites (blank) and mudstones (dashed) with some iron formations. The Central Rand Group is dominated by fluvial arkoses and subarkoses (now metamorphosed) (blank) and conglomerates (circles) that are interbedded with marine transgressive units (e.g. mudstones — dashed). The Klipriviersberg Group comprises basalts (V's), while the Platberg Group comprises: mudflow diamictites (triangles); fan-delta sandstones (circles); mudstones (dashes).

of fault zones which had a reverse oblique-slip component during the deposition of the Central Rand Group (Myers *et al.*, 1993). This inversion occurred in a manner similar to the inversion of Cretaceous Laramide reverse faults of the western USA (Myers *et al.*, 1993) during Basin and Range Tertiary extension (e.g. Sales, 1983).

In understanding the history of the end phase of the Witwatersrand Basin, it is most important to appreciate the changes which took place during the deposition of the Witwatersrand Supergroup. This can be indicated by contrasting the style of tectonics and sedimentation of the lower part of the West Rand Group with those of the upper part of the Central Rand Group.

Sedimentation and tectonics of the lower part of the West Rand Group and its correlates

The West Rand Group sediments were deposited well beyond what is generally understood to be the confines of the main Witwatersrand Basin (Fig. 1) which came into being only in the latter half of Witwatersrand Supergroup sedimentation. Exploration drilling and geomagnetic imagery (Corner *et al.*, 1986) have shown that outliers are developed well to the north and southeast of the present basin (Witwatersrand Basin *sensu stricto*). The Godwan Group in the eastern Transvaal has long been correlated with the Witwatersrand Supergroup (Button, 1977) and it has recently been correlated with the Pongola Sequence (Myers, 1991). The Pongola Sequence of northern Natal and eastern Transvaal was correlated with the Witwatersrand Supergroup by Truter and Rossouw (1955), but subsequent Rb/Sr data argued against this correlation. The Mozaan Group, representing the upper half of the Pongola Sequence, is, however, very similar to the upper part of the West Rand Group in terms of cyclic style and facies development. It has recently been correlated with the Witwatersrand Supergroup on the basis of a review of single zircon U–Pb dates and sequence stratigraphical arguments by Beukes and Cairncross (1991). Combining these stratigraphical equivalents in Fig. 3 it can be seen that the mainly marine

sequence was spread over a wide area of the craton.

Sedimentary facies analysis of the West Rand Group by Eriksson *et al.* (1981) led to their interpretation of quartz arenites and mudstones deposited in shoreline, shoreface and subtidal settings varying from meso- to macro-tidal. Coarsening-upward cycles can be identified within the quartz arenite units. Figure 4 shows such cycles in the Orange Grove Quartzite Formation at the base of the West Rand Group. These are interpreted as progradational parasequences. Within this sequence the most distal deposits are ironstones, interpreted as counterparts of Phanerozoic pelagic deposits (in accordance with the arguments of Eriksson, 1982), which are analogous to the marine condensed sequences of Phanerozoic depositional systems (e.g. Van Wagoner *et al.*, 1988).

The laterally extensive deposition of the West Rand Group has been interpreted to have taken place in two different tectonic settings by different authors. One school of thought proposes a foreland basin (Burke *et al.*, 1986) perhaps analogous to the Cretaceous seaway of North America (e.g. Spearing, 1975). On the other hand, Clendenin *et al.* (1988) suggest that the epicontinental style of sedimentation reflects a phase of thermal cooling, following the development of an extensional rift basin which received the Dominion Group volcanics and sediments, a situation perhaps analogous to the Cretaceous to Recent North Sea Basin (Sclater &

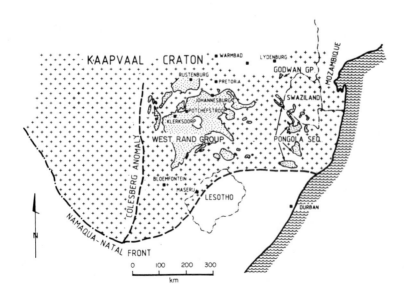

Fig. 3. The lateral extent of the West Rand Group and its equivalents. The Colesberg anomaly is a magnetic anomaly which marks the position of syn- to post-Central Rand Group middle crustal uplift (Corner *et al.*, 1986), causing erosion of West Rand Group to the west of this line.

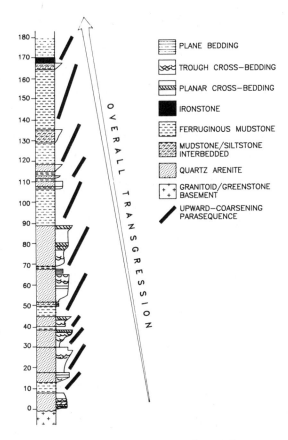

Fig. 4. Measured section through the Orange Grove Quartzite at the base of the Witwatersrand Supergroup, typifying early West Rand Group sedimentation. Coarsening-upward parasequences are developed in an overall transgressive systems tract, which ultimately leads to pelagic sediments (equivalent to marine condensed sequences of the Phanerozoic rock record) at the base of the uppermost parasequence.

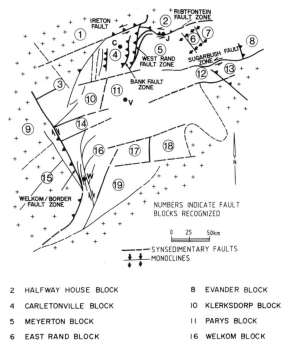

2	HALFWAY HOUSE BLOCK	8	EVANDER BLOCK
4	CARLETONVILLE BLOCK	10	KLERKSDORP BLOCK
5	MEYERTON BLOCK	11	PARYS BLOCK
6	EAST RAND BLOCK	16	WELKOM BLOCK

Fig. 5. A reconstruction of the structures which controlled sedimentation during the later stages of the Witwatersrand Basin. Individual fault-bounded blocks are numbered. (J, Johannesburg; W, Welkom; V, Vredefort; C, Carletonville.) Triangular tags indicate the up-thrown side of reverse faults or thrusts during Witwatersrand sedimentation. Arrows indicate sense of lateral movement.

Christie, 1980). I interpret the lower part of the West Rand Group as an epicontinental shelf-platform sequence deposited during thermal cooling. However, the incoming of fluvial sediments higher in the West Rand Group may represent the start of the conversion of the shelf to a foreland basin, as discussed in more detail later in this paper.

Sedimentation and tectonics of the upper part of the Central Rand Group

Through time Witwatersrand Supergroup sedimentation became much more localized in a 'shrinking' basin (Vos, 1975; Pretorius, 1976), which became triangular in shape (Pretorius, 1986). The northern and western margins of this basin are defined by faults (Fig. 5) associated with monoclines dipping into the basin as was originally recognized by Brock and Pretorious (1964). Recent papers (Stanistreet *et al.*, 1986; Winter, 1987; Myers *et al.*, 1990; Myers *et al.*, 1993) have shown, however, that these faults were related to compressional tectonics and are not extensional faults as previously envisaged. Evidence increasingly points to the fact that they were reverse oblique-slip faults during sedimentation (e.g. Stanistreet *et al.*, 1986; Myers *et al.*, 1993).

The east–west trending Rietfontein/Ireton Fault Zone along the northern margin of the basin is paralleled further south by other fault zones(Fig. 5), which appear to terminate against the western bounding fault (Welkom Line). Connecting faults between these east–west faults are generally north-northwest-trending and these define individual synsedimentary fault blocks which rotated and inclined independently during compression in a manner analogous to Cretaceous Laramide faultblocks of the western USA (Myers *et al.*, 1990, 1993).

This block-fault tectonic style was associated with sedimentation that was dominated by braided fluvial processes (e.g. Minter, 1978), which operated both on fluvial plains (e.g. Nami, 1983) and, at the other extreme, more localized stream-dominated alluvial fans (e.g. Kingsley, 1987). Pretorius (1976) was the first to suggest that fan sedimentation was important in the basin fill. Braided, fluvially dominated fans and braid deltas (Minter, 1978; McPherson *et al.*, 1987) or coalesced fan bajadas (Vos, 1975) otherwise predominated during Upper Central Rand Group sedimentation. Turner (1979) was the first to realize that these fans passed proximally into sediment veneered rock pediments and Tucker and Viljoen (1986) also indicate the proximal passage into rock pediments of the fluvial deposits of the Turffontein Subgroup. Distally the fluvial system interacted with a marine water body (Verrezen, 1987) at least as late as during the deposition of the lowermost Turffontein Subgroup (Holland, 1990).

Figure 6 shows a block diagram which portrays the sedimentation style across a faulted margin of the Witwatersrand Basin towards the basin centre. The proximal erosional pediment passed into a

fluvial bajada which interacted with a marine water body in the basin centre. Large upward-fining unconformity-bound packages resulted from phases of marginal uplift followed by basin subsidence (Stanistreet *et al.*, 1988; Stanistreet & McCarthy 1991). A borehole core correlation programme undertaken by the Sedimentology Division of the Geological Society in 1984 correlated such upward-fining packages all the way around the northern and western margin of the basin (reported by Camden-Smith *et al.*, 1986). Holland *et al.* (1990) and Holland (1990) have traced large- (*c.* 1100 m thick) and medium- (*c.* 100s m thick) scale fining-upward packages on a variety of scales into the middle of the basin, which is exposed by the much later emplacement of the Vredefort Dome (McCarthy *et al.*, 1986). The bajada must therefore have been of major proportions extending at least 350 km around the basin margin and 90 km into the basin centre at maximum extent (Fig. 7).

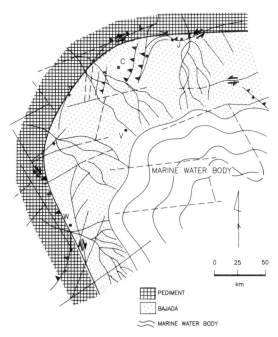

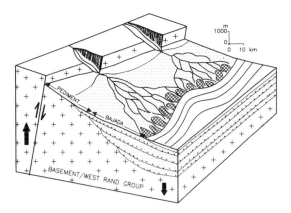

Fig. 6. Block diagram showing schematically the relationship between the faulted margins of the basin and fluvial depositional systems which interacted with a central marine water body during the deposition of the upper half of the Central Rand Group.

Fig. 7. Schematic palaeogeographical maps of the Witwatersrand Basin late in its history. (J, Johannesburg; C, Carletonville; V, Vredefort; W, Welkom.) At this stage the Rietfontein and Welkom fault zones were major controls on the pediment/bajada boundary. (See Fig. 6 for location.)

COLLISIONAL TECTONIC ESCAPE PROCESSES RELATED TO THE END PHASE OF THE WITWATERSRAND BASIN

Referring to the synsedimentary tectonic reconstruction for the upper part of the Central Rand Group in Fig. 5, the conjugate nature of the left-lateral, oblique-slip, east-trending, fault zones and the right-lateral, oblique-slip northwest-trending, fault zones seem to have developed under NE–SW-directed compression (see also Stanistreet *et al.*, 1986). This is confirmed by the trends of synsedimentary folding in the basin, which almost exclusively trends northwest (e.g. Antrobus & Whiteside, 1964; Camden-Smith & Stear, 1986). The resultant movement of the basement between the marginal faults was that of being 'squeezed-out' of the Kaapvaal Craton in a southeasterly direction (Fig. 8). McCarthy *et al.* (1990) have shown that a similar effect was operating during the time of extrusion of the lower half of the Klipriviersberg basalts.

Figure 1 shows the other lineaments that participated in this tectonic escape phenomenon, probably with more enhanced strike-slip motion. The Palala Shear Zone, which marks the southern margin of the Limpopo Belt, is documented as having operated as a left-lateral zone of shear (van Reenen *et al.*, 1987; McCourt & Vearncombe, 1992) and the Thabazimbi Fault Zone is thought to have acted in

this fashion (Clendenin *et al.*, 1988). The Colesburg Line (Fig. 1) indicates where major upthrow of deep crust has occurred (Corner *et al.*, 1986), probably in association with tectonic lineaments such as the Amalia Line and others parallel to it (Fig. 1) which would have acted in concert with the Border Structure (Welkom Fault Zone). Most of these lineaments can be seen to represent reactivations of older fault and shear zones associated with the end phases of greenstone belt evolution (Fig. 1).

Tectonic escape of continental crust over adjacent oceanic crust is associated with collision of a continent with another continent, microcontinent or volcanic arc. It was a process first suggested by Molnar and Tapponier (1975) for the area of Asia behind the Himalayan Orogen in which wedges of continental crust are escaping eastward away from the indenting India subcontinent.

A similar tectonic escape has been recognized where Arabia is colliding with Europe and a large area of Turkey and the Aegean is escaping westward into the Mediterranean Sea where it overrides oceanic crust (McKenzie, 1978). The continental crust underneath the Aegean Sea is experiencing 50% extension because of the escape and this is accompanied by the Neogene emplacement of basic magmas (Burke & Sengör, 1986). Up until now, however, no totally satisfactory example of the tectonic escape process having developed due to ancient continental collisions has been proposed.

THE TECTONICS AND SEDIMENTATION OF THE WITWATERSRAND BASIN COMPARED WITH THE MARACAIBO BASIN

In the two examples of modern tectonic escape mentioned above no sedimentary basin has been formed comparable with the formation of the Witwatersrand Basin. In northern South America, however, Burke and Sengö (1986) have suggested that the Maracaibo Basin has developed as a result of the collision of the Panama or Central America Arc with South America (Fig. 9). The escape of the Bonaire Block into the Caribbean Sea is accompanied by extension and subsidence which controls the position of the basin. Subsidence has localized the fresh to brackish modern Lake Maracaibo (Sarmiento & Kirkby, 1962); in the past, however, equivalent fully marine water bodies have occupied

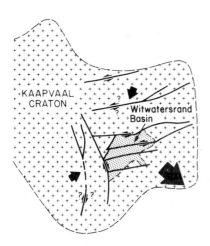

Fig. 8. The envisaged tectonic escape southeastward out of the Kaapvaal Craton of the 'block' on top of which the Witwatersrand Basin developed.

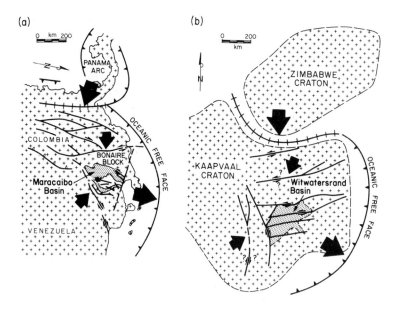

Fig. 9. (a) The origin of the Maracaibo Basin as a result of the tectonic escape of the Bonaire Block from the South American continent due to the collision of the Panama Arc (after Burke & Sengör, 1986). (b) The envisaged analogous development of the Witwatersrand Basin as a result of tectonic escape associated with the collision of the Zimbabwe Craton with the Kaapvaal Craton at the same scale. (NB (a) is rotated by 90° in order to facilitate comparison.)

the basin (Blaser & White, 1984). These centrally located water bodies are at present and have in the past been fringed by a fluvial apron on top of which meandering rivers operate (Hyne *et al.*, 1979a, b) (Fig. 10). Fluvial sediment is reworked at the shoreline either in association with the outbuilding of the Catatumbo Delta (Hyne *et al.*, 1979a, b), or along siliciclastic shorefaces dominated by counterclockwise coastal currents.

The bajada passes proximally into an uplifted hinterland in which erosive processes predominate and in which pediments have been developed in the past (Blaser & White, 1984). The uplifted areas to the west and east of the basin are controlled by major faults such as the left-lateral Santa Marta Fault and the right-lateral Bocono Fault (Rod, 1956) which control the outward escape of the Bonaire Block (Burke & Sengör, 1986). The basin-fill sedimentary sequence attains a thickness of 3500 m within the basin (Mencher *et al.*, 1953) but this thins down to zero as one moves onto the uplifted hinterlands to west and east (e.g. Link, 1952). This thinning is accompanied by major phases of erosion and overstepping as one moves to west and east (Mencher *et al.*, 1953). Marine and paralic sedimentation has predominated within the basin whereas fluvial sedimentation has predominated along the western and eastern margins (Blaser & White, 1984). Marine sedimentation was dominant early in the basin history (Late Cretaceous and

earliest Tertiary) but with basin confinement fluvial sedimentation has dominated in the latest stages (Quaternary). In the southern portion of the basin it can be seen (Fig. 10) that major block-faulting of the basement is interacting with the present-day sedimentation and there is borehole evidence that this has also happened in earlier stages of basin history (Mencher *et al.*, 1953). Associated with this block-faulting, through-going, left-lateral, oblique-slip faults are active internally to the basin at present (Naylor *et al.*, 1986) and also affect modern sedimentation.

All these features pertaining to the Maracaibo Basin are similar to features described above from the Archaean Witwatersrand Basin at a late stage in its history. Figure 9 also compares the two basins in terms of scale and in their relative positions with respect to the collisional events which controlled them. In the case of the Archaean example it was the collision of the Zimbabwe Craton which is thought to have activated the tectonic escape. In Fig. 10 the South American continent has been rotated by 90° and has been drawn at the same scale in order to facilitate the comparison. England and McKenzie (1981) have modelled tectonic escape and conclude that the size of the escaping block should be at least of the same order of magnitude as the colliding continental body. This is the case in both examples (Fig. 10). The proposal in Fig. 9a of an ocean–continent subduction zone to accommo-

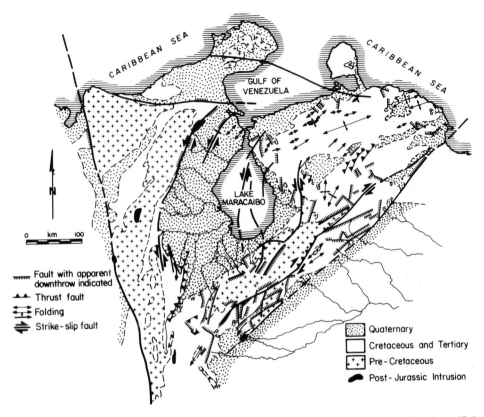

Fig. 10. Geology of the Maracaibo Basin. (Chief sources: Bucher, 1950; Ministry of Mines and Petroleum (Columbia), 1944; Naylor *et al.*, 1986.)

date the escape of continental crust southeastwards is supported by recent findings of Robb and Meyer (1991). They recognize a north–south line of geochemically distinct granites parallel and close to this eastern margin of the Kaapvaal Craton (Mpageni, Mbabane, Ngwempisi and Sicunusa Granites) which are dated at 2740 ± 15 Ma (Mpageni Granite, U–Pb single zircon date, Kamo *et al.*, 1990) and 2691 ± 4 Ma (Mbabane Granite, U–Pb single zircon date, Tegtmeyer & Kröner, 1987). These are I-type granites (Robb & Meyer, 1991) which span the tectonic escape phase proposed and which therefore may well reflect intrusions generated by subduction.

One major difference between the Maracaibo and Witwatersrand Basins is the style of fluvial sedimentation which operated within them. Meandering systems operate in the modern example, braided systems in the Archaean example. This difference may be ascribed to the lush tropical vegetation

which interacts with the fluvial systems in the modern Maracaibo Basin and which would not have been a factor in the Precambrian equivalent. The general absence of land plants prior to the Devonian Periodonly allowed the evolution of broad non-confined braidplains generally (e.g. Schumm, 1968; Long 1978; Fuller, 1985).

CENTRAL RAND GROUP SEDIMENTATION AND TECTONICS PRIOR TO THE COLLISION OF THE ZIMBABWE CRATON

Prior to the main phase of crustal thickening in the Limpopo Belt, representing continent–continent collision (e.g. Light, 1982), compressive events affected the shrinking Witwatersrand Basin, as evidenced by the compressive block-fault tectonics which affected the lower half of the Central Rand

Group (Myers *et al.*, 1990) and the upper portion of the West Rand Group. These compressive events caused basement uplift and successive rejuvenations of conglomerate sedimentation, developed at the base of fining-upward unconformity-bounded sequences traceable throughout the basin (see Stanistreet & McCarthy, 1991). At least two major uplift events of this type affected the Kaapvaal Craton during this period. The radical effect that these events had on sedimentary facies and dispersal patterns in the Witwatersrand Basin suggests that these were events similar to, but of a lower magnitude than the ultimate collision of the Zimbabwe microcontinent. In view of the proposed convergence history at this stage I suggest that these may have been caused by collision of exotic microplates or oceanic prominences, such as minor volcanic arcs or oceanic plateaux, with the Kaapvaal Craton, prior to the ultimate continental collision. In this regard McCourt (1991) and McCourt and Vearncombe (1992) have already suggested that the Central Zone of the Limpopo Belt may represent an exotic block within the collisional zone, although they time the emplacement of this *after* the collision of the Zimbabwe and Kaapvaal Cratons. Given the lack of age constraints on the timing of the emplacement, this could be more simply modelled as an exotic terrane which was incorporated *prior* to the collision of the Zimbabwe Craton.

A modern analogue exists of the beginning stages of such a conversion of a shelf into a foreland basin, as envisaged for the upper West Rand Group, an analogue associated with arc collision. Taiwan represents the orogeny associated with the collision of the Luzon Arc with the Eurasian continent in which a distal Miocene shelf sequence has been transformed into a fold-thrust belt (Ernst & Jahn, 1987). This has shed Plio-Pleistocene fan sediments westwards off the newly developing orogenic front to overlie these earlier shelf sequences, a process which continues today.

SUMMARY OF THE TECTONIC EVOLUTIONARY CONTEXT OF WITWATERSRAND TECTONIC ESCAPE

Figure 11 summarizes the four proposed stages in the tectonic evolution of the Kaapvaal Craton during the Late Archaean. An initial epicontinental marine stage (Fig. 11a), probably representing thermal cooling following the extensional rift basin which accepted Dominion Group volcanics and sediments, was succeeded by a stage (Fig. 11b) in which subduction along the northwest margin of the Kaapvaal Craton (at that time part of Vaalbara) caused the development of a foreland basin. The next stage (Fig. 11c) involved initial collision of the Zimbabwe Craton (microcontinent) with the Kaapvaal Craton causing tectonic escape which controlled the position of the end-stage Witwatersrand Basin (*sensu stricto*). Aspects of the emplacement of the Klipriviersberg Group lavas (McCarthy *et al.*, 1990) suggest that they were extruded into this basinal setting, in a fashion similar to the emplacement of Neogene basic magmas into the presently tectonically escaping Aegean block of the eastern Mediterranean. The final stage shows the extinction of the Witwatersrand Basin because of chronic indentation of the Zimbabwe Craton into the Kaapvaal Craton, which resulted in the development of a new impactogenal extensional rift (Fig. 1d), the Lichtenburg Basin, a development previously proposed by Burke *et al.* (1985).

CONCLUSIONS

During its end phase the Witwatersrand Basin was localized and controlled by tectonic escape effects in the Kaapvaal Craton associated with the compression and indentation caused by the collision between the Kaapvaal and Zimbabwe Cratons that resulted in the Limpopo Belt. This indentation subsequently caused the development of an extensional impactogenal rift basin, the Lichtenburg Basin, which developed orthogonally to the Limpopo belt. The extension represented by the Lichtenburg Basin inverted previous syn-Witwatersrand reverse oblique-slip fault zones which then developed as normal and normal oblique-slip fault structures.

The end-stage Witwatersrand Basin has a modern analogue in the form of the Maracaibo Basin, which is developing on top of the tectonically escaping Bonaire Block as a result of the collision between the Panama Arc and the South American continent. Both basins are characterized by fault-controlled fluvial aprons which surround a water body that is brackish or marine in nature. Both basins contain synsedimentary tectonic features that include both folding, which developed parallel with the orogenic front, and block-faulting.

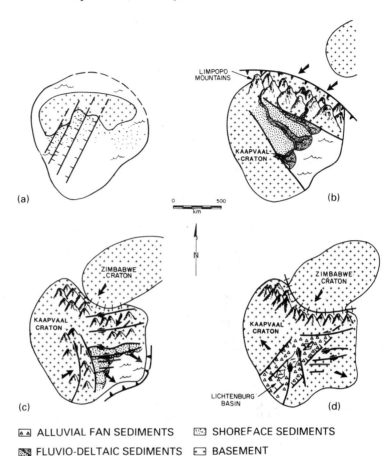

Fig. 11. Stages in the evolution of the Witwatersrand Basin: (a) the West Rand Group is deposited in an epicontinental marine shelf/platform setting over the previous Dominion extensional basin (orientation of faulting based on Clendenin *et al.*, 1990); (b) subduction under the northwest margin of the craton causes a foreland basin development; (c) initial collision causes tectonic escape and formation of the Witwatersrand Basin *sensu stricto*; (d) ultimate indentation causes an impactogenal rift, the Lichtenburg Basin.

ALLUVIAL FAN SEDIMENTS   SHOREFACE SEDIMENTS

FLUVIO-DELTAIC SEDIMENTS   BASEMENT

On the Kaapvaal Craton the tectonic escape phase was preceded by the deposition initially of an epicontinental marine sequence, the lowermost part of the Witwatersrand Supergroup. This sequence developed during thermal cooling following a rift phase represented by the formation of the Dominion Group. A foreland basin stage then developed which may have been caused by the collision and incorporation of exotic terranes and oceanic prominences into the area of the Limpopo Belt prior to the ultimate continent–continent collision of the Zimbabwe Craton with the Kaapvaal Craton.

ACKNOWLEDGEMENTS

The author would like to thank Guy Charlesworth, Terence McCarthy, Russell Myers and Laurie Robb for many discussions which have helped the development of ideas reflected in this paper. Special thanks go to Louis Nicolaysen, who directed the author's attention to new ideas on South American tectonics; Mike Meyer and Laurie Robb who kindly pointed out new data they have on granites near Barberton; and José van Harreveldt who conducted a thorough literature search on available sources on the Maracaibo Basin. Judy Wilmot and Lynda Whitfield kindly typed and drew diagrams, respectively, under very trying circumstances and Mark Hudson undertook photographic work. I thank Ken Eriksson and Tom Dreyer for their detailed and useful comments on a previous manuscript.

REFERENCES

ANTROBUS, E.S.A. & WHITESIDE, H.C.M. (1964) The geology of certain mines on the East Rand. In: *The Geology of some Ore Deposits of Southern Africa, 1* (Ed. Haughton, S.H.) eol. Soc. S. Afr., Johannesburg, pp. 125–160.

ARMSTRONG, R.A., COMPSTON, W., RETIEF, E.A. & WELKE, H.J. (1986) Ages and isotopic evolution of the Ventersdorp Volcanics. *Geocongress '86 Ext. Abstracts.* Geol. Soc. S. Afr., Johannesburg, pp. 89–92.

ARMSTRONG, R.A., RETIEF, E., COMPSTON, W. & WILLIAMS, I.S. (1990) Geochronological constraints on the evolution of the Witwatersrand Basin, as deduced from single zircon U/Pb ion microprobe studies. *Geocongress '90 Ext. Abstracts. Geol. Soc. S. Afr., Johannesburg, pp. 24–27.*

BEUKES, N. & CAIRNCROSS, B. (1991) A lithostratigraphic–sedimentological reference profile for the Late Archaean Mozaan Group, Pongola Sequence: applications to sequence stratigraphy and correlation with the Witwatersrand Supergroup. *S. Afr. J. Geol.* **94**, 44–69.

BLASER, R. & WHITE, C. (1984) Source-rock and carbonication study, Maracaibo Basin, Venezuela. In: *Petroleum Geochemistry and Basin Evaluation* (Ed. Weeks, L.G.) Mem. Am. Assoc. Petrol. Geol. **35**, 229–252.

BROCK, B.B. & PRETORIUS, D.A. (1964) Rand Basin sedimentation and tectonics. In: *Geology of some Ore Deposits of Southern Africa, 1* (Ed. Haughton, S.H.) Geol. Soc. S. Afr., Johannesburg, pp. **1**, 549–599.

BUCHER, W.H. (1950) (Compiler) Geologic-tectonic map of the United States of Venezuela Scale 1:1 000 000. Geological Society of America, Washington.

BURKE, K., KIDD, W.S.F. & KUSKY, T.M. (1985) Is the Ventersdorp Rift System of Southern Africa related to a continental collision between the Kaapvaal and Zimbabwe Cratons at 2.64 Ma ago? *Tectonophysics* **115**, 1–24.

BURKE, K., KIDD, W.S.F. & KUSKY, T.M. (1986) Archaean foreland basin tectonics in the Witwatersrand, South Africa. *Tectonics* **5**, 439–456.

BURKE, K. & SENGÖR, C. (1986) Tectonic escape in the evolution of the continental crust. In: *Reflection Seismology: the Continental Crust* (Eds Barazangi, M. & Brown, L.) Geodynam. Ser. Am. Geophys. Union., Washington **14**, 41–53.

BUTTON, A. (1976) Transvaal and Hammersley Basins — review of basin development and mineral deposits. *Miner. Sci. Eng.* **8**, 262–293.

BUTTON, A. (1977) Correlation of the Godwan Formation based on stratigraphic trends in the Witwatersrand Basin. *Inf. Circ. Econ. Geol. Res. Unit.* **109**, 7 pp.

CAMDEN-SMITH, P. & STEAR, W. (1986) Relationship between syndepositional structural elements and sedimentational patterns on East Rand Proprietary Mines Limited. *Geocongress '86 Ext. Abstracts.* Geol. Soc. S. Afr., Johannesburg, pp. 441–446.

CAMDEN-SMITH, P., STEWART, B., TAINTON, S., THERON, F., WOOD, A. & KARPETA, P. (1986) The stratigraphy of the Witwatersrand Sequence. *Exc. Guidebook Geocongress '86*, Geol. Soc. S. Afr., Johannesburg, pp. 104

CHENEY, E.S. (1990) Evolution of the "south western" continental margin of Vaalbara. *Ext. Abstracts Geocongress '90, 23rd Congr. Geol. Soc. S. Afr. Univ. Cape Town*, pp. 88–91.

CHENEY, E.S., ROERING, C. & STETTLER, E. (1988) Vaalbara. *Ext. Abstracts Geocongress '88, 22nd Congr. Geol. Soc. S. Afr. Univ. Natal*, pp. 85–88.

CLENDENIN, C.W., CHARLESWORTH, E.G. & MASKE, S. (1988)

Tectonic style and mechanism of Early Proterozoic successor basin development, southern Africa. *Tectonophysics* **156**, 275–291.

CLENDENIN, C.W., CHARLESWORTH, E.G. & MASKE, S. (1990) Structural styles of fault inversion influencing the Witwatersrand Supergroup: examples from Oukop, northeast of Klerksdorp. *S. Afr. J. Geol.* **93**, 202–210.

CORNER, B., DURRHEIM, R.J. & NICOLAYSEN, L.O. (1986) The structural framework of the Witwatersrand Basin as revealed by gravity and aeromagnetic data. *Geocongress '86. Ext. Abstracts.* Geol. Soc. S. Afr., Johannesburg, pp. 27–30.

DEWEY, J.F. & BURKE, K.C.A. (1973) Tibetan, Variscan and Precambrian basement reactivation: products of continental collision. *J. Geol.* **81**, 683–692.

ENGLAND, P. & McKENZIE, D.P. (1981) A thin viscous sheet model for continental deformation. *Geophys. J.R. Astron. Soc.* **70**, 295–321.

ERIKSSON, K.A. (1982) Archaean iron-formations: analogues of Holocene pelagic sediments. *Abstract, 11th Congress on sedimentology.* Int. Assoc. Sedimentol. Hamilton, Ontario, p. 2.

ERIKSSON, K.A., TURNER, B.R. & VOS, R.G. (1981) Evidence of tidal processes from the lower part of the Witwatersrand Supergroup, South Africa. *Sediment. Geol.* **29**, 309–325.

ERNST, W.G. & JAHN, B.M. (1987) Crustal accretion and metamorphism in Taiwan, a post-Palaeozoic mobile belt. *Philos. Trans. R. Soc. London* **A321**, 129–161.

FULLER, A.O. (1985) A contribution to the conceptual modelling of pre-Devonian fluvial systems. *S. Afr. J. Geol.* **88**, 189–194.

HOLLAND, M.J. (1990) A palaeoenvironmental analysis of the Turffontein Subgroup (Witwatersrand Supergroup) around the Vredefort Dome. Unpubl. M.Sc. thesis, Univ of the Witwatersrand, 256 pp.

HOLLAND, M.J., STANISTREET, I.G. & McCARTHY, T.S. (1990) Tectonic control on the deposition of the Turffontein Subgroup around the Vredefort Dome. *S. Afr. J. Geol.* **93**, 158–168.

HYNE, N.J., COOPER, W.A. & PARKE, A.D. (1979a) Stratigraphy of intermontane, lacustrine delta, Catatumbo river, Lake Maracaibo, Venezuela. *Bull. Am. Assoc. Petrol. Geol.* **63**, 2042–2057.

HYNE, N.J., LAIDIG, L.W. & COOPER, W.A. (1979b) Pro-delta sedimentation on a lacustrine delta by clay mineral flocculation. *J. Sediment. Petrol.* **49**, 1209–1216.

KAMO, S.L., DAVIS, D.W. & DE WIT, M.J. (1990) U–Pb geochronology of Archaean plutonism in the Barberton region, South Africa: 800 m.y. of crustal evolution. *Geol. Soc. Austr. Abstr.* 27–53.

KINGSLEY, C.S. (1987) Facies changes from fluvial conglomerate to braided sandstone of the Early Proterozoic Eldorado Formation, Welkom Goldfield, South Africa In: *Developments in Fluvial Sedimentology* (Eds Ethridge, F.G., Flores, R.M. & Harvey, M.D.) Spec. Publ. Soc. Econ. Palaeontol. Mineral. **39**, 361–370.

LIGHT, M.P.R. (1982) The Limpopo Mobile Belt: A result of continental collision. *Tectonics* **1**, 325–342.

LINK, W.K. (1952) Significance of oil and gas seeps in world oil exploration. *Bull. Am. Assoc. Petrol. Geol.* **36**, 1505–1540.

LONG, D.G.F. (1978) Proterozoic stream deposits: some problems of recognition and interpretation of ancient sandy fluvial systems In: *Fluvial Sedimentology* (Ed. Miall, A.D.) Mem. Can. Soc. Petrol. Geol. **5**, 313–341.

MAYER, J.J. & ALBAT, H.M. (1988) The tectono-sedimentary setting of the area of the Vredefort Structure during deposition of the upper quartzite member of the Hospital Hill Subgroup. *S. Afr. J. Geol.* **91**, 239–247.

McCARTHY, T.S., CHARLESWORTH, E.G. & STANISTREET, I.G. (1986) Post-Transvaal structures across the northern portion of the Witwatersrand Basin. *Trans. Geol. Soc. S. Afr.* **89**, 311–323.

McCARTHY, T.S., McCALLUM, K., MYERS, R.E. & LINTON, P. (1990) Stress states along the northern margin of the Witwatersrand Basin during Klipriviersberg Group volcanism. *S. Afr. J. Geol.* **93**, 245–260.

McCOURT, S. (1991) Structure of the Limpopo Belt and adjacent granitoid-greenstone terranes: a review of Late Archaean crustal evolution in southern Africa. *Terra Abstracts, Suppl. 3 to Terra Nova* **3**, 22.

McCOURT, S. & VEARNCOMBE, J.R. (1987) Shear zones bounding the central zone of the Limpopo Mobile Belt, southern Africa. *J. Struct. Geol.* **9**, 127–137.

McCOURT, S. & VEARNCOMBE, J.R. (1992) Shear zones of the Limpopo Belt and adjacent granitoid-greenstone terraces: implications for Late Archaean collision tectonics in southern Africa. *Precamb. Res.* **55**, 553–570.

McKENZIE, D.P. (1978) Active tectonics of the Mediterranean region. *Geophys. J. R. Astron. Soc.* **55**, 217–254.

McPHERSON, J.G., SHANMUGAM, G. & MOIOLA, R.J. (1987) Fan-deltas and braid deltas: Varieties of coarse-grained deltas. *Geol. Soc. Am. Bull.* **99**, 331–340.

MENCHER, E., FICHTER, H.J., RENZ, H.H., WALLIS, W.E., PATTERSON, J.M. & ROBIE, R.H. (1953) Geology of Venezuela and its oil fields. *Bull. Am. Assoc. Petrol. Geol.* **37**, 690–777.

Ministry of Mines and Petroleum (Columbia) (1944) General geological map of the Republic of Columbia scale 1:2,000,000. Bogota.

MINTER, W.E.L. (1978) A sedimentological synthesis of placer gold, uranium and pyrite concentrations in Proterozoic Witwatersrand sediments. In: *Fluvial Sedimentology* (Ed. Miall, A.D.) Can. Soc. Petrol. Geol. **5**, 801–829.

MOLNAR, P. & TAPPONIER, P. (1975) Cenozoic tectonics of Asia: effects of a continental collision. *Science* **189**, 419–426.

MYERS, R.E. (1991) The geology of the Godwan Basin, eastern Transvaal. Unpublished PhD thesis, University of the Witwatersrand, 320 pp.

MYERS, R.E., McCARTHY, T.S. & STANISTREET, I.G. (1990) A tectono-sedimentary reconstruction of the development and evolution of the Witwatersrand Basin, with particular emphasis on the Central Rand Group. *S. Afr. J. Geol.* **93**, 180–201.

MYERS, R.E., STANISTREET, I.G. & McCARTHY, T.S. (1993) Two-stage basement fault-block deformation in the development of the Witwatersrand Goldfields, South Africa. In: *Characterization and Comparison of Ancient (Precambrian–Mesozoic) Continental Margins* (Eds Bartholomew, M.J., Hyndman, D.W., Mogk, D.W. & Mason, R.) *Proc. 8th Int. Conf. on Basement Tectonics at Butte, Montana, USA.* D. Reidel Publishing Co., Dordrecht, Holland.

NAMI, M. (1983) Gold distribution in relation to depositional processes in the Proterozoic Carbon Leader placer Witwatersrand, South Africa. In: *Modern and Ancient Fluvial Systems* (Eds Collinson, J.D. & Levin, J.) Int. Assoc. Sedimentol. Spec Publ. **6**, 563–575.

NAYLOR, M.A., MANDL, G. & SIJPESTEIJN, C.H.K. (1986) Fault geometries in basement-induced wrench faulting under different initial stress states. *J. Struct. Geol.* **8**, 737–752.

PRETORIUS, D.A. (1976) The nature of the Witwatersrand gold–uranium deposits. In: *Handbook of Stratabound and Stratiform Ore Deposits* (Ed. Wolf, K.H.) **7**, 29–88.

PRETORIUS, D.A. (1986) The geometry of the Witwatersrand Basin and its structural setting in the Kaapvaal Craton. *Geocongress '86 Ext. Abstracts,* Geol. Soc. S. Afr., Johannesburg, pp. 47–52.

ROBB, L.J. & MEYER, M. (1991) Uranium-bearing accessory minerals in contrasting high-Ca and low-Ca granites from the Archaean Barberton Mountain Land. In: *Primary Radioactive Minerals* (Ed. Augustinius, S.S.) Theophrastus Publications, Athens, 417 pp.

ROD, E. (1956) Strike-slip faults of northern Venezuela. *Bull. Am. Assoc. Petrol. Geol.* **40**, 457–476.

SALES, J.K. (1983) Collapse of Rocky Mountain basement uplifts. *Rocky Mountain Assoc. Geol.* 79–97.

SARMIENTO, R. & KIRBY, R.A. (1962) Recent sediments of Lake Maracaibo *J. Sediment. Petrol.* **32**, 698–724.

SCHUMM, A.A. (1968) Speculations concerning palaeohydrologic controls of terrestrial sedimentation. *Geol. Soc. Am. Bull.* **79**, 1573–1588.

SCLATER, K.G. & CHRISTIE, P.A.F. (1980) Continental stretching: an explanation of the post mid-Cretaceous subsidence of the central North Sea basin. *J. Geophys. Res.* **85**, 3711–3739.

SPEARING, D.R. (1975) Shallow marine sands. In: *Depositional Environments as Interpreted from Primary Sedimentary Structures and Stratification Sequences.* (Eds Harms, J.C., Southard, J.B., Spearing, D.R. & Walker, R.G.) Soc. Econ. Palaeontol. Mineral. Short Course **2**, 103–132.

STANISTREET, I.G. & McCARTHY, T.S. (1991) Changing tectono-sedimentary scenarios relevant to the development of the Witwatersrand Basin. *J. Afr. Earth Sci.* **13**, 65–81.

STANISTREET, I.G., MARTIN, D. McB., SPENCER, R. & BENEKE, D. (1988) The importance of diamictites in the understanding of the tectonics and sedimentation of the Witwatersrand basin. *Ext. Abstracts Geocongress '88, 22nd Congr. Geol. Soc. S. Afr., Univ. Natal,* pp. 607–610.

STANISTREET, I.G., McCARTHY, T.S., CHARLESWORTH, E.G., MYERS, R.E. & ARMSTRONG, R.A. (1986) Pre-Transvaal wrench tectonics along the northern margin of the Witwatersrand Basin, South Africa. *Tectonophysics* **131**, 53–74.

TAINTON, S. & MEYER, F.M. (1990) The stratigraphy and sedimentology of the Promise Formation of the Witwatersrand Supergroup in the western Transvaal. *S. Afr. J. Geol.* **93**, 103–117.

TANKARD, A.J., JACKSON, M.P.A., ERIKSSON, K.A., HOBDAY, D.K., HUNTER, D.R. & MINTER, W.E.L. (1982) *Crustal Evolution of Southern Africa: 3.8 Billion Years of Earth History.* Springer-Verlag, Heidelberg, 523 pp.

TAPPONIER, P., PELTZER, G., LE DAIN, A.Y., ARMIJO, R. & COBBOLD, P. (1982) Propagating extrusion tectonics in Asia: new insights from simple experiments with plasticine. *Geology* **10**, 611–616.

TEGTMEYER, A.R. & KRÖNER, A. (1987) U-Pb zircon ages bearing on the nature of Early Archaean greenstone belt evolution, Barberton Mountain Land, southern Africa. *Precamb. Res.* **36**, 1–20.

TRUTER, F.C. & ROSSOUW, P.J. (1955) Geological map of the Union of South Africa 1:1 000 000. Geological Survey, Pretoria.

TUCKER, R.F. & VILJOEN, R.P. (1986) The geology of the West Rand Goldfield, with special reference to the southern limb. In: *Mineral Deposits of Southern Africa* (Eds. Anhaeusser, C.R. & Maske, S.) **1**, 649–688.

TURNER, P.A. (1979) The planar auriferous Witwatersrand conglomerates: pediment mantles of the Proterozoic. *Geocongress '79 Absracts. Geol. Soc. S. Afr., Port Elizabeth*, pp. 359–375.

VAN REENEN, D.D., BARTON, J.M., ROERING, C., SMIT, C.A. & VAN SCHALKWYK, J.F. (1987) Deep crustal response to continental collision: the Limpopo belt of southern Africa. *Geology* **15**, 11–14.

VAN WAGONER, J.C. POSAMENTIER, H.W., MITCHUM, R.M. *et al.* (1988) An overview of the fundamentals of sequence stratigraphy and key definitions. In: *Sea-level Changes: an Integrated Approach* (Eds Wilgus, C.K., Hastings, B.S., Posamentier, H.W., Kendall, St.C.C.G., Ross, C.A. & van Wagoner, J.C.) *Spec. Publ. Soc. Econ. Geol.Mineral.* **42**, 39–45.

VERREZEN, L. (1987) Sedimentology of the Vaal Reef palaeoplacer in the western portion of Vaal Reefs mine. Unpublished MSc thesis, Rand Afrikaans Univ., 194 pp.

VOS, R.G. (1975) An alluvial plain and lacustrine model for the Precambrian Witwatersrand deposits of South Africa, *J. Sediment. Petrol,* **45**, 480–493.

WALLMACH, T. & MEYER, F.M. (1990) A petrogenetic grid for metamorphosed aluminous Witwatersrand Shales. *S. Afr. J. Geol* **93**, 93–102.

WINDLEY, B.F. (1984) *The Evolving Continents,* 2nd Edn. John Wiley, London 399 pp.

WINTER, H. DE LA REY (1987) A cratonic foreland model for Witwatersrand Basin development in a continental back-arc, plate tectonic setting. *S. Afr. J. Geol.* **90**, 409–427.

Spec. Publs Int. Ass. Sediment. (1993) **20**, 337–397

Tectonically induced clastic sediment diversion and the origin of thick, widespread coal beds (Appalachian and Williston basins, USA)

E.S. BELT

Department of Geology, Amherst College, Amherst, MA 10002, USA

ABSTRACT

When clastic sediments are pervasive in a fresh-water depositional setting upgradient of eustatic influence, tectonically induced drainage diversion may be required to divert fluvial channels away from the basin so that regionally widespread peat deposits can develop. Two different settings for coal beds have been examined: the Appalachian foreland basin (Middle and early Upper Pennsylvanian strata), and the Williston intracratonic basin lower Palaeocene strata). Strata studied from the foreland basin accumulated within a few tens of kilometres of the Appalachian compressional orogen. Strata studied from the intracratonic basin lay hundreds of kilometres from the Bighorn and Black Hills source areas.

In each basin, a stratigraphical interval comprising coarse clastics sediments consists of well-defined trunk and subsidiary channel-belts that are coeval with lenticular coals in the adjacent floodbasins. The clastic-rich interval alternates in the strata with a thinner interval of regionally widespread coal and mudrock. Five regionally extensive intervals containing mineable coals with mud partings were studied within the non-marine facies of both basins. In both study areas, palaeogeographical maps show thick siliciclastic-dominated sequences with 10–30-km spacing between trunk drainages in a direction perpendicular to palaeoslope. This drainage spacing contrasts with the more than 75-km (Appalachian basin) to more than 100-km (Williston basin) spacing of trunk drainages mapped within the thinner intervals containing widespread coal deposits.

Previous models of the origin of the widespread coal and mud-rich intervals in the Appalachian basin depend either on autogenic processes (delta-lobe switching, avulsion, peat doming) or on allogenic processes (glacio-eustasy, climatic fluctuation, and lithospheric flexure). These models account for neither the change in the spacing of trunk drainages, nor the high water table during both the coarse clastic-rich intervals and the coal and mud-rich intervals, nor the apparent simultaneous accumulation of thick peat through many thousands of square kilometres within a few tens of kilometres of a tectonically active source area. Previous models of the origin of widespread coal and mud-rich intervals in the Williston basin depend either on autogenic processes (avulsion, peat doming) or on allogenic processes (glacio-eustasy, ratchet tectonics). None of these is likely to cause the widespread coal and mud-rich intervals examined.

Different drainage diversion models are proposed for the origin of widespread coal intervals in the two basins. A fold-salient and blind-thrust drainage diversion model accounts for the development of five regionally extensive Middle and early Upper Pennsylvanian coal beds that today abut the Allegheny Front. The depositional basin was flooded with coarse clastic sediments during the times when the thrust- induc d-topographic ridges were breached by streams.

In the Williston basin, source area tectonics had little or no influence on the change in drainage spacing during the development of widespread coal intervals. Palaeogeographical maps for the coarse clastic sediment interval above and below these coal/mud intervals show from 90° to 180° changes in the direction of the regional palaeoslope. Changes in the position of regional depocentres, when coupled with the period of time during which trunk channel-belts became deranged prior to re-establishment in the direction of the new palaeoslope, account for the development of widespread coal intervals. These conclusions are consistent with stream drainage diversion that resulted from tectonically induced basin topography, rather than from eustatic, climatic, or ratchet tectonic controls.

INTRODUCTION

Coal geologists and sedimentologists in recent years have debated the origin of widespread coal beds within thick siliciclastic sequences. Most would concur that ephemeral, lenticular coal beds are due largely to autogenic* processes controlled by sediment dispersal mechanisms, and that extremely widespread coal zones (those that extend for thousands to hundreds of thousands of square kilometre are influenced chiefly by allogenic processes. Delta-lobe switching, crevasse-splay development and the doming effect of peat bogs are the chief autogenic mechanisms considered by previous workers. Eustasy, climatic change, and tectonism are the chief allogenic mechanisms previously considered.

Some previous workers argue that autogenic processes dominate the origin of all coal beds. They believe that coal facies, regardless of how they might correlate, actually lens out every few kilometres. For example, the Upper Kittanning coal bed is shown as a series of discrete lenses at about the same stratigraphical position (Ferm & Staub, 1984, fig. 3). This *ad hoc* assumption of lenticularity is based on the obvious discontinuous occurrence of present-day peat deposits. As a consequence, it has long been held that Appalachian basin coal zones formed at sites that either bordered coastal lagoons and adjacent coastal plains, or formed in lower delta plains and in river floodplains (Ferm, 1970; Donaldson, 1974). Each of these types of autogenic environment is so commonly clastic-rich that coal beds thicker than a few centimetres are unlikely (McCabe, 1984; Kosters *et al.*, 1987; Scott, 1989) unless the conditions are appropriate for domed peat deposits (McCabe, 1984). In order for any autogenic hypothesis to be valid, syndepositional channel-belts must be demonstrated to have been coeval with the peat deposits.

Underground mining in western Maryland, and adjacent regions of West Virginia (Belt & Lyons, 1989) demonstrates that the Upper Freeport coal bed and zone extend more than 75 km parallel to the Allegheny Front (shown on Fig. 1). Sporadic distribution of these coals (Lyons *et al.*, 1984) is most commonly due not to alluvial ridge deposits coeval with the coal, but to erosion from younger

*The terms *allogenic* and *autogenic* rather than allocyclic and autocyclic are used in order to avoid the notion that cyclicity is present *de facto* in any of the facies herein described.

downcutting channels. However, in southwestern Pennsylvania coarse channel-belts have been shown to coexist with the widespread Upper Freeport coal deposits (Sholes *et al.*, 1979; Skema *et al.*, 1982). Because an entire coal-bearing zone is so much more widespread than the intervening siliciclastic facies, some type of allogenic mechanism must be involved when postulating its origin. However there is considerable debate over which of the many allogenic possibilities is most influential.

Some authors subscribe to a single factor allogenic model, that is, a model in which one basin-wide mechanism dominates not only the origin of the coal-rich interval but also the clastic-rich interval.

The most popular of the allogenic concepts is eustasy. Glacio-eustasy has been recently considered the most important single factor controlling the origin of widespread coal beds and their inter-bedded siliciclastic facies in the Appalachian basin (Busch & Rollins, 1984; Busch & West, 1987). Eustasy without glaciation (Chervan & Jacob, 1985) is a popular model for the Williston basin. The hypothesis of Busch and Rollins (1984) and Busch and West (1987) is probably untenable because of the lack of any marine deposits within 15–20 m stratigraphically above the top of the Upper Freeport coal bed and zone throughout western Maryland and central West Virginia (Belt & Lyons, 1989), eastern Ohio (Glen Merrill, written communication, 1987), northeastern Kentucky (Carlson, 1965), and southwestern Pennsylvania (Busch, 1984). Although Scanlon (1991) claims that, in central western Maryland, marine facies lie directly above the Upper Freeport coal zone (upper sub-facies A), marine fossils have not been found in those strata, (Belt & Lyons, 1989). The model that the Upper Freeport coal deposits formed on a lower delta plain (Ferm, 1970; Donaldson, 1974; Ferm & Staub, 1984) is also unlikely because the deposits above and below that coal zone are non-marine in origin and the channel-belts do not form a distributary pattern (Fig. 2) (Belt & Lyons, 1989).

Likewise, glacio-eustasy is not likely to be the cause of any coal beds in the strata studied in the Williston basin because continental glaciation has not been reported for the Late Cretaceous and Palaeocene. In addition, tectonically driven types of global sea-level change are unlikely to have controlled widespread coal deposition in nearly all of the Palaeocene facies because of the near absence of any marine deposits (Belt *et al.*, 1992).

Other previous models for the origin of wide-

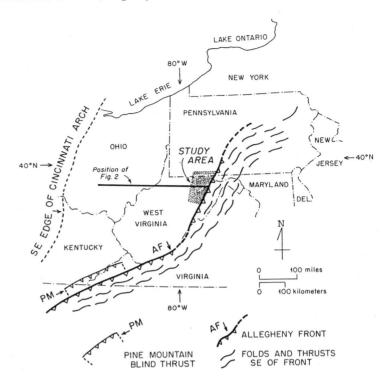

Fig. 1. Geological setting of the study area in the Appalachian basin.

spread facies concentrate on a mechanism of litho-spheric flexure. Heller *et al.* (1988) and Flemings and Jordan (1990) argue that lithospheric flexure is the most important control producing clastic-rich *vs* clastic-poor deposits within the Appalachian basin. Flemings and Jordan (1990) also argue that fine-grained clastic deposits formed in deep water dur-ing the thrust-loaded periods of downwarping. Coarse clastic deposits are considered to be absent in the basin centre and distal part of the foredeep but can be present in proximal regions. Although none of the above authors specifically address the origin of widespread coal deposits, lithospheric flexure clearly ought to have occurred during the time of peat accumulation in the Appalachian basin, but the question remains whether it was responsible for the origin of those coals. However, an entirely different allogenic mechanism is re-quired for similar coal deposits in the Williston basin where lithospheric flexure did not take place (Ahern & Mrkvicka, 1984; Fowler & Nisbet, 1985).

Both the Williston intracratonic basin and the Appalachian foreland basin are examined in this report to attempt to determine possible causes for the production of widespread coal beds in those two

basins (Table 1). Because the coal-bearing facies in each basin can be shown to have been beyond the influence of eustasy (Belt & Lyons, 1989; Wise *et al.*, 1991; Belt *et al.*, 1992), other allogenic factors, such as tectonism and climate change, that exclude eustatic effects become important considerations.

The influence of tectonism on the production and exclusion of siliciclastic sediment from a peat-accumulating basin has been considered infre-quently in recent publications, although the concept was widely understood from at least the late 19th century (Dawson, 1878, pp. 138–141). The exam-ples of tectonic influences on sediment diversion refer simply to the clastic material, not to peat deposits (Hirst & Nichols, 1986; Schmitt & Steidt-mann, 1990). There are two exceptions. One is a case where tectonic influences were applied to the origin of widespread coal deposits in a foreland basin (Richardson *et al.*, 1986). The other involved drainage diversion within the Williston deposi-tional basin (Winczewski & Groenewold, 1982).

In this paper, a drainage diversion/tectonic model is considered as an additional allogenic mechanism for developing widespread coal zones interbedded within thick siliciclastic sequences. It

Table 1. Comparison of central Appalachian and Williston basin settings for widespread coal deposits

Basin	Stratigraphical sequence	Tectonic setting	Depositional setting	Source of sediment	Stability of global climate	Palaeolatitude	Lithospheric flexure	Causes for siliciclastic diversion
Appalachian	Allegheny Formation and Lower Conemaugh Group	Foreland basin (compressional tectonics)	Alluvial plain with no eustatic influence	Appalachian orogen within 100 km of depositional basin	Unstable over short time spans due to active southern hemisphere glaciation; gradual long term changes	Equatorial region	Occurred but not of significant influence on facies studied	Peat swamps develop due to diversion of sediment from fold salients to embayments in orogenic zone
Williston	Fort Union Formation	Intracratonic basin (inversion tectonics)	Alluvial plain with eustatic influence possible at only two brief intervals	Laramide uplifts in Wyoming and S. Dakota over 100 km from depositional setting	Gradual change over long time spans; short glacio-eustatic changes unlikely	12° to 13°N of the equator	Did not occur	Peat swamps develop when drainages are diverted by the shifting of depocentre site from one region to another

may act aloneor in conjunction with other allogenic and autogenic mechanisms.

NORTHERN APPALACHIAN BASIN

The northern Appalachian basin study area lies west of the Allegheny Front in Garrett County, Maryland (a triangular-shaped part, west of Cumberland, MD) and portions of West Virginia south and southeast of Maryland (Fig. 1). Middle and early Upper Pennsylvanian strata there have been studied in the Upper Potomac coal field, the Lower Youghiogheny coal field (Jacobsen & Lyons, 1985), and in the Castleman coal field (Lyons *et al.*, 1985). Supplementary information was derived from the Upper Youghiogheny and Georges Creek coal fields in Garrett County as well as from some of the coal fields in southwestern Pennsylvania. (For location of these coalfields, see Fig. 4a.)

The interval examined in detail (Fig. 2) includes the strata between the Upper Freeport coal bed and the Brush Creek coal bed (Belt & Lyons, 1989; Wise *et al.*, 1991). The depositional details developed from this study interval have been extrapolated into southwestern Pennsylvania, eastern Ohio, northeastern Kentucky, and central West Virginia, using the Upper Freeport coal zone as the horizon of correlation (Fig. 2). Furthermore these depositional

details appear to be applicable to equivalent types of facies found through much of the Allegheny Formation below the Upper Freeport coal bed, including the sequence from the Lower Kittanning coal bed to the Brush Creek coal bed (Fig. 2).

Two contrasting facies (A and B, Fig. 3) occur in the stratigraphical interval between the Upper Freeport coal zone and the Brush Creek coal bed. Facies A is subdivided into three subfacies. The middle of the three, the Upper Freeport coal zone, is the most conspicuous. The Upper Freeport coal bed and coal zone consist of a regionally widespread coal and mudrock. Where the Upper Freeport coal bed contains the fewest and thinnest partings, it averages 2.7 m thick. It contains many thin mudrock partings, the middle parting usually being less than half a metre thick. In two localized areas, the middle parting reaches a thickness of 10 m in the southern part of the Upper Potomac coal field and this increases to 20 m for a distance of 18 km to the northeast. This thick parting probably represents a time of ponded water conditions within the coal swamp, as the region of the thick parting is roughly circular in plan. Other mudrock partings within the Upper Freeport coal zone are shoestring in geometry and sinuous and probably represent muddy channelways within the swamp.

The other two subfacies of interval A lie above and below the Upper Freeport coal zone. The lower

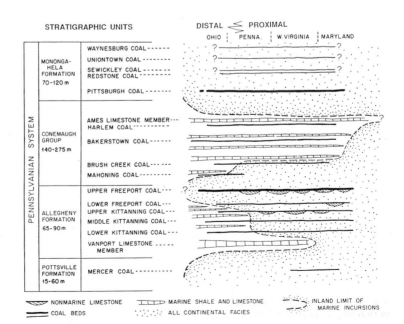

Fig. 2. Pennsylvanian facies of central Appalachian basin along line of section shown in Fig. 1. Stratigraphical interval examined in this report lies between Lower Kittanning coal bed and the Brush Creek coal bed. (Modified from Wise *et al.*, 1991, fig. 1.)

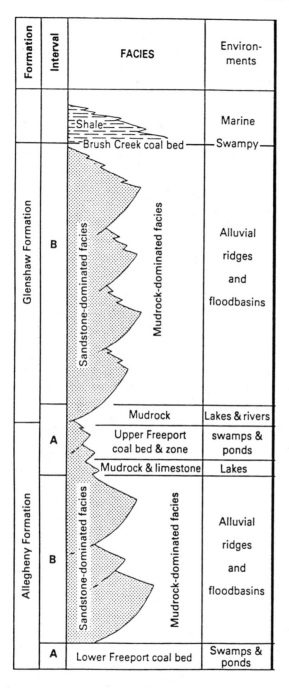

Formation	Interval	FACIES	Environments

Fig. 3. Depositional facies below and above the Upper Freeport coal zone. Facies interval A consists of fine-grained lake, lake-margin and coal-swamp deposits. Facies interval B consists of coarse sandy meander-belt deposits with laterally equivalent fine-grained floodplain deposits of crevasse lobe, lake, and lenticular coal swamp origin. (Modified from Belt & Lyons, 1989, fig. 14.)

subfacies has been called the Upper Freeport limestone. It consists of mudrock and limestone overlain by a seat earth associated with the overlying Upper Freeport coal zone. The mudrock is dark grey and contains fossil roots. It probably resulted from deposition during flood events but subsequently became an incipient palaeosol. The limestone is nodular, brecciated in places and yet laminated. Fossil roots developed in it after carbonate deposition ceased. It is probably of lacustrine origin. These observations on the Upper Freeport limestone are consistent with those of Weedman (1990) in southwestern Pennsylvania. All fine-grained limestone and mudrock units beneath the Upper Freeport coal zone are extremely widespread, though locally coarse sandstones and pebbly sandstones are found directly beneath the seat earth of the coal. Such sandstones are of channel-belt origin and probably represent the dying channel phases within the underlying facies B interval.

The Upper Freeport coal zone is overlain by mudrock and fine-grained sandstones of both lacustrine and fluvial origin (upper unit of facies A). The mudrock is well laminated and also contains flaser and linsen beds of rippled very fine sandstone with mud drapes. Escape and dwelling burrows of low invertebrate diversity are present locally. This low-diversity fauna plus the well preserved fronds and seeds of various types of ferns suggests a freshwater rather than marine depositional environment (P.C. Lyons, US Geological Survey, personal communication, 1987).

Facies B, above facies A in the study interval, consists of two subfacies: one is coarse-grained, and the other is fine-grained. The coarse-grained subfacies consists of coarse- to medium-grained cross-bedded sandstone and pebbly sandstone. These occur in well-defined multistoreyed meander-belts. Each storey contains trough and tabular cross-beds, and where the tops have not been eroded by younger storeys, the sandstones fine upward into current-rippled sandstone with mudrock interbeds (Belt & Lyons, 1989, figs 5 & 8). Clay plugs are locally present, and the dispersion of palaeocurrent vectors suggests that the palaeochannels had high to moderate sinuosity. This complex of coarse sandstones is referred to the Mahoning Sandstone Member. It is of fluvial origin within the study area. Where the sand grains have not been altered diagenetically, the Mahoning Sandstone contains rock fragments of sedimentary, metamorphic and plutonic origin (J.T. O'Connor, US Geological Survey, personal communication, 1988).

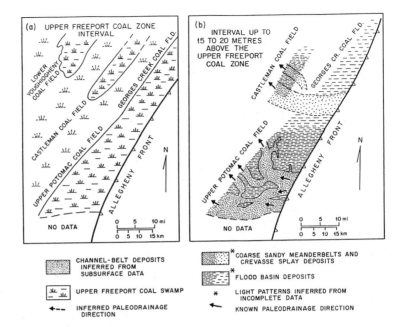

Fig. 4. Palaeogeographical maps superimposed on outlines of the various coal fields: (a) shows interval A and (b) shows interval B as defined in Fig. 3. (Modified from Wise *et al.*, 1991, fig. 2.)

Mahoning Sandstone units greater than 3 m thick were used (Belt & Lyons, 1989) to outline trunk and subsidiary palaeodrainage in the study area. A 20-m-thick interval of strata above the Upper Freeport coal zone was selected for the map shown in Fig. 4b. Channel-belts were defined, and the wide ones (some up to 10 km) contain the greater numbers of storeys. These are probably the longer-lived trunk systems. Their margins indicate a west-to-northwest orientation. This direction of the trunk palaeochannels (Belt & Lyons, 1989) is nearly perpendicular to the Allegheny Front (Fig. 1), which today forms the western boundary of the Valley and Ridge Province. These palaeodrainages flowed directly away from the Appalachian orogen and toward the Middle and Late Pennsylvanian sea, rather than turning axially down the foreland basin as is commonly the case (Richardson *et al.*, 1986).

The trunk meander-belts contain up to five stacked storeys suggesting that subsidence in the region was fairly slow. This contrasts with the channel-belts in the Williston basin which rarely show more than two storeys (Belt *et al.*, 1992) and where the subsidence rate was higher than in the Appalachian basin.

The fine-grained subfacies of facies B interval were deposited in the floodbasins that lie between the coarse sandy meander-belts. These deposits consist of mudrock and fine-grained sandstones of

lacustrine, palaeosol, swamp, and crevasse-lobe origin. Sinuous mudrock-filled channels with lateral accretion deposits of siltstone and claystone are interpreted as the result of anastomosed channels (Belt & Lyons, 1989, figs 10 & 11).

The most obvious deposits of the floodbasin are those of crevasse-lobe origin which comprise coarsening-up sequences of mudrock and fine- to medium-grained sandstone (Belt & Lyons, 1989, fig. 9). The lobes built out from the trunk channel-belts into the floodbasins. Thin lenticular coal beds are occasionally found at the top of the crevasse lobes.

Regional extrapolation

Because the Upper Freeport coal zone extends uninterrupted without coeval channel-belt breaks to the north into the Castleman and Youghiogheny coal fields, we conclude that the Upper Freeport swamp extended far further than 75 km parallel to the Allegheny Front (Wise *et al.*, 1991); probably for more than 150 km parallel to the Allegheny Front (Fig. 1).

Figure 4 shows a comparison between the A and B type of facies intervals found from the Upper Freeport level to 20 m above the Upper Freeport coal bed and coal zone. Figure 4a shows that the coal beds within interval A have been correlated parallel to the Allegheny Front for 75 km. Throughout the Maryland and West Virginia extent of the Upper Freeport coal zone, channel-belts of the kind found in interval B cannot be proven to have been coeval with the Upper Freeport coal zone (Fig. 4a). However, coarse sandstone channel-belts occur in Pennsylvania to the north of the study area (Sholes et al., 1979; Skema et al., 1982; Ferm & Staub, 1984; Stanton et al., 1986). Coarse standstones coeval with the Upper Freeport coal bed and coal zone preclude the climatic model of Cecil (1990) which relies on complete trapping of siliciclastics in the mountains during the humid (i.e. regional swamp) part of the climate cycle. In contrast to interval A, the trunk palaeodrainages in interval B occur in positions less than 30 km apart (Fig. 4b). Because the streams flowed directly away from the Allegheny Front, they presumably represent direct-line drainage away from the Appalachian orogen.

Discussion

Climatic fluctuation has been cited recently to explain the presence of widespread coal zones in the Appalachian basin (Cecil, 1990). This concept was first proposed by Wanless and Shepard (1936). Some aspects of Cecil's model are reasonable because, during Middle and early Late Pennsylvanian time, southern hemisphere glaciation caused many repetitions of climatic change (Veevers & Powell, 1987). Even though such change may not have been as pronounced in equatorial regions, such as the central Appalachian foreland basin (Table 1), as in higher latitudes, it might still have been recorded in the fossil flora.

Unfortunately, there is no evidence of Middle and early Late Pennsylvanian cycles of land-plant change (Phillips et al., 1985; Lyons & Darrah, 1989; Winston, 1990) consistent with a 0.5-My interval of time estimated (based on Lippolt et al., 1984) for the clastic deposition between each of the Kittanning and Freeport coal horizons. The megafloral evidence currently available indicates that gradual change took place throughout the time interval represented by Allegheny and Conemaugh strata.

The climatic model of Cecil (1990) thus relies more on assumed fluctuation in rainfall and sediment production, than on palaeobotanic relationships. During the time of widespread coal swamps, such as Upper Freeport, Cecil argues that higher rainfall caused a greater growth of vegetation in the mountains than occurred during drier times. This vegetation purportedly decreased the production of sediment from the mountainous regions, and hence coal swamps could develop in the depositional basin. The presence of wide, coarse siliciclastic channel-belts coeval with the Upper Freeport coal zone would seriously weaken the Cecil (1990) argument. As mentioned above, at least two such channel-belts have been found in southwestern Pennsylvania (Sholes et al., 1979; Skema et al., 1982).

Lithospheric flexure might be a viable allogenic control for at least some of the Middle and early Upper Pennsylvanian facies in the central Appalachian basin. In the clastic facies B interval, the coarseness of the channel-belt deposits suggests a proximal depositional setting. The coal/mudrock facies A intervals also are unlikely to represent the thrust-loading events, even though a good modern example of thrust-loaded peat and lake deposits exists today in Bangladesh (Coates et al., 1991). there, peat and lacustrine sediment is accumulating in a 100-km-wide zone within the tectonically depressed Sylhet basin abutting the ramp/thrust fault of the Shillong Massif (D.A. Coates, US Geological Survey, personal communication, 1991). The facies interval A of the study area is not considered to have resulted from thrust loading because water depths of both facies A and B (floodbasin phase) are essentially the same (Belt & Lyons, 1989). The facies do not indicate a deepening event but rather an increase and lessening of the supply of clastic sediment from the source area. Because there are no marine deposits above any Kittanning or Freeport coal deposit in the study area, both facies A and B intervals were probably deposited in a proximal depositional setting.

Each of the Kittanning and Freeport coal deposits represent very rapid tropical peat growth rates (approximately 1.0 mm/year or slightly greater) that outstripped the rapid thrust-loaded subsidence rates that ranged from 0.5 mm/year to 1.0 mm/year (McCabe, 1991, fig. 4). In the proximal depositional setting with no deepening events recorded, low-ash peat swamps are unlikely to develop unless a further refinement is added, namely exclusion of most clastic sediments by drainage diversion in the source area. Because both blind-thrust ridges and lithospheric loading are combined in the foreland basin compressive setting, the clastic-sediment diversion mechanism is perhaps the preferable controlling mechanism on the development of the peat swamps discussed here because regionally developed peat occurred about every 0.5 My, a rate much too high to be accounted for solely by lithospheric flexure. Thrust-loading events of 2 My duration are believed to have occured in the Appalachian basin (Flemings & Jordan, 1990).

ANTICLINAL RIDGE DIVERSION MODEL

Recent bedrock maps southeast of the Allegheny Front (Evans, 1989; Ferrill & Dunne, 1989; Wilson, 1989) indicate many zones of thin-skinned thrusting (Fig. 1). These are fold duplexes, including blind-thrusts, which might have formed topographic ridges during Middle and early Late Pennsylvanian time. If these topographic ridges diverted river systems then the trunk palaeodrainage could have been diverted many tens of kilometres away from some parts of the northern central Appalachian basin. This diversion would have focused the trunk channel-belts into a part of the Appalachian basin that would lie hundreds of kilometres to the northeast or southwest from the next trunk drainage system.

The fold salients and re-entrants of the present-day Zagros Mountains (Oberlander, 1965) have effectively diverted the main drainage of that area by 600 km into a 20-km-wide re-entrant. The modern topography was produced by thin-skinned folds whose deeply eroded analogues are found southeast of the Allegheny Front adjacent to the study area (Ferrill & Dunne, 1989). Blind-thrust ridges are seen today in other areas of the Appalachians, for example in the Sequatchee anticline and Pine Mountain thrusts (see PM, Fig. 1) southwest of the study area.

Recent work by Pashin and Carroll (1993) invokes drainage diversion that controlled facies in the Black Warrior coal basin of Alabama, but not by tectonics associated with the Sequatchee anticline. They propose a thrust ridge (the Birmingham anticlinorium) that blocked drainage during the early Pennsylvanian between the Black Warrior basin and the Cahaba basin to the southeast. The Cahaba rapidly subsided and trapped northwest-flowing sediment southeast of the Birmingham ridge, thus allowing many different types of facies (including coal) to develop in the larger Black Warrior basin to the northwest.

Thus during the Middle and early Late Pennsylvanian, various types of fold-induced topography could have produced a series of ridges between the main source of clastic sediments and the adjacent foreland basin. This type of topography would then divert the streams and their clastic loads into the embayments and away from the salients (Fig. 5a). Large, shallow lakes and areas of swamps that formed in the depositional basin were protected from coarse clastic sediments by the fold-salient ridges. Occasionally mud and very fine sand was brought into the basin by erosion of an adjacent fold ridge. Anastomosed streams fed the many ponds and lakes in the peat swamp. The mud that was deposited formed the numerous partings found in the coal bed including the thicker ones of the Upper Freeport coal zone (Belt & Lyons, 1989). Once erosion had breached the fold ridges (Fig. 5b),

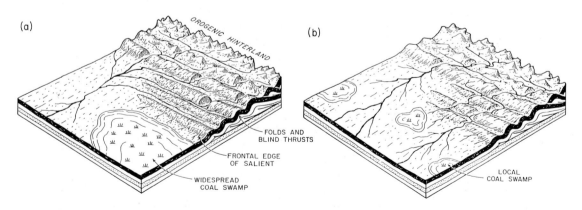

Fig. 5. (a) Proposed model for the origin of the peat and lacustrine interval A. (b) Proposed model for the origin of the coarse clastic depositional interval B. (Modified from Wise *et al.*, 1991, fig. 3.)

coarse siliciclastic material would be free to enter the portion of the basin that was previously protected. These siliciclastic inputs formed various types of channel-belts. Only thin, lenticular coal beds could form in the floodbasins between the trunk channel-belts.

The thickness and geographical distribution of each widespread coal bed in the Middle and early Late Pennsylvanian are different. The three Kittanning and two Freeport coal zones considered in this text (Fig. 2) differ significantly in regional extent, thickness, and type and number of partings. The model of fold-salient growth and destruction can account for such variation. Different rates of fold-ridge growth and erosional breakdown would control the thickness of the peat that formed in its lee. Furthermore, the length of the fold salient along depositional strike would control the regional extent of the resulting coal bed.

WILLISTON BASIN

Widespread coal beds and coal zones are also found within Palaeocene strata of the Williston basin (Fig. 6, Canadian portion not shown). The tectonic

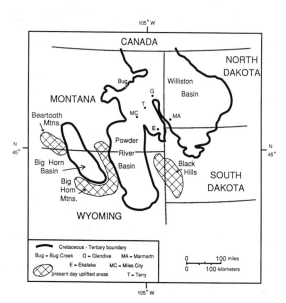

Fig. 6. USA portion of the Williston basin. Study area in Montana lies within the triangle described by MC, MA and E. Study area in North Dakota lies 30 km north and northeast of MA. (Modified from Belt *et al.*, 1992, fig. 1.)

setting and distance to the source of siliciclastic sediment in this basin contrasts markedly with the Appalachian basin, although the depositional environments are similar. The sediments in both basins were deposited on vast alluvial plains.

The Williston is an intracratonic basin that subsided sporadically from Early Cambrian to Early Tertiary times due to periods of thermal contraction (Ahern & Mrkvicka, 1984; Fowler & Nisbet, 1985). During the Middle and Late Mesozoic, a foreland basin formed along the western edge of the North American craton. Compressional forces directed to the east during the Late Cretaceous produced Sevier-type, thin-skinned overthrusting in western Wyoming and western Montana. Continued compression towards the east ultimately produced Laramide-type, thick-skinned basement uplifts in southwestern Montana, Wyoming, northwestern South Dakota, and central Colorado (Cross, 1986; Beck *et al.*, 1988; Dickinson *et al.*, 1988). These uplifts are high-angle reverse faults and/or normal faults where the basement and adjacent sediment within the basin are contiguous.

The Beartooth, Bighorn and Black Hills massifs (Fig. 6) are the uplifts that supplied sediment to the western Williston basin. These became important source areas for Early Palaeocene sediments that were transported by the aggrading streams of the alluvial plain (Belt *et al.*, 1984, 1992). When both the Bighorn and the Black Hills massifs had become sufficiently uplifted to define both sides of the Powder River basin (Fig. 6), fluvial systems were focused northward and northeastward into the Williston basin, although at times the drainage was reversed and flowed southward out of the Powder River basin, as earlier noted by Dickinson *et al.* (1988). During the latest Cretaceous, prior to the development of these three uplifts, most of the alluvial sediment deposited in the Williston basin originated in the overthrust belt of western Montana and Alberta.

Interval studied

The stratigraphical interval analysed includes the Late Cretaceous (Maastrichtian/Lancian) Hell Creek Formation and the Palaeocene Fort Union Formation (Fig. 7). Within the Fort Union Formation, facies of the Tullock, Lebo and Tongue River Members were examined in southeastern Montana (Belt *et al.*, 1992), and facies of the Ludlow, and Tongue River Member were examined in south-

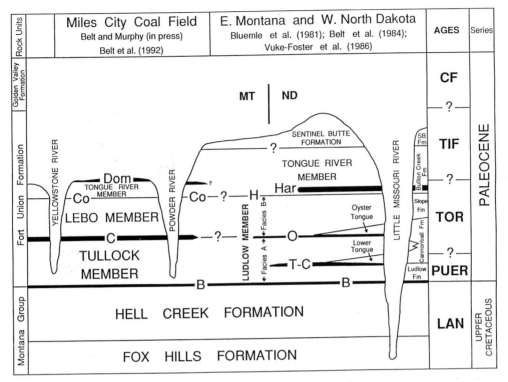

Fig. 7. Stratigraphical succession and correlation of facies discussed in the report. Lower Tongue and Oyster Tongue of the marine Cannonball Formation are brackish-water tongues in western North Dakota. All other facies are non-marine. Regionally widespread coal beds and coal zones are shown by heavy lines. Coal zones: C, C-coal zone; Co, Contact coal zone; Dom, Dominy coal zone; B, Boundary coal zone, T-C, T-Cross coal zone, O, Oyster coal bed; Har, Harmon coal zone. Mammal ages: LAN, Lancian; PUER, Puercan; TOR, Torrejonian; TIF, Tiffanian; CF, Clarkforkian. States: MT, Montana; ND, North Dakota. (Modified from Belt *et al.*, 1992, fig. 2.)

western North Dakota (Belt *et al.*, 1984). Palaeogeographical maps of the Hell Creek Formation were made for two districts near Ekalaka, Montana (Fig. 6, location E; Belt & Murphy, in press). Observations on palaeochannel directions were made near Glendive, Montana (Fig. 6, location G) and between Miles City and Ekalaka, Montana (Fig. 6, locations MC and E, respectively).

Most Lancian to Clarkforkian facies in Montana are of fluviolacustrine origin. These facies could not be of deltaic origin (as claimed by Cherven & Jacob, 1985) because tributaries enter the channel-belts and distributaries are not shown on the palaeogeographical maps (Belt *et al.*, 1992). In western North Dakota, however, two marine tongues of the Cannonball Member are evidence of at least some Puercan and Torrejonian delta facies. But all of the channel-belt patterns associated with those tongues show strictly fluvial patterns as well (Belt *et al.*,

1984). Only one coal bed, the Oyster coal bed (Fig. 7), is associated with one of the marine tongues, and might be of eustatic origin. Farther west, the C-coal zone is clearly of non-marine origin (Belt *et al.*, 1992, fig. 10). The fluvial aspect of the majority of facies is especially exhibited in the large Upper Ludlow and Tongue River meander-belts (Belt *et al.*, 1984). At this stratigraphical level, the Cannonball Sea, if it still existed, may have been several thousands of kilometres to the northeast, possibly in Canada, although evidence there is scanty (Broughton, 1979).

The Middle and early Upper Pennsylvanian and the Palaeocene facies discussed in this report are both fluvial in origin; however, their sedimentological styles differ considerably. The narrow channel-belt widths, the low number of storeys within a channel-belt, and the nearly complete vertical sequence of facies in individual channel deposits indicate that the subsidence rate in the Williston basin during the Early Palaeocene was more rapid than in the central Appalachian basin during the Middle and early Late Pennsylvanian. In the Williston basin widespread coal beds and coal zones are found only in Palaeocene strata. Any coal beds from the Late Cretaceous Hell Creek Formation are thin and lenticular. In contrast, the thick widespread coal beds of the Tongue River Member of western North Dakota, southeastern Montana, and northeastern Wyoming are extensively mined. These lie mainly in the Powder River basin (Fig. 6). Correlation of each of these coal horizons (Fig. 7) has been based more on lithostratigraphy than on biostratigraphy because age dating of Lancian and Palaeocene strata in the area, while locally excellent, is still incomplete (Archibald *et al.*, 1987; Archibald & Lofgren, 1990).

The coal beds and zones shown on Fig. 7 are but a small sample of the widespread coal horizons found in the Williston basin (Groenewold *et al.*, 1979; Cole & Sholes, 1980; Sholes *et al.*, 1982; Daly *et al.*, 1985; Sholes *et al.*, 1989; McLellan & Olive, 1991; Daly *et al.*, in press). Although some of these coal deposits lie hundreds to thousands of kilometres from the sources of clastic sediment, many of those found in southeastern Montana and in the Powder River basin lie within 100 km of these areas.

The coal beds and zones discussed here range in age from the boundary between the Hell Creek Formation and the Fort Union Formation (Boundary coal zone, Fig. 7) to Clarkforkian. Because these coal zones cover huge areas that greatly exceed the size of the floodbasins in the underlying and overlying clastic facies, previous views of their origin require consideration.

Previous models on the origin of Williston basin coal beds

Cherven and Jacob's (1985) concept of eustatic control on the origin of all coal beds in the Williston basin needs to be critically examined. As mentioned earlier, the Oyster coal bed (Fig. 7), is the only widespread one that might be of eustatic origin. Cherven & Jacob's (1985) claim for any others is untenable. The T-Cross coal bed, lying below the lower marine tongue (Fig. 7) lenses out within the Little Missouri study area (Belt *et al.*, 1984). Coal beds younger than the Oyster coal bed and older than the T-Cross coal bed were deposited in Williston basin facies that, at least south of Canada,

contained no marine deposits.

Three possible options for their origin are as follows.

1 Widespread swamps developed by the filling of large lakes. This is not unreasonable because many of the widespread coal beds either overlie, or are interbedded with, laminated muds that are likely to be of lacustrine origin. However, the origin of such widespread lakes would then need to be explained.

2 Each coal zone resulted from a regional climatic shift to more humid conditions; a shift to a drier climate brought a flood of clastic material into the basin. This is an unlikely hypothesis because ponded water facies are common in the floodbasins within the interval of clastic facies above and below the interval containing the widespread coal zone, thus indicating a consistent (non-cyclic) humid climate. These climatic conditions are further supported by the assemblages of plant species (Hickey, 1977, 1980; Johnson, 1982, 1989).

3 Source area relief was eroded down during the time when each widespread coal zone developed, hence clastic sediments would not have been available to the depositional basin (Kent, 1986). This is termed a ratchet tectonic model. The difficulty with this concept is that not only degradation, but also renewed uplift, would have to occur simultaneously in both the Bighorn and the Black Hills source areas. Either of the two source areas alone could have produced all of the accumulated clastic sediment.

If ratchet tectonics had occurred, the stratigraphical record should contain an unconformity for each uplift event. One unconformity has been discovered on the northwest margin of the Black Hills. Preliminary results show it to have preceded the Late Torrejonian and to have occurred after the Early Torrejonian (unpublished data by S.M. Vuke-Foster, Montana Bureau of Mines and the author, 1991). Published information on uplifts associated with either of these source areas indicates only one uplift event for each source area during the entire Palaeocene (McLellan *et al.*, 1990; Whipkey *et al.*, 1991). Thus the concept that cyclic or periodic source area uplift and erosion events would produce widespread coal zones in various districts throughout the Williston basin seems unlikely.

Exclusion of clastic sediments from the Palaeocene depositional basin (McCabe, 1984, 1987, 1991) is the most viable concept for the production of widespread coal zones in the Williston basin (Belt *et al.*(1992). Eustasy, source-area tectonics, and climatic shift are not likely models to accomplish such a marked reduction of clastic deposits over wide areas. The model of palaeodrainage switching introduced below is consistent with the known tectonic history of the Williston basin and is based on maps showing directions of channel-belts from the Miles City coal field (Belt *et al.*, 1992) and, nearly 100 km to the east, from the Little Missouri River study area (Belt *et al.*, 1984).

Facies studied

Strata in the Williston basin can be subdivided into clastic-rich intervals and coal-rich intervals just as in the central Appalachian basin. Lebo Member clastic-rich intervals contain channel-belts, typically with well-defined levees, and floodbasins. The flood-basin facies contains crevasse lobe, lacustrine and, in rare instances, anastomosed channel deposits. In contrast, the clastic-rich intervals of the Tongue River Member contain few or no leves deposits, and the floodbasins are typically the site of ponded water.

In both Lebo and Tongue River facies, lenticular coals developed in the floodbasin. Those from the Lebo generally developed farthest from the alluvial ridges, whereas those in the Tongue River occur on thin crevasse lobes that passed into standing water on the floodplain. This style of lenticular coal contrasts markedly with the widespread coal beds and zones found in the Palaeocene strata.

Widespread coal beds and coal zones in the Williston basin developed within discrete stratigraphical intervals. Like the Upper Freeport coal bed and coal zone of the central Appalachians, clastic sediments are associated with these coals, but in contrast no major channel-belts have been established to be coeval with any one of the zones. Compared with the Upper Freeport coal zone, the Palaeocene coal zones can contain very small sandy channe-belts between the bounding coals of the zone (Belt *et al.*, 1992, fig. 10). These channel sand deposits are all fine-grained and are noticeably different in scale from the trunk channel-belts in the clastic-rich interval of strata above and below the coal zone.

Evidence of palaeodrainage direction changes

Fifteen local palaeogeographical maps were made at different stratigraphical horizons in Lebo and Tongue River facies in the Miles City coal field, although only eight of these have been published (Belt *et al.*, 1992). Twelve local palaeogeographical maps were made at different stratigraphical horizons in the Ludlow and Tongue River facies in the Little Missouri study area, and seven of these were published (Belt *et al.*, 1984).

Figure 8 shows more complete palaeogeographical coverage than published earlier, including Hell Creek Formation maps (Belt & Murphy, in press), and some observations on the Tullock Member along the Powder River between Miles City and Ekalaka (Fig. 6). It also shows that Palaeocene palaeodrainages have shifted in time yet maintained a consistent trend over wide regions. This has been demonstrated throughout nearly 400 m of Late Cretaceous and Palaeocene strata (approximately 8 My) in both eastern Montana and western North Dakota. These palaeodrainages are also shown on regional palaeographical maps (Figs 9–12). The conclusions of this regional synthesis are as follows.

Regional trunk palaeodrainages flowed to the south, southeast, and the east during deposition (approximately 3 My) of Hell Creek and Tullock/Lower Ludlow sediments (Fig. 9). At this time, neither the Bighorns nor the Black Hills had yet been uplifted (Belt *et al.*, 1992). The Bighorns became uplifted at about 65 Ma, which is just after or at the K/T boundary (Giegengack *et al.*, 1988, personal communication, 1989; Whipkey *et al.*, 1991), yet the palaeodrainages to the north of them persisted to flow to the southeast (Fig. 10). These Tullock rivers entered the Cannonball sea presumably in central South Dakota. At the position of the C-coal zone (southeastern Montana) and of the Oyster coal bed (southwestern North Dakota), the Black Hills possibly became uplifted. This contrasts with the views of McLellan *et al.* (1990) who reported uplift at about the K/T boundary.

The Mid-Torrejonian uplift of the Black Hills created a major shift in palaeodrainage directions throughout the western Williston basin (Figs 8, 10 & 11). During the time of Lebo/Upper Ludlow deposition (approximately 2 My) the palaeodrainage direction was to the northeast into the Cannonball sea, which was retreating into Canada. The early Tongue River sedimentation in western Williston basin showed less regional consistency of palaeodrainage directions, though the strata contain many more and thicker widespread coal zones. Some of the earliest directions are to the southwest (Figs 8 & 12).

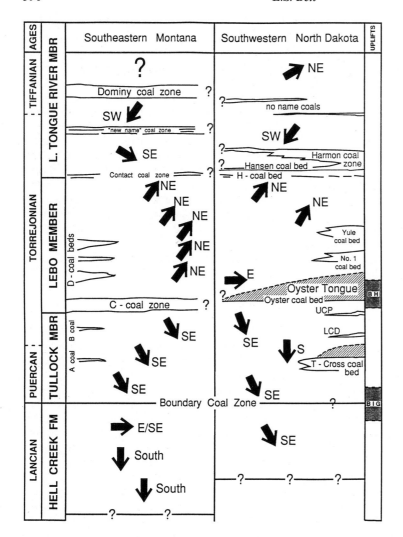

Fig. 8. Heavy arrows indicate average palaeodrainage directions based on a palaeogeographical map (Belt et al., 1992; Belt and Murphy, in press) at the stratigraphical position shown. (Modified from Belt et al., 1992, fig. 21.)

DEPOCENTRE SHIFTING MODEL

Nemec (1988), in a recent article on coal bed correlation within subsiding basins, has argued that peat swamp development can be controlled by intrabasinal differential subsidence. Unfortunately, Nemec did not give any example to support his hypothesis. The Williston basin model suggested by us (Belt et al., 1992) appears to be such a case. Our conclusions rest mainly on two types of evidence: (1) the presence of intervals of siliciclastic fluvial and lacustrine facies that are punctuated by an interval dominated by widespread coal beds; and (2) trunk palaeodrainages in the siliciclastic intervals above the coal zone diverge by 90° to 180° in

trend from the palaeodrainage direction below the coal zone.

The primary requirement for concluding a tectonic control is to establish that the coal beds and coal zones could not have been formed simply on floodplains between large channel-belts. Thus the spacing between palaeodrainage channels in the clastic-sediment interval must be less than the minimal aerial extent of the coal zone. Fortunately the palaeodrainage spacing can be determined from several of the palaeogeographical maps, both in eastern Montana and in North Dakota (Belt et al., 1984, 1992). The results of those studies indicate that the trunk drainage spacing, parallel to dip of the regional palaeogradient, changes from a few

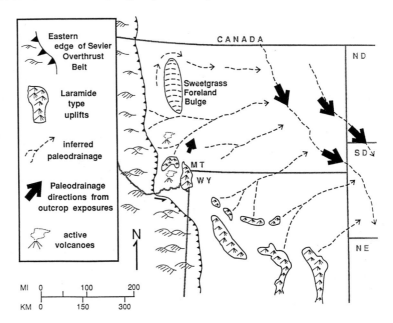

Fig. 9. Hell Creek (Lancian) palaeogeography. Heavy arrows are palaeodrainage directions based on field work by the author and by Fastovsky and Dott (1986) and Johnson *et al.* (1989). Note the absence of Beartooth, Bighorn, and Black Hills basement-cored Laramide uplifts.

tens of kilometres (in the clastic interval) to over 100 km (in the coal zone). A summary is shown in Fig. 8. Note that the palaeodrainage direction, and hence the direction of maximum differential subsidence below both the C-coal zone (Montana) and the

Oyster coal zone (North Dakota), lies to the southeast, whereas the palaeodrainage direction above those two coal zones lies to the northeast. Other palaeodrainage changes associated with widespread coal zones are also noted on Fig. 8, although the

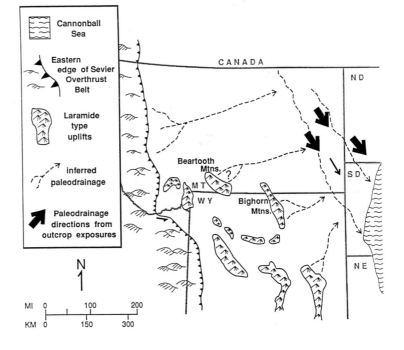

Fig. 10. Tullock and Lower Ludlow (Puercan and Early Torrejonian) palaeogeography. Heavy arrows represent the data collected by the author (Belt *et al.*, 1992). Note the presence of the Cannonball Sea. The Bighorns and possibly the Beartooths were uplifted and acted as source areas. The absence of the Black Hills is based on the palaeodrainage directions that trend directly towards and through the position of the uplift. (Modified from Belt *et al.*, 1992, fig. 24.)

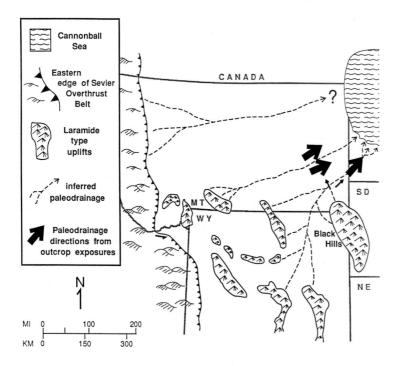

Fig. 11. Lebo and Upper Ludlow (Middle Torrejonian) palaeogeography. Heavy arrows represent the data collected by the author (Belt *et al.*, 1984, 1992). Note that the Black Hills were uplifted, and that the regional palaeodrainage direction is northeast. This coincides with the retreat of the Cannonball Sea into Saskatchewan, Canada, followed by the progradation of the Ludlow delta in the same direction (Belt *et al.*, 1984). (Modified from Belt *et al.*, 1992, fig. 25.)

most widespread and best dated coal zone, the Boundary coal zone, is not associated with any palaeodirection changes within the Williston basin. Perhaps in this case, a climatic threshold was passed from dry and warm to more humid and temperate (Johnson *et al.*, 1989). With the exception of the C-coal zone and Oyster coal bed, most of these drainage changes are not related to uplift in the source area. Because these widespread coal beds are so far removed (in excess of 150 km) from the Black

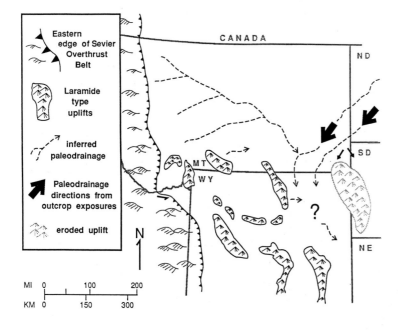

Fig. 12. Lower Tongue River (Late Torrejonian) palaeogeography. Heavy arrows represent the data collected by the author (Belt *et al.*, 1984, 1992). Palaeodrainage directions seem to suggest that depocentres C and possibly D (see Fig. 13) were controlling drainage directions.

Hills and the Bighorn uplifts, the cause for palaeodrainage shifting observed in the Miles City study area (Belt *et al.*, 1992) and in the Little Missouri River study area (Belt *et al.*, 1984) must be sought within the Williston basin rather than in the regions of the basement uplift that lay in Wyoming and South Dakota (Fig. 6).

Differential subsidence might have caused the shifting of the largest depocentres, which, in turn, would have established sand-filled channel-belts that became beheaded. Beheading would have produced a time of deranged drainage within an area during which little or no sand would have entered the region. During this time, channel-belt re-adjustment would take place until the developing depocentre began to attract sandy channel-belts and the new drainage direction became established. Peat swamps, if optimal edaphic conditions prevailed, would develop in the deranged region during the time of channel re-adjustment.

The locations of the various postulated depocentres are shown in Fig. 13. The location at the largest Late Lancian to Early Torrejonian depocentre is shown at position A (which also might have been

situated to the south in Nebraska). Depocentre A seemed to consistently attract all the drainages external to the Laramide intramontaine basins. Some of these channel-belts may have originated in the overthrust belt of central Alberta. Neither the Late Lancian to Early Puercan uplift of the Bighorns (Fig. 8) nor the gradual northward movement of the Cannonball Sea (Fig. 10) seem to have deflected the southeast palaeodrainage directions.

The uplift of the Black Hills in Mid-Torrejonian time effectively blocked the earlier southeast flowing drainages (Fig. 11). At this time, the region of dominant subsidence shifted from locality A to locality B (Fig. 13), the region which had been a depocentre in the Williston basin during much of the Palaeozoic (Fowler & Nisbet, 1985). During the same period the Cannonball Sea began to leave the Williston basin (Fig. 11). From Late Torrejonian and Early Tiffanian times shifts of depocentre position seemed to have been more frequent and less confined within the Williston basin. Regionally widespread, thick (up to 10 m) peat beds became more abundant. These changes in style are consistent with the Tongue River depositional setting, which contained widespread lakes on the floodplains (Belt *et al.*, 1992). The sediments were deposited in an area with low regional gradient. Slight increase in differential compactional and/or tectonic subsidence would thus cause highly divergent palaeodrainage changes. Those depocentres with greater differential subsidence would have attracted the major palaeodrainages. In one case the drainages flowed toward the southwest (Fig. 8). The depocentres may have been on either the Bighorn or Powder River basin (Fig. 13, localities D and C, respectively). If the Powder River basin had attracted siliciclastic deposits during early Tongue River time, the outflow of sediment from that basin would have been to the south, as suggested by Dickinson *et al.* (1988).

The Bighorn and the Black Hills basement-cored, thick-skinned Laramide thrusts resulted from varying stresses that were transmitted into and through the Williston basin cratonic crust to the northeast (Cloetingh, 1988). The uplift of the Bighorns caused no changes in palaeodrainage suggesting that no shift in depocentre occurred (Figs 9 & 10). In contrast, the Mid-Torrejonian Black Hills uplift caused major shifting of palaeodrainages (Fig. 8) and resulted in the widespread development of the C-coal and Oyster coal beds.

Varying stress patterns due to Laramide compres-

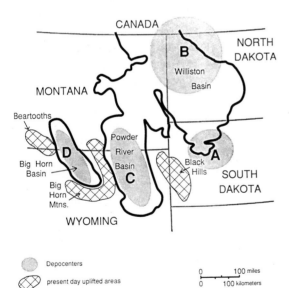

Fig. 13. Depocentres that influenced Late Cretaceous and Early Palaeocene palaeodrainage directions. Depocentre A was of greatest significance during the Lancian, Puercan and Early Torrejonian. Depocentre B was of greatest significance during the Middle Torrejonian. Depocentre C and possibly D were significant during the Late Torrejonian and Early Tiffanian. (Modified from Belt *et al.*, 1992, fig. 23.)

sive forces directed to the northeast along the southwestern margin of the Williston basin crust seem to account for the shifting of depocentres with varying intensity during the 8-My depositional record. This, in turn, caused the diversion of palaeodrainages with the consequent production of widespread peat-forming swamps.

CONCLUSIONS

Facies in the Appalachian and Williston basins were deposited by similar non-marine processes in different tectonic settings. The successions are subdivided into two equally widespread types of stratigraphical intervals. Intervals dominated by coarse channel-belt and fine floodbasin siliciclastic, sediments alternate with those dominated by coal, mudrock and/or freshwater limestone. Palaeogeographical maps of the siliciclastic-rich intervals show 10- to 30-km spacing between trunk channel-belts. This contrasts with the drainage spacing in the coal and mudrock-rich intervals of more than 75 km in the Appalachian basin and more than 100 km in the Williston basin. These relationships suggest allogenic mechanisms were responsible for the origin of the coal and mudrock-rich interval.

The most important allogenic mechanisms considered by previous workers have been eustasy, climatic fluctuation, and lithospheric flexure. Eustasy was found to be an unlikely controlling factor for the origin of the ten widespread coal zones considered in this paper because none are overlain (or underlain) by marine deposits. Climatic fluctuation is an unknown factor because of the lack of palaeobotanical evidence for temperature/humidity changes in the order of 0.5-My intervals. Lithospheric flexure can be ruled out in the Williston basin, and although present in the Appalachian basin, the facies studied do not record evidence of deepening pulses on a 0.5-My cycle. Drainage diversion is preferable because both the clastic sediment-rich and the coal and mudrock-rich intervals indicate similar shallow-water lacustrine and swamp conditions. Two models of drainage diversion of clastic sediment have been proposed (Wise et al., 1991; Belt et al., 1992). In the central Appalachian model, the diversion takes place in the source area due to compressional folds that produce topographic uplift in the drainage basins of the eroding streams. This, in turn, caused marked changes in the distances between trunk

drainages depositing within the basin. Similar effects can be seen in the Zagros Mountains today (Oberlander, 1965). As the topography is eroded in the intervals between thrust phases, the depositing streams in the basin resume a normal distance between the trunk alluvial ridges.

In the Williston basin model, the diversion of clastic material takes place within the depositional basin. Subcrustal interplate stresses (Cloetingh, 1988) caused the major depocentres to shift from one region to another. Palaeodrainages followed these shifts and, during the period of stream capture, large areas of deranged drainage were produced. Because high water-table conditions generally pertained in the Palaeocene facies, regionally extensive coal swamps developed in the deranged regions. Once the new palaeodrainage direction became established, the influx of coarse clastic material precluded further development of regionally extensive coal swamps.

These models allow for varying geometries of the regionally extensive coal swamps because the timing and extent of the tectonic influence varied in the two basins. This is consistent both with the different thicknesses and with the regional extent of the coal deposits.

ACKNOWLEDGEMENTS

George deV. Klein, Paul C. Lyons, and Gary T. Nichols critically reviewed the manuscript; their comments have been very helpful. Other discussions with Don Wise, Peter Crowley, and Bill Read helped to focus the arguments. Funding for these projects has been from the US Geological Survey (Coal Branches of both Denver, CO and Reston, VA), the North Dakota Geological Survey, the National Science Foundation (ROA grants), the W. M. Keck Foundation, an Amherst College fund for faculty research, and from personal resources. Marie Litterer drafted Figs 1–5; Emily H. Belt crafted the remainder by computer.

REFERENCES

AHERN, J.L. & MRKVICKA, S.R. (1984) A mechanical and thermal model for the evolution of the Williston Basin. *Tectonics* **3**, 79–102.
ARCHIBALD, J.D. & LOFGREN, D.L. (1990) Mammalian zonation near the Cretaceous–Tertiary boundary. In: *Dawn of the Age of Mammals in the Northern part of the Rocky*

Mountain Interior, North America (Eds Bown, T.M. & Rose, K.D.) Geol. Soc. Am. Spec. Pap. **243**, 31–50.

ARCHIBALD, J.D., GINGERICH, P.D., LINDSAY, E.H., CLEMENS, W.A., KRAUSE, D.W. & ROSE, K.D. (1987) First North American land mammal ages of the Cenozoic Era. In: *Cenozoic Mammals of North America* (Ed. Woodburne, M.O.) University of California Press, Ch. 3, 24–76.

BECK, R.A., VONDRA, C.F., FILKINS, J.E. & OLANDER, J.D. (1988) Syntectonic sedimentation and Laramide basement thrusting, Cordilleran foreland; timing of deformation. In: *Interaction of the Rocky Mountain Foreland and the Cordilleran Thrust Belt* (Eds Schmidt, C.J. & Perry, W.J.) Geol. Soc. Am. Mem. **171**, 465–487.

BELT, E.S. & LYONS, P.C. (1989) A thrust-ridge paleodepositional model for the Upper Freeport coal bed and associated clastic facies, Upper Potomac coal field, Appalachian Basin, U.S.A. In: *Peat and Coal. Origin, Facies, and Depositional Models* (Eds Lyons, P.C. & Alpern, B) *Int. J. Coal Geol.* **12**, 293–328.

BELT, E.S. & MURPHY, D.A. (in press) Fluvial facies and tectonic setting of the Hell Creek Formation near Ekalaka, southeastern Montana. In: *Magnetostratigraphy, Biostratigraphy, and Lithostratigraphy Spanning the Cretaceous–Tertiary Boundary in Eastern Montana* (Eds Butler, R.F. & Archibald, J.D.) University of Wyoming, Contributions to Geology.

BELT, E.S. FLORES, R.M., WARWICK, P.D., CONWAY, K.M., JOHNSON, K.R. & WASKOWITZ, R.S. (1984) Relationship of fluviodeltaic facies to coal deposition in the Lower Fort Union Formation (Palaeocene), south-western North Dakota. In: *Sedimentology of Coal and Coal-bearing Sequences* (Eds Rahmani, R.A. & Flores, R.M.) Int. Assoc. Sedimontol. Spec. Publ. **7**, 177–195.

BELT, E.S., SAKIMOTO, S.E.H. & ROCKWELL, B.W. (1992) A drainage diversion hypothesis for the origin of widespread coal beds in the Williston basin; examples from Paleocene strata, eastern Montana. In: *Coal Geology of Montana* (Eds Sholes, M.A. & Vuke-Foster, S.M.) Spec. Publ. Montana Bur. Mines Geol. **102**, 21–60.

BROUGHTON, P.L. (1979) Origin of coal basins by salt solution tectonics in western Canada. PhD thesis, University of Cambridge, England, 274 pp.

BUSCH, R.M. (1984) Stratigraphic analysis of Pennsylvanian rocks using a hierarchy of transgressive–regressive units. PhD thesis, University of Pittsburgh, PA, 427 pp.

BUSCH, R.M. & ROLLINS, H.B. (1984) Correlation of Carboniferous strata using a hierarchy of transgressive–regressive units. *Geology* **12**, 471–474.

BUSCH R.M. & WEST, R.R. (1987) Hierarchical genetic stratigraphy: a framework for paleoceanography. *Paleoceanography* **2**, 141–164.

CARLSON, J.E. (1965) Geology of the Rush Quadrangle, Kentucky. US Geological Survey, Map GQ–408, with marginal notes.

CECIL, C.B. (1990) Paleoclimate controls on stratigraphic repetition of chemical and siliciclastic rocks. *Geology.* **18**, 533–536.

CHERVEN, V.B. & JACOB, A.G. (1985) Evolution of Paleogene depositional systems, Williston Basin, in response to global sea level changes. In: *Cenozoic Paleogeography of the West-Central United States* (Eds Flores, R.M. & Kaplan, S.S.) Rocky Mountain Sect. Soc. Econ. Palae-

ontol. Mineral. Rocky Mountain Paleogeography Symposium, no. 3, pp. 127–170.

CLOETINGH, S. (1988) Intraplate stresses: a new element in basin analysis. In: *New Perspectives in Basin Analysis* (Eds Kleinspehn, K.L. & Paola, C.) Springer-Verlag, New York, pp. 205–230.

COATES, D.A., WHITNEY, J.W., SAWATZKY, D.L. & ALAM, A.K.M.K. (1991) Holocene deformation in the Ganges–Brahmaputra delta: a factor in flood distribution in Bangladesh. *Fall Meeting Program and Abstracts*, 9–13 December 1991, Am. Geophys. Union, 72, no. 44, 501.

COLE, G.A. & SHOLES, M.A. (1980) *Geology of the Anderson and Dietz Coal Beds, Big Horn County, Montana.* Montana Bureau of Mines and Geology, Geologic Map 14 and text 23 pp.

CROSS, T.A. (1986) Tectonic controls of foreland basin subsidence and Laramide style deformation, western United States. In: *Foreland Basins* (Eds Allen, P. & Homewood, P.) Int. Assoc. Sedimentol. Spec. Publ. **8**, 15–39.

DALY, D.J., GROENEWOLD, G.H. & SCHMIT, C.R. (1985) Paleoenvironments of the Paleocene Sentinel Butte Formation, Knife River area, west–central North Dakota. In: *Cenozoic Paleogeography of West-Central United States* (Eds Flores, R.M. & Kaplan, S.S.) Rock Mountain Sect. Soc. Econ. Paleontol. Mineral., Rocky Mountain Paleogeography Symposium, no. 5, pp. 171–185.

DALY, D.J. , SHOLES, M.A. & SCHMIT, C.R. (in press) Geology and depositional environments of the coal-bearing strata of the Fort Union lignite region. In: *Geology and Utilization of Fort Union Lignites* (Eds Finkelman, R., Tewalt, S. & Daly, D.J.) Energy and Environmental Research Centre, Grand Forks, North Dakota.

DAWSON, J.W. (1878) *Acadian Geology: the Geological Structure, Organic Remains and Mineral Resources of Nova Scotia, New Brunswick, and Prince Edward Island,* 3rd Edn. Macmillan, London, 694 pp.

DICKINSON, W.R., KLUTE, M.A., HAYES, M.J. *et al.* (1988) Paleogeographic and paleotectonic setting of Laramide sedimentary basins in the central Rocky Mountain region. *Geol. Soc. Am. Bull.* **100**, 1023–1039.

DONALDSON, A.C. (1974) Pennsylvanian sedimentation of central Appalachians. In: *Carboniferous of the Southeastern United States* (Ed. Briggs, G.) Geol. Soc. Am. Spec. Pap. **148**, 47–78.

EVANS, M.A. (1989) The structural geometry and evolution of foreland thrust systems, northern Virginia *Geol. Soc. Am. Bull.* **101**, 339–354.

FASTOVSKY, D.E. & DOTT, R.H. JR. (1986) Sedimentology, stratigraphy, and extinctions during the Cretaceous–Paleogene transition at Bug Creek, Montana. *Geology* **14**, 279–282.

FERM, J.C. (1970) Allegheny deltaic deposits. In: *Deltaic Sedimentation Modern and Ancient* (Ed. Morgan, J.P.) Soc. Econ. Paleontol. Mineral. Spec. Publ. **15**, 246–255.

FERM, J.C. & STAUB, J.R. (1984) Depositional controls of mineable coal bodies. In: *Sedimentology of Coal and Coal-bearing Sequences* (Eds Rahmani, R.A. & Flores, R.M.) Int. Assoc. Sedimentol. Spec. Publ. **7**, 275–289.

FERRILL, D.A. & DUNNE, W.M. (1989) Cover deformation above a blind duplex: an example from West Virginia, U.S.A. *J. Struct. Geol.* **11**, 421–431.

FLEMINGS, P.B. & JORDAN, T.E. (1990) Stratigraphic modeling of foreland basins: Interpreting thrust deformation and lithosphere rheology. *Geology* **18**, 430–434.

FOWLER, C.M.R. & NISBET, E.G. (1985) The subsidence of the Williston Basin. *Can. J. Earth Sci.* **22**, 408–415

GIEGENGACK, R., OMAR, G.I., LUTZ, T.M. & JOHNSON, K.R. (1988) Tectono-thermal history of the Bighorn Basin and adjacent highlands, Montana – Wyoming. *Geol. Soc. Am. Abstr. Programs* **20**, 6, 416.

GROENEWOLD, G.H., HEMISH, L.A., CHERRY, J.A., REHM, B.W., MEYER, G.N. & WINCZEWSKI, L.M. (1979) Geology and geohydrology of the Knife River basin and adjacent areas of west–central North Dakota. *North Dakota Geol. Surv. Rep. Investigation* **64**, 402 pp.

HELLER, P.L., ANGEVINE, C.L., WINSLOW, N.S. & PAOLA, C. (1988) Two-phase stratigraphic model of foreland-basin sequences. *Geology* **16**, 501–504.

HICKEY, L.J. (1977) Stratigraphy and paleobotany of the Golden Valley Formation (Early Tertiary) of western North Dakota. *Geol. Soc. Am. Mem.* **150**, 183 pp.

HICKEY, L.J. (1980) Paleocene stratigraphy and flora of the Clark's Fork basin. In: *Early Cenozoic Paleontology and Stratigraphy of the Bighorn basin, Wyoming,* (Ed. Gingerich, P.D.) University of Michigan, Papers on Paleontology, no. 24, 33–49.

HIRST, J.P.P. & NICHOLS, G.J. (1986) Thrust tectonic controls on Miocene alluvial distribution patterns, southern Pyrenees. In: *Foreland Basins* (Eds Allen, p. A. & Homewood, P.) Int. Assoc. Sedimentol. Spec. Publ. **8**, 369–394.

JACOBSEN, E.F. & LYONS, P.C. (1985) *Coal Geology of the Lower Youghiogheny Coal Field, Garrett County, Maryland.* US Geological Survey, Coal Investigation Map, C-101, with marginal notes.

JOHNSON, K.R. (1982) Paleobotany and sedimentology of some lower Ludlow strata, Fort Union Formation, southwestern North Dakota. Senior honors thesis, Amherst College, Amherst, MA, 159 pp.

JOHNSON, K.R. (1989) A high-resolution megafloral biostratigraphy spanning the Cretaceous–Tertiary boundary in the northern Great Plains. PhD dissertion, Yale University, New Haven, CT, 556 pp.

JOHNSON, K.R., NICHOLS, D.J., ATTREP, M. JR. & ORTH, C.J. (1989) High-resolution leaf-fossil record spanning the Cretaceous/Tertiary boundary. *Nature* **340**, 708–711.

KENT, B. (1986) Evolution of thick coal deposits in the Powder River basin, northeastern Wyoming. In: *Paleoenvironmental and Tectonic Controls in Coal-forming Basins of the United States* (Eds Lyons, P.C. & Rice, C.L.) Geol. Soc. Am. Spec. Pap. **210**, 105–122.

KOSTERS, E.C., CHMURA, G.L. & BAILEY, A (1987) Sedimentary and botanical factors influencing peat accumulation in the Mississippi delta. *J. Geol. Soc. London* **144**, 423–434.

LIPPOLT, H.J., HESS, J.C. & BURGER, K. (1984) Isotopische alter von pyroklastischen sanidinen aus kaolinkohlentonsteinen als korrelationsmarken für das mitteleuropäische Oberkarbon: *Fortschr. Geolo. Rheinl. Westfalen* **32**, 119–150.

LYONS, P.C. & DARRAH, W.C. (1989) Earliest conifers of North America: upland and/or paleoclimatic indicators? *Palaois* **4**, 480–486.

LYONS, P.C., JACOBSEN, E.F. & FLORES, R.M. (1984) Paleoen-

vironmental control of accumulation and quality of Upper Freeport coal bed (Allegheny Formation, Middle Pennsylvanian), Castleman coal field, Maryland (Abstract). *Am. Assoc. Petrol. Geol. Bull.* **68**, 1924.

LYONS, P.C., JACOBSEN, E.F. & SCOTT, B.K. (1985) *Coal Geology of the Castleman Coal Field, Garrett County, Maryland.* US Geological Survey, Coal Investigation Map C-98, with marginal notes.

McCABE, P.J. (1984) Depositional environments of coal and coal-bearing strata. In: *Sedimentology of Coal and Coal-bearing Sequences* (Eds Rahmani, R.A. & Flores, R.M.) *Int. Assoc. Sedimentol. Spec. Publ.* **7**, 13–42.

McCABE, P.J. (1987) Facies studies of coal and coal-bearing strata. In: *Coal and Coal-bearing Strata: Recent Advances* (Ed. Scott, A.C.) Geol. Soc. London Spec. Publ. **32**, 51–66.

McCABE, P.J. (1991) Tectonic controls on coal accumulation. *Soc. Géol. France Bull.* **162**, 277–282.

McLELLAN, M.W. & OLIVE, W.W. (1991) *Geologic Map of the Elk Ridge Quadrangle, Powder River County, Montana.* US Geological Survey, Miscellaneous Investigations Series, Map I-2140, with marginal notes.

McLELLAN, M.W., BIEWICK, L.R.H., MOLNIA, C.L. & PIERCE, F.W. (1990) *Cross Sections Showing the Reconstructed Stratigraphic Framework of Paleocene Rocks and Coal Beds in the Northern and Central Powder River Basin, Montana and Wyoming.* US Geological Survey, Miscellaneous Investigations Series, Map I-1959-A, cross section and map with marginal notes.

NEMEC, W. (1988) *Coal correlations and intrabasinal* subsidence: a new perspective. In: *New Perspectives in Basin Analysis* (Eds Kleinspehn, K.L. & Paola, C.) Springer-Verlag, New York, pp. 161–188.

OBERLANDER, T. (1965) *The Zagros Streams, a New Interpretation of Transverse Drainage in an Orogenic Zone.* Syracuse University, New York, Syracuse Geographical Series, no. 1, 168 pp.

PASHIN, J.C. & CARROLL, R.E. (1993) Origin of the Pottsville Formation (Lower Pennsylvanian) in the Cahaba synclinorium of Alabama: genesis of coalbed reservoirs in a synsedimentary foreland thrust system: *Proceedings of the 1993 International Coalbed Methane Symposium,* pp. 623–637.

PHILLIPS, T.L., PEPPERS, R.A. & DIMICHELE, W.A. (1985) Stratigraphic and interregional changes in Pennsylvanian coal-swamp vegetation: environmental inferences. *Int. J. Coal Geol.* **5**, 43–109.

RICHARDSON, R.J.H., STROBL, R.S., McCABE, P.J. & MACDONALD, D.E. (1986) Coal in a foreland basin: the Ardley coal zone, an Alberta example. *Program with Abstracts, Geological Association of Canada; Mineralogical Association of Canada; Canadian Geophysical Union, Joint Annual Meeting, February 1986,* pp. 118–119.

SCANLON, M.W. (1991) Depositional modelling and coal mineability — the Upper Freeport seam, northwest Virginia and western Maryland. MSc thesis, Institute for Mining and Minerals Research, University of Kentucky, Lexington, 75 pp.

SCHMITT, J.G. & STEIDTMANN, J.R. (1990) Interior ramp-supported uplifts: implications for sediment provenance in foreland basins. *Geol. Soc. Am. Bull.* **102**, 494–501.

SCOTT, A.C. (1989) Deltaic coals: an ecological and palaeobotanical perspective. In: *Deltas: Sites and Traps for Fossil Fuels* (Eds Pickering, K.T. & Whateley, M.K.G.) Geol. Soc. London Spec. Publ. **41**, 309–316.

Sholes, M.A., Edmunds, W.E. & Skema, V.W., (1979) *The Economic Geology of the Upper Freeport Coal in the New Stanton Area of Westmoreland County, Pennsylvania: a Model for Coal Exploration.* Pennsylvania Topographic and Geologic Survey, Mineral Resource Report **75**, 51 pp.

Sholes, M.A., Cole, G.A., Fine, D., Daniel, J. & Matson, R.E. (1982) *Sedimentological Controls on Distribution and Quality of Knobloch, Anderson, and Dietz Coals in Parts of Southeastern Montana.*, US Department of the Interior, Office of Surface Mining, OSM publication, Grant G 519 5026, unpublished report, 89 pp.

Sholes, M.A., Vuke-Foster, S.M. & Derkey, P.D. (1989) *Coal Stratigraphy and Correlation in the Glendive 30 × 60-minute Quadrangle, Eastern Montana and Adjacent North Dakota.* Montana Bureau of Mines and Geology, Geologic Map 49, text 9 pp.

Skema, V.W., Sholes, M.A. & Edmunds, W.E. (1982) The *Economic Geology of the Upper Freeport Coal in Northeastern Greene County, Pennsylvania.* Pennsylvania Topographic and Geological Survey, Mineral Resource Report 76, 51 pp.

Stanton, R.W., Cecil, C.B., Pierce, B.S., Ruppert, L.F. & Dulong, F.T. (1986) *Geologic Processes Affecting the Quality of the Upper Freeport Coal Bed, West–Central Pennsylvania.* US Geological Survey Open-File Report 86-173, 22 pp.

Veevers, J.J. & Powell, C. McA. (1987) Late Paleozoic glacial episodes in Gondwanaland reflected in transgressive–regressive depositional sequences in Euramerica. *Geol. Soc. Am. Bull.* **98**, 475–487.

Wanless, H.R. & Shepard, F.P. (1936) Sea level and climatic changes related to late Paleozoic cycles. *Geol. Soc. Am. Bull.* **47**, 1177–1206.

Weedman, S.D. (1990) Freshwater limestones of the Allegheny Group. *Pennsylvania Geol.* **21**, 9–16.

Whipkey, C.E., Cavaroc, V.V. & Flores, R.M. (1991) Uplift of the Bighorn Mountains, Wyoming and Montana — a sandstone provenance study. *US Geological Survey Bulletin 1917*, Chapter D, 20 pp.

Wilson, T.H. (1989) Geophysical studies of large blind thrust, Valley and Ridge province, central Appalachians. *Am. Assoc. Petrol. Geol. Bull.* **73**, 276–288.

Winczewski, L.M & Groenewold, G.H. (1982) A tectonic–fluvial model for Paleocene coal-bearing sediments, Williston basin, southwestern North Dakota. In: *Fifth Symposium on the Geology of Rocky Mountain coal.* (Ed. Gurgel, K.D.) *Utah Geol. Mineral Surv. Bull.* **118**, 76–88.

Winston, R.B. (1990) Implications of paleobotany of Pennsylvanian-age coal of the central Appalachian basin for climate and coal-bed development. *Geol. Soc. Am. Bull.* **102**, 1720–1726.

Wise, D.U., Belt, E.S. & Lyons, P.C. (1991) Clastic diversion by fold salients and blind-thrust ridges in coal-swamp development. *Geology* **19**, 514–517.

Spec. Publs Int. Ass. Sediment. (1993) **20**, 399–414

Plio-Pleistocene outer arc basins in southern Central America (Osa Peninsula, Costa Rica)

H. VON EYNATTEN,* H. KRAWINKEL†
and J. WINSEMANN‡

**Institut für Geowissenschaften, Johannes Gutenberg Universität,
Saarstraße 21, D-55099 Mainz, Germany;
†Institut für Geologie und Paläontologie, Universität Stuttgart,
Herdweg 51, D-70174 Stuttgart 1, Germany; and
‡Institut für Geologie, Universität Hamburg,
Bundesstraße, D–20146 Hamburg, Germany*

ABSTRACT

The Osa Peninsula forms part of the outer fore-arc area of the southern Central American isthmus near the triple junction of the Cocos, Nazca and Caribbean plates. Increasing northward drift rates at the eastern part of the Cocos Plate since Early Pliocene times has resulted in oblique subduction between the Cocos and Caribbean plates, producing trench-parallel dextral strike-slip faults in the fore-arc area. The Plio-Pleistocene sedimentary succession of the Osa Peninsula was deposited in two fault-bounded basins.

1 In the southern part of the Peninsula sedimentation was dominated by coarse-grained mass flow deposits, characteristic of coarse-grained delta and submarine fan depositional systems. Clast composition shows exclusive recycling of outer arc material. The overall retrogradational stacking patterns display clear deepening-upward trends, while high sedimentation rates continue. Therefore most probably strong basinal subsidence occurred, owing to the formation of a 'stepped fan system' seaward of a cliff-lined faulted slope.

2 In the eastern part of the Peninsula, Early Pliocene deposits start with a small coarse-grained deltaic section, overlain by 430 m of shelf deposits with intercalated tempestites. Aggradational stacking patterns as well as high sedimentation rates reflect high subsidence rates within an open marine embayment. Facies architecture and sedimentation style suggest deposition in a pull-apart basin. This basin opened within the outer fore-arc area in a trench-parallel direction.

The Osa arc area has been recently affected by strong uplift and arcward tilting caused by the subduction of the aseismic Cocos Ridge, which started 1 My ago. These subduction-related processes have superimposed the former strike-slip basin geometry. Major faults are now mainly perpendicular to the trench, following the direction of the Cocos Ridge.

INTRODUCTION

The Osa Peninsula forms part of the outer fore-arc area of southern Central America. The peninsula is located near the triple junction of the Cocos, Nazca and Caribbean plates (Fig. 1), where the Cocos Ridge is currently being subducted (Corrigan *et al.*, 1990).

A stratigraphical overview of the Osa Peninsula is given in Fig. 2. The basement consists of basalts (mid-ocean ridge basalts, island arc tholeiites, within-plate basalts) and radiolarites (Baumgartner *et al.*, 1984; Tournon, 1984; Berrangé & Thorpe, 1988; Appel, 1990). These basement rocks are overlain by Tertiary (Palaeocene to ?Oligocene) hemipelagic limestones, tuffs and volcaniclastic sediments (Azema *et al.*, 1981; Hein *et al.*, 1983;

Lew, 1983), which have been interpreted as trench-slope deposits (Christof, 1990; Seyfried *et al.*, 1991) or as an accretionary complex (Baumgartner, 1990; Winsemann, 1992), respectively. These deep-water deposits are bounded by an angular unconformity.

The following Plio-Pleistocene deep and shallow-marine siliciclastic sediments have been informally defined as the 'Punta la Chancha Formation' (Lew, 1983). Berrangé (1989) included these sediments in his 'Osa. Group' detailed facies analysis of the 'Punta La Chancha Formation' is given below.

The uppermost section consists of Late Pleistocene to Recent alluvial and coastal sediments ('Puerto Jimenez Group', Berrangé, 1989), which unconformably overlie the Plio-Pleistocene succession. The formation of this unconformity is related to the subduction of the Cocos Ridge, which started 1 My ago (Lonsdale & Klitgord, 1978) and has caused strong uplift and arcward tilting of the entire region (Lew, 1983; Seyfried *et al.*, 1991).

Previous investigations of the Osa area (e.g. Lew, 1983; Obando, 1986; Berrangé & Thorpe, 1988; Berrangé, 1989; Kriz, 1990) concentrated on tectonics, stratigraphy and regional mapping. The aim of this study is the development of a depositional model of the Plio-Pleistocene outer arc basins, based on: (1) a detailed facies analysis of the siliciclastic deposits; (2) an evaluation of the regional tectonic setting; and (3) a compilation of the plate tectonic data.

PLATE TECTONIC SETTING

The Neogene plate-tectonic processes within the study area were mainly controlled by the generation of the spreading centre between the Cocos and Nazca plates (Galapagos Rift) during the Early Miocene and the formation of the Cocos and Carnegie aseismic ridges by the Galapagos hot spot (Hey, 1977; Lonsdale & Klitgord, 1987; see Fig. 1). The Cocos Ridge reached the Middle America Trench off Panama about 10 My ago. Subduction ceased and the Coiba Ridge was cut off from the Cocos Ridge by the Coiba Transform Fracture Zone. This dextral transform fault then built the eastern boundary of the Cocos plate and later jumped stepwise westwards to its present position (Panama Fracture Zone) (Lowrie *et al.*, 1979). Since 1 My the Cocos Ridge has been subducted beneath the Osa and Burica Peninsulas, which now may behave as detached blocks due to decoupling from the arc (Adamek *et al.*, 1987).

Seismic data (Burbach *et al.*, 1984) indicate a new westward shift of the Panama Fracture Zone, resulting in an incipient fracture zone west of the Osa Peninsula, which will again cut off the Cocos Ridge. Recently published data of plate kinematics (DeMets *et al.*, 1990) give evidence for an oblique subduction in the Middle America Trench offshore from the study area (Fig. 1). Further evidence for oblique subduction is: (1) significantly higher spreading rates in the Costa Rica Rift in comparison to the Galapagos Rift (Hey *et al.*, 1977; see Fig. 1) and (2) additional north–south-directed spreading in the Panama Fracture Zone (Lowrie *et al.*, 1979). This should result in a largely northward directed movement of the Cocos plate at its eastern boundary since at least Early Pliocene times. Strike-slip fault planes are expected to develop arcward of oblique subduction zones (Dewey, 1980), due to stress-trajectory patterns in the upper plate (Fitch, 1972; in Mann *et al.*, 1983). A similar situation exists in southern Panama where the observed

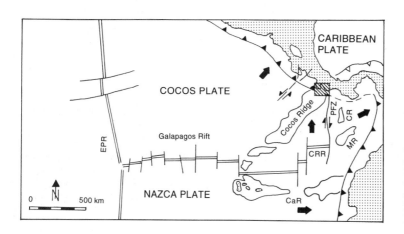

Fig. 1. Recent plate tectonic situation in the central East Pacific (after Berrangé & Thorpe, 1988; MacKay & Moore, 1990; Seyfried *et al.*, 1991). ▨, land; ▨, study area; ▲, Middle and South America Trench; →, relative plate movement. Abreviations: CaR, Carnegie Ridge; CR, Coiba Ridge; CRR, Costa Rica Rift; EPR, East Pacific Ridge; MR, Malpelo Ridge; PFZ, Panama Fracture Zone.

sinistral strike-slip faults (in Coiba Island and the Azuero Peninsula) most probably indicate oblique subduction (MacKay & Moore, 1990; Mann & Corrigan, 1990; Silver *et al.*, 1990).

REGIONAL TECTONIC FRAMEWORK

Analysis of aerial photographs and field data of the study area shows two major fault systems (Lew, 1983; Berrangé, 1989; Eynatten, 1991a, b), which consist of (1) trench-parallel (northwest–southeast striking) dextral strike-slip faults, and (2) trench-perpendicular (southwest–northeast striking) normal faults with a minor strike-slip component (Fig. 3).

Time relations between the two fault directions are given by the following observations: (1) the morphology and the drainage pattern of river systems follow trench perpendicular faults; (2) the eastern coastline configuration as well as the repetition of Plio-Pleistocene strata is controlled by trench-perpendicular, post-sedimentary normal faults; (3) north–south striking, synsedimentary normal faults in Plio-Pleistocene strata are related to extensional movements of trench-parallel dextral strike-slip faults (Christie-Blick & Biddle, 1985); and (4) lineations in Late Pleistocene to Recent alluvial plains are restricted to a trench-perpendicular direction (Eynatten, 1991b). Summarizing, the dextral, trench-parallel, strike-slip faults display evidence of synsedimentary deformation during Plio-Pleistocene times, while the normal faults perpendicular to the trench formed after the deposition of Plio-Pleistocene strata. We interpret the dextral strike-slip patterns of the Osa/Golfo Dulce area as arc-parallel, convergent-margin transcurrent faults (cf. Miall, 1990) due to oblique subduction. This structural regime was most probably initiated in Early Pliocene times and is today superceded by the subduction of the Cocos Ridge. The recent stress field is dominated by trench-perpendicular normal faults, striking in the direction of the Cocos Ridge. These faults are related to extension and uplift of the fore-arc area caused by the subduction of the Cocos Ridge (Corrigan *et al.*, 1990).

SEDIMENTARY FACIES AND DEPOSITIONAL ENVIRONMENT

The Plio-Pleistocene sedimentary succession of the

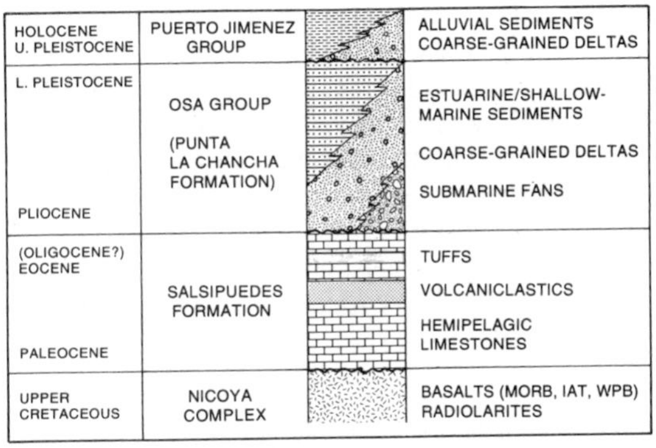

HOLOCENE U. PLEISTOCENE	PUERTO JIMENEZ GROUP		ALLUVIAL SEDIMENTS COARSE-GRAINED DELTAS
L. PLEISTOCENE	OSA GROUP (PUNTA LA CHANCHA FORMATION)		ESTUARINE/SHALLOW-MARINE SEDIMENTS
			COARSE-GRAINED DELTAS
PLIOCENE			SUBMARINE FANS
(OLIGOCENE?) EOCENE	SALSIPUEDES FORMATION		TUFFS
			VOLCANICLASTICS
PALEOCENE			HEMIPELAGIC LIMESTONES
UPPER CRETACEOUS	NICOYA COMPLEX		BASALTS (MORB, IAT, WPB) RADIOLARITES

Fig. 2. Stratigraphical overview and major depositional environments of the Osa Peninsula, Costa Rica. Data compiled from Azema *et al.* (1981), Lew (1983), Tournon (1984), Berrangé & Thorpe (1988), Berrangé (1989), Appel (1990), Baumgartner (1990), Christof (1990) and Eynatten (1991a).

Osa Peninsula ('Punta la Chancha Formation', see Fig. 2) crops out at three major locations along the southern and eastern coast of the Osa Peninsula: (1) the 'Punta la Chancha Section', (2) the 'Punta Río Claro Section' and (3) the 'Playa Sombrero Section' (see Fig. 3). The sediments consist of semiconsolidated conglomerates, sandstones and siltstones, displaying a fully marine fauna (Lew, 1983; Eynatten, 1991a). Clast composition of conglomerates and sandstones generally reflects re-

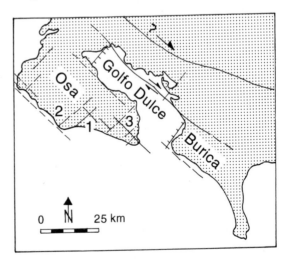

Fig. 3. Location map of the Osa area. Dotted areas indicate land. Numbers indicate locations of measured sections: 1, 'Punta la Chancha Section'; 2, 'Punta Río Claro Section 2'; 3, 'Playa Sombrero Section'. Tectonic data have been compiled after Lew (1983), Berrangé (1989), Corrigan *et al.* (1990) and Eynatten (1991b).

cycling of the underlying hemipelagic limestones and basaltic basement (Fig. 2).

Lew (1983) and Berrangé (1989) report a Middle to Upper Pliocene age (N20, N21) for the 'Punta la Chancha Formation'. Biostratigraphical data from the 'Playa Sombrero Section' (location 3 in Fig. 3), however, indicate Lower Pliocene (N18/N19, P. Sprechmann, personal communication, 1987) to Pleistocene ages (N22, M. Bolli, personal communication, 1990).

Section 1: 'Punta la Chancha'

The 'Punta la Chancha Section' is located on the southern coast of the Osa Peninsula (location 1 in Fig. 3). The section is 230 m thick and can be subdivided into three major facies associations, which correspond to different depositional subenvironments of a submarine fan (Fig. 4).

Observation

The lower part of the section (0–128 m; facies association 1, FA 1, see Fig. 4) is dominated by coarse-grained, graded to graded–stratified pebbly sandstones and, less common, intercalated lenses of conglomerates. Pebbly sandstone beds range in thickness from 0.1 to 1 m and contain a few cobbles

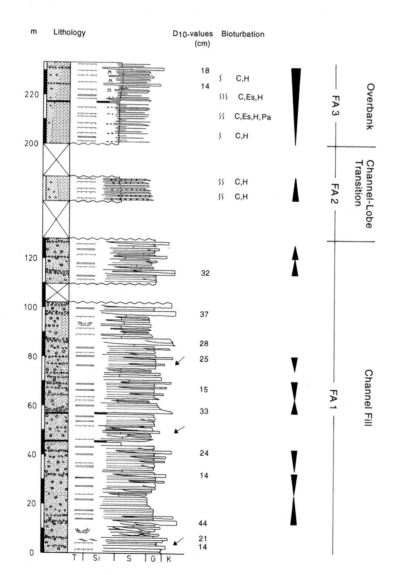

Fig. 4. (*Above and opposite.*)'Punta la Chancha Section', displaying different environments of a submarine fan system. Note complex pattern of coarsening- and fining-upward trends. For location see Fig. 3. (Modified after Eynatten *et al.*, 1992.)

and boulders, up to 32 cm in diameter. Basal contacts show erosional features. Most beds are laterally persistant at outcrop scale (about 50 m). Pebbles are usually subangular to subrounded. Conglomerates are up to a few metres thick and laterally pinch out at outcrop scale. Clast size varies between 5 and 69 cm (maximum D_{10} value is 44 cm). The matrix (up to 50%) consists of coarse sand and granules. These conglomerates are commonly disorganized (some are slightly normally graded) and display erosional basal contacts. The a(p)a(i)-imbrication of clasts in both conglomerates and pebbly sandstones indicate palaeocurrent directions from NE to SW. A few intercalated fine-grained beds (silt to fine sand) are up to 0.5 m in thickness and contain about 1% granules. Despite some coarsening- and fining-upward cycles (Fig. 4), the lower part of the 'Punta la Chancha Section' displays no significant trends in grain size, clast composition or bed thickness.

The middle part of the 'Punta la Chancha Section' (128–200 m; facies association 2, FA 2, see Fig. 4) is poorly exposed. The sediments consist mainly of 0.1–0.3-m-thick beds of normally graded coarse-grained sandstones and pebbly sandstones. Pebble size is up to 2 cm. Some of these beds pass upwards into laminated sandstones. Intercalated siltstone and sandstone beds are between 5 and 15 cm in thickness and show no visible sedimentary structures but a moderate bioturbation; with *Chondrites* ichnospecies (isp). and *Helminthopsis* isp.

Shell fragments are common in the fine-grained beds. The pebbly sandstone beds have strong erosive bases; rip-up clasts up to 10 cm in diameter are often concentrated at the base. The upper part of this section shows a fining-upward trend combined with thickening-upward of fine-grained beds.

The upper part of the 'Punta la Chancha Section' (200–230 m; facies association 3, FA 3, see Fig. 4) consists of 5–15-cm-thick siltstone beds with intercalations of very thin-bedded coarse-grained sandstones. Siltstones are commonly structureless. Some beds show horizontal laminations or fading ripples of fine sand which are mostly concentrated at the basal contact with coarse-grained layers. The content of sand in siltstones is up to 10% and most probably results from homogenization through bioturbation by *Chondrites* isp. and *Helminthopsis* isp. *Palaeophycos* isp. and a few biogenic escape structures have also been recognized. Wood fragments and small shell fragments are seen, but are not common. The thin-bedded sandstone layers have erosional bases, rip-up clasts and normal grading. Upsection they commonly pass into horizontal laminated medium to coarse-grained sandstones. Both siltstone and sandstone layers are laterally persistent at outcrop scale. Upsection the coarse-grained beds increase in thickness and commonly consist of pebbly sandstones. These beds show channelized features.

Interpretation

The textural characteristics of the pebbly sandstone facies types in the lower part (FA 1) of the 'Punta la Chancha Section' indicate depositional processes of high density turbidity currents (R_{1-3}, cf. Lowe, 1982; $A_{2.5}$–$A_{2.8}$, cf. Pickering *et al.*, 1986). The disorganized to normal graded conglomerates (facies types $A_{1.1}$ or $A_{2.3}$, cf. Pickering *et al.*, 1986) reflect deposition from sandy debris flow deposits (Ricci Lucchi, 1985), 'cohesionless' debris flow deposits (Nemec & Steel, 1984; Postma, 1986) or basal inertia flows (Postma *et al.*, 1988). The associated structureless, unbioturbated fine-grained beds are interpreted as having been deposited by muddy debris flows ('cohesive' debris flows, cf. Nemec & Steel, 1984).

In the sand-dominated facies association (FA 2) in the middle part of the section, grain sizes and bed thicknesses are clearly smaller than in FA 1. Channels are absent, but coarse-grained beds display highly erosive basal features. Facies types are mainly attributed to $C_{2.1}$ (Pickering *et al.*, 1986),

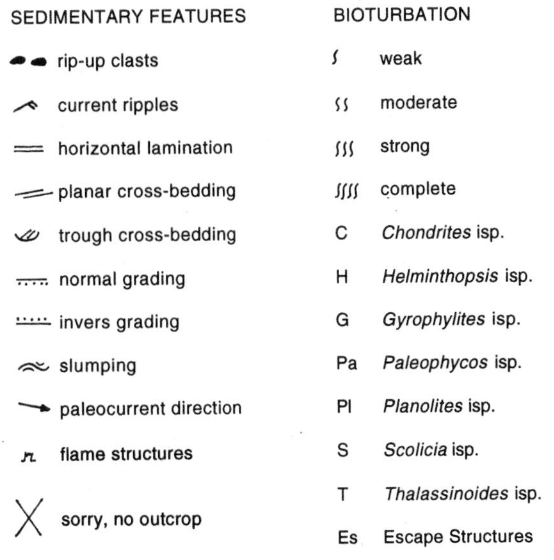

SEDIMENTARY FEATURES

- rip-up clasts
- current ripples
- horizontal lamination
- planar cross-bedding
- trough cross-bedding
- normal grading
- invers grading
- slumping
- paleocurrent direction
- flame structures
- sorry, no outcrop

BIOTURBATION

ſ	weak
ſſ	moderate
ſſſ	strong
ſſſſ	complete
C	*Chondrites* isp.
H	*Helminthopsis* isp.
G	*Gyrophylites* isp.
Pa	*Paleophycos* isp.
Pl	*Planolites* isp.
S	*Scolicia* isp.
T	*Thalassinoides* isp.
Es	Escape Structures

Fig. 4. Key.

suggesting deposition from high-density turbidity currents. The finer-grained and structureless beds most probably reflect rapid suspension settling.

The mud-dominated upper part of the 'Punta La Chancha Section' (FA 3) is characterized by Bouma T_{A-E} divisions (facies $C_{2.2}/C_{2.3}$, cf. Pickering *et al.*, 1986), indicating deposition from turbidity currents of low and high density (see also Stow, 1986). The upsection intercalated pebbly sandstones correspond to facies types $A_{1.4}$ and $A_{2.7}$ (cf. Pickering *et al.*, 1986), resulting from deposition from sandy debris flows or high density turbidity currents.

Lithology, sedimentology and depositional processes of the different facies associations (FA 1 to FA 3) of the 'Punta la Chancha Section' suggest three subenvironments of a submarine fan system (e.g. Mutti & Ricci Lucchi, 1972; Mutti & Normark, 1987; Shanmugam & Moiola, 1988). The coarse-grained channel-fill deposits of FA 1 represent a submarine canyon environment of the slope/upper fan part of a submarine fan. Alternating coarsening- and fining-upward trends result from the frequent migration of channels. The dominantly fine-grained facies (Bouma T_{A-E}) of FA 3 can be interpreted either as deposits of submarine lobes or as overbank areas. The channelized coarse-grained facies of the upper part of FA 3 narrows down the interpretation to distributary channels with associated overbank deposits. They are common in middle fan regimes but have not been reported from ancient depositional lobes (Shanmugam & Moiola, 1988). Suprafan lobes usually display a more sandy lithology, but are also known from middle fan settings (Shanmugam & Moiola, 1991). Owing to the coarsening-upward trend in FA 3 a prograding or laterally migrating middle fan distributary channel system is favoured. To a great extent FA 2 corresponds to FA 3, but facies C types are more sand-dominated and show highly erosional features. This may be best explained by the 'hydraulic jump' in a channel-lobe transition zone (Mutti & Normark, 1987).

The determination of bathymetric levels in submarine fans and subaqueous delta deposits is still questionable: sedimentological data are usually not diagnostic and the study of microfauna assemblages is often confusing because of the high degree of resedimentation. Ichnology provides a better method for determining bathymetry (e.g. Seilacher, 1967; Stow, 1986). The moderately bioturbated sediments of FA 2 and of FA 3 contain evidence of *Chondrites* isp. and *Helminthopsis* isp. Although

Chondrites isp. is known from all bathymetric levels, it is often referred to the *Zoophycos* ichnofacies, especially if not associated with shallow-water ichnospecies ((Bromley & Ekdale, 1984; Bromley & Ekdale, 1986; Ekdale *et al.*, 1984; Bromley, 1990). *Helminthopsis* isp. is grouped into the *Nereites* ichnofacies (Seilacher, 1967, in: Stow, 1986; Frey & Pemberton, 1984). Both above-mentioned ichnofacies of Seilacher indicate deeper bathymetric levels (slope, deep sea), whereas the *Nereites* ichnofacies is not generally deeper than *Zoophycos* ichnofacies (R.G. Bromley, personal communication, 1989). In accordance with the tiering principle (Bromley & Ekdale, 1986; Bromley, 1990), deep-tier ichnospecies (e.g. *Chondrites* isp., *Helminthopsis* isp.) should destroy the burrows of shallow-tier ichnospecies, under conditions of high bioturbation rate and continuous sedimentation. In our case, however, bioturbation is not complete and sedimentation rates depend on individual events (debris flows, turbidity currents). Therefore, shallow-tier bioturbation should be partially preserved. Since there is a complete absence of shallow-tier ichnospecies, the *Chondrites* isp. and *Helminthopsis* isp. ichnofacies must be primary, indicating a deeper depositional environment.

Summarizing, the 'Punta la Chancha Section' is interpreted as representing different stages of submarine fan development. Upsection, the transition of upper to mid-fan environments accompanied by a high sedimentation rate (more than 100 mm/1000 years; Lew, 1983; Eynatten, 1991a) reflect a high relative rate of subsidence.

Section 2: 'Punta Río Claro'

The 'Punta Río Claro Section' is located at the southern coast of the Osa Peninsula (location 2 in Fig. 3). It consists of 270 m of submarine coarse-grained delta deposits. Three different subenvironments can be distinguished, all of which show small and large fining- and thinning-upward cycles (Fig. 5).

Observation

The lowermost part of Section 2 (0–14 m; facies association 4, FA 4, see Fig. 5) is dominated by coarse-grained pebbly sandstones and conglomerates (maximum clast size 57 cm; maximum D_{10} value 33 cm). Beds are commonly about 1 m in thickness and pinch out laterally after a few tens of

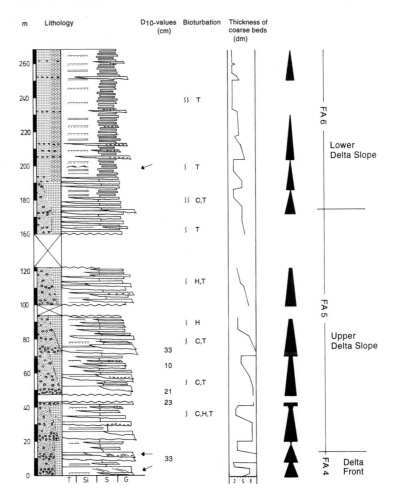

Fig. 5. 'Punta Río Claro Section', displaying small- and large-scale fining- and thinning-upward cycles from delta front to lower delta slope deposits of a coarse-grained delta. For location see Fig. 3; for key see Fig. 4. (Modified after Eynatten *et al.*, 1992.)

metres. Basal contacts show no distinct erosional features. These coarse-grained lenticular bodies contain neither wood and shell fragments nor bioturbation. Pebbly sandstones and conglomerates show planar cross-bedding, indicating lateral accretion in a westerly/southwesterly direction. Intercalated 5–20-cm-thick massive sandstones (fine- to medium-grained sand) contain wood and shell fragments. The sparse bioturbation is restricted to the surface regions of these beds. The sandstones are laterally persistent at outcrop scale (more than 100 m). A few of these beds display faint horizontal lamination or normal grading.

The middle part of the section (14–175 m, facies association 5, FA 5, see Fig. 5) is made up of alternating pebbly sandstones, conglomerates, sandstones, and siltstones, which build up four fining-

and thinning-upward cycles, each about 25 m in thickness (note outcrop gap between 122 and 160 m). Pebbly sandstones and conglomerates are commonly disorganized and show 0.5–1.2-m-thick, smooth lenticular beds. They extend laterally for up to 40 m. Clast size varies between 1 and 10 cm, but some boulders reach 62 cm in diameter (maximum D_{10}-value 33 cm). The degree of roundness (subangular to subrounded) and sorting is lower than in the basal part of the section. Occasionally observed a(p)a(i)-imbrications suggest transport from a northeasterly to southwesterly direction. Wood fragments (maximum size 20 cm) are common, but shell fragments and bioturbation are rare. The siltstone- and sandstone beds (fine-grained sand) are up to 30 cm in thickness and are generally structureless. Some display weak grading or faint horizontal lamination. Wood and shell fragments are common and bioturbation is dominated by *Thalassinoides* isp., *Chondrites* isp. and *Helminthopsis* isp. These siltstones and fine-grained sandstones alternate with medium- to coarse-grained sandstones with erosive bases. Grading and horizontal laminations are common and large clasts up to 30 cm in diameter occur occasionally. Organic material is less common in these sandstones than in the bioturbated siltstones.

The upper part of section 2 (175–270 m; facies association 6, FA 6, see Fig. 5) consists of coarse-grained sandstones and pebbly sandstones, which alternate with finer-grained siltstone and sandstone beds. The latter are similar to those of FA 5, despite a lack of *Helminthopsis* isp. in the ichnofacies. The coarse-grained sandstones and pebbly sandstone beds are 0.2–0.4 m thick and commonly grade upwards into horizontally laminated medium-grained sandstones. Pebble size is up to 2 cm in diameter. Bed contacts are erosional and rip-up clasts are concentrated at the bases. Some beds show complete amalgamation. Large clasts up to 90 cm in diameter are seen. Occasionally current ripples occur at the top of the sandstone beds, indicating transport in a southwesterly direction. Coarse-grained pebbly sandstones and conglomerates are rare and consist of lenticular disorganized beds up to 1 m thick.

Interpretation

The facies associations 5 and 6 are dominated by two main facies-types: (1) channelized, disorganized pebbly sandstones and conglomerates (facies types $A_{1.1}$ and $A_{1.4}$, cf. Pickering *et al.*, 1986), and (2) laterally more persistent, graded sandstones and fine-grained pebbly sandstones (facies types $C_{2.1}$/$C_{2.2}$, cf. Pickering *et al.*, 1986; S_3, cf. Lowe, 1982). The disorganized facies types show a(p)a(i)-imbrication and were most probably deposited from sandy/cohesionless debris flows or inertia flows. The finer-grained, graded sandstones and pebbly sandstones display typical Bouma T_{A-C} divisions and were deposited from high density turbidity currents.

Depositional models proposed for coarse-grained mass flow deposits (e.g. Nemec *et al.*, 1980, 1984; Clifton, 1984; Porębski, 1984) suggest a proximal environment for the coarse-grained, thick-bedded and disorganized debris flow deposits of FA 5. In contrast the normally-graded turbidites of FA 6 suggest a more distal facies. The large outsize clasts

of FA 6 (up to 90 cm) are therefore interpreted as sliding boulders (Postma, 1984; Prior *et al.*, 1984). It is possible to correlate directly the mass flow events of FA 5 and FA 6 if it is assumed that deposition occurred from basal inertia flows in proximal fan areas and from laterally or distally associated turbidity currents further down the slope (Postma *et al.*, 1988).

The intercalated, bioturbated, fine-grained sandstones and siltstones of FA 5 and 6 represent the normal sedimentation from suspension fall-out. Some of the coarser-grained, graded and/or horizontally laminated beds which are less bioturbated may be related either to low density turbidity currents or to buoyant plumes ('hypopycnal flow suspension fall-out', e.g. Wright, 1985).

Facies association 4 is the coarsest-grained part of the 'Punta Río Claro Section'. Fine-grained beds occur only infrequently. This is typical of a more proximal environment. The planar cross-bedded sets suggest deposition from debris fall avalanching. In this context, the massive sandstones most probably represent a rapid fall-out of suspension load from buoyant plumes.

The sedimentary succession of the 'Punta Río Claro Section' does not allow an unambiguous interpretation of the depositional environment. The occurrence of channelized mass-flow deposits and turbidites may reflect deposition either on a steep delta slope (e.g. Wescott & Ethridge, 1983; Postma, 1984; Surlyk, 1984; Prior & Bornhold, 1989) or in a submarine fan environment (e.g. Surlyk, 1984; Mutti & Normark, 1987; Shanmugam & Moiola, 1988). However, of the 'Punta Río Claro' deposits (Section 2), the following characteristics suggest deposition in the submarine part of a coarse-grained delta system.

1 *The dominance of laminar mass-flow deposits in channels.* The channelized facies types in FA 5 are interpreted as debris flow deposits. Associated fine-grained suspension fall-out sediments are only partly eroded. Conversely, the channelized facies types of the 'Punta la Chancha Section (Section 1) suggest the highly erosive and turbulent (FA 1) conditions of a submarine fan-canyon system. The minor erosional processes of FA 5 (Section 2) are more typical of steep coarse-grained delta slopes (e.g. Massari, 1984; Prior & Bornhold, 1988, 1989; Nemec, 1990b).

2 *The lack of overbank deposits.* Submarine overbank deposits are associated with feeder channels or distributary channels in submarine fan systems.

The 'Punta Río Claro Section' shows no overbank deposits, whereas FA 3 in the 'Punta la Chancha Section' (Section 1) comprises a series of overbank deposits.

3 *The high content of large wood fragments.* High wood contents in marine sediments are obviously an indicator of nearshore environments. Resedimentation processes must be considered, because they are able to transport wood fragments far down the slope. In the course of resedimentation, however, frequency and size of wood fragments should both decrease (as seen in the 'Punta la Chancha Section').

4 *A trace fossil association indicating a more shallow environment.* Compared with the 'Punta la Chancha' ichnofacies (Section 1), the 'Punta Río Claro' ichnofacies exhibits less frequent *Helminthopsis* isp. and *Thalassinoides* isp., but *Chondrites* isp. have become dominant. This gives further reason to assume a shallower environment for the 'Punta Río Claro Section'.

None of these arguments is definitive, but they do narrow down the interpretation to a more shallow coarse-grained deltaic environment. The most proximal FA 4 is dominated by debris fall avalanching processes, and can therefore be interpreted as mouth bar deposits or submarine gravel bars (cf. Prior & Bornhold, 1989). The lack of wave-reworking processes, apart from a better degree of roundness and sorting, might be explained by rapid subsidence (Kleinspehn *et al.*, 1984).

Summarizing, the 'Punta Río Claro Section' represents a deepening-upward sequence from delta front to lower delta slope in a retrogradational coarse-grained delta. This, together with sedimentation rates in excess of 100 mm/1000 years (Lew, 1983; Eynatten, 1991a), suggests rapid subsidence. Using the classification scheme of Ethridge and Wescott (1984), the term 'slope-type' (fan-) delta or 'coarse-grained delta' according to Nemec (1990a) is the most appropriate description of the palaeoenvironment. A more detailed classification cannot be given because of the lack of a subaerial feeder system. However, a 'debris cone type' delta or a 'gravitationally modified Gilbert-type' delta seems most reasonable, both reflecting steep gradient deep-water deltas (Postma, 1990).

Section 3: 'Playa Sombrero'

The 'Playa Sombrero Section' (Fig. 6) is located at the eastern coast of the Osa Peninsula (location 3 in Fig. 3). The section comprises a 470-m-thick succession of siltstones with intercalated sandy tempestites, which were deposited in an open marine embayment.

Observation

Section 3 starts with a 37-m-thick fining-upward sequence of pebbly sandstones, coarse-grained sandstones and sandy siltstones. Slump structures are common. Decrease in grain size is accompanied by increase in bioturbation. These sediments gradually pass upwards into a 430-m-thick succession of centimetre- to decimetre-thick interbeds of siltstones and fine-grained sandstones.

The siltstones contain no sedimentary structures since they are strongly bioturbated with *Chondrites* isp., *Thalassinoides* isp., *Helminthopsis* isp., *Scolicia* isp., *Gyrophyllites* isp., *Paleophycos* isp., and *Planolites* isp. Additional characteristics of this succession are abundant wood fragments (with sizes ranging from a few millimetres up to 90 cm in length) and a microfauna of fully marine benthic and planktonic foraminifera (H.M. Bolli & H.P. Luterbacher, personal communication, 1990). There is little evidence of a macrofauna. The thickness of siltstone beds varies between 10 and 35 cm.

These sandstone beds commonly exhibit horizontal lamination, current ripples, climbing ripples, convolute bedding, and biogenic escape structures. Thicknesses vary between a few millimetres and 15 cm. Foresets of current ripples suggest transport directions from northwest to southeast. In parts of the section, concretionary horizons are common, but only in the sandier beds. Bioturbation is restricted to the uppermost part of these beds and obviously reflects 'after-event communities' (cf. Seilacher, 1980), which are characterized by *Scolicia* isp. and *Palaeophycos* isp. The curves in Fig. 6 clearly demonstrate the interrelation between bioturbation, thickness of siltstone beds, and frequency of concretions. Thick-bedded siltstones correlate with a low degree of bioturbation and frequent concretions. Conversely, thin-bedded siltstones show a high degree of bioturbation and lack of concretions.

Interpretation

Two bathymetric indicators are important in the interpretation of the depositional environment for the 'Playa Sombrero Section'.

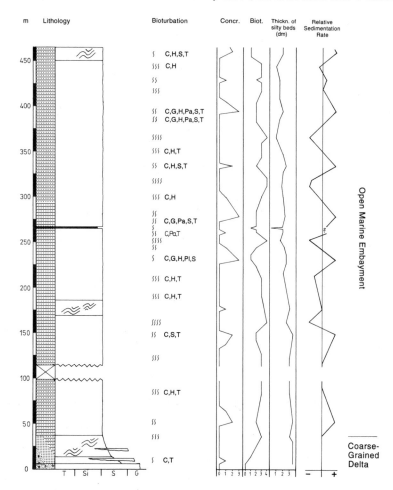

Fig. 6. 'Playa Sombrero Section', displaying the transition from basal coarse-grained delta deposits to sediments of an open marine embayment. The curves on the right show correlation between abundance of concretions (0, no concretions; 3, concretions very common), degree of bioturbation (0, no bioturbation; 4, completely bioturbated), thickness of fine-grained beds, and relative sedimentation rates (– , low; + , high). For location see Fig. 3; for key see Fig. 4. (Modified after Eynatten *et al.*, 1992.)

1 The lack of wave-induced sedimentary structures (e.g. oscillation ripples, hummocky cross-stratification) in the fully marine succession suggests sedimentation below storm-weather wave-base.

2 The high degree of bioturbation and ichnodiversity requires a nutrient- and oxygen-rich environment. In addition, the frequency of *Thalassinoides* isp., in spite of 'deep tier' overprint by *Chondrites* isp. and *Helminthopsis* isp. (Bromley & Ekdale, 1986), indicates a shallow water (shelf) environment (*Cruziana* ichnofacies, c.f. Seilacher, 1967; Stow, 1986; Ekdale & Bromley, 1984).

The occurrence of *Helminthopsis* isp. and *Chondrites* isp., associated with shallow-water ichnospecies, is known from environments characterized by stillwater conditions within the *Cruziana*-ichnofacies (Ekdale & Bromley 1984; Bromley,

1990, p. 241). Summarizing, deposition probably occurred in a shallow marine environment but below storm-weather wave-base.

Some of the sandstone beds display smooth undulating basal contacts and current ripples, followed by laminated beds of the upper flow regime, again current ripples and/or climbing ripples, and fine-grained laminated beds of the lower flow regime. These features are typical of a succession deposited from waxing flow conditions, flow peak (upper flow regime) and following waning flow conditions similar to those postulated for storm events lasting a few days (Morton, 1981). The sandstone beds show neither erosional features nor grading as generally expected in deposits from turbidity currents. The sandstone beds are therefore interpreted as tempestites. Assuming a total time of deposition of about 3 Ma (Early Pliocene to Early

Pleistocene), the tempestites occur with an average frequency of 1900 years. These '1000 year storm events' have a preservation potential of almost 100% when deposited below storm-weather wave-base (Hamblin & Walter, 1979; Hobday & Morton, 1984).

The siltstones should represent the suspension fall-out sedimentation during normal (fair weather) conditions. The up to 10% content of sand is most probably caused by the biological homogenization of deposits from small storm events. The average sedimentation rate of siltstones amounts to more than 100 mm/1000 years (Eynatten, 1991a). This value is about three times higher than the average sedimentation rate known for modern and ancient shelf environments (Füchtbauer, 1988, p. 860). A possible explanation for the high sedimentation rates is a local supply from a fluvial feeder system characterized by a high suspension load. The occurrence of terrestrial organic matter within the siltstones supports this assumption. The high sedimentation rate with no obvious environmental changes during deposition indicates a rapid and steady subsidence, most probably accompanied by a relative coastal transgression. Taking into account these observations and assumptions, and in view of the existence of brackish fauna assemblages northwestward of the section (Berrangé, 1989), we postulate a 'sediment bypassing' estuarine environment (Nichols & Biggs, 1985) landward of the section. Observed directions of palaeocurrents indicate transport from northwest to southeast. Therefore, we assume that the 'Playa Sombrero Section' was formed within a storm-dominated, hydrodynamically autonomous, open marine embayment within the outer arc.

The coarse-grained debris flow and/or turbidity current deposits of the lowermost part of the section are interpreted as coarse-grained deltaic deposits. Due to the sparse pattern of outcrops a more detailed facies analysis cannot be given. These sediments most probably indicate fault-controlled deposits within basal or marginal areas of the embayment. Comparable deposits, which are clearly fault controlled, crop out on the opposite shore of the Golfo Dulce (Eynatten, 1991b).

DISCUSSION

The characteristics of the Plio-Pleistocene siliciclastic sediments of the Osa Peninsula are:

1 rapid lateral facies changes;
2 high sedimentation rates in excess of 100 mm/1000 year;
3 fault-controlled coarse-grained deposits with evidence of synsedimentary tectonic movement; and
4 deposition within small isolated basins.

The observed sedimentary features of all three measured sections suggest a high rate of subsidence. All of these features are characteristic of sedimentary basins in strike-slip/transcurrent fault settings (Mitchell & Reading, 1986; Allen & Allen, 1990; Miall, 1990). This is consistent with the regional and plate-tectonic data. The Plio-Pleistocene depositional systems of the Osa Peninsula were deposited in at least two different tectonically controlled basins.

1 The 'Punta la Chancha-Section' and the 'Punta Rio Claro Section' to the south of the Peninsula are interpreted as representing submarine fan and coarse-grained delta deposits, respectively. High sedimentation rates and deepening-upward cycles in both sections indicate high rates of subsidence suggestive of a fault-controlled basin. Observed palaeocurrent directions give evidence for transport of recycled outer arc material from northeast to southwest, perpendicular to the strike-slip fault system. The sediment was transported down the northwest–southeast striking Middle America Trench and deposited in trench slope basins. In this highly active tectonic setting, complete facies sequences of deltas and submarine fans are unlikely to occur (Underwood & Bachmann, 1982).

These are the most likely interpretations of the depositional environments. However, some questions remain; e.g. was the submarine fan formed as part of a faulted prodelta setting? Is the distinction between real deltas, steep debris cones (which are included in the coarse-grained delta scheme of Postma, 1990) and conical deep-water deltas (Nemec, 1990b) somewhat irrelevant for the interpretation of the tectono-sedimentary setting? In addition, submarine canyons may occur elsewhere on a steep trench slope (Underwood & Bachmann, 1982), and the sedimentological distinction between steep coarse-grained delta slopes and submarine fan slopes still remains a problem. Integrating our data with these unanswered questions, we present a model of a 'stepped fan system' developed at the trench-parallel faulted slope of the Middle America Trench (Fig. 7B). This model incorporates a poorly developed delta front (FA 4), an upper and a lower delta slope (FA 5/FA 6), a fault-controlled

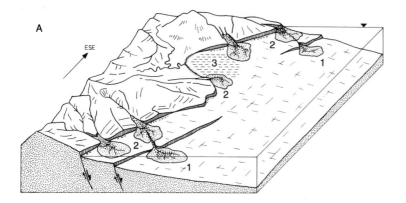

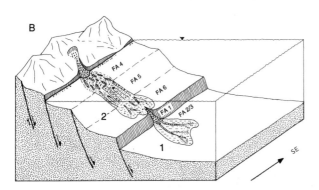

Fig. 7. Depositional model of the Plio-Pleistocene sedimentary environments. Numbers refer to depositional systems: 1, submarine fan; 2, coarse-grained delta; 3, open marine embayment. (A) Integrative model of trench slope sedimentation (coarse-grained deltas, submarine fans) and initial opening of an open marine embayment along trench-parallel strike-slip/dip-slip faults (modified after Eynatten *et al.*, 1992). (B) Detailed model of a 'stepped fan system' of coarse-grained deltas and submarine fans at the trench slope. Facies associations (FA 1 to FA 6) correlate to Figs 4 and 5.

submarine canyon (FA 1) as well as a distal submarine fan (middle fan settings of FA 2 and FA 3). The trench-parallel faults have a proposed lateral (strike-slip) displacement as well as a half-graben-like, dip-slip component to explain the high subsidence rates (Fig. 7B). This basin configuration is comparable to the situation during the Neogene in northern Costa Rica, where several fault-angle basins (cf. Ballance, 1980) existed seaward of the outer arc (Seyfried *et al.*, 1991).

2 In the eastern part of the Peninsula the shallow-marine succession of the 'Playa Sombrero-Section' reaches a thickness of almost 500 m. The lack of environmental change and high sedimentation rates suggest steady subsidence, creating space for almost 500 m of sediment in 3 Ma. Owing to the sedimentary patterns and dynamics of these sediments, we assume deposition in a coastal embayment (Figs 7A & 8). Basal and marginal, coarse-grained, mass-flow deposits suggest a fault-bordered embayment. Considering a strike-slip setting in the outer fore-arc area, the model of a pull-apart basin fits the data

best. Rhythmic variations of sediment input are most probably related to variations in tectonic activity and/or to climatic changes. Although a correlation of the phases of higher sediment supply with the lowstand phases of 3rd-order cycles (cf. Haq *et al.*, 1988) also seems possible, we favour a tectonic control in this strongly subsiding pull-apart setting.

Summarizing the development of sedimentary environments and depositional systems, the regional tectonic framework, and the plate tectonic setting we suggest a three-stage model of the Early Pliocene to Recent tectono-sedimentary history of the Osa area.

1 *Early Pliocene* (Fig. 8, 3.5 Ma): During the latest Miocene to earliest Pliocene subsidence was initiated along trench-parallel (northwest–southeast striking) strike-slip/dip-slip faults due to oblique subduction of the Cocos plate offshore of the study area. Cliff-lined coasts and coarse-grained deltas developed at the seaward edge of the outer arc. In the eastern part, a pull-apart basin (marine embayment) was initiated with marginal coarse-grained deltas and shallow-marine deposits, which were strongly influenced by sediment input from an estuarine river mouth (Fig. 7A).

2 *Late Pliocene to early Pleistocene* (Fig. 8, 2 Ma): During this time, the major part of the 'Punta la Chancha Formation' was deposited. At the trench

slope, strike-slip/dip-slip faults led to the formation of a stepped fan system of coarse-grained deltas and submarine fans (Fig. 7B). The embayment grew with increasing lateral displacement and formed a pull-apart basin within the outer arc.

3 *Late Pleistocene to Recent* (Fig. 8, 0.5 Ma): About 1 My ago the subduction of the Cocos Ridge began and caused a tectonic reorganization of the entire region. The Osa and Burica Peninsulas of the outer fore-arc area underwent strong uplift and arcward tilting. Major normal faults were formed perpendicular to the trench, most probably due to the updoming of the outer arc and trench-parallel extension (Corrigan *et al.*, 1990).

In this context it should be pointed out that the recent Golfo Dulce and the former pull-apart basin, which is now uplifted and tilted, are different basins. The modern active subsidence at the arcward boundary of the Golfo Dulce (Fischer, 1980) may be better explained by the subduction of an aseismic ridge (Burbach *et al.*, 1984), leading to the subsidence of the inner fore-arc (Cross & Pilger, 1982), rather than by movements along still active strike-slip faults as proposed by Berrangé and Thorpe (1988).

ACKNOWLEDGEMENTS

This study was supported by the German Research Foundation (Se 490/1-1; Se 490/1-2; Se 490/1-3). The help and cooperation of H.M. Bolli and H.P. Luterbacher (dating of Foraminifera) and R.G. Bromley (determination of trace fossils) is gratefully acknowledged. We would like to thank R. Gaupp, K. Hinkelbein, W. Jacoby, T. Reischmann and G. Wörner for valuable discussions. R.J. Steel, L. Frostick and two anonymous reviewers provided many helpful suggestions on the manuscript. And without the help and continuous support of H. Seyfried, this study could not have been achieved.

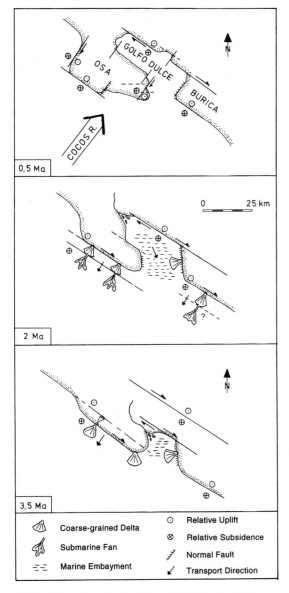

Fig. 8. Three-stage model of the early Pliocene to Recent tectono-sedimentary history of the Osa area. For further explanations see text.

REFERENCES

ADAMEK, S., TAJIMA, F. & WIENS, A. (1987) Seismic rupture associated with subduction of the Cocos Ridge. *Tectonics* **6**, 757–774.

ADAMEK, S., FROHLICH, C. & PENNINGTON, W.D. (1988) Seismicity of the Caribbean–Nazca boundary: constraints on microplate tectonics of the Panama region. *J. Geophys. Res.* **93**, 2053–2075.

ALLEN, P.A. & ALLEN, J.R. (1990) *Basin Analysis.* Blackwell Scientific Publications, Oxford.

APPEL, H. (1990) *Geochemie und K/Ar-Datierung an Magmatiten in Costa Rica, Zentralamerika.* Unpublished Diplomarbeit, Universität Mainz, Germany.

AZEMA, J., GLACON, G. & TOURNON, J. (1981) Nouvelles données sur le Paléocène á foraminifères planctonique de la bordure pacifique du Costa Rica (Amérique Centrale). *C. R. Somm. Soc. Geol. France.* **3**, 85–88.

BALLANCE, P.F. (1980) Models of sediment distribution in non-marine and shallow marine environments in oblique-slip fault zones. In: *Sedimentation in Oblique-slip Mobile Zones* (Eds Ballance, P.F. & Reading, H.G.) Int. Ass. Sediment. Spec. Publ. **4**, 229–236.

BAUMGARTNER, P.O. (1990) Mesozoic and Tertiary arcs, seamounts and accretionary terranes of Costa Rica (Central America). *Abstracts with Programs, Annual Meeting of the Geological Society of America, 1990,* A338, No. 06007, Dallas.

BAUMGARTNER, P.O., MORA, C., BUTTERLIN, J. *et al.* (1984) Sedimentación y paleogeografía del Cretacico y Cenozoico del litoral Pacifico de Costa Rica. *Rev. Geol. Am. Central* **1**, 57–136.

BERRANGÉ, J.P. (1989) The Osa Group: an auriferous Pliocene sedimentary unit from the Osa Peninsula, southern Costa Rica. *Rev. Geol. Am. Central* **10**, 67–93.

BERRANGÉ, J.P. & THORPE, R.S. (1988) The geology, geochemistry and emplacement of the Cretaceous–Tertiary ophiolitic Nicoya-Complex of the Osa Peninsula, southern Costa Rica. *Tectonophysics* **147**, 193–220.

BROMLEY, R.G. (1990) *Trace Fossils.* Unwin Hyman, London, 271 pp.

BROMLEY, R.G. & EKDALE, A.A. (1984) *Chondrites*: a trace fossil indicator of anoxia in sediments. *Science Reprint Series* **224**, 872–874.

BROMLEY, R.G. & EKDALE, A.A. (1986) Composite ichnofabric and tiering of burrows. *Geol. Mag.* **123**, 59–65.

BURBACH, G.V., FROHLICH, C., PENNINGTON, W.D. & MATUMOTO, T. (1984) Seismicity and tectonics of the subducted Cocos plate. *J. Geophys. Res.* **89**, 7719–7735.

CHRISTIE-BLICK, N. & BIDDLE, K.T. (1985) Deformation and basin formation along strike-slip faults. In: *Strike-slip Deformation, Basin Formation, and Sedimentation* (Eds Biddle, K.T. & Christie-Blick, N.) Soc. Econ. Petrol. Mineral. Spec. Publ. **37**, 1–34.

CHRISTOF, W. (1990) Fazieskartierung und sedimentologische Untersuchungen der paläogenen Salsipuedes-Formation auf der Halbinsel Osa, Costa Rica. Unpublished Diplomarbeit Universität, Stuttgart, 68 pp.

CLIFTON, H.E. (1984) Sedimentation units in stratified resedimented conglomerate, Paleocene submarine canyon fill, Point Lobos, California. In: *Sedimentology of Gravels and Conglomerates* (Eds Koster, E.H. & Steel, R.J.) Can. Soc. Petrol. Geol. Mem. **10**, 429–441.

CORRIGAN, J., MANN, P. & INGLE, J.C. JR. (1990) Forearc response to subduction of the Cocos Ridge, Panamá–Costa Rica. *Geol. Soc. Am. Bull.* **102**, 628–652.

CROSS, T.A. & PILGER, R.H. (1982) Controls of subduction geometry, location of magmatic arcs, and tectonics of arc and back-arc regions. *Geol. Soc. Am. Bull.* **93**, 545–562.

DEMETS, C., GORDON, R.G., ARGUS, D. & STEIN, S. (1990) Current plate motions. *Geophys. J. Intern.* **101**, 425–478.

DEWEY, J.F. (1980): Episodicity, sequence, and style at convergent plate boundaries. In: *The Continental Crust and its Mineral Deposits* (Ed. Srangway, D.W.). Geol. Ass. Can. Spec. Paper **20**, 553–573.

EKDALE, A.A. & BROMLEY, R.G. (1984) Comparative ichnology of shelf-sea and deep sea chalk. *J. Paleont.* **58**, 322–332.

EKDALE, A.A., BROMLEY, R.G. & PEMBERTON, S.G. (1984) Ichnology, the use of trace fossils in sedimentology and stratigraphy. *Soc. Econ. Paleont. Mineral., Short Course Notes* **15**, Tulsa.

ETHRIDGE, F.G. & WESCOTT, W.A. (1984) Tectonic setting, recognition and hydrocarbon reservoir potential of fan-delta deposits. In: *Sedimentology of Gravels and Conglomerates* (Eds Koster, E.H. & Steel, R.J.) Can. Soc. Petrol. Geol. Mem. **10**, 217–235.

EYNATTEN, H. VON (1991a) *Das Plio-/Pleistozän der Halbinsel Osa (Costa Rica, Mittelamerika).* Unpublished Diplomarbeit, Universität Mainz, 90 pp.

EYNATTEN, H. VON (1991b) *Geologische Kartierung des Südostens der Halbinsel Osa sowie der Westspitze der Halbinsel Burica (Costa Rica, Mittelamerika).* Unpublished Diplom-Kartierung, Universität Mainz, 53 pp.

EYNATTEN, H. VON, KRAWINKEL, H. & WINSEMANN, J. (1991) Sedimentation history and geodynamic significance of Plio-Pleistocene shallow and deep marine forearc sediments. *Zbl. Geol. Paläont.* (Teie I, 6, 1479–1492.)

FISCHER, R. (1980) Recent tectonic movements of the Costa Rican Pacific coast. *Tectonophysics* **70**, T25–T33.

FREY, R.W. & PEMBERTON, S.G. (1984) Trace fossil facies models. In: *Facies Models,* 2nd Edn (Ed. Walker, R.G.) Geoscience Canada Reprint Series **1**, 189–208, Toronto.

FÜCHTBAUER, H. (1988) Konglomerate und Breccien. In: *Sedimente und Sedimentgesteine, 4. Auflage* (Ed. Füchtbauer, H.). E. Schweizerbart, Stuttgart, pp. 69–96.

HAMBLIN, A.P. & WALKER, R.G. (1979) Storm dominated shallow marine deposits: the Fernie-Kootenay (Jurassic) transition, southern Rocky Mountains. *Can. J. Earth Sci.* **16**, 1673–1690.

HAQ, B.U., HARDENBOL, J. & VAIL, P.R. (1988) Cenozoic chronostratigraphy and eustatic cycles. In: *Sea Level Changes, an Integrated Approach* (Eds Wilgus, C.K., Hastings, B.S., Kendall, C.G.St.C., Posamentier, H.W., Ross, C.A. & van Wagoner, J.C.) Soc. Econ. Paleont. Mineral. Spec. Publ. **42**, 71–108.

HEIN, J.R., KUIJPERS, E.P., DENYER, P. & SLINEY, R.E. (1983) Petrology and geochemistry of Cretaceous and Paleogene cherts from western Costa Rica. In: *Siliceous Deposits on the Pacific Region* (Eds Iijima, A., Hein, J.R. & Siever, R.) Develop. Sedimentol. **36**, 143–175.

HEY, R. (1977) Tectonic evolution of the Cocos–Nazca spreading center. *Geol. Soc. Am. Bull.* **88**, 1404–1420.

HEY, R., JOHNSON, G.L. & LOWRIE, A. (1977) Recent plate motions in the Galapagos area. *Geol. Soc. Am. Bull.* **88**, 1385–1403.

HOBDAY, D.K. & MORTON, R.A. (1984) Lower Cretaceous shelf storm deposits, northeast Texas. In: *Siliciclastic Shelf Sediments* (Eds Tillman, R.W. & Siemers, C.T.) Soc. Econ. Paleont. Mineral. Spec. Publ. **34**, 205–213.

KLEINSPEHN, K.L., STEEL, R.J., JOHANNESSEN, E. & NETLAND, A. (1984) Conglomeratic fan-delta sequences, Late Carboniferous — Early Permian, western Spitsbergen. In: *Sedimentology of Gravels and Conglomerates* (Eds

Koster, E.H. & Steel, R.J.) Can. Soc. Petrol. Geol. Mem. **10**, 279–294.

KRIZ, S.J. (1990) Tectonic evolution and origin of the Golfo Dulce gold placers in southern Costa Rica. *Rev. Geol. Am. Central* **11**, 27–40.

LEW, L. (1983) The Geology of the Osa Peninsula, Costa Rica: observations and speculations about the evolution of part of the outer arc of the Southern Central American Orogen. Unpublished MSc thesis, Pennsylvania State University, 128 pp.

LONSDALE, P. & KLITGORD, K.D. (1978) Structure and tectonic history of the eastern Panama Basin. *Geol. Soc. Am. Bull.* **89**, 981–999.

LOWE, D.R. (1982) Sediment gravity flows: II. Depositional models with special reference to the deposits of high-density turbidity currents. *J. Sediment. Petrol.* **52**, 279–297.

LOWRIE, A., AITKEN, T., GRIM, P. & McRANEY, L. (1979) Fossil spreading center and faults within the Panamá Fracture Zone. *Mar. Geophys. Res.* **4**, 153–166.

MacKAY, M.E. & MOORE, G.F. (1990) Variation in deformation of the South Panamá accretionary prism: response to oblique subduction and trench sediment variation. *Tectonics* **9**, 683–698.

MANN, P. & CORRIGAN, J. (1990) Model for late Neogene deformation in Panama. *Geology* **18**, 558–562.

MANN, P., HEMPTON, M.R., BRADLEY, D.C. & BURKE, K. (1983) Development of pull-apart basins. *J. Geol.* **91**, 529–554.

MASSARI, F. (1984) Resedimented conglomerates of a Miocene fan-delta complex, southern Alps, Italy. In: *Sedimentology of Gravels and Conglomerates* (Eds Koster, E.H. & Steel, R.J.) Can. Soc. Petrol. Geol. Mem. **10**, 259–278.

MIALL, A.D. (1990) *Principles of Sedimentary Basin Analysis*, 2nd Edn. Springer, New York, 668 pp.

MITCHELL, A.H.G. & READING, H.G. (1986) Sedimentation and tectonics. In: *Sedimentary Environments and Facies*, 2nd Edn (Ed. Reading, H.G.) Blackwell Scientific, Oxford, pp. 471–519.

MORTON, R.A. (1981) Formation of storm deposits by wind-forced currents in the Gulf of Mexico and the North Sea. In: *Holocene Marine Sedimentation in the North Sea Basin* (Eds Nio, S.D., Schüttenhelm, R.T.E. & van Weering, Tj.C.E.) Int. Ass. Sediment. Spec. Publ. **5**, 385–396.

MUTTI, E. & NORMARK, W.R. (1987) Comparing examples of modern and ancient turbidite systems: problems and concepts. In: *Marine Clastic Sedimentology* (Eds Leggett, J.K. & Zuffa, G.G.) Graham & Trotman, London, pp. 1–38.

MUTTI, E. & RICCI LUCCHI, F. (1972) Le torbiditi dell' Appennino settentrionale: introduzione all' analisi di facies. *Soc. Geol. Ital. Mem.* **11**, 161–199.

NEMEC, W. (1990a) Deltas — remarks on terminology and classification. In: *Coarse-grained Deltas* (Eds Colella, A. & Prior, D.B.) Int. Ass. Sediment. Spec. Publ. **10**, 3–12.

NEMEC, W. (1990b) Aspects of sediment movement on steep delta slopes. In: *Coarse-grained Deltas* (Eds Colella, A. & Prior, D.B.) *Intern. Ass. Sediment. Spec. Publ.* **10**, 29–73.

NEMEC, W. & STEEL, R.J. (1984) Alluvial and coastal conglomerates: their significant features and some com-

ments on gravelly mass-flow deposits. In: *Sedimentology of Gravels and Conglomerates* (Eds Koster, E.H. & Steel, R.J.) Can. Soc. Petrol. Geol. Mem. **10**, 1–31.

NEMEC, W., POREBSKI, S.J. & STEEL, R.J. (1980) Texture and structure of resedimented conglomerates: examples from Ksiaz Formation (Famennian–Toarnaisian), southwestern Poland. *Sedimentology* **27**, 519–538.

NEMEC, W., STEEL, R.J., POREBSKI, S.J. & SPINNANGR, A. (1984) Domba conglomerate, Devonian, Norway: process and lateral variability in a mass flow-dominated, lacustrine fan delta. In: *Sedimentology of Gravels and Conglomerates* (Eds Koster, E.H. & Steel, R.J.) Can. Soc. Petrol. Geol. Mem. **10**, 295–320.

NICHOLS, N.M. & BIGGS, R.B. (1985) Estuaries. In: *Coastal Sedimentary Environments*, 2nd Edn (Ed. Davis, R.A.) Springer, New York, pp.77–186.

OBANDO, J.A. (1986) Sedimentoligía y tectonica del Cretácico y Paleógeno de la region de Golfito, Península de Burica y Península de Osa, Provincia de Puntarenas, Costa Rica. Unpublished Tesis de Licenciatura, Univ. de Costa Rica, 211 pp.

PICKERING, K.T., STOW, D.A.V., WATSON, M.P. & HISCOTT, R.N. (1986) Deep-water facies, processes and models: a review and classification scheme for modern and ancient sediments. *Earth Sci. Rev.* **23**, 75–174.

POREBSKI, S.J. (1984) Clast size and bed thickness trends in resedimented conglomerates: Example from a Devonian fan-delta succession, southwest Poland. In: *Sedimentology of Gravels and Conglomerates* (Eds Koster, E.H. & Steel, R.J.) Can. Soc. Petrol. Geol. Mem. **10**, 399–411.

POSAMENTIER, H.W. & VAIL, P.R. (1988) Eustatic controls on clastic deposition II — sequence and system tract models. In: *Sea Level Changes, an Integrated Approach* (Eds Wilgus, C.K., Hastings, B.S., Kendall, C.G.St.C., Posamentier, H.W., Ross, C.A. & van Wagoner, J.C.) Soc. Econ. Paleont. Mineral Spec. Publ. **42**, 125–154.

POSTMA, G. (1984) Mass flow conglomerates in a submarine canyon: Abrioja fan-delta, Pliocene, southeast Spain. In: *Sedimentology of Gravels and Conglomerates* (Eds Koster, E.H. & Steel, R.J.) Can. Soc. Petrol. Geol. Mem. **10**, 237–258.

POSTMA, G. (1986) Classification for sediment gravity-flow deposits based on flow conditions during sedimentation. *Geology* **14**, 291–294.

POSTMA, G. (1990) Depositional architecture and facies of river and fan deltas: a synthesis. In: *Coarse-grained Deltas* (Eds Colella, A. & Prior, D.B.) Int. Ass. Sediment. Spec. Publ. **10**, 13–27.

POSTMA, G., NEMEC, W. & KLEINSPEHN, K.L. (1988) Large floating clasts in turbidites: a mechanism for their implacement. *Sediment. Geol.* **58**, 47–61.

PRIOR, D.B. & BORNHOLD, B.D. (1988) Submarine morphology and processes of fjord fan deltas and related high gradient systems: modern examples from British Columbia. In: *Fan Deltas: Sedimentology and Tectonic Setting* (Eds Nemec, W. & Steel, R.J.) Blackie & Son, London, pp. 125–143.

PRIOR, D.B. & BORNHOLD, B.D. (1989) Submarine sedimentation on a developing Holocene fan delta. *Sedimentology* **36**, 1053–1076.

PRIOR, D.B., BORNHOLD, B.D. & JOHNS, M.W. (1984) Depositional characteristics of a submarine debris flow. *J. Geol.* **92**, 707–727.

Ricci Lucchi, F. (1985) Influence of transport processes and basin geometry on sand composition. In: *Provenance of Arenites* (Ed. Zuffa, G.G.) Reidel, Dordrecht, pp. 19–45.

Seilacher, A. (1967) Bathymetry of trace fossils. *Mar. Geol.* **5**, 413–428.

Seilacher, A. (1980) Storm deposits as a tool in facies analysis: II. Sandy tempestites. *Int. Ass. Sediment. 1st Europ. Mtg., Abstract-Band,* 47–49, Bochum.

Seyfried, H., Amann, H., Astorga, A. *et al.* (1991) Anatomy of an evolving island arc: tectonic and eustatic control in the south Central American forearc area. In: *Sedimentation Tectonics and Eustasy* (Ed. MacDonald, D.I.M.) Int. Ass. Sediment. Spec. Publ. **12**, 217–240.

Shanmugam, G. & Moiola, R.J. (1988) Submarine fans: characteristics, models, classification and reservoir potential. *Earth Sci. Rev.* **24**, 383–428.

Shanmugam, G. & Moiola, R.J. (1991) Types of submarine fan lobes: models and implications. *Am. Ass. Petrol. Geol. Bull.* **75**, 156–179.

Silver, E.A., Reed, D.L., Tagudin, J.E. & Heil, D.J. (1990) Implications of the north and south Panama thrust belts for the origin of the Panama orocline. *Tectonics* **9**, 261–281.

Stow, D.A.V. (1986) Deep clastic seas. In: *Sedimentary Environments and Facies,* 2nd Edn (Ed. Reading, H.G.) Blackwell Scientific, Oxford, pp. 399–444.

Surlyk, F. (1984) Fan-delta to submarine fan conglomerates of the Volgian–Valanginian Wollaston foreland group, East Greenland. In: *Sedimentology of Gravels and Conglomerates* (Eds Koster, E.H. & Steel, R.J.) Can. Soc. Petrol. Geol. Mem. **10**, 359–382.

Tournon, J. (1984) Magmatismes du Mesozoique a L'actuel en Amerique Centrale: L'example de Costa Rica, des ophiolites aux andesites. Unpublished Dissertation, Université Pierre et Marie Curie, Paris, 335 pp.

Underwood, M.B. & Bachmann, S.B. (1982) Sedimentary facies associations within subduction complexes. In: *Trench Forearc Geology* (Ed. Leggett, J.K.) Geol. Soc. London Spec. Publ. **10**, 537–550.

Wescott, W.A. & Ethridge, F.G. (1983) Eocene fan delta-submarine fan deposition in the Wagwater Trough, east-central Jamaica. *Sedimentology* **30**, 235–247.

Winsemann, J. (1992) Tiefwasser-Sedimentationsprozesse und -produkte in den Forearc-Becken des mittelamerikanischen Inselbogensystems: eine sequenzstratigraphische Analyse. *Profil* **2**, 1–218.

Wright, L.D. (1985) River deltas. In: *Coastal Sedimentary Environments,* 2nd Edn (Ed. Davis, R.A.) Springer, New York, pp. 1–76.

Spec. Publs Int. Ass. Sediment. (1993) **20**, 415–465

Mesozoic–Tertiary sedimentary and tectonic evolution of Neotethyan carbonate platforms, margins and small ocean basins in the Antalya Complex, southwest Turkey

A.H.F. ROBERTSON

*Department of Geology and Geophysics, University of Edinburgh,
Edinburgh EH9 3JW, Scotland*

ABSTRACT

The allochthonous Antalya Complex ('Antalya nappes') in the Isparta angle of southwest Turkey exemplifies important sedimentary and tectonic processes, related to the interaction of carbonate platforms, margins and small ocean basins in the Eastern Mediterranean Neotethyan area. The complex is mainly composed of Mesozoic sedimentary, igneous and minor metamorphic rocks of mostly microcontinental and oceanic origin. Individual areas are discussed in turn. The *western area* illustrates an original eastward transition from a large carbonate platform (Bey Dağları), across a passive margin, to an oceanic basin, with one or more off-margin carbonate platforms. The *northwestern area* is restored as a small oceanic basin (Isparta Çay unit), between a continuation of the large Bey Dağları carbonate platform and a smaller carbonate platfrom to the east (Davras Dağ). The *northern area* preserves remnants of several other carbonate platfroms, with contrasting Mesozoic depositional histories, either shallow-water (Kaymaz Dağ), or more deeply submerged (Barla Dağ). The *northeastern area*, uniquely, preserves a lithological transition from a carbonate platform to the northeast (Anamas Dağ), across its former passive margin to an oceanic basin located to the southwest. The *eastern* and *southeastern areas* preserve a westward transition from another large carbonate platform (Karacahisar), across an ocean strand, to a large off-margin carbonate platform (Sütçüler–Dulup unit). Finally, the previously much discussed *southwestern area* (Antalya–Kumluca) preserves a palaeogeographically complex passive margin, influenced by Neotethys strike-slip movement prior to tectonic emplacement.

Following stable shelf deposition along the northern margin of Gondwana in Late Permian time, rifting in the Early Triassic was followed by final continental break-up and sea-floor spreading in the Late Triassic (Carnian–Norian) to form a mosaic of continental slivers, volcanic seamounts and small carbonate platforms within the Neotethyan ocean. Oceanic crust was also generated in the Late Cretaceous, possibly in a supra-subduction zone setting. Regional compression began in the latest Cretaceous (Maastrichtian), and led to subduction–accretion, as evidenced by volcanic–sedimentary mélange. The deep-water passive margins of the carbonate platforms were also deformed and thrust-imbricated. Emplacement directions were generally outwards, away from the Isparta angle in all areas. Suturing was completed during Late Palaeocene–Early Eocene, with collision and imbrication of the carbonate platforms. In the Late Eocene, the northeast area was rethrust towards the southwest, related to suturing of a separate Neotethyan strand further northeast. The suture zone in the Isparta angle area then subsided to form an Oligo-Miocene clastic basin (Aksu basin). During the Early Miocene (Burdigalian) the apex of the Isparta angle subsided as a foredeep ahead of thrust sheets advancing from a separate Neotethyan ocean basin system to the northwest (Lycian nappes). During Late Miocene (pre-Messinian), the eastern limb of the Isparta angle experienced westward and southward thrusting and reverse faulting. Crustal extension and strike-slip faulting exploited the Isparta angle suture zone in the Plio-Quaternary, with development of the north-south Kovada graben and its offshore continuation into Antalya Bay.

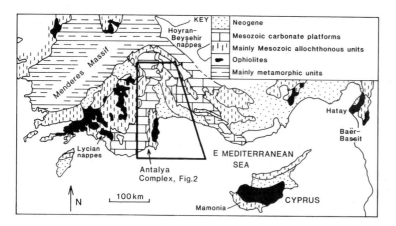

Fig. 1. Outline tectonic map of the Eastern Mediterranean, showing the location of the Antalya Complex, southwest Turkey in relation to the main tectono-stratigraphical units in the region.

INTRODUCTION

The purpose of this paper is to synthesize the Mesozoic–Early Tertiary sedimentary and tectonic evolution of the allochthonous, Neotethyan Antalya Complex in southwest Turkey (Fig. 1), based on new data from the Isparta angle, or Courbure d'Isparta (Blumenthal, 1960–63), within the Tauride Mountains (Fig. 2). This is a classic area for study of processes of microplate interaction within the Neotethyan ocean, including rifting, passive margin and deep-sea deposition, ocean crust genesis, subduction–accretion and collision. Conclusions derived from this area have considerable implications for the wider Tethyan region, in the Eastern Mediterranean and elsewhere.

HISTORY OF RESEARCH

The Antalya area was first investigated by a French group, culminating in tectonic synthesis of southwest Turkey as a whole (Brunn *et al.*, 1971; Dumont *et al.*, 1972; Marcoux *et al.*, 1975; Monod, 1976; Delaune-Mayere *et al.*, 1977; Poisson, 1977; Gutnic *et al.*, 1979). More recently, a British group studied the southwest and northeastern areas of outcrop, as summarized by Robertson and Woodcock (1984) and by Waldron (1984a, b, c; Fig. 3). Extensive fieldwork was also carried out by the Turkish Geological Survey (MTA), centred on Kumluca in the southwest (Şenel, 1980) and Isparta, further north (Figs 2 & 3). Other notable contributions concerned the stratigraphy of the Tauride units (Marcoux, 1970, 1974, 1976, 1977); the petrology and structure of the ophiolite (Juteau,

1975; Juteau *et al.*, 1977; Reuber, 1984); fossil and radiometric age dating (Thuizat *et al.*, 1981; Yilmaz, 1981, 1984; Yilmaz *et al.*, 1981) and Miocene foreland basin sedimentation (Hayward, 1982, 1984; Hayward & Robertson, 1982). However, the present paper is the first to synthesize the Isparta angle area as a whole, based on first-hand fieldwork.

TECTONIC SETTING

The allochthonous Antalya units are a structurally complex association of Palaeozoic, Mesozoic and Early Tertiary sedimentary, igneous and areally minor metamorphic rocks. These units typically structurally overlie large carbonate platform units, known as the 'Tauride autochthons'. The structurally lower Antalya units mainly comprise Mesozoic basinal sediments, mafic extrusives and ophiolitic rocks. Structurally higher units are mainly composed of Mesozoic shallow-water carbonates. In some areas, however, this simple structural order breaks down, for example, where basinal and platform units are structurally interleaved, or where renewed thrusting has disrupted the structural order after initial tectonic emplacement.

Simplified cross-sections of the Antalya units are shown in Fig. 4 and the overall chronology of the main structural events identified is summarized and interpreted in Table 1.

CONTROVERSIES

The Antalya units were first treated as a vast,

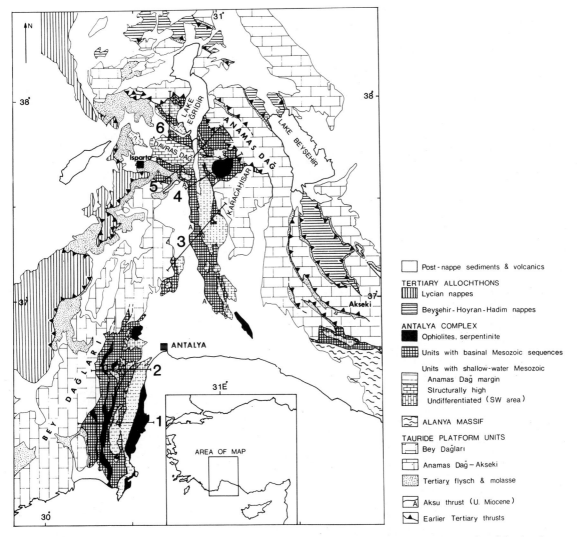

Fig. 2. Tectonic map of the Isparta angle, showing the relationships of the various allochthonous units of the Antalya Complex to the relatively autochthonous Tauride carbonate platforms. (Modified after Waldron, 1984b.)

far-travelled, pile of thrust sheets, the traditional 'lower', 'middle' and 'upper' 'Antalya nappes' (Lefevre, 1967; Brunn, 1974; Brunn *et al.*, 1971; Gutnic *et al.*, 1979). The 'lower nappe' comprised 'proximal' sedimentary units, the 'middle nappe' more 'distal' sediments, volcanics and ophiolitic rocks, and the 'upper nappe' was mainly shallow-water carbonates.

Opinions were divided as to whether these 'nappes' were thrust from the south (Monod, 1977; Poisson, 1977), or from the north (Ricou *et al.*, 1974, 1975, 1979). In the southerly derived model,

the Antalya 'nappes' were thrust a substantial distance (i.e. tens to hundreds of kilometres) from a continental margin and oceanic basin to the south ('Pamphylian Basin'), presumably located in the present Antalya Bay area or further south. In this interpretation the carbonate platforms on either side of the Isparta angle (e.g. Bey Dağları and Anamas Dağ; Fig. 2) simply join up beneath the Antalya 'nappes' as a single regionally extensive Tauride carbonate platform. The present outcrop pattern then resulted only from complicated thrusting, while the initial palaeogeography was essen-

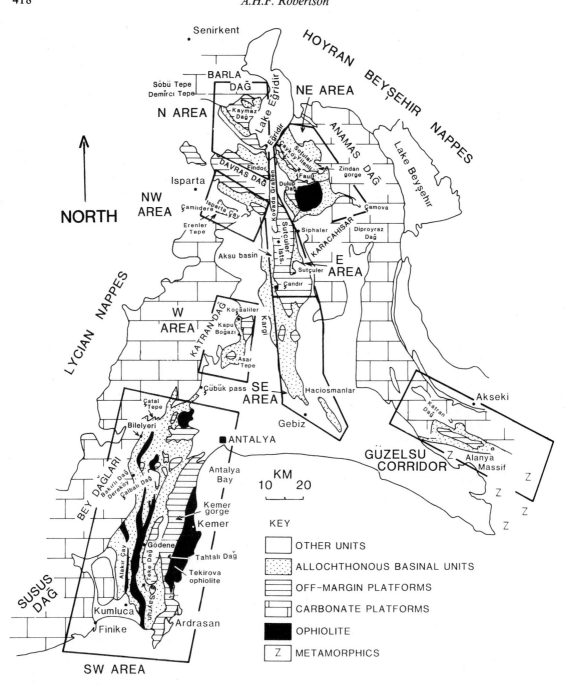

Fig. 3. Outline map showing the division of the Antalya Complex into the specific areas discussed in this paper. Places shown include those not marked on more detailed maps.

tially simple (Monod, 1976). The northerly derived model was similar, except that the Antalya 'nappes' were derived from a Neotethyan basin located far to the north, in central or northern Turkey (Ricou *et al.*, 1974, 1975, 1979). The carbonates of the 'upper nappe' could then represent either remnants of the northern margin of the sutured Neotethys and/or oceanic build-ups within a northerly Neotethys.

Objections to both the far-travelled hypotheses were set out in detail by Robertson and Woodcock (1984), based on stratigraphical, sedimentological and structural evidence mainly from the southwest outcrop area (Kumluca–Antalya; Fig. 3). Woodcock and Robertson (1977b) argued in favour of a less far-travelled hypothesis, with derivation from a Neotethyan basin now preserved as a suture zone essentially within the Isparta angle area. These authors introduced the term *Antalya Complex*, in preference to 'Antalya nappes', since most of the features unifying the Antalya units are lithological and stratigraphical, not merely structural. In addition: (1) the individual 'nappes' are often composite (e.g. sedimentary and igneous units of the 'middle Antalya nappe'); (2) sedimentary continuity can be demonstrated locally between different tectonic 'nappes' (e.g. in the northeast outcrop area; Wal-

dron, 1984a, b, c); (3) high-angle, rather than low-angle tectonics dominate at least the classic southwest outcrop area, where the tripartite nappe system breaks down entirely.

Robertson and Woodcock (1980, 1982) interpreted the southwest area of the Antalya Complex (Fig. 3) as a palaeogeographically complex passive margin adjacent to a large Mesozoic carbonate platform, the Bey Dağları. The 'upper units' (e.g. Çalbalı Dağ; Fig. 3) originated as off-margin continental slivers, capped by Mesozoic carbonate platforms. Emplacement in Late Cretaceous to Early Tertiary was achieved by a combination of thrusting and strike-slip faulting.

Based on detailed mapping of the northwest outcrop area (east of Isparta; Fig. 2) Waldron (1984b) proposed a concept of the Isparta angle as a mosaic of small carbonate platforms, separated by one, or several, Neotethyan oceanic strands within the Isparta angle area. A major objective of the work reported here was to test and extend this hypothesis, by study of the whole area.

Poisson (1984) accepted that the Isparta angle was the site of a deep marine basin, bordered by opposing carbonate platforms (i.e. Bey Dağları–Anamas Dağ/Karacahisar; Fig. 2). The 'upper Anta-

Table 1. Summary of the main structural events in the Antalya Complex and overall interpretations. Based on the literature and new data presented in this paper

	Period/Epoch	Structural event	Interpretation
1.6	Pliocene/Pleistocene	Extension and/or strike-slip, Kovada graben	Reactivation of Antaya Complex suture zone
5	Miocene	Compression in southwest, west, east and southeastern areas	Tightening of suture due to advanced collision elsewhere
23	Oligocene	Extension	Post-suture Aksu clastic basin
35	Eocene	Northeast to southwest thrusting in northeastern area only Northwest to southeast thrusting in north area	Rethrusting due to collisions further north
50	Palaeocene	Large-scale overthrusting	Collision and suturing of platforms and basins
65	Cretaceous U L	Extensional faulting, subsidence	Supra-subduction zone spreading within Neotethys
145	Jurassic U M L	Extensional faulting, volcanism Gentle subsidence	Renewed sea-floor spreading Relative stability
208	Triassic U M L	Extensional faulting, volcanism Crustal extensions minor volcanism	Initial sea-floor spreading Rifting of Neotethys

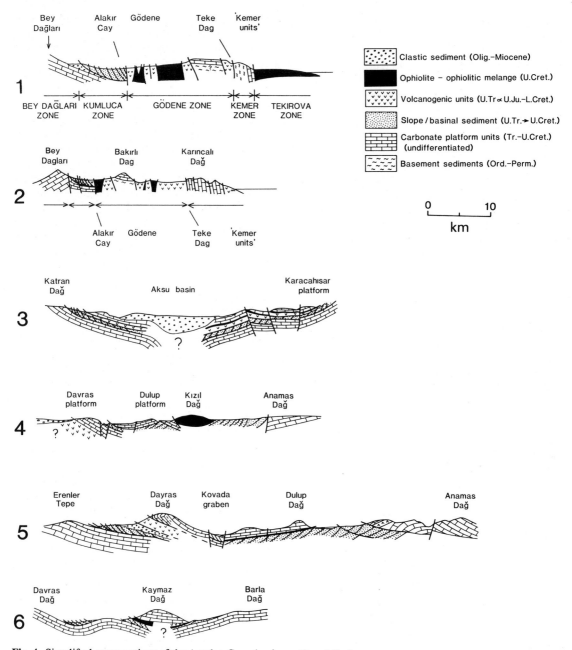

Fig. 4. Simplified cross-sections of the Antalya Complex in southwest Turkey.

lya nappe', however, was not explained in this hypothesis. However, in his thesis Poisson (1977) suggested that the upper nappe was an out-of-sequence thrust sheet, resulting from rethrusting of the carbonate platform basement after initial overthrusting by the 'Antalya nappes'.

Recently, Marcoux *et al.* (1989) have returned to the earlier concept of thrusting of the Antalya 'nappes' from a single Neotethyan ocean basin, located far to the north. However, as discussed by Robertson *et al.* (1991), these authors did not show that small-scale structural evidence of southward

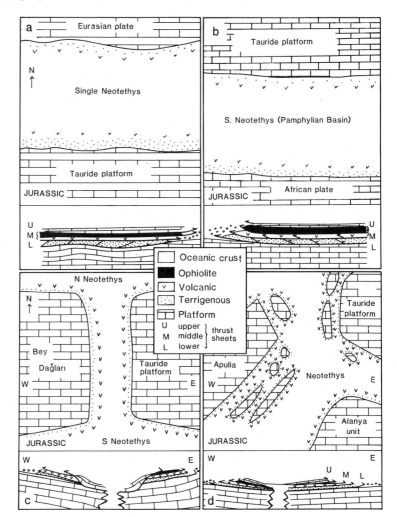

Fig. 5. Alternative tectonic interpretations of the Antalya units, based loosely on published concepts: (a) thrust from the north from a single Neotethyan ocean basin (Ricou *et al.*, 1974, 1975, 1979; Marcoux *et al.*, 1989); (b) thrust from a southerly Neotethyan ocean basin (Dumont *et al.*, 1972; Monod, 1976, 1977); (c) thrust from a deep marine basin within the Isparta angle (Poisson, 1984) — upper carbonate limestone units emplaced by out-of-sequence thrusting (Poisson, 1977); (d) an origin as a palaeogeographically complicated small ocean basin within the Isparta angle; units were thrust a relatively short distance outwards onto adjacent carbonate platforms (Robertson & Woodcock, 1980, 1982, 1984; Waldron, 1981, 1984a, b; present paper).

displacements within the Antalya units relates to the critical, initial phase of Late Cretaceous–Early Tertiary tectonic emplacement, rather than to later post-suturing deformation. Also, north of Isparta (e.g. Sübü Tepe and Demirci Tepe; Fig. 3) a succession of virtually undeformed Upper Cretaceous pelagic carbonates passes, without any depositional break, into red, hemipelagic carbonates, interbedded with thin (less than 0.25 m thick) calciturbidites of upper Paleocene–lower Eocene age (S. Yalkinçaya, personal communication, 1985). Tectonic disruption and thrusting of allochthons through this area from the northwest during this time interval, as in Marcoux *et al.*'s (1989) hypothesis is thus precluded.

In summary, the alternative tectonic models for the emplacement of the Antalya units are shown in Fig. 5. Controversy clearly persists and, therefore there is now an urgent need to take account of evidence from the whole of the Antalya Complex outcrop.

METHODOLOGY AND TECTONOSTRATIGRAPHY

The geology of the Antalya area is complicated and it is thus essential to describe and interpret each area individually before attempting an overall palinspastic reconstruction and synthesis. For this purpose, the Antalya Complex is divided into the following areas (Fig. 3): (1) *west area* (Kocaaliler); (2) *northwest area* (Isparta Çay–Davras Dağ); (3) *north area* (Davras Dağ–Kaymaz Dağ–Barla Dağ); (4) *northeast area* (Lake Eğridir–Anamas Dağ–Dulup Dağ); (5) *east area* (Karacahisar–Sütçüler); and (6) the *southeast area* (Gebiz–Serik). Finally, (7) the *southwest area*, which has been discussed at length elsewhere (e.g. Robertson & Woodcock, 1984), will be summarized briefly. Antalya Complex outcrops further east (Güzelsu corridor; Monod, 1977) are excluded, although new radiolarian data are included in the Appendix.

Attempts to analyse complex orogenic segments similar to the Isparta angle must draw on data from a range of disciplines within the earth sciences. The most important of these are as follows.

1 *Regional mapping.* From this it is clear that the concept of low-angle thrust nappes applies well to either side of the Isparta angle, but breaks down in the southwest area, where steeply dipping thrust sheets dominate.

2 *Gross stratigraphy.* The overall age of each of the units has implications for the timing of tectonic events, particularly rifting, spreading and emplacement. For example, radiolarites were previously thought to be entirely Upper Triassic (e.g. Poisson, 1977), but are now known to span a much longer time interval (see Appendix). Also, the age of 'flysch' successions stratigraphically overlying Mesozoic carbonate platforms constrains tectonic emplacement.

3 *Sedimentology and facies analysis.* This has the potential to identify distinct depositional settings, notably the existence of slope and basinal units. Allochthonous slope and basinal units of the Antalya Complex can be identified, and related to a number of Mesozoic carbonate platform units within the Isparta angle.

4 *Structural kinematic studies.* These reveal the relative emplacement directions in each area. Detailed studies of the southwest (Woodcock & Robertson, 1982) and northeast (Waldron, 1981, 1984a) Antalya Complex areas have already shown that initial emplacement was essentially towards the neighbouring, relatively autochthonous, carbonate platforms during Late Cretaceous–Early Eocene time, a conclusion borne out by reconnaissance structural studies of the other areas during this work. In other words, the Antalya oceanic units were generally thrust eastwards and westwards, away from an ocean basin located within the present Isparta angle suture zone.

5 *Basalt 'immobile' element studies.* Such studies show that the Antalya basalts of both Upper Triassic and Upper Jurassic–Lower Cretaceous age originated in a range of within-plate to mid-ocean ridge-type (MORB) settings (Robertson & Waldron, 1992) and support the existence of Neotethyan oceanic basins as early as Late Triassic in this area.

NOMENCLATURE

Formal stratigraphical units were previously erected for the northeast (Waldron, 1984c) and southwest areas (Robertson and Woodcock, 1981a, b, c). However, to avoid a proliferation of names, informal stratigraphical terms are adopted here. The overall stratigraphy of the main carbonate platform and basinal successions is shown in Fig. 6a and b.

Much debate has surrounded the application of tectonic terms in the Antalya units.

Antalya Nappes versus Antalya Complex. A thrust nappe is defined as 'an allochthonous tectonic sheet which has moved along a thrust fault' (McClay, 1981). A complex is defined is 'an assemblage of rocks of any age that have been folded together, intricately mixed, inverted, or otherwise complicated' (American Geological Institute, 1961). The term 'complex' is retained here as the definition fits the Antalya units much better than that of a thrust nappe. Note that the term Antalya Complex is purely descriptive and specifies no deformation mechanisms.

Mélange is used here as a purely descriptive field term for a pervasively mixed unit, irrespective of whether mixing took place by sedimentary, or tectonic processes, or both. Many sedimentary mélanges were generated by mass flow processes (e.g. debris flows).

Broken formation, a field term for thrust units that are so disrupted that they are intermediate between intact thrust sheets and mélange.

DISCUSSION OF INDIVIDUAL AREAS

Each of the areas within the Anatlya Complex is discussed below in turn, working clockwise a round the Isparta angle, beginning with the western area.

Western area

Traced northwards from the large southwestern area of outcrop (Antalya–Kumluca; Fig. 3), the Antalya Complex slope and basinal units disappear beneath Plio-Pleistocene sediments and are represented only by small outcrops of mélange and debris flows ('olistostromes'), faulted against the Bey Dağları carbonate platform succession (e.g. at Çübük Pass; Fig. 3). Similar units, however, reappear 20 km further north in the *western area* (Fig. 3).

Structural setting

In the western area, the Antalya Complex is thrust over a local, relatively autochthonous carbonate platform unit, the Katran Dağ (Figs 3, 7 & 8), an unbroken extension of the Bey Dağları northwards from the southwestern area. The basal thrust of the allochthonous units dips eastwards at a moderate angle (Fig. 8a). Above are three thrust-bounded units, roughly corresponding to the traditional 'middle' and 'upper' 'Antalya nappes'. The lower structural levels are sedimentary rocks, taking the form of coherent thrust sheets, dismembered thrust sheets, broken formation and mélange. Thrust-sheet imbrication and the geometry of occasional medium-scale folds indicates a generally westward emplacement towards the relatively autochthonous Katran Dağ carbonate platform. Above, there are thrust sheets of Upper Triassic basaltic lava, interbedded with deep-sea sediments (Fig. 8a). The thrust at the base of this volcanic-dominated unit cuts obliquely across the underlying sedimentary thrust sheets (southeast of Kocaaliler; Fig. 7). Between the sedimentary and volcanic units is a 60-m-thick zone of sheared serpentinite, mafic lava, quartzose turbiditic sandstones, limestone with *Halobia* and radiolarian cherts, individually forming thrust wedges up to several metres thick. Above the basalts, a moderately steep eastwards-dipping thrust brings in a higher unit, mainly composed of shallow-water carbonates (e.g. Gökin Tepe; Fig. 8a). Similar platform carbonates are exposed along strike further to the south (e.g. Alisivri Tepe; Kevke Tepe; Fig. 6a; Poisson, 1977). To the east, pillowed flows and radiolarites are locally exposed beneath the carbonate thrust sheets (along the western slopes of the Aksu River). Eastwards again, the Antalya Complex is unconformably overlain by Miocene and Plio-Quaternary clastic sediments (Fig. 3).

Lithologies

Autochthonous carbonate platform. Poisson (1977) recognised a succession of Lower Jurassic to Cenomanian age in the relatively autochthonous Katran Dağ carbonate platform (e.g. Kapu Boğazı; Fig. 3). This is overlain by pelagic carbonates of Upper Turonian, Coniacian and Danian age and then, further west by ophiolite-derived olistostromes of Palaeocene–lower Eocene age (e.g. Bademağacı; Poisson, 1977). By contrast, the succession in the western part of the Katran Dağ carbonate platform continues through Eocene pelagic limestones (Middle Lutetian), Lower Miocene limestones (Aquitanian Karabayır Formation) and flysch (Burdigalian–Lower Langhian Karakuş Tepe Formation).

The upper part of the succession along the eastern margin of the Katran Dağ carbonate platform was re-examined in this study, with the aim of shedding light on the emplacement history of the Antalya Complex. The upper surface of Cenomanian platform carbonates, largely rudist limestones, is fissured and karstic and then overlain by 40 m of medium-bedded micritic limestones, with abundant Upper Cretaceous planktonic foraminifera (e.g. *Globotruncana*). The highest stratigraphical levels of this succession contain silt and pebbly conglomerates, composed of reworked clasts of neritic and planktonic foraminiferal limestone, with *Globotruncana*, basalt, radiolarite and rare quartz. Locally (e.g. Karadağ area, Upper Cretaceous pelagic limestones are absent and allochthonous Antalya Complex units overlie several hundred metres of massive mega-breccias with metre-sized clasts, conglomerates, grey and white biomicrites and biosparites. Fossils in these coarse clastic sediments include large pelecypods, gastropods and rudists. They are overlain by tens of metres of more regularly stratified marls, calciturbidites and limestone debris flows with clasts up to 2.5 m in size.

Further west, sedimentation on the Bey Dağları (Fig. 6a) continued into Early Tertiary time, until the arrival of ophiolite-bearing debris flows during the Palaeocene–Early Eocene (e.g. near Bademağacı; Poisson, 1977).

Sedimentary sheets. Thick, intact successions were measured in structurally overlying sedimentary thrust sheets (Fig. 8b & c, log 1). At the base there are '*Halobia* limestones' and marls, of Upper Triassic age. The succession continues upwards through

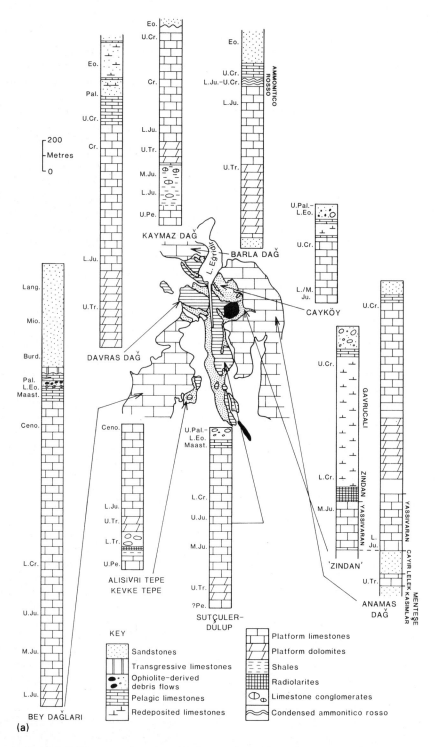

Fig. 6. (*above and opposite.*) Summary of stratigraphical successions. (a) Platform and platform edge units within the Isparta angle area. Sources of data: mainly Poisson, 1977; Gutnic *et al.*, 1979; this study. The southwestern area is excluded (see Robertson & Woodcock, 1982).

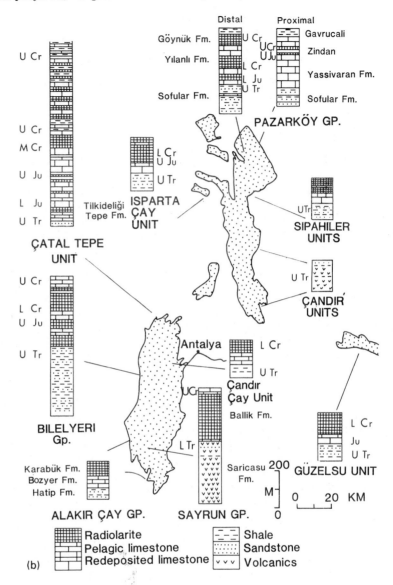

Fig. 6. (*Continued*) (b) Slope and basinal sedimentary successions of the Antalya Complex as a whole.

(b)

weakly-bedded, to massive, sugary marbles, that are interbedded with thin-to medium-bedded calciturbidites, calcareous shales and black chert, of replacement origin. The succession culminates in 100 m of limestone conglomerates, with clasts up to 1.5 m in diameter. Similar conglomerates crop out extensively, as fragmentary thrust sheets further north in the area (e.g. near Kocaaliler; Fig. 7). The clasts in the conglomerates are mainly shallow-water carbonates, including micrite, pelmicrite, benthic foraminifera, shell fragments, echinoderm

plates and algal-coated grains, often with micritic envelopes. Terrigenous material is absent.

The above units are structurally overlain by contrasting thrust sheets of more 'distal' sedimentary rocks. The lithologies are mostly turbiditic quartzose sandstones, rich in plant material at the base, passing upwards into thin- to medium-bedded calcilutites ('*Halobia* limestones'), followed by reddish shales, ribbon radiolarites and silicified calciturbidites (Fig. 8b, log 2). The radiolarians found in these beds include a form known in the Albian

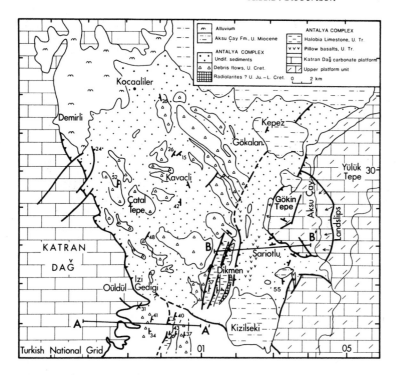

Fig. 7. Detailed map of the northern part of the western area. Note the dominance of limestone breccias and olistostromes thrust over the eastern margin of the relatively autochthonous Katran Dağ carbonate platform. Mapping by MTA, with sedimentological and structural data from this study.

(Mid- Cretaceous) (E.A. Pessagno, personal communication, 1984).

Further north, the highest structural levels of the thrust sheets of sedimentary rocks are dominated by pelagic carbonates, with Upper Cretaceous planktonic foraminifera (e.g. *Globotruncana*). There are also intercalations of ophiolite-derived debris flows (e.g. southwest of Kocaaliler; Fig. 7), which contain tectonic inclusions of radiolarite, serpentinite and basic lava.

Volcanic units. The structurally higher levels of the western area are dominated by pillow lavas, with subordinate massive basalts up to several hundred metres thick (Fig. 8b, log 3). There are also depositional intercalations of red radiolarite and pink ferruginous micritic limestone several metres thick containing Upper Triassic *Halobia* shell fragments.

Upper carbonate units. Structurally above are thrust units of shallow-water carbonates of Upper Permian to Upper Cretaceous (Cenomanian) age (i.e. of the 'upper Antalya nappe'; Poisson, 1977). These limestones are unconformably overlain by Miocene coarse clastic sediments to the northeast (e.g. north and northeast of Yülük Tepe; Fig. 7). Upper

Permian fossiliferous, argillaceous limestones in the upper carbonte units are conformably overlain by Lower Triassic (Skythian) pelecypod- and gastropod-bearing, vermicular limestones and dolomites. Elsewhere, Poisson (1977) identified Upper Permian limestone clasts in a Skythian-aged matrix (Kızılsekı breccias). The succession then passes into Upper Triassic to Cenomanian platform carbonates, terminating in pelagic carbonates.

Interpretation

The structural base of the succession, the Katran Dağ carbonate platform, is correlated with the regionally extensive Bey Dağları carbonate platform (Fig. 9a). The eastern margin of this platform is not exposed, however, and back reef, carbonate flat and lagoonal facies dominate (Poisson, 1977). After the Cenomanian, the platform was tilted, eroded, then subsided and was overlain by pelagic carbonates. This was followed by input of clastic silt that was presumably eroded from slope and basinal units of the Antalya Complex during its initial stages of emplacement in Maastrichtian time. In some areas, the platform was faulted and underwent mass wasting at this stage, giving rise to debris flows and megabreccias, mainly composed of neritic limestone.

The structurally lower, sedimentary sheets of the Antalya Complex are interpreted as deep-water facies laid down, from the Late Triassic onwards, near the base-of-slope of a contemporaneous carbonate platform, which is assumed to be the regionally extensive Bey Dağları unit. The Upper Cretaceous conglomerates at the top of the succession are similar to those in the adjacent, relatively autochthonous, carbonate platform succession (Katran Dağ E margin) and represent an important depositional link between these two units (Fig. 9b). As the carbonate platform was deformed, possibly in response to flexural loading, mass wasting took place and debris was shed eastwards into a deep-water, base-of-slope setting, where it was later detached and thrust back over the platform.

The structurally higher thrust sheets of sedimentary rocks are interpreted as distal, basin-plain deposits. Upper Triassic quartzose sandstone turbidites were derived from an unexposed metamorphic basement. The Halobia limestones represent periplatform oozes that were shed eastwards from newly generated carbonate platforms. The ophiolite debris in the overlying Late Cretaceous pelagic carbonates shows that this area was overthrust by Neotethyan oceanic crust, although substantial ophiolite thrust sheets are not preserved. The ophiolitic debris reached as far west as the inferred base-of-slope of the Katran Dağ carbonate platform during Maastrichtian/Palaeocene time, however, the allochthonous units were not thrust over the Bey Dağları carbonate platform until the Late Palaeocene–Early Eocene, when ophiolitic debris flows appeared within the *in situ* successions.

The structurally higher basic volcanics and associated pelagic and hemipelagic sediments originated further out in the Neotethyan ocean basin within the Isparta angle area to the east. The basalts analysed from this unit are of within-plate type, mid-ocean ridge-type (MORB) and depleted MORB type (Robertson, 1990; Robertson & Waldron, 1990). The basal contact of this unit was an important thrust surface along which distal, oceanic units were transported over more proximal, marginal units.

Following rifting in the Early Triassic (Fig. 9a), the structurally overlying shallow-water limestone unit ('upper Antalya nappe') is interpreted as a carbonate build-up which developed within the Neotethyan ocean basin to the east. The thickness

was considerably less than that of the subjacent Bey Dağları carbonate platform (Katran Dağ; Fig. 6a). This applies to all the structurally upper carbonate platform units in the Isparta angle. A possible explanation is that these units originated within the Neotethyan ocean, which remained magmatically active and thus buoyant, topographically high areas where carbonate deposition was limited, whereas marginal areas cooled earlier and subsided further allowing thicker platform successions to accumulate. A hiatus in the Cenomanian in the western area, as elsewhere, was followed by pelagic deposition. Uncertainties remain as to whether the various upper limestone sheets in the western area formed part of one, or several, build-ups, and how far these units extend eastwards beneath the Neogene–Quaternary sediment cover. The inferred emplacement direction is shown in Fig. 9b.

Northwest area

Structural setting

The northwestern area of the Antalya Complex (Fig. 3) is bordered to the west by the northwards extension of the Bey Dağları carbonate platform, Erenler Tepe and includes the well-known Isparta Çay sedimentary units (Allasinaz *et al.*, 1974) and the Davras Dağ carbonate massif to the east (Fig. 3).

In the west, the local carbonate platform (Erenler Tepe, Fig. 10) is disconformably overlain by an important sedimentary mélange, the 'Çamlıdere olistostrome' (Poisson, 1977; Fig. 10a & b). This unit is tectonically overlain first by thrust sheets of deep-sea sedimentary rocks (Isparta Çay unit). Unconformably above are Lower Miocene (Aquitanian) shallow-water limestones, up to several tens of metres thick (Poisson, 1977), passing upwards into quartzose and ophiolite-derived turbidites of Burdigalian age (Poisson, 1977; Fig. 10a & b). To the east, the Miocene sediments are then overthrust by a volcanic–sedimentary mélange, of c. 4 km structural thickness. This, in turn, is structurally overlain to the east by an inverted succession of calcareous flysch of Early Tertiary age. Eastwards again, these sediments pass structurally upwards (but stratigraphically downwards) into Mesozoic platform carbonates of the Davras Dağ carbonate platform (Fig. 10a). Along its southern margin the Davras Dağ platform carbonates are thrust over volcanic–sedimentary mélange of Palaeocene–Early Eocene

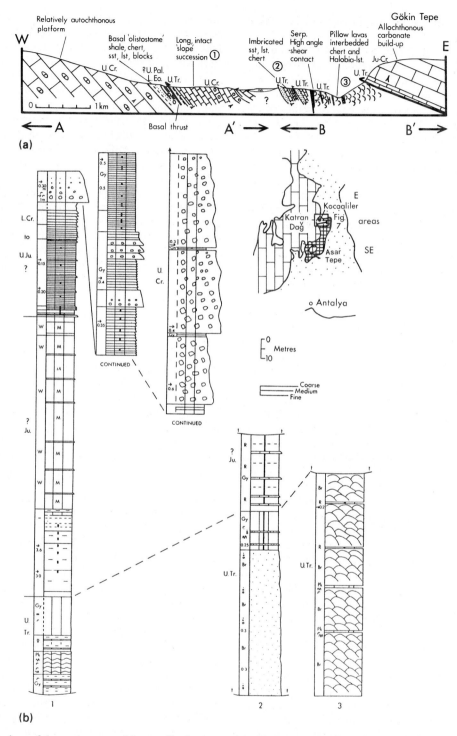

Fig. 8. Geology of the western area: (a) generalized cross-section; (b) measured sedimentary logs of thrust successions along the eastern margin of the relatively autochthonous Katran Dağ carbonate platform. More distal slope and basinal units are present in successively higher sheets. The locations are marked on the generalized cross-section above. (c) Key to the logs in Figs 8 and 12.

age, which in turn, is thrust over Miocene clastic sediments (east of Yukarıgökdere; Fig. 18).

Lithologies

Carbonate platform. Typical platform carbonates, of probable Turonian age form the base of the local succession of Erenler Tepe (Fig. 10a & b). They are depositionally overlain by 60–80 m of thin- to medium-bedded calciturbidites, interbedded with pelagic limestones, dated by Poisson (1977) as 'Lower Senonian'. The calciturbidites exhibit small-scale sedimentary structures, including grading, cross-lamination, convolute lamination and local burrowing. Petrographical studies reveal micritic intraclasts, shell fragments, calcite-replaced radiolarian tests, rare benthic foraminifera and scattered quartz grains. The highest levels of the succession

include more argillaceous intercalations several metres thick with detrital quartz grains, polycrystalline quartz and muscovite, but no ophiolite-derived material.

Sedimentary mélange. The highest levels of the autochthonous platform succession pass over a sharp, but clearly sedimentary, contact into a sedimentary mélange ('Çamlıdere olistostrome'), some 300 m thick (Fig. 10a & b). This unit contains planktonic foraminifera of Upper Palaeocene–Lower Eocene age within a sparse matrix. The mélange mainly exhibits a 'block-against-block' fabric. Many of the blocks are elongate and lithologies comprise Mesozoic basinal sediments, poorly cemented unsorted breccias, crudely stratified coarse sandstones and chalky marls. Disrupted intercalations of pelagic carbonate can be traced several

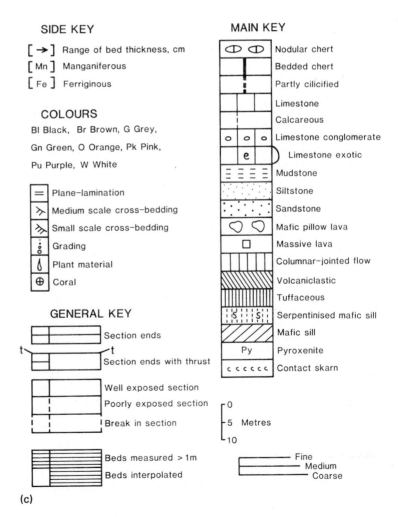

Fig. 8. *(Continued)* (c)

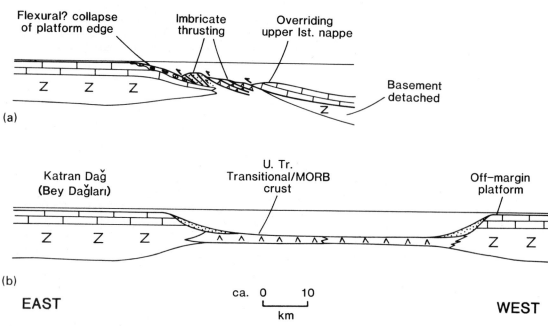

Fig. 9. Reconstruction of the western area of the Antalya Complex. (a) Upper Cretaceous–Palaeocene: during closure, prior to suturing. (b) Upper Triassic–Lower Cretaceous: passive margin phase.

hundred metres laterally, mainly in the lower levels of the mélange. The clasts in the breccias are highly angular and are mainly composed of radiolarian chert. In thin section the matrix of the mélange is mostly angular fragments of radiolarian chert, often replaced by calcite, pelletal limestone (with rhombs of diagenetic dolomite) and scattered quartz grains. The highest levels of the mélange consist of sheared, varicoloured, siliceous mudstone with scattered, small, rounded chalk pebbles. In places, the sedimentary mélange is separated from overlying Antalya Complex units by a zone of disrupted radiolarian chert up to 10 m thick.

Deep-water sediment thrust sheets. The structurally overlying sedimentary units are well exposed along a tributary of the Isparta Çay (Isparta River; Fig. 10a & b). Detailed mapping and new radiolarian determinations (see Appendix) establish that the succession is repeated by thrusts on a scale of tens to several hundreds of metres (Fig. 11). Taken together the dip of the thrust sheets, the presence of local outcrop-scale folds and the geometry of small duplex structures all indicate thrust movement towards the southwest. Previously this unit was treated as an over 1-km-thick, single stratigraphical succession (Allasinaz *et al.*, 1974; Poisson, 1977).

The stratigraphy within individual thrust sheets typically begins with poorly exposed, medium- to thick-bedded, medium-grained turbiditic quartzose sandstones, in beds up to 1.8 m thick, often rich in plant material (Fig. 12). Some intervals are markedly channelized, with individual channels up to 10 m wide. Channel axes contain conglomerate lags with numerous rounded pebbles of black chert up to 15 cm size formed by the replacement of carbonate. In some successions, sandstone is largely absent and bedded chert predominates (Fig. 12, log 2b). Intercalated mudstones contain illite, smectite and traces of chlorite (Allasinaz *et al.*, 1974).

The quartzose sandstones are conformably overlain by grey, medium-bedded calcilutites, with shaley partings and nodular replacement cherts ('*Halobia* limestones'). The carbonate consists of micrite with numerous radiolarians replaced by calcite and some terrigenous silt. In addition, intercalations of fine- to medium-grained calciturbidites up to 1.5 m thick contain micritic intraclasts, pellets, small benthic foraminifera, echinoderm plates, shells and scattered quartz grains. Several intercalations of rubbly limestone debris flows up to 16 m thick are dominated by shallow-water derived carbonates (Fig. 12, log 4a). Allasinaz *et al.* (1974) reported a rich fauna, including ostracods, sponge spicules, the ammonite

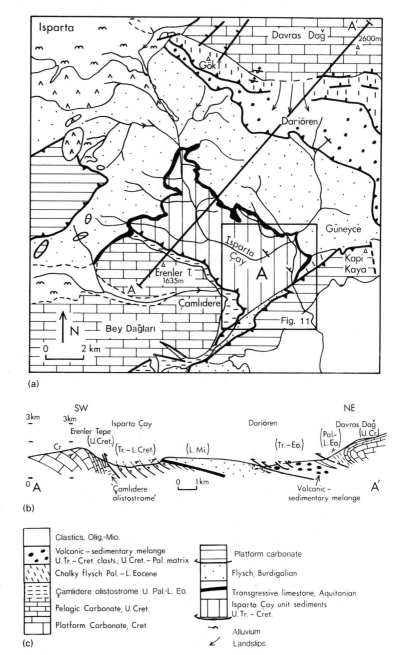

Fig. 10. Geology of the northwestern area of the Antalya Complex (Fig. 3). (a) Geological map of the northern part of the northwestern area, based on Poisson (1977), with additional data from this study. The box shows the area of the more detailed map in Fig. 11. (b) Structural cross-section of the area. (c) Key.

Auloceras and seven species of *Halobia*, confirming a Carnian–Norian (Upper Triassic) age.

In some thrust sheets, mainly the structurally lower ones, the succession continues upwards into non-calcareous radiolarian sediments via a several-metres-thick transitional unit of thinly-bedded, pink, calciturbidites and siliceous mudstones (Fig. 12, log 7). The overlying successions mainly comprise ribbon radiolarian cherts and red, non-calcareous, siliceous mudstones intercalated with fine- to medium-grained, turbiditic, packstones and grainstones up to tens of centimetres thick. Original

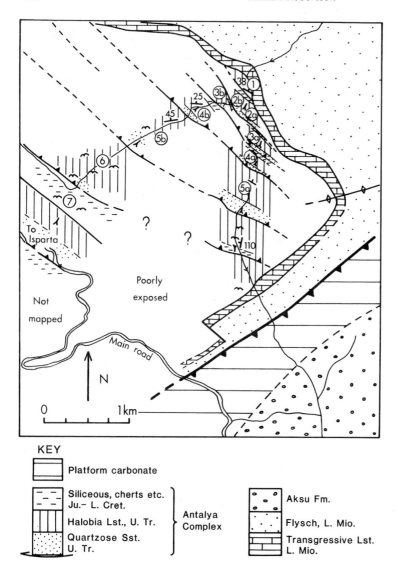

KEY

| | Platform carbonate |

	Siliceous, cherts etc. Ju.– L. Cret.
	Halobia Lst., U. Tr.	Antalya Complex
	Quartzose Sst. U. Tr.

	Aksu Fm.
	Flysch, L. Mio.
	Transgressive Lst. L. Mio.

Fig. 11. Structural sketch map of imbricated thrust sheets in the Isparta Çay, northwestern area. See Fig. 10 for location of map. Numbers in circles refer to the individual thrust sheets. Measured sedimentary logs are shown in Fig. 12.

carbonate sediments are now almost entirely silicified. Optical examination of thin sections reveals pellets, ooids, oncolites, gastropods, bivalves, rare benthic foraminifera, echinoderm plates, calcareous algae and scattered quartz grains. E. A. Pessagno (personal communication, 1984) interpreted the fauna as Upper Jurassic–Lower Cretaceous in age (see Appendix). Thus, the successions in the individual thrust sheets span a time interval at least from Upper Triassic (Carnian–Norian) to Lower Cretaceous.

Volcanic–sedimentary unit. Further east, the sedi-

mentary thrust sheets described above are unconformably overlain by shallow-water limestone of Aquitanian age (Poisson, 1977; Fig. 10a & b). These limestones pass transitionally upwards into strongly sheared successions of medium-bedded terrigenous turbidites, interbedded with mudstones of Burdigalian age (Poisson, 1977). This unit is, in turn, overridden by volcanic–sedimentary mélange along an eastward-dipping thrust (e.g. near Darioren; Fig. 10b).

The base of the volcanic–sedimentary mélange is locally marked by a several-metres-thick and 10-m-long, thrust-bounded sliver of schistose amphibo-

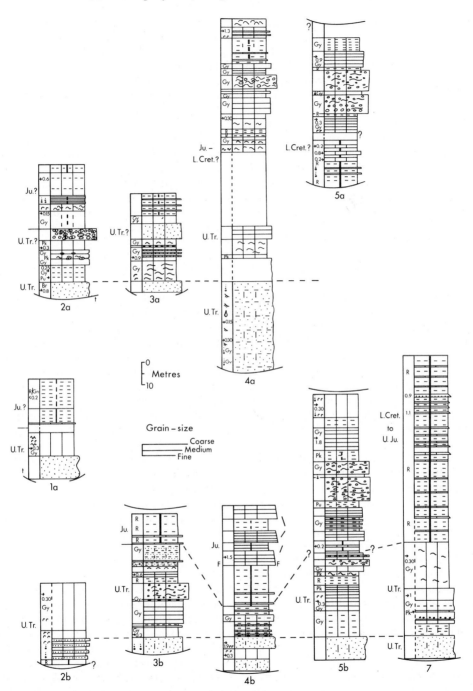

Fig. 12. Measured sedimentary logs of thrust sheets within the Isparta Çay unit, northwestern area. Fossil evidence indicates Upper Triassic to Lower Cretaceous ages. The numbers refer to the individual thrust sheets shown in Fig. 11. See Fig. 8c for key.

lite, mainly comprising green hornblende, quartz, sphene, plagioclase, opaque minerals and chlorite.

The overlying volcanic–sedimentary mélange (Fig. 10a & b) is dominated by pillow lava, massive basalt and hyaloclastite. Several relatively intact lava sheets up to tens of metres thick contain depositional intercalations of ferruginous and/or manganiferous radiolarian chert up to several metres thick (e.g. northeast of Darioren; Fig. 10a). The mélange includes disrupted sheets of Upper Triassic '*Halobia* limestones', up to 20 m long, and thick-bedded calciturbidites and debris flows with shallow-water derived limestone clasts. Locally (southwest of Darioren; Fig. 10a), three sub-vertically dipping sedimentary sheets are located within the middle part of the mélange. These sheets are 80 m thick and can be traced up to 1 km laterally. The succession in each sheet begins with turbiditic quartzose sandstones, followed by '*Halobia* limestones', and then by red radiolarian cherts dated as Upper Jurassic–Lower Cretaceous (see Appendix). Elsewhere, structurally higher levels of the mélange include debris flows, interbedded with chalky flysch containing planktonic foraminifera of Upper Palaeocene–Lower Eocene age. The debris flows range from polymict, with clasts of basalt, '*Halobia* limestone', radiolarite and sandstone, to nearly monomict, composed, for example, of lava, micaceous sandstone, and/or serpentinite.

The volcanic–sedimentary mélange, mostly sheared debris flows, can be traced southeastwards around the margins of the Davras Dağ and is last seen cropping out along the western margin of the neotectonic Kovada graben (e.g. near Yukarıgökdere; Figs 3 & 18).

In summary, the volcanic–sedimentary unit mainly consists of Upper Triassic subaqueous basalts and Upper Triassic–Cretaceous deep-water sediments. Basalts analysed are within-plate type (WPB), mid-ocean ridge type (MORB) and depleted MORB type (Robertson, 1990; Robertson & Waldron, 1990). The mélange matrix comprises debris flows, volcaniclastic sandstones and chalky flysch, of Upper Palaeocene–Lower Eocene age.

Davras Dağ carbonate platform. The succession in the structurally overlying Davras Dağ carbonate platform (Fig. 6a) begins with Upper Triassic dolomites and is then overlain, in turn, by Jurassic–Lower Cretaceous platform carbonates and then by Upper Cretaceous pelagic carbonates. The succession passes first into Palaeocene, red, hemipelagic

limestones and then into *Nummulites*-bearing flysch of Lower Eocene age (Gutnic, 1977; Gutnic *et al.*, 1979). This succession is critical as it effectively rules out any thrusting over the Davras Dağ platform during Maastrichtian–Early Eocene, when the first emplacement of Antalya Complex units was taking place elsewhere.

The *Nummulites*-rich flysch is particularly well exposed in a large west-facing, recumbent fold along the southwestern margin of the Davras Dağ carbonate platform (Fig. 10b), where it comprises turbiditic sandstones, calciturbidites, mudstones, hemipelagic carbonates and pelagic chalk. Thicker-bedded intercalations include amalgamated, channelized debris flows, up to 10 m thick, that pass upwards into cross-bedded sandstone and then into silty chalk, within individual depositional units. Complete A–E Bouma sequences of turbidites (Bouma, 1962) are also common. Most of the clasts in the debris flows are moderately well rounded, composed mainly of mafic lava, chert and *Nummulites*-bearing limestone. In thin section, the sandstone is compositionally heterogeneous and includes quartz, polycrystalline quartz, siltstone, biomicrite and radiolarian chert. Bioclasts comprise planktonic foraminifera, *Nummulites*, and shell fragments. Igneous grains include basic lava, devitrified volcanic glass, pyroxene, plagioclase and rare serpentinite. The calciturbidites are packed with planktonic foraminifera, often broken, together with large shell fragments, recrystallized radiolarians, terrigenous grains, minor chlorite and pyrite.

Traced southeastwards around the margins of the Davras Dağ massif, similar flysch is locally isoclinally folded, with folds facing mainly towards the southwest.

Interpretation

Moving from southwest to northeast across the northwest area each of the units is restored as follows. During the Late Cretaceous (post-Turonian) the local Bey Dağları carbonate platform (Erenler Tepe) subsided and was overlain by pelagic carbonates and calciturbidites, followed by input of terrigenous silt, presumably eroded from slope and basinal units of the Antalya Complex during the initial stages of emplacement (Fig. 13b). The structurally overlying Çamlıdere 'olistostrome' formed by mass wasting of sedimentary units of the Antalya Complex during final emplacement onto the carbonate platform in Late Palaeocene–Early Eocene time (Fig. 13c). However, ophiolitic material did not reach the thrust front in this area.

The overlying thrust sheets of Upper Triassic–Lower Cretaceous sediments (Isparta Çay unit) accumulated as channelized facies near the base of a carbonate platform slope, interpreted as the Bey Dağları to the west (Fig. 13a). The Upper Triassic 'Halobia limestone' originated as periplatform ooze, interbedded with calciturbidites and debris flows, derived from the neighbouring (Bey Dağları) carbonate platform. The supply of shallow-water carbonate allochems decreased after the Triassic, but proximity to the carbonate platform is still indicated by the presence of shallow-water derived calciturbidites, within bedded chert successions of Upper Jurassic–Lower Cretaceous age. During the initial stages of emplacement (Maastrichtian–Early Palaeocene?), the sedimentary sheets were thrust southwestwards, forming an imbricate fan (Fig. 13b).

The blocks in the volcanic–sedimentary mélange comprise deep-marine basalts and sediments. Upper Triassic 'Halobia limestone', quartzose sandstones and radiolarian cherts were all interbedded with pillow lavas, as in the southwest area of the Antalya Complex (Gödene Zone; Robertson & Woodcock, 1981c; Fig. 13a). The mélange is believed to have been assembled by processes of tectonic accretion related to subduction of Neotethyan oceanic crust. During this accretion, lavas readily disaggregated to form debris flows, while more competent 'Halobia limestones' either survived as intact blocks or became broken locally. Deeper, plutonic ophiolitic rocks were subducted and are preserved in this area only as sheared serpentinite along thrust planes. The amphibolite sliver at the base of the volcanic–sedimentary unit could be interpreted as a fragment of a subophiolite metamorphic sole (e.g. Woodcock & Robertson, 1977a) formed in contact with a hot overthrusting ophiolite thrust sheet not now exposed in the area.

The Eocene *Nummulites*-rich flysch overlying the Davras Dağ carbonate platform (Fig. 10) was derived from basinal volcanic and sedimentary units of the Antalya Complex and a coeval carbonate shelf. The flysch possibly accumulated in a thrust-top basin, developed as a carbonate platform unit and was thrust westwards over a collapsed Neotethyan basin to the east (Fig. 13c) during Late Palaeocene–Early Eocene times.

The subsequent, early Miocene transgression of the Antalya Complex thrust sheets of deep-sea sediments (Isparta Çay unit) was possibly related to a global eustatic sea-level high (Haq *et al.*, 1987). Overlying this, Burdigalian flysch is seen as a flexural foreland basin, related to regional south-eastwards transport of thrust sheets into the area (the Lycian nappes) in post-Langhian time (Fig. 13d; Hayward, 1982). This was associated with southwestwards thrusting and folding of the Davras Dağ carbonate platform and cover during the Late Miocene (pre-Messinian Aksu phase; Poisson, 1977; Akbulut, 1977).

Northern area

Structural setting

This area includes the northern margin of the Davras Dağ platform, two other carbonate platform units further north, the Kaymaz Dağ and Barla Dağ; and associated deep-sea sedimentary and ophiolitic units (Fig. 14). The Kaymaz Dağ is mapped as a large thrust sheet located above both the Davras Dağ and the Barla Dağ carbonate platform units (Fig. 15b). Despite some stratigraphical differences the Barla Dağ is tentatively correlated under Lake Eğridir with the Anamas Dağ carbonate platform to the east (Gutnic, 1977; Waldron, 1981). The Barla Dağ and the Davras Dağ are assumed to have been separate carbonate platforms, although this cannot be proved because of the lack of exposure. In the northwest of the area the Barla Dağ was locally thrust over the Kaymaz Dağ following deposition of Eocene transgressive limestones (Fig. 15c).

Lithologies

Davras Dağ carbonate platform. The highest stratigraphical levels of the Davras Dağ platform succession, well exposed along the northeast margin of the outcrop (e.g. Findos; Figs 14 & 15a), are *Globotruncana*-bearing pelagic carbonates of Maastrichtian age interbedded with redeposited terrigenous-free shallow-water carbonates (Gutnic, 1977).

Rudist-rich limestones and overlying pelagic limestones of Upper Cretaceous age also crop out in an elongate isolated exposure further north (e.g. Kara Tepe; Fig. 14), which is interpreted as a continuation of the Davras Dağ carbonate platform beneath a syncline of Antalya Complex slope and basinal units. The carbonate platform succession is exposed in the southeast in a neotectonic fault scarp adjacent to Lake Eğridir (Sivri Tepe; Fig. 14). A

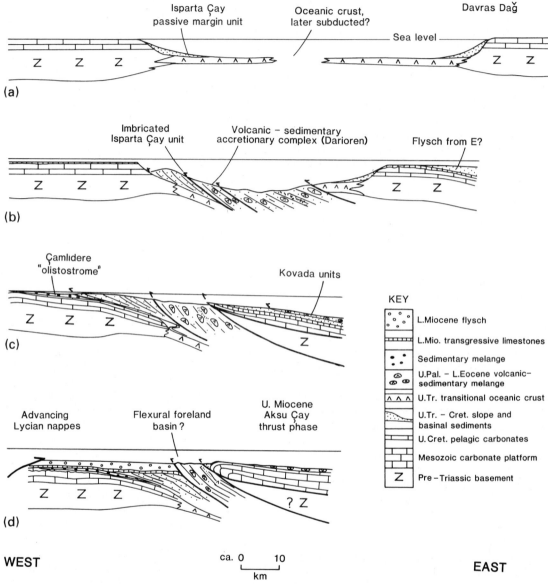

Fig. 13. Reconstruction of the northwestern area of the Antalya Complex. See text for explanation. (a) Late Triassic–Early Cretaceous, passive margin phase; (b) Maastrichtian–Early Palaeocene, during initial closure involving subduction– accretion; (c) Late Palaeocene–Early Eocene, during suturing; (d) Miocene, during rethrusting and folding.

Lower Cretaceous succession (Barremian–Aptian) of shallow-water platform carbonates is overlain there by 40 m of more thinly-bedded (up to 30 cm thick) calciturbidites, with micritic intraclasts and shaley partings. Petrographical study of the calciturbidites reveals grains of micrite and intrasparite, with planktonic and benthic foraminifera (e.g. miliolids), echinoderm plates and small quartz grains. Further northwest (Pembeli; Fig. 14), the highest levels of the northward continuation of the Davras

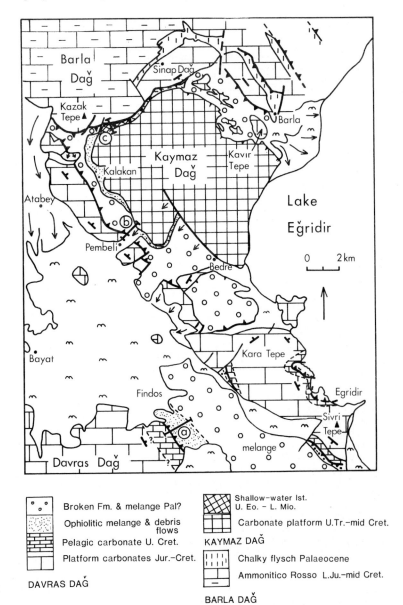

Fig. 14. Geological map of the northern area of the Antalya Complex. Modified after Gutnic (1977). Note locations of cross-sections (a), (b), (c) shown in Fig. 15.

Legend:

- Broken Fm. & melange Pal?
- Ophiolitic melange & debris flows
- Pelagic carbonate U. Cret.
- Platform carbonates Jur.–Cret.

DAVRAS DAĞ

- Shallow-water lst. U. Eo. – L. Mio.
- Carbonate platform U.Tr.–mid Cret.

KAYMAZ DAĞ

- Chalky flysch Palaeocene
- Ammonitico Rosso L.Ju.–mid Cret.

BARLA DAĞ

Dağ carbonate platform are medium- to thick-bedded pelagic limestones, with numerous well preserved *Globotruncana*.

Ophiolite-derived debris flows. Along the northeast margin of the Davras Dağ carbonate platform the Upper Cretaceous pelagic carbonates mentioned above are conformably overlain by a debris-flow (mélange) rich in ophiolite-derived debris. Locally

(e.g. Findos; Fig. 14), the base of this unit is calcareous mudstone, alternating with calciturbidites, up to 0.6 m thick. These, in turn, are overlain by chalky debris flows containing angular to subrounded clasts of limestone, chert, serpentinite, gabbro and lava. The highest levels of the sedimentary mélange are unstratified and contain numerous blocks of cherty mudstone, pelagic limestone and disrupted sheets of red ribbon radiolarites up to

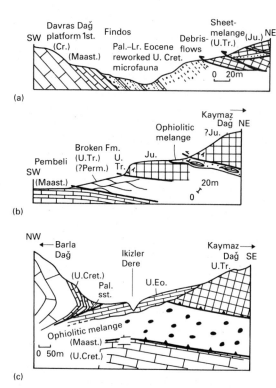

(a)

(b)

(c)

Fig. 15. Local cross-sections of the northern area of the Antalya Complex. (a) Submergence of the relatively allochthonous Davras Dağ carbonate platform in the Late Cretaceous, followed by deposition of ophiolite-derived debris flows, then overthrusting by dismembered thrust sheets (broken formation), mainly composed of Triassic platform slope facies; near Findos. (b) Overthrusting by the Kaymaz Dağ carbonate platform over ophiolitic mélange, sedimentary broken formation, interpreted as the imbricated edge of the Kaymaz Dağ carbonate platform; near Pembeli. (c) The Kaymaz Dağ carbonate platform is thrust over ophiolitic mélange, followed by Eocene trangression and later folding; near Kalakan. (Modified after Gutnic, 1977.)

10 m long. A few clasts are very well rounded and appear to be water-worn. Petrographical examination has revealed the following pelagic grains: micrite, planktonic foraminifera and radiolarian shells replaced by calcite. Shallow-water derived material includes benthic foraminifera, echinoderm plates, shells, oncolites, polyzoans and calcareous algae. Sedimentary lithoclasts comprise radiolarian-rich micritic limestone, radiolarian chert and quartzose sandstone. Igneous-derived grains include basalt, dolerite, serpentinite and clinopyroxene crystals. The matrix includes sparse very small planktonic

foraminifera of Palaeocene–Lower Eocene age.

Elsewhere (Sivri Tepe; Fig. 14), the base of the succession above the pelagic chalks is massive, coarse-grained, soft-weathering sandstone, almost entirely composed of serpentinite, with scattered grains of quartz, plagioclase, ferromagnesian minerals and opaque minerals.

Sedimentary mélange and broken formation. Sedimentary mélange and broken formation are exposed at structurally higher levels (e.g. southwest of Sivri Tepe; Figs 14 & 18). 'Olistoliths' include turbiditic quartzose sandstone, '*Halobia* limestone', radiolarite, mafic lava, gabbro and serpentinite, all up to several metres in diameter. An up to 1-km-thick, structurally higher, unit includes disrupted blocks of limestone of Upper Permian age (Ulsalaner, 1945). Mesozoic shallow-water and slope-derived carbonates and small ophiolitic blocks (e.g. gabbro, 1 m in diameter) also occur within a poorly exposed matrix of pelagic chalk, ophiolite-derived sandstone, medium-bedded bioclastic turbidites and debris flows. Microscopic examination of the sandstones reveals pelagic (planktonic foraminifera) and neritic carbonates (redeposited calcarenite, shells, small oncolites, echinoderm plates) and basinal sediments, including radiolarite, siltstone and quartzose sandstone. Igneous constituents include basalt, dolerite, serpentinite and crystals of clinopyroxene and plagioclase. Dolomite and pyrite are also present. Once again only tiny pelagic foraminifera are present, of Palaeocene–Lower Eocene? age.

Northeast of Findos (Fig. 14), the broken formation dominates. This comprises medium-bedded calciturbidites, containing shallow-water derived allochems, calcilutites and red chert of replacement origin. These rocks were correlated with the 'Antalya nappes' by Gutnic (1977) and include units of Upper Triassic and Lower Jurassic age.

Ophiolite-derived tectonic mélange. This unit structurally underlies the Kaymaz Dağ carbonate platform (Fig. 14), where it is 80–150 m thick (e.g. at Pembeli; Fig. 15b). The mélange dips northwards at a moderate angle and is composed of angular blocks of fresh gabbro, partly serpentinized peridotite, pyroxenite, mafic lava, red ribbon radiolarian chert and limestone, in a matrix of sheared serpentinite. The blocks range from several metres, up to 20 m in diameter. Several tectonic inclusions of Upper Triassic limestones of shallow-water and/or slope facies are present within the mélange (Fig. 15b). Elsewhere (near Bedre; Fig. 14), similar ophiolitic mélange is overlain tectonically by several metres of partly recrystallized planktonic foraminiferal chalk and then by the Kamaz Dağ carbonate platform thrust sheet.

Kaymaz Dağ and Barla Dağ carbonate platforms. The structurally overlying Kamaz Dağ carbonate platform to the north exposes a thick, still poorly dated, Lower Triassic to Upper Cretaceous succession (Gutnic, 1977; Fig. 6a). At the base yellow-brown calcareous mudstones of Lower Triassic age contain blocks of Upper Permian fossiliferous limestone (Bedre Formation; Gutnic, 1977). Further north relatively thin successions of Upper Triassic to Cretaceous age, including Upper Triassic Hallstatt facies limestones, are exposed in small klippen (e.g. Kavır Tepe, 1 km south of Barla; Fig. 14; Gutnic *et al.*, 1979). In the north the Kaymaz Dağ thrust sheet is underlain by mélange (Complexe de Barla), including blocks of quartzose sandstone, 'Halobia limestone' and radiolarian chert in a matrix of ophiolite-derived, turbiditic sediments ('wildflysch'). Along its northwest margin (at Kalakan; Figs 14 & 15c), the thrust contact between the Kaymaz Dağ carbonate platform and underlying ophiolitic mélange is sealed and unconformably overlain by *Nummulites*-rich limestones of Upper Eocene age (Gutnic, 1977).

In summary, the Kaymaz Dağ carbonate platform rests on the mélange with a low-angle thrust contact both to the north and to the south and is therefore interpreted as a huge klippe (Gutnic, 1977; Fig. 14). Along its northwestern margin, however, the Davras Dağ was overthrust by the Barla Dağ carbonate platform (Fig. 15c), probably during Late Eocene time.

Further north the succession in the large Barla Dağ carbonate platform unit (Fig. 6a) begins with sandstones overlain, in turn, by Upper Triassic (Norian) dolomites and Lower Jurassic platform limestones (Yassivaran Limestone). This is followed by a distinctive condensed unit 20–40 m thick, composed of red, nodular, pelagic Ammonitico Rosso ranging in age from Mid-Jurassic to Mid-Cretaceous. The succession continues upwards into Upper Cretaceous, red pelagic limestones and then passes into chalky flsych of Palaeocene–Eocene age (Gutnic, 1977; Gutnic *et al.*, 1979). Successions are regionally folded and verge towards the south (e.g. at Kalakan; Fig. 15c). Further north,

still within the Barla Dağ platform unit, contrasting, less condensed, Mesozoic successions are exposed (Senirkent Limestones) including distinctive limestone breccias. Upper Cretaceous pelagic limestones are overlain by distinctive, red Palaeocene hemipelagic sediments and then by Eocene *Nummulites*-bearing flysch again without ophiolitic intercalations. The relationship with the more northeasterly Kir Dağ carbonate platform has not been studied.

Interpretation

The Kaymaz Dağ is viewed as another, separate carbonate platform located to the north of the Davras Dağ carbonate platform within the Neotethys ocean (Fig. 16a). Upper Permian limestones, emplaced over the Davras Dağ platform in the south (south of Sivri Tepe), are seen as remnants of the pre-rift basement of the Kaymaz Dağ platform (Gutnic *et al.*, 1979). Initial rifting in the Early Triassic is recorded by the Upper Permian blocks in a muddy matrix (Bedre Formation). Remnants of slope and base-of-slope units of the former passive margin of the Kaymaz Dağ carbonate platform (e.g. 'Halobia limestones') are preserved in the various mélange and broken formation units. An inferred oceanic strand between the Davras Dağ and the Kaymaz Dağ carbonate platforms is now represented by ophiolitic mélange. One possibility is that the Kaymaz Dağ carbonate platform originally joined up with other, structurally high, carbonate platform units, particularly the Dulup–Sütçüler carbonate platform to the southeast of Lake Eğridir (see below).

The Barla Dağ is interpreted as another carbonate platform developed further north. The Lower Jurassic limestones (Yassiveren Limestones) of the Barla Dağ succession can be correlated with similar facies in the Anamas Dağ carbonate platform succession east of Lake Eğridir (Gutnic, 1977; Waldron, 1984b). However, condensed Ammonitico Rosso is not present in the Anamas Dağ succession (Fig. 6a). It is assumed here that the Anamas Dağ and Barla Dağ carbonate platforms were essentially contiguous but were separated by submarine normal faults during the Jurassic and Cretaceous. The Barla Dağ platform subsided after the Late Triassic, while the Anamas Dağ continued to undergo shallow-water carbonate platform deposition.

During the Early–Mid-Jurassic the Barla Dağ carbonate platform subsided along extensional

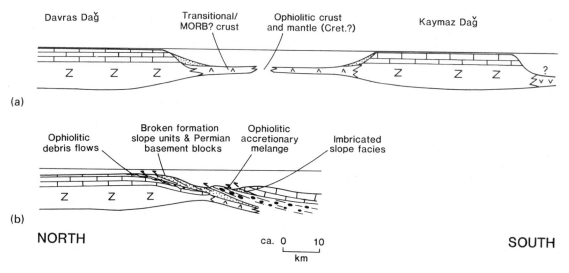

Fig. 16. Reconstruction of the northern area of the Antalya Complex. See text for explanation. (a) Late Triassic–Early Cretaceous, passive margin; (b) Late Cretaceous–Palaeocene, during initial collision.

faults and then underwent condensed pelagic deposition. Submarine faulting along the margins of this unit is recorded in the deposition of limestone breccias further north (Senirkent unit; Fig. 3).

During the Late Cretaceous the inferred, small, oceanic basin between the Davras Dağ and Kaymaz Dağ carbonate platforms began to close. The Davras Dağ carbonate platform subsided during the Maastrichtian, followed by deposition of pelagic carbonate, and then debris flows derived from the emplacing Antalya Complex in Palaeocene–Early Eocene? time. Ultramafic ophiolitic rocks were eroded and serpentinite-rich clastics were shed onto the by then submerged platform (e.g. Sivri Tepe). Disrupted sheets of '*Halobia*-limestone', calciturbidites and radiolarian sediments are interpreted as emplaced remnants of the former passive carbonate margins (Davras Dağ and/or Kaymaz Dağ; Fig. 16a). Ophiolite-derived mélange, as described above, is interpreted as the accreted remnants of oceanic crust and mantle derived from the Neotethyan basin (Fig. 16b). The Kaymaz Dağ platform formed an isolated platform within the ocean basin. It was then emplaced as a huge thrust sheet, both southwards over the Davras Dağ and northwards over the Barla Dağ carbonate platforms. The emplacement apparently took place in Palaeocene–Early Eocene times. Long distance overthrusting over the Davras Dağ and Barla Dağ carbonate platforms, either from the north or from the south

(present coordinates), are effectively precluded as successions in both of these units continue unbroken up into the Eocene without intercalated ophiolitic units. Late stage, local, thrusting of the Barla Dağ over the Kaymaz Dağ carbonate platform was probably related to Late Eocene, generally southward-directed compression that is most apparent in the northeast area (see below). In summary, the north area preserves remnants of three carbonate platforms, the Davras Dağ, the Kaymaz Dağ and the Barla Dağ.

Northeast area

The northeast area of the Antalya Complex extends westwards from the relatively autochthonous Anamas Dağ, Mesozoic carbonate platform unit as far as the neotectonic Kovada graben and also includes the Dulup Dağ carbonate platform and the Kızıl Dağ ophiolite (Figs 2 & 3). Much of this area was mapped in detail by Waldron (1981; 1984a, b, c).

Structural setting

The southwestern margin of the Anamas Dağ carbonate platform succession experienced multiple stages of thrusting and high-angle faulting (e.g. southwest of Sorkun Yayla; Fig. 17). A key point, established by Waldron (1984a, b), is that platform successions of the Anamas Dağ platform interior

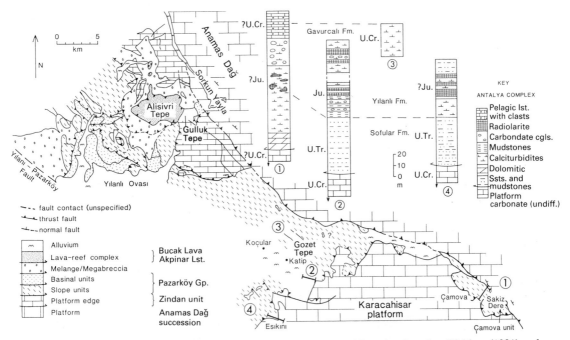

Fig. 17. Simplified geological map of the northeastern area of the Antalya Complex, based on Waldron (1981) and Dumont (1976). Measured logs of successions in the southeastern part of the area are shown above. Detailed logs of the west of the area are given in Waldron (1984b, c; see also Figs 3 & 4).

pass laterally into platform edge facies exposed along the southwestern margin of the Anamas Dağ (Zindan succession of Dumont, 1976; Fig. 4). Towards the southwest, both of these units are thrust over more distal slope and basinal successions of Upper Triassic to Lower Cretaceous age (Pazarköy Group; Waldron, 1984c; Figs 6 & 17). This thrusting apparently took place during the Late Eocene and was related to the emplacement of the Hoyran–Beyşehir–Hadim nappes onto the northeastern margin of the Anamas Dağ carbonate platform (Monod, 1977).

The basinal sediments of the Pazarköy Group (Fig. 5) are repeated by numerous thrusts. The outcrop is split by a major northwest–southeast trending high-angle fault, the Yılanlı-Pazarköy Fault (Fig. 17). Southwestward of this fault, six main sedimentary thrust sheets are exposed. Facing directions of medium-scale folds and duplex geometries indicate thrusting *towards the northeast*, i.e. towards the Anamas Dağ carbonate platform (Waldron, 1984a, b). In detail, the thust geometry is complex and some degree of local out-of-sequence thrusting can be recognized (e.g. Çamgöl thrust; Waldron, 1981). The youngest sediments that pre-

date thrusting are Upper Cretaceous–Palaeocene, ophiolite-derived debris flows (Göynük Formation). West of the Pazarköy-Yılanlı Fault the basinal sediments structurally overlie an elongate, north–south trending sliver of Late Mesozoic platform carbonates, the Çayköy unit (Fig. 3).

Further west again, additional basinal thrust sheets of the Pazarköy Group are interleaved with, and structurally overlain by, thick sheets of Mesozoic platform carbonates represented by the Dulup Dağ and Hudulca Tepe (Waldron, 1981; Fig. 18). The Dulup Dağ platform carbonates continue southwards without any break in outcrop into a much larger unit here termed the *Sütcüler-Dulup thrust sheet*. The north margin of the Dulup Dağ carbonate platform is underlain structurally by ophiolite-derived debris flows, mélange and megabreccias, of Maastrichtian–Palaeocene age (Göynük Formation). Further southeast, the Dulup Dağ carbonates are in high-angle contact with the deeper (plutonic) levels of an undated, possibly Late Cretaceous, ophiolite, the Kızıl Dağ of Eğridir (Juteau, 1975; Fig. 18). Detailed mapping suggests the Dulup Dağ is the structurally higher unit (Waldron, 1981).

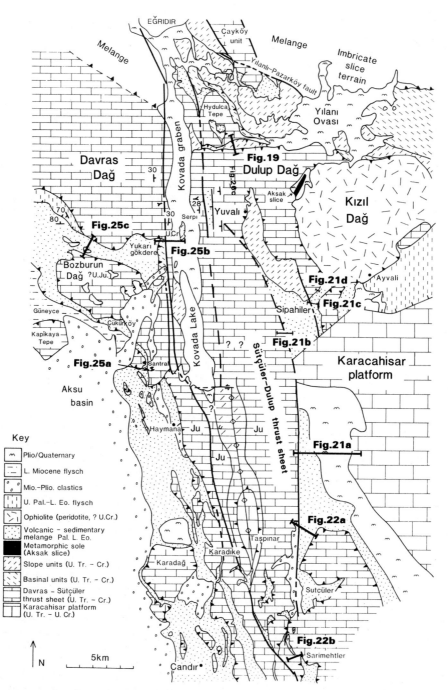

Fig. 18. Simplified geological map of the apex of the Isparta angle, including parts of the northwest, northeast and eastern areas of the Antalya Complex. The geology is based on Akbulut (1977), unpublished MTA mapping, Waldron (1981) and this study.

Proximal basinal sediments of Late Triassic–Late Cretaceous age also occur along the southwestern margin of the Anamas Dağ (Fig. 17), southeast of the area mapped by Waldron (1984a). Further south still similar basinal sediments are thrust over the northern margin of another major Mesozoic carbonate platform unit, the Karacahisar massif (Fig. 17). A thin tectonic slice of Mesozoic proximal deep-water sediments, the Çamova unit (Dumont, 1976), is located at the junction of the Anamas Dağ and the Karacahisar carbonate platform units to the southeast (Fig. 17).

Lithologies

Anamas Dağ carbonate platform. The Anamas Dağ unit in the east exhibits a fully developed Mesozoic succession (Dumont, 1976; Fig. 6), beginning with shallow-marine, terrigenous mudstones and sandstones (Upper Norian Kasımlar Formation) overlain by Upper Triassic dolomitic sediments (Menteşe Dolomite) and then by shallow-water limestones (Leylek Limestone). The platform succession is then interrupted by a distinctive interval of red and orange quartzose sandstones, conglomerates, siltstones, mudstones and pisolithic ironstones, the Çayır Formation (Monod & Akay, 1984). Clasts in this unit were derived from a psammitic metamorphic terrain (schist and quartzite), a platform carbonate succession (Upper Permian neritic fossiliferous limestones), and from more basinal sediments (Triassic? chert). The source area is thought to be Palaeozoic basement and Early Triassic rift units in the Lake Beyşehir area further east (Brunn *et al.*, 1971). Above, there are limestones of Early Jurassic age (Yassıveren Limestone) and then shallow-water platform carbonates and finally Late Cretaceous pelagic limestones and Eocene flysch (Dumont, 1976; Fig. 6). Palaeocene facies are apparently absent.

Westerly carbonate platform units. Platform carbonates also crop out further west as the structurally low Çayköy unit and the structurally higher Dulup Dağ unit (Fig. 6),

The Çayköy unit succession is dated as Lower Jurassic to Maastrichtian (Waldron, 1981; Fig. 6) and includes massive neritic limestones, rudist limestones and, at the top, pelagic limestones with replacement chert interbedded with calciturbidites up to 0.9 m thick. The platform succession is depositionally overlain by sheared debris flows of

Maastrichtian–Palaeocene age containing blocks of all the Antalya Complex sedimentary and igneous rocks (Göynük Formation) and also broken formation composed of relatively proximal slope sediments.

Further west again is the clearly overriding Sütçüler–Dulup thrust sheet encompassing complete Upper Triassic to Upper Cretaceous successions of platform carbonates (Fig. 6). The base of the succession is made up of Upper Triassic dolomites (Kovada Dolomite) overlain by Jurassic and Lower Cretaceous platform limestones (Dumont & Kerey, 1975a). The highest levels of the Dulup Dağ carbonate platform succession are exposed locally in tilted fault blocks along the eastern margin of the neotectonic Kovada graben (e.g. west of Yuvalı and near Serpi; Fig. 18). Steeply westward-dipping successions of typical Upper Cretaceous pelagic limestones containing *Globotruncana* are depositionally overlain by 80 m of debris flows with clasts including chalk, serpentinite, radiolarite, pink and grey chert and minor lava.

Easterly carbonate platform edge and slope units.

Platform edge units (Pazarköy Group) are exposed along the southwest margin of the Anamas Dağ platform and comprise Upper Triassic quartzose sandstones (Sofular Formation), Lower Jurassic proximal, redeposited carbonates (Yassıveren Formation), Upper Jurassic–Lower Cretaceous thinly-bedded limestones and chert (Zindan Formation), and Upper Cretaceous redeposited and pelagic limestones (Gavurcalı Formation; Dumont *et al.*, 1980; Fig. 6). Distinctive yellow siltstones and mudstones near the base of the Yassıveren Formation can be correlated with the Çayır Formation (Fig. 6) of the Anamas Dağ platform succession further east (Waldron, 1984b). Waldron (1981) also reported that the Upper Triassic, proximal facies of the Pazarköy Group are associated with alkaline lavas (Bucak Lavas) of geochemically within-plate type. These, in turn, are depositionally overlain by small patch reefs (Akpınar Tepe Member; Fig. 6). In the north, these units form the crest of a major ridge (Alisivri Tepe; Fig. 17), underlain by sandstones of the Sofular Formation and by tectonic mélange. Further south (Göllük Tepe; Fig. 17) reef limestones of the Akpınar Tepe Member are depositionally underlain by massive dolomites correlated with the Menteşe Dolomite of the Anamas Dağ succession.

Further southwest, along the Anamas Dağ margin in the Koçular–Katip area (Fig. 17), the Anamas Dağ platform succession and adjacent units also experienced Eocene southwestwards thrusting over proximal base-of-slope facies correlated with the Pazarköy Group (Fig. 17). Small klippen of Anamas Dağ carbonate platform successions commonly overlie slope and basinal sediments. The thrust plane is marked by tectonic breccia with shattered chert and sandstone. Above this, the Anamas Dağ platform succession locally comprises Upper Triassic mudstones and sandstones (Kasımlar Formation) with *in situ* coral patch reefs up to 70 m in diameter (Tilkideliği Member) followed by platform carbonates (Menteşe Dolomite and Leylek Limestone).

Basinal sediments. More distal units (Pazarköy Group; Fig. 6) are exposed beneath the southwesterly-directed thrust sheets of Anamas Dağ carbonate platform margin units. Rare slumps are directed towards the south and southwest (Waldron, 1981). Successions begin with Upper Triassic turbiditic sandstones (Sofular Formation), which are overlain by a distinctive interval of pink pelagic limestone (Kirazlar Tepe Member) and then by Lower and Middle Jurassic mixed, redeposited carbonate-chert successions (Yılanlı Formation) including pelletoidal grainstones, packstones, intraformational conglomerates and ribbon radiolarites. Above come pelagic limestones and Upper Cretaceous–Palaeocene ophiolite-derived sediments (Göynük Formation; Fig. 6). Similar basinal successions are exposed in thrust sheets located west of the Yılanlı–Pazarköy fault (Fig. 17). The structurally higher of these sheets, immediately below the overriding Dulup Dağ thrust sheet, is characterized by unusually well developed Upper Triassic 'Halobia limestones' (Akdoğan Member). Similar facies are also interthrust with the Dulup Dağ and Hudulca Tepe carbonate platform units (Yuvalı Group; Fig. 18).

The Dulup Dağ platform unit is structurally underlain by broken formation, mélange and debris flows (Göynük Formation; Fig. 19). Clasts in the debris flows include basalt, peridotite, chert and limestone. The larger slices of basalt are mid-ocean ridge-type tholeiites, dated as Upper Jurassic–Lower Cretaceous (Havutlu Lavas) using radiolarians in associated cherts (Waldron, 1981). In the southeast, the Kızıl Dağ ophiolite is dominantly harzburgite tectonite, with minor dykes and lenses

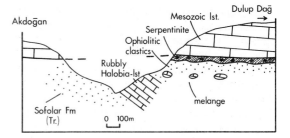

Fig. 19. Cross-section across the north margin of the Dulup Dağ carbonate platform, showing the underlying ophiolite-derived mélange and slope and basinal sediments of the Antalya Complex. See Fig. 18 for location.

of pyroxenite, dunite and chromite (Juteau, 1975; Fig. 18). The stretching lineation is orientated east–west (Juteau, 1975). The Kızıl Dağ ophiolite is locally bordered to the northwest by small slices of poorly exposed greenschists (Aksak unit), for which the Havutlu Lava unit is the most obvious protolith (Fig. 18).

Platform junction zone. Along the southwestern margin of the Anamas Dağ (Koçular-Katip area; Fig. 17), slope and basinal units of the Antalya Complex are dominated by imbricated, folded sedimentary successions beginning with Upper Triassic quartzose sandstone turbidites correlated with the Sofular Formation. In places these sediments are interbedded with calcilutites and cherty debris flows (e.g. Gozet Tepe; Fig. 17, log 3) and are overlain by thick-bedded, redeposited limestones. Possible lateral equivalents include thick-bedded limestone breccias up to 80 m thick, with angular blocks up to 1 m in diameter. Stratigraphically higher levels include Upper Jurassic–Lower Cretaceous? cherts 60 m thick correlated with the Zindan Formation and Upper Cretaceous successions of redeposited rudist limestones and pelagic carbonates.

Further south, similar basinal sediments are thrust over the major Karacahisar carbonate platform. The highest levels of this platform succession comprise 30 m of pink, thin-bedded, pelagic carbonates (e.g. at Esikini; Fig. 17). The structurally overlying basinal sediments are thrust imbricated and can be correlated with the Pazarköy Group (Fig. 17, logs 2 & 4). Successions begin locally with Upper Triassic sandstones and shales, passing upwards into 15 m of thick-bedded calciturbidites with beds up to 2 m thick and with coral debris and

'*Halobia* limestones' correlated with the Sofular Formation. These are then overlain by siliceous and redeposited limestone facies correlated with the Yılanlı Formation.

An elongate wedge of basinal facies is tectonically intercalated between the Karacahisar and Anamas Dağ platforms (Çamova unit; Dumont, 1976; Fig. 17). The contact with the Anamas Dağ platform to the north is faulted. However, the contact with the Karacahisar platform to the south is an open folded, low-angle, thrust. Where exposed (Sakız Dere; Fig. 17, log 1), the succession in the Camova unit begins with recrystallized dolomites, overlain by 65 m of (Upper Triassic?) massive, relatively fine-grained limestone, with occasional interbeds, exhibiting parallel lamination. Rare bedded chert is of replacement origin. Above this comes thinner-bedded dark grey porcellaneous limestones 8-m thick, with abundant black replacement chert (Upper Jurassic–Lower Cretaceous?). There is then 25 m of limestone conglomerates. The succession ends with a 15-m-thick unit of Upper Cretaceous, *Globotruncana*-bearing, pink pelagic carbonates with interbeds of thinly-bedded, medium-grained calciturbidites and rare, thin (tens of centimetres), sedimentary breccias.

Interpretation

Following Triassic rifting, the Anamas Dağ carbonate platform was bordered by a fault-controlled margin, with a deep basin to the southwest (present coordinates; Fig. 20). The basinal Pazarköy Group is assumed to have been underlain by Late Triassic transitional (i.e. marginal) oceanic crust. The most proximal (easterly) unit of this inferred volcanic basement is (Waldron, 1981) the Bucak Lavas and overlying patch reefs. Structurally underlying units (Pararköy Group) are interpreted as more distal sediments that accumulated in a basin floored by oceanic crust, located between the Anamas Dağ to the northeast and a separate Dulup Dağ platform to the southwest. The thick Upper Triassic '*Halobia* limestones' of the structurally higher sheets in the west are interpreted as periplatform ooze that accumulated near the base-of-slope of the Dulup Dağ carbonate platform. Smaller basins also existed within this platform in which similar '*Halobia* oozes' also accumulated.

The relationship between the Çayköy carbonate platform unit and the much larger Davras Dağ carbonate platform to the west is problematic, mainly because of lack of exposure across the

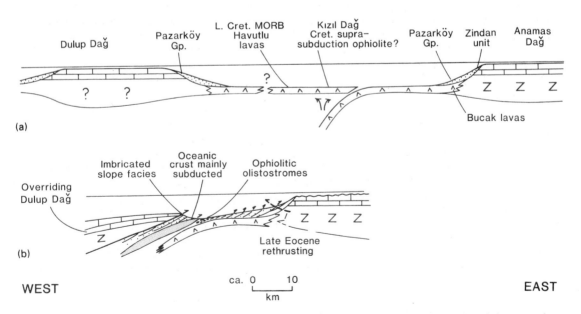

Fig. 20. Reconstruction of the northeastern area of the Antalya Complex. The harzburgitic Kızıl Dağ is interpreted as a remnant of Late Cretaceous(?) oceanic mantle, possibly from a supra-subduction zone setting. (a) Late Triassic–Late Cretaceous, during inferred supra-subduction zone spreading; (b) Late Cretaceous–Palaeocene, during collision and suturing.

neotectonic Kovada graben. The Çayköy unit was either an eastward extension of the Davras Dağ platform, or a small separate carbonate platform. Much hinges on the direction of emplacement of mélange and broken formation overlying the western margin of the Çayköy unit. Eastward emplacement, as inferred by Waldron (1981), apparently would preclude correlation with the Davras Dağ platform, since the sedimentary cover of this unit extends up into the Eocene, post-dating the Palaeocene–Lower Eocene? emplacement of the basinal Antalya Complex units above the Çayköy platform. However, the Kovada graben may mark the site of significant neotectonic strike-slip faulting and thus the Davras Dağ platform may not have blocked eastward emplacement of basinal units over the Dulup Dağ. In summary, the Çayköy carbonate platform unit is tentatively correlated with the Davras Dağ carbonate platform to the west.

During the Mesozoic the Karacahisar platform was bordered by a passive margin to the northwest, now preserved as overthrust proximal sediments of the Antalya Complex (Pazarköy Group). However, the Karacahisar and Anamas platforms were separated by a rift, in which proximal redeposited carbonate, pelagic carbonates and anoxic sediments accumulated. This rift is assumed to have been relatively narrow (ten kilometres at most) in view of the absence of volcanic or ophiolitic units.

During Late Jurassic–Lower Cretaceous time, spreading apparently took place within the ocean basin, marked by extrusion of tholeiitic basalts (Havutlu Lava) which are dated by intercalated radiolarites (Waldron, 1981). The Kızıl Dağ ophiolite is assumed to have formed in the Late Cretaceous, coeval with ophiolites in the southwest area (Tekirova Zone; see later). In the northeast area, genesis possibly took place above a subduction zone, in view of the dominantly harzburgitic nature of the Kızıl Dağ (e.g. Pearce *et al.*, 1982). If so, the MORB extrusives (Havutlu Lava) cannot have been co-magmatic with the Kızıl Dağ peridotite. The thrust-juxtaposed Aksak metamorphic slice (Fig. 18) possibly represents a metamorphic sole unit, related to thrusting of oceanic crust (i.e. Havutlu Lava) below the young, still hot, Kızıl Dağ ophiolite (e.g. Woodcock & Robertson, 1977a).

During the Maastrichtian–Palaeocene the northeast area was compressed and slope and basin sediments were imbricated and thrust northeast towards the Anamas Dağ carbonate platform slope (Zindan unit). Restoration of the thrust sheets assuming 'in-sequence' displacement implies that both the MORB extrusives and the Kızıl Dağ ophiolite were derived from an ocean basin located between the Anamas and Karacahisar carbonate platforms to the east and the Dulup–Sütçüler platform in the west (Fig. 20b).

One problem with this interpretation is that ophiolitic material is present between the overlying Dulup carbonate platform thrust sheet and proximal facies beneath, which are interpreted as slope sediments of the Dulup carbonate platform. Waldron (1981, 1984b) therefore proposed that the Kızıl Dağ ophiolite was first thrust northeastwards, shedding ophiolite debris over the Dulup Dağ platform onto the carbonate platform slope to the northeast. This was followed by emplacement of the Kızıl Dağ ophiolite into the basin further northeast of the Davras Dağ platform. Late-stage, out-of-sequence thrusting finally carried the Dulup carbonate platform over the Kızıl Dağ ophiolite. During this study it became clear that ophiolitic units are also present beneath the Sütçüler-Dulup carbonate platform thrust sheet further south, usually located directly below the basal thrust. Structurally underlying slope units are interpreted as imbricated marginal units of the relatively autochthonous Karacahisar carbonate platform to the east. On balance, therefore, it seems likely that the Kızıl Dağ ophiolite originated in a Neotethyan basin sited between the Anamas Dağ /Karacahisar carbonate platforms to the east and the Dulup–Sütçüler carbonate platform to the west.

In either interpretation, the restriction of the intact Kızıl Dağ ophiolite to a small area could reflect preferential preservation within an original oceanic embayment, where the northern margin of the Karacahisar platform was offset northeastwards by a strike-slip? carbonate margin, before joining up with the northwest–southeast trending, orthogonally rifted, Anamas Dağ carbonate margin (Fig. 2).

Eastern area

The eastern area of the Antalya Complex is defined as the eastern limb of the Isparta angle bordering the Karacahisar carbonate platform, from the Dulup Dağ carbonate platform in the north southwards to the vicinity of Çandır (Fig. 3).

Structural setting

The Mesozoic Karacahisar carbonate platform is structurally overlain by slope and basinal sedimentary and volcanogenic units of the Antalya Complex. The overall steeper dip of thrust sheets with respect to the relatively autochthonous carbonate platform basement (Karacahisar), combined with evidence from small-scale thrust duplexes in the highest levels of the carbonate platform, indicate emplacement varying from eastwards to northeast-wards (e.g. north of Sütçüler, Fig. 18). The slope and basinal units are, in turn, structurally overlain by the southward continuation of the Dulup car-bonate platform, the Sütçüler Limestones of Akbulut (1977) Figs 21. Towards the east, the Sütçüler–Dulup carbonate platform thrust sheet was emplaced over the Karacahisar platform and later faulted with little or no exposure of the slope and basinal units of the Antalya Complex (e.g. north of Sütçüler, Fig. 18). The Sütçüler–Dulup thrust unit is internally folded on a large scale (e.g. near Karadiken, Fig. 18) and was dissected by both reverse faults and thrusts related to the Late Mi-ocene Aksu phase and by neotectonic normal faults. In the west of the area, the Sütçüler–Dulup thrust sheet is represented by numerous small klippen (e.g. near Haymana; Fig. 18). These units are transgres-sively overlain by, and/or faulted against, Pliocene? conglomerates (e.g. near Haymana, Fig. 18; Akbu-lut, 1977). Slope and basinal units of the Antalya Complex are also well exposed in windows through the Sütçüler–Dulup thrust unit further east (e.g. near Çandır; Fig. 18). Towards the west (Fig. 18), the Antalya Complex as a whole is unconformably overlain by Late Miocene clastics (Aksu Formation; Akbulut, 1977).

Lithologies

Karacahisar carbonate platform. In the east, the Upper Triassic interval of the Karacahisar carbon-ate platform includes a huge reef build-up, the Diproyraz Dağ (Fig. 3). In contrast to the Anamas Dağ carbonate platform, the end-Triassic/basal Ju-rassic quartzose clastic unit (Çayır Formation) is absent (Dumont, 1976). Instead a an Upper Trias-sic dolomitic interval (Menteşe Dolomite) is over-lain unconformably by shallow-water limestones, passing without a break into Jurassic and Creta-ceous platform carbonates (Akkilise Limestone; Fig. 6). The upper surface of the Cenomanian inter-val is marked by sparite-cemented erosional brec-cias and in turn disconformably overlain by Upper Cretaceous pelagic limestones (Esekini Limestones; Akbulut, 1977).

Where they are exposed locally the highest strati-graphical levels of the Karacahisar carbonate plat-

form succession comprise shallow-water limestones overlain by 35 m of Upper Cretaceous thin-bedded pelagic carbonates with shaley and pelagic interbeds rich in replacement chert. The succession ends with 9 m of thicker-bedded bioclastic turbidites, soft-weathering sandstones, marl and mudstone, over-lain by debris flows derived from the slope and basinal units of the Antalya Complex. The upper-most sandstones contain scattered radiolarians and *Globotruncana*, shell fragments and grains of radio-larian chert, micrite and basalt, set in a fine-grained calcareous matrix.

Slope and basinal units are also exposed struc-turally below the Sütçüler–Dulup thrust sheet (Taşpınar area; Fig. 18). At the base of the exposed succession, thin- to medium-bedded Upper Creta-ceous pelagic limestones and minor calciturbidites (up to 15 m thick) pass depositionally upwards into several tens of metres of ophiolite-derived sand-stones, interbedded with chalky debris flows with clasts of limestone, chert, gabbro, basalt, peridotite and lenses of sheared serpentinite up to 8 m thick. This unit is inferred to represent a window into a westward extension of the upper levels of the Karacahisar carbonate platform, although an origin as deeper level imbricates of the Sütçüler–Dulup allochthonous carbonate platform unit cannot be ruled out.

Basinal sediments. In the northeast of the area (southwest of Sipahiler; Fig. 18), the Karacahisar carbonate platform is trangressively overlain by Plio-Quaternary limestone conglomerates (Akbulut, 1977; Fig. 21a). However, further south, slope and basinal sediments of the Antalya Complex are exposed, thrust over, and/or faulted against, the Karacahisar carbonate platform succession (Fig. 22a). These units range from severely disrupted thrust sheets to broken formation and debris flows with a pelagic carbonate matrix. The clasts are composed of relatively 'proximal' base-of-slope facies and include Upper Triassic thick-bedded quartzose turbidites, '*Halobia* limestones' and silty radiolarian cherts (Sipahiler units, Fig. 6b). This unit is tectonically overlain by more intact thrust sheets of sedimentary rocks, mainly Upper Triassic quartzose sandstone, '*Halobia* limestones' and radi-olarian chert. Structurally higher thrust sheets in-clude a 25-m-thick limestone, 40 m of thick-bedded turbiditic sandstones, and serpentinite-derived de-bris flows (Fig. 21c & d).

Further east (near Ayvalı; Fig. 18), disrupted sedimentary thrust sheets as described above are

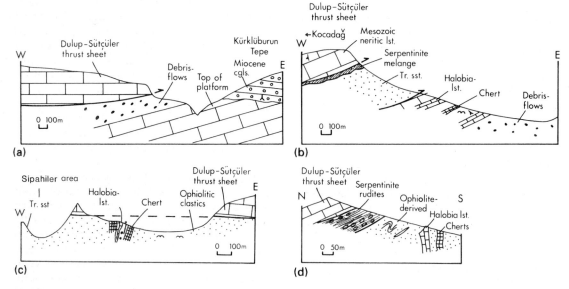

Fig. 21. Local cross-sections in the northern part of the eastern area of the Antalya Complex. See Fig. 18 for locations. (a) General cross-section, showing the Dulup–Sütçüler unit thrust over slope and basinal sediments of the Antalya Complex; (b) local cross-section, showing the imbricated nature of the slope and basinal units below the Dulup–Sütçüler thrust sheet; (c) illustrates the overthrust nature of the Dulup–Sütçüler carbonate platform unit; (d) local cross-section, showing the presence of ophiolite-derived conglomerates below the Dulup–Sütçüler thrust sheet.

structurally overlain by a 60-m-thick interval of very deformed, white, pelagic chalk, ophiolite-derived sandstone and limestone debris flows, with occasional disrupted sheets of, pink, micritic limestones and ribbon chert. The sandstones contain *Globotruncana*, large shell fragments and calcareous algae. There are also sedimentary lithoclasts of quartzose sandstone and siltstone, pelagic and neritic limestone, and radiolarian chert. Igneous lithoclasts are altered basalt, dolerite, zeolite, plagioclase, pyroxene and serpentinite. Metamorphic lithoclasts are represented by quart–albite–epidote schist. Other constituents are fresh biotite crystals and opaque mineral grains.

Volcanic–sedimentary mélange. Structurally higher units are well exposed in the southern part of the area (i.e. Fig. 22a & b). The tectonic contact between the lower sedimentary unit (described above) and an upper volcanic–sedimentary unit is marked by stringers of detached blocks, up to 10 m in size, composed of recrystallized white limestone, with chalky partings, containing poorly preserved small planktonic foraminifera of Palaeocene–Lower Eocene? age. In this area the volcanic–sedimentary unit consists mainly of dismembered thrust sheets

of pillow lava and massive lava up to 100 m thick, with little interlava sediment. However, elsewhere (e.g. near Sütçüler; Fig. 18), the volcanic–sedimentary unit is mainly mélange, consisting of blocks of pillow and massive lava. Flows are both feldspar-phyric and aphyric and vary greatly in vesicularity. The more competent massive lavas and '*Halobia* limestones' make up most of the larger blocks (up to 4 m in diameter). The matrix is volcaniclastic and highly sheared.

Volcanic–sedimentary mélange is also widely exposed beneath the western margin of the overriding Sütçüler–Dulup thrust unit (e.g. north and south of Haymana; Fig. 18), where it is cut by several strands of sheared serpentinite up to 10 m thick. Poorly stratified, matrix-supported, debris flows and turbiditic, polymict sandstones dominate. The debris flows comprise angular, to subangular clasts of mafic extrusives, diabase, gabbro and serpentinized peridotite, up to 10 cm in diameter. The largest blocks are mainly '*Halobia* limestones' and red ribbon radiolarites up to 8 m across. There are also several metre-thick intervals of white, finely laminated pelagic chalk containing calcite-replaced radiolarians and small planktonic foraminifera of Palaeocene–Lower Eocene(?) age.

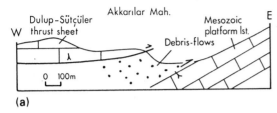

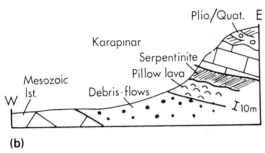

Fig. 22. Local cross-sections of the southern part of the eastern area of the Antalya Complex. See Fig. 18 for locations. (a) The Sütçüler Limestone above forms part of a regionally extensive flat thrust sheet, over ophiolitic debris flows and dismembered thrust sheets of Mesozoic slope and basinal sediments. (b) Here, the Sütçüler Limestone thrust sheet overlies ophiolite-derived debris flows. Note also the presence of pillow lavas and the Miocene cover facies.

The volcanic–sedimentary mélange is locally overlain structurally by steeply-dipping thrust sheets of turbiditic sandstones and bioclastic limestones. For example, south of Sütçüler (Fig. 18), medium- to thick-bedded turbiditic sandstones, with pale grey mudstone partings, are overlain by several tens of metres of bioclastic limestones and conglomerates, rich in intraformational mud chips. Turbiditic sandstones up to 40 m thick are exposed further north at the same structural level, where they are interbedded with calcarenites and limestone conglomerates. These sandstones comprise mixtures of sedimentary (quartz, quartzite), igneous (perthite, plagioclase) and carbonate grains (pellets, pisoliths, ooids, echinoderm plates). Plutonic ophiolite-derived grains are, by contrast, virtually absent.

Ophiolitic units. Ophiolitic thrust sheets and debris flows (excluding the volcanic–sedimentary mélange, described above) structurally overlie all the units described above and are located immediately below the Sütçüler–Dulup thrust sheet. South of Sütçüler, volcanic–sedimentary mélange is overlain

by a 20 m-thick unit of sheared lava with scattered limestone blocks, and then by a sheared and brecciated serpentinite 20 m thick (Fig. 22b). The shear fabric is subhorizontal. Ophiolitic rocks are also exposed in windows through the Sütçüler–Dulup thrust unit further west (near Çandır; Fig. 18), and include ophiolitic debris flows, often monomict and varying from mainly basaltic, to gabbroic, or rich in serpentinite.

Further north (northeast of Sipahiler; Fig. 18) important 60–80-m = thick unit of serpentinite conglomerates can be traced laterally for up to 1 km (Fig. 21d). Entrained within these conglomerates are blocks of white crystalline limestone, up to 0.8 m in diameter and elongate inclusions of less recrystallized pelagic chalk up to 2.5 m long by 0.8 m thick. The importance of these particular serpentinite conglomerates is that they are located at essentially the same structural level as the Kızıl Dağ peridotite further north (see above). These units are also lithologically similar to ophiolitic units mapped by Waldron along the northern margin of the Dulup Dağ carbonate platform thrust sheet in the northeast area.

Upper carbonate unit; Sütçüler Limestones. Akbulut (1977) erected a 1245-m-thick composite succession in the Sütçüler Limestones, beginning with Upper Triassic '*Halobia* limestones' passing in turn into Upper Traissic to Liassic shallow-water limestones (up to 390 m), then Mid–Upper Jurassic limestones (295 m), Cretaceous neritic limestone (up to 460 m) and finally 50 m of Upper Cretaceous pelagic limestones (Fig. 6a). Fine-grained, ophiolite-derived, sediments are found locally in the highest levels of the Sütçüler Limestones (Akbulut, 1977). The existence of stratigraphical breaks within the Upper Triassic–Upper Jurassic and Jurassic–Mid Cretaceous intervals was also suggested by Akbulut (1977), but based on limited biostratigraphical data.

Of note are basal sediments of the Sütçüler unit, exposed immediately above the basal thrust in the large window near Taşpınar (Fig. 18). There, several tens of metres of red, orange and greenish mudstones, siltstones and orthoquartzites are overlain by medium- to thick-bedded hemipelagic calcilutites and calciturbidites ('*Halobia* limestones'), which become progressively coarser-grained and thicker-bedded upwards and then pass transitionally into shallow-water limestones of Lower Jurassic age. Elsewhere radiolarite is exposed in the base of the Sütçüler unit (e.g. 3 km southeast of Çandır;

Fig. 18), interstratified with pink ferruginous calcilutites and again passing upwards into shallow-water platform carbonates. More commonly, the basal thrust of the Sütçüler–Dulup thrust sheet is located at a higher stratigraphical level, within Lower Jurassic–Lower Cretaceous shallow-water platform carbonates.

Interpretation

The Karacahisar carbonate platform exhibits a typical Tauride carbonate platform evolution with post-Cenomanian subsidence (Fig. 23a). Unlike the Anamas Dağ carbonate platform further north, clastics of the Çayır Formation, of end-Triassic to basal Jurassic age, are absent possibly because of isolation from source areas further northeast (i.e. Beyşehir area; Fig. 3) by the intra-platform basin discussed above (Çamova unit). The sedimentary units structurally overlying the Karacahisar carbonate platform are envisaged as remnants of the passive margins of this carbonate platform, emplaced by a combination of thrusting and gravity sliding into a pelagic carbonate-receiving foredeep (Fig. 23b). The presence of ophiolite-derived debris in the highest levels of the carbonate platform succession shows that oceanic crust was already being emplaced during the Maastrichtian at least onto the structurally higher carbonate platform units.

The volcanic–sedimentary unit represents more distally-derived debris flows that were finally emplaced in Late Palaeocene–Early Eocene? times (Fig. 23b). The main source was Upper Triassic deep-water basaltic lavas and associated sediments (e.g. '*Halobia* limestone'). Several basalts analysed were found to be of within-plate type (Robertson & Waldron, 1990). Mass wasting, submarine erosion and volcaniclastic deposition took place, again in a deep-water pelagic carbonate-depositing setting. Dismembered ophiolitic rocks, including serpentinite debris flows, represent more distal (i.e. westerly derived) Late Cretaceous(?) oceanic crust and mantle, emplaced following mass wasting and gravitational reworking (Fig. 23b).

Deposition of the relatively thin Sütçüler–Dulup carbonate platform succession was apparently interrupted by hiatuses which may have been tectonically controlled. The lowest exposed unit records variable Upper Triassic quartzose clastic deposition, overlain by periplatform '*Halobia* ooze', and then by shallow-water carbonates. The Dulup Dağ carbonate platform was constructed to the north on an inferred horst of Upper Permian limestones. Further south the Sütçüler Limestone platform developed on a submerged basement of unknown composition which was overlain by Upper Triassic Hallstatt facies. Assuming 'in-sequence' thrusting, Neotethyan oceanic crust, represented by the ophi-

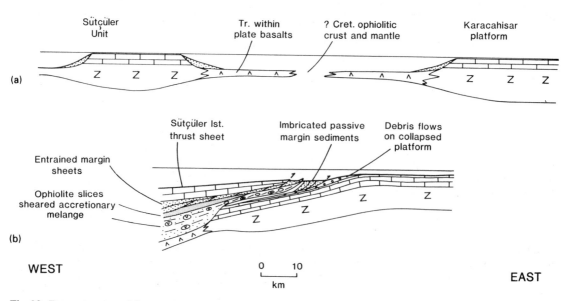

Fig. 23. Reconstruction of the eastern area of the Antalya Complex. (a) Late Triassic–Early Cretaceous, passive margin phase; (b) Late Cretaceous–Early Eocene, during collision.

olitic units, was located between the Karacahisar carbonate platform to the east and the Sütçüler platform unit to the west. Upper, during Late Cretaceous–Early Eocene times, the Sütçüler–Dulup carbonate platform was detached from its basement and emplaced eastwards over all other units.

Southeast area

The southeast area of the Antalya Complex (Fig. 3) is effectively the southward extension of the units described above (Fig. 18), although there are some differences in structural style.

Structural setting

The structural trend of the Antalya Complex curves towards the east and then disappears beneath Plio-Quaternary sediments of the Antalya plain. The southern extension of the Karacahisar carbonate platform is transgressively overlain by thick Miocene sediments (Aksu Formation) and then further east by the north–south trending Kırkkavak basin (Dumont & Kerey, 1975b), thus obscuring the inferred thrust contact between slope/basinal units of the Antalya Complex and the adjacent Karacahisar carbonate platform. In this area, the Antalya Complex is cut by numerous northeast–southwest trending reverse faults and thrusts related to the Upper Miocene Aksu phase (Poisson, 1977). Numerous klippen and/or fault-bounded units of Mesozoic neritic limestone up to several kilometres across are correlated with the Sütçüler–Davras thrust unit further north. The area is also by affected neotectonic normal faults, stepping down to the west.

Lithologies

The slope and basinal units of the Antalya Complex in this area are dominated by lava–sedimentary mélange and broken formation. The main lithologies are Upper Triassic quartzose sandstones and mafic lavas, associated with 'Halobia limestones' and ribbon radiolarites.

The higher structural levels of the mélange are cut by moderately- to steeply-dipping thrust slivers of sheared serpentinite, up to several hundred metres wide and several kilometres long. The mélange adjacent to these serpentinites is also sheared. Entrained within the serpentinites are small thrust

wedges of ophiolite-derived sandstone and white pelagic chalk, containing small planktonic foraminifera of Palaeocene–Lower Eocene(?) age in a micritic matrix. Elsewhere the serpentinite strands incorporate blocks of sheared gabbro, ophiolite-derived debris flows, sandstone, mudstone and pelagic chalk.

Within the dismembered, overriding limestone thrust sheet, local successions can be reconstructed: for example, Upper Triassic, pink 'Halobia limestones', passing conformably upwards into Jurassic shallow-water platform limestones (Fig. 24), as in the Sütçüler unit further north. In most cases, the base of this unit is marked by sheared serpentinite. However, locally in the extreme southeast of the area intact pillow lava successions are interbedded with turbiditic ribbon radiolarite and quartzose sandstone. These extrusives are overthrust by pink 'Halobia limestone' which is, in turn, overlain by neritic limestones.

In the extreme southeast, near Gebiz (Fig. 3), the volcanic–sedimentary unit is in high-angle fault contact with fossiliferous limestones and mudstones in which A. Racey (personal communication, 1985) identified *Lepidocyclina* (*Eulepidina* and *Nephrolepidina*) of Middle Oligocene to Early Miocene age, a primitive *Miogypsina* of? Lower Miocene age, also *Amphistegina* sp. and *Operculina* sp. The age of this succession is interpreted as

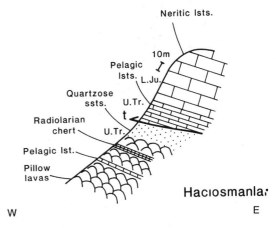

Fig. 24. Local cross-section of the southeastern area. Upper Triassic pillow lavas are interbedded with radiolarites and 'Halobia limestones', and overlain by quartzose sandstones. Upper Triassic pelagic limestones follow, then Jurassic neritic limestones. The contact at the base of the limestones is probably a thrust. See Fig. 3 for location.

Lower Miocene, suggesting correlation with the Aquitanian limestones of the northwestern area (see above). This succession is overlain structurally by a 200-m-thick, intact thrust sheet of Upper Triassic sediments containing *Halobia*. In detail, poorly exposed quartzose sandstones are overlain by thin- to medium-bedded calcilutites and grey radiolarian micrites interbedded with fine- to-medium-grained bioclastic calciturbidites exhibiting numerous intraformational slump horizons. Shaley and siliceous sediments dominate the higher levels of the succession, which is unconformably overlain by ophiolite-derived conglomerates of the Upper Miocene? Aksu Formation (e.g. at Karakütük). These Mesozoic sediments are correlated with sedimentary thrust sheets below the Sütçüler–Dulup thrust unit further north in the eastern area.

Interpretation

During the Mesozoic, the eastern area was bordered by the Karacahisar carbonate platform. The volcanic–sedimentary mélange and overriding limestone thrust sheets preserve remnants of neotethyan oceanic crust and the southwards extension of the Sütçüler–Dulup carbonate platform. Locally exposed slope/basinal sedimentary sheets could represent the imbricated, overthrust eastern passive margin of this carbonate platform. Emplacement of the volcanic–sedimentary mélange again involved large-scale submarine erosion and redeposition in a pelagic, carbonate-depositing, basin. Serpentinite was sheared into the structural succession along with other lithologies, in a manner reminiscent of the Gödene Zone in the southwestern area of the Antalya Complex, where large-scale strike-slip faulting is inferred (Robertson & Woodcook, 1980; see below).

Apex of the Isparta angle

The discussion of each of the units around the Isparta angle given above allows remaining questions about the relationship of units exposed on both sides of the neotectonic Kovada graben (Fig. 3) to be considered. In particular, were the Davras Dağ and Sütçüler–Dulup thrust sheets originally a single, continuous carbonate platform, or were they two separate platforms? High-angle neotectonic faults, recent alluvium and Kovada Lake obscure critical contact relationships except in the south, near the Kovada hydroelectric plant (Santral 1; Figs

18 & 25a). There, thin-bedded, Upper Cretaceous, pelagic carbonates and thin- to medium-bedded calciturbidites are disconformably overlain by 3 m of sheared pelagic carbonates and chalky debris flows including clasts of serpentinite, gabbro, lava and chert. Despite strong neotectonic faulting, this succession can be traced northwards into successions of the Davras Dağ carbonate platform unit. This unit is overlain structurally by a limestone klippe correlated with the Sütçüler–Dulup thrust sheet (3 km southwest of Santral 1). The local succession is pink nodular calcilutites and 'Halobia limestones' of inferred Upper Triassic age. Further east, a klippe of massive limestone, also correlated with the Sütçüler–Dulup thrust sheet, was dated as Jurassic? by Akbulut (1977; his locality 14), while an isolated klippe further south (his locality 15) yielded a Cretaceous age. In summary, the Sütçüler–Dulup thrust unit *overlies structurally* the Davras Dağ carbonate platform.

In the southern part of the Kovada graben, the Davras Dağ carbonate platform sediments, described above, are overlain structurally to the east by east-dipping, imbricated sedimentary sheets. Similar rocks crop out as small, isolated hills further north within the graben (Figs 18 & 25b). These successions are dominated by medium- to thick-bedded, proximal, Upper Triassic slope facies 'Halobia limestones' and ribbon radiolarites. Lithologically they resemble, for example, thrust sheets imbricated beneath the northern margin of the Dulup Dağ and Hudulca Tepe (see above). The Kovada units are interpreted as slope facies of the original eastern margin of the Davras Dağ carbonate platform.

Along the southeastern margin of the Davras Dağ, Akbulut (1977) correlated a thrust sheet of mainly Upper Jurassic platform carbonates with the Sütçüler Limestones (e.g. Bozburun Dağ; Asar Tepe; Fig. 18). This implied structural continuity of the carbonate platforms across the Kovada graben. However, mapping indicates that this overriding limestone unit is an extension of the large Davras Dağ carbonate platform thrust sheet (Figs 18 & 25c). Beneath, there is a volcanic–sedimentary unit, dominated by polymict, ophiolite-derived debris flows, broken formation, coarse-grained volcaniclastic sandstones and disrupted sheets of 'Halobia limestone' up to 40 m thick (Fig. 25c). Further southeast (near Yukarıgökdere; Fig. 18), similar mélange is overlain structurally by sheared serpentinite, highly deformed pillow basalt, ribbon radio-

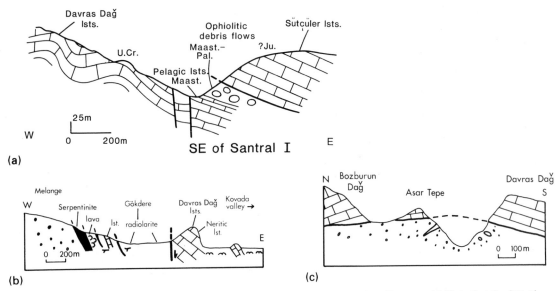

Fig. 25. Local cross-sections across the apex of the Isparta angle in the Kovada valley area. (a) Note that the Sütçüler Limestones structurally overlie the Davras Dağ carbonate platform unit, near the hydroelectric plant (Santral 1); (b) the Davras Dağ carbonate platform unit is thrust over volcanic–sedimentary mélange. Limestone units exposed further south (Asar Tepe and Bozburun Dağ) are klippes related to the same thrust sheet and not part of the Sütçüler Limestones (cf. Akbulut, 1977); (c) downfaulted Davras Dağ carbonate platform, adjacent to the neotectonic Kovada graben. The Davras Dağ in this area is structurally underlain by thrust sheets of slope and basinal sediments and ophiolite-derived mélange. Further south (near Cukurköyü), the entire stack is seen to be thrust over Miocene sediments.

larites, disrupted sheets of mudstones, radiolarites, calciturbidites and limestone debris flows (Fig. 25b). Radiolarites from this locality were dated as Upper Jurassic–Lower Cretaceous and Early Cretaceous (Upper Hauterivian–Lower Albian). Further south (Fig. 18) more coherent successions include Upper Triassic quartzose sandstones, 'Halobia limestones' and radiolarian cherts. All of these units structurally underlie the Davras Dağ carbonate platform thrust sheet and are correlated with volcanic–sedimentary mélange beneath the southwest margin of the Davras Dağ (Fig. 10a). They are interpreted as derived from the Mesozoic oceanic strand between the Davras Dağ carbonate platform to the east and the Bey Dağları carbonate platform to the west.

In summary, the Davras Dağ and Sütçüler–Dulup thrust sheets existed as two separate carbonate platforms within the Neotethys ocean (Fig. 26a). Basinal sediments and ophiolites were thrust, apparently westwards, over the eastern edge of the Davras Dağ platform during Maastrichtian–Early Eocene times. Simultaneously, the inferred Mesozoic small ocean basin to the west collapsed and was overridden by the Davras Dağ carbonate platform (Fig. 13). The Davras Dağ thrust sheet and the underlying basinal units were folded and thrust westwards over Miocene sediments during the Upper Miocene Aksu phase. The Sütçüler–Dulup thrust sheet further east underwent high-angle reverse faulting and thrusting during this time, followed by neotectonic faulting to form the Kovada graben.

Southwest area: summary

Before proceeding to an overall interpretation of the Antalya Complex, published information on the classic southwest area must be summarized. A number of tectonic units, termed zones, are recognized (Woodcock & Robertson, 1977b, 1982; Fig. 4, parts 1 & 2). These are tectono-stratigraphical units separated by steeply dipping faults and, from west to east, are as follows.

1 The *Bey Dağları Zone* comprises a relatively autochthonous carbonate platform, the Bey Dağları (Poisson, 1977), ranging from Lower Jurassic (e.g. Katran Dağ) to Lower Tertiary in age and locally

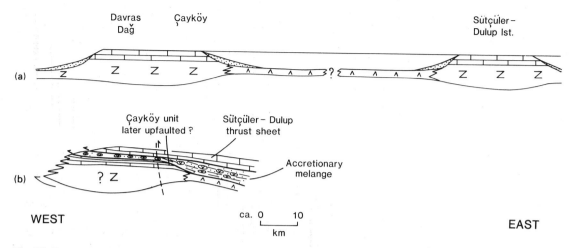

Fig. 26. Reconstruction of the relationship between the Davras Dağ and Sütçüler units on either side of the neotectonic Kovada graben; (a) Late Triassic–Early Cretaceous, passive margin phase; (b) Late Cretaceous–Palaeocene following suturing.

overlain by Miocene ophiolite-derived clastics (northwest of Kumluca; Hayward, 1982; Hayward & Robertson, 1982).

2 The *Kumluca Zone* is an imbricate fan of deep-water Neotethyan slope and basinal sediments of Upper Triassic to Upper Cretaceous age (Robertson, 1981; Robertson & Woodcock, 1981a, b)

3 The *Gödene Zone* is a thick succession of mainly Upper Triassic mafic extrusives (Juteau, 1970, 1975) and associated deep-water sediments, cut by high-angle, anastomosing, strands of sheared serpentinite (Robertson & Woodcock, 1980; Woodcock & Robertson, 1982). Tectonic discrimination of Upper Triassic basalts using 'immobile' major and trace-elements (Pearce *et al.*, 1982) indicates within-plate, MORB and depleted MORB-type settings (Robertson, 1990; Robertson & Waldron, 1990)). The Gödene Zone also includes Upper Palaeozoic basement lithologies (e.g. Bakırlı Dağ) and Mesozoic carbonate platform rocks (e.g. Çalbalı Dağ; Fig. 3).

4 The *Kemer Zone* is dominated by Palaeozoic (Ordovician–Permian) basement sedimentary units, depositionally overlain by Mesozoic carbonate platform facies (e.g. 'Kemer units'; Figs 3 & 4).

5 Lastly, the *Tekirova Zone* is the deeper, plutonic levels of an Upper Cretaceous ophiolite (Juteau, 1975; Juteau *et al.*, 1977; Reuber, 1984; Yilmaz, 1981, 1984), unconformably overlain by ophiolite-derived clastics (Robertson & Woodcock, 1982: Lagabrielle *et al.*, 1986).

Interpretation

Robertson and Woodcock (1982) and Poisson (1984) interpreted the Bey Dağları as a Mesozoic carbonate platform bordering a small, southerly Neotethyan ocean basin. The Kumluca Zone (Çatal Tepe unit; Alakır Çay and Bilelyeri Groups; Fig. 5) was identified as the thrust-imbricated passive margin of the Bey Dağları carbonate platform. Robertson and Woodcock (1982) and Yilmaz (1984) related the Upper Triassic basaltic pillow lavas (Juteau, 1977; e.g. Sayrun) of the Gödene Zone to rifting and continental break-up, a conclusion supported by new geochemical data on Upper Triassic basalts (Robertson & Waldron, 1990). A sizeable continental sliver was rifted off to form the Palaeozoic basement of the Kemer Zone carbonate platform. Other smaller carbonate platform units were also constructed on continental basement (Bakırlı Dağ) and/or volcanic seamounts (Tekke Dağ?). In the north, at least one carbonate platform (Çalbalı Dağ/Bakırlı Dağ) was located not far from the margin (tens of kilometres at most, Robertson & Woodcock, 1981b). Similar, rifted-off, continental crust slivers are known from the Dibba Zone, northern Oman Mountains, an area interpreted as a strike-slip influenced passive margin segment of Gondwana (Robertson *et al.*, 1990). There are also marked similarities with modern Atlantic transform rifted margins (e.g. eastern equatorial Atlantic) particularly in the presence of rifted outer margin basement ridges (Mascle & Blarez, 1987; Mascle *et al.*, 1988).

An oceanic strand of variable width thus existed between the Bey Dağları carbonate platform and several off-margin carbonate build-ups within Neotethys to the east. This basin is now preserved only as highly deformed ophiolitic units within the Gödene Zone. Dated slivers of greenschist and amphibolites imply that this basin was first tectonically disrupted during the Late Cretaceous (c. 78–73 Ma; Yilmaz, 1984).

Reuber (1984) interpreted the Tekirova ophiolite further east as Late Cretaceous oceanic crust and mantle generated adjacent to a sinistrally slipping oceanic fracture zone. The Tekirova ophiolite was already uplifted and deeply eroded by the Maastrichtian, but without admixed terrigenous sediment, probably because it was then still located within Neotethys (Robertson & Woodcock, 1982).

Further west, westward thrust imbrication of distal basinal sediments (Kumluca Zone) towards the Bey Dağları carbonate platform took place during the Maastrichtian(?). However, unbroken carbonate deposition continued on the Bey Dağları platform during Upper Cretaceous–Early Tertiary times, showing that overthrusting did not take place. Further inferred strike-slip and westward compression occurred during the Early Tertiary, when the eastward margin of the Bey Dağları subsided and was overlain by nummulitic flysch (Hayward & Robertson, 1982). The complex was finally emplaced by relatively short distance (tens of kilometres at most) westward thrusting over a small clastic foreland basin in the Late Miocene.

In the past the strike-slip model (Fig. 27) for the southwestern area has led to several misunderstandings. First, for some it implied that the Antalya units in the southwest area were relatively autochthonous, which is far from being the case. Secondly, the strike-slip model was wrongly assumed to apply to the whole of the Isparta angle, despite that fact that Waldron (1981, 1984a, b) had described contrasting structural styles in the northeastern area. Most of the inferred strike-slip faulting place within the Neotethyan ocean, may have taken away from the large bordering carbonate platforms. Units in the Isparta angle underwent pervasive deformation during suturing of carbonate platforms prior to Middle Eocene time, which would have destroyed any evidence of oceanic strike-slip displacements in this area. By contrast, the southwestern area still

JURASSIC–LR. CRETACEOUS

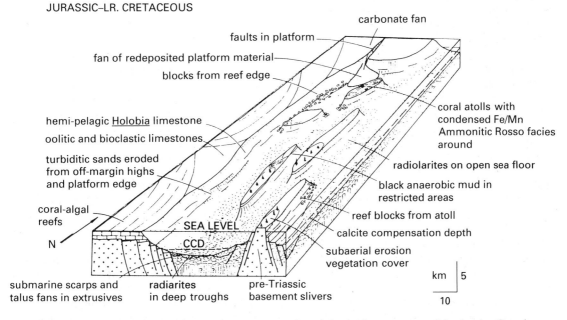

Fig. 27. Block diagram illustrating the inferred tectonic evolution of the southwestern area of the Antalya Complex. Note: the exact size, position and number of the off-margin platforms remains conjectural. Emplacement in the Late Cretaceous–Early Tertiary was achieved mainly by strike-slip and thrusting in this area. CCD, calcite compensation depth. (After Robertson & Woodcock, 1980.)

remains in an essentially pre-suture setting, with Neotethyan oceanic crust still out to the southeast in the Eastern Mediterranean. In summary, the southwestern area of the Antalya Complex records the remnants of a palaeogeographically complex, small ocean basin and its margins.

RECONSTRUCTION AND HISTORY OF PLATFORMS AND BASINS

The palaeogeographical and tectonic history of each area of the Antalya Complex is summarized in Table 2. The palaeogeography inferred for the Mid-Eocene soon after suturing of the Isparta angle is shown in Fig. 28. Early Eocene is the time of final emplacement, as documented by the presence of ophiolitic debris flows within the *in situ* succession of the Bey Dağları platform (western area, near Badamağaci; Poisson, 1977). During the Neogene, the Bey Dağları carbonate platform underwent some 30° of anticlockwise rotation (Kissel & Poisson, 1987; Morris, 1990; Morris & Robertson, 1993). Possible rotations of carbonate platform units along the eastern margin of the Isparta angle, however, remain largely unknown. The reconstructions of the Isparta angle area shown in and Fig. 30 takes account of known palaeomagnetic rotations and also assumes that the area was tectonically assembled, mainly by 'in-sequence', piggy-back, thrusting.

The main uncertainty is the relative widths of the oceanic strands between the carbonate platforms (since these were almost entirely subducted). However, it is assumed, for reasons discussed earlier, that oceanic crust lay between the Anamas Dağ–Karacahisar carbonate platforms and the Sütçüler–Dulup carbonate platform.

The Antalya Complex developed as a mosaic of small carbonate platforms within the Neotethys Figs 29 & 30). The north margin of Gondwana in this area was first rifted in the Early Triassic (Skythian), followed by renewed extension and final continental break-up in the Late Triassic (Carnian–Norian). In places, break-up faults cut across earlier rifts and carbonate platforms became established on earlier fault-zones and basins (e.g. Kaymaz Dağ; Sütçüler unit). Pelagic Hallstatt facies was locally overlain depositionally by Jurassic shallow-water carbonates (e.g. eastern, southweastern and northwestern areas). Within-plate, transitional and MORB-type lavas were erupted in the Late Triassic

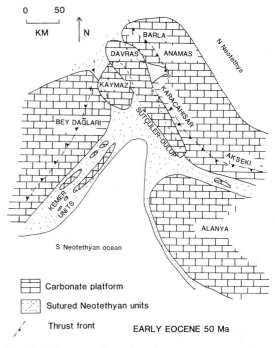

Fig. 28. Palaeotectonic setting of the Antalya Complex after suturing of the Isparta angle was completed in the Early Eocene, but before Late Eocene and Late Miocene deformation in the north. The southwestern area remained in a pre-collisional setting with Neotethys still to the south.

(Robertson, 1990; Robertson & Waldron, 1990). These lavas are geochemically very similar to those, for example, of Oman (Haybi Complex; Searle *et al.*, 1980), another area of Neotethyan continental break-up (e.g. Robertson & Searle, 1990). Passive margins began to develop around the carbonate platforms from the Late Triassic onwards. Deeper-water slope and basinal areas had subsided below the carbonate compensation depth (CCD) by Early Jurassic times. On one large rifted platform (Barla Dağ), sedimentation failed to keep pace with subsidence and condensed Ammonitico Rosso accumulated on a submerged platform from Mid-Jurassic to Mid-Cretaceous times. Further crustal extension, in Late Jurassic–Early Cretaceous times, was marked by sediment redeposition and hydrothermal activity. Tholeiitic volcanism of Upper Jurassic–Early Cretaceous age, at least in the northeast and southwestern areas (Waldron, 1981; Yilmaz, 1984) is suggestive of renewed spreading.

By the Upper Cretaceous, Africa and Eurasia

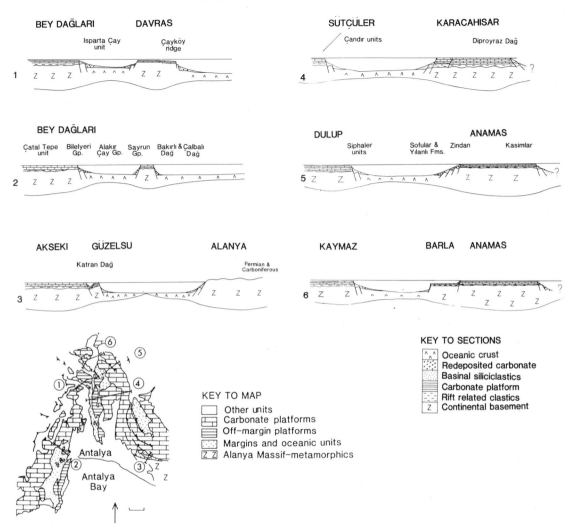

Fig. 29. Sketch cross-sections illustrating the inferred Mesozoic palaeogeography of the Isparta angle Neotethyan area during the passive margin phase (Late Triassic–Early Cretaceous).

were converging in the Eastern Mediterranean area (Livermore & Smith, 1984). A possible manifestation of this in the Antalya area was genesis of the harzburgitic Kızıl Dağ ophiolite (northeastern area), possibly above a subduction zone. Also, the Late Cretaceous Tekirova ophiolite (southwestern area) formed near an oceanic transform fault (Reuber, 1984). Subduction of old, dense (Late Triassic?) oceanic crust and mantle could have led to regional crustal extension and this may have been the driving force of subsidence of the carbonate platforms after Cenomanian times. Alternatively, eustatic sea-

level rise may have been the main control (Haq *et al.*, 1987).

With further convergence in the Late Cretaceous, the Neotethyan oceanic crust in the Isparta angle area was progressively subducted, while overlying volcanics and sediments were detached and partly preserved within accretionary prisms. Marginal sedimentary sheets were deformed into imbricate fans (e.g. Isparta Çay units). Thrusting was towards the west in the southwest, west and northwestern areas (after removing palaeomagnetic rotations), but towards the east in the northeast and eastern

Table 2. Summary of data and interpretations for each area of the antalya Complex Based on the literature and new data presented in this paper see Fig. 3 for locations of the areas

Age	Area	
	Western area	Northwest area
QUATERNARY	Uplift and erosion, terracing	Uplift, erosion, local acidic volcanism (south of Isparta)
PLIOCENE	Deposition on Antalya Plain to southeast	Clastic basin to south
MIOCENE		Westwards-thrusting and folding of platform and basinal units; Lycian nappes thrust from northwest
Tortonian ————	Subsidence and clastic deposition in Aksu basin to east	
Burdiglian ————		
Aquitanian ————		Deepening flysch basin, flexural subsidence; compression from shallow-water carbonate transgression over deformed Antalya Complex sediments (Isparta Çay)
OLIGOCENE	? Erosion	
EOCENE		
PALAEOCENE	Final westwards emplacement onto platform	Final thrusting of Antalya Complex
		Calcareous flysch on Davras Dağ platform: accretionary mélange to west
CRETACEOUS	Turbidites and debris flows from advancing margin and oceanic crustal units: slicing of margin sediments and thrusting of oceanic crust	Imbrication and westwards thrusting of basinal sediments (Isparta Çay)
Senonian ————		
Cenomanian ————	Disconformity, then subsidence of platform	Subsidence of Bey Dağları and Davras Dağ platforms: pelagic deposition
Lower ————	Stable carbonate platform deposition on Bey Dağları platform (Katran Dağ) to west. similar, but more condensed, deposition on upper unit platform to east (Alisivri Tepe)	Stable carbonate platform deposition on both Bey Dağları (Erenler Tepe) to west and separate Davras Dağ platform
JURASSIC		
Upper ————		
Middle ————	Inferred, stable deep-sea deposition on passive margins and marginal (transitional) oceanic crust to east	Channelized calciturbidites within siliceous basinal deposits to east of Bey Dağları platform (Isparta Çay)
Lower ————		
TRIASSIC		
Upper ————	Passive margins develop, following continental break-up and initial spreading	Proximal base-of-slope facies of Bey Dağları preserved in Isparta Çay unit
Norian –		
Carnian –		
Middle ————	Platform deposition established on off-margin platform	Quartzose turbidites and periplatform *Halobia* ooze; oceanic crust to east (Darioren)
Ladinian –		
Anisian –		
Lower Skythian ————	Rifting and reworking of Permian basement; unstable shelf deposition on newly formed, off-margin carbonate platform to east	
PERMIAN		
Upper ————	Muddy shelf carbonates, locally exposed in upper unit to east	

Continued

Area		
Northern area	**Northeast area**	**Eastern and southeast area**
Uplift erosion; lake sedimentation (Lake Eğridir)	Uplift–erosion extensive lake sedimentation (Lake Eğridir)	Uplift, erosion, Plio-Quaternary clastic deposition in Antalya basin to southwest
Southwards-directed compression and folding in Upper Miocene?	West-directed faulting and compression, increases southwards	Reverse faulting, local west-directed thrusting. Messinian evaporite, then clastic deposition in remnant basin to southwest
	Erosion	
L. Eocene–Miocene nummulitic cover sediments in Barla Dağ and southwards thrusting of Kaymaz Dağ over Davras Dağ and basinal units		Initial transgression; large foraminifera
	Southwest-directed thrusting of Anamas platform: reimbrication of emplaced Antalya Complex thrust sheets	Erosion?
Chalky clastic flysch on Barla Dağ; flexural foreland basin? Southwards thrusting and emplacement of ophiolite, marginal units and basinal sediments onto Davras Dağ platform, mainly olistostromes	Deposition of Anamas Dağ platform ends with chalky flysch; flexural basin?	Accretionary mélange, then thrusting of Sütçülers Limestone eastwards as major out-of-sequence thrust sheet; strike-slip faulting in southeast
	Imbrication of Anamas Dağ passive margin northeastwards; ophiolitic debris flows and mélange thrust over Davras Dağ margin (and Çayköy unit); thrusting of Dulup Dağ platform over its marginal facies	Initial slicing of marginal and basinal units bordering Karacahisar and Dulup–Sütçüler platforms, Cen.; flooding of Karachisar and Dulup–Sütçüler platforms, U. Cretaceous genesis of oceanic crust and mantle (Kızıl Dağ)
Drowning of Davras Dağ and Kaymaz Dağ platforms; genesis of oceanic crust between Davras Dağ and Kaymaz Dağ platforms?		
U. Triassic–Cenomanian: stable carbonate platform deposition on both Davras Dağ and Kaymaz Dağ platforms; M. Jurassic–L. Cretaceous condensed Ammonitico Rosso on submerged Barla Dağ platform to north	U. Triassic–Cenomanian: stable shallow-water carbonate deposition on Anamas Dağ platform in northeast; separate development of smaller platforms to weat (Hudulca Tepe). Larger platform sited to west (Dulup Dağ) with own marginal/slope facies; continued passive margin and basinal deposition (Pazarköy Gp)	U. Triassic–Cenomanian: stable shallow-water carbonate deposition on Anamas Dağ; more interrupted deposition on Dulup–Sütçüler platform Passive margins border Karacahisar and Dulup–Sütçüler platforms
Proximal base-of-slope facies (e.g. <u>Halobia</u> limestones) accumulated along north margin of Davras Dağ platform and south margin of Kaymaz Dağ platform; U. Triassic marginal crust formed between these two platforms; now preserved as blocks in mélange (e.g. Findos); rifting of Barla Dağ platform to north	Red clastics in Anamas Dağ; rifting of Neotethyan basins to northeast? Rift-related deposition ofclastics (Kasimlar Fm); evaporitic deposition (Mentesçe Fm) on Anamas Dağ platform; rifting and development of proximal (Zindan Gp) and more distal slope sediments (Pazarköy Gp). Within-plate type volcanics with carbonate build-ups, near passive margin edge (Bucak Lava)	Ammonitico Rosso, well developed at base of Sütçüler unit Proximal base-of-slope, quartzose turbidites and peri-platform ooze around margins of platforms pass basinwards into pillow basalts and deep-sea sediments; lavas of within-plate type
Limestones, preserved as blocks in broken formation southwest of Egridir; remnant of pre-rift basement		

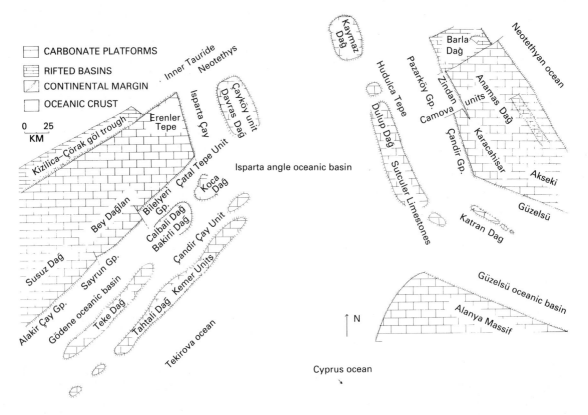

Fig. 30. Palaeogeographical sketch illustrating the inferred Isparta angle Neotethyan ocean during the Upper Jurassic–Early Cretaceous; i.e. after Upper Triassic spreading, but prior to genesis of ophiolite in the Upper Cretaceous and subsequent deformation. The reconstruction takes account of 30° of anticlockwise, neotectonic rotation of the Bey Dağları. Note that this is not intended as an accurate palinspastic reconstruction, but illustrates the most likely setting suggested by the available data.

areas, i.e. outwards from the Isparta angle. During convergence the adjacent platform margins were uplifted in response to crustal flexure and/or faulting, followed by collapse to form foredeeps. This is similar to events which occurred during the emplacement of the Late Cretaceous Oman ophiolite (Robertson, 1987). During the Early Tertiary, the Isparta angle was tightened and deformed slope and basinal units were thrust further onto adjacent carbonate platforms creating more new foredeeps during Palaeocene–Early Eocene times. The former intra-oceanic platforms were also finally assembled, by collision during this time. *Nummulites*-bearing flysch accumulated in thrust-top basins during the later stages of collision in Early Eocene times (e.g. Davras Dağ). By contrast, the southwestern area remained in a largely pre-collisional setting and underwent mainly strike-slip related deformation in

Late Cretaceous–Early Tertiary times, prior to final emplacement over the Bey Dağları carbonate platform during the Miocene.

Tectonic assembly of the mosaic of carbonate platforms in the Isparta angle was complete by Early Eocene time (Fig. 28). Subsequently, the northeastern area was again thrust southwestwards, an event that is thought to relate to suturing of another Neotethyan ocean basin further northeast, the 'inner Tauride ocean' of Sengör and Yilmaz (1981). Soon afterwards the Isparta angle subsided giving rise to the Oligo-Miocene clastic Aksu basin (Fig. 3; Poisson *et al.*, 1983; Akay *et al.*, 1985). Northerly and easterly land areas of the Isparta angle were transgressed by shallow-water limestone in the Early Miocene (Aquitanian). Regional compression then further tightened the Isparta angle. Turbidites accumulated in a 'flexural' foreland basin in the Burdigalian, followed by thrusting of the Lycian nappes into the apex of the Isparta angle from the northwest in post-Langhian–pre-Tortonian time (Poisson, 1977). The southwestern area of the Antalya Complex was also thrust westwards over Lower Miocene ophiolite-derived sediments. In Upper Miocene (pre-Messinian) times, the Antalya Complex and the Aksu Formation were deformed by west- then south-directed folding, reverse faulting and local thrusting (Aksu phase; Akay *et al.*, 1985). An open marine basin still persisted in the south until Late Pliocene? times. During the Late Pliocene/ Early Quaternary? the whole area was uplifted, associated with reactivation of crustal weakness zones to form the Kovada graben and its southwards offshore extension into Antalya Bay.

In conclusion, the sedimentary and tectonic history of the Antalya Complex can be taken to exemplify the evolution of palaeogeographically complex areas of Neotethys, or other small ocean basin systems like the modern Caribbean (Pindell, 1985; Pindell & Barrett, 1990). Similarly complicated Neotethyan areas include the Pindos ocean in Greece, Albania and Yugoslavia and the Vardar ocean basin in northeast Greece, as summarized in Fig. 31. On a regional scale the Isparta angle oceanic basin appears to have separated two microcontinents rifted from Gondwana, Apulia to the west and the Tauride carbonate platforms to the east.

ACKNOWLEDGEMENTS

I thank the MTA (Maden Tetkik ve Arama Enstitüsü), particularly S. Yalkinçaya for logistical support. The work was funded by a NERC research grant. Helpful comments on the manuscript were made by John Waldron, Lynne Frostick and an anonymous referee. Emile Pessagno is thanked for providing valuable radiolarian determinations.

APPENDIX

Radiolaria determined by E. A. Pessagno, University of Texas, Dallas, Texas.

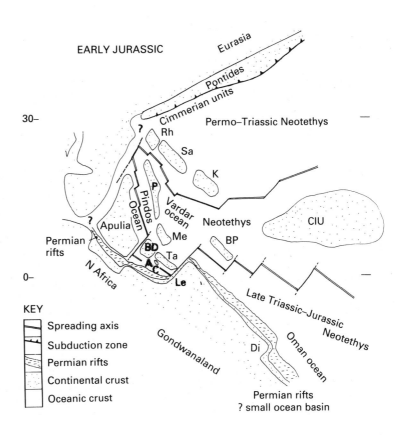

Fig. 31. Palaeotectonic sketch of the Eastern Mediterranean Neotethys during the Middle Jurassic. This time slice followed break-up in the Late Triassic, but preceded initial ocean basin deformation in the Greek area during late Middle/Late Jurassic time. A, Alanya Massif; BD, Bey Dağlari platform; BP, Bitlis-Pütürge units; C, Cyprus; CIU, Central Iranian units; Di, Dibba Zone; K, Kirşehir Massif; Le, Levant; Me, Menderes Massif; P, Pelagonian Zone; Rh, Rhodope Zone; Sa, Sakarya unit. (From Robertson *et al.*, 1991.)

Western area

T/83/169: 1 km south of Kocaaliler
Zifondium sp. (Unnamed form; occurs in Albian).
Archaeodictyomitra sp.

Northwestern area

T/83/2: volcanic–sedimentary mélange, Darioren,
Isparta Çay, Fig. 10.
Mirifusus sp. aff. *M. guadalupensis* Pessagno
Biostratigraphic determination: Zone 2 to Zone 5;
possibly Zone 2 to lower Zone 3 *sensu* Pessagno,
Blome, and Longoria. Late Jurassic (Kimmeridg-
ian) to Early Cretaceous (Upper Valanginian to
Lower Hauterivian); possibly Kimmeridgian to
Lower Tithonian.

T/83/19: locality as above.
Mirifusus?
Archaeodictyomitra sp.
Ristola sp.
Archaeospongoprunum sp.
Tripocyclia sp.
Biostratigraphical determination. Superzonel 1,
Zone 1A to Zone 5; possibly Zone 2 to Zone 5.
Aalenian to Lower Hauterivian; possibly Kim-
meridgian to Lower Hauterivian.

T/83/28: Isparta Çay unit, Isparta CCay: locality as
above.
Hsumm maxwelli Pessagno.
E. ptyctum Riedel and Sanfilippo.
Mirifusus sp.
Angulobracchia sp.
Biostratigraphical determination. Zone 2 to Zone 3.
Upper Jurassic: Kimmeridgian to Lower Titho-
nian.

T/83/117: Yukarigökdere, Kovada graben.
Archaeodictyomitra sp.
Cecrops septemporatus (Parona).
Biostratigraphical determination. Zone 5, Subzone
5C. Early Cretaceous: Upper Valaginian to Lower
Hauterivian.

T/83/119: locality as above.
Holocryptocanium sp.
Thanarla sp. aff. *elegantissima* (Cita).
Archaeodictyomitra lacrimula (Foreman).
Pseudodictyomitra carpatica (Lozyniak).
Pseudodictyomitra (?) *lilyae* (Tan Sin Hok).

Preaconocaryomma (?) sp.
Biostratigraphical determination: Zone 6 to Zone 8;
probably Zone 7. Early Cretaceous: Upper Hauterivian to Upper Albian; probably Lower Albian. The absence of *Pantenellium* Pressagno and
Thanaria conica (Aliev) is significnat.

T/83/120: locality as above.
Mirifusus bailevi Pessagno.
Mirifusus sp.
Ristola sp.
Trupocyclia sp.
Vallupus hopsoni Pressagno & Blome
Podobursa sp.
Podocapsa amphitreptera Foreman.
Biostratigraphical determination: Zone 4. Upper
Jurassic: Upper Tithonian. Note: Zones 2, 3 and
4, redefined by Pessagno, Blome and Longoria.

Güzelsu Corridor

T/83/198: imbricated sheets below south margin of
Katran Dağ, Güzelsu Corridor.
Xitus sp.
Ristola sp. cf. *Boesii* (Parona).
Mirifusus n. sp.
Pseudodictyomitra carpatica (Lozyniak).
Pantanellium corriganesis Pessagno.
Pantanellium sp. (5 +).
Thanaria conica (Aliev).
Podobursa sp.
Crucella sp.
Emiluvia sp.
Alievium helenae Schaaf.
Aconiotyle sp. cf. *umbilicata* Foreman.
Biostratigraphical determination: Zone 5, Subzone
5C. Early Cretaceous: Upper Valanginian/Lower
Hauterivian.

T/83/199: (As locality 198).
Pseudodictyomitra carpatica (Lozyniak).
Thanarla conica (Aliev).
Biostratigraphical determination: Zone 5, Subzone
5C to Zone 6. Early Cretaceous: Upper Valangin-
ian to Upper Aptian.

REFERENCES

American Geological Institute (1961) *Dictionary of Geological Terms. Dolphin Books, New York, 545 pp.*
AKAY, E. UYSAL, S., POISSON, A., CRAVATTE, J. & MULLER, C.

(1985) Stratigraphy of the Antalya Neogene Basin (in Turkish). *Bull. Geol. Soc. Turkey* **28**, 105–119.

AKBULUT, A. (1977) Etude géologique d'une partie du Taurus occidentale au sud d'Eğridir (Turquie). Thèse Univ. Paris Sud, Orsay **203**.

ALLASINAZ, A., GUTNIC, M. & POISSON, A. (1974) La formation de l'Isparta Çay calcaires à Halobies, grés à plantes, et radiolarites d'age Carnien(?) Norien (Taurides — Region d'Isparta — Turquie). *Sch. Erdwiss, Komm. Oster Akad.* **2**, 11–21.

BLUMENTHAL, M. (1960–1963) Le system structural du Taurus sud-anatolien. In: *Livre à mémoire de Professor P. Fallot*, Mém. Soc. Géol Fr., 1. **2**, 611–662.

BOUMA, A.H. (1962) *The Sedimentology of some Flysch Deposits.* Elsevier, Amsterdam.

BRUNN, J. H. (1974) Le problème de l'origine des nappes et leurs translations dans les Taurides occidentales. *Bull. Soc. Géol. Fr.* **16**, 101–106.

BRUNN, J.H., DUMONT, J. F., GRACIANSKY, P.C., DE GUTNIC, M., JUTEAU, T., MARCOUX, J., MONOD, O. & POISSON, A. (1971) Outline of the Geology of the Western Taurides. In: *Geology and History of Turkey* (Ed. Campbell, A.S.) Petrol Explor. Soc. Libya, Tripoli, pp. 225–252.

DELAUNE-MAYERE, M., MARCOUX, J., PARROT, J. F. & POISSON, A. (1977) Modèle d'évolution mésozoique de la paléomarge téthysienne au niveau des nappes radiolaritiques et ophiloitiques du Taurus lycien, d'Antalya et du Baër-Bassit. In: *Structural History of the Mediterranean Basins* (Eds Biju-Duval, B. & Montadert, L.) Technip, Paris, pp. 79–94.

DUMONT, J.F. (1976) Etudes géologiques dans les Taurides Occidentales: Les formations paléozoiques et mésozoiques de la coupole de Karacahisar (Province d'Isparta, Turquie). Thesis Univ. Paris-Sud, Orsay.

DUMONT, J.F. & KEREY, E. (1975a) Eğridir gölü güneyinin temel jeolojik etüdü (Basic geological study of the southern lake of Eğridir). *Bull. Geol. Soc. Turkey* **18**, 169–174.

DUMONT, J.F. & KEREY, E. (1975b) L'accident de Kirkkavak, un decrochment majeur dans le Taurus occidental: *Bull. Soc. Géol. France* **16**, 1071–1073.

DUMONT, J.F., MARCOUX, J., MONOD, O. & POISSON, A. (1972) Le Trias des Taurides occidentales (Turquie). Définition du bassin pamphylien: un nouveau domaine à ophiolites à la marge externes de la chain taurique. *Z. Deutsch Geol. Ges.*, **123**, 385–409.

DUMONT, J.F., UYSAL, S. & MONOD, O. (1980) La série de Zindan: un élément de liaison entre plateforme et bassin à l'Est d'Isparta (Taurides Occidentales, Turquie). *Bull. Soc. Géol. France* (7)**22**, 225–232.

GUTNIC, M. (1977) Géologie du Taurus pisidien au Nord d'Isparta (Turquie). Principaux résultats extraits des notes de M. Gutnic entre 1964 et 1971 par O. Monod. *Trav. Lab. de Géol. Hist.*, Univ. Paris-Sud, Orsay, 130 pp.

GUTNIC, M., MONOD, O., POISSON, A. & DUMONT, J. F. (1979) Géologie des Taurides occidentales (Turquie). *Mém. Soc. Géol. France* **137**, 1–112.

HAQ, B.U., HARDENBOL, J. & VAIL, P.R. (1987) Chronology of fluctuating sea levels since the Triassic. *Science* **235**, 1156–1167.

HAYWARD, A.B. (1982) Türkiye 'nin güneybatısındaki Bey Dağları ve Susuz Dağ masiflerinde Miyosen yaşlı kır-

ıntıı tortulların stratigrafisi. *Bull. Géol. Soc. Turkey* **25** (2), 81–89.

HAYWARD, A.B. (1984) Miocene clastic sedimentation related to emplacement of the Lycian Nappes and Antalya complex, S.W. Turkey. In: *Geological Evolution of the Eastern Mediterranean.* (Eds, Dixon, J.E. & Robertson, A.H.F.) Spec. Publ. Geol. Soc. London **17**, 287–300.

HAYWARD, A.B. & ROBERTSON, A.H.F. (1982) Direction of ophiolite emplacement inferred from Tertiary and Cretaceous sediments of an adjacent autochthon; the Bey Dağları, S.W. Turkey. *Bull. Geol. Soc. Am.* **93**, 68–75.

JUTEAU, T. (1970) Petrogenèse des ophiolites des Nappes d'Antalya (Taurus Lycien Oriental). Leur liaison avec une phase d'expansion oceanique active au Trias Supérieur. *Sci. Terre.* **15**, 265–288.

JUTEAU, T. (1975) Les ophiolites des nappes d'Antalya (Taurides occidentales, Turquie). Thèse, Doct. Sci., *Mém. Sci. Terre Nancy*, **32**, 1–692.

JUTEAU, T., NICHOLAS, A., DUBESSY, J., FRUCHARD, J.C. & BOUCHEZ, J.L. (1977) Structural relationships in the Antalya ophiolite complex, Turkey: possible model for an oceanic ridge. *Bull. Geol. Soc. Am.* **88**, 1748.

KISSEL, C. & POISSON, A. (1987) Etude palaéomagnetique des formations cénozoiques des Bey Dağları (Taurus occidentales — Turquie). *C. R. Acad. Sci., Paris* **304**, 343–348.

LAGABRIELLE, Y., WHITECHURCH, H., MARCOUX, J. *et al.* (1986) Obduction-related ophiolitic polymict breccias covering the ophiolites of Antalya (Southwestern Turkey). *Geology* **14**, 734–737.

LIVERMORE, R.A. & SMITH, A.G. (1984) Some boundary conditions for the evolution of the Mediterranean region. In: *Geological Evolution of the Mediterranean Basin* (Eds Stanley, D.J. & Wezel, F.C.) Springer-Verlag, Berlin, pp. 83–110.

LEFEVRE, R. (1967) Un nouvel élément de la géologie du Taurus lycien: les nappes d'Antalya (Turquie). *C. R. Acad. Sci. Paris* **165**, 1365–1368.

MCCLAY, K.R. (1981) What is a thrust? What is a nappe? In: *Thrust and Nappe Tectonics* (Eds McClay, K.R. & Price, M.J. Spec. Publ. Geol. Soc. London **9**, 1–5.

MARCOUX, J. (1970) Age carnien des termes éffusives du cortège ophiolites des Nappes d'Antalya (Taurus Lycien oriental, Turquie). *C. R. Acad. Sci. Paris* **271**, 185–267.

MARCOUX, J. (1974) Alpin type Triassic of the upper Antalya nappe (western Taurides, Turkey). In: *Die Stratigraphie der alpin-mediterranean Trias* (Ed. Zapfe, H.) Vienna pp. 145–146.

MARCOUX, J. (1976) Les séries triasiques des nappes à radiolarites et ophiolites d'Antalya (Turquie): homologies et signification probable. *Bull. Soc. Géol. Fr.* **18**, 511–512.

MARCOUX, J. (1977) Geological sections of the Antalya region. In: *Western Taurus Excursion Geological Guidebook* (Ed Guvenç, T.) VI Colloquium on the Geology of Aegean Regions, Izmir, 1977.

MARCOUX, J., POISSON, A. & DUMONT, J.F. (1975). Le domaine d'Antalya, témoin de la fracturation de la plateforme africaine au cours du Trias. *Bull. Soc. Géol. Fr.* **16**, 116–127.

MARCOUX, J., RICOU, L.E., BURG, J.P. & BRUN, J.P. (1989) Shear-sense criteria in the Alanya thrust system (southwestern Turkey): evidence for southward emplacement. *Tectonophysics* **161**, 81–91.

MASCLE, J & BLAREZ, E. (1987) Evidence for transform margin evolution from the Ivory Coast — Ghana Continental margin. *Nature* **326**, 378–381.

MASCLE, J., BLAREZ, E. & MARIAHS, M. (1988) The shallow structures of the Guinea and Ivory Coast — Ghana transform margins: their bearing on the Equatorial Atlantic Mesozoic evolution. *Tectonophysics* **155**, 193–209

MONOD, O. (1976) La "courbure d'Isparta": une mosaique de blocs autochtones surmontés de nappes composites à la jonction de l'arc hellénique et de l'arc taurique. *Bull. Soc. Géol. Fr.* **18**, 521–532.

MONOD, O. (1977) Récherches géologiques dans le Taurus occidental au sud de Beyşehir (Turquie). *Thèse de Doct. Sci. Univ. Paris-Sud, Orsay, 442 pp.*

MONOD, O. & AKAY, E. (1984) Evidence for a Upper Triassic–Early Jurassic orogenic event in the Taurides. In: *The Geological Evolution of the Eastern Mediterranean* (Eds Dixon, J.E. & Robertson, A.H.F.) Spec. Publ. Geol. Soc. London **17**, 113–128.

MORRIS, A. (1990) Palaeomagnetic studies of the Mesozoic–Tertiary tectonic evolution of Cyprus, Turkey, Greece. Unpublished PhD thesis, University of Edinburgh.

MORRIS, A. & ROBERTSON, A.H.F. (1993) Miocene remagnetisation of Mesozoic Antalya complex units in the Isparta angle, SW Turkey. *Tectonphysics* **220**, 243–266.

PEARCE, J.A., LIPPARD, S.J. & ROBERTS S. (1982) Characteristics and tectonic significance of supra-subduction zone ophiolites. In: *Marginal Basin Geology* (Eds Kokelaar, B.P. & Howells, M.F.) Spec. Publ. Geol. Soc. London **16**, 77–98.

PINDELL, J.L. (1985) Alleghenian reconstruction and subsequent evolution of the Gulf of Mexico, and proto-Caribbean. *Tectonics* **4**, 1–39.

PINDELL, J.L. & BARRETT, J.F. (1990) Geological evolution of the Caribbean region: a plate-tectonic perspective In: *Decade of North American Geology, 1 Caribbean Region*, **H**, (Eds Case, J.E. & Dengo, G.) Geological Society of America, Boulder, Co.

POISSON, A. (1977) Récherches géologues dans les Taurides occidentales (Turquie). *Thèse Doc. Sci., Univ. Paris-Sud, Orsay, 1–795.*

POISSON, A. (1984) The extension of the Ionian trough into southwestern Turkey. In: *Geological Evolution of the Eastern Mediterranean* (Eds Dixon, J.E. & Robertson, A.H.F.) Spec. Publ. Geol. Soc. London **17**, 241–250.

POISSON, A., AKAY, E., CRAVATTE, J., MULLER, C. & UYSAL, S. (1983) Données nouvelles sur la chronologie de mise en place des nappes d'Antalya au centre de l'angle d'Isparta (Taurides occidentales, Turquie). *C.R. Acad. Sci. Paris* **296**, 923–926.

REUBER, I. (1984) Mylonitic ductile shear zones within tectonics and cumulates as evidence for an oceanic transform fault in the Antalya ophiolite, SW Turkey. In: *The Geological Evolution of the Eastern Mediterranean* (Eds Dixon, J.E. & Robertson, A.H.F.) Spec. Publ. Geol. Soc. London **17**, 319–334.

RICOU, L.E., ARGYRIADIS, I. & LEFEVRE, R. (1974) Proposition d'une origine interne pour les nappes d'Antalya et le massif d'Alanya (Taurides occidentales, Turquie). *Bull. Soc. Géol. Fr.* (7) **16**, 107–111.

RICOU, L.E., ARGYRIADIS, I. & MARCOUX, J. (1975) L'axe calcaire du Taurus, un alignement de fenëtres arabo-africaines sous des nappes radiolaritiques, ophiolitiques et métamorphiques. *Bull. Soc. Géol. Fr.* (7) **17**, 1024–1043.

RICOU, L.E., MARCOUX, J. & POISSON, A. (1979) L'allochtonie des Bey Dağları orientaux. Reconstruction palinspastique des Taurides occidentales. *Bull. Soc. Gél. Fr.* (7) **21**, 125–134.

ROBERTSON, A.H.F. (1981) Metallogenesis on a Mesozoic passive continental margin, Antalya Complex, S.W. Turkey. *Earth Planet Sci. Lett.* **54**, 323–345.

ROBERTSON, A.H.F. (1987) The transition from an Upper Cretaceous foreland basin related to ophiolite emplacement in the Oman Mountains. *Geol. Soc. Am. Bull.* **99**, 633–653.

ROBERTSON, A.H.F. (1990) Microplate tectonics and evolution of the Mesozoic Tertiary Isparta angle, SW Turkey. *IESCA, 1990, International Earth Sciences Congress on Aegean Regions*, 1–6th Oct. 1990, Izmir, Turkey, Abstracts.

ROBERTSON, A.H.F. & SEARLE M.P. (1990) The northern Oman Tethyan continental margin: stratigraphy, structure, concepts and controversies. In: *The Geology and Tectonics of the Oman Region* (Eds Robertson, A.H.F., Searle, M.P. & Ries, A.C.) Spec. Publ. Geol. Soc. London **49**, 3–26.

ROBERTSON, A.H.F. & WALDRON, J.W.F. (1990) In: Geochemistry and tectonic setting of Upper Triassic and Upper Jurassic–Early Cretaceous basaltic extrusives from the Antalya Complex, SW. Turkey. (Eds Sarasin, M.Y. & Eronat, A.H. *Proc. Internat. Aegean Geology, IESCA, Izmir* **2**, 279–299.

ROBERTSON, A.H.F. & WOODCOCK, N.H. (1980) Strike-slip related sedimentation in the Antalya Complex, S.W. Turkey. In: *Sedimentatiuon in Oblique-slip Mobile Zones* (Eds Ballance, P.F. & Reading, H.G.) Int. Assoc. Sedimentol. Spec. Publ. **4**, 127–145.

ROBERTSON, A.H.F. & WOODCOCK, N.H. (1981a) Alakır Çay Group, Antalya Complex, S.W. Turkey: deposition on a Mesozoic passive carbonate margin. *Sediment. Geol.* **30**, 95–131.

ROBERTSON, A.H.F. & WOODCOCK, N.H. (1981b) Bilelyeri Group, Antalya Complex, S.W. Turkey: deposition on a Mesozoic passive continental margin. *Sedimentology* **28**, 381–399.

ROBERTSON, A.H.F. & WOODCOCK, N.H. (1981c) Gödene Zone, Antalya Complex, S.W. Turkey: volcanism and sedimentation on Mesozoic marginal oceanic crust. *Geol. Rundsch.* **70**, 1177–1214.

ROBERTSON, A.H.F. & WOODCOCK, N.H. (1982) Sedimentary history of the south-western segment of the Mesozoic–Tertiary Antalya continental margin, south-western Turkey. *Ecolgae Geol. Helv.* **75**, 517–562.

ROBERTSON, A.H.F. & WOODCOCK, N.H. (1984) Segment of the Antalya Complex, Turkey as a Mesozoic–Tertiary continental margin. In: *The Geological Evolution of the Eastern Mediterranean* (Eds Dixon, J.E., & Robertson, A.H.F.) Spec. Publ. Geol. Soc. London **17**, 251–272.

ROBERTSON, A.H.F., BLOME, C.D., COOPER, D.W.J., KEMP, A.E.S. & SEARLE, M.P. (1990) Evolution of the Arabian continental margin the Dibba Zone, Northern Oman Mountains. In: *The Geology and Tectonics of the Oman*

Region (Eds Robertson, A.H.F., Searle, M.P. & Ries, A.C.) Spec. Publ. Geol. Soc. London **49**, 251–284.

ROBERTSON, A.H.F., CLIFT, P.D., DEGNAN, P.J. & JONES, G. (1991) Palaeogeographic and palaeotectonic evolution of the Eastern Mediterranean Neotethys. *Palaeoclimol. Palaeogeogr. Palaeocecol* **87**, 289–343.

SEARLE, M.P., LIPPARD, S.J., SMEWING, J.D. & REX, D.C. (1980) Volcanic rocks beneath the Semail ophiolite nappe in the northern Oman Mountains and their significance in the Mesozoic evolution of Tethys. *J. Geol. Soc. London* **137**, 589–604.

ŞENEL, Y.M. (1980) Teke Toroslari Güneydogusunun Jeolojisi, Finike–Kumluca–Kemer (Antalya). Unpublished report, M.T.A. Ankara, Turkey, 106 pp.

ŞENGÖR, A.M.C. & Yilmaz, Y. (1981) Tethyan evolution of Turkey: a plate tectonic approach. *Tectonophysics* **75**, 181–241.

THUIZAT, R., WHITECHURCH, H., MONTIGNY, R. & JUTEAU, T. (1981) K–Ar dating of some infra-ophiolitic metamorphic soles from the Eastern Mediterranean: new evidence for oceanic thrustings before obduction. *Earth Planet. Sci. Lett.* **52**, 302–310.

ULSALANER, C. (1945) A preliminary description of the Carboniferous and Devonian fauna discovered in the western Taurides. *Bull. Min. Res. Exp. Inst. Turkey* **25**, 599–603.

WALDRON, J.W.H. (1981) Mesozoic sedimentary and tectonic evolution of the northeast Antalya Complex, Eğridir, S.W. Turkey. Unpublished PhD thesis, University of Edinburgh, 239 pp.

WALDRON, J.W.H. (1984a) Structural history of the Antalya Complex in the "Isparta Angle", S.W. Turkey, In: *Geological Evolution of the Eastern Mediterranean* (Eds Dixon, J.E. & Robertson, A.H.F.) Spec. Publ. Geol. Soc. London **17**, 273–286.

WALDRON, J.W.H. (1984b) Evolution of carbonate platforms on a margin of the Neotethys ocean: Isparta angle, southwestern Turkey. *Ecolgae Geol. Helv.* **77**, 553–582.

WALDRON, J.W.H. (1984c) Antalya Karmaşığı Kuzeydoğu Uza niminin Isparta Bölgesindeki Stratigrafisi ve Sedimanter Evrimi. *M.T.A. Bull.* **97–98**, 1–20.

WOODCOCK, N.H. & ROBERTSON, A.H.F. (1977a) Origins of some ophiolite-related metamorphic rocks of the 'Tethyan' belt. *Geology* **3**, 373–376.

WOODCOCK, N.H. & ROBERTSON, A.H.F. (1977b) Imbricate thrust belt tectonics and sedimentation as a guide to emplacement of part of the Antalya Complex, S.W. Turkey. In: *Sixth Coloquium on the Geology of the Aegean Region.* (Eds Izdar, E. & Nakoman, E.) Piri Reis Int. Contrib. Ser. Publ. no. 2, 661–670.

WOODCOCK, N.H. & ROBERTSON, A.H.F. (1982) Wrench and thrust tectonics along a Mesozoic–Cenozoic continental margin: Antalya Complex, S.W. Turkey. *J. Geol. Soc. London* **139**, 147–163.

YILMAZ, P.O. (1981) Geology of the Antalya Complex, S.W. Turkey. Unpublished PhD thesis, University of Texas. Univ. Microfilms Int. Michigan U.S.A., 194 pp.

YILMAZ, P.O. (1984) Fossil and K–Ar dating for the age of the Antalya Complex, S.W. Turkey. In: *The Geological Evolution of the Eastern Mediterranean* (Eds Dixon, J.E. & Robertson, A.H.F.) Spec. Publ. Geol. Soc. London **17**, 335–348.

YILMAZ, P.O., MAXWELL, J.C. & MUEHLBERGER, W.R. (1981) Antalya kompleksinin yapısal evrimi ve doğu Akdeniz deki' yeri. *Yerbilileri* **7**, 119–127.

Spec. Publs Int. Ass. Sediment. (1993) **20**, 467–480

Growth of a slope ridge and its control on sedimentation: Paola slope basin (eastern Tyrrhenian margin)

A. ARGNANI *and* F. TRINCARDI

Istituto per la Geologia Marina — CNR, Via Zamboni 65, 40127 Bologna, Italy

ABSTRACT

The Tyrrhenian Sea is a Neogene back-arc basin originated within the Africa–Europe continental collision zone and is characterized by the presence of several slope basins. Paola Basin is one of these slope basins and is located on the eastern margin of the Tyrrhenian Sea. It is small (70×20 km), confined, and bounded seawards by a composite slope ridge. The sector of the eastern Tyrrhenian margin where Paola Basin is located presents the thickest Plio-Quaternary sedimentary succession among the peri-Tyrrhenian basins because of its proximity to the rapidly uplifting Calabrian Arc. The sedimentary evolution of this area is broadly divided into three stages by the development of a sedimentary ridge.

The unit deposited during the first stage represents a thick basin-margin wedge which prograded during the Pliocene. This wedge was deformed by a regional contractional episode and detached from its basement. As a result, a sediment ridge was constructed and a slope basin was defined. The second stage unit originated during ridge growth and shows wedging of reflectors.

The infilling of the slope basin represents the last stage of evolution and shows a greater variety of seismic facies compared with the two previous stages. A suite of sedimentary bodies with facies typical of coarse-grained deposits occurs at the base of slope and indicates multiple proximal sediment sources along the basin margin that shifted in position through time. Channelized and mounded deposits, as well as ponded distal turbidites, fill morphological depressions while, at the same time, morphological highs are mantled by thick sedimentary drapes. This variability in facies reflects the complex interaction of the steepening of the basin margin, due to contraction, of the topographical confinement exerted by the ridge, and of the onset of Quaternary sea-level fluctuations.

INTRODUCTION

Slope ridges and basins are fairly common features along various types of continental margins. They exert a major control on the physiography and the sedimentary style of the margin where they occur. In fact, the growth of slope ridges and basins affects in various ways the gradient of the slope, the sediment delivery routes, and the topography of depositional areas. However, the sediment fill of slope basins records their evolution and can be used to retrace the depositional history of the margin.

This paper aims at presenting the sedimentary evolution of the Paola slope basin, which originated along part of the margin of a young back-arc basin, the Tyrrhenian Basin (Fig. 1). This back-arc basin is one of the extensional regions that originated in the

Mediterranean area during the convergence between Africa and Europe (Horvath & Berckhemer, 1982).

Back-arc-related extensional tectonics gave rise to the Tyrrhenian rifting during the Late Tortonian (Malinverno & Ryan, 1986; Trincardi & Zitellini, 1987; Sartori, 1990). Since then, extension continued until Late Pleistocene, propagating southeastward and creating two areas with an extremely thinned crust, possibly floored by oceanic crust (Kastens *et al.*, 1988; Kastens & Mascle, 1990). All along its margins, the Tyrrhenian Sea is surrounded by slope basins showing different structural style and sedimentary architecture, the so-called peri-Tyrrhenian basins (Fabbri *et al.*, 1981). Some of

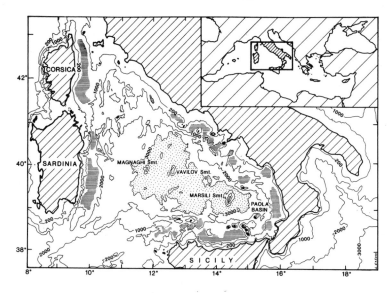

Fig. 1. Bathymetry of the eastern Tyrrhenian Sea margin and location of Paola slope basin. Dashed pattern indicates the major slope basins existing on the margin. Dotted pattern refers to the Tyrrhenian Sea bathyal plain.

these slope basins originated by block-faulting while others are bounded by magmatic intrusions or volcanic structures.

Among the peri-Tyrrhenian basins, Paola Basin (Fig. 1) presents the thickest Plio-Quaternary sedimentary fill (up to 4800 m; Barone *et al.*, 1982). The basin lies adjacent to the rapidly uplifted Calabrian Arc composed of stacked slices of continental crystalline basement, ophiolites and metamorphosed sedimentary rocks thrust onto the southern margin of the Alpine–Apennine system

(Amodio-Morelli *et al.*, 1976; Knott, 1987; Dietrich, 1988). The physiography of the basin (Fig. 2) is characterized by a very steep shelf (0°50′) and upper slope (up to 6°) well over the world average (0°07′ and 4° respectively; Shepard, 1963) and by the presence of a system of slope ridges separating the basin from a less steep, lower slope (Gallignani, 1982; Canu & Trincardi, 1989; Argnani & Trincardi, 1990). At the southern end of the basin the ridge system fades out into a broad terrace that is deeply incised by the Angitola Canyon, a major

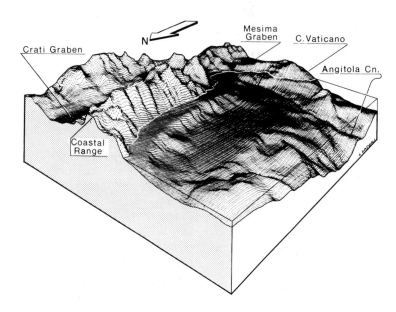

Fig. 2. Mesh diagram of the bathymetry of Paola Basin and of the topographical relief of the adjacent mainland. The system of slope ridges bordering the basin is evident as is the deep Angitola Canyon. The most important tectonic features dissecting the Calabrian Arc are also shown.

canyon that connects the shelf to the Tyrrhenian basin plain. This area is the offshore continuation of the Catanzaro Graben and presents structural trends running west-southwest–east-northeast.

Based on a 2500-km network of mostly single channel sparker profiles, the stratigraphical and structural evolution of this segment of the eastern Tyrrhenian margin (ETM) is outlined. In addition, the sedimentary architecture of Paola slope basin is presented. Sparker profiles (1 kJ) were fired every 2 s, recorded with a 2-s sweep and band-pass filtered typically between 200 and 600 Hz (Trincardi *et al.*, 1988). The sparker profiles were run jointly with 3.5-kHz echo-sounding profiles and navigation was provided by a Loran C hyperbolic positioning system. In particular, the effects of the growth of the slope ridge system on the sedimentary style will be discussed.

EVOLUTION OF THE MARGIN AND GROWTH OF THE SLOPE BASIN

Seismic data

Through interpretation of seismic profiles the gross geometry of the margin has been identified. Profile MS-1 (Fig. 3) runs across the basin where the system of slope ridges is best developed, and provides a typical picture of this margin (Argnani & Trincardi, 1990). The main features to be observed are:
1 the marked landward-dipping reflector (labelled Y) that corresponds to the Messinian unconformity and represents the substrate of the sedimentary wedge of the margin;
2 the topography of the ridge system, still well-expressed on the sea floor, that is unrelated to the trend of the basin substrate and that confines upslope the flat area of Paola Basin; and
3 the external shape and internal reflection characters of the three seismic units identified.

The lower unit is poorly imaged but shows a pattern of almost parallel reflections that appears conformable with the upper boundary and downlapping onto the lower one. Its thickness distribution suggests that its external shape, now broadly folded towards the basin margin, was originally that of a landward-thickening wedge. The intermediate unit is mostly confined underneath Paola Basin where it displays an upward-concave lens shape. Its internal reflections converge both towards the slope ridges and the basin margin (Fig. 3). Finally, the upper unit, with its almost horizontal reflections onlapping onto slope ridges and the basin margin, reflects the filling of the slope basin. The details of this sedimentary filling are the object of the following sections.

Age of units

No data are available to constrain the age of the seismic units because of the lack of wells in the study area. A rough estimate can be made, however, based on the average sedimentation rate (0.9 mm/ year). Such a rate has been obtained from the thickest part of the sedimentary succession (4300 m) considering that the sedimentary fill covers the whole of the Plio-Quaternary (time span of 4.8 My). In this way we obtained a Pliocene age for

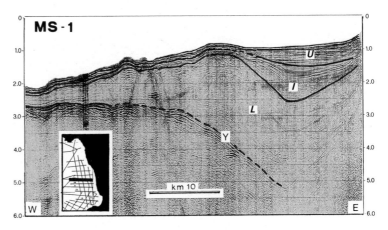

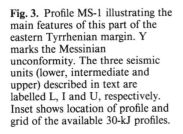

Fig. 3. Profile MS-1 illustrating the main features of this part of the eastern Tyrrhenian margin. Y marks the Messinian unconformity. The three seismic units (lower, intermediate and upper) described in text are labelled L, I and U, respectively. Inset shows location of profile and grid of the available 30-kJ profiles.

the lower unit, an early Pleistocene age for the intermediate unit and a Middle Pleistocene to present age for the upper unit. Therefore, the intermediate and upper units deposited during the formation of the adjacent Marsili Basin (Kastens & Mascle, 1990). As sedimentation rate is a function of relief of the source area, and knowing that the uplift rate of the Calabrian Arc increased during the Pleistocene (Ghisetti, 1981), we believe that the ages attributed to the three seismic units are the oldest possible.

Recent extension

A situation similar to that seen on profile MS-1 is observed all along the basin (Gallignani, 1982; Canu & Trincardi, 1989; Argnani & Trincardi, 1990). In some places, however, extensional faults post-date and cut through the ridge system and affect also the upper unit (Fig. 4). These faults are mostly developed along the lower slope.

Evidence of contraction

The trend of the slope ridges is shifted with respect to the axis of the basin substrate. This suggests that there is some sort of decoupling within the sedimentary succession, although the poor resolution of the

seismic data does not allow image any clear structure. Given the lack of diapiric features, we believe that the ridges are most likely shale-cored anticlines due to contraction (Argnani & Trincardi, 1990). Similar features have been reported from the Gulf of Mexico where they originated because of gravitational spreading of soft sediments and where they are connected with extensional structures usually in the form of growth faults (Winker & Edwards, 1983). No such extensional faults have been observed in this segment of the eastern Tyrrhenian margin. Furthermore, the wedging of internal reflections within the intermediate unit, which was deposited during the growth of the ridge system, suggests that the basin margin too was relatively uplifted at the same time. It is during this span of time that the lower unit was synclinally folded and that the slope basin originated. Furthermore, the broad terrace bounding the basin to the south originated because of uplift and inversion of a previous local depocentre (Canu & Trincardi, 1989); the timing of this uplift coincided with the growth of the ridges. It follows that the contraction responsible for the growth of the slope ridges affected a wide segment of the eastern Tyrrhenian margin (Argnani & Trincardi, 1990). The late extensional episode is connected with Quaternary subsidence documented for the adjacent Marsili Basin and only marginally affected Paola Basin.

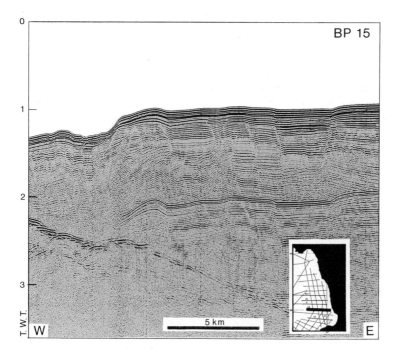

Fig. 4. Detail of profile BP 15 showing extensional faults affecting and post-dating the sedimentary slope ridge. Inset shows location of profile and grid of other 30-kJ profiles.

Evolution of the margin

The evolution of this part of the eastern Tyrrhenian margin is schematically represented in Fig. 5. Stage I (Fig. 5a) was characterized by deposition of a sedimentary wedge along the young Tyrrhenian margin during the Pliocene. A regional contractional episode affected the margin, possibly in the early Pleistocene, and produced a roughly east–west-oriented shortening (Fig. 5b). Within the sedimentary prism one or more décollement surfaces

originated giving rise to a system of anticlines. Such anticlines are best developed in areas where a thick sediment pile accumulated during Stage I and formed the slope ridges confining Paola Basin. During the final stage (Fig. 5c) the slope basin was progressively filled and only partly affected by extensional faulting that mostly occurred on the lower slope.

The contractional episode affected the sedimentary history of the margin in various ways. Firstly, it created the slope basin and its topographical confinement; secondly, it steepened the slope, in particular the upper slope, and the shelf up to values much larger than the average along the Tyrrhenian margins; and thirdly, it contributed to the elevation of the adjacent sediment source area. All these effects are recorded, to some extent, in the sedimentary fill of the slope basin, which will be addressed in the following section.

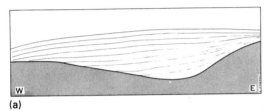

(a)

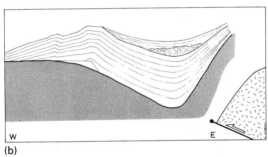

(b)

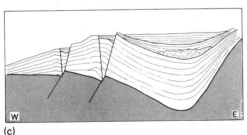

(c)

Fig. 5. Stratigraphical and structural evolution of the part of the eastern Tyrrhenian margin where Paola Basin is located. Stage I (a) depicts a sedimentary wedge of the extensional margin not affected by synsedimentary tectonics and corresponding to the lower unit of profile MS-1. Approximate age interval covers the whole of the Pliocene. During Stage II (b) the margin suffered a regional contraction that resulted in a build-up of the ridge system and in creation of the slope basin. Estimated time interval is Early Pleistocene. Stage III (c): contraction stops, slope basin is progressively filled up and extensional faulting affects the lower part of the margin.

PAOLA BASIN SEDIMENTARY FILL

In this section we examine the effects of the growth of the slope ridges on the lateral distribution, internal organization and stacking pattern of depositional elements derived by turbiditic and mass-transport processes.

Because of its steep gradient and its proximity to the clastic sources, the upper slope is an area of degradation and/or sediment bypass where slump and slide scars, often of great lateral extent and size are common, as well as shallow erosional gullies (Canu & Trincardi, 1989). Small levee wedges grow at the base of the upper slope on the basinward termination of the gully systems.

Sedimentation in the basin appears, in general, to have been dominated by the presence of large amounts of fine-grained sediment, despite the fact that the source region is located close to the basin and dominated by short and high-relief drainage systems. The large amount of fine-grained sediment reaching the basin is indicated by the deposition of thick draping units that mantle the morphological relief. Facies and geometries indicative of coarser-grained sediment are restricted to the base-of-slope area. Sand content decreases rapidly away from the base of the upper slope and directly reflects the location of sediment entry points. This character is expressed on 3.5-kHz echo-sounding profiles by a systematic distal increase in acoustic penetration

(Gallignani, 1982), and on 1-kJ sparker profiles by a concurrent decrease in reflections amplitude and complexity of reflections geometry (Trincardi *et al.*, 1990). The area of deposition can be subdivided into two major zones: the base of slope, characterized by a dominance of high-amplitude discontinuous reflections taken to indicate larger amounts of sand, and the distal confined basin, where reflections show good lateral continuity and lower amplitudes. The location of the transition between these two areas shifted through time and primarily reflects changes in the volume of coarse-grained sediments delivered to the basin.

Base-of-slope area

Base-of-slope deposits show, on seismic reflection profiles, a dominance of high-amplitude discontinuous reflections associated with channelised and/or mounded external geometry. In this section we describe briefly the seismic and morphological character of the most significant depositional elements that occurred in the base-of-slope area, where they are thick enough to be resolved. The recognition of basin-wide key reflectors allowed us to establish the relative timing of the base-of-slope depositional elements or stacks of elements (Trincardi *et al.*,

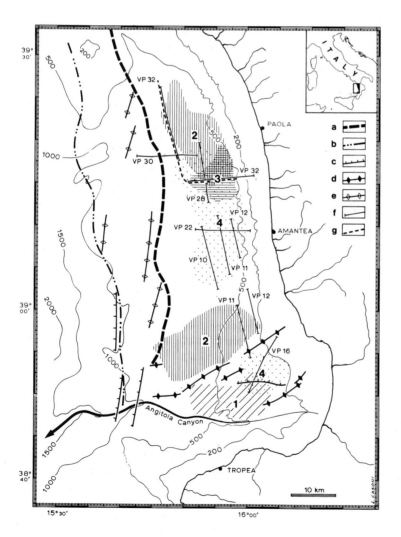

Fig. 6. Location and relative timing of major base-of-slope deposits. (Modified after Canu & Trincardi, 1989.) Numbers indicate relative ages from older to younger. (a) Western limit of basin; (b) axis of basement high; (c) extensional faults; (d) fold axis in the area affected by basin inversion; (e) slope ridges; (f) location of 1-kJ profiles shown in Figs 7–16.

1990). The location and timing of the most-developed base-of-slope deposits are schematically shown in Fig. 6; it appears that, within the upper unit, deposition was controlled by numerous entry points active along the margin.

1 *Depositional mounds* showing bidirectional downlap of internal reflections are located in the area of major change in gradient at the base of the upper slope (Fig. 7). Overlapping mounded reflection packages have a convex upper surface and a non-erosive base. They are small-sized (typically a few km^2), with and have high relief on the sea floor. Numerous shallow channels dissect the mounds in their proximal termination (Fig. 8a). In a few kilometres from the upper slope, channels disappear, reflection continuity increases and the convexity of the upper surface becomes less prominent (Figs 8b & 8c). A profile perpendicular to the basin axis and crossing the same mounded base-of-slope deposits summarises the above observations showing the rapid change in internal geometry of a stack of mounds toward the basin center (Fig. 9). Depositional mounds commonly occur in turbidite sys-

tems on the downslope termination of major feeding channels and represent the locus where the highest amount of sand accumulates (Mitchum, 1985; Bouma *et al.*, 1989).

2 *Channel-levee complexes* are characterized by low-amplitude reflections converging with a low-angle away from the channnel axis (Fig. 11). Channels are typically small-sized and are probably ephemeral features (Fig. 8a); exceptions are found in areas influenced by feeding systems of larger size and/or more persistent through time (Fig. 10); in these cases levee wedges several tens of metres thick are observed. Profile VP 12 (Fig. 11) crosses the upper slope close to the apex of the channel-levee complex of Fig. 10. A complicated pattern of reflections is given by the growth of multiple levee wedges and by their internal deformation induced or influenced by differential compaction of sediments. Channel-levee complexes tend to show two stages of evolution; the first is dominated by the deposition of a 'core' of coarse sediment characterized by high-amplitude discontinuous reflections encased in low-amplitude continuous reflections packages. This stage is accompanied by the growth of the levee wedges. The following stage is dominated by the progressive smoothing of the relief of the levee by the filling of the channel with high-amplitude continuous onlapping reflections. Comparable channel-levee complexes are found in the Quaternary record on continental margins under the influence of major rivers; examples from the Mediterranean are the Rhone fan (Droz & Bellaiche, 1985), the Ebro margin (O'Connell *et al.*, 1987; Field & Gardner, 1990), and, although on a smaller scale, the Crati fan (Ricci Lucchi *et al.*, 1983–84).

3 *Mass-transport deposits* are characterized by a hummocky upper surface and chaotic to transparent internal reflection patterns (Figs 8c & 12). Relief on the upper surface of such deposits is higher in the proximal areas where rotated slump blocks can be encountered (Fig. 12, base of slope). Examples of possible mudflows of basin-wide extent within Paola Basin are discussed in Gallignani (1982). The largest slide deposit within the upper sequence is about 220 km^2 in extent; it shows evidence of remoulding of sea-floor sediment during and following the slide emplacement as well as a component of 'upslope' transport against the morphological barrier produced by the pre-existing slope ridge (Trincardi & Normark, 1989; Fig. 8c). Mass-transport deposits represent a significant

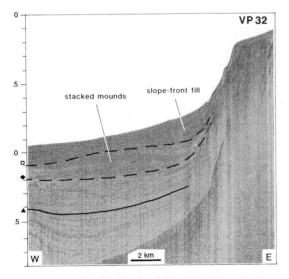

Fig. 7. One-kilojoule sparker profile VP 32 (location in Fig. 6) perpendicular to the basin axis across the base-of-slope area. Note the updip convergence of reflections of the intermediate unit (below reflector marked with triangle toward the upper slope as well as the growing slope ridge. Within the filling sequence, base-of-slope mounds are onlapped, on their basinward termination, by distal confined deposits characterized by more continuous reflection packages. The triangle, diamond and square mark key reflectors.

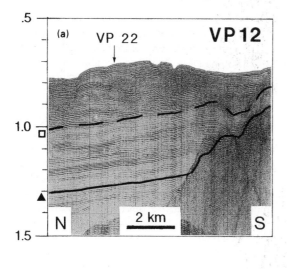

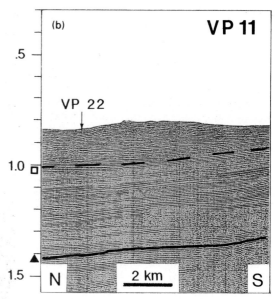

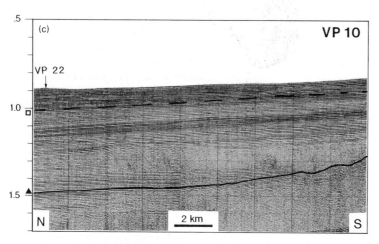

Fig. 8. One-kilojoule sparker profiles (VP 10–12) cutting a base-of-slope mound (location in Fig. 6), from proximal (a) to distal (c). Note the decrease in complexity of reflection packages away from the source. The proximal portion of the deposit is dissected by channels connected to the gully system occurring on the upper slope. Symbols as in Fig. 7.

contribution to the basin fill, as observed in other slope basins such as those of the California Borderland (Field, 1981; Nardin *et al.*, 1982) or those in the Gulf of Mexico (Bouma, 1981).

Base-of-slope depositional elements, where best developed, tend to show repetitive stacking patterns (Canu & Trincardi, 1989): mounded deposits develop on basin floor plane-parallel deposits and are covered, in their landward termination by younger channel-levee complexes. When abandoned, channel-levee complexes are draped and partly filled by deposits characterized by continuous plane-parallel reflection packages and tend to decrease their mor-

phological expression on the sea floor. Mass-transport deposits have highly variable volumes and lateral extents, and their vertical distribution within the filling sequence appears more random than in the case of the other elements.

Distal confined basin

This area is characterized by plane-parallel reflections of good lateral continuity and low amplitude (Fig. 13). Reflection packages of this kind onlap against the slope ridge as well as the more proximal base-of-slope deposits indicating confinement and

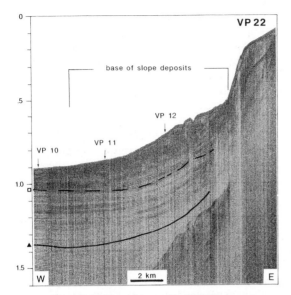

Fig. 9. One-kilojoule sparker profile (VP 22) perpendicular to those of Fig. 8 and to the regional contours (location in Fig. 6). The lower part of the profile shows evidence of reflection convergence towards the upper slope. Symbols as in Fig. 7.

a significant component of longitudinal disperal along the axis of the basin (Figs 13 & 14). With respect to those observed in the base-of-slope area, distal confined deposits reflect a higher content of fine sediments probably deposited by diluted turbidity flows. Interpretation of 3.5-kHz echo-

sounding profiles however, shows evidence, of mass-transport deposits of basin-wide extent but with thickness at the limit of the instrument resolution (Fig. 12). These deposits may contribute significantly to sedimentation in the distal and confined parts of the basin and may be undetectable in the seismic pattern typical of fine-grained turbidites.

A complex pattern of reflections onlapping against and draping onto the slope ridge results from morphological confinement of gravity flows (mostly fine-grained 'distal' turbidites). Gravity flow deposits tend to fill topographical depressions, but the large amount of fine sediments delivered to the basin contributes to the preservation of topographical relief through hemipelagic draping (Figs 12–14).

DISCUSSION

During the Quaternary the ETM underwent an episode of contraction which caused the development of a system of slope ridges and slope basins (Argnani & Trincardi, 1990). The slope ridges which define the basins have variable origins and times of formation; in general, however, the filling sequences that accumulated after the growth of such barriers became more complex from north to south. This observation probably reflects the decreasing distance from the Apennine chain and the increasing uplift rates that characterize the chain along the

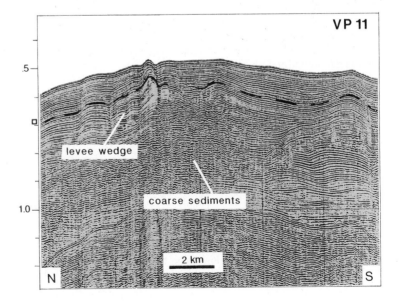

Fig. 10. Particulars of channel-levee complex. One-kilojoule sparker profile (VP 11) located in Fig. 6. Note the asymmetry of the levees, the right-hand one being more developed, and the coarse-grained deposits characterizing the early stage of development of the system. Square as in Fig. 7.

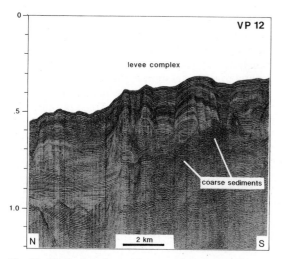

Fig. 11. Coalescing channel-levee complexes on the upper slope affected by shallow deformation probably induced by differential compaction. One-kilojoule sparker profile (VP 12). For location see Fig. 6.

Calabrian Arc (up to 0.8 mm/year for the southern sector of the arc; Ghisetti, 1981).

The change in style of deposition induced by the growth of the slope ridges is particularly evident in the case of Paola Basin, where the average sedimentation rate through the Plio-Quaternary was high and the distance from the uplifting chain minimal. Tectonic deformation influenced the growth of depositional systems in at least two ways: firstly, it affected the source area by controlling lateral variation in the rates of uplift, as well as the arrangement of drainage systems; and secondly, it affected the area of deposition by inducing changes through time of basin size and geometry, as well as causing the overall 'immature' physiography of the margin.

Based on a gross estimate of its age, it appears that the upper unit, which represents the filling of the slope basin, was concurrent with glacially-induced Quaternary sea-level fluctuations. The overall physiography of the margin, and the activity of multiple point sources suggest that the effects of sea-level fluctuations are probably less evident than in cases where the turbidite systems were fed by major rivers and located on mature passive margins. In this tectonically active setting, the signals of the topographically-enhanced sedimentary dynamics interfere with the eustatic signature. The influence of Quaternary sea-level changes on bathyal-plain deposition was documented in the Vavilov and Marsili basins by the ODP Leg 107 results (Robertson et al., 1990). It is suggested that sediment remains trapped in the peri-Tyrrhenian basins and shelf areas during relative high stands and is transported downslope during relative sea-level falls and low stands (Robertson et al., 1988). Features like the steep and narrow shelf and the steep slope make it unlikely that there was storage of large volumes of sediment on the shelf during relative sea-level high stands and the successive removal of sediment to the basin during relative sea-level fall and low stands (Ricci Lucchi, 1985). This fact is likely to cause distortions in the classic systems tracts arrangement expected during sea-level fluctuations (Posamentier & Vail, 1988), although it is difficult to unravel the interplay between sedimentary dynamics and eustasy.

Many parameters control the growth of fan and non-fan systems along continental margins (Stow et al., 1983–84; Mutti & Normark, 1987). Among these the volume, frequency and grain size of sedimentary inputs, the size and the gradients of the receiving basin and the confinement effects exerted

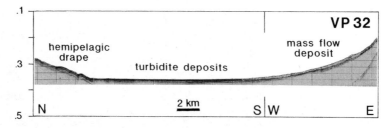

Fig. 12. 3.5-kHz echo-sounding profile located in the northern portion of the basin (Fig. 6). Transparent deposit with hummocky upper surface on the base of slope is interpreted as having resulted from mass-transport processes. Note that strong returns accompanied by low energy penetration (coarser deposits) onlap against the morphological barrier, which grows through the accumulation of finer-grained draping deposits. This profile was shot along with the sparker profile shown in Figs 7 and 14.

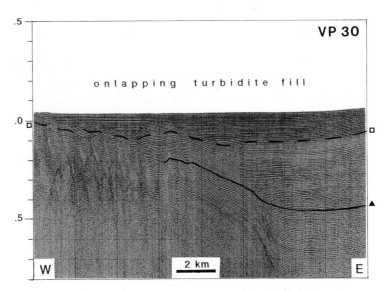

Fig. 13. Profile VP 30 (1-kJ sparker) crossing the basin (location in Fig. 6) and showing the diffuse faulting that affects the crest of a slope ridge buried by deposits of the upper unit. Distal confined deposits are characterized by laterally-continuous, low-amplitude, plane-parallel reflections. These deposits interfinger with draping deposits on the morphological highs through an area of alternating onlapping terminations and draping reflections. Distal deposits also onlap onto the depositional relief that dominates the base-of-slope area. Symbols as in Fig. 7.

on gravity flows by the sea-floor morphology are considered to be relevant ones. Given the rather small size of the Paola slope basin the effect of confinement upon gravity flows is expected to be maximized. However, because of the steep gradients, the narrow shelf and the many small rivers forming the drainage system, it is likely that sediment input came from small rather than large volume entry points. As a consequence, large and confined base-of-slope deposits are not the rule.

The lateral extent of each of the depositional

elements described above depends on the volume of sediment that characterizes each depositional event (e.g. basin-wide slide or mass-transport deposit versus local failure at the base of the upper slope) as well as the persistence through time of any given feeding system. Within Paola Basin, most of the base-of-slope deposits are constructed by depositional elements of small size; as a consequence their geometry is affected less severely by the growth of the slope ridges (Figs 7 & 9).

An exception to this general observation is in the

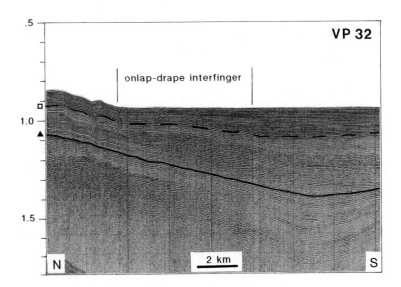

Fig. 14. Distal confined deposits seen on a 1-kJ sparker profile (VP 32) parallel to the axis of the basin. Location in Fig. 6, symbols as in Fig. 7.

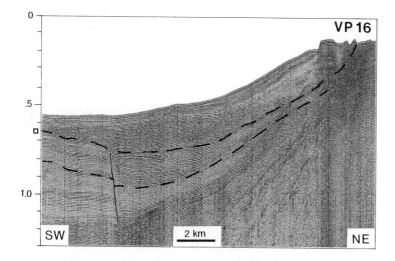

Fig. 15. One-kilojoule sparker profile (VP 16) from the southern portion of the study area (location in Fig. 6) showing base-of-slope deposits stacked within a morphologically confined area. In this case deposition of coarser-grained sediments filled the morphological low and partly buried the nearby morphological high. Square as in Fig. 7.

southern part of the study area where deformation related to the activity of the Catanzaro graben structure originated a localized sub-basin less than 20 km^2 in extent (Fig. 15). The basin is no longer evident on a bathymetric map of the present-day sea floor. It was filled with base-of-slope deposits, arranged in a complex pattern of stacked mounds. Two episodes of filling can be distinguished on the basis of different styles of deposition. A lower part, characterized by mostly continuous reflections, mimics the onlapping/draping patterns observed against the slope ridge system in the rest of the basin (Fig. 15). A thick sequence of plane-parallel reflections, due to draping of the distal high, maintains

the relief along the edge of the basin and thus the conditions favourable to induce ponding of successive deposits. The upper part is dominated by mounded deposits which gradually fill the morphological low and then 'prograde' onto the distal high; draping by fine-grained deposits during this stage cannot compete with coarser-grained sedimentation and fails to maintain the ridge morphology (Fig. 15).

Base-of-slope deposits tend to compensate pre-existing morphological lows induced on the sea floor by the occurrence of older base-of-slope deposits (mostly depositional mounds). This kind of confinement of sediment gravity flows is exerted by

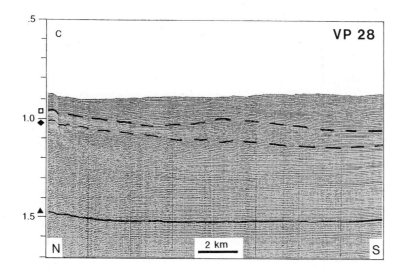

Fig. 16. Compensation cycles within base-of-slope deposits. Morphological lows formed between depositional mounds and controlled the placement of younger base-of-slope deposits. One-kilojoule Sparker profile (VP 28) located in Fig. 6. Symbols as in Fig. 7.

older depositional relief on the sea floor and takes place on a smaller scale but is equally important in controlling distribution, thickness and morphology of some of the base-of-slope deposits (Fig. 16).

CONCLUSIONS

Paola Basin is that sector of the eastern Tyrrhenian margin where the thickest sedimentary section was deposited during the Plio-Quaternary. Three stages of development of the margin were recorded by the three major depositional sequences of the Plio-Quaternary section. The lower unit is a thick sedimentary wedge deposited during Stage-I extension and subsidence of the margin. The intermediate unit was deposited during a regional episode of tectonic contraction (Stage II) which affected the margin during the early Pleistocene. The effects of this process are more evident in the areas where a thick section accumulated during Stage I and resulted in the growth of the slope ridges on the margin. As a consequence, a slope basin originated and the upper unit represents the filling of such a basin (Stage III). The sedimentary fill of this basin has been investigated in detail, using high-resolution seismic data, in an attempt to understand the relationships between tectonics and sedimentation. Increased uplift rates on the mainland and steepening of topographical gradients favoured an increase in sedimentation rates and promoted a high variability of depositional geometries. During Stage III, the basin was fed by multiple point sources derived from the unstable uplifting margin. These sources were ephemeral in nature and correspond to the numerous drainage systems presently dissecting the Calabrian Arc.

Coarser-grained, base-of-slope, depositional elements, in particular, do not appear to have been subject to recurrent confinement throughout the entire sequence because of their small volume relative to the size of the basin. Depositional elements typical of submarine fans, such as channels, levee wedges and channel-mouth mounds, can be recognized. These have morphological expression on the sea floor and wedge out before reaching the slope ridge. The distribution and internal geometry of distal finer-grained deposits show a more evident confinement against morphological features. Such deposits result from diversion of flows along the basin axis induced by the slope ridges and other tectonic barriers of more local extent. Finally, the

great amounts of fine-grained sediment delivered to this small topographically-confined basin leads to a tendency to preserve the morphological highs by the deposition of thick packages of hemipelagic, draped sediment. This process competes with the gradual infill of morphological lows by a suite of gravity-driven, base-of-slope and distal confined deposits.

ACKNOWLEDGEMENTS

The authors acknowledge Alastair Robertson and Ron Steel for their thoughtful reviews of the manuscript and Luciano Casoni for preparing the figures. This is contribution no. 933 of the Instituto per la Geologia Marina.

REFERENCES

AMODIO-MORELLI, L., BONARDI, G., COLONNA, V. *et al.*, (1976) L'Arco Calabro-Peloritano nell'orogene Apenninico-Maghrebide. *Mem. Soc. Geol. It.* **17**, 1–60.

ARGNANI, A. & TRINCARDI, F. (1990) Paola slope basin: evidence of regional contraction on the Eastern Tyrrhenian Margin. *Mem. Soc. Geol. Ital.* **44**, 93.

BARONE, A., FABBRI, A., ROSSI, S. & SARTORI, R. (1982) Geological structure and evolution of the marine areas adjacent to the Calabrian Arc. *Earth Evolution Sci.* **3**, 207–221.

BOUMA, A. (1981) Depositional sequences in clastic continental slope deposits, Gulf of Mexico. *Geo-Mar. Lett.* **1**, 115–121.

BOUMA, A.H., COLEMAN, J.M., STELTING, C.E. & KOHL, B. (1989) Influence of relative sea level change on the construction of the Mississippi Fan. *Geo-Mar. Lett.* **9**, 161–170.

CANU, M. & TRINCARDI, F. (1989) Controllo eustatico e tettonico sui sistemi deposizionali nel bacino di Paola (Plio-Quaternario), margine tirrenico orientale. *Geo. Geol.* **51/2**, 41–61.

DIETRICH, D. (1988) Sense of overthrust shear in the Alpine nappes of Calabria (Southern Italy). *J. Struct. Geol.* **10**, 373–381.

DROZ, L. & BELLAICHE, G. (1985) Rhone deep-sea fan: morpho-structure and growth pattern. *Am. Ass. Petrol. Geol. Bull.* **69/1**, 460–479.

FABBRI, A., GALLIGNANI, P. & ZITELLINI, N. (1981) Geologic evolution of the peri-Tyrrhenian sedimentary basins. In: *Sedimentary Basins of Mediterranean Margins* (Ed. Wezel, F.C.) Tecnoprint, Bologna, pp. 101–126.

FIELD, M.E. (1981) Sediment mass transport in basins: control and patterns. In: *Depositional Systems of Active Continental Margin Basins* (Eds Douglas, R.G., Colburn, I.P. & Gorsline, D.S.) Soc. Econ. Paleont. Miner. Pacific Section, Short Course Notes, pp. 61–83.

FIELD, M.E. & GARDNER, J.V. (1990) Pliocene–Pleistocene growth of the Rio Ebro margin, northeast Spain: A prograding-slope model. *Geol. Soc. Am. Bull.* **102**: 721–733.

GALLIGNANI, P. (1982) Recent sedimentation processes on the Calabrian continental shelf and slope (Tyrrhenian Sea, Italy). *Oceanol. Acta* **5**, 493–500.

GHISETTI, F. (1981) Evoluzione neotettonica dei principali sistemi di faglie della Calabria Centrale. *Boll. Soc. Geol. Ital.* **98**, 387–430.

HORVATH, F. & BERCKHEMER, H. (1982) Mediterranean back arc basins. In: *Alpine–Mediterranean Geodynamics* (Eds Berckhemer, H. & Hsu, K.) Am. Geophys. Union Geodyn. Ser. **7**, 141–173.

KASTENS, K. & MASCLE, J. (1990) The geological evolution of the Tyrrhenian Sea: an introduction to the scientific results of ODP Leg 107. In: *Proc. ODP, Sci. Results, 107* (Eds Kastens, K.A., Mascle, J. *et al.*) College Station, TX, pp. 3–26.

KASTENS, K., MASCLE, J., AUROUX, C. *et al.* (1988) ODP Leg 107 in the Tyrrhenian Sea: insights into passive margin and back-arc basin evolution. *Geol. Soc. Am. Bull.* **100**, 1140–1156.

KNOTT, S.D. (1987) The Liguride Complex of Southern Italy — a Cretaceous to Paleogene accretionary wedge. *Tectonophysics* **142**, 217–226.

MALINVERNO, A. & RYAN, W.B.F. (1986) Extension in the Tyrrhenian Sea and shortening in the Apennines as a result of arc migration driven by sinking of the lithosphere. *Tectonics* **5**, 227–245.

MITCHUM, JR, R.M. (1985) Seismic stratigraphic expression of submarine fans. In: *Seismic Stratigraphy II: an Integrated Approach to Hydrocarbon Exploration* (Eds Berg, O.R. & Woolverton, D.G.) Am. Ass. Petrol. Geol. Mem. **39**, 117–138.

MUTTI, E. & NORMARK, W.R. (1987) Comparing examples of modern and ancient turbidite systems: problems and concepts. In: *Marine Clastic Sedimentology* (Eds Leggett, J.K. & Zuffa, G.G.) 1–38.

NARDIN, T.R., EDWARDS, B.D. & GORSLINE, D.S. (1982) Santa Cruz basin, California borderland: dominance of slope processes in basin sedimentation. In: Geology of Continental Slopes (Eds Doyle, L.J. & Pilkey, O.H.) *Soc. Econ. Paleont. Miner. Spec. Publ.* **27**, 209–221.

O'CONNELL, S., RYAN, W.B.F. & NORMARK, W.R. (1987) Modes of development of slope canyon and their relation to channel and levee features on the Ebro sediment apron off-shore northeastern. Spain. *Marl. Petrol. Geol.* **4**, 308–319.

POSAMENTIER, H.W. & VAIL, P.R. (1988) Eustatic controls on clastic deposition II — sequence and system tract models. In: *Sea-level Change — an Integrated Approach* (Eds Wilgus, C.K., Hastings, B.S., Kendall, C.G.St.C., Posamentier, H.W., Ross, C.A. & Van Wagoner, J.C.) Soc. Econ. Paleont. Miner. Spec. Publ. **42**, 125–154.

RICCI LUCCHI, F. (1985) Influence of transport processes and basin geometry on sand composition. In: *Provenance of Arenites* (Ed. Zuffa, G.G.), NATO ASI Series, Reidel Publishing, Dordresht, pp. 19–45.

RICCI LUCCHI, F., COLELLA, A., GABBIANELLI, G., ROSSI, S. & NORMARK, W.R. (1983–1984) The Crati Submarine Fan, Ionion Sea. *Geo-Mar. Lett.* **3**, 71–77.

ROBERTSON, A., HIEKE, W., MASCLE, G. *et al.* (1990) Summary and synthesis of Late Miocene to recent sedimentary and paleoceanographic evolution of the Tyrrhenian Sea, Western Mediterranean: Leg 107 of the ODP. In: *Proc. ODP, Sci. Results, 107* (Eds Kastens, K.A., Mascle, J. *et al.*), College Station, TX, pp. 639–663.

SARTORI, R. (1990) The main results of ODP Leg 107 in the frame of Neogene to Recent geology of peri-Tyrrhenian areas. In: *Proc. OPD, Sci. Results, 107* (Eds Kastens, K.A., Mascle, J. *et al.*), College Station, TX, pp. 715–730.

SHEPARD, F.P. (1963) *Submarine Geology*, 2nd Edn. Harper and Row, New York, 557 pp.

STOW, D.A.V., HOWELL, D.G. & NELSON, C.H. (1983–1984) Sedimentary, tectonic and sea level controls on submarine fan and slope-apron turbidite system. *Geo-Mar. Lett.* **3**, 57–64.

TRINCARDI, F. & NORMARK, W.R. (1989) Suvero Pleistocene slide, Paola basin, Southern Italy. *Mar. Petrol. Geol.* **6**, 324–335.

TRINCARDI, F. & ZITELLINI, N. (1987) The rifting of the Tyrrhenian basin. *Geo-Mar. Lett.* **7**, 1–6.

TRINCARDI, F., CIPOLLI, M., FERRETTI, P. *et al.* (1988) Slope basin evolution on the Eastern Tyrrhenian Margin: preliminary report. *Giorn. Geol.* **42**, 1–16.

TRINCARDI, F., CANU, M., FIELD, M.E. & NORMARK, W.R. (1990) Seismic facies and basin history related to sea level cycles and tectonism: Paola slope basin, Eastern Tyrrhenian Margin. *Am. Ass. Petrol. Geol. Annual Convention, June 2–6, 1990*, p. 184.

WINKER, C.D. & EDWARDS, M.B. (1983) Unstable progradational clastic shelf margins. In: *The Shelfbreak: Critical Interface on Continental Margins* (Eds Stanley, D.J. & Moore, G.T.) Soc. Econ. Paleont. Miner. Spec. Publ. **33**, 139–157.

Spec. Publs Int. Ass. Sediment. (1993) **20**, 481–497

Sedimentation and tectonics of the Khomas Hochland accretionary prism, along a Late Proterozoic active continental margin, Damara Sequence, central Namibia

P.A. KUKLA* *and* I.G. STANISTREET

Department of Geology, University of the Witwatersrand,
Private Bag 3, Johannesburg 2050, RSA

ABSTRACT

The Khomas Trough in central Namibia is one of the basins that was filled during sedimentation of the Late Proterozoic Damara Sequence (750–570 Ma). The basin that was filled by a thick sequence, now preserved as metagreywackes and pelites of the Kuiseb Formation, which were subsequently multiply deformed and thrusted during the Damaran Orogeny (570–450 Ma). Minor lithologies included are graphite schists, calc-silicates and scapolite schists. The intercalated Matchless Amphibolite contains tholeiitic metavolcanic rocks, including pillow lavas and breccias, as well as ultramafic lithologies, metagabbroic lenses, graphite schists, cherts, and a marble unit. Large exotic blocks of serpentinite were tectonically emplaced into the thrust pile.

Original sedimentary structures, the vertical facies distribution of progradational and retrogradational cycles, and the lateral extent of major sedimentary units indicate that major parts of the sedimentary sequence have been deposited as an elongate deep-sea trench fan. The regional structural pattern is characterized by several phases of coaxial deformation with folds verging consistently to the southeast. The structural regime is markedly heterogeneous and is associated with thrusting which developed elongate thrust slices traceable laterally for at least 150 km. The thermal evolution comprises a prolonged amphibolite-facies metamorphism with the peak occurring late in the deformational history. These features are explained in a tectono-sedimentary model which involves the evolution of a Late Proterozoic accretionary prism, named the Khomas Hochland accretionary prism, within a convergent continental margin setting.

Rift initiation took place along old tectonic weakness zones between the Congo and Kalahari Cratons. This was followed by the opening of the Khomas Sea and the development of mature oceanic shelves on the two adjacent cratons. Regional geological constraints indicate that the Khomas Sea opened as a relatively small, gulf-type oceanic basin. A passive continental margin was maintained on the southern craton during closure of the Khomas Sea, which occurred at a northwest dipping subduction zone along the northern Andean-type active margin. Submarine fan sediments of the elongate trench, some pelagic sediments and pillow lavas of the oceanic crustal sequence were scraped off during accretionary prism evolution. Ultimately continental collision resulted in: (1) obduction of the accretionary prism onto the Kalahari Craton, involving the emplacement of serpentinite pods together with basement slices; (2) strong southeast vergent folding and thrusting; (3) extensive shortening; (4) prograde regional metamorphism through crustal thickening; and (5) widespread granitic plutonism.

INTRODUCTION

The tectono-sedimentary evolution of active continental margins is still enigmatic in many orogens of pre-Palaeozoic age. This is mainly because of repeated tectono-thermal rejuvenations which result in complex overprint patterns and obliteration of previous structures in old orogenic belts. These processes are accompanied by intense deformation

---

*Present address: Shell Research BV, KSEPL, PO Box 60, 2280 AB Rijswijk, The Netherlands.

and metamorphism. Within the metasedimentary sequences of such orogens it is common that there is a lack of biostratigraphical control and of the delicate sedimentary structures which are essential for palaeoenvironmental interpretations, caused respectively by absence of macrofossils, metamorphic recrystallization and penetrative deformation. Lithologies are furthermore mostly monotonous, without marker horizons, and the sedimentary successions have usually been severely disrupted by subsequent phases of folding and thrusting. The uplift history of the orogens may be long-lasting, reaching the high temperature–low pressure peak of prograde metamorphism after the main deformation events. Age dating to determine the timing of deposition is therefore often inconclusive in such metasedimentary rocks.

The latter factors also apply partly to the Late Proterozoic Damara Orogen in Namibia (Fig. 1), particularly its inland branch. It represents a well-preserved fold-and-thrust belt which has passed through stages of sediment accumulation, subsidence and orogenesis and which has been overprinted only slightly by subsequent geotectonic processes.

Palaeoenvironmental interpretations in the southern Damara Orogen have previously been based upon lithological considerations and not facies analysis and these have been combined with some structural, petrological and geochemical information to infer strongly opposing models of crustal evolution. Initially aulacogen-related models have been proposed by Martin and Porada (1977) and an oceanic subduction model by Blaine (1977). The reader is directed to Miller (1983a) for an overview.

During our studies in the Khomas Trough of the southern Damara Orogen (Fig. 1) we have discovered a variety of sedimentary structures which are well preserved despite the amphibolite facies metamorphic grade (P.A. Kukla *et al.*, 1990). These have led us to undertake, for the first time, a detailed sedimentary facies analysis and taken together with structural and petrological studies these help to constrain the setting of the Khomas Trough as a whole. This paper synthesizes the results of our investigations and proposes a new model for the geotectonic evolution of the inland branch of the Damara Orogen. The main study area is a north–south trending, 80-km-long river section through the Khomas Hochland about 120 km west of Windhoek (Fig. 1). Important structural and sedimentological markers and sequences have also been mapped laterally (Fig. 2).

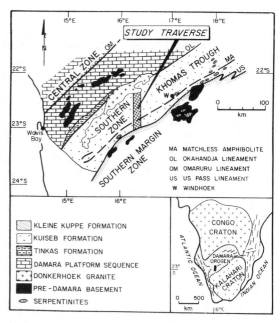

Fig. 1. Location of the study traverse and tectonostratigraphical zones in the inland branch of the Damara Orogen. Note position of the Okahandja (OL) and Us Pass (US) Lineaments and the Matchless Amphibolite (MA). (Modified after Miller *et al.*, 1983.)

GEOLOGICAL SETTING AND ROCK RELATIONSHIPS OF THE SOUTHERN DAMARA OROGEN

The evolution of the Damara Orogen has been summarized by Miller (1983a) who reviews various igneous and metamorphic dating programmes. Initial intracontinental rifting is best dated by volcanic extrusives at 750 ± 65 Ma (U–Pb zircon). Post-tectonic intrusions are dated at about 505 Ma (C. Kukla *et al.*, 1990) and biotites are dated at about 450 Ma having been previously interpreted as representing a cooling age (Miller, 1983b).

The Khomas Trough in central Namibia forms part of the inland branch (Fig. 1) of the orogen. It comprises mainly multiply deformed metagreywackes and pelites of the Kuiseb Formation. Minor lithologies include graphite schists, calc-silicates, and scapolite schists. The Matchless Amphibolite is structurally emplaced within the metasedimentary sequence (Fig. 1). Alpine-type ultramafic bodies occur as exotic blocks in the southern portion of the Khomas Trough (Barnes, 1982). A subdivision of

Fig. 2. Landsat image of the central and western Khomas Trough showing marker units and major thrust zones. Note their lateral extent. (Reproduced with permission of the Satellite Applications Centre of the CSIR.)

the Khomas Trough has been undertaken by Kukla *et al.* (1988) on the basis of structural and sedimentological markers such as graphite schists, scapolite schists, thick pelitic units, metamorphosed carbonates (marble and tremolite schist) and major zones of high strain which represent thrust zones. These sedimentological and structural features may be traced on the Landsat image, which emphasizes the

extreme lateral persistence of the marker horizons (Fig. 2).

Porada and Wittig (1983) have established that, towards the north, the Kuiseb Formation of the Khomas Trough interfingers with calcareous turbidites of the Tinkas Formation, a lateral equivalent of the platform carbonates of the Karibib Formation (Fig. 3). The latter is overlain by meta-

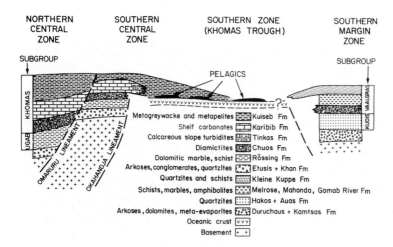

Fig. 3. Rock relationships of the Khomas Trough and adjacent areas in the Damara Orogen. Stratigraphical contacts are emphasized in their palinspastic context. Note the development of a carbonate platform sequence in the Central Zone, which was later replaced by clastic terrigenous sediments, and of a clastic platform/margin sequence to the south of the Khomas Trough. Correlations and terminology of the Southern Margin Zone and the southern and northern Central Zone are after Hoffmann (1983), Henry *et al.* (1990) and Badenhorst (1987) respectively.

greywackes and pelites of the Kuiseb Formation. South of the trough the allochthonous thrusted platform/margin sequence of the Southern Margin Zone (Hoffmann, 1983) comprises conglomerates, quartzites, glacio-marine diamictites, pelitic schists, amphibolites, ultramafic pods and basement inliers (Fig. 3).

SUBMARINE FAN SEDIMENTATION

Sedimentary patterns and facies analysis

Rock profiles have been measured in detail on a variety of scales from centimetres upwards. The primary nature of the layering in which the above lithotypes are organized is confirmed by sharp bases and tops of beds, graded beds and various erosional features. Graded beds are more quartz- and feldspar-rich at the base and more mica-rich at the top, representing a metamorphosed compositional grading. Sedimentary structures preserved include trough cross-lamination, load structures, scour surfaces, flute casts, channels, rip-up clasts, and clastic dykes, documented in P.A. Kukla *et al.* (1990). A high proportion of the metagreywacke units, displaying sedimentary structures that are arranged in Bouma sequences, show that the sedimentary protoliths were deposited by low-density and high-density turbidity currents. Palaeocurrent determinations on flute casts and sets of trough cross-laminae show a remarkable uniformity throughout the traverse, with orientations indicating current flow from the northeast to the southwest (230–250°).

The preservation of original sedimentary features has led to the definition of turbiditic and pelagic facies types in the Kuiseb schists using attributes such as bedding style, vertical bed thickness distribution, variations in the psammite/pelite ratio and sedimentary structures (Fig. 4). Facies classes B, C, D, E and G according to the turbidite facies classification of Pickering *et al.* (1986) and Mutti and Ricci Lucchi (1975) are present. A summary and comparison of turbidite facies is shown in Fig. 4 and the facies classification of Pickering *et al.* (1986) will be followed in this paper.

Facies class B beds (Fig. 4) comprise (1) structureless metagreywackes (psammites) which are usually 10 cm–3 m thick; some beds reach 5 m (facies B1.1). Depositional processes assigned to this facies are high-density turbulent flows (Lowe, 1982). The psammites are similar to facies S_3 sands of Lowe (1982), who interpreted deposition from high-concentration turbidity currents by suspension sedimentation. More abundant are psammites (2) which contain sedimentary structures and show bed thicknesses ranging from 10 cm to about 3 m, mostly being in the order of 70 cm. In the 10–70-cm-thick range, cross-lamination in the upper few centimetres as well as load casts and flute casts at the base are commonly developed. Also, erosional channels and scours, sandstone dykes and rip-up clasts were observed. Channels preserved on outcrop scale show a maximum depth of 3 m and a width of 5 m. This facies compares best with facies B2 of Mutti and Ricci Lucchi (1975) and might indicate the transition from high-density suspension flow to residual low-density currents (Lowe,

THIS STUDY Facies Types		Pickering et al. (1986) Facies Class	Mutti & Ricci Lucchi (1975) Facies
	medium– to very thick–bedded psammites a)massive, structureless b)massive, "organized" c)parallel–stratified d)cross–stratified	B 1.1 – B 2.1 B 2.2	A1 or B1 B2 B1 –
	graded psammite/pelite couplets a)very thick– to thick–bedded b)medium–bedded c)thin–bedded d)very thick– to thick–bedded, mud–dominated	C 2.1 C 2.2 C 2.3 C 2.4	C2 D2 D1 –
	thin–laminated metasiltstones	D 2.1	D1
	thin– to very thick–bedded pelites a)structureless b)organized"	E 1 or G E 2 or G	thin– to medium–bedded D or G

Fig. 4. Summary of the facies classification of this study compared with other studies. The classification of Pickering *et al.* (1986) is used in the text.

1982). Further facies are (3) parallel-stratified psammites (facies B2.1), which may be interpreted as traction carpet deposits (Hiscott & Middleton, 1979; Lowe, 1982), and (4) cross-stratified psammites (facies B2.2). Such discrete cross-stratified layers are attributed to tractional processes beneath dilute turbidity currents or bottom currents (Lowe, 1982; Pickering *et al.*, 1986).

The psammite–pelite couplets of *facies class C* (Fig. 4) show bed thicknesses from 1 cm to more than 2 m and a wide range of psammite/pelite ratios from >10:1 to 1:10. Generally, graded layering characterizes this facies and sedimentary structures within Bouma sequences are abundant. Calcareous lithologies which are now preserved as calc-silicates may form part of individual turbidites. They occur preferentially at the base of massive facies B and C psammites and also are finely dispersed within the troughs of cross-laminated facies C psammites. High-concentration turbidity currents with rapid settling seem to have been the predominant depositional process for the very thick- to thin-bedded graded layers (facies C2.1 to C2.3) but low-concentration turbidity currents produce thin-bedded units and decreasing psammite/pelite ratios (Piper, 1978; Walker, 1978; Stow & Bowen, 1980). Mud-dominated, very thick- to thick-bedded graded layers (facies C2.4; Fig. 4) are 50 cm to more than 2 m thick and locally show spectacular sand filled dykes as well as load casts. Pickering and Hiscott (1985) attribute this unusual facies to large-

volume, high-concentration turbidity currents confined within small basins.

The thinly-laminated siltstone *facies class D* is only developed in the central part of the study traverse. Sedimentary units comprising this facies are typically 20 cm to 350 cm thick with individual bed thicknesses ranging from 1 mm to about 7 cm. The majority of the beds show normal grading and turbidite sequences comprise mostly E2 and E3 divisions of Piper (1978). Bases often show micro-loading and flame structures. Cross-stratification is occasionally developed in the lower quartz-rich parts of layers (E1 division of Piper, 1978). Thin-laminated siltstones are hydrodynamically interpreted to have formed by low-concentration turbidity currents (Piper, 1978; Stow & Piper, 1984). Pickering (1984) attributed the thin laminations to discrete (waning) flow events with an accompanying depositional sorting process as described by Stow and Bowen (1978).

The thin- to very thick-bedded pelite *facies class E* comprises pelitic layers which are not interrupted much by graded psammite/pelite couplets (facies E1). The thickness varies from about 20 cm to several metres. No sedimentary structures have been found. This facies may also be compared with the E3 division of Piper (1978). Graded, laminated pelites with bed thicknesses of 10 cm to more than 10 m belong to facies E2 (Fig. 4). It may not be excluded, however, that some of the lithologies may also be classified as facies G according to Pickering et al. (1986). Fine-grained sediment may be generally deposited as turbidites, contourites and as a result of pelagic–hemipelagic settling. It has been pointed out by Stow and Piper (1984) that these processes strongly interact and overlap during both transport and deposition. Structureless muds may be interpreted as resulting from nepheloid settling through the water column to be finally reworked by deep-sea bottom currents (Gorsline, 1984). This facies may also result from ponding of thick, dilute turbidity currents as suggested by Pickering et al. (1986). Piper (1978) points out that the internal organization of muds is largely dependent on the deceleration of the turbidite flow concerned. Thin facies E2 turbidites may therefore be deposited through grain-by-grain settling or aggregate settling within low-concentration turbidity currents (Pickering et al., 1986).

Pelagic facies encountered in the Khomas Trough are graphitic schists, pelitic units and some calcareous lithologies. Graphite schists are usually between 1 and 10 m thick. Stable isotope data obtained from several graphite schists demonstrate clearly their biogenic origin (Kukla, 1990a). The interpretation of the calcareous lithologies, such as intercalated calc-silicates and a marble unit close to the Matchless Amphibolite, is more problematic due to diagenetic and metamorphic processes which may involve element and isotope mobilities. Whilst most of the calc-silicate lithologies are interpreted to have been deposited by resedimentation processes (see above), isotope signatures of the up to 2-m-thick marble unit indicate a hydrothermal origin of the carbon (Kukla, 1990a). This marble is therefore interpreted as a chemogenic pelagic sediment (facies G) which has been deposited in a pelagic setting through the accumulation of exhalatively derived carbonate. It is also likely that, in addition to the turbiditic pelites described, a considerable percentage of the pelites, especially the thick-bedded units, may in fact represent non-turbiditic hemipelagic and pelagic deposits. These are, however, difficult to prove and to discriminate in metamorphic rocks. The carbon-rich pelagic facies described (graphite schists) are here attributed to facies G 'biogenic oozes, hemipelagites and chemogenic sediments' of Pickering et al. (1986).

Facies sequences and sedimentary cycles

The facies described above are developed as facies sequences throughout the Khomas Trough. A representative set of measured sections is presented in Fig. 5. This shows that vertical facies sequences mostly involve very thick-bedded (better developed towards the north) to thin-bedded (better developed towards the south) facies B layers which are associated with facies C and partly facies E at the northern margin of the Khomas Trough (top row of measured sections in Fig. 5). Facies B, C, D and partly E are developed in the centre (centre row of measured sections in Fig. 5); and facies E predominates in the south (base row of measured sections in Fig. 5). In the south and centre, pelagic facies G graphite schists are associated with turbiditic facies B and C. The turbiditic carbonate facies is distributed throughout the trough. Generally, the bed thickness of psammites and pelites decreases from a maximum of about 5 m in the north to a maximum of 20 cm in the south and Fig. 5 also shows an increase in the development of pelitic sequences southwards. An exception is a limited area just south of the confluence of the Koam and Amsas Rivers in the south where thick metagreywacke sequences are developed (Fig. 5).

Further analyses of measured sections has shown that facies sequences are organized within well-defined sedimentary cycles which are developed on all scales. Major cycles (> 500 m thick) are visible on the Landsat image of the area (Fig. 2), which reflects broad changes from psammite-rich to pelite-rich units. The cycles may be traced laterally for at least 150 km and their distribution along the study traverse is shown in Fig. 6.

Small-scale cycles (Fig. 7a) and medium-scale cycles (Fig. 7b) vary in thickness from 2 to 30 m and from 200 to 500 m respectively. These are nested within the large-scale cycles (Fig. 6). The northern side of the Khomas Trough comprises two major thinning-upward sequences, the southern of which includes minor thickening-upward cycles (Fig. 6). Major parts of the Khomas Trough contain nested thickening-upward cycles, but these are superseded by non-cyclic successions in the south (Fig. 6), with only a thin unit of thickening-upward cycles incorporated.

Since the Kuiseb schists are strongly deformed and metamorphosed, a considerable amount of internal folding is present within the sedimentary cycles. Notwithstanding this strong but widely co-axial deformation, sequential lithofacies changes have been shown to occur on all scales. Therefore the cycles are regarded as coherent stratigraphical units which extend for more than 150 km laterally.

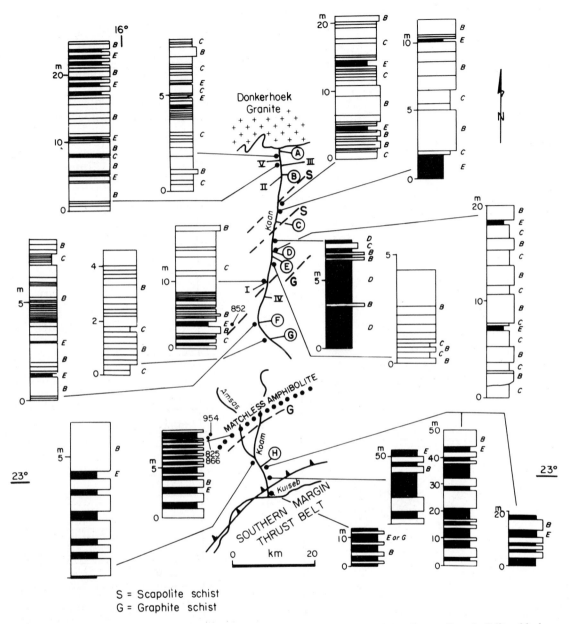

S = Scapolite schist
G = Graphite schist

Fig. 5. A selection of detailed measured sections within the Kuiseb Formation of the Khomas Trough. Pelites, black shading; psammites, blank. Facies nomenclature in accordance with Pickering *et al.* (1986). Respective facies are specified on top of the individual measured section in cases of facies consistency throughout the section.

Facies associations and the depositional setting of an elongate submarine fan

The facies and facies models which have been documented above are interpreted to indicate elongate submarine fan and basin plain sedimentary environments. The main reasons for interpreting major proportions of the Kuiseb sediments to have been deposited as an elongate submarine fan are:

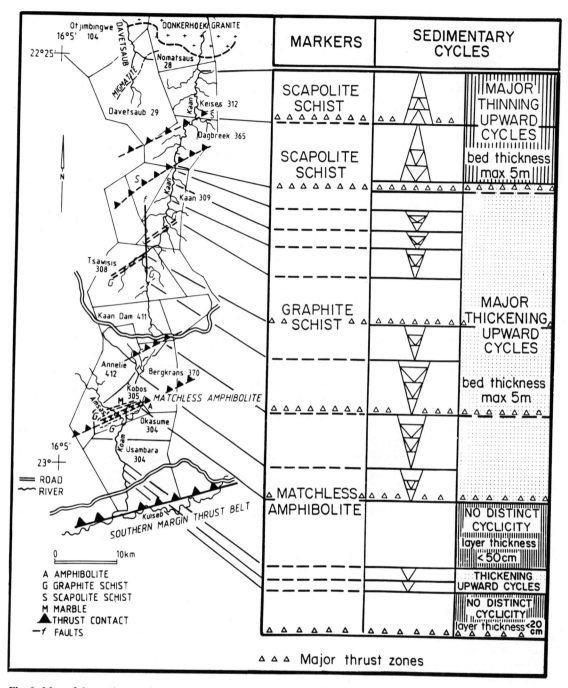

Fig. 6. Map of the study area showing the position of large- and medium-scale sedimentary cycles in relation to marker units and major thrusts. Large-scale cycles and marker units may be traced laterally for more than 150 km on the Landsat image (Fig. 2).

Fig. 7. Photographs showing sedimentary cycles. (a) Small-scale cycle (3.5 m thick) on the farm Tsawisis 308. White bars represent individual bed thicknesses. Note hammer for scale. (b) Medium-scale (*c.* 200 m) thickening-upward cycles etched in the landscape on the farm Dagbreek 365.

(a)

(b)

1 progradational and retrogradational cyclic as well as non-cyclic gravity-flow deposits are traceable for more than 150 km laterally;

2 consistent palaeocurrent patterns show south-west directed sediment transport;

3 size of sedimentary cycles;

4 uniformity of lithologies and grain sizes;

5 development of channelized and non-channelized deposits;

6 'proximal' to 'distal' (Walker, 1976) facies changes within the Khomas Trough.

In favour of basin plain environments are:

1 the intercalation of thin-bedded turbidites with pelagic deposits;

2 the overall decreasing bed thickness of turbidites towards the southern margin of the Khomas Trough;

3 the increasing thickness of large-scale pelitic successions up to a 100-m scale towards the south.

Interpreting submarine fan facies associations from 'proximal' to 'distal' has limitations. Various authors have put emphasis on the problems arising from the comparison of modern fan settings with those of ancient fans (Normark *et al.*, 1983/1984; Shanmugam *et al.*, 1988). Ancient fans are mostly interpreted from vertical facies sequences, which are dependent on their degree of preservation. The latter usually limits the detection of large-scale morphological features which would otherwise help to confirm a fan shape.

The distribution and organization of facies and vertical facies sequences, together with cyclic patterns are none the less interpreted to document the following submarine fan depositional subenvironments in the Khomas Trough (Fig. 8). Middle fan deposition is indicated in the north by the dominance of thick-bedded over thin-bedded turbidites comprising mainly facies B and C within an overall thinning-upward trend; facies D and E are subordinate (Fig. 5). Small, scale channels, incomplete Bouma sequences, and largely symmetrical, irregular sequences of thinning- and thickening-upward cycles, however, do indicate the widespread occurrence of interchannel deposits (Nelson & Nilsen, 1974, 1984; Stow, 1986).

Outer fan deposition is recorded in the centre of the trough with facies associations confirming mostly outer fan lobe and fan fringe deposits. Generally this area comprises intercalated facies B, C and D, and occasional thick facies E pelites. Characteristics of the outer fan lobe deposits are: (1) non-channelized thickening-upward 'megasequences' (10–30 m thick); (2) small-scale thickening-upward compensation cycles (up to 5 m thick), minor thinning-upward cycles and widespread symmetrical cycles; (3) a uniformity of palaeocurrents; and (4) high psammite/pelite ratios. The presence of facies D also in the centre of the Khomas Trough indicates intercalated fan fringe deposits (Nelson & Nilsen, 1984). These contain thin-bedded turbidite sequences which are organized into repetitive small-scale thickening-upward cycles (2–5 m thick)

comprising mostly facies D associated with C and some facies B.

Basin plain deposits are developed from just north of the Matchless Amphibolite towards the south (Fig. 8). The sequences, which are typically non-cyclic, are further characterized by the systematic increase in the amount and bed thickness of pelites and graphite schists, and the decreasing thickness of turbidite beds which are intercalated with facies E and G hemipelagic and pelagic muds. A narrow area comprising thickening-upward cycles with facies B and C turbidites is intercalated within this basin plain succession. This might be explained either by structural interposition or more probably by an episode of outbuilding of a lobe sequence into such distal environments, similar to one of the basin plain lobes described by Pickering (1981) from the Late Proterozoic Kongsfjord turbidite system and by Ricci Lucchi and Valmori (1980) from the Italian Apennines.

Since there is no evidence for coarse-grained, conglomerate facies A and slump deposits throughout the study traverse, it is inferred that slope and inner fan environments have not been preserved or exposed at the present erosion level. Palaeocurrent evidence predicts that these should be developed in areas to the east now covered by Tertiary sediments (Fig. 1).

In conclusion, the lithofacies, the facies associations, the uniform palaeocurrents and the long distance correlation of sedimentary sequences suggest that an elongate submarine fan has been operating in the basin separating the Congo and Kalahari Cratons. 'Proximal' to 'distal' relation-

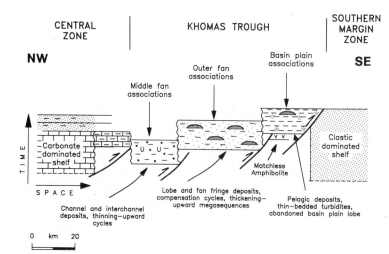

Fig. 8. Map and schematic profile across the Khomas Trough showing the inferred submarine fan depositional palaeosubenvironments and their palinspastic relationship. Facies were developed laterally to one another, but have also been thrust on top of one another during continuing deposition.

ships occur from north to south confirming the development of various middle fan, outer fan and basin plain environments through time.

THE MATCHLESS AMPHIBOLITE

The Matchless Amphibolite is a 350-km-long and up to 3-km-wide (Miller, 1983b) sequence of deformed tholeiitic oceanic metavolcanics (Finnemore, 1978; Breitkopf & Maiden, 1987). It further includes pillow lavas, breccias, metagabbroic lenses (Fig. 9) and elsewhere ultramafic lithologies (Barnes, 1982). These are intercalated with pelagic sediments such as pelitic schists, graphite schists, small amounts of cherts, and a 2-m-thick marble unit (Fig. 9). Massive sulphide (copper) ore deposits are aligned along the Matchless Amphibolite. In parts of the sequence graded layering is an important way-up indicator. The interposition of various lithologies within the Matchless Amphibolite is associated with the development of high- and low-strain zones. Throughout the sequence, high strain

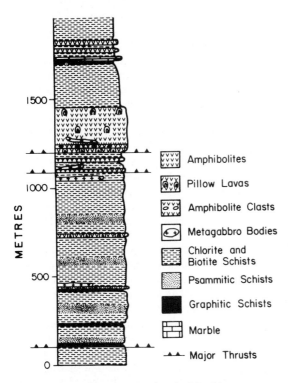

Fig. 9. Measured section showing the Matchless Amphibolite sequence in the study traverse (Annelie 412 farm).

has been taken up preferentially by units of low competence such as pelites and carbonate-replaced volcanics. Some pillow lavas have preserved the original spheroidal pillow shapes, whereas others have been extremely elongated. Other structural features include thrusts with an associated down-dip lineation, shear zones and boudinage structures which indicate layer-parallel extension. The particular depositional and structural styles within the Matchless sequence are illustrated in Fig. 9. Structural emplacement is indicated in the lower parts of the sequence by the interposition of thinly layered amphibolites, pelitic and graphitic schists, with metagabbro lenses and, elsewhere, ultramafic pods (Barnes, 1982). Upper parts of the sequence show a more coherent relationship between carbonate-rich amphibolites and units of graphite schist and marble. Geochemical analyses of the metabasalts indicate N-type MORB compositions (Finnemore, 1978; Miller, 1983c; Kukla, 1990a) although Breitkopf and Maiden (1987) have found within-plate basalt compositions in the western Khomas Trough.

The massive sulphide deposits in the vicinity of the Matchless Amphibolite have been interpreted as having been formed close to a sea-floor spreading centre (Killick, 1983; Klemd *et al.*, 1987; Breitkopf & Maiden, 1988). The lack of recognition of a complete ophiolitic sequence and of mélange deposits has been used to argue against the oceanic crustal origin of the Matchless Amphibolite (Martin, 1983). Others have interpreted the MORB compositions of the metabasalts as indicating their origin in a mid-ocean ridge type setting (Kasch, 1983; Miller, 1983c). This study has shown, however, that the sequence comprises pillow lavas (with N-type MORB characteristics), pelagic sediments, and metagabbros. Its association with ultramafic bodies is additional evidence supporting an interpretation that the Matchless Amphibolite represents the upper portion of an oceanic crustal sequence.

EVIDENCE FOR AN IMBRICATE THRUST PILE

Preserved sedimentary layering and sedimentary structures provide a good stratigraphical control for the delineation of fold structures. Several phases of deformation have been discerned, of which the initial three phases show coaxial folding (Kukla *et al.*, 1988). Fold axes are subhorizontal and consistently have a northeasterly trend, with folds verging

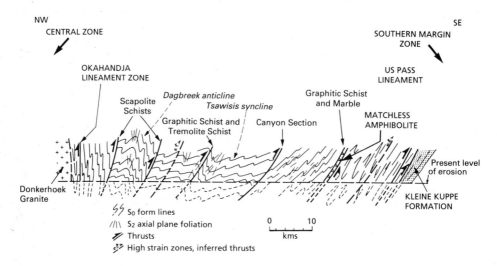

Fig. 10. Simplified structural profile across the western Khomas Trough showing the organization of the sequence into distinct imbricate thrust slices relative to the major D_2 phase of folding.

towards the southeast. The first deformation overturned parts of the sequence and was associated with thrusting. This was followed by open to tight folding which overprinted the earlier overturned folds to develop downward-facing structures in parts of the sequence (Kukla *et al.*, 1989). This second deformation has regional significance and was associated with a major phase of thrusting and faulting and the shearing-out of the steep limbs of some of the F_2 folds. A schematic structural profile along the study traverse is shown in Fig. 10, indicating the change in fold style across the Khomas Trough. This shows a steepening of the structures towards the north and the position of major thrust zones which are associated with most pelitic portions of the sequence. The third deformation is characterized by intensified shearing with the development of a pronounced penetrative cleavage and folds within high-strain shear zones. These shear zones, situated in thick pelitic sequences, acted as conduits for intensive fluid migration, as evidenced by the development of the scapolite schists along the shear zones in the northern Khomas Trough (Fig. 2). These units have been interpreted to have originated from the impregnation of Na- and Cl-rich fluids along veins and fractures (Kukla, 1990a, b) which broadly followed the major thrust dislocations (Fig. 10). In contrast to the earlier deformation phases, the D_4 phase is represented by

crenulation cleavages. Late, right-lateral strike-slip displacement along the northern margin of the Khomas Trough (Okahandja Lineament) is evident from the (post-D_4) clockwise rotation of previous structural elements.

The strain distribution throughout the area is distinctly heterogeneous. Narrow high-strain zones, several hundreds of metres wide, show a remarkable development of F_3 folds and the S_3 cleavage. Earlier fabrics are largely overprinted and original sedimentary structures are completely obliterated in these zones. The major shear zones are developed in thick pelitic units (Fig. 10) and are the ultimate expression of the heterogeneous distribution of strain. Minimum shortening values between 50 and 80% have been based on strain analyses from folds, pillow lavas, and calc-silicate spindles by Hälbich (1977), Sawyer (1981) and Kukla (1990a).

On the basis of the structural data we interpret the Kuiseb Formation in the Khomas Trough as representing a sequence of metasediments and amphibolites which has been telescoped by trusting. This has developed imbricate, elongate thrust slices with a lateral extent of more than 150 km. The individual slices are floored by thick pelitic units. It is noteworthy that the structural evolution is one of progressive deformation, emphasized by the inheritance of the major thrusts from one phase to the next (Kukla *et al.*, 1989).

ASPECTS OF THE THERMAL EVOLUTION

Metamorphic isograds mapped by Hoffer (1977) describe a concentric distribution around an area close to Walvis Bay on the Atlantic coast (Fig. 1) where partial melting of Damaran pelitic lithologies occurs. Sawyer (1981), Hartmann *et al.* (1983) and Kasch (1983) have subsequently confirmed that metamorphic temperatures increase and pressures decrease inward towards the Central Zone. The highest pressures obtained so far are 10 kbar, recorded from the southern Khomas Trough by Kasch (1983). In the study area metamorphic index minerals in pelitic lithologies have grown from syn-D_2 to post-D_3. Estimates of *P–T* based on microprobe analyses of post-D_3 mineral phases show 520–570°C at 4–7 kbar in the southern Khomas Trough; 530–600°C at 2 kbar in the centre, and 590–630°C at 4–6 kbar in the north (Kukla, 1990a). At the northern margin of the trough, migmatization occurs suggesting a temperature peak of regional metamorphism of about 660°C. The post-tectonic Donkerhoek Granite (Fig. 2) has also intruded along the northern margin of the Khomas Trough.

Barrovian-type regional metamorphism has thus occurred in this part as well as the remainder of the Damara Orogen late in its deformational history. Lower to upper amphibolite facies conditions are developed and *P–T* Estimates suggest that different crustal levels are exposed in the southern, central and northern part of the Khomas Trough.

THE KHOMAS HOCHLAND ACCRETIONARY PRISM

Previous discussions on the geodynamic evolution of the inland branch of the Damara Orogen have concentrated mainly on whether or not oceanic crust formed to be subsequently subducted during continental convergence. Aulacogen-related models include those of Martin and Porada (1977) and Porada (1983) and the delamination model of Kröner (1982). Ocean floor subduction models include the subduction of a wide ocean (Hartnady, 1978; Kasch, 1979; Barnes & Sawyer, 1980), the subduction of a narrow, Red Sea-type ocean (Miller, 1983b) and small ocean basins associated with strike-slip shear movements (Downing & Coward, 1981). For the Khomas Trough, Blaine (1977) and Miller *et al.* (1983) have suggested a fore-arc setting. Downing and Coward (1981) and Hoffmann (1983) speculated upon an accretionary wedge setting for the Khomas Trough. We think that our research discriminates between the various possibilities and we propose that the sediments of the Khomas Trough developed as an accretionary prism of Late Proterozoic age evolving in a series of stages.

The striking sedimentological and structural asymmetries within the inland branch have recently led Henry *et al.* (1990) to relate the early rift phase to the development of two low-angle detachment systems. This ultimately resulted in the evolution of the Khomas Sea between the Kalahari and Congo Cratons. Rifting ceased with the onset of spreading in the Khomas Sea and the formation of oceanic

Fig. 11. Depositional palaeo-environments during early continental convergence. The Kuiseb Formation metasediments were deposited within an elongate trench fan. The greywackes and muds were derived from a volcanic arc to the east of the exposed Khomas Trough. Pelagic and hemi-pelagic sedimentation dominates the basin plain. In this model carbonates associated with the Matchless Amphibolite accumulate as chemogenic sediments in the pelagic environment close to the mid-ocean ridge. Continental platform and slope sedimentation, partly in submarine fans, occurs on the Congo and Kalahari Cratons on either side of the Khomas Sea.

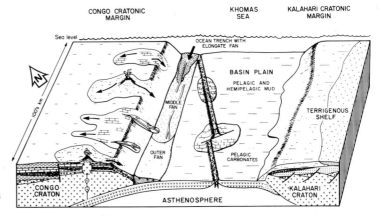

crust is implied by the MORB compositions of the Matchless metabasalts. Tectonic shortening indicates that the widest opening of the Khomas Sea was 500 km minimum. During the opening of the Khomas Sea, stable carbonate shelf conditions occurred in the north on the Congo Craton whilst in the south, on the Kalahari Craton, terrigenoclastic platform/margin sedimentation prevailed (Fig. 3).

Northwest orientated subduction was initiated along the northern continental margin and the convergence was associated with the development of an oceanic trench which contained the elongate submarine fan system. A major source area is indicated towards the northeast. Intercalated graphite schists, carbonates and thick pelites confirm that fan sedimentation and pelagic sedimentation were contemporaneous within the basin. The inferred palaeoenvironmental setting during early B-subduction is shown in Fig. 11. To the north of the Khomas Sea, a magmatic arc is situated close to the trench. The carbonates on the Congo cratonic shelf were superseded during continued subduction by clastic terrigenous psammites and pelites of the Kuiseb Formation, the bulk of which probably entered the platform from the east as indicated by

the measured sections of Badenhorst (1987). The southern terrigenous platform of the Kalahari Craton remained stable with local fan sedimentation occurring (Porada & Wittig, 1983). Hartnady et al. (1985) have suggested from tectonic slip-vector data that a possible Euler pole for the clockwise rotation of the Kalahari Craton during closure of the Khomas Sea was situated in north-eastern Zambia.

During subduction, the accretionary prism evolved through the offscraping of submarine fan lithologies from the descending oceanic crustal slab (Fig. 12a). The dominance of metasediments within the present-day Khomas Trough indicates a high sediment input into the oceanic trench with subsequent preferential accretion of clastic sediment instead of oceanic basalts and pelagics. This particular setting is also thought to account for the initiation and maintenance of shallow subduction within the Khomas Sea gulf which is thought to have comprised young buoyant oceanic crust (Stanistreet et al., 1991). The prism backstop was located at the northern margin and we suggest that this was an early expression of the Okahandja Lineament. Early thrusting and folding occurred

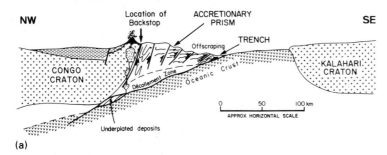

(a)

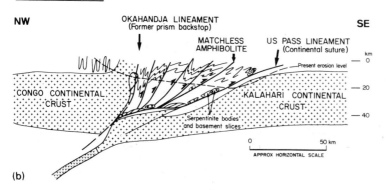

(b)

Fig. 12. (a) Tectonic setting of the southern Damara Orogen during late convergence of the Congo and Kalahari Cratons. The oceanic trench contains the elongate submarine fan and the accretionary prism evolves through the offscraping of sediments and pillow lavas from the descending slab. Early folding and thrusting occurs within the prism. (b) Collision between the Congo and Kalahari Cratons is accompanied by strong south-east vergent folding and thrusting and emplacement of ultramafic pods together with basement slices.

within the prism which is represented by the D_1 phase of deformation in the Khomas Trough.

The collision-phase (Fig. 12b) involved strong southeast vergent folding and thrusting with the development of isoclinal and open folds (D_2 deformation). Scapolitization of the wall rocks of the thrusts indicates that fluids were channelized into thrust and fault zones, a process described by Moore (1989), and they participated during the development of the shear zones in a manner similar to that described by Peacock (1987).

Finally rotation of structural elements occurred close to the Okahandja Lineament during late dextral strike-slip movements. This lineament was also partly the site of post-tectonic granite intrusions (Fig. 12b) at about 505 Ma (C. Kukla *et al.*, 1990). The final collision, involving the continental A-subduction of the Kalahari beneath the Congo Craton, led to the post-tectonic peak of high-*T*–low-*P* metamorphism in this part of the orogen. Differential uplift is thought to account for the superimposition of higher pressure and lower pressure sequences developed in the Khomas Trough.

In this model the Matchless Amphibolite represents uppermost oceanic crust (pillow lavas and pelagics) which has been sliced off an oceanic crustal topographic high (oceanic plateau, aseismic ridge or mid-ocean ridge) to be incorporated into the accretionary prism. During collision and obduction of the accretionary prism, Alpine-type ultramafic blocks were structurally emplaced together with basement slices along the northern margin of the Kalahari Craton (Figs 1 & 12b). To explain the intercalation of metabasalts, massive sulphide ore deposits and the turbidites of the elongate trench fan the analogy of the southern Peru–Chile trench applies. The interaction of newly formed, hot, young buoyant oceanic crust with trench sediments is also a factor where the Chile rise is being shallowly subducted beneath South America (Cross & Pilger, 1982).

ANCIENT AND MODERN ANALOGUES

Lothologies, structural styles, sedimentation patterns, time sequence of events and the size of the Khomas Hochland accretionary prism are most reminiscent of the Lower Palaeozoic Southern Upland accretionary prism (e.g. Leggett *et al.*, 1979; Kemp, 1987) in view of:

1 series of imbricate thrust slices steepening towards the backstop (Southern Uplands Fault here compared with the Okahandja Lineament);
2 extreme lateral persistence (>150 km) of stratigraphical marker horizons and thrust slices;
3 the predominance of lateral trench-fill turbidite sedimentation; and
4 no mélange deposits are recognized.

The Chugach terrain in Alaska has been explained as a Cretaceous trench-fill deposit (Nilsen & Zuffa, 1982; Sample & Moore, 1987). Broad similarities with the Khomas Trough exist in terms of: (1) linear extent (2000 km long, 100 km wide); (2) contained lithologies; (3) late-phase granite intrusions towards the fore-arc; and (4) structural style comprising coaxial deformation 'seawards' and non-coaxial deformation with rotation and strike-slip faulting 'landwards'.

The time sequence of events during deformation of the Shumagin region of the modern Aleutian trench (Lewis *et al.*, 1988) can be compared with the Khomas Trough involving successively early folding and thrust faulting, thrust faulting and strike-slip faulting.

ACKNOWLEDGEMENTS

We are grateful to the Geological Survey of Namibia, the Deutsche Forschungsgemeinschaft and the Foundation for Research Development for funding this research. This study has benefited from discussions with E.G. Charlesworth and G. Henry (Johannesburg); C. Kukla, M. Okrusch, H. Häussinger and B. Bühn (Würzburg); R. McG. Miller, R. Swart and K.H. Hoffmann (Windhoek); and with A. Bouma (Louisiana) and K. Eriksson (Virginia). L. Whitfield and M. Hudson are thanked for their preparation of figures. We thank Jutta Winsemann (Stuttgart) and an anonymous referee for their valuable comments.

REFERENCES

BADENHORST, F. (1987) Lithostratigraphy of the Damara Sequence in the Omaruru area of the northern Central Zone of the Damara Orogen and a proposed correlation across the Omaruru Lineament. *Comm. Geol. Surv. S.W.Afr./Namibia* 3, 3–8.

BARNES, S.-J. (1982) Serpentinites in central South West Africa/Namibia — a reconnaissance study. *Mem. Geol. Surv. S.W.Afr./Namibia* 8, 90 pp.

BARNES, S.-J & SAWYER, E.W. (1980) An alternative model for the Damara Mobile Belt. Ocean crust subduction and continental convergence. *Precambr. Res.* **13**, 297–336.

BLAINE, J.L. (1977) Tectonic evolution of the Waldau Ridge structure and the Okahandja Lineament in part of the Central Damara Orogen, west of Okahandja, South West Africa. *Bull. Precambr. Res. Unit Univ. Cape Town* **21**, 99 pp.

BREITKOPF, J.H. & MAIDEN, K.J. (1987) Geochemical patterns of metabasites in the southern part of the Damara Orogen, SWA/Namibia: applicability to the recognition of tectonic environment. In: *Geochemistry and Mineralization of Proterozoic Suites* (Eds Pharaoh, T.C., Beckinsale, R.D. & Rickard, D.) Geol. Soc. London Spec. Publ. **33**, 355–361.

BREITKOPF, J.H. & MAIDEN, K.J. (1988) Tectonic setting of the Matchless Belt pyritic copper deposits, Namibia. *Econ. Geol.* **83**, 710–723.

CROSS, T.A. & PILGER, R.H. JR. (1982) Controls on subduction geometry, location of magmatic arcs, and tectonics of arc and back-arc regions. *Geol. Soc. Am. Bull.* **93**, 545–562.

DOWNING, K.N. & COWARD, M.P. (1981) The Okahandja Lineament and its significance for Damaran tectonics in Namibia. *Geol. Rundsch.* **70**, 972–1000.

FINNEMORE, S.H. (1978) The geochemistry and origin of the Matchless amphibolite belt, Windhoek district, South West Africa. *Geol. Soc. S. Afr. Spec. Publ.* **4**, 433–477.

GORSLINE, D.S. (1984) A review of fine-grained sediment origins, transport and deposition. In: *Fine-grained Sediments: Deep-water Processes and Facies* (Eds Stow, D.A.V. & Piper, D.J.W.) Geol. Soc. London Spec. Publ. **15**, 17–34.

HÄLBICH, I.W. (1977) Structure and tectonics along the southern margin of the Damara mobile belt, South West Africa. *Ann. Univ. Stellenbosch, Ser. ÄA1(Geol)* **2**, 149–247.

HARTMANN, O., HOFFER, E. & HAACK, U. (1983) Regional metamorphism in the Damara Orogen: interaction of crustal motion and heat transfer. In: *Evolution of the Damara Orogen of South West Africa/Namibia* (Ed. Miller, R.McG.) Geol. Soc. S. Afr. Spec. Publ. **11**, 233–241.

HARTNADY, C.J. (1978) The stratigraphy and structure of the Naukluft nappe complex. *14th and 15th Annual Rep. Precambr. Res. Unit.* Univ. Cape Town, 163–170.

HARTNADY, C.J.H., JOUBERT, P. & STOWE, C. (1985) Proterozoic crustal evolution in southwestern Africa. *Episodes* **8**, 236–244.

HENRY, G., CLENDENIN, C.W., STANISTREET, I.G. & MAIDEN, K.J. (1990) A multiple detachment model for the early rifting stage of the Late Proterozoic Damara Orogen in Namibia. *Geology* **18**, 67–71.

HISCOTT, R.N. & MIDDLETON, G.V. (1979) Depositional mechanics of thick-bedded sandstones at the base of a submarine slope, Tourelle Formation (Lower Ordovician), Quebec, Canada. *Soc. Econ. Paleont. Miner. Spec. Publs* **27**, 307–326.

HOFFER, E. (1977) Petrologische Untersuchungen zur Regionalmetamorphose Al-reicher Metapelite im südlichen Damara-Orogen (Südwest-Afrika). Habilitationsschrift (unpubl.), Univ. Göttingen 150 pp.

HOFFMANN, K.H. (1983) Lithostratigraphy and facies of the Swakop Group of the southern Damara Belt, SWA/Namibia. In: *Evolution of the Damara Orogen of South*

West Africa/Namibia (Ed. Miller, R.McG.) Geol. Soc. S. Afr. Spec. Publ. **11**, 43–63.

KASCH, K.W. (1979) A continental collision model for the tectono-thermal evolution of the (southern) Damara belt. *16th Annual Rep. Precambr. Res. Unit.* Univ. Cape Town, pp. 101–107.

KASCH, K.W. (1983) Tectonothermal evolution of the southern Damara Orogen. In: *Evolution of the Damara Orogen of South West Africa/Namibia* (Ed. Miller, R.McG.) Geol. Soc. S. Afr. Spec. Publ. **11**, 255–265.

KEMP, A.E.S. (1987) Tectonic development of the Southern Belt of the Southern Uplands accretionary complex. *J. Geol. Soc. London* **144**, 827–838.

KILLICK, A.M. (1983) Sulphide mineralization at Gorob and its genetic relationship to the Matchless Member, Damara Sequence, SWA/Namibia. In: *Evolution of the Damara Orogen of South West Africa/Namibia* (Ed. Miller, R.McG.) Geol. Soc. S. Afr. Spec. Publ. **11**, 381–384.

KLEMD, R., MAIDEN K.J. & OKRUSCH, M. (1987) The Matchless copper deposit SWA/Namibia: a deformed and metamorphosed massive sulfide deposit. *Econ. Geol.* **82**, 163–171.

KRÖNER, A. (1982) Rb–Sr geochronology and tectonic evolution of the Pan-African Damara Belt of Namibia, southwestern Africa. *Am. J. Sci.* **282**, 1471–1507.

KUKLA, C., KRAMM, U., OKRUSCH, M. & GRAUERT, B. (1990) Isotope studies and their implication for the metamorphic processes in the Kuiseb Formation of the southern Damara Orogen, Namibia. *23rd Earth Science Congress, University Cape Town, Ext. abstract*, pp. 305–308.

KUKLA, P.A. (1990a) Tectonics and sedimentation of a late Proterozoic convergent continental margin, Khomas Hochland, central Namibia. PhD thesis (unpubl.), University of the Witwatersrand, Johannesburg, 287 pp.

KUKLA, P.A. (1990b) Unusual scapolitization of Late Proterozoic deep-sea fan sequences in an amphibolite grade metamorphic terrain, Damara Orogen, Central Namibia. *Europ. J. Mineralogy* **2**, Beih. 1 (abstract volume), 145–146.

KUKLA, P.A., OPITZ, C., STANISTREET, I.G. & CHARLESWORTH, E.G. (1988) New aspects of the sedimentology and structure of the Kuiseb Formation in the western Khomas Trough, Damara Orogen, SWA/Namibia. *Comm. Geol. Surv. Namibia* **4**, 33–42.

KUKLA, P.A., CHARLESWORTH, E.G., STANISTREET, I.G. & KUKLA, C. (1989) Downward-facing structures in the Khomas Trough of the Damara Orogen, Namibia. *Comm. Geol. Surv. Namibia* **5**, 53–57.

KUKLA, P.A., KUKLA, C., STANISTREET, I.G. & OKRUSCH, M. (1990) Unusual preservation of sedimentary structures in sillimanite-bearing metaturbidites of the Damara Orogen, Namibia. *J. Geol.* **98**, 91–99.

LEGGETT, J.K., McKERROW, W.S. & EALES, M.H. (1979) The Southern Uplands of Scotland; a Lower Palaeozoic accretionary prism. *J. Geol. Soc. London* **136**, 755–770.

LEWIS, S.D., LADD, J.W. & BRUNS, T.R. (1988) Structural development of an accretionary prism by thrust and strike-slip faulting: Shumagin region, Aleutian Trench. *Geol. Soc. Am. Bull.* **100**, 767–782.

LOWE, D.R. (1982) Sediment gravity flows: II. Depositional models with special reference to the deposits of high density turbidity currents. *J. Sediment. Petrol.* **52**, 279–297.

MARTIN, H. (1983) Overview of the geosynclinal, structural and metamorphic development of the Intracontinental Branch of the Damara Orogen. In: *Intracontinental Fold Belts — Case Studies in the Variscan Belt of Europe and the Damara Belt in Namibia* (Eds Martin, H. & Eder, F.W.) Springer-Verlag, Berlin–New York, pp. 611–654.

MARTIN, H. & PORADA, H. (1977) The intracratonic branch of the Damara orogen in South West Africa. I. Discussion of the geodynamic models. II. Discussion of relationships with the Pan-African Mobile Belt system. *Precambr. Res.* **5**, 311–357.

MILLER, R.McG. (1983a) Evolution of the Damara Orogen of South West Africa/Namibia (Ed.). *Geol. Soc. S. Afr. Spec. Publ.* **11**, Johannesburg, 515 pp.

MILLER, R.McG. (1983b) The Pan-African Damara Orogen of South West Africa/Namibia. In: *Evolution of the Damara Orogen of South West Africa/Namibia* (Ed. Miller, R.McG.) Geol. Soc. S. Afr. Spec. Publ. **11**, 431–515.

MILLER, R.McG. (1983c) Tectonic implications of the contrasting geochemistry of Damaran mafic volcanic rocks, South West Africa/Namibia. In: *Evolution of the Damara Orogen of South West Africa/Namibia* (Ed. Miller, R.McG.). Geol. Soc. S. Afr. Spec. Publ. **11**, 115–139.

MILLER, R.McG., BARNES, S.-J. & BALKWILL, G. (1983) Possible active margin deposits within the southern Damara Orogen: the Kuiseb Formation between Okahandja and Windhoek. In: *Evolution of the Damara Orogen of South West Africa/Namibia* (Ed. Miller, R.McG.) Geol. Soc. S. Afr. Spec. Publ. **11**, 73–88.

MOORE, J.C. (1989) Tectonics and hydrogeology of accretionary prisms: role of the décollement zone. *J. Struct. Geol.* **11**, 95–106.

MUTTI, E. & RICCI LUCCHI, F. (1975) Turbidite facies and facies associations. In: *Examples of Turbidite Facies and Associations from Selected Formations of the Northern Apennines* (Eds Mutti, E., Parea, G.C., Ricci Lucchi, F. *et al.*), Field Trip Guidebook A-11, 9th International Association of Sedimentologists Congr., Nice, pp. 21–36.

NELSON, C.H. & NILSEN, T.H. (1974) Depositional trends of modern and ancient deep-sea fans. In: *Modern and Ancient Geosynclinal Sedimentation* (Eds Dott, R.H. & Shaver, R.H.) Soc. Econ. Paleont. Miner. Spec. Publ. **19**, 54–76.

NELSON, C.H. & NILSEN, T.H. (1984) Modern and ancient deep-sea fan sedimentation. *Soc. Econ. Paleont. Miner. Short course*, no. 14, 404 pp.

NILSEN, T.H. & ZUFFA, G.G. (1982) The Chugach Terrane, a Cretaceous trench-fill deposit, southern Alaska. In: *Trench–Forearc Geology* (Ed. Leggett, J.K.) Geol. Soc. London Spec. Publ. **10**, 213–227.

NORMARK, W.R., MUTTI, E. & BOUMA, A.H. (1983/1984) Problems in turbidite research: a need for COMFAN, *Geo-Marine Lett.* **3**, 53–56.

PEACOCK, S.M. (1987) Thermal effects of metamorphic fluids in subduction zones. *Geology* **15**, 1057–1060.

PICKERING, K.T. (1981) Two types of outer fan lobe sequence, from the late Precambrian Kongsfjord Formation submarine fan, Finnmark, North Norway. *J. Sediment. Petrol.* **367**, 77–104.

PICKERING, K.T. (1984) Facies, facies associations and sediment transport/depositional processes in a late Precambrian upper basin-slope/pro-delta, Finnmark, N.

Norway. In: *Fine-grained Sediments: Deep-water Processes and Facies* (Eds Stow, D.A.V. & Piper, D.J.W.) Geol. Soc. London Spec. Publ. **15**, 343–362.

PICKERING, K.T. & HISCOTT, R.N. (1985) Contained (reflected) turbidity currents from the Middle Ordovician Cloridorme Formation, Quebec, Canada: an alternative to the antidune hypothesis. *Sedimentology* **32**, 373–394.

PICKERING, K.T., STOW, D.A.V., WATSON, M.P. & HISCOTT, R.N. (1986) Deep-water facies, processes and models: a review and classification scheme for modern and ancient sediments. *Earth-Sci. Rev.* **23**, 75–174.

PIPER, D.J.W. (1978) Turbidite muds and silts on deep-sea fans and abyssal plains. In: *Sedimentation in Submarine Canyons, Fans, and Trenches* (Eds Stanley, D.J. & Kelling, G.H.) Dowden, Hutchinson and Ross Inc., Stroudsburg, Pennsylvania, pp. 163–176.

PORADA, H. (1983) Geodynamic model for the geosynclinal development of the Damara Orogen, Namibia, South West Africa. In: *Intracontinental Fold Belts — Case Studies in the Variscan Belt of Europe and the Damara Belt in Namibia* (Eds Martin, H. & Eder, F.W.) Springer-Verlag, Berlin–New York, pp. 503–541.

PORADA, H. & WITTIG, R. (1983) Turbidites in the Damara Orogen. In: *Intracontinental Fold Belts — Case Studies in the Variscan Belt of Europe and the Damara Belt in Namibia* (Eds Martin, H. & Eder, F.W.) Springer-Verlag, Berlin–New York, pp. 543–576.

RICCI LUCCHI, F. & VALMORI, E. (1980) Basin-wide turbidites in a Miocene, over-supplied deep-sea plain: a geometrical analysis. *Sedimentology* **27**, 241–270.

SAMPLE, J.C. & MOORE, J.C. (1987) Structural style and kinematics of an underplated slate belt, Kodiak and adjacent islands, Alaska. *Geol. Soc. Am. Bull.* **99**, 7–20.

SAWYER, E.W. (1981) Damaran structural and metamorphic geology of an area southeast of Walvis Bay, SWA/Namibia. *Geol. Surv. S.W.Afr./Namibia Mem.* **7**, 94 pp.

SHANMUGAM, G., MOIOLA, R.J., MCPHERSON, J.G. & O'CONNELL, S. (1988) Comparison of turbidite facies associations in modern passive-margin Mississippi fan with ancient active-margin fans. *Sediment. Geol.* **58**, 63–77.

STANISTREET, I.G., KUKLA, P.A. & HENRY, G. (1991) Sedimentary basinal responses to a late Precambrian Wilson cycle: the Damara Orogen and Nama Foreland, Namibia. *Afr. J. Earth Sci.* **13**, 141–156.

STOW, D.A.V. (1986) Deep clastic seas. In: *Sedimentary Environments and Facies* (Ed. Reading, H.G.) Blackwell Scientific Publications, Oxford, pp. 399–444.

STOW, D.A.V. & BOWEN, A.J. (1978) Origin of lamination in deep-sea fine-grained sediments. *Nature* **274**, 324–328.

STOW, D.A.V. & BOWEN, A.J. (1980) A physical model for the transport and sorting of fine-grained sediment by turbidity currents. *Sedimentology* **27**, 31–46.

STOW, D.A.V. & PIPER, D.J.W. (1984) Deep-water fine-grained sediments: facies models. In: *Fine-grained Sediments: Deep-water Processes and Facies* (Eds Stow, D.A.V. & Piper, D.J.W.) Geol. Soc. London Spec. Publ. **15**, 611–646.

WALKER, R.G. (1976) Facies models, 2. Turbidites and associated coarse clastic deposits. *Geosci. Can.* **3**, 25–36.

WALKER, R.G. (1978) Deep-water sandstone facies and ancient submarine fans; models for exploration for stratigraphic traps. *Am. Assoc. Petrol. Geol. Bull.* **62**, 932–966.

Index